中国农业科学院
兰州畜牧与兽药研究所科技论文集
（2017）

中国农业科学院兰州畜牧与兽药研究所　主编

中国农业科学技术出版社

图书在版编目（CIP）数据

中国农业科学院兰州畜牧与兽药研究所科技论文集.2017 / 中国农业科学院兰州畜牧与兽药研究所主编.—北京：中国农业科学技术出版社，2020.6
ISBN 978-7-5116-4716-0

Ⅰ.①中…　Ⅱ.①中…　Ⅲ.①畜牧学-文集②兽医学-文集　Ⅳ.①S8-53

中国版本图书馆 CIP 数据核字（2020）第 068860 号

责任编辑　闫庆健　努尔古丽·阿布里哈衣尔
责任校对　马广洋

出 版 者　中国农业科学技术出版社
北京市中关村南大街 12 号　邮编：100081
电　　话　(010)82106632(编辑室)　(010)82109704(发行部)
(010)82109703(读者服务部)
传　　真　(010) 82106625
网　　址　http://www.castp.cn
经 销 者　各地新华书店
印 刷 者　北京富泰印刷有限责任公司
开　　本　880 mm×1 230 mm　1/16
印　　张　35.5　彩插　16 面
字　　数　1 170 千字
版　　次　2020 年 6 月第 1 版　2020 年 6 月第 1 次印刷
定　　价　160.00 元

《中国农业科学院兰州畜牧与兽药研究所科技论文集（2017）》

编委会

前　言

近年来，在中国农业科学院科技创新工程的引领下，研究所的科研水平快速提升。我所科研人员和管理人员不但有工作上的热情，更有对工作认识上的高度和对学科理解上的深度。他们在紧张繁忙的实践活动中，笔耕不辍，将自己的研究成果写成论文。这不单是科研人员和管理人员的工作总结、过程记录，更是他们智慧的结晶，最终成为研究所的一笔宝贵财富。

为了珍惜这笔财富，加强优秀论文的交流与传播，营造更加浓厚的学术氛围，促进科研水平和管理水平的提升，切实推进研究所的科技创新，科技管理处搜集了 2017 年研究所科研人员公开发表的论文编印成《中国农业科学院兰州畜牧与兽药研究所科技论文集》第六卷，共 110 篇。由于时间仓促，可能还有论文未能收录，敬希鉴谅！

编者

2020 年 6 月

目　录

IL-6 Promotes FSH-Induced VEGF Expression Through JAK/STAT3 Signaling Pathway in Bovine Granulosa Cells

Meng YANG, Lei WANG, Xurong WANG, Xuezhi WANG,
Zhiqiang YANG, Jianxi LI

(Engineering & Technology Research Center of Traditional Chinese Veterinary Medicine of Gansu Province, Lanzhou Institute of Husbandry and Pharmaceutical Sciences, Chinese Academy of Agricultural Sciences, Lanzhou, China)

Abstract: Background/Aims: Vascular endothelial growth factor (VEGF) has been demonstrated to playa pivotal role in the regulation of angiogenesis in ovarian follicular development, particularly during the preovulatory period. Although numerous studies have shown that interleukin-6 (IL-6) is one of the major inducing factors that regulate the expression of VEGF in non-ovarian cells, whether it involved in regulating the expression of VEGF in normal ovarian granulose cells is still unknown. The aim of this study was to elucidate the mechanisms underlying the effect of IL-6 on FSH-induced VEGF expression in bovine granulosa cells derived from large follicles. Methods: VEGF mRNA expression in granulosa cells after IL-6 with/without inhibitors treatment was analyzed by RT-qPCR. Phosphorylation levels of ERK1/2 and STAT3 proteins induced by IL-6 were analyzed by western blotting. The protein levels produced by granulosa cells were detected by ELISA. Results: High concentration of IL-6 (10ng/mL) can significantly up-regulate FSH-induced VEGF gene and protein expression levels in granulosa cells, and also promote the VEGF upstream regulators HIF-1α and COX2 mRNA expression. VEGF expression levels were significantly decreased after specifically blocking HIF-1α and COX2 by using inhibitors. The up-regulation effect of IL-6 on FSH-induced VEGF expression in granulosa cells mainly through activating the JAK/STAT3 signaling pathway, which can be impaired by JAK inhibitors. Conclusion: IL-6 can promote FSH-induced VEGF expression in granulosa cells, which is mainly achieved by increasing the expression of HIF-1α and COX2. This promoting effect is mediated by activating the JAK/STAT3 pathway. Moreover, there may be a synergistic relationship between FSH and IL-6 in the regulation of VEGF expression.

Key words: IL-6; VEGF; Granulosa cells; JAK/STAT3 pathway; Angiogenesis

1 INTRODUCTION

Female fertility depends on the fully development of ovarian follicles until ovulation, which is a continuous process of multi-factor participation.This process is accompanied b a reciprocal cycle of angiogenesis, particularly during the preovulatory period[1].Newly formed blood vessels can promote the transport of oxygen, nutrients and hormones to the ovaries, and also ensure that different hormones can transfer to target cells[2].A large number of regulatory factors involved in angiogenesis have been identified and characterized.In particular, vascular endothelial growth factor (VEGF) has been demonstrated to play a pivot role in the regulation of angiogenesis.VEGF is a specific mitogen for vascular endothelial cells that can directly induce vascular endothelial cell proliferation and formation of blood vessels[3].In addition to endothelial cells, VEGF gene and protein are also expressed in ovarian granulosa cells, granulosa-lutein cells, thecal cells and thecal-lutein cells[4-6]. It has been found that changes in VEGF expression and its distribution in the ovary are closely related to follicular development and ovulation[7-9].

In addition to regulating the biological function of follicular angiogenesis, VEGF can also increase vascular permeability and promote estrogen and prostaglandin synthesis in ovarian granulosa cells[1,10].It is not only involved in the normal physiological processes of ovarian activity, but also involved in the occurrence and development of many ovulatory disorders[11-13].Numerous factors have been found to regulate VEGF expression, such as hypoxia-inducible factors, cytokines, growth factors and gonadotropins[1,11,14-20]. Although numerous studies have shown that inflammatory cytokines are one of the major inducing factors that regulate the expression of VEGF in non-ovarian cells[21-27], whether inflammatory cytokines are involved in regulating FSH-induced VEGF expression in ovarian follicles are still unknown.

IL-6 produced by a variety of cell types, including ovariangranulosa cells[28,29].IL-6 exerts the biological function by binding to the IL-6 receptor consisting of IL-6Rα and a second protein on the cell membrane.IL-6 first binds to IL-6Rα, which subsequently associates with a gp130 dimer.The dimerized gp130 triggers activation of the ERK1/2 and JAK/STAT3 signaling pathways[30]. Many studies have found that IL-6 is involved in the angiogenesis process under physiological and pathological conditions, and this effect may be achieved by promoting the expression of VEGF[31,32]. Therefore, we speculate that IL-6 may also regulate the expression of VEGF in normal granulosa cells.However, IL-6 is a pleiotropic cytokine that exerts different functions in different microenvironments.Moreover, the biological functions of IL-6 in different cells are also different[33]. It remains unclear whether IL-6 can modulate FSH-induced VEGF expression in ovarian granulosa cells.

In the present study, we examined the effects of IL-6 on FSH-induced VEGF mRNA and protein expression levels and regulatory mechanisms in bovine granulosa cells.The results showed that IL-6 could promote FSH-induced VEGF expression mRNA and protein in granulosa cells, which was mainly achieved by up-regulating HIF-1α and COX2 expression.In addition, we further found that the JAK/STAT3 signaling pathway participates in this regulatory process.

2 MATERIALS AND METHODS

2.1 Antibodies and reagents

Recombinant bovine IL-6 was purchased from Kingfisher Biotech.U0126, AG490, BAY87-2243, YM155 and NS398 were purchased from Selleck Chemicals.ERK1/2 antibody and phospho-ERK1/2 antibody were purchased from Cell Signaling Technology. Anti-STAT3 antibody and phospho-STAT3 antibody were obtained from LifeSpan BioSciences.β-actin antibody was purchased from Abcam.

2.2 Granulosa cell cultures

Bovine ovaries were collected from a local abattoir and brought to the laboratory within 2h after euthanasia.The separation and culture of granulosa cells were carried out according to our previous report[34].Briefly, healthy follicles greater from 8mm to 17mm in diameter were separated after washed with sterile PBS three times.Ovarian follicles were cut in half, and then the follicle walls were washed with serum-free cell culture medium.Granulosa cells were collected by briefly centrifuge and washed three times with the culture medium.The cells were counted and assessed for viability using trypan blue staining.Granulosa cells were cultured in DMEM/F12 (Gibco, USA) supplemented with 1g/L BSA (Sigma, USA), 1% nonessential amino acids (Gibco, USA), 1% insulin-transferrin-selenium (Gibco, USA) in a humidified atmosphere under 5% CO_2 at 37℃.

2.3 Extraction of total RNA and preparation ofcDNA

Thegranulosa cells were cultured in 6-well plates containing 1×10^6 viable cells in 2ml cell culture medium.Total RNA was extracted from granulosa cells using Trizol reagent (Invitrogen, USA) according to the manufacturer's instructions.cDNA was generated from 1μg of total RNA by using the PrimeScript RT reagent kit (Takara, China).Briefly, 1μg of total RNA was mixed with 2μl of 5×g DNA Eraser Buffer, 1μl of gDNA Eraser.RNase Free dH_2O was complemented to 10μl, and then incubated at room temperature for 5min to remove genomic DNA.Followed by addition of 10μl of the reverse transcription reaction solution to carry out the reverse transcription reaction and the resulting cDNA was stored at-20℃ for subsequent RT-qPCR.

2.4 Quantitative real-time PCR

Quantitative real-time PCR was performed in a total volume of 25μl containing 12.5μl SYBR premix Ex Taq Ⅱ, 1μl forward primer (10μM), 1μl reverse primer (10μM).The primers involved in the experiment were synthesized by Beijing Genomics Institute.The sequences of all primers used in this work are as follows: IL-6 (forward): 5′-ATGCTTCCAATCTGGGTTCAATC-3′, IL-6 (reverse): 5′-ACTCGTTCTGGAGGTAGTCCAGGTA-3′; VEGF (forward): 5′-CCCACGAAGTGGTGAAGTTCA-3′, VEGF (reverse): 5′-CCACCAGGGTCTCGATGG-3′; HIF-1α (forward): 5′-CCATTTTCCACTCAGGACAC-3′, HIF-1α (reverse): 5′-AATTCATCACTGGTGGCTGT-3′; COX2 (forward): 5′-CCTTTAAGGCTTACCTACTCACCAG-3′, COX2 (reverse): TGTCAGTGTAGCACATCCAG-3′; Survivin (forward): 5′-CCTGGCAGCTCTCTCAAG-3′, Survivin (reverse): 5′-TAAGTAGGCCAACACGAAAG-3′; FSHR (forward): 5′-AGC-

CCCTTGTCACAACTCTATGTC - 3′, FSHR (reverse): 5′ - GTTCCTCACCGTGAGGTAGATGT - 3′; GAPDH (forward): 5′ - GATGGTGAAGGTCGGAGTGAAC - 3′, GAPDH (reverse): 5′ - GTCATTGATGGCGACGATGT - 3′. Relative quantifications of mRNA were performed using the $2^{-\Delta\Delta CT}$ comparative method. RNase free dH_2O was used as the negative control reaction.

2.5 SDS/PAGE and western blotting analyses

Granulosa cells were cultured in 60-mm dishes containing 2.5×10^6 viable cells and incubated with IL-6 and inhibitors when the cells reached subconfluence. The cells were washed with precooled PBS and then lysed with M-PER Mammalian Protein Extraction Reagent (Thermo Scientific). The protein concentrations were determined by the BCA protein assay after the cell lysate centrifugation. The lysate (20μg) was then resolved on 12% SDS/PAGE gels and electrophoretically transferred to a PVDF membrane. After blocking, the primary antibody was incubated with the protein antigen transferred to the PVDF membrane. After sufficient binding, secondary antibodies were combined with corresponding primary antibody. The immune complexes were visualized by the enhanced chemiluminescence reaction.

2.6 Measurement of VEGF production bygranulosa cells

Thegranulosa cells were cultured in 24-well plates containing 1×10^5 viable cells in 1ml cell culture medium. The supernatant was collected and centrifuged at 3h after the IL-6 and FSH with or without inhibitors treatment. These samples were stored at -80℃ until assayed. VEGF secreted by granulosa cells was detected by bovine VEGF ELISA reagent set (GenWay Biotech, USA). Absorbance was measured at 450nm using a microplate reader. The concentration of VEGF in the samples was calculated by comparison with the standard curve.

2.7 Data analysis

Data represent means±SEM from at least three independent experiments. Comparisons between groups were performed by one-way ANOVA. The significance of the differences between the mean value of the control group and each treated group was determined using the Tukey test. A value of $P<0.05$ was considered significant.

3 RESULTS

3.1 IL-6 promotes FSH-induced VEGF expression ingranulosa cells

Many studies have found that IL-6 can promote VEGF expression in non-granulosa cells and granulosa cell lines. However, whether it regulates VEGF expression in normal granulose cells is still unknown. Here, we examined the effect of different concentrations of IL-6 (0.1-10ng/mL) on FSH-induced VEGF expression in granulosa cells. The results showed that FSH-induced VEGF mRNA expression levels were significantly increased after treatment with any concentration of IL-6. It was also found that higher concentration of IL-6 (10ng/mL) had a more pronounced effect on FSH-induce VEGF mRNA expression than lower concentration (0.1-3ng/mL) (Fig. 1A). The change trend of VEGF protein production in culture supernatant was consistent with mRNA expression in granulosa cells after IL-6 treatment (Fig.3A).

In addition, we also analyzed FSH-induced VEGF expression ingranulosa cells at different time (1.5, 3, 6 and 12h) with the same concentration of IL-6 (10ng/mL) treatment. The results showed that FSH-induced VEGF mRNA expression levels were significantly increased at 1.5h and 3h after IL-6 treatment compared with the control group (1.5h) (Fig.1B).

3.2 IL-6 promotes FSH-induced HIF-1α and COX2 mRNA expression ingranulosa cells

Previous studies have shown that HIF-1α, COX2 andsurvivin are important regulators the expression of VEGF.However, little is known about whether these factors are involved the regulation process of IL-6 on FSH-induced VEGF expression in granulosa cells.We examined the effect of IL-6 on FSH-induced HIF-1α, COX2 and survivin mRNA expression in granulosa cells.As shown in Fig.2A, the FSH-induced HIF-1α mRNA expression levels were significantly higher than that of the control group after any concentration of IL-6 (0.1-10ng/mL) treatment, while COX2 mRNA expression level was increased only at 1ng/mL (Fig.2B).Although the expression levels of survivin mRNA was significantly higher from that of the control group after treatment with different concentrations of IL-6, there was no significant difference compared with the FSH treatment alone (Fig. 2C).Thus, IL-6 does not up-regulate FSH-induced survivin mRNA expression in granulosa cells.

3.3 HIF-1α and COX2 are involved in the regulation of IL-6 on FSH-induced VEGF expression in granulosa cells

In order to further demonstrate the important role of HIF-1α and COX2 in IL-6 promotes FSH-induced VEGF expression in granulosa cells, we treated cells with HIF-1α, COX2 and Survivin inhibitors to specific blocking HIF-1α, COX2 and survivin, respectively.As shown in Fig. 2D, the expression levels of VEGF mRNA were significantly lower in the HIF-1α and COX2 inhibitor-treated groups than in the FSH+IL-6 group.However, the expression level of VEGF mRNA was not significantly altered after survivin inhibitor treatment. The change trend of VEGF protein in culture supernatant was consistent with mRNA expression in granulose cells after inhibitors treatment (Fig.3B).

3.4 IL-6 promotesphosphorylation of ERK1/2 and STAT3 proteins in granulosa cells

IL-6 activates ERK1/2 and JAK/STAT3 signaling pathways bydimerization of gp130 after binding to IL-6Rα on the ovarian granulosa cells membrane.In order to clarify the regulation mechanisms of IL-6 on VEGF expression in granulosa cells, we used western blotting to detect the phosphorylation levels change of ERK1/2 and STAT3 proteins at different time points (5, 15, 30 and 60min) after IL-6 treatment.The results showed that IL-6 could promote FSH-induced phosphorylation of ERK1/2 and STAT3 protein in granulosa cells (Fig. 4).

3.5 The regulation of IL-6 on FSH-induced VEGF expression ingranulosa cells requires the involvement of JAK/STAT3 signaling pathway

In order to furtherelucidate the regulatory mechanism of IL-6 on FSH-induced VEGF

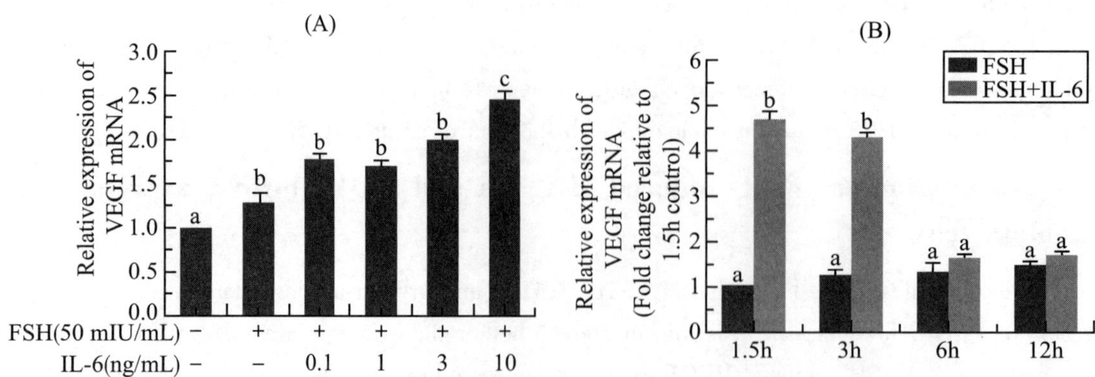

Fig. 1. IL−6 promotes FSH−induced VEGF mRNA expression ingranulosa cells.

Note: (A): Granulosa cells were treated with FSH (50mIU/mL) and different concentrations of IL−6 (0.1−10ng/mL) for 3h.The VEGF mRNA expression levels were analyzed using RT−qPCR. (B): Granulosa cells were treated with FSH (50mIU/mL) or/and the same concentration of IL−6 (10ng/mL), and mRNA expression levels were analyzed at 1.5, 3, 6 and 12h using RT−qPCR, respectively.Results were expressed as the mean±SEM of at least 3 independent experiments.Values without a common letter were significantly different ($P<0.05$).

expression granulosa cells, we used ERK1/2 and JAK inhibitors to specifically block the corresponding signaling pathways.As shown in Fig. 5, the expression levels of VEGF mRNA in AG490 (JAK inhibitors) group was significantly lower than that in control group (FSH+IL−group).However, the expression of VEGF mRNA was not significantly changed after U0126 (ERK1/2 inhibitors) treatment.

3.6 Gonadotropin promotes IL−6 mRNA expression in granulosa cells

There are complex interrelationships between steroid hormones, cytokines, growth factors and gonadotropin−derived gonadotropins in the local follicles during follicular development and ovulation. In order to clarify the interaction between FSH and IL−6 regulating the expression of VEGF in granulosa cells, we investigate the effect of gonadotropin on the expression of IL−6 in granulosa cells.As shown in Fig. 6A, Gonadotropin, including FSH and LH, can promote the expression of IL−6 mRNA, whereas IL−6 can also increase the expression of FSHR mRNA in granulosa cells (Fig. 6B).

4 DISCUSSION

In this study, we found that IL−6 can promote the FSH−induced VEGF expression in bovine granulosa cells, which is mainly achieved by increasing the expression of HIF − 1α and COX2. VEGF mRNA and protein expression levels were significantly reduced after specifically blocking HIF−1α and COX2 expression by inhibitors.We further found that JAK/STAT3 signaling pathway involved in the regulation of IL−6 on FSH−induced VEGF expression in granulosa cells.

VEGF is not only involved in the occurrence and development of many ovarian diseases, but also plays an important role in the physiological processes, such as follicular development, ovulation and corpus luteum formation[1,2].Although a large number of studies have confirmed that inflammatory cytokines play a pivotal role in these biological events in follicular development and ovulation, IL−6 as a major type of inflammatory cytokine, whether it is involved in the regulation of

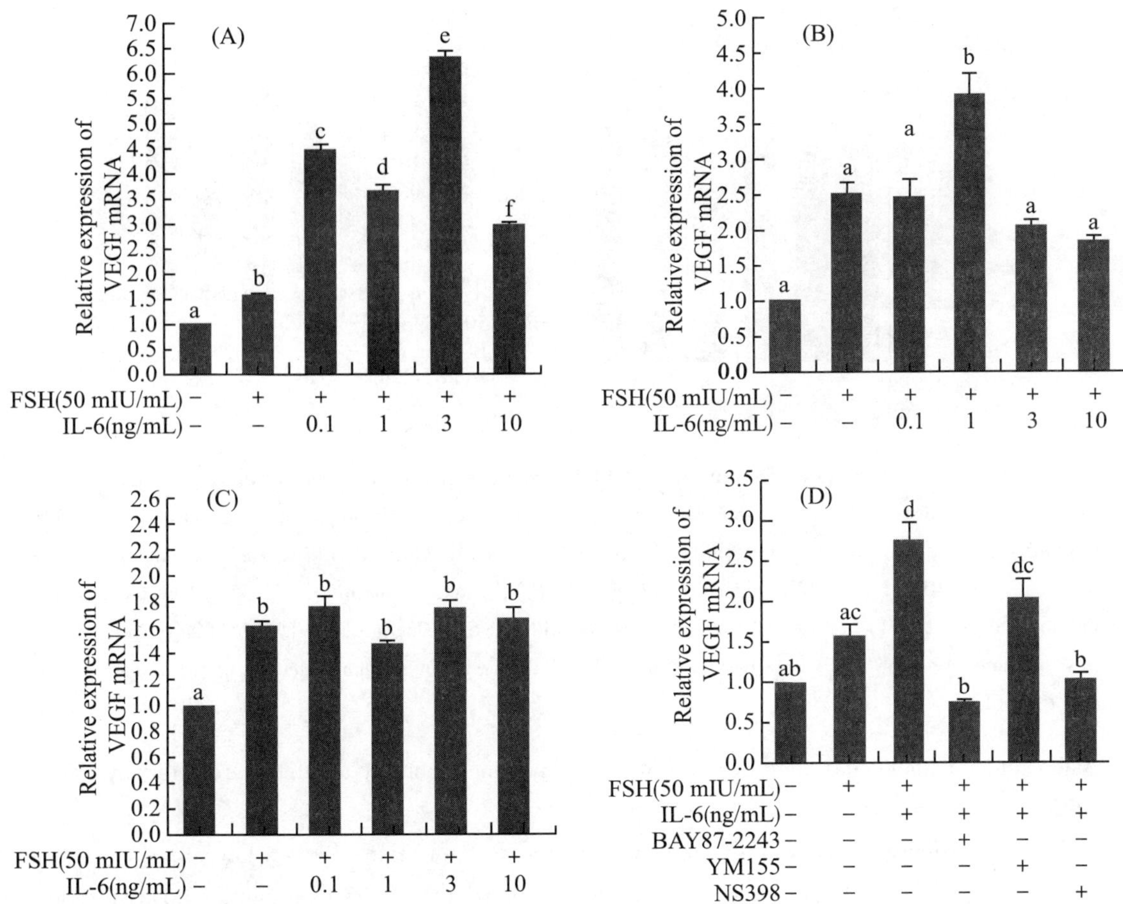

Fig. 2. IL-6 promotes FSH-induced HIF-1α and COX2 mRNA expression ingranulosa cells.

Note: (A), (B) and (C) Granulosa cells were treated with FSH (50mIU/mL) and different concentrations of IL-6 (0.1-10ng/mL) for 3h. The mRNA expression levels of HIF-1α, COX2 and Survivin were analyzed using RT-qPCR, respectively. (D) Granulosa cells were pretreated with HIF-1α inhibitor BAY87-2243 (10μM), Survivin inhibitor YM155 (10μM) or COX2 inhibitor NS398 (10μM) for 1h. Cells were then treated with FSH (50mIU/mL) and/or IL-6 (10ng/mL) for 3h. VEGF mRNA levels were measured by RT-qPCR. Results were expressed as the mean±SEM of at least 3 independent experiments. Values without a common letter were significantly different ($P<0.05$).

VEGF expression in granulosa cells is unclear. Here, we investigated the regulatory role of IL-6 on FSH-induced VEGF expression in normal granulosa cells. We found that high concentrations of IL-6 can promote FSH-induced VEGF mRNA and protein expression. This is consistent with the results of the study on granulose cell lines[11]. Together with previous studies, IL-6 and FSH may have synergistic effect in regulating VEGF expression. In order to further clarify this synergistic effect, we investigated the IL-6 and FSH receptor expression in granulosa cells derived from large follicles after treatment with gonadotropin and IL-6, respectively. The results show that FSH can promote the expression of IL-6 mRNA, whereas IL-6 can also increase the expression of FSHR mRNA in granulosa cells. This result demonstrates that it is possible to amplify the regulatory effect of FSH or IL-6 on cells after IL-6 combined with FSH treatment. It also suggests that IL-6 no only regulates VEGF ex-

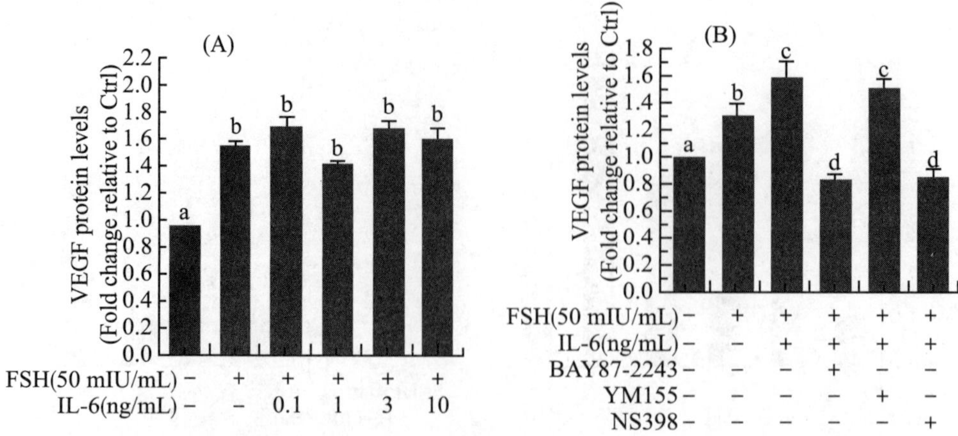

Fig. 3. IL−6 promotes FSH−induced VEGF protein expression levels ingranulosa cells.

Note: (A) Granulosa cells were treated with FSH (50mIU/mL) and different concentrations of IL−6 (0.1−10ng/mL) for 3h, and protein expression levels of VEGF were analyzed using ELISA. (B) Granulosa cells were pretreated with HIF−1α inhibitor BAY87−2243 (10μM), Survivin inhibitor YM155 (10μM) or COX2 inhibitor NS398 (10μM) for 1h.Cells were then treated with IL−6 (10ng/mL) and FSH (50mIU/mL) for 3h. VEGF protein expression levels were measured by ELISA.Values without a common letter were significantly different ($P<0.05$).

pression under pathological conditions, but also plays an important regulatory role in physiological conditions.

The expression of VEGF is regulated by a variety of factors.HIF−1α is a key promoter of VEGF expression in pathological conditions[35−37].Our study shows that IL−6 can significantly increase the FSH−induced HIF−1α expression level in normal granulosa cells.In addition, we also found that IL−6 can also promote the FSH−induced COX2 expression level, which is another important gene that regulates VEGF expression in the follicular development and ovulation.It can be speculated that the regulatory effect of IL−6 on FSH−induced VEGF expression may be achieved by inducing the expression of HIF−1α and COX2. In order to confirm this inference, we used BAY87−2243 (HIF−1α inhibitor) and NS398 (COX2 inhibitor) to specifically block HIF−1α and COX2, and found that IL−6 promotes FSH−induced VEGF mRNA expression was significantly decreased.Taken together, the above studies have shown that IL−6 mainly promotes FSH−induced VEGF expression in granulosa cells by inducing the expression of HIF−1α and COX2.

Forgranulosa cells, IL−6 regulates the expression of target genes by binding to IL−6 receptors and initiating ERK1/2 and JAK/STAT3 signaling pathways. In order to elucidate the regulatory mechanism of IL−6 promotes FSH−induced VEGF expression in granulosa cells, we used inhibitors to specifically block the corresponding signaling pathway.The results show that FSH−induced VEGF mRNA expression levels were decreased significantly after AG490 (JAK inhibitor) treatment.However, the expression of VEGF mRNA was not significantly changed after U0126 (ERK1/2 inhibitors) treatment.In addition, we also used western blotting to detect changes in intracellular phosphorylation levels after IL−6 treatment.The results showed that IL−6 can significantly increase FSH−induced phosphorylation levels ERK1/2 and STAT3 proteins in granulosa cells. The results once

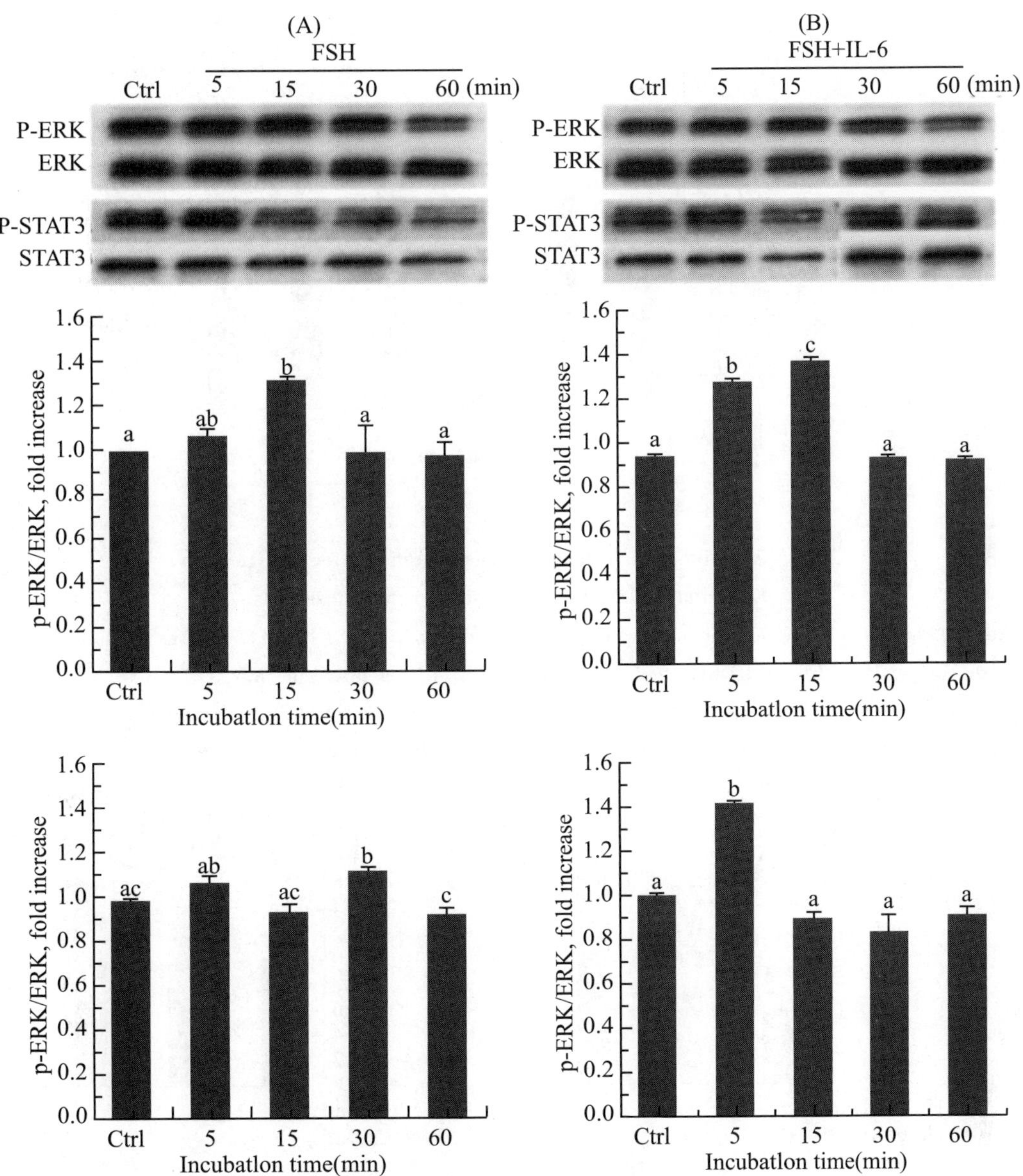

Fig. 4. IL-6 promotes FSH-induced phosphorylation of ERK1/2 and STAT3 proteins in granulose cells.

Note: (A) The subconfluent granulose cells were exposed to FSH (50mIU/mL) for 5, 15, 30 and 60min. Whole-cell lysates were used for western blotting to detect levels of phosphorylated ERK1/2 (p-ERK1/2) and phosphorylated STAT3 (p-STAT3). (B) the granulose cells were exposed to FSH (50mIU/mL) and IL-6 (10ng/mL) for 5, 15, 30 and 60min. Whole-cell lysates were used for western blotting to detect levels of phosphorylated ERK1/2 (p-ERK1/2) and phosphorylated STAT3 (p-STAT3). The detection of ERK1/2 and STAT3 proteins served as a loading control. The phosphorylation level of the protein is expressed as the ratio of phosphorylated protein to the total protein. Results were expressed as the mean±SEM of at least 3 independent experiments. Values without a common letter were significantly different ($P<0.05$).

again demonstrated the synergistic effect between IL-6 and FSH regulating VEGF expression. All of the above studies have shown that JAK/STAT3 signaling pathway is involved in the regulation of

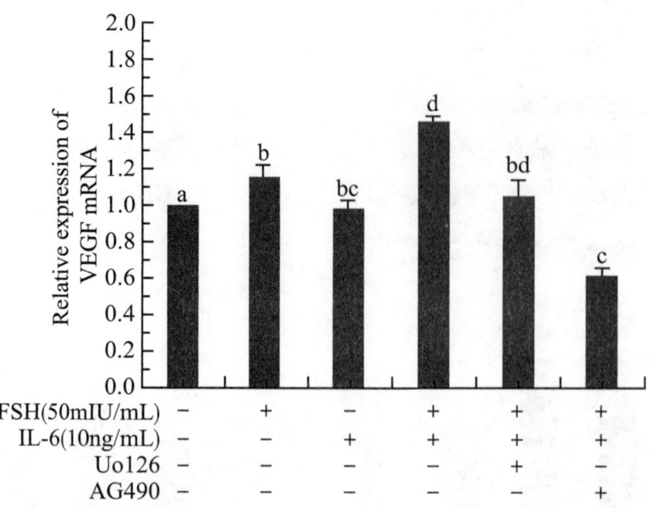

Fig. 5. JAK inhibitor attenuates thepromotive effect of IL−6 on FSH−induced VEGF expression in granulosa cells.

Note: Granulosa cells were pretreated with ERK1/2 inhibitor U0126 (10μM) or JAK inhibitor AG490 (10μM) for 1h and the cells were then treated with IL−6 (10ng/mL) for 3h. VEGF mRNA levels were measured by RT−qPCR. Results were expressed as the mean±SEM of at least 3 independent experiments. Values without a common letter were significantly different ($P<0.05$).

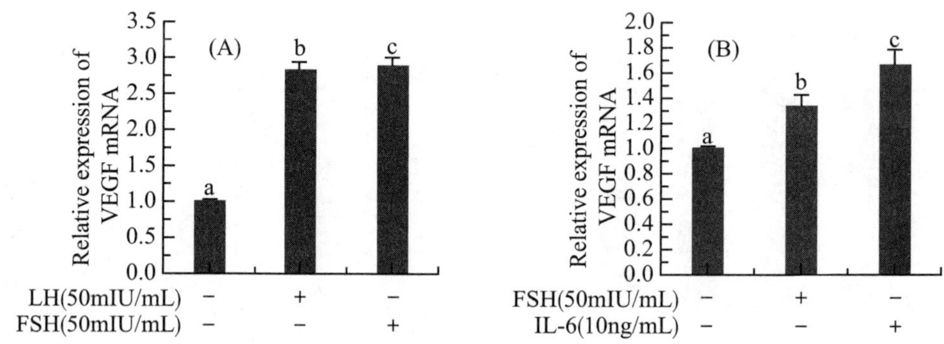

Fig. 6. FSH and LH promote IL−6 mRNA expression ingranulosa cells.

Note: (A) Granulosa cells were treated with FSH (50mIU/mL) or LH (50mIU/mL) for 3h. The IL−6 mRNA expression levels were analyzed using RT−qPCR. (B) Granulosa cells were treated with FSH (50mIU/mL) or/and IL−6 (10ng/mL) for 24h, and FSHR mRNA expression levels were analyzed using RT−qPCR. Results were expressed as the mean±SEM of at least 3 independent experiments. Values without a common letter were significantly different ($P<0.05$).

VEGF expression in granulosa cells. This is consistent with the results of the study in the granulose cell lines[11,37]. In addition to the ERK1/2 pathway, many studies have shown that the JAK/STAT3 pathway plays a key role in different induction factors. Moreover, the biological functions of granulosa cells will be inhibited after specific blocking of the JAK/STAT3 signaling pathway. Our study combined in previous studies shows that JAK/STAT3 pathway plays an important role in regulating follicular development and ovulation, including promoting VEGF gene expression.

Numerous studies have confirmed that inflammatory cytokines play an important regulatory role in ovarian activity[38-40].In particular, the discovery and application of specific inhibitors and gene knockout animals have more substantiated the previous speculation about the involvement of inflammatory cytokines in follicular development and ovulation. However, the previous and our present study mainly focused on a single type of inflammatory cytokines.The study still lacks the role of cytokine network in ovarian cycle.At present, the biological effects of cytokines in regulating follicular development and ovulation have been recognized, but the mechanisms of the overall regulatory effect in the ovarian cycle remains to be studied.

5 CONCLUSION

Our study confirms that IL-6 can promote FSH-induced VEGF synthesis in bovine granulosa cells by increasing the expression of HIF-1α and COX2. The up-regulation effect of IL-6 on FSH-induced VEGF expression in bovine granulosa cells mainly through activating the JAK/STAT3 signaling pathway.It is not only conducive to clarify the intrinsic mechanism of IL-6 in regulating ovarian activities, but also help to elucidate the possible mechanism of the occurrence of ovulatory disorders.

ACKNOWLEDGEMENTS

This work was supported by the Dairy Industry Technology and Systems Projects (CARS-37), Agro-scientific Research in the Public Interest (201303040-01) and Chinese Central Government for Basic Scientific Research Operations in Commonweal Research Institutes (1610322014004).

DISCLOSURE STATEMENT

All authors declare that they have no Disclosure Statement.

REFERENCES OMITTED

(发表于《Cell Physiol Biochem》, 院选 SCI, IF: 5.103)

IL-1α Up-Regulates IL-6 Expression in Bovine Granulosa Cells via MAPKs and NF-κB Signaling Pathways

Meng YANG, Xurong WANG, Lei WANG,
Xuezhi WANG, Zhiqiang YANG, Jianxi LI

(Lanzhou Institute of Husbandry and Pharmaceutical Sciences, Chinese Academy of Agricultural Sciences, Lanzhou, China)

Abstract: Background/Aims: IL-6 is one of the main cytokines in regulating ovarian follicular development and ovulation. However, the factors that regulate IL-6 expression in follicles are still unclear. The aim of this study was to elucidate the mechanisms underlying the effect of IL-1α on IL-6 expression in granulosa cells. Methods: IL-6 expression after IL-1α with/without inhibitors treatment was analyzed by RT-qPCR and ELISA. The phosphorylation of proteins induced by IL-1α was analyzed by western blot. The intracellular cAMP level was assayed by immunoassay kit. Results: IL-1α has a dose-dependent effect on IL-6 expression in granulosa cells. This promoting effect can be significantly attenuated by Erk, c-Jun, p38 and IκB proteins inhibitors, respectively. Moreover, the phosphorylation levels of Erk, c-Jun, p38 and IκBα proteins were significantly increased after IL-1α treatment. In addition, we also found that IL-1α not only reversed the cAMP attenuated IL-6 expression, but also increased IL-1α mRNA expression in granulosa cells. Conclusion: The regulation of IL-1α on IL-6 expression is mediated by activation of MAPKs and NF-κB signaling pathways. Moreover, IL-1α may regulate the ovulation-related genes expression in granulosa cells by an autocrine and/or paracrine manner.

Key words: IL-6; Granulosa cells; IL-1α; MAPKs pathway; NF-κB pathway

1 INTRODUCTION

Female fertility depends on the normal development and growth of ovarian follicles and eventually leads to ovulation. These processes are precisely regulated by many genes, including follicle stimulating hormone, luteinizing hormone and follicle derived steroid hormones[1,2]. In addition, many local factors, including inflammatory cytokines play an important regulatory role in the ovarian cycle, especially in ovulation[3-5]. Moreover, it is also involved in the occurrence and development of ovulatory disorder infertility[6-8].Inflammatory cytokines not only play a biological effect by themselves, but also have a complex regulatory relationship between each other, both at the gene level and the protein level.

IL-6 is produced by various types of cell, such as leukocytes, keratinocytes, endothelial cells, fibroblasts, and some tumor cells. It often displays hormone-like characteristics that affect homeostatic processes in many systems, including the reproductive system[3,9,10]. Similar to the effects in the inflammatory response, IL-6 also plays a dual role in regulation of follicular development and ovulation. It not only suppresses steroidogenesis induced by FSH[11,12], but also promotes the LH receptor expression[13] and cumulus-oocyte complexes expansion[14]. Although previous studies have confirmed that IL-6 and its receptors expressed in granulosa cells[15-17], the factors that regulate IL-6 expression in ovarian follicles are still unknown.

IL-1 is another important type of inflammatory cytokines, mainly from the activated monocytes/macrophages and epithelial and endothelial cells. It induced expression of pro-inflammatory cytokines and chemokines through activating mitogen-activated protein kinase (MAPKs) and nuclear factor-kappa B (NF-κB) signaling pathways by binding to IL-1 receptor type 1 (IL-1R1) on the cell membrane[18-20]. IL-1 is also one of the most important cytokines in local ovarian follicles[21]. It involved in several ovulation associated events such as synthesis of proteases, regulation nitric oxide and prostaglandin production[22]. On the other hand, previous studies has also found that hCG-induced ovulation rates were decreased in rats after treatment with IL-1 receptor antagonist (IL-1Ra)[23,24]. Although IL-1α and IL-1β are encoded by different genes, they exert similar biological effects by binding to the same receptor (IL-1R1). However, IL-1α is more able to rapidly initiate the inflammatory cytokine expression both in the mature and precursor forms when binding to IL-1R1, and cause the aseptic inflammation, which has many similarities with ovulation[25]. It remains unclear whether IL-1α can modulate the expression of IL-6 in granulose cells.

In this study, we examined the effects of IL-1α on IL-6 mRNA expression and protein production in granulosa cells. Moreover, the mechanisms of its action were also studied. The results showed that IL-1α could up-regulates the IL-6 expression via MAPKs and NF-κB signaling pathways. The regulatory effects of IL-1α on ovarian granulosa cells may be achieved through an autocrine and/or paracrine manner.

2 MATERIALS AND METHODS

2.1 Antibodies and reagents

Recombinant bovine IL-1α was purchased from Kingfisher Biotech. U0126, SP600125, SB203580 and BAY11-7082 were obtained from Beyotime Biotechnology. Erk antibody, p-Erk antibody, c-Jun antibody, p-c-Jun antibody, IκBα antibody and p-IκBα antibody were purchased from Cell Signaling Technology. p38 antibody, p-p38 antibody and β-actin antibody were purchased from LifeSpan BioSciences. Forskolin were purchased from Abcam.

2.2 Granulosa cell cultures

Bovine ovaries were collected from a local abattoir and brought to the laboratory in saline at 30-

35℃ in 2h after euthanasia. Healthy follicles greater than 8mm in diameter were separated after sterile PBS three times. Ovarian follicles were cut in half, and then the follicle walls were washed with serum-free cell culture medium. Granulosa cells were collected by briefly centrifuge and washed three times with the culture medium. The cells were counted and assessed for viability using trypan blue staining. Granulosa cells were cultured in Dulbecco's Modified Eagle Medium/Nutrient Mixture F-12 (DMEM/F12; Gibco, USA) supplemented with 1g/L bovine serum albumin (BSA; Sigma, USA), 1% nonessential amino acids (Gibco, USA), 1% insulin-transferrin-selenium (ITS; Gibco, USA) in a humidified atmosphere under 5% CO_2 at 37℃.

2.3 Quantitative real-time PCR

Thegranulosa cells were cultured in 6-well plates containing 1×10^6 viable cells in 2ml cell culture medium. The fresh medium was replaced 24h before inhibitors and IL-1α treatment. Total RNA was extracted from granulosa cells using Trizol reagent (Invitrogen, USA) according to the manufacturer's instructions. cDNA was generated from 1μg of total RNA by using the PrimeScript RT reagent kit (Takara, China). Real-time PCR was performed in a total volume of 25μl containing 12.5μl SYBR premix Ex Taq Ⅱ, 1μl forward primer (10μM), 1μl reverse primer (10μM). The primers involved in the experiment were synthesized by Beijing Genomics Institute. The sequences of all primers used in this work are as follows: IL-6 (forward, ATGCTTCCAATCTGGGTTCAATC-3'; reverse, 5'-ACTCGTTCTGGAGGTAGCCAGGTA-3'), IL-1α (forward, 5'-CCTCTCTCT-CAATCAGAAGTCC-3'; reverse, 5'-CCACCATCACCACATTCTCC-3'), GAPDH (forward, 5'-GATGGTGAAGGTCGGAGTGAAC-3'; reverse, 5'-GTCATTGATGGCGACGATGT-3'). Relative quantifications of mRNA were performed using the $2^{-\Delta\Delta CT}$ comparative method. RNase free dH_2O was used as the negative control reaction.

2.4 Measurement of IL-6 secretion bygranulosa cells

Granulosa cells were cultured in 24-well plates containing 1×10^5 viable cells in 1ml cell culture medium supernatant was collected and centrifuged at 24h after the IL-1α with or without inhibitors treatment. These samples were stored at-80℃ until assayed. IL-6 secreted by granulosa cells was measured by ELISA kit (Abcam, UK). Absorbance was measured at 450nm using a microplate reader. The concentration of IL-6 in the samples was calculated by comparison with the standard curve.

2.5 SDS/PAGE and western blot analyses

Granulosa cells were cultured in 60mm dishes and incubated with IL-1α when the cells reached subconfluence. The cells were washed with pre-cooled PBS and then lysed with RIPA lysis buffer. The protein concentrations were determined by the BCA protein assay after the cell lysate centrifugation. The lysate (20μg) was then resolved on 12% SDS/PAGE gels and electrophoretically transferred to a polyvylidene difluoride (PVDF) membrane. After blocking, the primary antibody was incubated with the protein antigen transferred to the PVDF membrane. After sufficient binding, secondary antibodies were combined with corresponding primary antibody. The immune complexes were visualized by the enhanced chemiluminescence reaction.

2.6 Detection of intracellular Cyclic AMP levels

Granulosa cells were treated with different concentrations of IL−1α when the cells reached subconfluence. Competitive immunoassay kit (Abcam, UK) was used to determine of intracellular Cyclic AMP (cAMP) level in cells. Briefly, removed the media from adherent cells and add enough 0.1M HCl to cover the bottom of the plate. The cells were incubated in 0.1M HCl for 10min at room temperature. Cell debris was removed by centrifugation after cells were uniform lysed. The supernatant were stored at −80℃ until assayed. Absorbance was measured at 405nm using a microplate reader. The intracellular cAMP level in the samples was calculated by comparison with the standard curve.

2.7 Data analysis

All data shown are means ± SEM from three independent experiments. Comparisons between groups were performed by one−way ANOVA. The significant differences between the control group and each treated group were determined using the Tukey test. A value of $P<0.05$ was considered significant.

3 RESULTS

3.1 IL−1α promotes IL−6 expression ingranulosa cells

To investigate the effect of IL−1α on IL−6 biosynthesis, we analyzed the mRNA expression and protein production of IL−6 in granulosa cells. Granulosa cells derived from large follicles were cultured 6−well plates and treated with different concentrations of IL−1α (0−10ng/mL). As shown in Fig. 1 A, IL−1α has a dose−dependent effect on IL−6 mRNA expression. The expression levels of IL−6 mRNA were significantly increased after treatment with higher concentrations of IL−1α (3ng/mL and 10ng/mL) compared with lower concentration (0−1ng/mL). The change trend of IL−6 protein production in culture supernatant was consistent with mRNA expression in granulosa cells after IL−1α treatment (Fig. 1C).

3.2 IL−1α induced IL−6 expression is attenuated by inhibitors ofMAPKs and NF−κB signaling pathways

It is known that the IL−1 can induce synthesis and secretion of inflammatory cytokines through MAPKs and NF−κB signaling pathways[16]. However, little is known about the biological function of IL−1α in local follicles. To identify the signaling pathways involved IL−1α up−regulation of IL−6 expression in granulosa cells, we analyzed IL−6 expression in the absence or presence of U0126, SP600125, SB203580 or BAY11−7082. As shown in Fig. 1 B and Fig. 1 D, the promoting effect of IL−1α on the expression of IL−6, both at the gene level and protein level in granulosa cells was significantly attenuated by the corresponding inhibitor of MAPKs and NF−κB signaling pathway compared with the control group (IL−1α at 10ng/mL).

3.3 IL−1α inducesphosphorylation of Erk, c−Jun, p38 and IκBα proteins

To further verified the role ofMAPKs and NF−κB signaling pathways in IL−1α up−regulation of IL−6 expression in granulosa cells, we used western blot analysis the protein phosphorylation induced by IL−1α (10ng/mL) at 5min and 15min. The results showed that the phosphorylation level of Erk after IL−1α treatment was significantly higher than the control group at 15min (Fig. 2A and Fig. 2E). We also found that the phosphorylation levels of c−Jun and p38 were significantly increased after exposure of IL−1α, either at 5min or 15min (Fig. 2B, 2C, 2F and 2G). In addition, the IκBα phosphorylation was significant increased at 15min after IL−1α treatment compared with the control group (Fig. 2D and Fig. 2H).

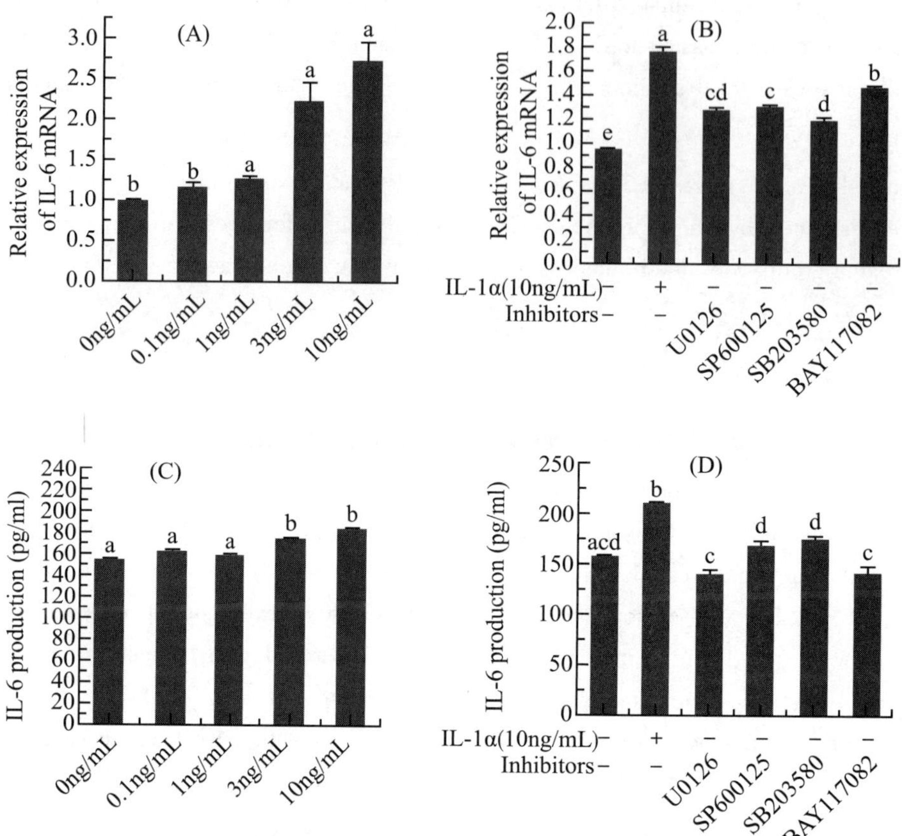

Fig. 1. Effects of IL−1α on IL−6 expression ingranulosa cells.

Note: (A) and (C), Granulosa cells were treated with different concentrations of IL−1α (0−10ng/mL), and the mRNA expression levels and protein production levels of IL−6 were analyzed at 3h or 24h using RT−qPCR or ELISA, respectively. (B) and (D), Granulosa cells were pretreated with Erk inhibitor U0126 (10μM), c−Jun inhibitor SP600125 (10μM), P38 inhibitor SB203580 (10μM) or IκBα inhibitor BAY11−7082 (10μM) for 1h. Cells were then treated with IL−1α (10ng/mL) for 3h or 24h. IL−6 mRNA and protein production levels were measured by RT−qPCR or ELISA, respectively. Results were expressed as the mean±SEM of at least 3 independent experiments. Values without a common letter were significantly different ($P<0.05$).

3.4 cAMP attenuated IL−6 expression is reversed by high concentration of IL−1α

It is known thatcAMP is involved in the regulation of many cell functions, for example, regulation of cytokine synthesis and secretion[26-28]. Here, we examined the intracellular cAMP levels of granulosa cells after stimulated by different concentrations of IL−1α (0−10ng/mL) for 1.5h and

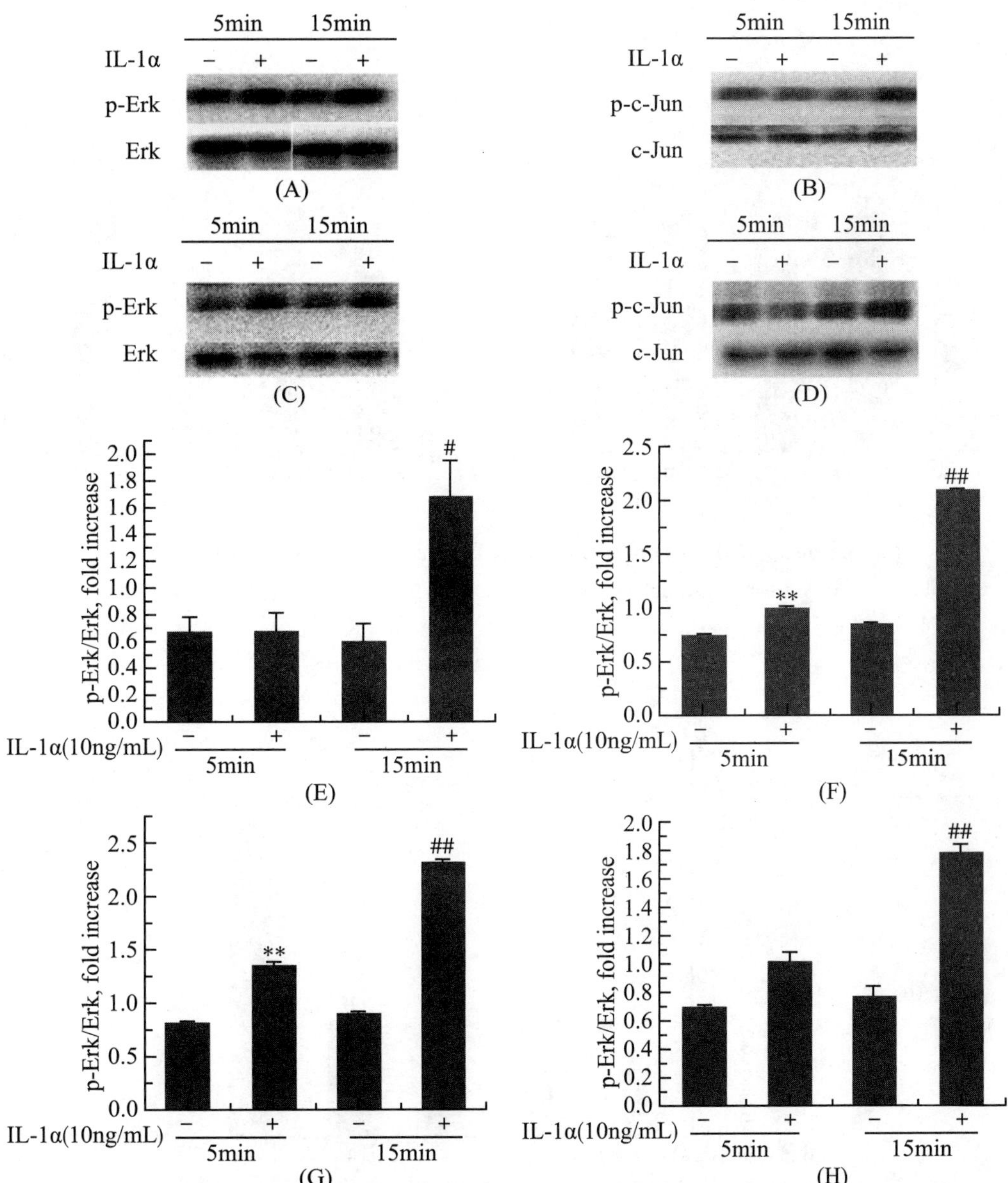

Fig. 2. Effects of IL-1α onphosphorylation of Erk, c-Jun, p38 and IκBα proteins in granulose cells.

Note: (A), (B), (C) and (D), The subconfluent granulose cells were exposed to IL-1α (10ng/mL) for 5 or 15min. Whole-cell lysates were used for Western blot to detect levels of phosphorylated Erk (p-Erk), phosphorylated c-Jun (p-c-Jun), -phosphorylated p38 (p-p38) and phosphorylated IκBα (p-IκBα). The detection of Erk, c-Jun, p38 and IκBα proteins served as a loading control. The phosphorylation level of the protein is expressed as the ratio of the phosphorylated protein to the total protein. Results were expressed as the mean±SEM of at least 3 independent experiments. * ($P < 0.05$) versus control group (5min). # ($P < 0.05$) versus control group (15min).

3h, respectively. As shown in Fig. 3A and Fig. 3B, no significant differences were observed in intracellular cAMP levels between the treatment groups, neither at 1.5h nor 3h. However, we still observed that the intracellular cAMP levels after IL-1α treatment remained higher than the control group (0ng/mL).

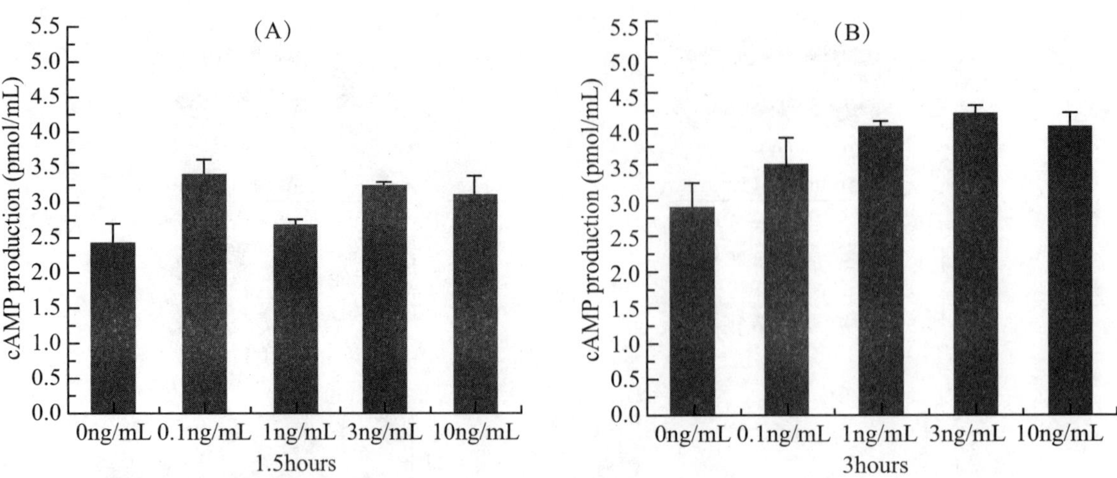

Fig. 3. Effects of IL−1α on intracellularcAMP production in granulosa cells.

Note: Granulosa cells were treated with different concentrations of IL−1α (0−10ng/mL), and the concentrations of intracellular cAMP were measured by ELISA at 1.5h (A) and 3h (B), respectively.Results were expressed as the mean±SEM.

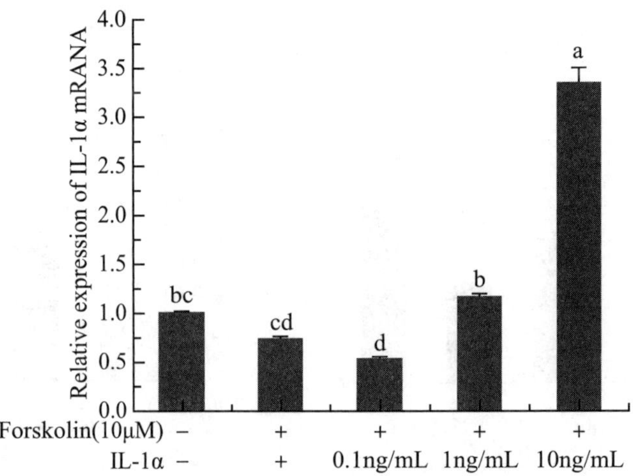

Fig. 4. Effects ofcAMP on IL−6 mRNA expression in granulosa cells.

Note: Granulosa cells were pretreated with Forskolin (10μM), a cAMP specific agonist, for 1h.Cells were then treated with different concentrations of IL−1α (0.1−10ng/mL) for 3h.IL−6 mRNA levels were measured by RT−qPCR.Results were expressed as the mean±SEM of at least 3 independent experiments.Values without a common letter were significantly different ($P<0.05$).

In order to further elucidate the effects ofcAMP on IL−6 mRNA expression in granulose cells, we analyzed IL−6 expression after treatment with Forskolin (10μM) and different concentrations of IL−1α (0.1, 1 and 10ng/mL).As shown in Fig.4, expression level of IL−6 was reduced by Forskolin treatment alone, although no significant difference in statistics.Moreover, the expression level of IL−6 was significantly decreased after treatment with Forskolin and low concentration of IL−1α

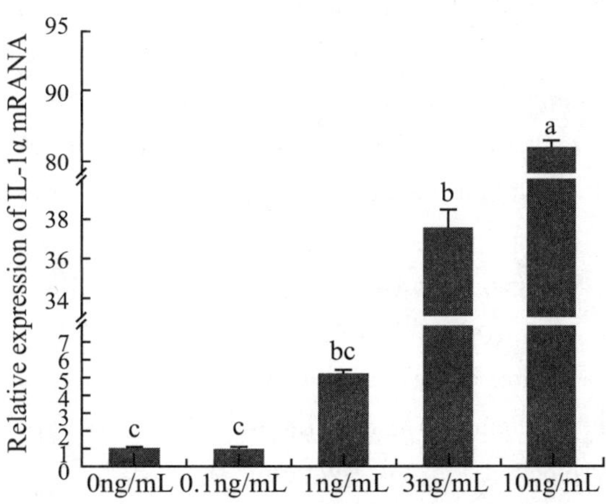

Fig. 5. Effects of IL−1α on IL−1α mRNA expression in granulosa cells.

Note: Granulosa cells were treated with different concentrations of IL−1α (0−10ng/mL), and the mRNA levels of IL−1α were analyzed at 3h using RT−qPCR. Results were expressed as the mean±SEM of at least 3 independent experiments. Values without a common letter were significantly different ($P<0.05$).

(0.1ng/mL) compared with the control group. But, this attenuating effect can be reversed by high concentrations of IL−1α (10ng/mL).

3.5 IL−1α promotes IL−1α mRNA expression ingranulosa cells

Many studies have found that the level of IL−1 in follicular fluid and serum sharply increased at preovulatory period, and plays an important regulatory role in ovulation−related gene expression[29-31]. Here, we used different concentrations of IL−1α (0−10ng/mL) on the granulosa cells from large follicle with the characteristics of pre−ovulation granulosa cells. The expression levels of IL−1α mRNA were analyzed after treatment with IL−1α 3h later. The results (Fig. 5) showed that IL−1α has a dose−dependent effect on IL−1α mRNA expression. Interestingly, a dramatic increased in IL−1α mRNA expression level was observed after treatment with of IL−1α at 10ng/mL.

4 DISCUSSION

Inflammatory cytokines is the main cytokine type involved in the regulation of ovarian granulosa cell biological functions. In this study, we discovered that IL−1α could up−regulate IL−6 expression in granulosa cells derived from large follicles. Moreover, our study further found that MAPKs and NF−κB signaling pathways involved in this regulation process.

Many studies have shown that cytokines, especially inflammatory cytokines involve in the regulation of follicular development and ovulation. Oocyte−cumulus complexes and granulosa cells were the main regulating objects of inflammatory cytokines[32]. Mutual induction of cytokine gene expression is an important feature of inflammatory cytokines. However, the factors that regulate the expression of IL−6 in the ovarian follicles are still unknown. In this study, we found that the expression levels of IL−6 mRNA and protein in granulosa cells were up−regulated by high concentrations of IL−1α. It indicated that IL−1α not only plays directly regulatory roles on the granulosa cells by itself, but also indirectly regulates the biological functions of cells through synthesis of other cytokines.In addition, we also found that IL−1α could up−regulate the expression of IL−1α mRNA in granulosa cells.Although IL−1α can be expressed by granulosa cells, it may also be synthesized by other cells in the ovary.Especially in the preovulatory period, a large number of immune cells are rapidly recruited to the follicles and play a crucial role as facilitators of ovulation[33-35].It suggests that IL−1α may regulate the ovulation−related genes expression in granulosa cells by an autocrine and/or paracrine manner.

IL−1 induces cytokine gene expression mainly by triggering MAPKs and NF−κB signaling pathways.MAPKs signaling pathways involved in the regulation of diverse responses, including female follicular development and ovulation.However, the functions of cytokines will vary with different microenvironment.To clarify the mechanisms underlying the effect of IL−1α on IL−6 expresion in granulosa cells, we specifically blocked the corresponding signaling pathways by using inhibitors. Our data show that up−regulation of IL−6 expression by IL−1α was attenuated by inhibitors of U0126, SP600125, and SB203580 respectively.In order to further confirm the effect of MAPKs pathways on IL−6 expression after IL−1α treatment, we detected phosphorylation of Erk, c−Jun and p38 proteins of MAPKs pathway by western blot.The results show that the phosphorylation of signaling proteins was significantly increased after IL−1α treatment compared with the control group.The above studies indicated that the regulation of IL − 1α on IL − 6 expression is mediated by activation of MAPKs signaling pathway. NF − κB signaling pathway is another important pathway for IL − 1 activation in the regulation of inflammatory cytokine gene expression.It not only plays the most important role in the immune system, but also regulates genes expression in many other systems.In this study, we found that IL−1α−induced IL−6 expression in granulosa cells can be attenuated by NF−κB inhibitors (BAY11 − 7082). We also found that the phosphorylation of IκBα protein was increased after exposure of IL−1α.It also indicated that NF−κB signaling pathway is also involved in the IL−1α−induced IL−6 expression in granulosa cells.

There isainteraction between cAMP and MAPKs signaling pathways in regulating gene expression[36].It has been reported that cAMP can regulate the biosynthesis of IL − 6 in many types of cells[37-39].Here, we observed that the intracellular cAMP levels after IL−1α treatment was higher than the control group, although no significant difference in statistics.In order to further clarify the regulatory role of cAMP in IL−1α−induced IL−6 expression, Forskolin, a cAMP specific agonist, was used to treat cells in combination with IL−1α.The results showed that cAMP can reduce the IL−6 mRNA expression, and this effect can be reversed by high concentrations of IL − 1α. Inhibitory effect of cAMP on IL−6 expression observed in the present study is consistent with previous study on lung fibroblasts[40].It suggests that the cAMP may play a negative role in IL−1α−induced IL−6 expression in ovarian granulosa cells.In addition, it is also imply that there is a regulatory relationship between IL−1α, IL−6 and cAMP in processes of the follicular development and ovulation.

In summary, our study demonstrates that IL−1α can promote IL−6 expression in bovine granulosa cells.This effect is mediated through the MAPKs and NF−κB signaling pathways. IL − 1α may play a regulatory role in granulosa cells through an autocrine and/or paracrine manner.

ACKNOWLEDGEMENTS

This work was supported by the Dairy Industry Technology and Systems Projects (CARS−37),

Agro-scientific Research in the Public Interest (201303040-01) and Chinese Central Government for Basic Scientific Research Operations in Commonweal Research Institutes (1610322014004).

DISCLOSURE STATEMENT

All authors declare that they have no conflict of interest.

REFERENCES OMITTED

(发表于《Cellular Physiology and Biochemistry》，院选 SCI，IF：5.103)

Ultrasound-assisted Extraction of Polysaccharides from *Rhododendron aganniphum*: Antioxidant Activity and Rheological Properties

Xiao GUO[1], Xiaofei SHANG[1*], Xuzheng ZHOU[1], Baotang ZHAO[2], Jiyu ZHANG[1**]

(1. Key Laboratory of New Animal Drug Project of Gansu Province, Key Laboratory of Veterinary Pharmaceutical Development of Ministry of Agriculture, Lanzhou Institute of Husbandry and Pharmaceutical Sciences, Chinese Academy of Agricultural Sciences, Lanzhou 730050, China
2. College of Food Science and Engineering of Gansu Agricultural University, Lanzhou 730070, China)

ABSTRACT: In this study, we aimed to optimize the extraction of polysaccharides from the leaves of *Rhododendron aganniphum* and investigate its rheological properties and antioxidant activity. After optimizing the operating parameters using a Box-Behnken design (BBD), the results showed that the optimal ultrasound-assisted extraction conditions were as follows: extraction temperature, 55℃; liquid-solid ratio, 25 : 1; extraction time, 2.2h; and ultrasound treatment power, 200W. The optimized experimental yield of polysaccharides by ultrasound-assisted extraction (PUAE) was 9.428%, higher than that obtained by hot water extraction (PHWE) for 12h at the same liquid-solid ratio and extraction temperature. In the in vitro antioxidant activity tests, PUAE had higher positive radical scavenging activity for hydroxyl, superoxide and 1, 1-diphenyl-2-picrylhydrazyl (DPPH) radicals than PHWE. However, PUAE and PHWE solutions had similar intermolecular interactions in the steady-shear flow and dynamic viscoelasticity tests, resulting in similar macroscopic behaviour. With respect to the apparent viscosity, storage modulus (G') and loss modulus (G'') of PUAE were lower at the same shear rate or angular frequency. All PUAE solutions exhibited non-Newtonian shear-thinning pseudoplastic behaviour that was accurately described by the Carreau model but was better fit by the power-law model at high shear rates ($\geq 1/s$), which demonstrated that the variation in the apparent viscosity dependence was greater at higher concentrations and shear rates. The G' and G'' of the solutions increased as the experimental frequency increased from 0.05 to 500 rad/s under all experimental concentrations, and the modulus crossover point decreased gradually with increasing PUAE concentration. The above results demon-

* First co-author. http: //dx.doi.org/10.1016Zj.ultsonch.2017.03.021, 1350-4177/

** Corresponding author at: E-mail addresses: gx_ 139417@163.com, infzjy@sina.com (J.Zhang).

strated that the ultrasound-assisted extraction methods gave a higher yield of polysaccharides from the leaves of R.agan-niphum with a shorter extraction time than the hot water extraction method, which could affect the apparent viscosity and dynamic viscoelasticity.PUAE presented good radical scavenging activity for DPPH, superoxide and hydroxyl radicals in vitro and could be used as a natural antioxidant in the food and medical industries.

Key words: Antioxidant activity; Rheological properties; *Rhododendron aganniphum*; Ultrasound-assisted extraction

1 INTRODUCTION

As one of the largest genera of vascular plants, *Rhododendron* L. (Ericaceae) comprises 8 subgenera with more than 850 species[1,2].The majority of the species grow in the Himalayan region, Southeast Asia and Malesia, with others distributed in North America, Europe and North-East Australia[3,4].The phytochemistry and bioactivity of this genus have been investigated, and many plants have been demonstrated to have significant biological activities, including anti-inflammatory, analgesic, anti-microbial, antidiabetic, antioxidant, insecticidal and cytotoxic activity[5,6]. However, *Rhododendron aganniphum*, one of the dominant species of the genus *Rhododendron* in the Tibet region, has not been studied.It grows at altitudes of 2 700-4 700m throughout China, Nepal, India, Bhutan and Sikkim, and as a folk medicine, the flowers and leaves are widely used by the local people to treat general body weakness, inflammation, and lung and skin disorders[7,8].

Plant polysaccharides have attracted considerable attention due to their multi-functional bioactivity, including antioxidant, immunomodulatory, antitumour, and hypoglycaemic properties[9,10] and have been found in some *Rhododendron* sp.[11,12].When dispersed in water, plant polysaccharides can form sols; therefore, they have been widely used as thickeners, gel agents, emulsifiers and stabilizers in the food, medical, cosmetic, chemical, oil drilling and other industries[13].As important antioxidants, polysaccharides isolated from some natural products can be explored as novel and natural functional or healthy foods to prevent oxidative damage in living organisms[14]. Rheological properties are a type of mechanical property describing the deformation and flow of material due to applied stress and strain.Polysaccharides are viscoelastic materials that exhibit liquid and solid characteristics simultaneously[15].The rheological behaviours of polysaccharides are the basis for their applications as functional foods, pharmaceutical and cosmetic industries. These rheological properties of polysaccharides in aqueous solutions affect various technological processes, such as heating, stirring, mixing, and filtering[16-18].Thus rheological data are essential for functional food product quality evaluations, engineering calculations and process design[19].

Because preserving the structure of polysaccharides during their isolation is crucial to maintain their bioactivity for the intended application, conventional methods, such as heating water to extract polysaccharides, have been gradually abandoned due to hydrolysis, ionization, or oxidation as a result of longer extraction times[20].Newer techniques such as microwave-assisted extraction, supercritical fluid extraction, ultrasound-assisted extraction (UAE) and supercritical fluid extraction have been developed to improve extraction by reducing energy and time while obtaining higher bioactivity[21,22]. Among these techniques, UAE is inexpensive, environmentally friendly, less time consuming and efficient[23,24], and has been used to extract polysaccharides from natural products[14,25-27]. Meanwhile, the secondary effect of this technology and more contact among analytes and sorbent led to rise the mass transfer by different mechanism correspond to microstreaming, micro-turbulence, acoustic waves and micro jets, and don't change significantly in equilibrium characteristics of the sorption/desorption system[28-30].

In this paper, UAE was used to extract polysaccharides from the leaves of *R. aganniphum*, and the extraction conditions were optimized using a Box - Behnken design (BBD). Then, the antioxidant activity and rheological characteristics of the polysaccharides obtained by ultrasound-assisted extraction (PUAE) were investigated, and as a comparable group, the antioxidant activity and rheological characteristics of the polysaccharides obtained by hot water extraction (PHWE) were also studied.

2 MATERIALS AND METHODS

2.1 Materials

Rhododendronag anniphum was collected from the north slope of Shergyla Mountain near the Linzhi region of Tibet (July 2014). It was authenticated by Zhen Xing, an associate professor at the Agricultural and Animal Husbandry College of Tibet University. A voucher specimen with accession number ZSY412 was submitted to the Herbarium of the Lanzhou Institute of Animal and Veterinary Pharmaceutics Sciences, Chinese Academy of Agricultural Science (Lanzhou, China). The fresh stems and leaves were air-dried, crushed and stored in plastic bags at room temperature.

2.2 Chemicals and reagents

1, 1-Diphenyl-2-picrylhydrazyl (DPPH), nitro blue tetrazolium (NBT), nicotinamide adenine dinucleotide disodium salt (NADH - 2Na), phenazine methosulfate (PMS) were purchased from Sigma Chemical Co. (St. Louis, MO, USA), and ethylene diamine tetraacetic acid (EDTA), vitamin E (VE), ascorbic acid (Vc), Tris-HCl, H_2O_2 and thiobarbituric acid (TBA) were purchased from Sinopharm Chemical Reagent Co. (Shanghai, China); all other reagents were analytical grade.

2.3 Instrumentation

An ultrasonic cleaner (KQ-250DE, Kunshan Ultrasonic Instruments Co., Ltd., China) at 45 kHz frequency and 250W power was used for the ultrasound - assisted extraction of polysaccharides from R. aganniphum. A DD - 5M centrifuge (Cence ©, China) was used to accelerate the phase separation. A R-215 rotary evaporator (BU CHI, Switzerland) and a Modulyo freeze dryer (Edwards, United Kingdom) were used for sample concentration and drying, respectively. An Anton Paar Physica MCR 301 rheometer (Anton Paar, GmbH, Germany) with parallel plate geometry (PP50, stator inner diameter = 50mm) and concentric cylinder geometry (CC27, stator inner diameter = 27mm) was used to measure rheological properties.

2.4 Ultrasound-assisted extraction of polysaccharides

2.4.1 UAE experiment

The extraction process was based on a previous method with modifications[31]. The dried leaf powder was refluxed with 85% ethanol to remove some coloured ingredients and small-molecule impurities. After removing the solvent, the pretreated samples were soaked in distilled water at room temperature for 6h and extracted twice in an ultrasound cleaner at 45 kHz. The extract was collected by filtration and concentrated using a rotary evaporator at 63℃ under vacuum. The concentrated extract was precipitated in four volumes of ethanol overnight at 4℃ and then centrifuged for 15min at 4 500 rpm. The resulting precipitates were washed sequentially with anhydrous ethanol and acetone and dried to afford crude polysaccharides named PUAE. Singlefactor experiments were performed for the ratio of water to raw material, extraction temperature, extraction time, and ultrasound power. One factor was changed while the other factors were held constant in each experiment, and each single-factor experiment was repeated thrice. The polysaccharide yield (%) was calculated using Eq. (1):

$$Y(\%) = \frac{W_1(g)}{W_0(g)} \times 100\% \tag{1}$$

where Y is the polysaccharide yield (%); W_1 is the amount of extracted polysaccharides (g); and W_0 is the dried sample weight (g).

2.4.2 Box-Behnken experimental design

Based on the results of the single-factor experiments (data not presented), a three-level Box-Behnken experimental design (BBD) with three factors was applied to determine the optimal ranges of the variables that significantly affected the extraction efficiency: ultrasound power (X_1: Power), extraction time (X_2: Time) and extraction temperature (X_3: Temperature). The design consisted of 17 experimental points in random order[30,32,33]. The experimental designs of the code and the levels of each factor were presented in Table 1. The significance of the model was evaluated by analysis of variance (ANOVA). The accuracy and general ability of the polynomial model were evaluated based on the determination coefficient (R^2) and adjusted coefficient of determination (R^2_{adj}). Several confirmation experiments were conducted to verify the experimental validity.

Table 1 Box-Behnken experimental results.

No.	X_1: power (W)	X_2: time (h)	X_3: temperature (℃)	Y: yield (%)[a]
1	(−1) 150	(−1) 1	(0) 55	8.012
2	(+1) 250	(−1) 1	(0) 55	8.059
3	(−1) 150	(+1) 3	(0) 55	8.272
4	(+1) 250	(+1) 3	(0) 55	8.501
5	(−1) 150	(0) 2	(−1) 50	7.643

(continued)

No.	X_1: power (W)	X_2: time (h)	X_3: temperature (℃)	Y: yield (%)[a]
6	(+1) 250	(0) 2	(−1) 50	7.713
7	(−1) 150	(0) 2	(+1) 60	7.831
8	(+1) 250	(0) 2	(+1) 60	8.142
9	(0) 200	(−1) 1	(−1) 50	6.594
10	(0) 200	(+1) 3	(−1) 50	8.341
11	(0) 200	(−1) 1	(+1) 60	7.942
12	(0) 200	(+1) 3	(+1) 60	7.873
13	(0) 200	(0) 2	(0) 55	9.459
14	(0) 200	(0) 2	(0) 55	9.361
15	(0) 200	(0) 2	(0) 55	9.653
16	(0) 200	(0) 2	(0) 55	9.304
17	(0) 200	(0) 2	(0) 55	9.252

[a] Mean of triplicate determination.

2.5 Hot water extraction

As a comparable test, hot water extraction (HWE) technology was used to extract polysaccharides from the leaves ofR.aganni-phum under the same conditions (ratio of water to material of 25: 1 w/v, and extraction temperature of 55℃) for 12h.

2.6 In vitro antioxidant activity

2.6.1 DPPH radical scavenging activity assay

The DPPH radical scavenging activity was evaluated using a previously reported method with minor modifications[34].Different concentrations of PUAE or PHWE (2.0mL; 0.02-2.0mg/mL) were incubated with a methanol solution of DPPH (2.0mL; 0.1mmol/L) at 37℃ for 30min in the dark.The absorbance was recorded at 517nm using distilled water as a control and vitamin E (VE) for comparison.The DPPH radical scavenging ability was calculated using Eq. (2):

$$\text{scavenging effect}(\%) = \left(1 - \frac{A_i - A_j}{A_0}\right)100\% \qquad (2)$$

A_0 is the absorbance of the DPPH solution without sample; A_i is the absorbance of the test sample mixed with DPPH; and A_j is the absorbance of the sample without DPPH.

2.6.2 Hydroxyl radical scavenging activity assay

The hydroxyl radical scavenging activities of PUAE and PHWE were determined according to previous work with minor modifications[35,36].Briefly, PUAE and PHWE were dissolved in deionized water at 0.02-2.0mg/mL.The sample solution (0.1mL) was mixed with 0.6mL of reaction buffer (20mM phosphate buffer (pH 7.4), 2.67mM deoxyribose, and 100mM EDTA), 0.05mL of

2.0mM Vc, 0.05mL of 10mM H_2O_2, and 0.2mL of 0.4mM ferric chloride and incubated for 15min at 37℃; then, 1mL of 1% TBA and 1mL of 2% trichloroacetic acid (TCA) were added to terminate the reaction. The mixture was boiled for 15min and then cooled to room temperature. Distilled water was used as a control, and ascorbic acid (V_C) was used for comparison. The absorbance of the mixture was measured at 532nm against a blank. The hydroxyl radical scavenging ability was calculated using Eq. (3):

$$\text{scavenging effect}(\%) = (1 - \frac{A}{A_0}) \times 100\% \tag{3}$$

where A_0 is the absorbance of the solution without the sample and A is the absorbance of the test sample mixed with the reaction solution.

2.6.3 Superoxide radical scavenging activity assay

The superoxide radical scavenging activities of PUAE and PHWE were determined according to a previously reported method with slight modifications[35,37]. Polysaccharides were dissolved in deionized water at 0.02–2.0mg/mL. Reaction mixtures containing 1mL of varying concentrations of sample, 1mL of Tris-HCl (16mM, pH 8.0), 1mL of NADH-2Na (338μM), 1mL of NBT (72μM) and 1mL of PMS (30μM) were incubated at room temperature for 5min, and the absorbance was measured at 560nm. Distilled water was used as a control, and V_C was used for comparison. The superoxide radical scavenging ability was calculated using Eq. (4):

$$\text{scavenging effect}(\%) = (1 - \frac{A}{A_0}) \times 100\% \tag{4}$$

where A_0 is the absorbance of the solution without sample and A is the absorbance of the test sample mixed with the reaction solution.

2.7 Rheological measurement

2.7.1 Sample preparation

Rheological measurements of PUAE and PHWE solutions were performed according to Wei et al[38] with minor modifications. PUAE (1, 5, 10 and 50mg/mL) and PHWE (1mg/mL) were dissolved in deionized water with magnetic stirring for 4h (approximately 130rad/min) at 25℃ and overnight at 4℃. The solutions were allowed to stand at room temperature for 1h before measurements to remove entrapped air and allow equilibration.

2.7.2 Steady-shear flowbehaviour

Steady-shear flow measurements of different concentrations of PUAE solutions (1–50mg/mL) and of PHWE (1mg/mL) were conducted at 25℃ and pH 7.0 using MCR 301 rheometer with PP50. The shear stress and the apparent viscosity were recorded as a function of the shear rate (0.01 to 100 /s). Cross [Eq. (5)], Carreau [Eq. (6)] and power-law [Eq. (7)] models[38-41] were used to fit the experimental data of the solutions at different concentrations:

$$\eta = \eta_\infty + \frac{(\eta_0 - \eta_\infty)}{[1 + (a\dot{\gamma})^d]} \tag{5}$$

$$\eta = \eta_\infty + \frac{(\eta_0 - \eta_\infty)}{[1 + (c\dot{\gamma})]^p} \tag{6}$$

$$\eta = k \times \dot{\gamma}^{n-1} \tag{7}$$

where η, η_0 and η_∞ are the apparent, zero shear and infinite shear rate apparent viscosities (Pa · s), respectively; $\dot{\gamma}$ is the shear rate (1/s); a and c are time constants (s); k is the consistency coefficient (Pa · s^n); d and p are rate indices (dimensionless); and n is the flow behaviour index (dimensionless).

2.7.3 Frequency sweep

Oscillatory dynamic frequency sweeping was used to simultaneously assess the storage modulus (G') and loss modulus (G'') of the viscoelastic properties of PUAE and PHWE solutions at various concentrations[42]. Before the tests, the linear viscoelastic range of the solutions was established using a strain sweep (0.01%–100%) at a constant frequency of 1Hz (data not presented). Then, a dynamic oscillatory frequency sweep of different concentrations of PUAE and PHWE at 25℃ and pH 7.0 was performed using MCR 301 rheometer with CC27 in controlled-strain mode. The angular frequency (ω) range was 500–0.05 rad/s with 0.1% strain amplitude.

2.8 Statistical analyses

All experiments were performed in triplicate, and the extraction data were subjected to analysis of variance for a completely random experimental design. One-way ANOVA and Student's t-test were performed to identify differences among means. Statistical significance was declared at $P <$ 0.05. Theantioxidation results were expressed as the mean±standard deviation (SD), and the rheological measurement data were analysed using RHEOPLUS/32 V3.40, the software provided by MCR301 rheometer and Origin 8.0.

3 RESULTS AND DISCUSSION

3.1 Optimization of the extraction conditions using BBD

The experiments shown in the design matrix inTable 1 were performed. Using multiple regression analysis, Design-Expert 7.0 software generated a second-order polynomial equation to express the relationship between the process variables and response. The final equation in terms of coded factors was

$$Y = 9.4 + 0.082X_1 + 0.3X_2 + 0.19X_3 + 0.044X_1X_2 + 0.06X_1X_3 - 0.45X_2X_3 - 0.53X_1^2 - 0.67X_2^2 - 1.05X_3^2 \tag{8}$$

where X_1 is the ultrasonic power (W); X_2 is the extraction time (min); and X_3 is the extraction temperature (℃).

The statistical significance of the second-order polynomial equation wasanalysed using an F-test, and ANOVA for the response surface quadratic polynomial model was performed using Design-Expert 7.0. As shown in Table 2, the fitness of the quadratic regression model was highly significant ($P<0.0001$)[31,32,43]. The lack of fit indicates the failure of a model to represent data in the experimental domain at points not included in the regression. In the present study, the lack of fit had a small F-value ($F=1.811$) and a large P-value ($P=0.286$), indicating that the lack of fit F-statistic was not significant[34]. Therefore, the model equation was adequate for predicting polysaccharide yield of the PUAE under all tested variable combinations. R^2 represents the goodness of fit of a model; the R^2 for the model (0.979) was close to 1.0, which represents a satisfactory correlation between the observed and predicted values. R^2_{adj} was 0.951, which meant that most of the variation in the polysaccharide yield could be predicted by the model[33,44]. The P-values were used to determine the significance of each coefficient, which in turn may reveal the pattern of interactions between variables. The smaller the value of P implied the more significant the corresponding coefficient. The linear term coefficients (X_2, X_3), quadratic term coefficients (X_1^2, X_2^2, X_3^2) and cross product coefficient (X_2, X_3) were significant ($P<0.05$). Therefore, the extraction temperature and extraction time significantly affected polysaccharide yield of the PUAE.

3.2 Analysis of response surfaces

The 3D response surfaces and 2D contour plots are graphical representations of regression equations. They provide a method to visualize the relationship between the response and experimental level of each variable and the type of interaction between two test variables. The shape of the contour plot, circular or elliptical, indicates whether the mutual interaction between variables is significant. A circular contour plot indicates that the interaction between the corresponding variables is negligible, whereas an elliptical contour plot indicates that the interaction is significant.

The three-dimensional surface plots obtained by plotting the response on the Z-axis against any two variables while maintaining the other variables at their zero level were presented in Fig. 1. The effects of different ultrasound power and extraction time on polysaccharide yield of the PUAE are shown in Fig. 1A and a for an extraction temperature and liquid-solid ratio fixed at 55℃ and 25 : 1, respectively. The polysaccharide yield increased as the ultrasound power increased from 150W to 200W and then decreased as the ultrasound power increased from 200W to 250W.

Fig. 1B and b showed the 3D response surface plot and contour plot for varying ultrasound power and extraction temperature at a fixed extraction time (2h) and liquid-solid ratio (25 : 1). The maxi-mum polysaccharide yield was achieved at an ultrasound power and extraction temperature of approximately 200W and 55℃, respectively. Figs. 1C and 1c showed polysaccharide yield of the PUAE as a function of extraction time and temperature at fixed ultrasound power (200W) and liquid-solid ratio (25 : 1); the polysaccharide yield increased rapidly as the ultrasound treatment time increased from 1hto 2.3h. However, the yield of polysaccharide decreased very slowly as the treatment time increased beyond 2.3h. The polysaccharide yield increased as the extraction temperature increased from 50℃ to 56.5℃ and then decreased slowly as the temperature increased from 56.5℃ to 60℃.

Table 2 Analysis of variance of the experimental results of the BBD.

Variables	SS[a]	DF[b]	MS[c]	F value	P value
Model	10.380	9	1.153	35.741	<0.0001
X_1	0.054	1	0.054	1.682	0.2361
X_2	0.712	1	0.712	21.979	0.0022
X_3	0.281	1	0.281	8.561	0.0221

(continued)

Variables	SS[a]	DF[b]	MS[c]	F value	P value
X_1X_2	7. 884E-003	1	7. 884E-003	0. 243	0. 6363
X_1X_3	0. 014	1	0. 014	0. 442	0. 5270
X_2X_3	0. 832	1	0. 832	25. 614	0. 0015
X_1^2	1. 167	1	1. 167	36. 281	0. 0005
X_2^2	1. 892	1	1. 892	58. 402	0. 0001
X_3^2	4. 641	1	4. 641	143. 638	<0. 0001
Residual	0. 234	7	0. 032		
Lack of fit	0. 131	3	0. 043	1. 811	0. 2857
Pure error	0. 096	4	0. 024		

[a] Sum of square.

[b] Degree of freedom.

[c] Mean square.

3. 3 Verification of the predictive model

The optimal values of the variables were determined using Design-Expert 7. 0 software as follows: extraction temperature, 55. 22℃; liquid-solid ratio, 25 : 1; ultrasound treatment time, 2. 2h; and ultrasound treatment power, 204. 47W. Under the optimal conditions, the maximum predicted polysaccharide yield of the PUAE was 9. 443%. Considering operating convenience, the optimal parameters were as follows: extraction temperature, 55℃; liquid-solid ratio, 25 : 1; ultrasound treatment time, 2. 2h; and ultrasound treatment power, 200W.

To ensure the predicted results were not biased toward practical values, experimental rechecking was performed using the optimal conditions. The mean value of (9. 428±0. 021)% (n=3) obtained from real experiments demonstrated the validity of the BBD experimental model. The good correlation between the experimental and predicted values confirmed that the response model accurately and adequately represented the PUAE extraction. The validation results revealed no significant differences between the experimental and predicted values, suggesting that the response model adequately reflected the expected optimization.

3. 4 Comparison of UAE with HWE

In the test, to evaluate the extraction efficiency of the UAE method that we developed, we compared and investigated the HWE method for extracting polysaccharides. Under similar extraction conditions, UAE (9. 428%) gave a higher yield of polysaccharides from the leaves ofR. aganniphum with a shorter extraction time than HWE (8. 451%), which indicated that the application of UAE had a positive effect on the polysaccharide yield.

3. 5 In vitro antioxidant activity

3. 5. 1 DPPH radical scavenging activity

The DPPH radical is a stable free radical that is widely used to evaluate the free radical scaven-

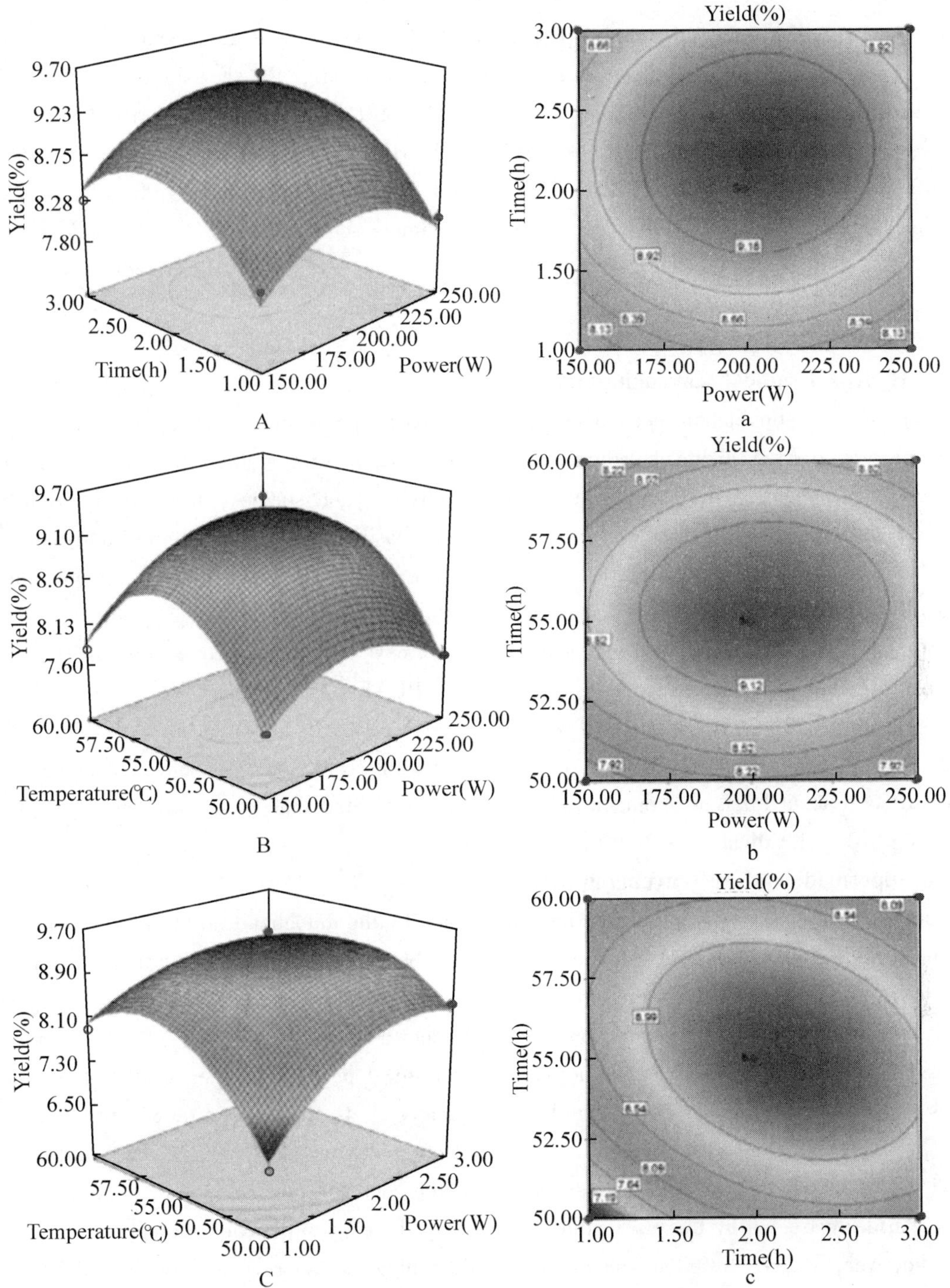

Fig. 1. Three-dimensional response surface and two-dimensional contour plots

A and a: ultrasound power and extraction time; B and b: ultrasound power and extraction temperature; C and c: extraction time and extraction temperature.

ging activity of natural compounds. When a DPPH methanol solution is reduced, the absorbance at 517nm decreases, and the solution changes from purple to light yellow[32,37]. Fig. 2a shows the scavenging activities of PUAE and PHWE on the DPPH radical compared with VE. In the range of 0. 02mg/mL to 2. 0mg/mL, all samples possessed DPPH radical scavenging capacity in a concentration-dependent manner. However, at the same concentrations, PUAE exhibited better DPPH radical scavenging activity than PHWE. At a concentration of 0. 2mg/mL, PUAE, PHWE, and VE scavenged 84. 139%, 52. 671%, and 94. 176% of DPPH radicals, respectively.

In addition, the scavenging effects of PUAE and PHWE all increased significantly with concentration increasing. The EC_{50} (the concentration of sample scavenging 50% of the radical) were 0. 041 and 0. 194mg/mL for PUAE and PHWE, respectively, and the DPPH radical scavenging capacity of VE was 56. 847% at 0. 02mg/mL. These results demonstrated that PUAE possessed moderate DPPH radical scavenging ability.

3. 5. 2 Hydroxyl radical scavenging activity

There are twoantioxidation mechanisms: suppression of the generation of hydroxyl radicals and scavenging of generated hydroxyl radicals[32,33]. In this paper, we studied the latter antioxidant mechanism, scavenging hydroxyl radicals. Among reactive oxygen species, the hydroxyl radical is one of the most potent oxidants. The hydroxyl radical can easily cross cell membranes and react with most biomacromolecules, and hydroxyl radicals can cause severe damage to adjacent biomolecules or cell death[33,34]. The hydroxyl radical scavenging activities of the three samples are shown in Fig. 2b. All samples showed an obvious scavenging effect on hydroxyl radical in a concentration-dependent manner. At 0. 2mg/mL, the scavenging activities of PUAE, PHWE and Vc were 39. 773%, 16. 581% and 61. 283%, respectively. The EC_{50} values of PUAE, PHWE and Vc were 0. 529, 0. 783 and 0. 092mg/mL, respectively, and compared with PHWE, PUAE showed excellent scavenging ability (80. 601%) at 0. 6mg/mL. These results indicated that PUAE was more effective at scavenging hydroxyl radical than PHWE.

3. 5. 3 Superoxide radical scavenging activity

Superoxide radicals, the initial free radicals formed by the mitochondrial electron transport system, must be scavenged because they can degrade to create other radicals, such as hydroxyl, singlet oxygen, and H_2O_2[35-37]. In the present study, superoxide radicals were generated using a PMS/NADH system to assay the reduction of NBT. As shown in Fig. 2c, PUAE, PHWE and V_C all exhibited superoxide radical scavenging activities in a dose-dependent manner at all tested concentrations, and the scavenging effects significantly increased as the concentration increased from 0. 02mg/ mL to 0. 2mg/mL. At 0. 2mg/mL, the scavenging activities of PUAE, PHWE and V_C were 84. 871%, 80. 296% and 96. 033%, respectively. At the concentration range of 0. 2-2mg/mL, the polysaccharides extracted by the two methods all displayed steady superoxide radical scavenging activity; however, PUAE exhibited greater superoxide radical scavenging activity than PHWE. The EC_{50} values of PUAE and PHWE were 0. 072 and 0. 114mg/mL, respectively, and the superoxide radical scavenging capacity of V_C was 53. 828% at 0. 02mg/mL. These results indicated that PUAE had observable superoxide radical scavenging activity.

3. 6 Rheological properties

3. 6. 1 Steady-shear flow properties

The flow curves of the aqueous solutions at different concentrations (1-50mg/mL) wereanal-

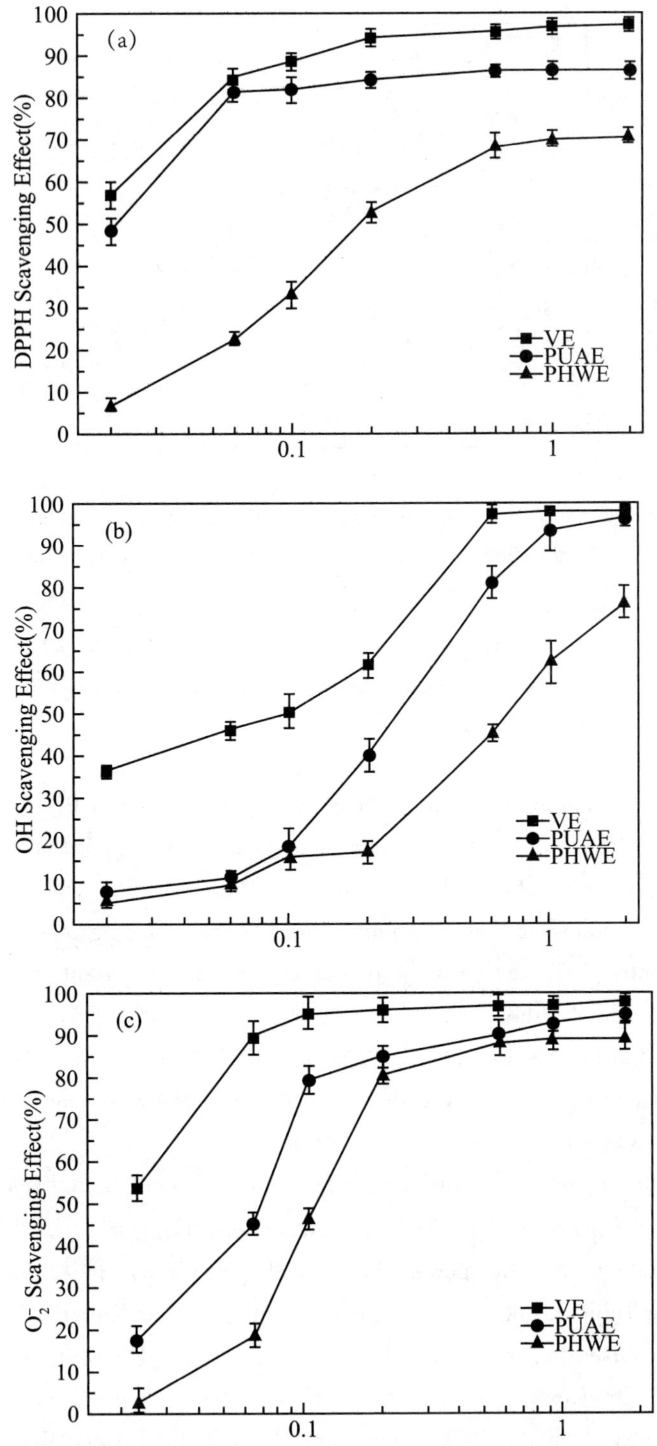

Fig. 2. Activity of PUAE and PHWE at different concentrations

Note: (a) DPPH radical scavenging activity of PUAE, PHWE and VE at different concentrations, (b) hydroxyl radical scavenging activity of PUAE, PHWE and Vc at different concentrations, (c) superoxide radical scavenging activity of PUAE, PHWE and Vc at different concentrations. Data are shown as the mean± SD (n=3).

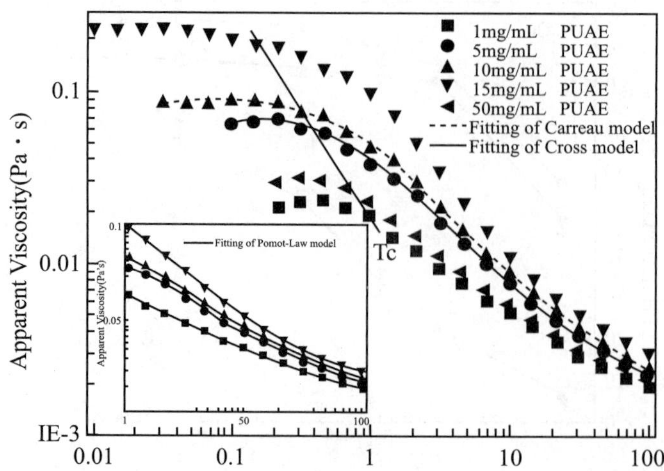

Fig. 3. Dependence of the apparent viscosity on the shear rate at different PUAE (1, 5, 10 and 50mg/mL) and PHWE (1mg/mL) concentrations.

Note: Dotted lines and solid lines represent the Cross model and Carreau model, respectively. Continuous lines represent the critical shear rate (γ_C) obtained by the Carreau model. The inset shows the flow curves of different PUAE concentrations when the shear rate is greater than 1/s and the Power-law model fitting of the apparent viscosity. Steady-shear flow behaviour were carried out at 25℃ and pH 7.0.

ysed using the steady-shear flow test over a shear rate (γ) range of 0.01-100/s. The apparent viscosities (η) of the PUAE and PHWE solutions are shown in Fig. 3. Both a Newtonian plateau and shear-thinning regime were observed in all solutions. The apparent viscosity remained constant in the Newtonian plateau, indicating that re-entanglement of the molecules occurs at the same rate as disentanglement. By contrast, in the shear-thinning region, the apparent viscosity decreased with increasing shear rate due to the decreased rate of entanglement[41,45-47]. Although Newtonian behaviour was observed at lower shear rates, PUAE and PHWE showed non-Newtonian pseudoplastic behaviour. The flow curve of PUAE was similar to that of PHWE at 1mg/mL, but the apparent viscosity of the former was lower at the same shear rate.

The zero shear rate apparent viscosity (η_0) and other parameters of PUAE solutions obtained by fitting the Cross model [Eq. (5)] and Carreau model [Eq. (6)][38,41,48] are shown in Tables 3 and 4, respectively, and the power-law model [Eq. (7)] results at high shear rates ($\geq$1/s) are shown in Table 5. Both regression coefficient (R^2) and relative deviation error (RE) as suitability index were used to assess model which is adequate in describing the rheological characteristics[38]. Compared with Cross model, good agreement between the experimental data and the Carreau model was observed for the PUAE solutions ($R^2 \geq 0.991$), and the fit of the power-law model was better at high shear rates ($R^2 \geq 0.993$) due to the two models had higher value of R^2 and lower value of RE. The η_0 increased from 0.024Pa · s to 0.230Pa · s, and c increased from 0.944s to 4.721s as the concentration increased from 1mg/mL to 50mg/mL.

A double logarithmic plot of g_0 from the Carreau model versus the PUAE concentration is shown in Fig. 4a. The linear relationship between the logarithmic η_0 (Logη_0) and logarithmic concentration

(Log C) (slope of 0.573) indicated that increasing the PUAE concentration increased the intermolecular connections; therefore, $\eta_0 \propto c^{0.573}$ in this study. The time constant (c) values increased in the shear-thinning region, indicating the dependence of the apparent viscosity on the shear rate[38]. The reciprocal of the time con-stant (1/c), which represents the critical shear rate ($\dot{\gamma}_C$), provides an indicator of the initiation shear rate for shear-thinning, that is, $\dot{\gamma}_C = 1/c$[49-51]. $\dot{\gamma}_C$ was concentration-dependent, i.e., solutions with higher concentrations underwent the transition at lower shear rates (Fig. 4b).

Table 3 Cross model parameters at different PUAE concentrations.

RABP (mg/mL)	Cross model fitting parameters					
	a (s)	d	η_0 (Pa.s)	η_∞ (Pa · s)	R^2	RE
1	0.509	0.590	0.024	0.003	0.986	0.0012
5	0.876	0.804	0.071	0.003	0.995	0.0024
10	0.877	0.808	0.088	0.003	0.996	0.0019
50	1.507	0.887	0.239	0.004	0.998	0.0031

Table 4 Carreau model parameters at different PUAE concentrations.

RABP (mg/mL)	Carreau model fitting parameters					
	c (s)	p	η_0 (Pa · s)	η_∞ (Pa · s)	R^2	RE
1	0.944	0.281	0.024	0.002	0.991	0.0005
5	1.794	0.394	0.068	0.003	0.997	0.0014
10	1.893	0.397	0.086	0.003	0.998	0.0014
50	4.721	0.443	0.230	0.005	0.994	0.0040

Table 5 Power-law model parameters for shear rates greater than 1/s at different PUAE concentrations.

RABP (mg/mL)	Power-law model fitting parameters			
	K (Pa · s^n)	n	R^2	RE
1	0.018	0.456	0.996	0.0005
5	0.039	0.329	0.993	0.0018
10	0.048	0.284	0.994	0.0028
50	0.102	0.077	0.997	0.0103

η/η_0 versus $\dot{\gamma}_C$ is plotted at different PUAE and PHWE concentrations in Fig. 5. The dimensionless forms were similar to a singleflow curve, and at very low shear rates, the flow curves approached unity[52]. The results indicated that PUAE and PHWE had similar intermolecular interactions, which was the origin of the similar macroscopic behaviour[53]. Similar behaviour has been re-

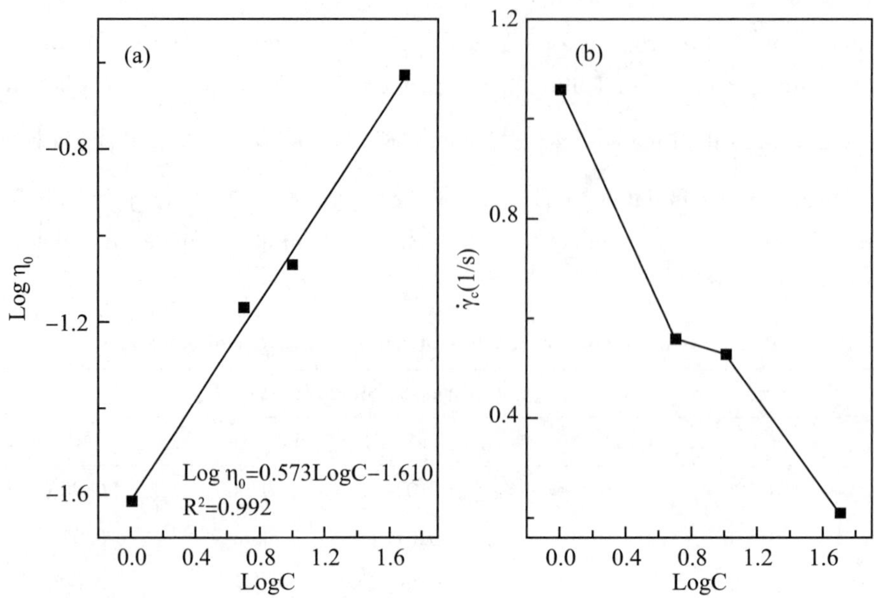

Fig. 4. Logarithmic concentration–dependence of the logarithmic zero shear viscosities (a) and linear critical shear rates (b) for the PUAE solutions.

ported for xanthan gum and polysaccharides from Adansonia digitata leaves[47,54]. The rate index p (0.281<p<0.443), is a measure of the extent of the dependence of the viscosity of the PUAE solution on the shear rate in the shear-thinning region. For Newtonian solutions, p = 0, but p approaches unity with increasing shear thinning[55]. The apparent viscosity was dependent on the concentration and shear rate, and it presented the positive manner to the concentration and the negative effect to the shear rate. These results were consistent with the power–law fitting results at high shear rates (Fig. 3, inset and Table 5), and Non–Newton was stronger due to reduction of n value with the increasing of concentration. K value has a positive relationship with apparent viscosity[56,57]. Table 5 showed that K value was increased, indicating the apparent viscosity was increased with concentration increasing.

3.6.2 Frequency sweep analysis

Fig. 6shows the frequency sweeps that were used to determine G′ and G″ of PUAE and PHWE aqueous solutions (1 to 50mg/mL). The behaviour was liquid–like at low frequencies (G″>G′) and solid–like at high frequencies (G′ >G″), producing a crossover in the modulus traces (G″= G′). These results are consistent with results reported for tara gum, fenugreek gum, locust bean gum, guar gum and the polysaccharides of Ulva lactuca and Adansonia digitata leaves[58,39,47]. The frequency sweeps of PUAE were similar to those of PHWE at 1mg/mL. Compared with PHWE, the PUAE solutions exhibited lower G′ and G″ at lower frequency and similar G′ and G″ at high frequency (Fig. 6, inset).

The G′ and G″ of PUAE solutions increased with increasing concentration and frequency, and the superposition of G′ and G″ was not related to the concentration at high frequency. The crossover points occurred at angular frequencies of 2.549, 2.372, 1.934 and 1.477 rad/s for 1, 5, 10 and

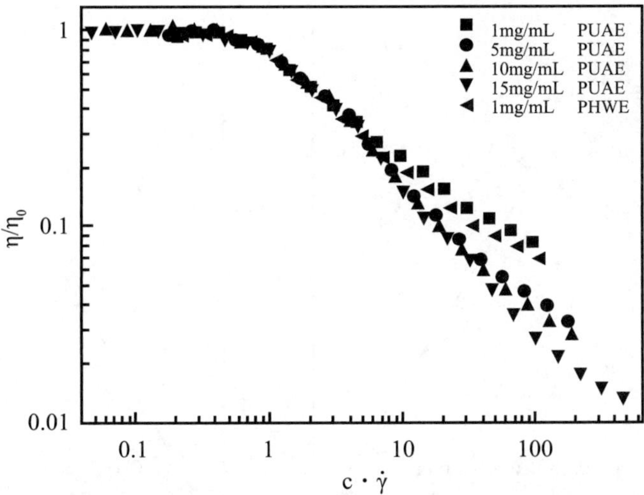

Fig. 5. Generalized flow curves of the dependence of the viscosity on the shear rate of the PUAE and PHWE solutions at different concentrations.

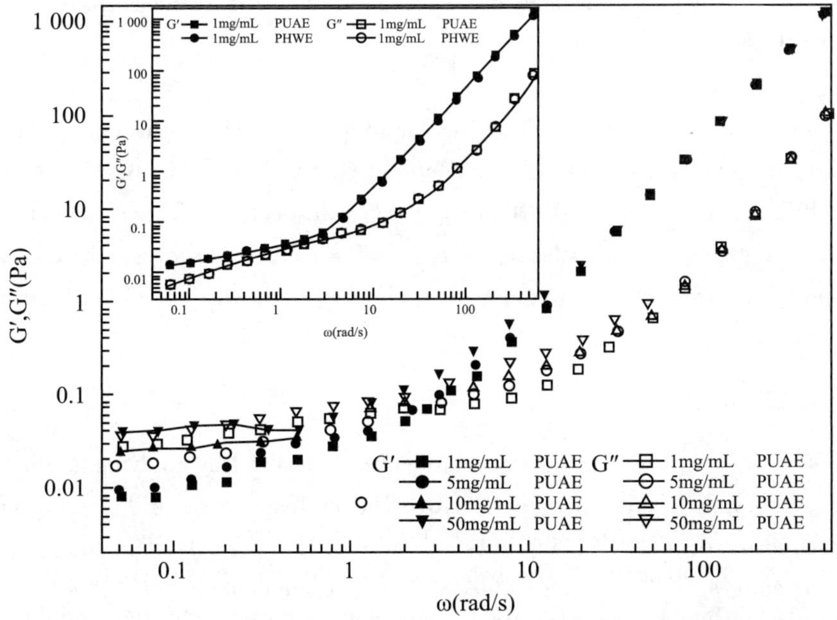

Fig. 6. Frequency sweeps performed at 0.1% strain to determine the storage modulus (G′, filled symbols) and loss modulus (G″, open symbols) of the PUAE solutions (1, 5, 10 and 50mg/mL).

Note: The inset shows the frequency sweeps of the PUAE and PHWE solutions at 1mg/mL. All measurements were made at 25℃.

50mg/mL, respectively, and the modulus ($G' = G''$) increased from 0.065 to 0.088 Pa as the concentration increased from 1 to 50mg/mL. For the PUAE solutions, the influence of the concentration on the crossover point was meaningful, and the frequency of the crossover points decreased with increasing RABP concentration. However, an opposite relationship was observed between the

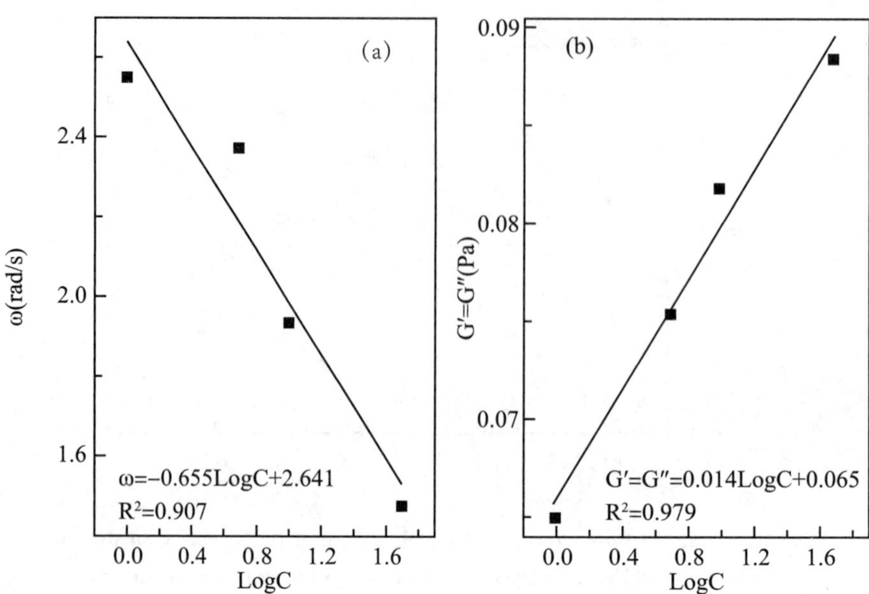

Fig.7. Dependence of the angular frequency (ω) (a) and the storage modulus (G′) and loss modulus (G″) (b) on the logarithmic concentration of the PUAE solutions when G′ = G″ in the frequency sweeps.

modulus (G′ = G″) and concentration. The angular frequency x and logarithm of the concentration were nearly linearly related (Fig. 7a). The modulus (G′ = G″) was identical (Fig. 7b), and the regression coefficients (R^2) were 0.907 and 0.979, respectively. The crossover of G′ and G″ values provides insights on the viscoelastic behaviour of a material. However, the crossover and variation of G′ and G″ occurred at high frequencies (Fig. 6), indicating weak gel behaviour of PUAE at these concentrations.

4 CONCLUSION

In this study, the extraction of polysaccharides was optimized by applying an ultrasound extraction technique according to an experimental BBD. The optimal conditions of ultrasound extraction were an ultrasound power of 200W, liquid-solid ratio of 25 : 1, extraction time of 2.2h, and an extraction temperature of 55℃, which yielded a maximum percentage (9.428%) higher than that of hot water extraction (8.451%). In addition, PUAE exhibited good radical scavenging activity for DPPH, superoxide and hydroxyl radicals in vitro, and ultrasound-assisted extraction methods affected the apparent viscosity and dynamic viscoelasticity. These results indicate that ultrasound-assisted extraction methods could be used to extract polysaccharides from Rhododendron aganniphum, and this paper also provided a theoretical basis for the application of PUAE in medicine and healthcare products.

ACKNOWLEDGEMENTS

This work was financed by the National Science and Technology Infrastructure Program

(2015BAD11B01) and the National Beef Cattle and Yak Industry Technology System of Modern Agriculture Construction Special Fund (CARS-38).

REFERENCES OMITTED

（发表于《Ultrasonics Sonochemistry》，院选 SCI，IF：4.556）

A Rapid and Simple HPLC-FLD Screening Method with QuEChERS as the Sample Treatment for the Simultaneous Monitoring of Nine Bisphenols in Milk

Lin XIONG[1,2*], Ping YAN[1,2,3], Min CHU[1,3], Ya-Qin GAO[1,2], Wei-Hong LI[1,2], Xiao-Ling YANG[1,2]

(1. Lanzhou Institute of Husbandry and Pharmaceutical Sciences, Chinese Academy of Agricultural Sciences, Lanzhou 730050 Gansu, China; 2. Laboratory of Quality & Safety Risk Assessment for Livestock Product (Lanzhou), Ministry of Agriculture, Lanzhou 730050 Gansu, China; 3. Key Laboratory for Yak Genetics, Breeding & Reproduction Engineering of Gansu Province, Lanzhou 730050 Gansu, China)

ARTICLEINFO: Chemical compounds studied in this article:

Bisphenol A (PubChem CID: 6623)

Bisphenol B (PubChem CID: 66166)

Bisphenol C (PubChem CID: 6620)

Bisphenol E (PubChem CID: 608116)

Bisphenol F Diglycidyl Ether (PubChem CID: 91511)

Bisphenol F bis (2, 3-dihydroxyprophy) Ether (PubChem CID: 3928015)

Bisphenol F Glycidyl 2, 3-Dihydroxypropyl Ether (PubChem CID: 71360307)

Bisphenol A Bis (2, 3-dihydroxyprophy) Ether (PubChem CID: 110678)

Bisphenol A diglycidyl ether (PubChem CID: 2286)

ABSTRACT: A specific, precise and accurate high performance liquid chromatographic (HPLC) analytical method with a fluorescence detector (FLD) was established for the simultaneous determination of nine bisphenols (BPs) in milk samples.Samples were extracted ultrasonically with acetonitrile and cleaned using the QuEChERS technique. Under the optimized conditions, good linearity was obtained for the nine BPs and the correlation coefficients (R^2) ranged from 0. 9942 to 0. 9997. Recovery values for the nine bisphenols in spiked samples were 75. 82-93. 86% with intra-day and inter-day relative standard deviations (RSDs) from 2. 6 to 11. 1%.The limits of detection (LODs) and quantification (LOQs) were 1. 0-3. 1μg/kg and 3. 5-9. 8μg/kg, respectively.The results demonstrated clearly that the approach developed

* Corresponding author.

provides reliable, simple, rapid and environmentally-friendly quantification and identification of nine bisphenols in a fatty matrix and could be used for monitoring bisphenols in milk.
Key words: Bisphenols; QuEChERS; HPLC-FLD; Milk

1 INTRODUCTION

Many compounds that are introduced into the environment by human activity can disrupt the endocrine system of higher life forms. The consequences of such disruption can be profound because of the crucial role that hormones play in controlling development. Since hormones affect reproduction and cellular development, and probably alter the risk of carcinogenesis, chronic environmental exposure may have a major impact on health. Bisphenol A (BPA), a high-volume chemical with 2.7 billion kilograms produced annually worldwide (Park et al., 2016), is used to make polycarbonate plastic and epoxy resins for linings in cans and pipes (Staniszewska, Nehring, & Mudrak-Cegiołka, 2016). It can be found, for example, in baby bottles, plastic drinking bottles, microwaveable food products, canned drinks and foods (Zimmers et al., 2014).

Milk is a nourishing food that supplies protein. During milking, transportation, production and drinking, milk inevitably comes into contact with cans, bottles and packing boxes (Niu, Zhang, Duan, Wu, & Shao, 2015). If these containers contain BPA, this hazardous substance can infiltrate the milk.

Environmental pollution is a serious environmental contamination problem and BPA can be found in water and soil from where it can migrate into milk via the food cycle. In more serious cases, BPA can lead to abnormalities of the human reproductive system (Mei, Deng, et al., 2013; Wang, Chen, & Bornehage, 2016). Studies have shown that BPA can decrease sperm count and fertility in males, and increase the rates of breast cancer as well as other hormone-linked diseases in women (Khalil et al., 2014).

Facing growing restrictions on the use of BPA in food contact materials, the plastic and canning industries are seeking alternatives that can replace BPA (Heffernan et al., 2016; Sun et al., 2014). Thus, over the past few years, new compounds have been designed to resemble the physicochemical properties of BPA, many of which belong to the same chemical family as p, p-bisphenols (Rocha et al., 2016). Among these structural analogues, bisphenol B (BPB), bisphenol C (BPC) and bisphenol E (BPE), bisphenol F diglycidyl ether (para-para-BFDGE), bisphenol F bis (2, 3-dihydroxyprophy) ether (BFDGE · $2H_2O$), bisphenol F glycidyl 2, 3-dihydroxypropyl ether (BFGDGE · H_2O), bisphenol A bis (2, 3-dihydroxyprophy) ether (BADGE · $2H_2O$) and bisphenol A diglycidyl ether (BADGE) are, apparently, the most important (Fattore, Russo, Barbato, Grumetto, & Albrizio, 2015). Unfortunately, these analogues have potentially harmful toxicological profiles (Schmidt, Kotnik, Trontelj, Knez, & Mašič, 2013; Švajger, Dolenc, & Jeras, 2016).

In the US, Food and Drug Administration (FDA) regulations no longer allow of BPA-based polycarbonate resins in baby bottles. The US Environmental Protection Agency (EPA) has established a BPA reference dose (RfD) of 50μg/kg/d, based on the lowest observable adverse effect level (LOAEL) (Vandenberg, Maffini, Sonnenschein, Rubin, & Soto, 2009). To reduce the risk of possible health problems associated with BPA, the tolerable daily intake for BPA was established as 10μg/kg body weight per day by the European Commission's Scientific Committee on Food (SCF) (European Commission., 2002). The European Commission (EC) established a specific migration limit (SML) for BPA of 0.6μg/g of food or food simulants in Commission European Journal of Communication, 2004/19/EC (Commission European Journal of Communication, 2004). The use of BFDGE has been forbidden (European European European Commission, 2005), and the SMLs fixed by the European Commission are 9.0mg/kg for BADGE and its hydroxyl derivatives and 1.0mg/kg for BADGE and its chlorinated derivatives (European Commission, 2005; European European European Commission, 2011).

Current detection methods forBPs are reliant on complex instruments. For example, gas chromatography-mass spectrometry (GC-MS) (Deceuninck et al., 2014), HPLC (Liu et al., 2013; Wu et al., 2014) and liquid chromatography-mass spectrometry (LC-MS) (Vitku et al., 2015). However, major drawback of LC-MS methodology is the charge competition observed in electrospray ionization, leading to questionable results especially where complex biological matrices (such as foodstuffs) are concerned (Gallart-Ayala, Moyano, & Galceran, 2007). GC-MS samples require some derivatization pretreatment prior to chromatographic separation (Deceuninck et al., 2014). More importantly, for non-specialized and primitive laboratories in developing countries, the price of GC-MS and LC-MS instruments is prohibitive. Expensive instrumentation, time-consuming sample pre-treatment and expert personnel for operation limit the practical application of these methods (Wang, Zeng, Wei, & Lin, 2006).

Molecular imprinting (MI) (Wu et al., 2015), enzyme linked immune sorbent assay (ELISA) (Mita et al., 2007), electrochemical analysis method (Xue et al., 2013) and biosensor methods (Mei, Qu, et al., 2013) have also been used for rapid detection of BPs. These methods avoid the drawbacks of instrument-based methods, and are highly sensitive and convenient, but only few BPs like BPA can be tested and multifarious BPs cannot be tested simultaneously. Apart from these methods, HPLC-FLD has great potential for simultaneous determination of multiple BPs due to its simple operation and low cost.

Anastassiades, Lehotay, Stajnbaher, and Schenck (2003) developed an analytical methodology for the simultaneous extraction and isolation of pesticides from food matrices. Analytical methodologies characterized as: quick, easy, cheap, effective, rugged and safe (QuEChERS) have undergone various modifications and enhancements since the introduction of this concept. As a "green" analytical approach (Capela, Homemn, Alves, & Santos, 2016), QuEChERS is used widely to determine pesticide due to its simplicity, low cost and relatively high efficiency (Pérez-Burgos et al., 2012), and has been applied for testing hazardous substance in foods of animal origin recently (Huertas-Pérez et al., 2016; Liu, Lin, & Fuh, 2016). However, as far as we know, there has been no report on the determination of BPs in milk using the QuEChERS method with pre-concentration for sample treatment.

The purpose of this study was to develop a rapid and simple method for the simultaneous determination of nine BPs (BPA, BPB, BPC, BPE, para-para-BFDGE, BFDGE · $2H_2O$,

BFGDGE · H_2O, BADGE · $2H_2O$ and BADGE) in milk. In the method developed, samples were prepared using the QuEChERS method and analysed with HPLC-FLD. The economic benefits and wide use of HPLC-FLD were combined with the rapid, convenient and environmentally-friendly QuEChERS method for sample pre-treatment. This method could be applied for routine laboratory analysis of nine BPs in large numbers of milk samples.

2 MATERIALS AND METHODS

2.1 Chemicals, reagents and solution

Analytical standards with high purity were used in the experiment. BPA (Purity 98.5%) and BPB (Purity 99.8%) were provided by Dr. Ehrenstorfer GmbH (Augsburg, Germany); BFDGE · $2H_2O$ (Purity 96%), para-para-BFDGE (Purity 90%), BFGDGE · H_2O (Purity 93.98%), BADGE · $2H_2O$ (Purity 93%) and BADGE (Purity 98%) were provided by Toronto Research Chemical, Inc. (Toronto, Canada); BPC (Purity 98.9%) and BPE (Purity 100%) were provided by AccuStand, Inc. (New Haven, USA). Single standard solutions were obtained by dissolving 10mg of each standard in methanol (HPLC grade) in a 100mL volumetric flask, and 1μg/mL mixed stock standard solutions of nine BPs were prepared by the diluting single standard in mobile phase. This was stored at -20℃ in refrigerator for up to 6months. The standard curve was based on six serially-diluted concentrations from the 1μg/mL mixed stock standard solutions. HPLC grade methanol and acetonitrile were obtained from Tedia Company, Inc. (Fairfield, USA). Formic acid was analytical grade and obtained from Sigma-Aldrich (Santa Clara, USA). Primary secondary amine (PSA) (40-63μm, ultraclean bulk for QuEChERS) and C_{18} (40-63μm, ultraclean bulk for QuEChERS) were obtained from CNW Technologies GmbH (Duesseldorf, Germany). All other chemicals were of analytical grade, unless stated otherwise, and purchased from Sinopharm (Shanghai, China); they were used without further purification.

2.2 Instrumentation and apparatus

Separation was performed using a Waters e2695 separation module with fluorescence detector (Milford, USA) and a Waters Xcharge C_{18} column (250mm×4.6mm, 5μm, 100Å) (Milford, USA). An electronic analytical balance BSA224S-CW accurate to 0.1mg (Gottingen, Germany), homogenizer BÜCHI Mixer B-400 (Flawil, Switzerland), vortex Mixer Labinco L24 (Breda, Netherlands), ultrasonic apparatus Scientz Biotechnology (Ningbo, China), shaker ZP-200 (Taicang, China), membrane filters 0.22μm Millipore Millex-GV (Massachusetts, USA), centrifugal tube Coring Centristar™ (Coring, USA), Milli-Qultrapure water system Millipore (Molsheim, France), centrifuge Omnifuge 2.ORS (Osterode, Germany) and nitrogen evaporators Peak Scientific N100DR (Glasgow, UK) were used for pretreatment.

2.3 Sample collection and preparation

Each milk sample (5±0.02g) obtained from farms in Lanzhou was placed in a polypropylene centrifuge tube (15mL). Formic acid (0.1%) acetonitrile (5mL) solution was added to the milk and mixture ultra-sonicated for 10min. One gram of anhydrous magnesium sulphate and 2.0g sodium chloride were added, and the mixture was immediately shaken for 5min, followed by centrifugation at 10, 000 rpm for 5min at 10℃.Then, the supernatant was transferred to a new tube containing 0. 1mg PSA, 0. 1mg C_{18} and 0. 25mg $MgSO_4$, vortexed for approximately 30s, and centrifuged at 10, 000 rpm for 5min at 10℃.The liquid was transferred to another tube, evaporated to dryness under nitrogen at 50℃ and reconstituted with 0. 1% formic acid/acetonitrile (55 : 45, v/v) solution. Finally, the solution was shaken and passed into an LC vial through a PTFE Millipore filter.

2. 4 LC operating conditions

The mobile phase, which was degassed and filtered before analysis, consisted of 0. 1% formic acid (A) and acetonitrile (B).The eluent flow rate was 1. 0mL/min with the mobile phase initially consisting of 55% A and 45% B, then decreased linearly to 45% A from 0. 0min to 7. 0min, held there for 17. 0min from 7. 0min to 24. 0min, and finally recovered to 55% A and held for 4. 0min. Fluorescence measurements were carried out at 215-245nm excitation wavelength and 295-315nm emission wavelength.The injection volume was 20μL and the temperature of the column oven was maintained at 35℃.

3 RESULTS AND DISCUSSION

3. 1 Optimization of developed method

3. 1. 1 Optimization of HPLC parameters

The correlation between composition of the mobile phase and resolution of the chromatogram was addressed.Parameters A using gradient elution and Parameters B (0. 1% formic acid: acetonitrile, 55 : 45, v/v) were set and compared.Chromatograms of samples spiked with mixed standard solutions of nine BPs at 50μg/kg were contrasted.Parameters A exhibited a high degree of separation and higher selectivity for analysing BPs and all nine were separated from baseline and exhibited nice peak shapes.With Parameters B, all baseline resolved peaks were obtained, but three peaks after 14. 0min were very smooth, which led to low sensitivity for BPC, para-para-BFDGE and BADGE. So, parameters A were selected for subsequent work.

3. 1. 2 Optimization of excitation wavelength and emission wavelength

The excitation and emission wavelengths were two factors decisive among the instrument parameters for sensitivity, so they were optimized in single factor experiments.Two experiments to addressing the relationship between peak height, and excitation and emission wave-lengths were designed. A standard mixture of the nine BPs (50μg/L) was filtered directly into a vial and analysed using different excitation and emission wavelengths. Firstly, the excitation wavelength was set at: 215, 220, 225, 230, 235, 240 or 245nm.Peak heights for the nine BPs were obtained and the chromatograms are shown in Fig. 1. The range of peak heights for the BPs was 4. 98 - 14. 69, 7. 77 - 24. 80, 10. 99-41. 04, 12. 28-52. 52, 10. 47-51. 96, 7. 22-38. 07 and (4. 55-23. 63) $\times 10^5$ mV.When the excitation wavelength was at 230nm, all nine chromatograms had the maximum peak height [(12. 28-52. 52) $\times 10^5$mV], and maximum sensitivity could be obtained.So, 230nm was chosen as optimal for excitation.

With the excitation wavelength at 230nm, the emission wavelength was set at 295, 300, 305, 310 or 315nm. The range of peak heights for the nine BPs was 8. 01 - 32. 41, 9. 92 -

39. 12, 10. 99−41. 04, 9. 67−38. 20 and (8. 22−31. 51) $\times10^5$mV. When the emission wavelength was 305nm, peak heights for the BPs were greatest [(10. 99−41. 04) $\times10^5$mV] and this wavelength (305nm) was selected for subsequent as optimal for emission. At these wavelengths, the range of peak heights for the nine BPs was at a maximum [(12. 73−56. 02) $\times10^5$ mV] and the best sensitivity was obtained.

3. 1. 3 Optimization of dosage of cleaning agent C_{18} and PSA

C_{18} can remove fat, pigments and vitamins in milk while PSA can remove the carbohydrate and organic acids. Amount of C_{18} and PSA in QuEChERS have a key role in the result, so these were optimized. Conditions for the experiment were kept the same, but the amounts of C_{18} and PSA (1 : 1, m: m) were varied (0. 025, 0. 05, 0. 1, 0. 2 and 0. 4g) before recovery from five spiked samples was determined. The results showed that 70. 1%−82. 4% recovery was achieved for the nine BPs at 0. 1g. When the amounts of C_{18} and PSA were at 0. 2g or 0. 4g, recovery was lower (66. 1%−80. 4%) because superfluous cleaning agent C_{18} and PSA can absorb some target compounds. When the amounts of C_{18} and PSA were less at 0. 1g, recovery was 65. 2%−80. 1% and did not exhibit any significant variation, but there were more peaks, which had a negative effect on sensitivity and accuracy. So, the amounts of C_{18} and PSA used in the final optimised method were both 0. 1g, which was best from an economic point−of−view as well as providing optimal efficiency.

3. 1. 4 Optimization of ultrasound extraction time

Methods for BP extraction include ultrasonic−assisted extraction (UAE), liquid−liquid extraction (LLE) (Yang, Guan, Yin, Shao, & Li, 2014) and liquid−solid extraction (LSE) (Geens, Neels, & Covaci, 2009; Schmidt, Müller, & Göen, 2013). In contrast, LLE and LSE need plenty of organic solution and take more time, so ultrasonic−assisted extraction (UAE) is the most popular due to its high performance. The dependence of accuracy on recovery for the nine BPs on extraction time was examined at 5, 10, 15, 20 and 25min as well as shaking (15min or 30min) (Fig. 2). When the extraction time was 5min, incomplete extraction led to minimum recovery (70. 1%−81. 2%) due to insufficient time. When the extraction time was 10min, recovery increased to 75. 7%−83. 6%. When the extraction time was 15min, recovery was 75. 7%−84. 1%, and the increase in accuracy was not remarkable. When the extraction time was increased to 20 and 25min, recovery was less (65. 6% − 82. 0%). Shaking for 15 or 30min produced recovery of 65. 2%−81. 2%. Compared with ultrasound extraction, traditional shaking was inefficient and time−consuming. Thus, 10min of ultrasound extraction was used in the optimised method.

3. 2 Purifying effect of QuEChERS method

The effect of C_{18} and PSA on impurities was assessed in three−groups of contrast experiment (Fig. 3): (A) blank milk sample prepared using QuEChERS, (B) blank milk sample not cleaned and (C) 10μg/kg spiked milk sample prepared using QuEChERS. Both the blank and spiked samples cleaned using QuEChERS contained fewer impurities; their chromatograms were smooth and concise, and no interfering peaks were observed at the retention times for the nine BPs.

3.3 The advantages of QuEChERS method compared with other methods

Purification of BPs is based on reversed SPE, solid-phase micro-extraction (SPME) and dispersive liquid-liquid microextraction (DLLE) (Cunha & Fernandes, 2013). These procedures are time-consuming, expensive and laborious, and often include many processing steps. SPE is the most common technique (Regueiro & Wenzl, 2015). Although traditional SPE sorbents have high capacity and can trap a wide range of analytes, they suffer from low efficiency due to their low selectivity toward specific target molecules (Sun et al., 2014). There is a demand for rapider, simpler and cheaper methods for pretreatment of complex matrices, such as milk, and QuEChERS meets this (Halle, Claparols, Garrigues, Franceschi-Messant, & Perez, 2015). The QuEChERS method developed was compared with the traditional method (Table 1). Only 5mL acetonitrile was used in the QuEChERS method developed, minimizing the volume of organic solvent used and developing a more environmentally-friendly approach. The QuEChERS method only needed 25min to prepare samples, and eight steps, the least of any alternative, meaning QuEChERS would reduce milk sample pre-treatment and minimize sources of error. Thus, the QuEChERS method developed was relatively quick, simple and environmentally-friendly.

3.4 Method validation

Quality parameters-linearity, accuracy, limit of detection (LOD), limit of quantification (LOQ), precision (repeatability and intermediate precision), and selectivity were evaluated. Bisphenol-free milk samples were used as blanks to verify method selectivity and, after spiking, confirm the assignment of peak identities and calculate linearity, precision and accuracy of the method.

3.4.1 Linearity

Specificity was tested byanalysing a spiked-milk sample. Linearity of the FLD chromatographic response was tested with six calibration points across the range 5-100μg/kg. The linear regression coefficients for the calibration curves are summarized in Table 2, which shows good results were achieved (R^2 0.9942-0.9997).

3.4.2 LODs and LOQs

Method sensitivity was evaluated based on limits of detection (LODs) and quantification (LOQs). LODs and LOQs were calculated

from the signal-to-noise ratio of chromatograms for milk samples (S/N = 3 for LODs and S/N = 10 for LOQs), as shown in Table 2. LODs for the nine BPs in milk were in the range 1.0-3.1μg/kg and LOQs were in the range 3.5-9.8μg/kg. LOQs for the method developed were less 0.6μg/g, therefore, according to the SML for BPA (0.6μg/g, Commission Directive 2004/19/EC) for food or food simulant and the SML (9.0mg/kg of food) for BADGE and its hydroxyl derivatives, adequate for routine monitoring of nine BPs in milk samples.

3.4.3 Recovery

To estimate of recovery, blank milk samples were fortified with the mixed standard solution at three concentrations (10, 50 and 100μg/kg) in intra-day (n = 18) and inter-day experiments (n=6). 18 spiked samples were analysed in the intra-day experiment (same day), and every day

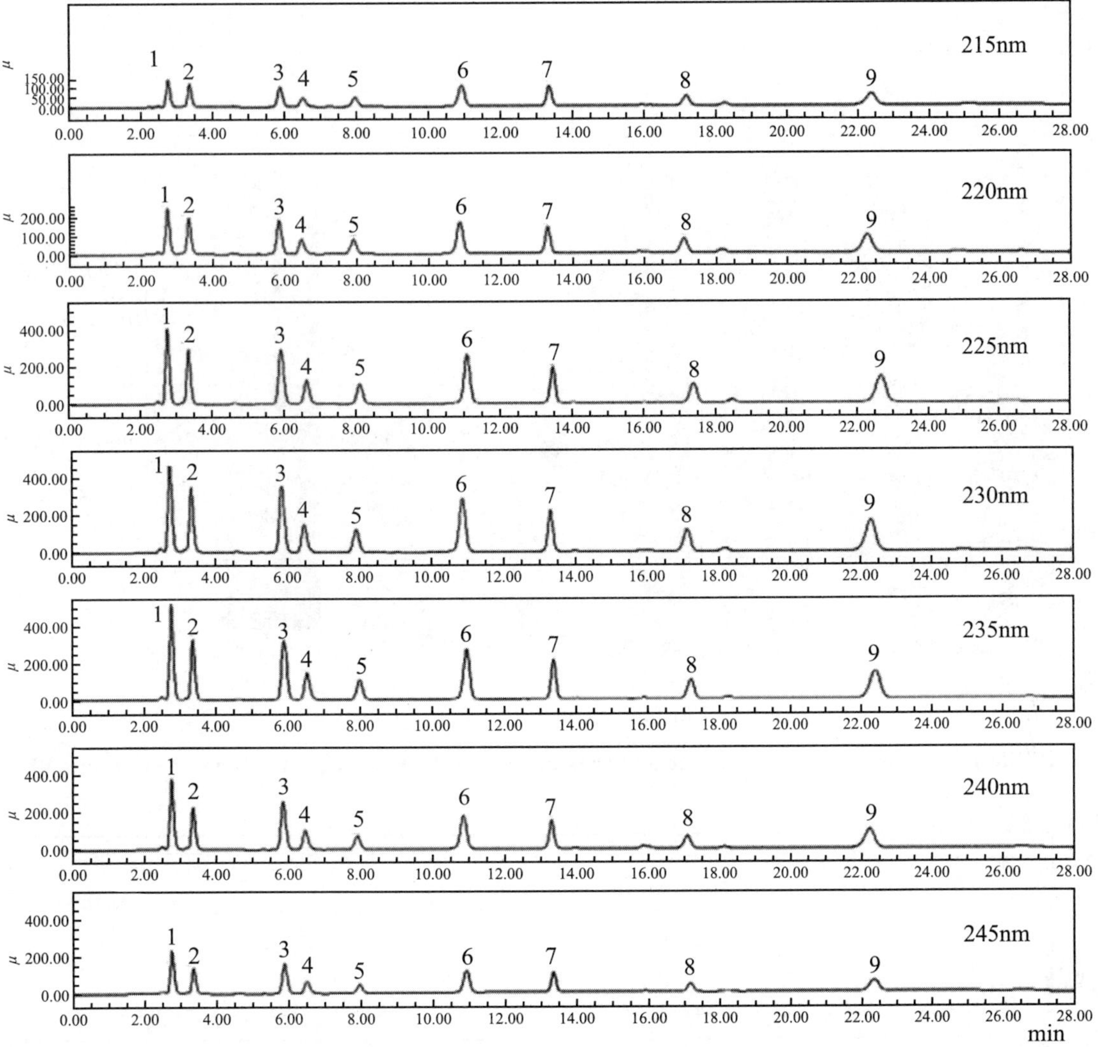

Fig. 1. Comparison of 7 chromatograms at 215–245nm excitation wavelength.

(1) BFDGE · $2H_2O$, (2) BADGE · $2H_2O$, (3) BFGDGE · H_2O, (4) BPE, (5) BPA, (6) BPB, (7) BPC, (8) para-para-BFDGE and (9) BADGE.

six spiked samples were analysed for three consecutive days.High recoveries were obtained using this simple method, based on QuEChERS extraction with C_{18} and PAS.The results indicated that recoveries for the nine BPs in milk were 75. 82%–93. 86% in the inter-day experiment and 77. 60%–89. 68% in intra-day experiment (Table 3).The results indicated that nine PBs in milk could be quantified using the method developed.

3. 4. 4 Precision

The precision, expressed as relative standard deviations (RSDs), was determined in the intra-and inter-day experiments.RSDs for the nine BPs in milk are shown in Table 3. The inter-and intra-day RSDs for the nine BPs in milk ranged from 2. 6% to 13. 0% and from 5. 5% to 11. 1%, respectively, at three concentrations (10, 50, and 100μg/kg).

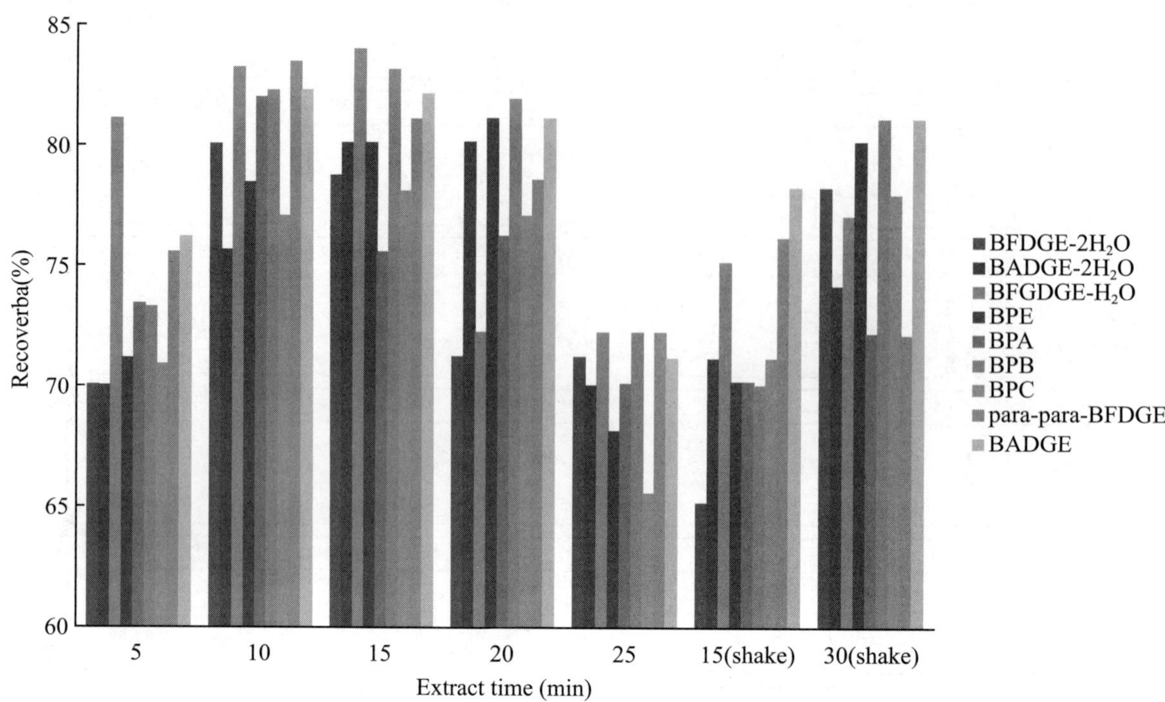

Fig. 2. Effect of ultrasound extraction time and shaking extraction time on the recovery rates of nine BPs.

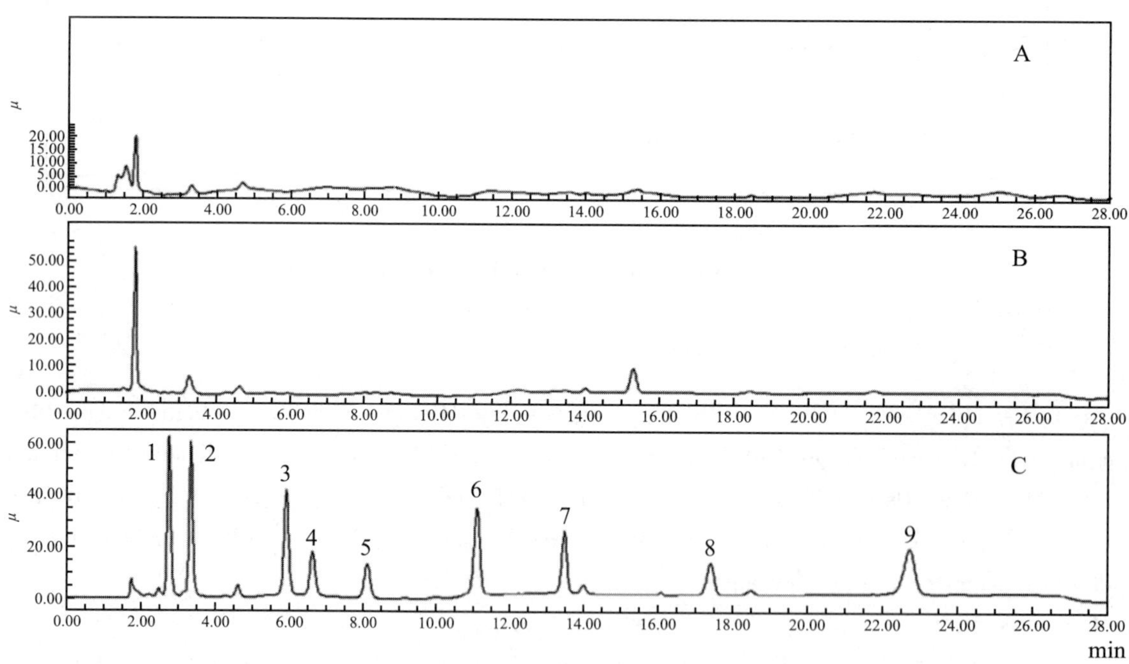

Fig. 3. Purifying effect of QuEChERS method on BPs in milk samples.

(1) BFDGE · $2H_2O$, (2) BADGE · $2H_2O$, (3) BFGDGE · H_2O, (4) BPE, (5) BPA, (6) BPB, (7) BPC, (8) para-para-BFDGE and (9) BADGE.

Table 1 The QuEChERS developed compared with the traditional methods.

Pretreatment method (matrix)	Extract organic solvent	Volume (mL)	Elapsed time (min)	Number of processes
LLE-SPE (Milk) (Zimmers et al., 2014)	hexane/acetonitrile	3+12	about 60	18
UAE-LLE-SPE (Juice) (Regueiro and Wenzl, 2015)	acetonitrile/methanol	5	about 45	9
LLE/LSE-SPE (milk etc.) (Deceuninck et al., 2014)	acetonitrile/cyclohexane	14+15	about 13	12
LLE-SPE (Milk) (Park et al., 2016)	methanol/hexane/LMT/acetonitrile	1. 5+4+2+3	about 60	11
LLE-DLLME (edible oils) (Liu etal., 2013)	n-hexane/methanol/	4+2	about 30	12
UAE-LLE-QuEChERS (Milk) (QuEChERS)	acetonitrile	5	about 25	8

Table 2 Linear range, regression coefficient (R^2), retention time, LOD and LOQ for BPs.

Compound	Retention time (min)	R^2	Linear range (μg/kg)	LOD (μg/kg)	LOQ (μg/kg)
BFDGE · $2H_2O$	2. 742	0. 9991	5-100	1. 0	3. 5
BADGE · $2H_2O$	3. 344	0. 9942	5-100	1. 3	3. 8
BFGDGE · H_2O	5. 892	0. 9979	5-100	1. 2	4. 1
BPE	6. 539	0. 9972	10-100	3. 0	9. 2
BPA	8. 006	0. 9992	10-100	3. 1	9. 8
BPB	10. 968	0. 9997	5-100	1. 5	5. 0
BPC	13. 367	0. 9959	10-100	1. 8	6. 2
para-para-BFDGE	17. 217	0. 9988	10-100	2. 5	8. 4
BADGE	22. 444	0. 9987	10-100	2. 3	7. 8

Table 3 Recovery rates and RSDs for nine BPs from milk samples obtained using HPLC-FLD.

Compound	Spiked level (mg/kg)	Intra-day Recovery (%) /RSDs (n=6)			Inter-day Recovery (%) /RSD (n=18)
		1 day	2 days	3 days	
BPA	10	79. 34 8. 6	78. 34 7. 5	79. 21 7. 3	77. 60 7. 0
	50	87. 05 8. 6	87. 91 11. 2	76. 26 10. 5	86. 80 9. 5
	100	84. 43 7. 6	81. 55 8. 2	81. 94 6. 3	86. 97 9. 6

(continued)

Compound	Spiked level (mg/kg)	Intra-day Recovery (%) /RSDs (n=6) 1 day	2 days	3 days	Inter-day Recovery (%) /RSD (n=18)
BPB	10	85. 55 11. 6	79. 05 7. 1	82. 33 5. 2	88. 19 10. 5
	50	82. 34 8. 0	90. 89 6. 9	79. 53 7. 0	85. 77 10. 7
	100	82. 74 8. 4	83. 44 2. 9	88. 34 2. 6	89. 68 10. 4
BPC	10	82. 34 9. 9	83. 11 5. 9	86. 68 5. 5	84. 91 9. 2
	50	76. 03 5. 6	89. 24 9. 0	80. 46 9. 3	86. 22 10. 0
	100	89. 28 7. 2	90. 13 3. 8	81. 12 10. 8	87. 32 9. 3
BPE	10	84. 91 9. 5	85. 49 6. 3	82. 80 5. 7	85. 49 11. 1
	50	80. 06 4. 9	90. 49 8. 4	83. 06 6. 1	87. 11 10. 2
	100	80. 21 4. 8	84. 63 7. 8	79. 85 7. 5	87. 69 11. 1
para-para-BFDGE	10	85. 99 10. 7	79. 82 5. 5	79. 83 9. 3	84. 59 10. 9
	50	85. 93 5. 51	93. 86 9. 4	75. 82 3. 6	86. 26 7. 2
	100	82. 32 4. 1	86. 49 7. 0	83. 83 8. 9	89. 39 8. 8
BFDGE-$2H_2O$	10	81. 87 9. 3	86. 96 5. 3	81. 30 10. 0	84. 59 10. 4
	50	82. 32 9. 2	89. 85 7. 8	85. 00 8. 1	86. 60 10. 4
	100	86. 79 5. 7	84. 62 13. 0	79. 96 6. 9	87. 79 9. 8

(continued)

Compound	Spiked level (mg/kg)	Intra-day Recovery (%) /RSDs (n=6)			Inter-day
		1 day	2 days	3 days	Recovery (%) /RSD (n=18)
$BFGDGEH_2O$	10	81. 25 8. 5	84. 28 6. 8	85. 74 8. 1	89. 31 9. 3
	50	82. 69 9. 9	85. 46 11. 5	83. 03 9. 6	88. 78 9. 2
	100	83. 04 8. 1	86. 87 7. 6	83. 61 7. 0	89. 47 7. 8
BADGE-$2H_2O$	10	93. 32 11. 2	82. 52 6. 7	80. 10 11. 5	87. 29 8. 7
	50	80. 81 7. 3	89. 519. 0	83. 50 7. 4	89. 19 10. 5
	100	87. 43 5. 8	85. 12 9. 7	79. 97 7. 9	78. 19 7. 9
BADGE	10	91. 80 11. 1	85. 20 5. 1	77. 92 9. 6	87. 61 7. 9
	50	90. 66 3. 1	89. 98 8. 5	85. 16 6. 2	81. 14 8. 6
	100	86. 44 6. 9	82. 34 11. 4	81. 89 8. 6	78. 44 5. 5

3. 5 Application on real samples

Milk samples were collected from different supermarkets and dairy farms in and around Lanzhou in China. A total of 50 milk samples were collected in 50mL polypropylene tubes and shipped on ice to the laboratory within 5h, and stored at−20℃ in refrigerator until analysis. The analytical method was applied successfully for the identification nine BPs in a total 50 milk samples. BPA was found in one at 13. 74 μg/kg using the method developed and verified at 14. 31μg/kg (n = 3) using the established LC−MS/MS method (Park et al., 2016). These were below permitted level (0. 6μg/g, Commission Directive 2004/19/EC). BADGE · $2H_2O$ was found in one sample at 15. 80μg/kg and BFDGE · $2H_2O$ was found in two at 16. 23 and 17. 82μg/kg, respectively. The amount of BADGE · $2H_2O$ was below permitted level (9. 0mg/kg, European Commission, 2005); the use of BFDGE has been forbidden in Europe (European Commission, 2005), but the limitation requirement of BFDGE has not been set in China. These three positive results were verified using the LC−MS/MS method according to Chinese national standard SN/T 3150—2012 (Chinese national standard, 2012); values were 16. 68, 15. 15 and 19. 22μg/kg, respectively. Detection rates for BPA, BADGE · $2H_2O$ and BFDGE · $2H_2O$ were 2, 2 and 4%, respectively. The relative tolerances of measurements using the method developed and verified method were 4. 06, 5. 42, 6. 88 and

7.56%, respectively, and less than 10%.These data further illustrates the reliability and accuracy of the method developed.

4 CONCLUSIONS

An HPLC-FLD method using QuEChERS for preconcentration was established for the simultaneous determination of nine BPs in milk, for the first time. Extraction and purification analysis of nine BPs in milk were examined in depth.Compared with methods published previously, the method developed in this study had many advantages, such as less organic solvent, inexpensive instrumentation, less manual operation and shorter sample preparation time. Good quality results, including recovery, precision, linearity, LOD and LOQ were achieved.Finally, milk samples collected from supermarkets and dairy farms were analysed to demonstrate the applicability of the method. Among nine BPs, BPA, BADGE · $2H_2O$ and BFDGE · $2H_2O$ were detected in some samples.The results were also verified using LC-MS/MS, and the reliability and accuracy of the method developed were verified.The method developed is suitable for rapid determination of trace amounts of BPs in milk in routine analysis and could be used in laboratories and by testing agencies.

ACKNOWLEDGEMENTS

This research was funded by research grants from the innovation project of Chinese academy of agricultural sciences (grant number: CAAS-ASTIP-2014-LIHPS-01), fund of scientific plan on Gansu province (grant number: 1606RJYA285) and special fund of Chinese academy of agricultural sciences (grant number: 1610322014014).The authors are grateful to the Ministry of Agriculture of the People's Republic of China for financial support.

CONFLICT OF INTEREST

The authors have no conflicts of interest.

REFERENCES OMITTED

(发表于《Food Chemistry》, 院选 SCI, IF: 4.529)

UPLC-Q-TOF/MS-based Metabonomic Studies on the Intervention Effects of Aspirin Eugenol Ester in Atherosclerosis Hamsters

Ning MA, Yajun YANG, Xiwang LIU, Xiaojun KONG,
Shihong LI, Zhe QIN, Zenghua JIAO, Jianyong LI

Based on the pro-drug principle, aspirin andeugenol were used to synthesize aspirin eugenol ester (AEE) by esterification reaction. In present study, the anti-atherosclerosis effects of AEE were investigated in hamsters with the utilization of metabonomic approach based on UPLC-Q-TOF/MS. Biochemical parameters and histopathological injures in stomach, liver and aorta were evaluated. In atherosclerotic hamster, oral administration of AEE normalized biochemical profile such as reducing TG, TCH and LDL, and significantly reduced body weight gain, alleviated hepatic steatosis and improved pathological lesions in aorta. Slight damages in stomach mucous were found in AEE group. Plasma and urine samples in control, model and AEE groups were scattered in the partial least squares-discriminate analysis (PLS-DA) score plots. Thirteen endogenous metabolites in plasma such as lysophosphatidylcholine (LysoPC), leucine and valine, and seventeen endogenous metabolites in urine such as citric acid, phenol sulphate and phenylacetylglycine were selected as potential biomarkers associated with atherosclerosis. They were considered to be in response to anti-atherosclerosis effects of AEE, mainly involved in glycerophospholipid metabolism, amino acid metabolism and energy metabolism. This study extended the understanding of endogenous alterations of atherosclerosis and offered insights into the pharmacodynamic activity of AEE.

As a growing health challenge in the world, atherosclerosis is a complex chronic disease characterized by the accumulation of lipids within arterial walls, dyslipidemia, endothelial dysfunction, and chronic inflammation[1]. Atherosclerosis can cause narrowing, hardening and even complete blockage of arteries, which is a major cause of mortality in patients with cardiovascular diseases (CVD). Eugenol is a light yellowish oily liquid extracted from certain essential oils such as clove oil, nutmeg, cinnamon, basil and bay leaf. Evidences from numerous studies have showed that eugenol possesses antibacterial, antiviral, antifungal, antioxidant, antithrombotic, antiparasitic, and anti-inflammatory properties[2-5]. In addition, eugenol could lower blood lipids to alleviate hyperlipemia, which is beneficial to the inhibitation of atherosclerosis[6-8]. Therefore, eugenol has gained the attraction from researchers in atherosclerosis prevention and treatment. However, irritation and vulnerablity to oxidation are the main constraints in its application, which are mainly caused by the phenolic hydroxyl group.

In ancientChina and Egypt, natural substance salicylic acid and its derivatives obtained from willow bark or leaves had been used to ease pain, fever and inflammation. Acetylsalicylic acid (aspirin), the best-known salicylic derivative, was synthesized by Felix Hoffman in 1897[9]. In addition to the anti-inflammatory, antipyretic and analgesic effects, aspirin is widely used to reduce the risk of CVD and certain cancers[10,11]. Moreover, increasing reports have demonstrated that aspirin has therapeutic effects on atherosclerosis[12,13]. However, the side effects of aspirin such as gastrointestinal damage limit its application[14].

Key Lab of New Animal Drug Project of Gansu Province, Key Lab of Veterinary Pharmaceutical Development, Ministry of Agriculture, Lanzhou Institute of Husbandry and Pharmaceutical Science of Chinese Academy of Agricultural Sciences, Lanzhou, China. Correspondence and requests for materials should be addressed to J.L. (email: lijy1971@ 163. com)

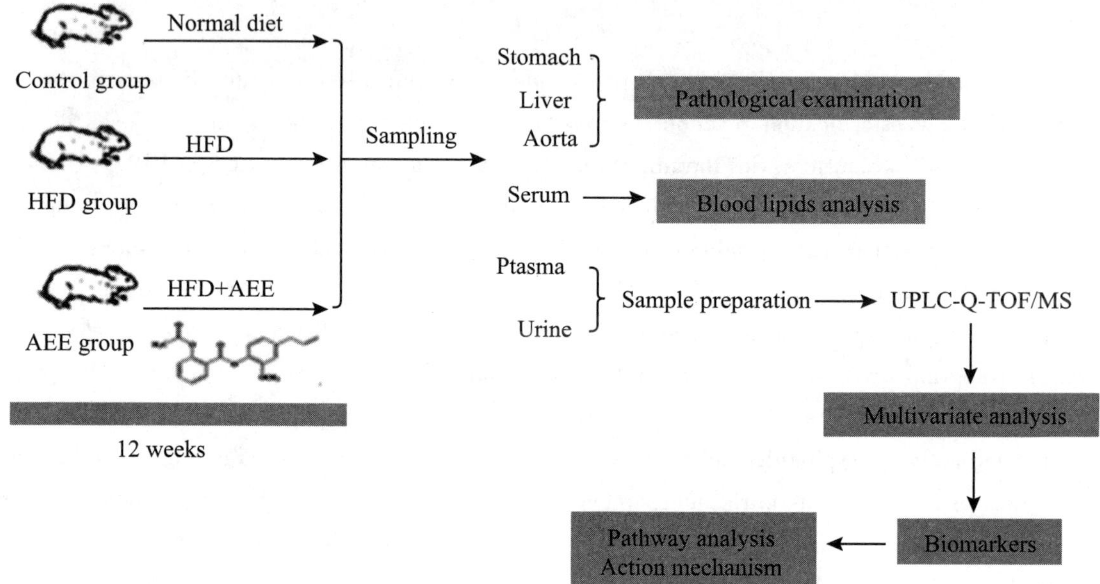

Figure 1. Schematic overview of the experiment design.

In order to increase the therapeutic effects of aspirin andeugenol on atherosclerosis and reduce their disadvantages, aspirin eugenol ester (AEE) was synthesized according to pro-drug principle[15]. AEE is a white and odorless crystal. In the chemical structure of AEE, carboxylic group from aspirin and hydroxyl group from eugenol were chemically masked to reduce the gastrointestinal side effects and improve structural stability[16]. The toxicological studies indicated that AEE was non-genotoxic in vitro or in vivo and its toxicity was obvious lower than its precursors, which suggested that AEE was a safe compound with good druggability[17,18]. Metabolism study proved that AEE was decomposed into salicylic acid and eugenol after administration, which could show their original activities and act synergistically[19]. Moreover, pharmacodynamics experiments showed AEE could regulate blood lipid levels in hyperlipidemic rats such as reducing triglycerides (TG), total cholesterol (TCH) and low density lipoprotein cholesterol (LDL)[20,21]. However, no data has been reported the effects of AEE on atherosclerosis.

With the comprehensive analysis of small molecules, metabonomics is a versatile tool to evaluate the toxicity or therapeutic effect of compounds from the understanding of the dynamic biochemical compositions. In present study, the atherosclerosis in Syrian golden hamster induced with high fat diet (HFD), an excellent model for studying atherosclerosis[22], was used to evaluate the effects of AEE on atherosclerosis with the application of UPLC-Q-TOF/MS-based plasma and urine metabonomics. Meanwhile, the body weight gain (BWG), blood biochemical indices, atherosclerosis index (AI) and the histopathological changes of aorta, stomach, and liver were also investigated in the study.

1 MATERIALS AND METHODS

Chemicals and reagents. AEE (transparent crystal, purity: 99.5% with RP-HPLC) was prepared in Lanzhou Institute of Husbandry and Pharmaceutical Sciences of Chinese Academy of Agricultural Science. MS-grade formic acid was supplied by TCI (Shanghai, China). Deionized water (18MΩ) was prepared with a Direct-Q©3 system (Millipore, USA). MS-grade acetonitrile was purchased from Thermo Fisher Scientific (USA). Carboxymethylcellulose sodium (CMC-Na) was supplied by Tianjin Chemical Reagent Company (Tianjin, China).

Animal experiment. A total of 30 male Syrian golden hamsters weighing 100-110g were purchased from Charles River Company (Vital River, Beijing, China). All animals were housed in facilities by group at a controlled relative humidity (45%-65%) and temperature (22±2℃). Hamster feed and drinking water were supplied ad libitum. All of the experimental protocols and procedures were approved by the Institutional Animal Care and Use Committee of Lanzhou Institute of Husbandry and Pharmaceutical Science of Chinese Academy of Agricultural Sciences (Approval No. NKMYD201601). Animal welfare and experimental procedures were performed strictly in accordance with the Guidelines for the Care and Use of Laboratory Animals issued by the US National Institutes of Health.

Study design. The experimental design was shown in Fig. 1. Hamsters were assigned into 3 groups (n=10): (1) control group, in which hamsters were fed with normal diet; (2) high fat diet (HFD) group, in which hamsters were fed with HFD; (3) AEE group, in which the hamsters were simultaneously fed with HDF and AEE (27mg/kg body weight). The normal diet (12.3% lipids, 63.3% carbohydrates and 24.4% proteins) was purchased from Keao Xieli Feed Co., Ltd (Beijing, China) and the atherogenic HFD (40% lipids, 43% carbohydrates and 17% proteins) was supplied by Research Diet, Inc. (product D12079B, New Brunswick, NJ).

In thesubchronic toxicity, the no-observed-adverse-effect level (NOAEL) of AEE was considered to be 50mg/kg/day 18. Meanwhile, in our previous study, five week treatment of AEE dosed at 18, 36, 54mg/kg can reduce the levels of TG, TCH and LDL in hyperlipidemic rats[23]. Based on the dose used in the former studies, the dose of AEE was selected as 27mg/kg in the present study. The study was conducted for 12 weeks and AEE suspensions were prepared in 0.5% CMC-Na. According to the individual body weight, hamsters in AEE group were intragastrically (i.g.) administered with AEE. For eliminating the effect of CMC-Na (vehicle), hamsters in control and HFD groups were treated with equal volume of CMC-Na as AEE group.

Sample collection. After fasted for 10–12 hours, hamsters from each treatment group were sacrificed under anesthesia induced with pentobarbital (intraperitoneal injection, 30mg/kg). Blood samples were collected from the heart into normal and heparin–treated vacuum tubes to prepare serum and plasma, respectively. Serum and plasma were obtained after centrifugation of blood (2500 rpm at 4℃ for 10min), and stored at –80℃ until analysis. Individual hamsters were placed in metabolic cages (1 per cage) to obtain 24–hour urine collections and be stored at –80℃ before analysis. The aorta was carefully isolated from hamster and extravascular fat tissue was removed. The tissues of aorta, liver and stomach were subsequently fixed in 4% formalin for pathological observations.

Measurement of biochemical parameters. Serum was analyzed using an automatic biochemistry analyzer (Erba XL–640, German). Biochemical parameters including total bilirubin (T–BIL), total protein (TP), albumin (ALB), alanine transaminase (ALT), aspartate aminotransferase (AST), lactate dehydrogenase (LDH), glucose (GLU), triglycerides (TG) and total cholesterol (TCH), low density lipoprotein cholesterol (LDL) and high density lipoprotein cholesterol (HDL), were analyzed in the experiment. The AI was calculated as followed: AI = (TCH–HDL) /HDL[24]. The kits for biochemistry analysis were provided by Ningbo Medical System Biotechnology Co., Ltd (Ningbo, China).

Histopathological examination. In order to investigate the histopathological changes, tissues of aorta, liver and stomach were formalin–fixed and paraffin embedded, sectioned, and stained with hematoxylin and eosin (HE) by using the standard protocol[25]. HE–stained sections were examined using a 13395H2X microscope (Leica, Germany). The morphometric analysis of the aorta images was carried out by Image–Pro Plus 6.0 software (Media Cybernetics, Bethesda, MD, USA). The stenosis ratio was calculated as percent of lumen area to total area of the aorta as previously described[26].

Sample preparation. The plasma samples were thawed at room temperature prior to analysis. Acetonitrile (400μL) was added to every 200μL plasma. After vigorous vortex–mixing for 1min and incubation for 10min, the mixture was centrifuged at 12 000g for 15min at 4℃ to precipitate the proteins. The supernatant was filtered through a 0.22μm nylon filter. An aliquot of 3μL sample was injected for analysis. Urine samples were thawed at room temperature, and then 600μL ice–cold methanol was added into 200μL of urine, vortex mixed and centrifuged at 13 000g for 15min at 4℃ to remove solid materials. The supernatant was also filtered through a 0.22μm nylon filter and an aliquot of 4μL was injected for analysis.

Data acquisition and processing. Metabonomics analysis was performed with an Agilent 1290 Infinity LC system coupled to an Agilent 6530 Accurate–mass Q–TOF mass spectrometer (Agilent, USA). Chromatographic separations of plasma and urine samples were performed on an Agilent ZORBAX SB–C18 threaded column (2.1×150mm, 1.8μm, Agilent Technologies, USA) maintained at 35℃. The mobile phase consisted of solvent A–water with 0.1% formic acid and solvent B–acetonitrile with 0.1% formic acid. The optimized gradient program was shown in Table S1. Flow rate of plasma sample was 0.3mL/min, and 0.35mL/min of urine sample. The post time was set to 5min for equilibration. Mass spectrometry was performed both in electrospray ionization in positive (ESI+) and negative (ESI–) ion modes. The fragment voltage was set at 135V and skimmer voltage was set at 65V.The capillary voltages were set at 4.0KV in positive mode and 3.5KV in negative mode, respectively.The drying gas flow (nitrogen) was set to 10 L/min at 350℃ and the nebulizer pressure was set at 45 psig.Data was collected in centroid mode from 50-1 000m/z using an extended dynamic model.

The raw MS data were firstly processed by Mass Hunter Qualitative Analysis software (Agilent technologies, USA) to converted to common data format (.mzData).The program XCMS was used for nonlinear alignment of the data in the time domain and automatic integration and extraction of the peak intensities.The parameters of the XCMS were default settings.The data were filtered by interquantile range and normalized to the total intensity for further multivariate data analysis.The obtained data were imported into SIMCA-P (version 13.0, Umetrics AB, Sweden) where principal component analysis (PCA) and partial least squares discriminant analysis (PLS-DA) were performed to data set analysis.The quality of PLS-DA models was described by R^2X, R^2Y, and Q^2 and its validity was evaluated by permutation testing (with 200 permutations).Variable importance in the projection (VIP>1) value of validated PLS-DA model and the P values of one-way ANOVA ($P<0.05$) were taken as the measurement indices for potential metabolites selecting.Identification of the metabolites was achieved through a mass-based search followed by manual verification.TOF-MS accurate mass value of the molecular ion of interest was searched against the METLIN and Human Metabolome Database (HMDB).Then, MS/MS analysis was carry out to confirm the structure of potential biomarkers by matching the masses of the fragments.The clustering analysis of the potential biomarkers and pathway analysis were performed on MetaboAnalyst 3.0 (http://www.metaboanalyst.ca/), and the metabolic pathway interpretation was performed using the KEGG database.

Statistical analysis.BWG, AI and blood biochemical parameters were expressed as mean±standard deviation (SD).The differences had been evaluated by one-way ANOVA with Fisher's least significant difference (LSD) test using the Statistical Package for Social Science program (SPSS 16.0, Chicago, IL, USA).Differences were considered significant at $P<0.05$.

2 RESULTS

AEE reduced body weight gain. Before the experiment, no difference was observed in body weight among the groups.Body weight gains (BWGs) of hamsters in different treatment groups were analyzed at the end of the study (Fig. 2a).After feeding with HFD for 12 weeks, the hamsters in HFD group had higher BWGs than that in the control ($P<0.01$).In comparison with the HFD group, BWGs were significantly decreased in AEE group ($P<0.01$).Significant difference was found between control and AEE groups ($P<0.01$), indicating the elevated BWGs induced by HFD were partly recovered by AEE treatment.

AEE ameliorated biochemical profile disorder.The results of serum biochemical parameters were shown in Table 1. After feeding with HFD for 12 weeks, the levels of biochemical parameters were all significantly increased in the serum of the HFD group than that in the control group ($P<0.01$). Compared with the HFD group, after AEE treatment, the levels of AST, HDL, LDL, TCH, TBIL, TP, LDH, ALB, GLU and TG were significantly reduced ($P<0.01$).No significant difference was observed in ALT between HFD and AEE groups.Biochemical parameters in AEE group including TBIL, TP, AST, LDH, TG, LDL and TCH showed no difference when compared with the control. However, the levels of ALB, ALT, GLU and HDL in AEE group were significant higher than those in the control ($P<0.01$). These results suggested that AEE reversed the disturbance in biochemical profile caused by HFD. According to the values of TCH, HDL and LDL, the AI was calculated in the study (Fig. 2b). The AI was markedly elevated in the HFD group in comparison with the control ($P<0.01$). In AEE group, the AI values were significantly reduced than those in the HFD group ($P<0.01$), and no statistical difference was observed between control group and AEE group, indicating the blood lipid levels were improved in hamsters with AEE treatment.

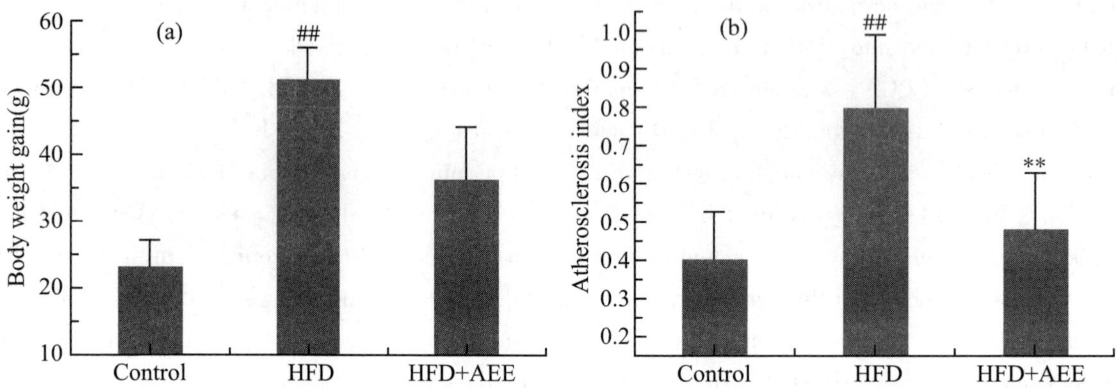

Figure 2. Effects of AEE on the body weight gain.

Note: (a) and atherosclerosis index (b) in atherosclerosis hamsters. Data were expressed as mean±standard deviation. HFD: high fat diet group; ## $P<0.01$ significant difference from the control group; ** $P<0.01$ significant difference from HFD group.

Table 1. Effects of AEE on serum biochemical indices in different treatment groups.

Variables	Control	HFD	AEE
TBIL (umol/L)	1.1±0.2	1.9±0.6##	1.1±0.3**
TP (g/L)	38±5	50±8##	40±4**
ALB (g/L)	18±3	28±4##	23±3** ##
ALT (U/L)	48±8	70±15##	64±12##
AST (U/L)	70±15	107±25##	59±17**
LDH (U/L)	143±40	222±49##	119±27**
GLU (mmol/L)	5.9±1.2	10.2±2.0##	8.9±1.8** ##
TG (mmol/L)	1.41±0.35	2.71±0.50##	1.09±0.28**
HDL (mmol/L)	0.99±0.22	1.96±0.06##	1.27±0.21** ##
LDL (mmol/L)	0.32±0.09	0.61±0.12##	0.29±0.08**
TCH (mmol/L)	1.73±0.37	3.48±0.52##	1.6±0.43**

Note: HFD: high fat diet; AEE: aspirin eugenol ester; ## $P<0.01$ significant difference from control group; ** $P<0.01$ significant difference from HFD.

AEE inhibited liver and atherosclerotic lesions. The results of the histopathological changes in the different groups were illustrated in Fig. 3. Histopathological examination of the liver in the control group showed that hepatocyte of hamster was normal and nuclear structure was clear, while significant morphological changes were observed in the HFD group. In the HFD group, liver sections showed that the hepatocyte had a large area with hydropic degeneration and some were found with cytolysis and fatty degenerations. In AEE treatment group, hepatocyte was nearly normal with the reduction of hydropic and fatty degenerations. From the above results, HFD consumption stimulated fat accumulation in hepatic cells and finally caused fatty liver in the atherosclerosis hamsters, whereas AEE reversed the HFD-induced liver injury. These results were in accordance with the results of serum biochemical parameters.

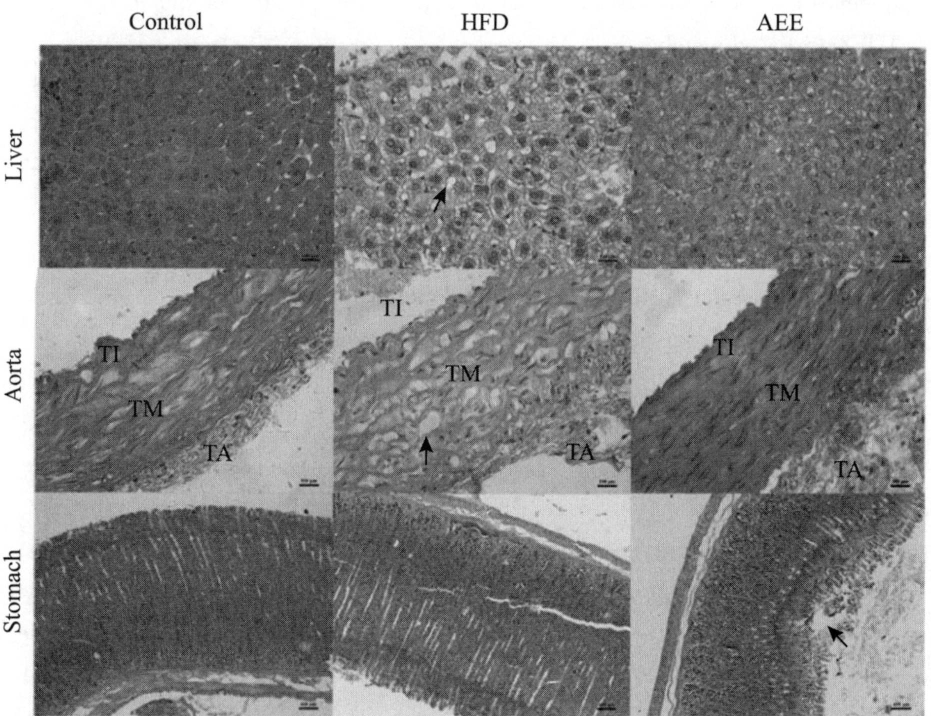

Figure 3. Representative photographs of hematoxylin-eosin (HE) staining of liver (×400), aorta (×400) and stomach (×100) in different treatment groups.

Note: TA: tunic adventitia; TM: tunic media; TI: tunic intima. Compared with the liver in control hamster, large fat droplets were observed in the liver of the hamsters fed with high fat diet; The structure of blood vessel in control group was integrated and the TI was smooth, whereas lots of foam cells, migration of smooth muscle cells and serious accumulation of fat were observed in HFD group; Hepatic steatosis and pathophysiologic changes of aorta were notably alleviated by AEE treatment. The slight necrosis and ecclasis of gastric mucosa were found in the AEE group. The typical pathological changes of liver, aorta and stomach were indicated by black arrows, respectively.

Gastrointestinal effects of oral administration of AEE were examined in the study. No damaging effects on the stomach were observed in control and HFD groups. After oral administration of AEE dosed at 27mg/kg for 12 weeks, stomach mucous membrane of hamster became uneven, and the evidence of slight degeneration, necrosis and ecclasis of mucosa epithelium could be found

(Fig. 3).

Nointimal or medial pathologic changes were observed in the aorta in control group (Fig. 3). Compared with the results in the control group, intimal thickening with foam cells and migration of smooth muscle cells were occurred in the HFD group. AEE caused a notable decrease of pathophysiologic changes of atherosclerosis induced by HFD. The percentage of lumen area to the cross-section of artery was also measured in the study. AEE treatment significantly ameliorated the aortic stenosis compared with the HFD group (Fig. 4, $P<0.01$). The mean value of lumen area to artery cross-section in AEE group was lower than that in the control, but there was no statistical difference between two groups. These results demonstrated that anti-atherosclerosis effects of AEE might be associated with the improvement of the aorta lesion.

Metabonomics analysis of plasma. In order to explore the possible action mechanisms of AEE, UPLC-Q-TOF/MS based metabonomic experiment was carried out. Representative total ion chromatograms (TICs) of the plasma samples analyzed by UPLC-Q-TOF/MS showed good separations and strong sensitivity of the established method (Fig. 51). PCA is an unsupervised multivariable statistical method to find out the metabolic distinction and the resulting data were displayed by score plots representing the distribution of samples in multivariate space. As indicated by the score plots in Fig. 5a and 5b, the plasma metabolic profiles in positive and negative modes of the control and HFD groups were clearly separated, which revealed that the perturbations of plasma metabolic profiles in HFD group were evident. Model parameter R2X representing the explanative ability of the model, were 0.627 and 0.625 in positive and negative modes, respectively, which showed the data can be highly elucidated by the two PCA models.

PLS-DA, a supervised multivariable statistical method, was conducted to further assess the influence of AEE on metabolic pattern. In PLS-DA analysis, the HFD and control groups were clearly separated, which was consistent with the found in PCA (Fig. 5c and 5d). Meanwhile, the metabolic profile of hamster in groups supplemented with AEE quite differed from the HFD group, indicating the disorders induced by HFD were ameliorated after AEE treatment. A clear separation among the control, HFD and AEE groups was observed in the score plots of the PLS-DA models. This distribution suggested that AEE treatments partially recovered the atherosclerosis status. The permutation test was performed to test the over-fitting of PLS-DA after modeling the data. Permutation tests generated the intercepts of $R^2=0.468$ and $Q^2=-0.373$ in positive mode and $R^2=0.50$, $Q^2=-0.508$ in negative mode (Fig. 5e and 5f), which demonstrated that the PLS-DA models were robust without overfitting.

VIP values concluding the contribution of the features for the model were employed to select the potential biomarkers. The candidate metabolites with VIP>1.0 and $P<0.05$ were considered as potential biomarkers. Following the threshold, 13 endogenous metabolites in plasma were selected, which may be related with how AEE influenced the development of atherosclerosis in hamster (Table 2). To fully and intuitively display the relationships and differences between samples, the selected biomarker data were analyzed using clustering heatmap (Fig. 52a). In the HFD group, the concentrations of lysophosphatidylcholine (LysoPC) (22 : 5), LysoPC (20 : 4), LysoPC (18 : 1), LysoPC (18 : 0), LysoPC (16 : 0), LysoPC (20 : 3), LysoPC (15 : 0),

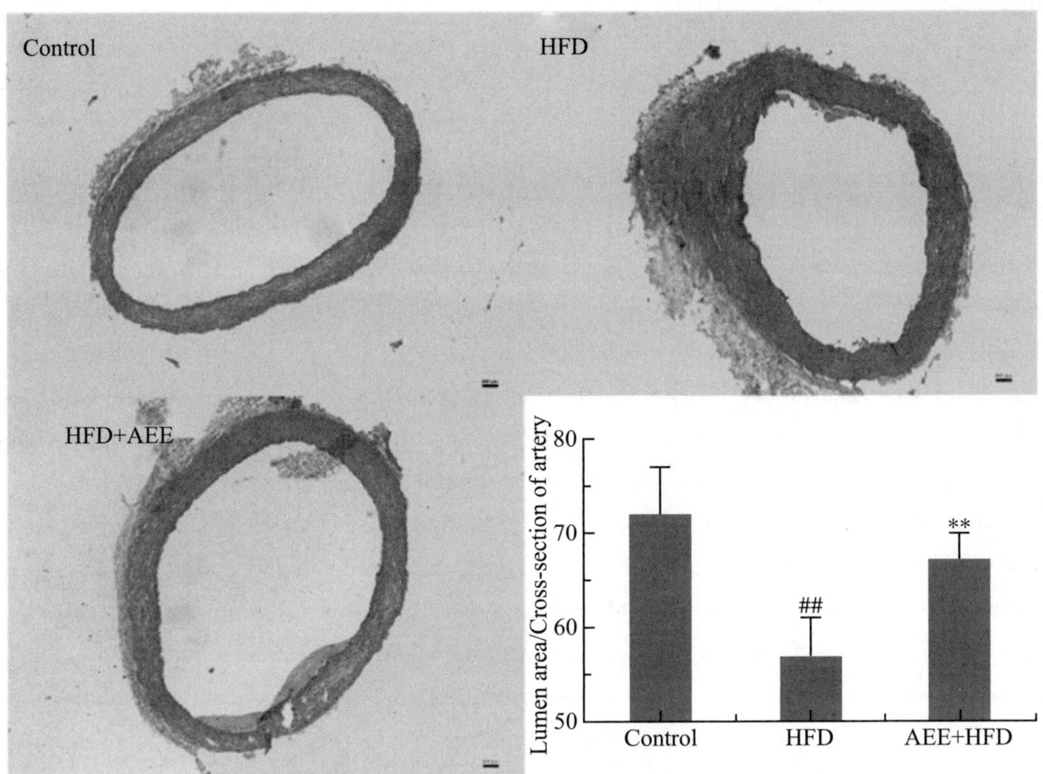

Figure 4. Representative cross-sections of aorta and image analysis of stenosis ratio in different group (×50).

Note: Compared with control group, blood vessel wall and lumina of aorta in HFD group became thicker and narrower (★). The stenosis ratio was expressed as percentage of lumen area to total area of the aorta. ## $P<0.01$ significant difference from the control; ** $P<0.01$ significant difference from HFD group.

LysoPC (14 : 0), LysoPC (16 : 1) and elaidic acid were significantly increased, while leucine, valine and docosahexaenoic acid (DHA) were decreased (Table 2). Interestingly, AEE treatment corrected and reversed the variations of the selected potential biomarkers such as LysoPC (20 : 4), leucine and valine. These results indicated that AEE treatment had regulation effects on selected metabolites.

Metabonomics analysis of urine. The results of urine metabolomics were similar to that of the plasma samples, suggesting the ameliorative effects of AEE in hamster with atherosclerosis. Typical TICs of urine extracts in positive and negative modes obtained from UPLC-Q-TOF/MS analysis were shown in Fig. 3S. PCA was used to globally understand the metabolic changes of control and model groups. PCA score plots in positive and negative modes (Fig. 6a and 6b) showed that there were significant deviations in atherosclerotic hamsters compared with the control. The samples in model group clustered away from those in the control group, indicating HFD had significant influence on metabolites in urine.

Score plots of PLS-DA models showed a clear separation among control, model and AEE groups (Fig. 6c and 6d). The cluster of the samples in model group was located far away from the control, suggesting the urine metabolic profile of atherosclerosis was different from the healthy controls. Urine metabolic profile of hamsters in AEE treated group fairly differed from the model group, which indicated that AEE improved deviations induced by HFD. These results were in accordance with the results of blood lipids analysis and pathological changes observation. The validation plot (Fig. 6e and 6f) strongly indicated that the original model was valid: the Q^2 regression line has a negative intercept (ESI+: $Q^2=-0.519$, ESI-: $Q^2=-0.345$), and all permuted R^2 values to the left of the intercept (ESI+: $R^2=0.468$, ESI-: $R^2=0.277$) were lower than the original point to the right.

With VIP>1.0 and $P<0.05$, 17 metabolites were selected as potential biomarkers associated with atherosclerosis in urine (Table 3). Heatmap of the metabolites in urine was shown in Fig. S3b. In comparison with the control, HFD significantly elevated the levels of citric acid, phenylglucuronide, phenol sulphate, phenylacetylgly-cine, p-Cresol glucuronide and acetylcysteine, while HFD reduced the pantothenic acid, hippuric acid, phenyllactic acid, azelaic acid, niacinamide, spermidine, DL-2-Aminooctanoic acid (DL-2-AC), leucine and riboflavin. AEE treatment showed a tendency of bringing altered metabolites to normal, such as the improvement of citric acid, phenylglucuronide, phenol sulphate and acetylcysteine. Potential biomarkers in urine were mainly involved in citrate cycle, amino acid metabolism and nicotinate and nicotinamide metabolism.

Pathway analysis. Metabonomics pathway analysis was carried out with MetaboAnalyst 3.0 to identify and visualize the most relevant metabolic pathways in hamster with atherosclerosis. The impact-value threshold was set to 0.05 and the pathway with impact-value above this threshold was filtered out. The summary of pathway analysis was shown in Table 2. Figure 7 showed that the pathways in response to atherosclerosis and AEE treatment were valine, leucine and isoleucine biosynthesis, glyoxylate and dicarboxylate metabolism, pantothenate and CoA biosynthesis, riboflavin metabolism, lysine degradation, nicotinate and nicotinamide metabolism and glycerophospholipid metabolism. These pathways were obviously disturbed by HFD administration, and could be acted as targets for AEE against atherosclerosis.

3 DISCUSSION

Hamsters are high sensitive to HFD which can elevate blood lipid levels and promote appreciable atherosclerosis in as little as 6 weeks. Like humans, the cholesteryl ester transfer protein (CETP) of hamster can transfer the cholesterol from HDL to LDL particles in plasma. Therefore, HFD-fed hamster is an invaluable and sensitive model for rapid establishment of atherosclerosis[27]. To our knowledge, this study was the first to explore the anti-atherosclerosis effects and mechanism of AEE in atherosclerosis hamster model. The histopathological results confirmed that the hamsters suffered severe atherosclerotic lesions in the aorta and hepatic damages after administration of HFD for 12 weeks. Notably, the histopathological changes of aorta and liver were obviously improved after AEE treatment, which proved the therapeutic effects of AEE on atherosclerosis. Furthermore, biochemical analysis and UPLC-Q-TOF/MS based metabonomic were applied to characterize the crucial parameters and metabolic pathways associated with AEE treatment.

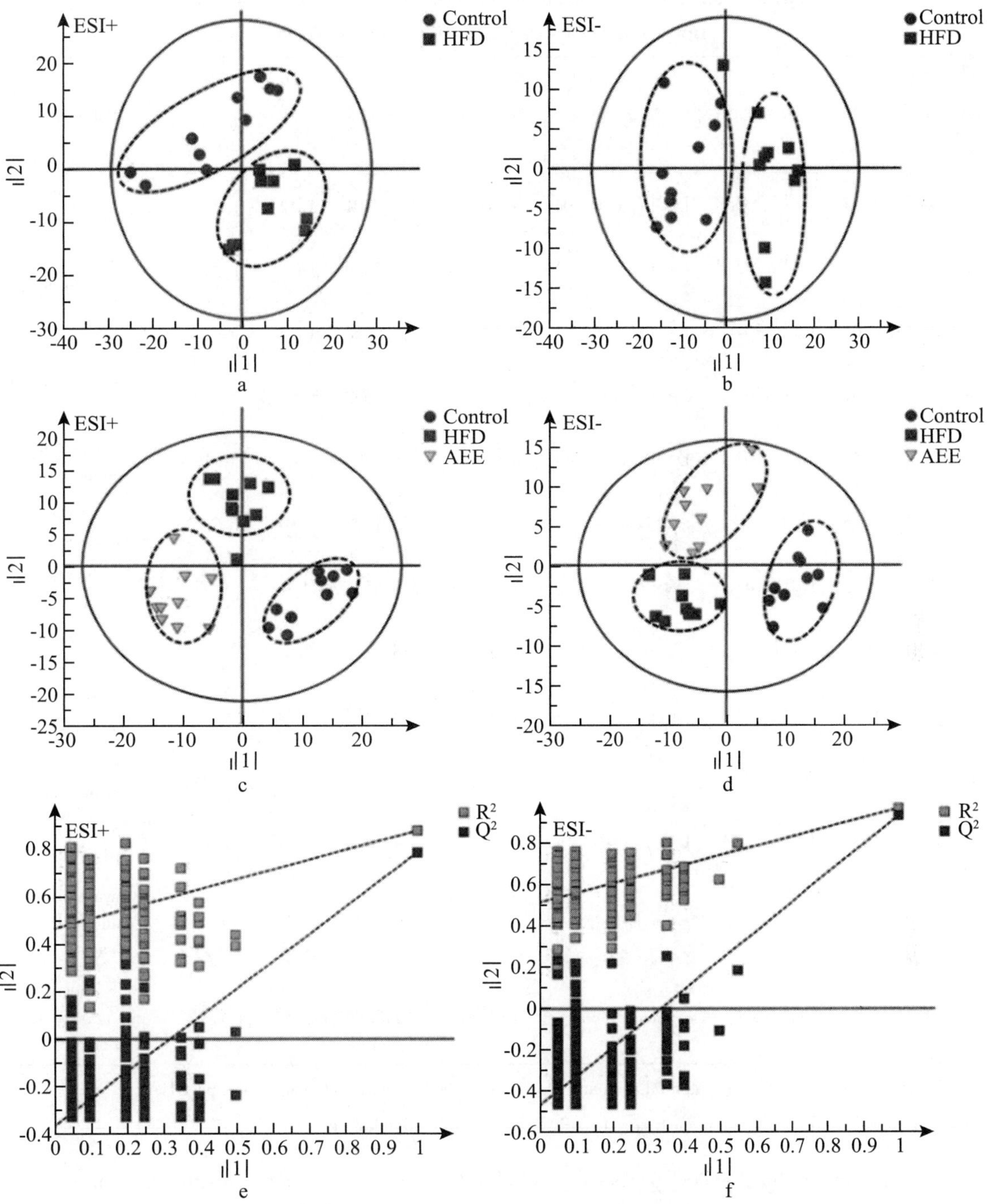

Figure 5. Multivariate data analyses of plasma based on UPLC-Q-TOF/MS analysis.

Note: (a, b) PCA score plot based on the plasma metabolic profiling of the control and atherosclerosis hamsters in positive and negative modes, ESI+: $R^2=0.627$, ESI−: $R^2=0.625$. (c, d) PLS-DA score plots of the control, HFD and AEE groups, ESI+: $R^2X=0.424$, $R^2Y=0.911$, $Q^2=0.813$; ESI−: $R^2X=0.613$, $R^2Y=0.917$, $Q^2=0.829$. (e, f) Permutation test of the PLS-DA models, ESI+: the intercepts of $R^2=0.468$ and $Q^2=-0.373$, ESI−: $R^2=0.50$, $Q^2=-0.508$.

Table 2 Summary of tag numbers based on the DGE data from Gansu Alpine fine wool sheep skin with different WFD

No.	RT	VIP	Formula	Metabolite	Adduction	m/z	Fold Change HFD/Control	Fold Change AEE/HFD	Pathway
1	9.97	5.94	$C_{24}H_{50}NO_7P$	LysoPC (16 : 0)	[M+H] +	496.3401	1.14*	1.24**	Glycerophospholipid metabolism
2	10.36	5.74	$C_{26}H_{52}NO_7P$	LysoPC (18 : 1)	[M+H] +	522.3560	2.07**	1.25**	Glycerophospholipid metabolism
3	12.15	4.01	$C_{26}H_{54}NO_7P$	LysoPC (18 : 0)	[M+H] +	524.3718	1.22*	1.18*	Glycerophospholipid metabolism
4	10.06	2.01	$C_{30}H_{52}NO_7P$	LysoPC (22 : 5)	[M+H] +	570.3559	5.59**	1.12	Glycerophospholipid metabolism
5	9.46	3.27	$C_{28}H_{50}NO_7P$	LysoPC (20 : 4)	[M+H] +	544.3403	1.79**	1.15	Glycerophospholipid metabolism
6	1.48	1.86	$C_6H_{13}NO_2$	Leucine	[M+H] +	132.1020	0.50*	1.68**	Amino acid metabolism
7	12.24	1.70	$C_{28}H_{52}NO_7P$	LysoPC (20 : 3)	[M+H] +	546.3534	1.06	1.97	Glycerophospholipid metabolism
8	1.19	1.58	$C_5H_{11}NO_2$	Valine	[M+H] +	118.0861	0.51**	2.22*	Amino acid metabolism
9	10.97	1.52	$C_{23}H_{48}NO_7P$	LysoPC (15 : 0)	[M+H] +	482.3357	1.00	8.44**	Glycerophospholipid metabolism
10	8.52	1.44	$C_{22}H_{46}NO_7P$	LysoPC (14 : 0)	[M+H] +	468.3090	8.54**	1.08	Glycerophospholipid metabolism
11	8.98	1.34	$C_{24}H_{48}NO_7P$	LysoPC (16 : 1)	[M+H] +	494.3244	1.70*	1.26	Glycerophospholipid metabolism
12	12.28	1.54	$C_{22}H_{32}O_2$	DHA	[M-H] -	327.2349	0.57**	0.75	Biosynthesis of unsaturated fatty acids
13	14.73	2.33	$C_{18}H_{34}O_2$	Elaidic acid	[M-H] -	281.2506	1.75**	1.1	Biosynthesis of unsaturated fatty acids

Table 2. Effects of AEE on potential biomarkers associated with atherosclerosis in plasma. DHA: docosahexaenoic acid; RT: retention time; LysoPC: lysophosphatidylcholine; * $P<0.05$, ** $P<0.01$.

The analysis of biochemical parameters is helpful to assess the general health status of animals. It was reported that there was a close association between atherosclerosis and lipid abnormalities, especially high levels of plasma LDL, TCH, and TG[28]. In this study, the elevated levels of LDL, TCH, HDL and TG in HFD group showed the metabolic disorder of lipids, which was also verified by the metabolomics analysis in the score plots. AI was calculated to evaluate the lipid-lowering effect of AEE. Recent studies have suggested that AI is a reliable index to access the relative contribution of lipids to the atherosclerosis. The decreased AI values in AEE group revealed that AEE could ameliorate blood lipid profile, which was consistent with our previous study[23]. The normalization of blood lipid profile could contribute to reducing the accumulation of fat, lipid and cholesterol in the aorta, which might be the reasons for the improved pathological results in the aorta in AEE group. As the primary source of energy for the body, GLU is transported from the intestines or liver to cells via the bloodstream. High concentrations of GLU in the HFD group might be caused by the increased levels of blood lipids and energy metabolism disorders[29]. Serum TBIL, TP, ALB, ALT, AST and LDH are important parameters of liver function. The increased levels of these parameters in the HFD group might indicate existing liver damage, which was confirmed by the pathological changes of liver tissue. With the AEE treatment, hamsters showed a reversible trend to normal levels in biochemical parameters and had a remarkable decrease of pathological changes in liver. It was suggested that 12-week AEE treatment was beneficial to improve the biochemical profile and pathological changes in atherosclerosis hamster. Based these results, it could be found that there was a mutual cause-and-effect relationship between pathological findings and biochemical parameters.

Aspirin can reduce the risk of CVD by its anti-inflammatory andantiplatelet effects via the irreversible acetylation of cyclooxygenases (COX) 1 and 2[30]. COX-1 is constitutively expressed in most tissues, and is the predominant form in gastric mucosa which is crucial for mucosal protection. COX-2 is absent under normal conditions, but elevated levels are found during inflammation[31]. The non-selective inhibitation of aspirin on both COX-1 and COX-2 is considered to be an underlying reason for the gastrointestinal side effects. Accumulated evidence indicates that eugenol displays antiulcer activities, in which eugenol can dose-dependently reduce gastric ulcers in rat gastric ulcer models[32-34]. Eugenol can also stimulate the synthesis of mucus, an important gastroprotective factor, which may be responsible for antiulcer activity. AEE is decomposed into salicylic acid and eugenol by the enzyme after administration, then salicylic acid and eugenol can play complementary roles to reduce gastrointestinal damage. Meanwhile, there is no direct contact of acidic group with gastric mucosa through masking the carboxyl of aspirin, which is simple and efficient way to reduce gastrointestinal side effects[35]. In our previous study, no lesion in stomach and duodenum was found in the rats fed with 50mg/kg AEE for 15 days[18]. However, slight pathological changes of gastric mucosa were observed in AEE group, which indicated there were some mild gastrointestinal side effects of AEE under the present experimental conditions. The pathological injury in gastric mucosa may be attributable to the animal species used in the experiment, long duration of the experiments (12 weeks) or the dosage of AEE (27mg/kg). Chen Yu et al. reported that oral administration of aspirin (200mg/kg) for seven days could induce severe mucosal damage in mice[36]. From the reasons of gastroprotective effects of eugenol and the disappearance of carboxyl group, it may be inferred that AEE has smaller gastrointestinal side effects than equal molar aspirin. Pathological change of gastric mucosa is a potential limitation in the application of AEE, which may be avoided by the appropriate control of dosage and administration time of AEE. Further studies are needed to reveal the underlying mechanism of gastrointestinal side effect in AEE treatment.

Metabolomics is a sensitive and effective approach for detecting biological responses by investigating the endogenous small molecule metabolites. Analysis of metabolite changes could provide valuable information, which is helpful to reveal the action mechanism of drug[37]. In this study, plasma and urine metabolomics analyses were applied to systematically evaluate the treatment effects of AEE in ath-

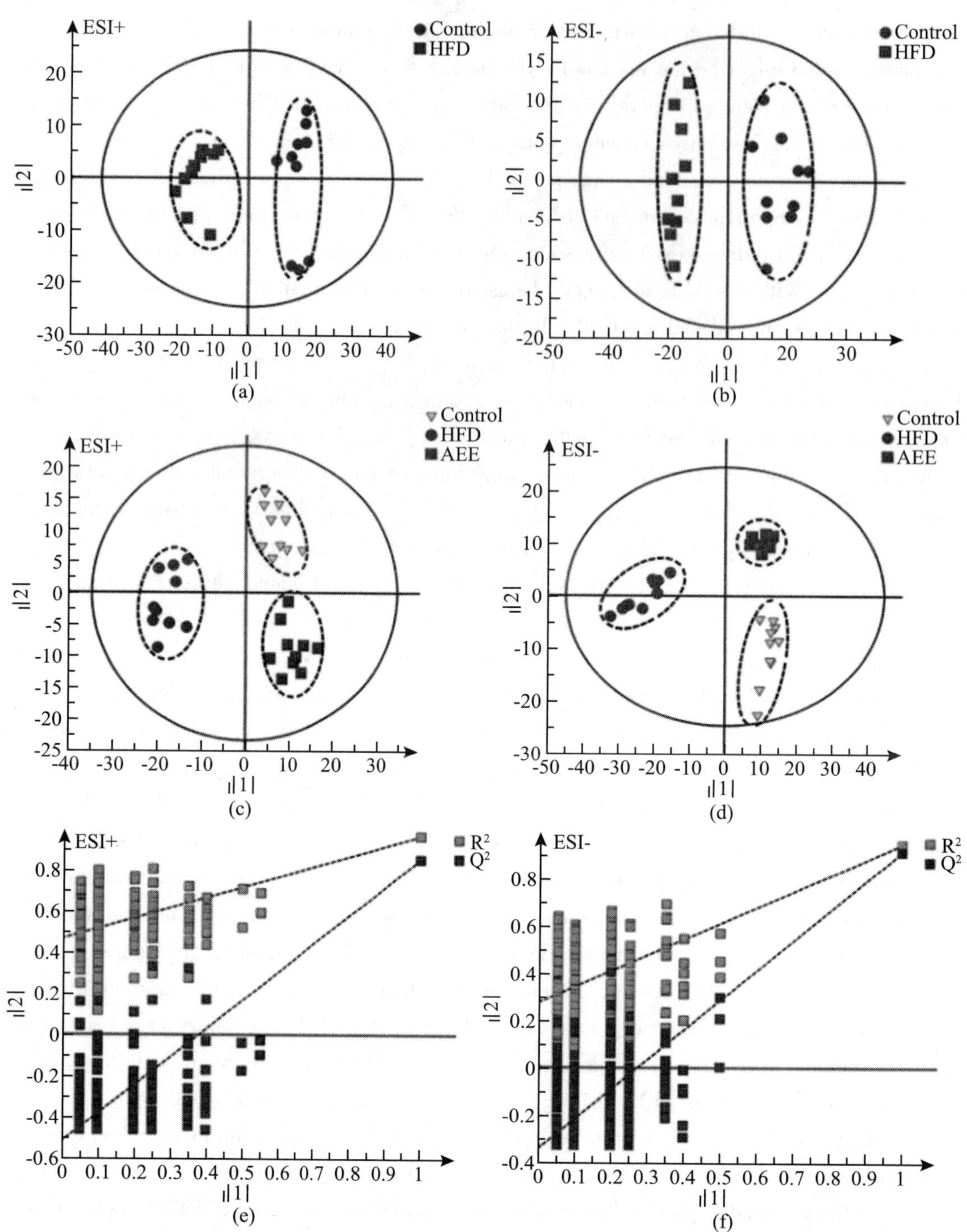

Figure 6. Multivariate data analyses of urine based on UPLC-Q-TOF/MS analysis.

Note: (a, b) PCA score plot based on the urine metabolic profiling of the control and atherosclerosis hamsters in positive and negative mode, ESI+: $R^2=0.535$, ESI−: $R^2=0.75$. (c, d) PLS-DA score plots of the control, HFD and AEE groups, ESI+: $R^2X=0.614$, $R^2Y=0.966$, $Q^2=0.884$; ESI−: $R^2X=0.7$, $R^2Y=0.96$, $Q^2=0.932$. (e, f) Permutation test of the PLS-DA models, ESI+: the intercepts of $R^2=0.468$ and $Q^2=-0.519$, ESI−: $R^2=0.277$, $Q^2=-0.345$.

erosclerosis hamsters. Multivariate data analysis indicated that the control group, HFD group, and AEE treatment group could be clearly distinguished from each other. Compared with the control, selected potential biomarkers associated with atherosclerosis exhibited differences in the HFD group, and be regulated by AEE treatment. Pathway analysis indicated that anti-atherosclerosis effects of AEE were mainly related with glycerophospholipid metabolism, amino acid metabolism, pantothenate and CoA biosynthesis, riboflavin metabolism, and biosynthesis of unsaturated fatty acids.

LysoPC is a major component of oxidized low-density lipoprotein, which plays functional roles in various diseases including diabetes, hyperlipidemia, atherosclerosis and cancer. It is generally believed that the increased LysoPCs can trigger inflammation and the autoimmune response, which may be related to the pathogenesis of atherosclerosis. In our study, LysoPCs were increased in hamster with HFD - induced atherosclerosis, which was good agreement with other reports 26. The relative content of LysoPCs recovered at different levels after AEE administration except LysoPC (20 : 3), LysoPC (15 : 0) and LysoPC (16 : 1). It is noteworthy that there are still controversies on the roles of LysoPCs in atherosclerosis. For example, some researchers reported that some kinds of LysoPCs had a strong inverse association with coronary artery disease[38,39]. After AEE treatment for 12 weeks, the abnormal levels of LysoPCs in atherosclerosis hamster were intervened. Perturbed glycerophospholipid metabolism suggested the complex physiological interplay during atherosclerosis progression. Further studies about the influence of AEE on LysoPCs are needed to elucidate its roles in atherosclerosis.

Table 3. Effects of AEE on potential biomarkers associated with atherosclerosis in urine. DL-2-AC: DL-2-Aminooctanoic acid, * $P<0.05$, ** $P<0.01$.

No.	RT	VIP	Formula	Metabolite	Adduction	m/z	Fold Change		Pathway
							HFD/ Control	AEE/ HFD	
1	1.16	1.30	$C_6H_8O_7$	Citric acid	M-H	191.0191	5.23**	0.73	Citrate cycle
2	4.23	1.58	$C_9H_{17}NO_5$	Pantothenic acid	M-H	218.1017	0.07**	0.74	beta-Alanine metabolism
3	6.05	2.88	$C_{12}H_{14}O_7$	Phenylglucuronide	M-H	269.0654	4.09**	0.93	
4	6.29	2.54	$C_6H_6O_4S$	Phenolsulphate	M-H	172.9899	2.54**	0.80*	
5	6.72	5.00	$C_9H_9NO_3$	Hippuric acid	M-H	178.0494	0.29**	0.93	Phenylalanine metabolism
6	7.09	2.68	$C_8H_7NO_4S$	Indoxyl sulfate	M-H	212.0007	0.77	0.81	Tryptophan metabolism
7	7.64	2.92	$C_{10}H_{11}NO_3$	Phenylacetylglycine	M-H	192.0650	1.26**	0.88*	Phenylalanine metabolism
8	8.22	3.75	$C_{13}H_{16}O_7$	p-Cresolglucuronide	M-H	283.0811	1.40**	0.88	
9	8.68	1.95	$C_9H_{10}O_3$	Phenyllactic acid	M-H	165.0542	0.25**	0.98	Tropane, piperidine and pyridine alkaloid biosynthesi

(continued)

No.	RT	VIP	Formula	Metabolite	Adduction	m/z	Fold Change		Pathway
							HFD/ Control	AEE/ HFD	
10	9.89	1.09	$C_9H_{16}O_4$	Azelaic acid	M-H	187.0960	0.38**	0.61**	
11	1.70	1.40	$C_6H_6N_2O$	Niacinamide	M+H	123.0555	0.20**	0.57	Nicotinate and nicotinamide metabolism
12	0.88	1.04	$C_7H_{19}N_3$	Spermidine	M+H	146.1652	0.33**	0.94	Arginine and prolinc metabolism
13	1.15	1.35	$C_6H_{11}NO_2$	Pipecolic acid	M+H	130.0863	0.34	1.67	Lysine degradation
14	1.14	1.08	$C_8H_{17}NO_2$	DL-2-AC	M+H	160.1331	0.27**	0.96	
15	1.87	1.26	$C_6H_{13}NO_2$	Leucine	M+H	132.1018	0.30*	0.71	Valine, leucine and isoleucine degradation
16	7.26	1.22	$C_{17}H_{20}N_4O_6$	Riboflavin	M+H	377.1467	0.68**	0.51**	Vitamin digestion and absorption
17	8.74	1.21	$C_5H_9NO_3S$	Acetylcysteine	M+H	164.0376	2.82**	0.86	

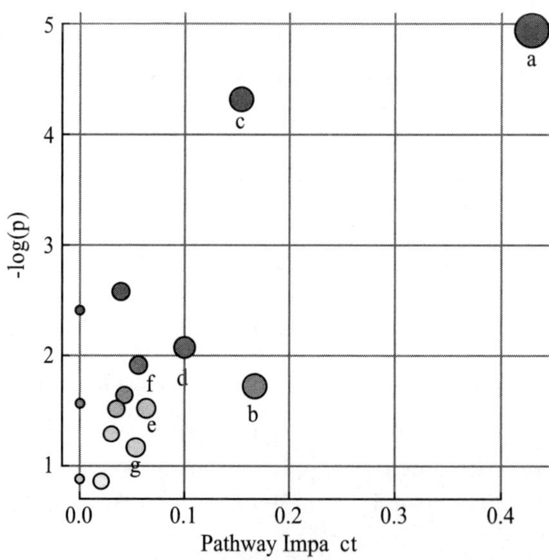

Figure 7. Possible disturbed metabolic pathways of the potential biomarkers for atherosclerosis and AEE treatment.

(a) Valine, leucine and isoleucine biosynthesis; (b) Glyoxylate and dicarboxylate metabolism; (c) Pantothenate and CoA biosynthesis; (d) Riboflavin metabolism; (e) Lysine degradation; (f) Nicotinate and nicotinamide metabolism; (g) Glycerophospholipid metabolism.

Diet supplementation withleucine not only plays key role in protein metabolism but also ameliorates lipid profile such as the reducing of TG, TCH and LDL. So the increased level of leucine in

AEE group was conducive to lowing blood lipids and treating atherosclerosis. According to the results, it was found that there was some relationship among metabolite level changes, pathological results and biochemical parameters.Song et al.reported that elevated levels of GLU, lipid accumulation and decreased level of valine implied energy metabolism impairment such as glycolysis inhibitation and the increase of fatty acid β-oxidation[40].In this study, the similar changes of GLU, blood lipids and valine were observed in HFD group.Citric acid, as a TCA cycle intermediate, was found to be higher in the HFD than that in the control.The increased free fatty acid oxidation and inhibitory TCA cycle might result in increasing the excretion of citric acid in urine[41].Valine and leucine can be used to produce succinyl-CoA by degradation reaction, and then enter into TCA cycle. AEE might greatly promote TCA cycle and attenuate energy metabolism impairment by ameliorating blood lipid profile, reducing GLU and citric acid, as well as elevating the level of valine and leucine. Hippuric acid is glycine conjugate of benzoic acid, which was involved into phenylalanine metabolism.It has been report that urinary hippuric acid was decreased in atherosclerosis rabbits, which might be related to the dietary intake and gut microbial metabolism[42,43]. Delaney J. et al. had reported that urinary phenylacetylglycine was raised in animals exhibiting abnormal phospholipids accumulation and might be a surrogate biomarker for phospholipidosis[44].Increased phenylacetylglycine caused by HFD consumption was restored by AEE treatment, suggesting the improvement of phospholipid metabolism and the reduction of phospholipid accumulation.

Most studies have demonstrated the positive effects of dietary DHA on cardiovascular health, that is, DHA can reduce inflammation and total body fat and attenuate dyslipidemia[45]. Azelaic acid, a nine carbon saturated aliphatic dicarboxylic acid, can inhibit atherosclerosis development and exert beneficial effect on hepatic key enzymes of carbohydrate metabolism[46,47]. In present study, we observed that the levels of DHA and azelaic acid in AEE group were lower than those in HFD group.The possible explanation for the results was that AEE increased the consumption of DHA and azelaic acid to produce inhibitory effect on inflammation and carbohydrate metabolism to against atherosclerosis development.Different studies have suggested a relationship between a high dietary intake of elaidic acid and increased risk of coronary artery disease through promoting lipid droplets accumulation[48].HFD significantly increased the level of elaidic acid, which might be one of the reasons to explain the hepatic steatosis.Results showed that AEE treatment improved liver fat pathological changes, whereas no significant difference of elaidic acid was observed between HFD and AEE groups.It was speculated that AEE might have other therapeutic mechanism on the liver pathological changes induced by HFD.

Pantothenic acid is a B-group vitamin important for lipid metabolism and coenzyme A synthesis.Previous study has shown that HFD could decrease the pantothenic acid level in urine through the suppression of TCA cycle[49].As a tryptophan metabolite, indoxyl sulfate is a circulating uremic toxin.It has been reported that indoxyl sulfate could stimulate glomerular sclerosis, increase systemic oxidative stress and induce endothelial dysfunction[50,51].In HFD group, the reduced excretion of indoxyl sulfate and pantothenic acid might be potential biomarkers of atherosclerosis, and that could accelerate the development of atherosclerosis.However, no significant difference of pantothenic acid and indoxyl sulfate was observed between HFD and AEE groups.Wei dong dai et al.has reported that p-Cresol glucuronide and riboflavin are identified as potential biomarkers in endothelial dysfunction rats[52]. Like indoxyl sulfate, p-Cresol glucuronide is also a uremic toxin, which can inhibit endothelial cells proliferation and wounded endothelium repairment[53]. AEE treatment showed favorable inhabitation of p-Cresol glucuronide, indicating that anti-atherosclerosis efficacy of AEE might ascribe to the protection of endothelial cells. Riboflavin plays key role in maintaining health, which is required in many biological process such as energy production, red blood cell formation and reproduction. Moreover, riboflavin uptake can counteract oxidative stress and minimize cardio-vascular risk[54]. Oxidative stress caused by HFD might increase riboflavin consumption to lead the reduction of riboflavin in the urine. AEE might aggravate riboflavin consumption to produce anti-atherosclerosis effects, resulting in further reduction of riboflavin in urine. Phenylglucuronide, phenol sulphate and acetylcysteine were increasingly excreted in the urines of the hamsters fed with HFD. They are usually served as antioxidants. In contrast to the healthy rats, increased phenylglucuronide, phenol sulphate and acetylcysteine in the urine of the atherosclerosis hamster suggested that more antioxidants were possibly produced to defense the increasing oxidative stress during pathological progression, and thereafter excreted massively in urine[55]. AEE treatment might inhibit the oxidative stress, and thus down-regulated levels of acetylcysteine and phenol sulphate. Niacinamide, spermidine and DL-2-AC were reduced in hamster with atherosclerosis, suggesting that there was a significant negative relationship between these metabolites and atherosclerosis. Very little is known about the mechanism of the reductions of these metabolites, and more studies are needed to explore their functional roles in atherosclerosis.

In the present work, the anti-atherosclerosis effect of AEE was confirmed by blood biochemistry, pathological examination and metabolomic analysis. Our results showed that AEE could significantly reduce HFD-induced body weight gains, normalize disturbed blood biochemistry, reduced excessive fat accumulation in hepatocyte and ameliorate pathological lesions of aorta. Furthermore, the PLS-DA score plots showed the complete distinction of HFD-induced atherosclerotic hamsters and AEE-treated hamsters. AEE effectively inhibited the metabolic alternations induced by atherosclerosis. Based on the metabonomic approach, the disturbed global metabolic profiling mainly associated with glycerophospholipid metabolism, amino acid metabolism, energy metabolism, riboflavin metabolism, and pantothenate and CoA biosynthesis was improved after AEE treatment. These findings demonstrated that UPLC-Q-TOF/MS-based metabonomic approach was a powerful tool to explore the underlying mechanism of atherosclerosis, and also be helpful for understanding the possible mechanism of AEE for anti-atherosclerosis.

REFERENCES OMITTED

（发表于《Scientific Reports》，院选 SCI，IF：4.529）

Gymnadenia conopsea (L.) R.Br.: A Systemic Review of the Ethnobotany, Phytochemistry, and Pharmacology of an Important Asian Folk Medicine

Xiaofei SHANG*, Xiao GUO*, Yu LIU, Hu PAN,
Xiaolou MIAO**, Jiyu ZHANG**

Abstract: Key Laboratory of New Animal Drug Project, Gansu Province, Key Laboratory of Veterinary Pharmaceutical Development of Ministry of Agriculture, Lanzhou Institute of Husbandry and Pharmaceutical Sciences of Chinese Academy of Agricultural Science, Lanzhou, China *Gymnadenia conopsea* (L.) R. Br. (Orchidaceae) is a perennial herbaceous orchid plant that grows widely throughout Europe and in temperate and subtropical zones of Asia. In China, its tuber has been used in traditional Chinese medicines, Tibetan medicines, Mongolian medicines and other ethnic medicines, and taken to treat numerous health conditions. The present paper provides a review of the traditional uses, phytochemistry, biological activities, and toxicology to highlight the future prospects of the plant. More than 120 chemical compounds have been isolated, and the primary components are glucosides, dihydrostilbenes, phenanthrenes, aromatic compounds, and other compounds. G. conopsea and its active constituents possess broad pharmacological properties, such as the tonifying effect, anti-oxidative activity, anti-viral activity, immunoregulatory, antianaphylaxis, antigastric ulcer, sedative, and hypnotic activities, etc. However, overexploitation combined with the habitat destruction has resulted in the rapid decrease of the resources of this plant, and the sustainable use of G. conopsea is necessary to study. Meanwhile, the toxicity of this plant had not been comprehensively studied, and the active constituents and the mechanisms of action of the tuber were still unclear. Further, studies on G. conopsea should lead to the development of scientific quality control and new drugs and therapies for various diseases; thus, its use and development require additional investigation.

Key words: Gymnadenia conopsea; Traditional medicine; Glucosides; Tonifying activity;

---

* Co-author first. Citation: Shang X, Guo X, Liu Y, Pan H, Mao X and Zhang J (2017) Gymnadenia conopsea (L.) R.Br.: A Systemic Review of the Ethnobotany, Phytochemistry, and Pharmacology of an Important Asian Folk Medicine. doi: 10.3389/fphar.2017.00024

** Corresponding author, E-mail: Xiaolou Miaomiaoxiaolou@caas.cn, Jiyu Zhang shangxf928@126.com

Anti-viral activity

Abbreviations: BSA, Bovine Serum Albumin; CAT, Catalase; DNA, Deoxyribonucleic Acid; DPPH, 1, 1-diphenyl-2-picrylhydrazyl; DTH, Delayed type hypersensitivity; HBsAg, Hepatitis B Surface Antigens; HBV, Hepatitis B virus; HDL-C, High Density Lipoprotein-Cholesterol; HE, Hemoglobin Electrophoresis; HIV-1, Human Immunodeficiency Virus-1; HPLC, High-performance Liquid Chromatography; HPLC-DAD-MS^n, High-performance Liquid Chromatography-diode Array Detection-tandem Mass Spectrometry; HPSEC-MALLS/RID, High Performance Size Exclusion Chromatography CoupledwithMulti-angle Laser LightScattering/refractive IndexDetector; IC50, 50% inhibition concentration; LD50, Median Lethal Dose; LDL-C, Low Density Lipoprotein-Cholesterol; M_W, Weight-average Molecular Weight; MDA, Malondialdehyde; MeOH, Methyl Alcohol; PACE, Polysaccharide Analysis by Carbohydrate Gel Electrophoresis; PCR, Polymerase Chain Reaction; SOD, superoxide dismutase; TC, Total Cholesterol; TCM, Traditional Chinese Medicine; TG, Triglyceride; WHO, World Health Organization.

1 INTRODUCTION

Gymnadenia conopsea (L.) R.Br. (Orchidaceae) is a perennial herbaceous flowering plant that is distributed from 200 to 4 700m altitude throughout northern Europe, including England, Ireland, Russia, etc., and temperate and subtropical zones in Asian countries, including Nepal, China, Japan, and the Korean peninsula (Commission of Flora Reipublicae Popularis Sinicae, 2004; http://frps.eflora.cn/frps).For thousands of years, due to the prominent effects on invigorating the spleen, nourishing the lungs and blood, regenerating body fluid, and controlling bleeding with astringents, the tubers of G.conopsea was ascribed as a reinforcing agent of traditional medicines in China.It has been primarily used to treat kidney asthenia, cough, and dyspnea induced by lung asthenia, consumption diseases, neurasthenia, chronic diarrhea, morbid leucorrhea, chronic hepatitis, and other diseases in some Asian countries (Chinese Materia Editorial Committee, State Chinese Medicine Administration Bureau, 2002). Because the contour of the tuber is similar to the palm of the human hand, the tuber was given the Chinese name Shou Zhang Shen, meaning "ginseng likes palm hands" (Figure 1).In 1977, the tuber of G.conopsea was listed in the Pharmacopeia of the People's Republic of China (Committee for the Pharmacopoeia of P. R.China, 1977).Now, it is widely used as a folk medicine and traditional health food by Tibetans, Mongolians, the Han people, and other ethnic groups in China.

Because of the marked therapeutic effects and nutritional actions, researchers have widely investigated the properties of the tuber of G.conopsea Modern pharmacological studies have shown that it possesses broad pharmacological properties and can be used in the following treatments: tonifying effect, antioxidative, anti-viral, gastric ulcer prevention, anti-aging, immunoregulatory, antianaphylaxis, sedative, hypnotic, etc.Most of these actions have closely matched traditional uses. The chemical compounds from this plant have also been extensively studied, and glucosides, toluylenes, dihydrostilbenes, phenanthrenes, aromatic compounds, and other compounds have been isolated and identified.

In this review, advances in the ethnobotanical, phytochemical, biological and pharmacological activities, and toxicology of G.conopsea are presented and critical assessment. And the data supports its use and exploitation in new drugs.

2 BOTANICAL DESCRIPTION

Commission of Flora Reipublicae Popularis Sinica According to the description by Meekers et al. (2012) and the Commission of Flora Reipublicae Popularis Sinicae (2004), Gymnadenia conopsea (L.) R.Br.is an apolycarpic, perennial, terrestrial, and fragrant orchid herb that belongs to the Gymnadenia genus of Orchidaceae family, and distributs at forests, grasslands, and waterlogged meadows from 200 to 4 700m altitude. It has about 69 synonyms of this species, but only Gymnadenia conopsea (L.) R.Br.is an accepted and approved name in the World. The stem is 20–60cm, erect, slim, terete, or angled above and leafy with 2–3 brown membranous sheaths at the base. The leaves are green with dimensions of (5.5–15) × (1–2.5) cm, and the lower leaves are erect to slightly spreading, more, or less narrowly oblong–lanceolate or linear–lanceolate, obtuse to subacute, and slightly hooded at the apex, entire, keeled, and folded and have 1–2 or more veins on each side of the midrib; the upper 2–3 leaves are smaller, lanceolate or bract–like, and taper to a fine point. The bracts are green and usually tend toward violet at the edges, and they are lanceolate and taper to a fine point at the apex. The raceme is 5.5–15cm long and ranges in color from pale pink to lilac (rarely white or bright magenta), and it is strongly scented with a flowering season from July to August. Inflorescence 11–26cm, slender; peduncle with one to a few scattered, lanceolate bracts 1.5–6cm; rachis 4–12cm, densely many flowered; floral bracts lanceolate, often longer than ovary and flower, apex long acuminate–caudate. And flowers fragrant, pink, rarely pinkish white; ovary 5–8mm including pedicel. Dorsal sepal broadly elliptic to broadly ovate–elliptic; lateral sepals reflexed, obliquely ovate, (4–5.5) × (3–4) mm, 3–veined, margin revolute, apex acute. Petals obliquely ovate–triangular, 3–veined, apex acute; lip spreading, broadly cuneate–obovate. There are two tubers with dimensions of (14–30) × (7.5–24) mm; the tubers are palmate lobed with thick segments that are tapering and obtuse, pressed together, and split halfway to the base into 3–6 lobes. The short and thick roots are sparse and grow horizontally or even toward the soil surface. The fruit has dimensions of 8.69.3× (2.6–2.7) mm, and they are erect with six ribs. The seeds have dimensions of 0.3× 0.1mm and are produced in large numbers (http://www.theplantlist.or g*; www.efloras.or g**; Figure 1).

3 TRADITIONAL USES

G.conopsea is widely distributed in northern Europe and certain Asian countries. Like some TCM, G.conopsea is not used in folk medicine, and just has been considered a fragrant orchid plant in some European countries. So, studies on the ethnopharmacology and clinical uses of the plant have been mainly focused on Asian countries, such as China, Nepal, and Japan.

---

* The Plant List, version 1 (Accessed 1 January, 2010).

** The Flora of China.

In China, G. conopsea is primarily distributed in Heilongjiang, Jilin, Liaoning, Hebei, Shanxi, Shan'xi, Gansu, Sichuan, and Yunnan Provinces and Inner Mongolia and Xizang Autonomous Regions at altitudes of 265-4 700m (Commission of Flora Reipublicae Popularis Sinicae, 2004). The tuber has been employed as a reinforcing agent of folk medicines and widely used as traditional Chinese medicine (Han national medicine), Tibetan medicine, Mongolian medicine, Baiyao (the Bai national medicine), Chaoyao (the Korean national medicine), and Naxiyao (the Naxi national medicine) to treat various diseases in China. The tuber has also been employed as a health care product with other medicines or food to improve the body and prevent illness, and could be made to tincture and galenical to treat impotence and the bronchial asthma in China and Russia, respectively (Mamedov and Craker, 2001; Matsuda et al., 2004; Gutierrez, 2010). But at the same time, because of the rare resource of this plant, the tuber of Gymnadenia crassinervis, Coeloglossum viride var. bracteatum, and Spiranthea lancea have been used as substitutes for G. conopsea in certain regions, such as Tibet region (Xie et al., 2005; Zi, 2008; Xue et al., 2009).

Figure 1. Photographs of *Gymnadenia conopsea* (L.) R.Br.and the tuber.

Note: We also thanks for the provider of Figure 1, Zhou Yao from http://www.plantphoto.cn.

As a traditional Tibetan medicine, the G. conopsea tuber has been widely used in the Tibetan region to treat lung disease and weakening by invigorating the kidney and moisturizing the lungs. The plant is known as "Wangla" and it has been recorded in the "Sibuyidian" the classical book of Tibetan medicine, since the eighth century. In Tibetan medicine, it could be used as a single medicine or as one composition mixed with other medicines to treat diseases. For example, after grinding to powder, it (30g) concoction with bee honey (40g), Rhizome Gastrodiae (30g), Radix Phlomii (30g), Herb Drosera peltat (30g), and Rhododendron parvifolium (30g) could be used to treat impotence, spermatorrhea, anemia, and insomnia for 3g twice per day (Chinese Materia Editorial Committee, State Chinese Medicine Administration Bureau, 2002). And according to the database of Tibetan prescriptions, out of

4500 traditional prescriptions in Tibet Autonomous Region, the G.conopsea tuber was used 104 times (rate of 2.3%), and 33 prescriptions were used to invigorate the body, strengthen the Yang, and lengthen human life; 26 prescriptions were used to treat kidney diseases; 12 prescriptions were used to treat gout and arthromyodynia diseases; 11 prescriptions were applied to treat lung disease; 7 prescriptions were used to treat eye diseases; and other prescriptions were used to treat parasitic diseases and additional diseases (Ji et al., 2009; Xue et al., 2009). Currently, the tuber of G.conopsea is combined with other medicines in various preparations to treat a number of diseases. Five preparations have been listed in the Chinese Pharmacopeia and approved by the State Food and Drug Administration of China of the People's Republic of China. Medicines such as "Shi Wei Shou Shen Powder" and "Fu Fang Shou Shen Wan" have been widely used to treat kidney asthenia, impotence and spermatorrhea, among other disorders (http://www.sfda.gov.cn, 2014; Table 1). Except the above effects on the clinic, the tuber of G.conopsea also could be used to treat hepatitis B by folk doctors only in the Tibetan region (Chinese Materia Editorial Committee, State Chinese Medicine Administration Bureau, 2002). At the same time, it has also been used as a common food item by local people in the Tibetan region, where it is usually cooked with vegetables and rice. For example, the tuber is used as an important ingredient in the famous dish "Shiguo Ji" (chicken cooked in a stone hotpot).

In the traditional Mongolian medicine, the tuber of G.conopsea is named "Erihaoteng" and has been historically recorded by many classical folk medicine books of the Inner Mongolia Autonomous Region. In additional, it has been widely used to treat kidney asthenia, lumbago and leg pain, light scurvy, spermatorrhea, and impotence (Gege et al., 2013). The tuber of G.conopsea and other traditional medicines have been combined into different preparations to treat various diseases. For example, the tuber and an additional 36 medicines are known as Shouzhangsheng-37 pills, which are used to treat kidney cold and asthenia, edema, tinnitus, spermatorrhea, impotence, stomach diseases, dyspepsia, and other diseases, and this preparation has been approved by the State Administration of Traditional Chinese Medicine of the People's Republic of China [http://www.sfda.gov.cn, 2014; (Si and Liu, 2013); Table 1]. Meanwhile, Wu et al. (2014) reported that in clinical practice of Mongolian medicine, after administrating orally two Shenzhujin pills (the tuber is the primary component) for three times per day for 21 days, the kidney deficiency of one patient (65 years) have been cured.

Furthermore, the tuber of G.conopsea is also used in Korean national medicine, Bai national medicine, and Naxi national medicine, where it is known as Yinyang Cao, Foshousheng, and Kaishelabei, respectively. However, it has been used as a reinforcing agent within all of these traditional medicines and used to treat similar diseases as previously indicated, such as invigorating the body, strengthening the Yang, etc. (Jia and Li, 2005). But up to now, except the experience-based uses of this plant, the relevant evidence-based clinical uses and the safety data scientifically are very rare and should be investigated further.

Table 1　Preparations in which Gymnadenia conopsea (L.) R.Br.was the primary component as listed in the Chinese Pharmacopeia and approved by the government *.

Preparation name	Main compositions	Usage
Shi Wei Shou Shen San	Tuber Gymnadenia, Cinnamomi Cortex, Myristicae Semen, Piperis Longi Fructus, Asparagi Radix, Granati Pericarpium, Canavaliae Semen, Carthami Flos, Moschus, Bear Gall.	Invigorating the kidney and treating spermatorrhea.
Fu Fang Shou Shen Wan	Tuber Gymnadenia, Polygonati Rhizoma, Cynomorii Herba, Chebulae Fructus, Rhizoma Mirabilis Himalaica, Asparagi Radix, Cordyceps, Tribuli Fructus, Rhizoma Przewalskia Tangutica, Herba Pleurospermum.	Warming the kidney and activating yang. Treating insufficiency of kidney - YANG, damage of essence, impotence, and spermatorrhea, etc.
Shou Shen Shen Bao Jiao Nang	Tuber Gymnadenia, Polygonati Rhizoma, Asparagi Radix, Folium Rhododendron Anthopogonoides, Cordyceps.	Warming the kidney and invigorating yin. Treating giddy dazzled, tinnitus, lumbar genu aching, and limp is faint induced by kidney asthenia.
Shou Zhang Shen Sanqi Wan	Tuber Gymnadenia, Calcite, Alpiniae Oxyphyliae Fructus, Granati Pericarpium, Zingiberis Rhizoma, Feces Trogopterori, Myristicae Semen, Piperis Longi Fructus, Gecko, Chebulae Fructus, Sal Ammoniac, Folium Rhododendron Anthopogonoides, etc.	Invigorating the kidney and strengthening yang. Treating kidney asthenia, edema, tinnitus, impotence, and spermatorrhea, etc.
Fu Fang Shou Shen Yi Zhi Jiao Nang	Tuber Gymnadenia, Polygoni Multiflori Radix Praeparata, Acanthopanacis Senticosi Radix, Polygonati Rhizoma, Angelicae Sinensis Radix, Asteragali Radix, Lycii Fructus, Schisandrae Chinensis Fructus, Corni Fructus, Rhizoma Polygala, Acori Tatarinowii Rhizoma, Paeoniae Radix Rubra.	Invigorating the kidney and liver, and treating the deffciency of liver and kidney, amnesia, palpitation, and insomnia induced by Qi and blood hemophthisis.

* Cited from "Chinese Pharmacopeia" and the Website: http: //www.sfda.gov.cn.

4　PHYTOCHEMISTRY

Approximately 129 compounds have been isolated and identified from G.conopsea to date, and most of compounds isolated from the tuber. Forty-nine glucosides compounds (one of the most important components) were identified, which could be divided into benzylester glucosides and other glucosides. Meanwhile, dihydrostilbenes, phenanthrenes, aromatic compounds, polysaccharides, and other compounds were also isolated and reported (Table 2; Figure 2). The different chemical compositions of G.conopsea provide the foundation for its different pharmacology activities.

4.1 Glucosides

As one of the most important components, glucosides are widely studied and separated from the tuber of G.conopsea.According to the structural conformation of the glucosides, they have been divided into benzylester glucosides and other glucosides.

4.2 Benzylester glucosides

Benzylester glucosides are produced by combining 2-isobutyl tartaric acid or 2-isobutyl hydroxysuccinic acid with 4-glycosylbenzyl alcohol. According to the differences in their organic acids, they are classified as (2R, 3S) -2-isobutyl tartaric acid derivates and (2R) -2-isobutyl hydroxysuccinic acid derivatives. Approximately 29 compounds have been isolated and identified. In 2006, Morikawa et al. (2006a, b) isolated the following compounds: gymnoside Ⅰ (1), gymnoside Ⅱ (2), gymnoside Ⅲ (3), gymnoside Ⅳ (4), gymnoside Ⅴ (5), gymnoside Ⅵ (6), gymnoside Ⅶ (7), gymnoside Ⅷ (8), gymnoside Ⅸ (9), gymnoside Ⅹ (10), loroglossin (11), dactylorhin A (12), dactylorhin B (13) and militarine (14). Meanwhile, (-) -4- [β-D-glucopyranosyl- (1→4) -β-D-glucopyranosyloxy] benzyl alcohol (15), (+) -4- [α-D-glucopyranosyl- (1→4) -β-D-glucopyranosyloxy] benzyl alcohol (16), (-) -4- [β-D-glucopyranosyl- (1→3) -β-D-glucopyranosyloxy] benzyl alcohol (17), (-) -4- [β-D-glucopyranosyl- (1→3) -β-D-glucopyranosyloxy] benzyl ethyl ether (18), (-) - (2R, 3S) -1- (4-β-D-glucopyranosyloxybenzyl) -2-O-β-D-glucopyranosy] -4- {4- [α-D-glucopyranosyl- (1→4) -β-D-glucopyranosyloxy] benzyl} -2-isobutyltartrate (19), (-) - (2R, 3S) -1- (4-β-D-glucopyranosyloxybenzyl) -2-O-β-D-glucopyranosyl-4- {4- [β-D-glucopyranosyl- (1→3) -β-D-glucopyranosyloxy] benzy l} -2-isobutyltartrate (20), (-) - (2R, 3S) -1- {4- [β-D-glucopyranosy l- (1→3) -β-D-glucopyranosyloxy] benzyl} -2-O-β-D-glucopyrano syl-4- (4-β-D-glucopyranosyloxybenzyl) -2-isobutyl-tartrate (21), (-) - (2R, 3S) -1- (4-β-D-glucopyranosyloxybenzyl) -4- {4- [β-D-glu copyranosyl- (1→6) -β-D-glucopyranosyloxy] benzyl} -2-isobutyl tartrate (22), (-) - (2R, 3S) -1- (4-β-D-glucopyranosyloxybenzyl) -4-methyl-2-isobutyltartrate (23) and (-) - (2R) -2-O-β-D-glu copyranosyl-4- (4-β-D-glucopyranosylbenzyl) -2-isobutyltartrate (24) were isolated from the tuber in 2008. And the further studies showed that at 10^{-5} M, the inhibition rates of the above compounds foracetylcholine esterase were <10%, and the positive drug donepezil gave an inhibition rate of 77. 2%.Then, the results presented that at 10^{-5} M, the inhibition rates of the above compounds formonoamine oxidase-B were <15. 2%, and the positive drug pargyline had an inhibition rate of 94. 5% (Zi et al., 2008). Meanwhile, dactylorhin E (25), coelovirins A (26), coelovirins B (27), coelovirins D (28), and coelovirins E (29) also were isolated (Zi, 2008).

4.3 Other glucosides

Other glucosides have also been isolated from the tuber of G.conopsea.In 2009, Yang identified 4-hydroxybenzyl-β-D-glucopyranoside (30), 4-methylphenyl-β-D-glucopyranoside (31), and4-hydroxyphenyl-β-D-glucopyranoside (32).The following compounds have also been isolated (Li et al., 2001; Morikawa et al., 2006b; Zi, 2008): 4-methoxymethylbenzyl-β-D-glucoside

(33), bis (4-hydroxybenzyl) -ether mono-β-D-glucopyranoside (34), 4- (β-D-glucopyranosyloxy) benzoic aldehyde (35), 4- (β-D-glucopyranosyloxy) benzyl ethyl ether (36), phenyl-β-D-glucopyranoside (37), 4-formylphenyl-β-D-glucopyranoside (38), benzyl-β-D-glucopyranoside (39), trans-ferulic acid-β-D-glucoside (40), cis-ferulic acid-β-D-glucoside (41), N^6 - (4 - hydroxybenzyl) adenine riboside (42), daucosterol (43), dioscin (44), dactylose A (45), dactylose B (46), n - butyl - P - D - fructopyranose (47), and thymidine (48).

Table 2 Compounds isolated fromGymnadenia conopsea (L.) R.Br. (the structure of the primary compounds are illustrated in Figure 2).

No.	Names	Parts	
1	Gymnoside Ⅰ	Tuber	
2	Gymnoside Ⅱ	Tuber	
3	Gymnoside Ⅲ	Tuber	
4	Gymnoside Ⅳ	Tuber	
5	Gymnoside Ⅴ	Tuber	
6	Gymnoside Ⅵ	Tuber	
7	Gymnoside Ⅶ	Tuber	
8	Gymnoside Ⅷ	Tuber	
9	Gymnoside Ⅸ	Tuber	
10	Gymnoside Ⅹ	Tuber	
11	Loroglossin	Tuber	
12	Dactylorhin A	Tuber	
13	Dactylorhin B	Tuber	
14	Militarine	Tuber	
15	(-) -4- [β-D-glucopyranosyl- (1→4) -β-D-glucopyranosyloxy] benzyl alcohol	Tuber	
16	(+) -4- [α-D-glucopyranosyl- (1→4) -β-D-glucopyranosyloxy] benzyl alcohol	Tuber	
17	(-) -4- [β-D-glucopyranosyl- (1→3) -β-D-glucopyranosyloxy] benzyl alcohol	Tuber	
18	(-) -4- [β-D-glucopyranosyl- (1→3) -β-D-glucopyranosyloxy] benzyl ethyl ether	Tuber	
19	(-) - (2R, 3S) -1- (4-β-D-glucopyranosyloxybenzyl) -2-O-β-D-glucopyranosy] -4- {4- [α-D-glucopyranosyl-Tuber (1-4) -β-D-glucopyranosyloxy] benzyl} -2-isobutyltartrate		
20	(-) - (2R, 3S) -1- (4-β-D-glucopyranosyloxybenzyl) -2-O-β-D-glucopyranosyl -4- {4- [β-D-glucopyranosyl- (1-3) -β-D-glucopyranosyloxy] benzyl} -2-isobutyltartrate	Tuber	Zi, 2008; Zi et al., 2008
21	(-) - (2R, 3S) -1- {4- [β-D-glucopyranosyl- (1→3) -β-D-glucopyranosyloxy]	Tuber	Zi, 2008; Zi et al., 2008

(continued)

No.	Names	Parts	
	benzyl} -2-O-β-D-glucopyranosyl-4- (4-β-D-glucopyranosyloxybenzyl) -2-isobutyltartrate		
22	(-) - (2R, 3S) -1- (4-β-D-glucopyranosyloxybenzyl) -4- {4- [β-D-glucopyranosyl- (1→6) -β-D-glucopyranosyloxy] benzyl} -2-isobutyltartrate	Tuber	Zi, 2008; Zi et al., 2008
23	(-) - (2R, 3S) -1- (4-β-D-glucopyranosyloxybenzyl) -4-methyl-2-isobutyltartrate	Tuber	Zi, 2008; Zi et al., 2008
24	(-) - (2R) -2-O-β-D-glucopyranosyl-4- (4-β-D-glucopyranosylbenzyl) -2-isobutyltartrate	Tuber	Zi, 2008; Zi et al., 2008
25	Dactylorhin E	Tuber	
26	Coelovirins A	Tuber	
27	Coelovirins B	Tuber	
28	Coelovirins D	Tuber	
29	Coelovirins E	Tuber	
30	4-Hydroxybenzyl-β-D-glucopyranoside	Tuber	
31	4-Methylphenyl-β-D-glucopyranoside	Tuber	
32	4-Hydroxyphenyl-β-D-glucopyranoside	Tuber	
33	4-Methoxymethylbenzyl-β-D-glucoside	Tuber	
34	bis (4-hydroxybenzyl) -ethermono-β-D-glucopyranoside	Tuber	
35	4- (β-D-glucopyranosyloxy) benzoic aldehyde	Tuber	
36	4- (β-D-glucopyranosyloxy) benzyl ethyl ether	Tuber	
37	Phenyl-β-D-glucopyranoside	Tuber	
38	4-Formylphenyl-β-D-glucopyranoside	Tuber	
39	Benzyl-β-D-glucopyranoside	Tuber	
40	trans-ferulic acid-β-D-glucoside	Tuber	
41	cis-ferulic acid-β-D-glucoside	Tuber	
42	N^6- (4-hydroxybenzyl) adenine riboside	Tuber	
43	Daucosterol	Tuber	
44	Dioscin	Tuber	
45	Dactylose A	Tuber	
46	Dactylose B	Tuber	
47	n-Butyl-β-D-fructopyranose	Tuber	
48	Thymidine	Tuber	Morikawa et al., 2006b
49	Batatasin III	Tuber	Matsuda et al., 2004
50	3′-O-methylbatatasin Ⅲ	Tuber	Matsuda et al., 2004

(continued)

No.	Names	Parts	
51	3′, 5-Dihydroxy-2-(4-hydroxybenzyl)-3-methoxy-bibenzyl	Tuber	Matsuda et al., 2004
52	3, 3′-Dihydroxy-2-(4-hydroxybenzyl)-5-methoxy-bibenzyl	Tuber	Matsuda et al., 2004
53	Gymconopin D	Tuber	Matsuda et al., 2004
54	3, 3′-Dihydroxy-2, 6-bis (4-hydroxybenzyl)-5-methoxybibenzyl	Tuber	Matsuda et al., 2004
55	5-O-methylbatatacin Ⅲ	Tuber	Yoshikawa et al., 2005
56	2-(4-Hydroxybenzyl)-3′-O-methylbatatacin Ⅲ	Tuber	Yoshikawa et al., 2005
57	Arundinin	Tuber	Yoshikawa et al., 2005
58	Arundin	Tuber	Yoshikawa et al., 2005
59	Bulbocodin C	Tuber	Yoshikawa et al., 2005
60	Bulbocodin D	Tuber	Yoshikawa et al., 2005
61	Gymconopin A	Tuber	Matsuda et al., 2004
62	Gymconopin B	Tuber	Yoshikawa et al., 2005
63	Gymconopin C	Tuber	Yoshikawa et al., 2005
64	1-(4-Hydroxybenzyl)-4-methoxy-9, 10-dihydrophenanthrene-2, 7-diol	Tuber	Matsuda et al., 2004
65	1-(4-Hydroxybenzyl)-4-methoxyphenanthrene-2, 7-diol	Tuber	Matsuda et al., 2004
66	Dihydrophenanthrene-4, 5-diol	Tuber	Matsuda et al., 2004
67	2-Methoxy-9, 10-4-methoxy-9, 10-dihydrophenanthrene-2, 7-diol	Tuber	Matsuda et al., 2004
68	Blestriarene A	Tuber	Matsuda et al., 2004
69	Blestriarene A	Tuber	Matsuda et al., 2004
70	Blestriarene A	Tuber	Matsuda et al., 2004
71	9, 10-Dihydro blestriarene B	Tuber	Matsuda et al., 2004
72	9, 10-9′, 10′-Dihydro blestriarene C	Tuber	Matsuda et al., 2004
73	4-Hydroxybenzyl alcohol	Tuber	Cai et al., 2006
74	4-Hydroxybenzyl aldehyde	Tuber	Cai et al., 2006
75	Pinoresinol	Tuber	Morikawa et al., 2006b
76	4-Hydroxybenzoic acid	Tuber	Yue et al., 2010
77	Vanillic acid	Tuber	Yue et al., 2010
78	Trans-p-coumaric acid	Tuber	Yue et al., 2010
79	Cis-p-coumaric acid	Tuber	Yue et al., 2010

(continued)

No.	Names	Parts	
80	4-Benzaldehyde	Tuber	Morikawa et al., 2006b
81	4-Hydroxybenzyl methyl ether	Tuber	Zi, 2008
82	3, 5-Methoxy benzaldehyde	Tuber	Morikawa et al., 2006b
83	4- [(4-Hydroxyphenyl) methoxy] -benzenemethanol	Tuber	Morikawa et al., 2006b
84	4, 4′-Dihydroxydiphenyl methane	Tuber	Morikawa et al., 2006b
85	Phenol	Tuber	Morikawa et al., 2006b
86	5-Hydroxymethyl furfural	Tuber	Morikawa et al., 2006b
87	1, 2-Dihydroxy benzene	Tuber	Morikawa et al., 2006b
88	2, 6-Dimethoxy phenol	Tuber	Morikawa et al., 2006b
89	Eugenol	Tuber	Morikawa et al., 2006b
90	4-Hydroxybenzene	Tuber	Morikawa et al., 2006b
91	4-Methoxy phenylpropanol	Tuber	Morikawa et al., 2006b
92	4-Ethoxy phenylpropanol	Tuber	Morikawa et al., 2006b
93	Contra-hydroxybenzyl dithioether	Tuber	Morikawa et al., 2006b
94	Syringol	Tuber	Morikawa et al., 2006b
95	Syringaldehyde	Tuber	Morikawa et al., 2006b
96	Gastrodin	Tuber	Yue et al., 2010
97	4-Hydroxybenzoic aldehydes	Tuber	Zi, 2008
98	4-Hydroxybenzoic acid	Tuber	Yue et al., 2010
99	3-Hydroxybenzoic acid	Tuber	Yue et al., 2010
100	4-Hydroxyisophthalic acid	Tuber	Yue et al., 2010
101	3-Methoxy-4-hydroxybenzoic acid	Tuber	Yue et al., 2010
102	Arctigenin	Tuber	Yue et al., 2010
103	erythro-buddlenol E	Tuber	Zi, 2008
104	Lappaol A	Tuber	Yue et al., 2010
105	Lappaol F	Tuber	Yue et al., 2010
106	2 - Hydroxy - 2 - (4 - hydroxyphenylmethyl) - 4 - methylcyclopent-4-en-1, 3-dione	Tuber	Zi, 2008
107	2-Hydroxy-3- (4-hydroxyphenyl) -4-hydroxymethylcyclo-pent-2-enone	Tuber	Zi, 2008
108	3, 3′ - Dihydroxy - 4 - (4 - hydroxybenzyl) - 5 - methoxy-bibenzyl	Tuber	Morikawa et al., 2006b
109	2 - C - (4 - hydroxybenzyl) - α - L - xylo - 3 - ketohexulofuranosono-1, 4-lactone	Tuber	Morikawa et al., 2006b

(continued)

No.	Names	Parts	
110	Arabinose	Tuber	Sun et al., 2010
111	Xylose	Tuber	Sun et al., 2010
112	Mannitose	Tuber	Sun et al., 2010
113	Galactose	Tuber	Sun et al., 2010
114	Glucose	Tuber	Sun et al., 2010
115	Secoxyloganin	Tuber	Zi, 2008
116	Quercitin-3, 7-di-O-β-D-glucopyranoside	Tuber	Zi, 2008
117	Kaempferol-3-P-glycosido-7-β-glycoside	Flower	Schonsiegel and Ingomar, 1969
118	Quercetin-3-P-glycosido-7-glycoside	Flower	Schonsiegel and Ingomar, 1969
119	Astragalin	Flower	Schonsiegel and Ingomar, 1969
120	Isoquercetin	Flower	Schonsiegel and Ingomar, 1969
121	Cyclo (L-Leu-L-Tyr)	Tuber	Zi, 2008
122	Cyclo (L-Leu-L-Pro)	Tuber	Zi, 2008
123	Cyclo (L-Val-L-Tyr)	Tuber	Zi, 2008
124	Cyclo (L-Ala-L-Phe)	Tuber	Zi, 2008
125	Thymidine	Tuber	Yue et al., 2010
126	β-sitoterol	Tuber	Yue et al., 2010
127	Octadecylselyl acid	Tuber	Yue et al., 2010
128	Di- (p-hydroxybenzyl) disulfide	Tuber	Li et al., 2007
129	4. 4′-Dihydroxybenzyl sulfoxide	Tuber	Li et al., 2007

4.4 Dihydrostilbenes

Dihydrostilbenes are other important components of G. conopsea. The basic parental nucleus of these compounds is dihydrogen alkene. The C-2, -3, -4, -5, -6 and-3′ positions of the basic nucleus areusually substituted. In 2004 and 2005, batatacin Ⅲ (49), 3′-O-methylbatatacin Ⅲ (50), 3′, 5-dihydroxy-2- (4-hydroxybenzyl) -3-methoxybibenzyl (51), 3, 3′-dihydroxy-2- (4-hydroxybenzyl) -5-methoxybibenzyl (52), gymconopin D (53), 3, 3′-dihydroxy-2, 6-bis (4-hydroxybenzyl) -5-methoxybibenzyl (54), 5-O-methylbatatacin Ⅲ (55), 2- (4-hydroxybenzyl) -3′-O-methylbatatacin Ⅲ (56), arundinin (57), arundin (58), bulbocodin C (59), bulbocodin D (60) were isolated and identified from the tuber, respectively (Matsuda et al., 2004; Yoshikawa et al., 2005).

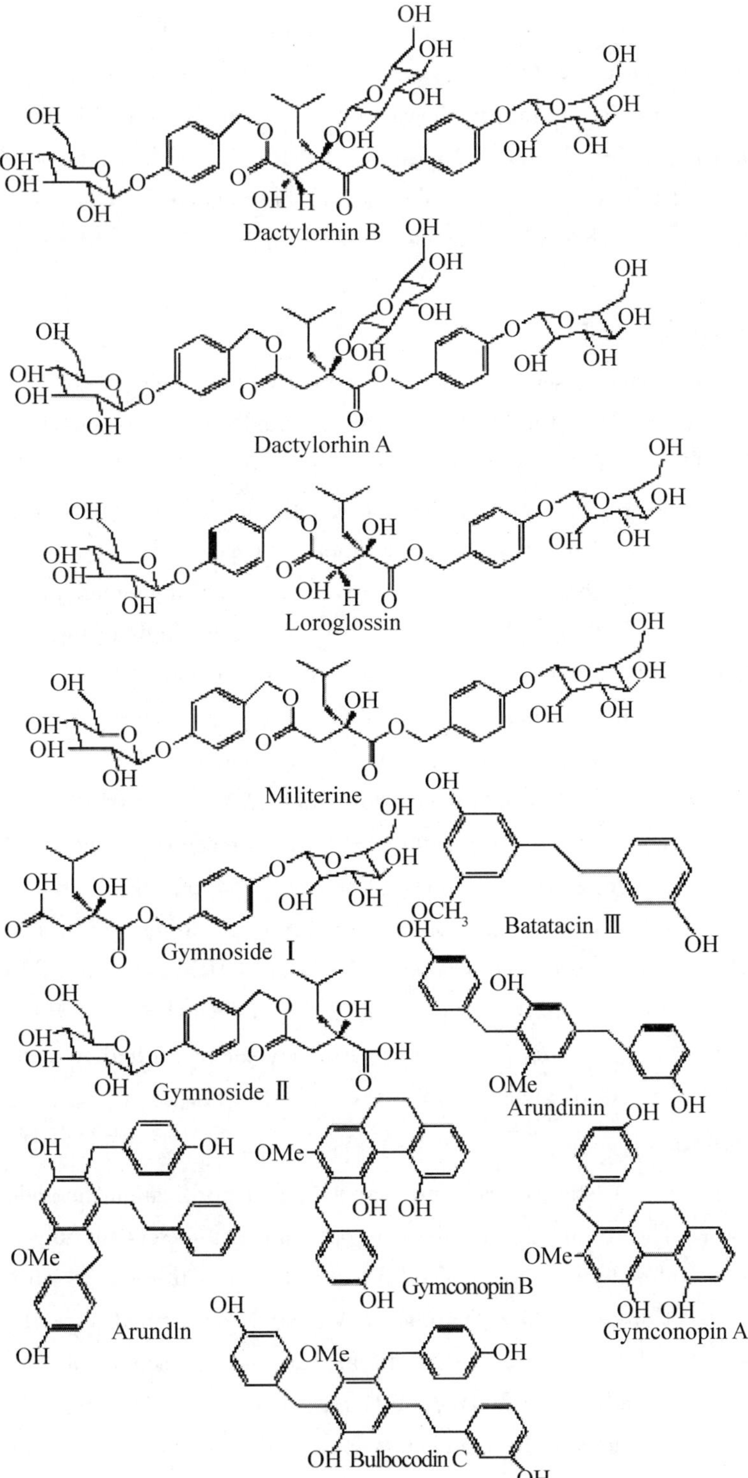

Figure 2. Chemical structures of the primary compounds from the tuber of Gymnadenia conopsea (L.) R.Br.

4.5 Phenanthrenes

A group of phenanthrene compounds was isolated from the tuber of G.conopsea and studied in recent decades, including gymconopin A (61), gymconopin B (62), gymconopin C (63), 1- (4-hydroxybenzyl) -4-methoxy-9, 10-dihydrophenanthrene-2, 7-diol (64), 1- (4-hydroxybenzyl) -4-methoxyphenanthrene-2, 7-diol (65), 2-methoxy-9, 10-dihydrophenanthrene-4, 5-diol (66), 4-methoxy-9, 10-dihydrophenanthrene-2, 7-diol (67), blestriarene A (68), blestriarene A (69), blestriarene A (70), 9, 10-dihydro blestriarene B (71), and 9, 10-9′, 10′-dihydro blestriarene C (72) (Matsuda et al., 2004; Yoshikawa et al., 2005).

4.6 Aromatic compounds

Approximately 20 aromatic compounds from G.conopsea have been studied, and most of them are phenol compounds.In 2006, Cai et al. (2006) isolated 4-hydroxybenzyl alcohol (73) and 4-hydroxybenzyl aldehyde (74) from the tuber. Then, pinoresinol (75), 4-hydroxybenzoic acid (76), vanillic acid (77), trans-p-coumaric acid (78), cis-p-coumaric acid (79), 4-benzaldehyde (80), 4-hydroxybenzyl methyl ether (81), 3, 5-methoxy benzaldehyde (82), 4-[(4-hydroxyphenyl) methoxy]-benzenemethanol (83), 4, 4′-dihydroxydiphenyl methane (84), phenol (85), 5-hydroxymethyl furfural (86), 1, 2-di-hydroxybenzene (87), 2, 6-dimethoxy phenol (88), eugenol (89), 4-hydroxybenzene (90), 4-methoxy phenylpropanol (91), 4-ethoxy phenylpropanol (92), contra-hydroxybenzyl dithioether (93), syringol (94), syringaldehyde (95), gastrodin (96), 4-hydroxybenzoic aldehydes (97), 4-hydroxybenzoic acid (98), 3-hydroxybenzoic acid (99), 4-hydroxyisophthalic acid (100), 3-methoxy-4-hydroxybenzoic acid (101), arctigenin (102), erythro-buddlenol E (103), lappaol A (104), lappaol F (105), 2-hydroxy-2-(4-hydroxyphenylmethyl) -4-methylcyclopent-4-en-1, 3-dione (106), 2-hydroxy-3-(4-hydroxyphenyl) -4-hydroxymethylcyclopent-2-enone (107), 3, 3′-dihydroxy-4- (4-hydroxybenzyl) -5-methoxybibenzyl (108), and 2-c- (4-hydroxybenzyl) -α-L-xylo-3-ketohexulofuranosono-1, 4-lactone (109) were isolated and identified from the tuber of G.conopsea (Morikawa et al., 2006b; Zi, 2008; Yue et al., 2010).

4.7 Polysaccharides

In 2009, Yang (2009) studied the four extraction and purification methods for the polysaccharides of G.conopsea.The results showed that the optimum parameters for solvent extraction were 80℃ for 2.5h, a ratio of material to water of 1∶30 with three repetitions, which yielded a polysaccharide extraction rate of up to 11.83%.The microwave extraction method were 250W for 20min, a ratio of material to water of 1∶30 with one repetition, and the extraction rate was up to 13.56%. The ultrasound extraction method was 240W, 60℃ for 10min, and three repetitions, and the extraction rate was 15.24%.Finally, the parameters for cellulose extraction were to add cellulose at 1% of the tuber concentration for 4h at 50℃ and pH 4.8, and the extraction rate was 16.58%. Thus, the cellulose method is the best among the four methods.Meanwhile, vapor phase chromatography was adopted to analyze the monosaccharides from the polysaccharides, and the results showed that the polysaccharides are composed of arabinose (110), xylose (111), mannitose (112), galactose (113), and glucose (114) (Sun et al., 2010). And in 2014, the water soluble polysaccharides from the tuber of G.conopsea collected at seven regions in China were investigated and compared using high performance size exclusion chromatography coupled with multi-angle laser light scattering/refractive index detector (HPSEC-MALLS/RID) and saccharide mapping based on polysaccharide analysis by carbohydrate gel electrophoresis (PACE), respectively. The results showed that the weight-average molecular weight (M_w) and the radius of gyration ($<S^2>z^{1/2}$) of polysaccharides were ranging from 4.46×10^5 to 7.41×10^5 Da and 73.3-94.2nm, respectively. And the exponent (v) values of $<S^2>z^{1/2}=kM_W^V$ were ranging from 0.36 to 0.42, which indicated that α-1, 4-and β-1, 3 (4) -glucosidic, α-1, 5-arabinosidic, β-1, 4-mannosidic and α-1, 4-D-galactosiduronic linkages existed in globular polysaccharides. The further studies results that the nitric oxide released from RAW 264.7 cells induced by polysaccharides were significantly affected by their α-1, 5-arabinosidic and β-1, 3 (4) -glucosidic, especially α-1, 4-D-galactosiduronic and β-1, 4-mannosidic linkages (Lin et al., 2015).

4.8 Other Compounds

From G.conopsea, six flavones were isolated from the flower, including secoxyloganin (115), quercitin-3, 7-di-O-β-D-glucopyranoside (116), kaempferol-3-β-glycosido-7-β-glycoside (117), and quercetin-3-β-glycoside-7-glycoside (118), astragalin (119) and isoquercetin (120) (Schonsiegel and Ingomar, 1969; Zi, 2008). Then, cyclo (L-Leu-L-Tyr) (121), cyclo (L-Leu-L-Pro) (122), cyclo (L-Val-L-Tyr) (123), cyclo (L-Ala-L-Phe) (124), thymidine (125), β-sitoterol (126), and octadecylselyl acid (127) were isolated (Yue et al., 2010; Zi et al., 2010). And di- (p-hydroxybenzyl) disulfide (128) and 4.4′-dihydroxybenzyl sulfoxide (129) also were identified (Li et al., 2007). Lan (2005) determined the trace element contents of the tuber of Gymnadenia conopsea (L.) R.Br., and the results showed that the Cu, Mn, Fe, Zn, Ni, Se, Cr, K, Ca, Mg, Cd, and Pb contents inμg/g were 7.6, 31.6, 60.3, 20.1, 2.1, 6.0, 2.0, 19, 000, 4100, 1800, 0.16, and 3.31, respectively.

5 PHARMACOLOGY

5.1 Tonifying Effect and Anti-Fatigue Activity

In China, the G.conopsea tuber has been used as a reinforcing agent for traditional medicines, and it has historically been used to treat kidney asthenia, lung asthenia, consumption diseases, neurasthenia, impotence, spermatorrhea, and other diseases. According to the theory of traditional Chinese medicine, kidney asthenia is related to weakness in the body. Thus, the tonifying activity (kidney-yang-tonifying activity) was the primary subject of study in China. The results showed that after administering orally 2 and 1g/kg the tuber for 10 days, the medicine relieved the symptoms in mice with yang deficiencies that had been induced by hydrocortisone (25mg/kg, i.g.), increased the body weight of kidney-yang-deficient mice (7.82 and 5.56g, $P>0.05$), prolonged the retention time on the rotating bar (19.57 ± 10.21min, $P<0.01$, and 5.90 ± 2.47min, $P<0.05$) and increased the kidney indexes of treated mice compared with those of the model group. And the increased weight and time on the rotating bar for the untreated control groups were 8.71g ($P>0.05$)

and 24.20±13.95min ($P<0.01$), respectively. Further, study showed that the tubers could also increase DNA synthesis in the spleen (48.89±18.68 and 46.63±13.80μg/g, $P<0.01$), kidneys (14.26±5.34 and 5.94±4.88μg/g, $P>0.05$), and liver (42.94±13.95μg/g and 42.02±17.69, $P<0.05$) of mice with yang deficiencies that were induced by hydrocortisone compared with the mice in the model group (27.25±13.18μg/g, $P<0.01$, 4.90±4.82μg/g, $P>0.05$, and 19.02±15.49μg/g). Theseresults suggested that the tuber has a significant tonifying effect on the kidney and strengthens the bodies of mice with kidney deficiencies that were induced by hydrocortisone (Lin, 2009).

In 2011, Zhao and Liu (2011) studied the anti-fatigue activity of the tuber of G.conopsea in mice. After administering orally (i.g.) aqueous extracts to mice (40, 20, and 10g/kg) for 6 days, the swimming times were markedly increased (1 441.6, 1 357.0, and 1 249.9 s, $P<0.01$) during the weight-bearing swimming tests in a dose-dependent manner. The swimming times of the control and positive groups (Radix ginseng, 20g/kg) were 491.1s and 1 685.9 s ($P<0.01$). According to the processing method of traditional Mongolia medicine in China, the tuber should be decocted in goat or cow milk before clinical use. After processing with a different method, Jin and Wang (2009) studied and compared the strengthening effects of the tuber of G.conopsea tonics. The results showed that compared with the control group (salinewater, 6.38±4.22min), a goat milk decoction (2g/kg.i.g.) could significantly increase the swimming time of mice (19.44 ± 12.6min, $P<0.01$); the unprocessed group (2g/kg) and cow's milk decoction group (2g/kg) had swimming times of 9.09±9.81min and 6.43±7.21min ($P>0.05$), respectively, but they did not show significant difference. All of the decoctions could increase the SOD activity and MDA content compared with that of the control group. Thus, the researchers indicated that after processing with goat milk, the tuber have a better effect with the strengthening tonics than unprocessed medicine (Table 4).

5.2 Anti-Oxidant Activity

The anti-oxidant activity of the tuber of G. conopsea was first comprehensive studied in 2006. The results showed that after isolating by Diaion HP-20 column chromatography (reverse-phase silica gel columnchromatography, Chromatorex ODSDM1020T, 100-200 mesh, Fuji Silysia Chemical, Ltd.), the MeOH-eluted fraction showed the radical scavenging activities for DPPH radical (SC_{50}=59.1μg/mL) and.O_2^- (IC_{50}=13.3μg/mL), but without the inhibitory activity on xanthine oxidase (IC_{50}>100μg/mL). Meanwhile, the radical scavenging activities for DPPH radical and O_2^- exhibiting inhibitory activity on formozan formation and xanthine oxidase of the acetone-eluted fraction were 55.4, 33.2, and 29.7μg/mL, respectively. But H_2O-eluted fraction didn't present any activities. Then, the anti-oxidant 13 compounds were studied further, some compounds showed the marked radical scavenging activities for DPPH radical and O_2^- exhibiting inhibitory activity on formozan formation and xanthine oxidase of 13 compounds (Morikawa et al., 2006b; Table 3). Meanwhile, the study simultaneously showed that at a concentration of 10^{-6} mol/L, arctigenin, lappaol A, and lappaol F have anti-oxidative activities that inhibit Fe^{2+}-cystine-induced lipid peroxidation in rat liver microsomes with inhibitory rates of 53, 59, and 52%, respectively, relative to that of vitamin E at 35% (Zi et al., 2008).

Table 3 Activities of specific compounds from *Gymnadenia conopsea* (L.) R.Br.

Compounds	Effects	In vitro test	References
Gymconopin A Gymconopin B 2–Methoxy–9, 10–dihydrophenanthrene–4, 5–diol 4–Methoxy–9, 10–dihydrophenanthrene–2, 7–diol 1– (4–Hydroxybenzyl) –4–methoxy–9, 10–dihydrophenanthrene–2, 7–diol 1– (4–Hydroxybenzyl) –4–methoxyphenanthrene–2, 7–diol Blestriarene A Batatacinlll 3′, 5–Dihydroxy–2– (4–hydroxybenzyl) –3–methoxy-bibenzyl3, 3′–Dihydroxy–2– (4–hydroxybenzyl) –5–methoxybibenzyl 3, 3′–Dihydroxy–2, 6–bis– (4–hydroxybenzyl) –5–methoxybibenzyl α–Tocopherol (+) –Catechin	Anti–oxidant activity	Radical scavenging activities for DPPH radical and. O_2^- exhibiting inhibitory activity on formozan formation and xanthine oxidase at 29. 2, 45. 8, and > 100μM; 33. 4, 21. 5, and >100μM; 31. 2μM, >100μM and null; 12. 7, 0. 95, and 44. 0μM; 8. 2, 0. 19, and 30. 5μM; 15. 7, 9. 4, and >100μM; 5. 8, 0. 27, and 4. 5μM; >40, 82. 8, and >100μM; >40, 9. 3, and 72. 9μM; >40, 13. 4, and 45. 1μM; >40, 13. 4, and 65. 2μM; 11. 0, null and null; 6. 0, 1. 5, and >10μM, respectively.	Morikawa et al., 2006b
Arctigenin Lappaol A Lappaol F	Anti–oxidant activity	Anti–oxidative activity inhibiting Fe^{2+}–cystine activity induced in rat liver microsomal lipid peroxidation with inhibitory rates of 53, 59, and 52%, compared with that of vitamin E at 35%.	Zi et al., 2008
(–) –4– [β–D–glucopyranosyl– (1β4) –β–D–glucopyranosyloxy] benzyl alcohol (–) –4– [β–D–glucopyranosyl– (1→3) –β–D–glucopyranosyloxy] benzyl ethyl ether (–) – (2R, 3S) –1– [4–βD–glucopyranosyloxybenzyl] –4–methyl–2–isobutyltartrate Cyclo [gly–L–S – (4 – hydroxybenzyl)] cys 2 – Hydroxy – 2 – (4 – hydroxyphenylmethyl) –4–methylcyclopent–4–en–1, 3–dione 2–Hydroxy–3– (4–hydroxyphenyl) –4–hydroxymethylcyclopent–2–enone Coelovirins E Dactylorhin E Dactylorhin B Militarine Gastrodin	Anti–HIV activity	At a concentration of 10^{-5} mol/L, the inhibition rates of these compounds against VSVC/HIV–luc model in 293 cell lines were 9. 0, 5. 0, 6. 2, 11. 9, 11. 3, 0. 6, 13. 3, 5. 1, 10. 0, 2. 4, and 0. 6%, respectively. Comparisons with the positive drugs zidovudine (10^{-7} mol/L, 85. 6%) and lamivudine (10^{-8} mol/L, 47. 4%) were performed.	Zi, 2008

In 2010, Li et al. (2010) screened and evaluated the aqueous extract, acidic aqueous extract (pH=3), 60% ethanol extract, 95% ethanol extract, n-butanol extract, n-butanol saturated by aqueous extract and ethyl acetate extracts of the tuber of G.conopsea for their radical scavenging capacity.The results showed that the radical scavenging capacity of the aqueous and acidic aqueous (pH=3) extracts were weak.The radical scavenging activities (IC_{50}) of the other extracts were 0. 1266, 0. 4537, 0. 3151, and 0. 1305mg/mL. Thus, the investigators indicated that the 95% ethanol extract was the most effective among the six extracts (Table 4).

5.3 Anti-Viral Activity

In 2008, Zi (2008) studied the anti-HIV activity of certain compounds from the tuber of G. conopsea at a concentration of 10^{-5}mol/L in relation to the drugs zidovudine (10^{-7}mol/L, 85. 6%) and lamivudine (10^{-8} mol/L, 47. 4%), and the inhibition rates of these compounds against the VSVC/HIV-luc model in 293 cell lines were 0. 6%-13. 3% (Table 3).

In Tibet, G.conopsea has primarily been used by local people to treat chronic hepatitis B.In 2002, Lu et al. (2002) studied the anti-HBV activity.The results showed that after treating the patient's serum that contained hepatitis B for 4h, the medicine (0. 03, 0. 06, and 0. 12mg/L) could inhibit eight hemagglutin units of HBsAg. Thus, they indicated that the tuber has medium anti-HBV activity. In 2003, Kimura et al. (2003) studied the inhibitory effect of Tibetan medicinal plants on viral polymerases.The results showed that at 100μg/mL, 28 species of 76 medicines presented the anti-RTase activity with the more than 70% inhibition rates. But after adding BSA to these drugs, except the fruit of Terminalia chebula and Areca catechu, other species didn't showthe anti-RTase activity, including G.conopsea (Kimura etal., 2003; Table 4).

5.4 Sedative and Hypnotic Activities

In 2009, Lin (2009) studied the sedative and hypnotic effects of the tuber of Gymnadeniae tubers at different dosages. In the sedative test, administering the tuber (2 and 1g/kg i. g.) for 30min, it could reduce the spontaneous activity of mice within 5min (6. 63±3. 18 and 5. 36±3. 44) at inhibitory rates of 37. 44% ($P<0.05$) and 45. 60% ($P<0.05$), and the control group (normal saline) and positive group (diazepam, 0. 004g/kg) exhibited inhibition of 10. 60±5. 77 and 3. 66±1. 47 (an inhibition of 64. 21%, $P<0.01$), respectively.The tubers could also decrease the frequency of upward-raising motion in mice forelimbswithin 5min, especiallyforthehigh dose group at 15. 42±11. 77 times ($P<0.05$).The upward-raising frequency of the control and positive groups were 22. 11±6. 92 and 9. 37±3. 65 ($P<0.01$), respectively.

In the mesmerism test, the medicine could prolong sleeping times at doses of 2 and 1g/kg i.g. (32. 94±14. 84min, $P<0.01$; and 28. 85±11. 28min, $P<0.05$, respectively) in mice and decrease the latency time (3. 00±0. 82min, and 3. 34±0. 52min, P>0. 05, respectively) induced by pentobarbital sodium over the threshold dose (50mg/kg) compared with mice in the control group (13. 96±8. 45 and 4. 80±1. 85min); the positive groups were 45. 31±12. 31min ($P<0.01$) and 1. 05±0. 31min ($P<0.05$).This treatment also increased the number of sleeping mice (18 and 18) and latency time (3. 11±1. 00min, $P<0.01$; and 2. 56±0. 49min, $P<0.05$), and it prolonged their sleeping times (28. 22 ± 10. 50min, $P<0.01$; and 20. 81 ± 9. 22min, $P<0.05$)

relative to that of the control group (7, 0.92±1.58 and 8.56±15.53min), which was induced by pentobarbital sodium at less than the threshold dose (45mg/kg), and the diazepam group (20, 6.35±0.82 and 38.22±3.84; $P<0.01$). These results indicated that the tuber of G.conopsea has marked sedative and hypnotic dose-dependent effects, and its mechanism of action requires further research (Table 4).

Table 4 The biological activities of *Gymnadenia conopsea* (L.) R.Br.

Pharmacological activities	Plant parts	Processing method/extract	Test system	Animal or test organism	Positive drug	References
Tonifying activity	Tuber	Powder, no extract	In vivo	Mice induced by hydrocortisone (25mg/kg, i.g.)	-	Lin, 2009
Anti-fatigue activity	Tuber	Aqueous extract	In vivo	Mice	Radix ginseng, 20g/kg	Zhao and Liu, 2011
	Tuber	Goat and cow milk decoction	In vivo	Mice	-	Jin and Wang, 2009
Anti-oxidant activity	Tuber	Methanol, aqueous and acetone-eluted fraction	In vitro	Radical scavenging assay	-	Morikawa et al., 2006b
	Tuber	Aqueous, acidic aqueous, 60% ethanol t, 95% ethanol, n-butanol, and n-butanol extracts	In vitro	Radical scavenging assay		Li et al., 2010
Anti-viral activity	Tuber	Aqueous extract	In vitro	The patient's serum with hepatitis B	-	Lu et al., 2002
Sedative and hypnotic activities	Tuber	Powder, no extract	In vivo	Mice	Diazepam	Lin, 2009
Preventing and treating gastric ulcers	Tuber	Powder, no extract	In vivo	Mice induced by a hydrochloride-ethanol solution	-	Lin, 2009
	Tuber	Powder, no extract	In vivo	Mice induced by a hydrochloride - ethanolsolution		Jiang et al., 2009
Immunoregulatory Activity	Tuber	The polysaccharide	In vivo	Mice	-	Shang et al., 2014
Anti-aging activity	Tuber	Shouzhangshen - 37 pill (the tuber of G.conopsea is the primary component of this pill)	In vivo	Mice induced by injecting D-galactose	Vitamin E	Si and Liu, 2013
Anti-hyperlipidemia activity	Tuber	Ethanol extracts	In vivo	Hyperlipidemia rats	Lovastatin	Zhang et al., 2013

(continued)

Pharmacological activities	Plant	Processing method/extract	Test	Animal or test organism	Positive drug	References
Antianaphylaxis activity	Tuber	Methanolic extract	In vivo	Mice	–	Matsuda et al., 2004
Anti-silicosis activity	Tuber	60% ethanol extract	In vivo	Rat lung interstitial fibrosis	–	Wang et al., 2007, 2008
	Tuber	Ethanol extract	In vivo	Rat lung interstitial fibrosis	–	Zeng et al., 2007
	Tuber	Ethanol extract	In vivo	Rat lung interstitial fibrosis	–	Chen, 2008

5.5 Preventing and *Treating Gastric Ulcers*

With normal saline as the control, the inhibitory effect of the tubers on gastric ulcers was induced by hydrochloride-ethanol solution (7.5 ml/kg for 3 days, i.g.).The macroscopic and pathological results showed that at doses of 2 and 1g/kg, the treatment could relieve the pathology index of gastric ulcers induced by hydrochloride-ethanol solution and decrease the gastric ulcer index (0.05±0.13cm and 0.22±0.38cm) at inhibitions of 88.86% ($P<0.01$) and 48.88%, respectively.The biochemical test showed that after treatment with Gymnadeniae tubers, the serum (9.30±4.18nmol/mL, $P<0.01$; and 13.71±3.89nmol/mL) and gastric (1.10±1.19nmol/mL, $P<0.05$; and 0.96±0.69nmol/mL) MDA contents could be decreased in rats.The gastric ulcer index, MDA content in the serum and gastric ulcer index of the model group were 0.44±0.18cm, 14.17±4.88 and 3.81±5.35nmol/mL, respectively.The above results suggest that the tubers exert a degree of inhibition on gastric ulcers that were induced by the hydrochloride-ethanol mixture (Lin, 2009).Jiang et al. (2009) also studied the beneficial effect of the tuber of G.conopsea on acute gastric ulcers in rats that were induced by hydrochloride-ethanol solution (7.5 ml/kg for 3 days).After drug administration orally (2 and 1g/kg) for 30min, the gastric ulcer areas were decreased with inhibitions of 88.86 ($P<0.01$) and 48.88 ($P<0.01$); the inhibition by a commonly prescribed drug (ranitidine, 1.95g/kg) was 91.12 ($P<0.01$).The pathology of the gastric ulcers was markedly improved by these drugs (Table 4).

5.6 Immunoregulatory activity

As a traditional folk medicine, the tuber of G.conopsea has been widely indicated to have good tonifying and immunoregulatory activities (Chinese Materia Editorial Committee, State Chinese Medicine Administration Bureau, 2002; Li et al., 2006).In 2014, Shang et al. (2014) first evaluated the immunoregulatory functions of the polysaccharide of the tuber of G.conopsea in mice.After administering orally the polysaccharide (100, 50, and 10μg/g) and distilled water for 28 days, the serum lysozyme content (104.8±7.8, 102.5±2.8, and 100.1±7.3, all $P<0.01$), thymus index (0.42±0.03, $P<0.01$; 0.35±0.01, $P<0.01$; and 0.25±0.02) andspleenindex (0.92±0.07, $P<0.01$; 0.85±0.04, $P<0.01$; and 0.79±0.10, $P<0.05$) weresignificantlyimproved relative to that of the control group (86.7±2.8mg/L, 0.22±0.02, and 0.72±0.05mg/g).

In the delayed type hypersensitivity (DTH) test, the drugs could promote the genesis of DTH and induceearswelling, especially for the high dose group (30.0±0.4mg, $P<0.01$) compared with the control group (15.6±0.5mg). The polysaccharides could significantly improve the phagocytic function ofmouse macrophages, and the phagocytosis rates were 69.6±1.3 ($P<0.01$), 58.6±0.5 ($P<0.05$) and 49.2±1.0 ($P<0.05$) compared with that of the control group at 40.8±0.7. These results suggest that the polysaccharide has a marked immunoregulatory function (Table 4).

5.7 Anti-Aging activity

In 2013, Si and Liu (2013) studied the anti-aging effects and mechanisms of the Shouzhangshen-37 pill on subacute aging in mice (the tuber of G.conopsea is the primary component of this pill). The subacute aging model mice were induced by injecting D-galactose (120mg/kg) once per day for 7 weeks. High (2.4g/kg), middle (1.2g/kg), and low (0.6g/kg) doses of Shouzhangshen-37 pills and a common drug (VE, 38.9mg/kg) were also administered i.g. for 7 weeks. After the treatment, the anti-aging effects and mechanisms were evaluated by observing learning and memory ability with step-through tests, the indexes of the brain, thymus, and spleen were calculated by measuring the SOD, CAT, and MDA contents of brain tissues and observing the pathomorphism changes in mouse cerebral tissue by HE coloration. The results showed that compared with the model group, the drugs could enhance the memory ability and decrease the number of mistakes within 300s (1.54 ± 0.53, $P<0.05$; 1.26 ± 0.42, $P<0.01$; and1.71 ± 0.82, $P<0.05$) compared with that of the model group (2.11±0.81) andVEgroup (1.26±0.42, $P<0.01$). Inaddition, thebraintissueindex (12.97 ± 1.89mg/g, $P<0.05$; 13.39 ± 1.39mg/g, $P<0.01$; and13.05±2.35mg/g, $P<0.05$), spleen (4.53±0.66mg/g, $P<0.05$; 4.76±0.82mg/g, $P<0.01$; and 4.58±0.65mg/g, $P<0.05$), and thymus (1.11±0.12mg/g, $P<0.05$; 1.21 ± 0.20mg/g, $P<0.01$; and1.14±0.25mg/g, $P<0.05$;) were improved compared with the model group (10.43±1.66, 3.63±0.55, and 0.91±0.18mg/g). The results of the biochemical test showed that the SOD activity (170.8±20.3U/mg, $P<0.05$; 180.6±14.7U/mg, $P<0.01$; and177.3±20.1U/mg, $P<0.05$) and CAT activity (6.51±0.22, 7.14±0.72, and 6.32± 0.23U/mg, all $P<0.05$) were increased and MDA content (3.21±0.98nmol/mg, $P<0.05$; 3.17±0.73nmol/mg, $P<0.01$; and 3.36±0.81nmol/mg, $P<0.05$) was decreased compared with that of the model group (156.9±14.5, 5.24±0.51U/mg, and 4.31±0.47nmol/mg). These results suggest that the Shouzhangshen-37 pill has an anti-aging effect in subacute aging mice. The mechanism may enhance the antioxidant activity in brain tissue, decrease the MDA content in brain tissue, protect the brain nerve cells, and improve memory ability in mice (Table 4).

5.8 Anti-Hyperlipidemia activity

In 2013, Zhang et al. (2013) studied the effects of the tuber of G.conopsea ethanol extracts on the blood lipids and liver function of experimental hyperlipidemia rats. The results showed that after treating the hyperlipidemia model rats that had been induced by high-fat diets with ethanol extracts (at 5, 2.5, and 1.25g/kg, i.g.), the TC and LDL-C serum contents of the model rats were not improved; however, the TG and HDL-C contents were decreased at 1.35, 0.89, 0.97mmol/L ($P<0.01$), and 0.49 ($P<0.05$), 0.53 and 0.49mmol/L ($P<0.05$) compared with the model group (2.43 and 0.64mmol/L), and the commonly prescribed drug lovastatin also showed markedly decreased TG and HDL-C content ($P<0.01$). In addition, the ALT and AST activities in the serum were also inhibited at10.71, 10.82, 8.21U/L ($P<0.05$), 6.97 ($P<0.01$), 13.60 ($P<0.05$), and 18.02U/L compared with that of the model group, and the commonly prescribed drug results were 8.53 ($P<0.05$) and 10.38 ($P<0.01$). These results suggest that the tuber could reduce blood lipids and protect the liver function in experimental hyperlipidemia rats (Table 4).

5.9 Antianaphylaxis activity

In 2004, Matsuda et al. (2004) studied the effects of the methanolic extract of the G. conopsea tuber. The results showed an anti-allergic effect on passive cutaneous anaphylaxis reactions in the ears of mice. The inhibitory effects of the principal constituents on β-hexosaminidase release, which acts as a marker of degranulation in RBL-2H3 cells, were examined, and five phenanthrenes (gymconopin B, 4-methoxy-9, 10-dihydrophenanthrene-2, 7-diol, 1-(4-hydroxybenzyl) -4-methoxyphenanthrene-2, 7-diol, 1-(4-hydroxybenzyl) -4-methoxy-9, 10-dihydrophenanthrene-2, 7-diol, blestriarene A) and six dihydrostilbenes (gymconopin D, batatasin Ⅲ, 3′-O-methylbatatasin Ⅲ, 3, 3′-dihydroxy-2-(4-hydroxybenzyl) -5-methoxybibenzyl, 3′, 5-dihydroxy-2-(4-hydroxybenzyl) -3-methoxybibenzyl, and 3, 3′-dihydroxy-2, 6-bis (4-hydroxybenzyl) -5-methoxybibenzyl) were found to inhibit antigen-induced degranulation by 65.5-99.4% at 100μM in RBL-2H3 cells (Table 4).

5.10 Anti-Silicosis activity

Silicosis is an important occupational disease caused by the inhalation of silica dust, and it is characterized by lung interstitial fibrosis. In 2007 and 2008, Wang et al. (2007, 2008) studied the effects of silica exposure on collagen synthesis in rat lungs and mechanisms of anti-oxidative stress by using a 60% ethanol extract of the tuber of G.conopsea. Silicotic animal models were established by surgically directing the tracheal instillation of silica into rat lungs. After administering the ethanol extract orally (8g/kg, per day), the rats were sacrificed and the samples were collected to assay the relative index at 7, 14, 21, 28, and 60 days. The results showed that the extract could reduce the lung/body weight ratio of rats (8.6, 6.99, $P<0.01$; 7.25, $P<0.05$; 7.97, $P<0.05$; and 9.75mg/g, $P<0.05$) relative to the model group (11.04, 9.28, 9.11, 11.98, and 13.91mg/g), and the extract could also improve pathological changes in the lung. This treatment could also ameliorate silica-induced pulmonary fibrosis by decreasing the type I and type Ⅲ collagen-positive area percentage in the lungs of rats exposed to silica at different time points. These studies also indicated that the ethanol extract could decrease the MDA content (4.78, $P<0.01$; 5.39, $P<0.05$; 5.48, $P<0.05$; 5.29, $P<0.05$; and 4.35mmol/L) and increase the SOD (302.67, 243.95, $P<0.01$; 293.38, $P<0.05$; 277.74, $P<0.05$; and 243.36 KU/L) and GPx (2 199.58, 2 359.34, $P<0.01$; 2 538.66, $P<0.01$; 2 422.41, $P<0.05$; and 2 298.04U/L) activities relative to that of the model group (6.43, 6.88, 6.98, 6.12, and 5.82mmol/L, 314.84, 183.62, 219.41, 218.44, and 226.50 KU/L, and 2 489.50, 1 015.34, 1 227.83, 1 814.90, and 1 867.38U/L). In 2007, Zeng et al. (2007) found that the ethanol extract could reduce the TNF-α integral optical density (86.86, 105.1, 122.09, 108, and 94.88, all $P<0.01$) of rat lungs compared with that of the model group (116.98, 140.1, 220.19, 140.9, and 110.9) at 7, 14, 21, 28, and 60 days, respectively. Before and after treating with a 60% ethanol extract, Chen (2008) studied the differential gene expression profiles of rat lung tissue in the early stage exposure to silica. After treating with the extract (8g/kg, per day, i.g.), there were 308 and 231 up-regulated and down-regulated genes among a total of 539 available genes compared with that of the model group. Up-regulated pathways might be associated with cell adhesion molecules, notch signaling pathways, and leukocyte transendothelial migration and cell communication. Down-regulated pathways might be related to genes in the linkage of complementary and coagulation cascades, hematopoietic cell lineages, urea cycle and Alzheimer's disease. The chip results were verified by using real-time PCR for SOD and HMOX genes with the same trends. Thus, the investigators indicated that the alcohol extract may inhibit silicosis progress during early exposure periods and attenuate airs acculitis and fibrosis of the lung. The results also showed that the ethanol extract (8g/kg, per day, i.g.) has the reverse effect on silicosis, and this process involved the cathepsin D precursor SEC14-like protein 3 and peroxiredoxin-1 (Chen, 2009; Table 4).

5.11 Other activities

In 2008, Zi (2008) studied the anti-cancer activity of certain compounds from the tuber of G.conopsea. The results showed that the IC_{50}-value of (-) -4- [β-D-glucopyranosyl- (1→4) -β-D-glucopyranosyloxy] benzyl alcohol against human hepatoma cancer cells (Bel7402) was 3.818μM compared with the commonly prescribed drug camptothecine at 12.5μM. At a concentration of 10^{-6} mol/L, gastrodin was active against serum deprivation-induced SH-SY5Y apoptosis.

At a dose of 5mg/kg (i.p.), dactylorhin B, coelovirin E and (2R, 3S) -1- (4-hydroxybenzyl) -2-hydroxy-4- (4-hydroxybenzyl) -2-isobutyltartrate could improve mouse memory deficits induced by scopolamine with a neuroprotective effect. Coelovirin E and (2R, 3S) -1- (4-hydroxybenzyl) -2-hydroxy-4- (4-hydroxybenzyl) -2-isobutyltartrate could extend the latent time for the step-down test in mice with improvements of 65 and 106%, respectively (Zi, 2008; Table 4).

6 PREPARATIONS AND QUALITATIVE AND QUANTITATIVE ANALYSIS

G.conopsea has been widely used as a folk medicine in the treatment or prevention of diseases for thousands of years in China, but it has not been listed in the Chinese pharmacopeia because there is a lack of scientific quality standards to control the quality of the tuber since 1977 (Committee for the Pharmacopoeia of P.R.China, 1977, 2000; Chinese Materia Editori al Committee, State Chinese Medicine Administration Bureau, 2002). Recently, experts have attempted to formulate the proper quality standard by analyzing the active compounds and determining the contents of plants from different habitats with different chromatographic equipment.

Now, as a strategy to control the quality of folk medicines, chemical fingerprint analyses have been accepted by the WHO (1991), State Food and Drug Administration of China (2000), and other authorities, which has been recognized as a rapid and reliable method of identifying and qualifying herbal medicines. Cai et al. (2006) first developed a high-performance liquid chromatography-diode array detection-tandem mass spectrometry (HPLC-DAD-MS^n) method for the chemical fingerprint analysis and rapid identification of major compounds in G.conopsea tubers. An HPLC separation was performed by using a linear gradient at room temperature (20℃) and a flow rate of 0.7ml/min with an Inertsil C_{18} ODS-3 column. The gradient elution started with a methanol: water mixture (20 : 80, v/v), and the methanol content was increased to 100% within 60min; the detection wavelength for fingerprint analysis was 270nm. Adenosine, 4-hydroxybenzyl alcohol, 4-hydroxybenzyl aldehyde, dactylorhin B, loroglossin, dactylorhin A, and militarine were well-separated and identified from the tuber of10 samples, which were collected from Sichuan, Qinghai, and Hebei Provinces and Tibet Autonomous Region of China and Nepal. Xue et al. (2013) also studied the fingerprint of G.conopsea by using HPLC. Their samples were separated on a Kromasil C_{18} (4.6×250mm, 5μm) column and gradient-eluted with a mixture of methanol and water (0.04% phosphoric acid) at a flow rate 0.7 ml/min; the detection wavelength was set at 222nm, and the column temperature was 30℃. The results showed that 13 common peaks were selected, and the method validation met the technical requirements of a fingerprint. The similarity of 12 samples from the different regions of the Xizang Autonomous Region and Qinghai Province was 0.904-0.989. The above two methods are simple, practical and reliable. The combined use of the two fingerprints confirmed the identification and quality assessment of G.conopsea. In 2009, after investigating the optimal extraction method, Li et al.firstly developed a HPLC method for the simultaneous quantification of five glucosyloxybenzyl 2-isobutylmalates in the tubers of G.conopsea, which was collected at five regions in China. The optimal extraction conditions was the direct reflux with 70% ethanol for 1h, and the compounds are separated on an Agilent Hydrosphere C18 (150×4.6mm i.d., 5μm) column using a mobile phase of acetonitrile-water including 0.3% acetic acid (adjusted with 36% acetic acid) with gradient elution at a flow rate of 1.0mL/min. Detection is set at a UV wavelength of 221.5nm. The recovery of the method is 97.7% - 101.0%, and linearity ($r > 0.9998$) is obtained for all the analytes. The results showed that the contents of dactylorhin B, dactylorhin E, loroglossin, dactylorhin A, and militarine were 1 638.28, 100.19, 279.91, 717.22, and 101.78mg/g (Lijiang City, Yunan Province), 665.82, 92.12, 248.15, 507.65, and 138.57mg/g (Weixian City, Hebei Province), 707.84, 85.64, 322.63, 705.70, and 137.30mg/g (Kangding City, Sichuan Province), 2 207.30, 204.46, 604.81, 2 213.39, and 228.04mg/g (Tibet Autonomous Region), and 4073.89, 156.81, 3060.63, 2230.67, and 563.56mg/g (Xining City, Qinghai Province). Yang et al. (2009) and Yang determined the contents of four marker compounds from the tubers of Gymnadenia conopsea (L.) R.Br.by HPLC, and the tubers were collected from 24 regions, including the Sichuan Province, Tibet Autonomous Region, Jilin Province, Heilongjiang Province, Qinhai Province, and other regions in China. At a detection wavelength of 222nm, the mobile phases of methanol (A) and water (B) were used in a gradient elution at 0.7mL/min. The initial condition was 15%-20% A in 0-10min, which increased to 45% A in 20min and 70% A in 35min. The results showed that the highest contents of dactylorhin A, dactylorhin B, loroglossin, and militarine from the rhizomes of24 different regions were 1.428 and 1.638mg/g from Lulang County in the Xizang region and 1.148 and 0.521mg/g from Jilin Province.In 2011, the gastrodin contents were assayed by HPLC, and the results showed that the highest content was 9.113mg/g in Tongde County of Qinghai Province, China, followed by 7.785mg/g (Beinagou region, Qinghai Province), 4.881mg/g (Huangnan Zeku region, Qinghai Province), 4.309mg/g (Guoluo region, Qinghai Province), 4.285mg/g (Batang Grassland, Qinghai Province), 3.895mg/g (Nangqian County, Qinghai Province), 3.719mg/g (Maixiu Forest Farm, Qinghai Province), 3.242mg/g (Jinyingtan, Qinghai Province), 3.024mg/g (Linzhi County, Tibet Autonomous Region), 3.103mg/g (Chengduo, Qinghai Province), 2.734mg/g (Bomi County, Tibet Autonomous Region), and 2.268mg/g (Beishan Forest Farm, Huzhu County, Qinghai Province; Xue et al., 2011).

Meanwhile, the content of four marker constituents of the tubers with the climatic factors were determined and analyzed by HPLC in 2014. The results showed that the samples collected from Milin County in the Tibet Autonomous Region had the highest gastrodin content at 2.725mg/g, and the Shennongjia region of Hubei Province had lowest content at 0.374mg/g.The samples collected from Kangding County of Sichuan Province, Aba region of Sichuan Province and Naqu County of the Tibet Autonomous Region had the highest contents of adenosine, 4-hydroxybenzyl alcohol and 4-hydroxybenzaldehyde at 1.485, 1.505, and 1.048mg/g, respectively.These findings suggest that the average temperature and annual average variations in monthly temperature change along with seasonal variations, and the highest temperature, variations in the scope of the annual temperature, average temperature of the most humid season, average temperature of the hottest season, annual rainfall, rainfall during the driest season, rainfall during the coldest season, and latitude and altitude of the habitat can have an effect on the active ingredient content and quality of the tuber of G. conopsea (Zhang et al., 2014).

The study results also indicated that the samples collected from Huzhu County of Qinghai Province had the highest contents of gastrodin (4.9242mg/g), dactylorhin A (8.2274mg/g), and militarine (7.4645mg/g; Yao and Lin, 2014).

7 TOXICITY

In 2007, Bai and Zheng (2007) evaluated the toxicity of the tuber of G.conopsea.In acute toxicity test, after administering the tuber (1.00, 2.15, 4.64, and 10g/kg, i.g.) to mice and rats, the behavior, and growth of the animals did not change after 2 weeks compared to the control group ($P>0.05$), and the LD_{50} was more than 10g/kg.The genotoxicity of the medicine was then studied, and the results showed that the rates ofbone marrow and sperm abnormalities were not changed compared with that of the control group, but the sperm abnormality rates of the group administered a common drug (cyclophosphamide, 30mg/kg) changed markedly ($P<0.01$).And after administering the tuber (1.67, 3.33, and 6.67g/kg, i.g) to rats for 30 days, the parameters of blood routine, biochemical indexes, organ coefficient, and organ pathology of rats haven't marked changed compared the control group ($P>0.05$).So they thought that the tuber didn't present the apparent toxicity.

8 CONCLUSIONS AND FUTURE PERSPECTIVES

As an important traditional medicine in China, the tuber of G.conopsea has been historically used in a number of clinical applications.However, lack of scientific quality standard to control the quality of the tuber has been hampered to the development of this plant.Up to now although investigators have begun studying and developing the quality of the tuber and its preparations by analyzing the marker compounds and determining the contents of plants from different habitats, a scientific method to control the quality of these products has not yet been developed and approved by government.Thus, a characteristic chemical and biological index should be established to monitor and evaluate quality of samples and maintain their clinical and pharmaceutical stability, and the method of accurately controlling the quality of the tuber in traditional medicine and preparations should be studied further.

With the excavation and abundant use of G.conopsea, as well as the over-grazing and disorder tourism resulted in the habitat destruction, the resources of this plant were rapid decreased.At the same time, because of the low reproductive capacity of this plant, once the resources and habitat have been destroyed, the species does not recover easily.In 2000, G.conopsea has been listed in the grade Π section of endangered species (Gesang and Gesang, 2010).Recently, investigators have begun studying and developing methods of cultivating this plant, such as tissue culture and artificial breeding, but this technology is still not applied in industry (Bao et al., 2008).Therefore, the sustainable use of G.conopsea is necessary to investigate.Of course, along with the decrease of resources, the studies on the aerial part should be paid more attention, especially for finding new active compounds and bioactivities.

In a word, phytochemical and pharmacological studies of G.conopsea have received great interest, and an increasing number of extracts and active compounds have been isolated that have demonstrated tonifying activity, anti-viral activity and immunoregulatory activity, among others.However, validating the correlations of the ethnomedicinal uses and pharmacological effects should be carried out further, and the toxicity of this plant also should be studied systematically.Meanwhile, the poor quality control, decreased resources, the increasing gap between more experience-based traditional uses and less evidence-based clinical trials for the tuber of G.conopsea has created challenges.Therefore, this plant should be studied and developed further, especially in the resource conservation.

AUTHOR CONTRIBUTIONS

XS and JZ conceived the review; XS, XG, XM wrote the manuscript; HP collected the literatures; and YL edited the manuscript.All the authors read and approved the final version of the manuscript.

ACKNOWLEDGMENTS

This work was financed by National Science and Technology Infrastructure Program of China (2015BAD11B01) and Key Technology R&D Program of Gansu Province (2016GS10130).The authors would also like to express their gratitude to Dr. Allan Grey at Lanzhou University for correcting the English in this paper. We also thanks for the provider of figure 1, Zhou Yao from http://www.plantphoto.cn.

REFERENCES OMITTED

（发表于《Frontiers in Pharmacology》，院选 SCI，IF：4.400）

UHPLC-Q-TOF/MS Based Plasma Metabolomics Reveals the Metabolic Perturbations by Manganese Exposure in Rat Models

Hui WANG[1,2], Zhiqi LIU[3], Shengyi WANG[2], Dongan CUI[2], Xinke ZHANG[1], Yongming LIU[2], Yihua ZHANG[1*]

(1. College of Veterinary Medicine, Northwest A & F University, Yangling 712100, China; 2. Lanzhou Institute of Husbandry and Pharmaceutical Sciences of Chinese Academy of Agricultural Sciences, Lanzhou 730050, China; 3. Institute of Agro-Products Processing Science and Technology, Chinese Academy of Agricultural Sciences, Beijing 100193, China)

Abstract: Although manganese (Mn) is an essential metal ion biological cofactor, high concentrations could potentially induce an accumulation in the brain and lead to manganism.However, there is no "gold standard" for manganism assessment due to a lack of objective biomarkers.We hypothesized that Mn-induced alterations are associated with metabolic responses to manganism.Here we use an untargeted metabolomics approach by performing ultra-high performance liquid chromatography coupled with quadrupole time-of-flight mass spectrometry (UHPLC-Q-TOF/MS) on control and Mn-treated rat plasma, to identify metabolic disruptions under high Mn exposure conditions.Sprague-Dawley rats had access to deionized drinking water that was either Mn-free or contained 200mg Mn per L for 5 weeks.Mn-exposure significantly increased liver Mn concentration in comparison with the control, and also resulted in extensive necrosis and dissolved nuclei, which suggested liver damage from hepatic histopathology.Principal component analysis readily distinguished the metabolomes between the control group and the Mn-treated group.Using multivariate and univariate analysis, Mn significantly altered the concentrations of 36 metabolites (12 metabolites showed a remarkable increase in number and 24 metabolites reduced significantly in concentration) in the plasma of the Mn-treated group.Major alterations were observed for purine metabolism, amino acid metabolism and fatty acid metabolism.These data provide metabolic evidence and putative biomarkers for the Mn-induced alterations in plasma metabolism.The targets of these metabolites have the potential to improve our understanding of cell-level Mn trafficking and homeostatic mechanisms.

Key words: Although manganese is an essential metal ion biological cofactor, exposure to high concentrations ofMn could potentially induce an accumulation in the brain and lead to manganism. However, there is no "gold standard" for manganism assessment due to a lack of objective biomarkers. This study uses an untargeted metabolomics approach, and the results provide metabolic evidence and putative biomarkers for Mn-induced alterations in plasma metabolism.

1 INTRODUCTION

Metals play a central role in biological systems due to their involvement in regulating numerous cellular processes.[1] All cells possess a battery of regulatory proteins that mediate the homeostasis of metal ions by regulating the expression of genes that encode metal transporters, intracellular chelators or other detoxification enzymes.[2] The importance of monitoring the effects of toxic metals on living organisms has increased progressively. Manganese (Mn) is an essential micronutrient, and is important in biochemical reactions that are involved in several significant physiological processes, including reproduction, development, energy metabolism, immune function, antioxidant defenses, etc.[3] But chronic or excessive Mn exposure is a cause of severe neurological dysfunction and may progressively extend the site of Mn deposition and toxicity from the globus pallidus to the entire area of the basal ganglia,[4] perturb the cellular redox potential, produce highly reactive hydroxyl radicals and cause striatal neurotoxicity that can be attributed to basal ganglia dysfunction.[5-7] Mn neurotoxicity is becoming a great public health concern due to a number of factors affecting an ever broader range of the population.[8] The understanding of Mn biology in cellular regulation is an area of expanding interest. However, the homeostatic mechanism for Mn at the cellular level is currently poorly understood.

Mn must be carefully regulated to ensure that good health is maintained. Impaired Mn homeostasis may alter the activity of Mn-dependent enzymes, alter the lipid and carbohydrate metabolism,[9] and alter the Mn-sensitive pathways.[3] Furthermore, Mn treatment of cells resulted in a decrease in cellular energy and the redox state,[10] which may contribute to its toxicity. However, the mechanism of Mn uptake regulation and the efflux of Mn into the cells is not well established, and the signalling pathways that are responsive to physiological Mn concentrations have not been described.[6] Early and reliable detection of the onset of manganism is a useful biomarker for the status of Mn and therefore can facilitate the prevention of manganism, but the investigation of a reliable biomarker for Mn exposure is still ongoing.[11,12] Determination of Mn-species in various body fluids such as plasma is a really powerful tool for Mn biomonitoring. Plasma is an integrative bio-fluid that is a preferred matrix due to the easy sampling procedure and the fact that it contains a great diversity of chemical structures, and there are vast differences in the dynamic range of the metabolites.[13,14] Understanding the targets of these metabolites may improve our understanding of cellular and intracellular Mn trafficking and the regulation of Mn homeostasis.

Omics technologies are valuable tools that can provide the means to identify biomarkers through measuring the biochemical changes associated with the mode of action at the level of DNA/RNA, proteins and metabolites.[15] Metabolomics is becoming an increasingly used tool for exhaustive studies of all metabolites contained in an organism.[16] It is known that external perturbations imposed on organisms can produce changes in their metabolites. These perturbations can be pathophysiological stimuli, environmental changes, nutritional stresses and genetic modification.[17,18] Therefore, metabolomics can be used to identify novel and potential metabolite markers, and to explore molecular mechanisms and the response of metabolic pathways to different perturbations. Metabolomics is a novel technology with great potential for metal toxicity research and provides a unique perspective on metal-induced changes in cellular metabolism.[19]

The janiform nature of Mn prompted us to investigate Mn speciation after exposure, since different Mn compounds might be transported or metabolized by different pathways. Here we assume that the metabolite profiles are associated with Mn induced pathophysiological processes. Therefore, to execute a large-scale detection of the metabolite features in plasma samples, untargeted plasma metabolomics based on ultra-high performance liquid chromatography coupled with quadrupole time-of-flight mass spectrometty (UHPLC-Q-TOF/MS) with high resolution, high throughput, and high sensitive technology was performed in this research.[20,21] An untargeted metabolomics workflow that acquires MS and MS/MS data sequentially was designed (Fig. 1). Quantitative information is extracted from MS datausing XCMS Online and metabolite features are simultaneously characterized by matching the MS/MS data to the METLIN database.[22] With the approach of metabolomics analysis, the versatile effects of Mn on different metabolic pathways in plasma and more detail on the mechanisms of action after Mn exposure could be observed.

2 MATERIALS AND METHODS

2.1 Standard solutions and reagents

$MnCl_2 \cdot 4H_2O$ (99.995%) used to prepare acute doses for oral administration was supplied by klamar (klamar, China). All solvents used for sample preparation were of optimal grade for mass spectrometry. Acetonitrile was purchased from Merck (Merck, 1499230-935), while methanoic acid and ammonium fluoride were supplied by Sigma-Aldrich. Ultra-high purity water was prepared by Millipore-Q SAS 67120MOLS HEIM (France).

2.2 Keeping and treatment of animals

The protocols of the study were approved by the Institutional Animal Care and Use Committee of the Lanzhou Institute of Husbandry and Pharmaceutics Sciences of Chinese Academy of Agricultural Sciences (animal use permit: SCXK20008-0013). Thirty SPF-grade male Sprague-Dawley rats weighing (148 ± 2) g were obtained from the Laboratory Animal Center of Lanzhou University (Lanzhou, China). All of the rats were randomly allocated to either the control or Mn-treated group (ten in the control group, and twenty in the Mn-treated group) and kept under constant temperature (20℃ ± 3℃) and humidity (60% ± 10%) in controlled rooms with a 12h/12h light/dark cycle. The rats of the control group received the AIN-93G diet and non-ionic water, and those in the Mn-treated group were fed the AIN-93G diet and Mn-rich water ($200mg \cdot L^{-1}$) for 5 weeks. The risk concentration of Mn in drinking water criteria of the WHO is $0.5mg \cdot L^{-1}$[23] Evidence in the literature suggests that chronic low-level Mn exposure is a major health concern.[24] The Mn-exposure concentration of $200mg \cdot L^{-1}$ has been used previously in our lab to study the neurological effects of a chronically low concentrated Mn intake by drinking water. The ingredient composition of the AIN - 93G diet is listed in Table 1 (35, 10 and 6mg · kg^{-1} of Fe, Mn, and Cu, respectively). All animals had free access to their food and water. All of the rats were acclimated to the room for 7 days. The feeding duration was 5 weeks until sacrifice. For euthanasia, animals were deeply narcotized with 5% isoflurane until total loss of consciousness. The body weight (BW) of all the rats in each group was measured prior to sacrifice. After cutting through the aorta abdominalis with a ceramic scalpel, blood was collected in Eppendorf (EP) tubes and centrifuged subsequently for 10min at 2000g and 4℃ to obtain EDTA-K2 plasma, which was stored at -80℃ until analysis. Furthermore, the livers of the rats were taken out and the liver weights (LW) were measured, then immediately immersed in 10% neutral buffered formalin for 24h. After fixation, the tissues were dehydrated through a gradient of alcohols, embedded in paraffin using conventional methods,[25] sectioned into 4mm thick slices, and mounted on glass microscope slides. The slices were stained with hematoxylin-eosin (H&E) and examined by a light microscope (Olympus × 51 model, Tokyo, Japan) equipped with a camera (Olympus 20, Tokyo, Japan).

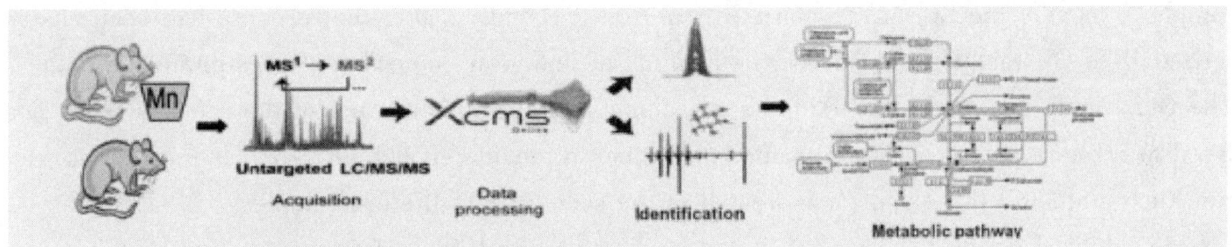

Fig. 1 Schematic workflow of the experimental approach using autonomous untargeted metabolomics.

Note: Rats were treated with toxicological Mn exposure conditions. Plasma was harvested for metabolomics profiling using untargeted metabolomics that acquires MS and MS/MS data sequentially. Quantitative information is extracted from the MS data using XCMS Online and identification followed by pathway correlation.

Table 1 The ingredient composition of the diet fed to rats

Ingredient	g · kg^{-1} diet
Corn starch	650
Casein	165
Soybean oil	90
Fiber	50
Mineral mix (AIN-93M-MX)	35
Vitamin mix (AIN-93-VX)	10

Abbreviations: AIN-93-MX, AIN for the laboratory animals' mineral mixtures; AIN-93-VX, AIN for the laboratory animals' vitamin mixtures.

2.3 Measurement of cholesterol and Mn content

The levels of total cholesterol (CHOL; No.F002-1, Nanjing Jiancheng Bioengineering Institute, Nanjing, China) in the plasma were measured according to the manufacturer's instructions.

Graphite furnace atomic absorption spectrometty (GFAAS; ZEEnit 700, Analytik Jena, Germany) was carried out to determine the Mn concentrations in the livers. The optimal operating conditions of GFAAS are described in our previous publication[26] and are carried out according to ref.[27].

2.4 Plasma samples and quality control sample preparation

Plasma samples collected from 30 male rats, and stored at -80℃, were used in this study. Sample treatment prior to analysis consisted of thawing at room temperature. Next, 100mL of sample was placed in EP tubes, and 400μL of methanol/acetonitrile (1 : 1, v/v) was added. The tubes were vortexed for 30s, incubated for 10min at -20℃, and then centrifuged (14 000 g, 15min, 4℃).

The supernatants were collected and dried with nitrogen, and then the lyophilized powder was stored at -80℃ prior to analysis.

In parallel to the preparation of the test samples, we prepared a bulk quality control (QC) sample. The QC samples served two purposes. The first purpose was to act as a regular quality control sample to monitor the LC-MS response in real-time. Secondly, after the response had been characterized, the QC samples were used as standards of unknown composition to calibrate the data.[28] The QC sample was made by mixing equal volumes (30μL) from each of the samples being analyzed to create a pooled sample of sufficient volume to provide enough QC samples for the analytical run. Each aliquot of this sample was treated in the same way as the test samples.

Lyophilized samples were reconstituted by dissolving in 100μL of solvent mixture containing water/acetonitrile (5 : 5, v/v). The samples were vortexed for 1min and centrifuged at 14 000g for 15min at 4℃. The supernatants were collected for UHPLC-Q-TOF/MS analysis.

2.5 Liquid chromatography conditions

Metabolomics analysis was performed with an Agilent 1290 Infinity LC ultra-high pressure liquid chromatograph (UHPLC) (Agilent, Palo Alto, USA) equipped with an electrospray ionization source operating in positive and negative ion modes.

For the metabolomics analysis, an ACQUITY UPLC HSS T3 column (2.1×100mm, 1.8mm, Waters MS Technologies, Manchester, UK) was used. The column was maintained at 25℃ and eluted at a flow rate of 300mL min^{-1}. The mobile phase of the positive ion mode was composed of A (0.1% formic acid in water) and B (0.1% formic acid in acetonitrile); the negative ion mode was composed of A (0.5mM ammonium fluoride in water) and B (acetonitrile). The process of linear gradient elution was as follows: 0-1.5min, 1% B; 1.5-13min, 1%-99% B; 13-16.5min, 99% B; 16.5-16.6min, 99-1% B; 16.6-20min, 1% B. The auto sampler was maintained at 4℃ and the injection volume was 2μL. The QC samples, prepared from the pooled plasma, were placed into the column at regular intervals in the analysis sequence (one QC after every 5 samples) in order to monitor the precision and stability of the method during its operation.

2.6 Q-TOF mass spectrometry conditions

The mass spectrometer (MS) used was a Triple TOF™ 5600 system (AB/Sciex, Foster City, USA) equipped with an electrospray (ESI) ionization source in positive (ESI^+) and negative

(ESI^-) ion modes. The MS properties were set as follows: scan range, m/z of 60 - 1 000 Da; product ion scan m/z range, 25-1000 Da; TOF MS scan accumulation time, 0.2s per spectra; product ion scan accumulation time, 0.05s per spectra; Ion Source Gas1 (Gas1), 40 psi; Ion Source Gas2 (Gas2), 80 psi; curtain gas (CUR), 30 psi; source temperature, 650℃; Ion Sapary Voltage Floating (ISVF), ±5 000 V; declustering potential (DP), ±60 V; collision energy, (35±15) eV. MS/MS data were acquired in the information dependent acquisition (IDA) mode and high sensitivity modes. The settings of IDA were as follows: exclude isotopes within 4 Da, candidate ions to monitor per cycle, 10. Meanwhile the QC sample was analyzed with every batch of five samples to monitor the system stability and data quality.

2.7 Data processing and statistical data analysis

For XCMS, the raw data files were first converted into the mzML format via ProteoWizard, and subsequently the converted files were imported into the XCMS software for nonlinear alignment in the time domain, automatic integration, and extraction of the peak intensities, with default parameter settings.[22] The data were subsequently processed using XCMS for peak alignment and data filtering. MetaboAnalyst 2.0 (http://www.metaboanalyst.ca) was used for the statistical analysis. Principle component analysis (PCA) and hierarchical clustering were performed for the unsupervised multivariate statistical analysis. Partial-least squares discrimination analysis (PLS-DA) was performed as a supervised method to identify the important variables with discriminative power. PLS-DA models were validated based on the multiple correlation coefficient (R^2) and cross-validated R^2 (Q^2) in cross-validation and permutation tests by applying 2000 iterations ($P>0.001$). The significance of the biomarkers was ranked using the variable importance in projection (VIP) score (>1) from the PLS-DA model. For the univariate analysis, candidate specific biomarkers were determined using Student's t-test statistics for comparison of the means between the control group and the Mn-treated group. $P<0.05$ was considered to be statistically significant.

2.8 Metabolites identification and pathway analysis

The Metlin database was used to identify potential specific biomarker candidates based on their MS signature and tandem mass spectrometry (MS/MS) spectra, as well as eventual contaminants. Identification of potential biomarkers was carried out by searching METLIN (http://metlin.scripps.edu/), HMDB (http://www.hmdb.ca/), KEGG (http://www.genome.jp/kegg/), MassBank (http://www.mass bank.jp/), LIPIDMAPS (http://www.lipidmaps.org/) and Chemspider (http://www.chemspider.com) using the exact molecular weights or the MS/MS fragmentation pattern data, and a literature search was conducted to identify the affected metabolic pathways and to facilitate further biological interpretation. Mass accuracy tolerance within 25 ppm was used as the mass window for the database search. For confirmation of the metabolite identities using an authentic chemical standard, the MS/MS fragmentation pattern of the chemical standard was compared with that of the candidate metabolite under the same LC-MS conditions to reveal any matching. In the case of unknown metabolites, molecular formulae were generated using Mass Profiler Professional (Agilent Technologies).

3 RESULTS AND DISCUSSION

3.1 Livers are susceptible to Mn

Mn treatment reduced the liver/body weight (LW/BW) ratios in the experiment rats (Fig. 2a). Consistently higher plasma CHOL levels were detected in the Mn-treated rats in comparison with the control rats (Fig. 2b). Furthermore, the Mn-treated rats also exhibited increasing liver Mn content at the end of the experiments (Fig. 2c); this observation showed that accumulation in the liver occurred in the Mn-exposed rats. As demonstrated in Fig. 2d, histology examinations of the hepatic tissue of the two groups depicted the damaged areas. More extensive necrosis and more dissolved nuclei were clearly observed in the Mn-treated group in comparison with the control group. Previous research demonstrated that the distribution of Mn to tissues by plasma is quick,[3] and that Mn is not only distributed in the basal ganglia structures,[29] but also in the liver where high concentrations are observed;[30] exposure to Mn showed gene expression changes associated with cholesterol homeostasis,[31] similar to our findings. However, Horning et al. reported that the nervous system is the primary target for excessive Mn, and that Mn exposure has not been linked with damage to the heart, kidney, liver, skin, blood, or stomach.[3] Thus, our research is the first demonstration that rats are vulnerable to Mn, as shown by the liver damage due to the excess Mn. This is likely due to the essential nature of micronutrient Mn in energy production and the high energy demands of the liver.

Plasma metabolic profiling by UHPLC-Q-TOF/MS

The stability of the analytical method is crucial for obtaining valid metabolomics data. To validate the system's performance during sample analysis, a pooled QC sample was applied that was a representative "mean" sample including all analytes used during the analysis.[32] QC samples were handled as real samples and inserted every five samples into the ESI positive or negative analysis batch to monitor the stability of the instrument. The similarity of the QCs included the peak shape, separation degrees, retention times, and intensity distribution of the metabolites involved in the profiles. The results indicated that the method was robust with good repeatability and stability, rather than the product of artefacts arising from technical errors, and was suitable for the measurement of the samples in this study.

A total of 17 917 molecular features were extracted from all the samples using XCMS software. Principal component analysis (PCA) was carried out using these molecular features on all the sample groups from the study including conditioning runs and quality control (QC) samples. The distribution of metabolic profiles for the QC samples in PCA can be seen in Fig. 3. All of the QC injections (Green) were clustered tightly in PCA space. The consistency of the repeated QC injections and reliable data quality across all the samples demonstrated the suitability of the method for metabolic profiling studies during the experiment.

A non-targeted metabolomics study was performed in order to analyze the effects on plasma metabolism in chronic Mn exposure in rats. UHPLC-MS is more suitable for metabolomics due to its higher resolution, peak capacity, and sensitivity compared to conventional HPLC-MS.[33] Plasma metabolic profiling was carried out under UHPLC-Q-TOF/MS conditions in the ESI positive and

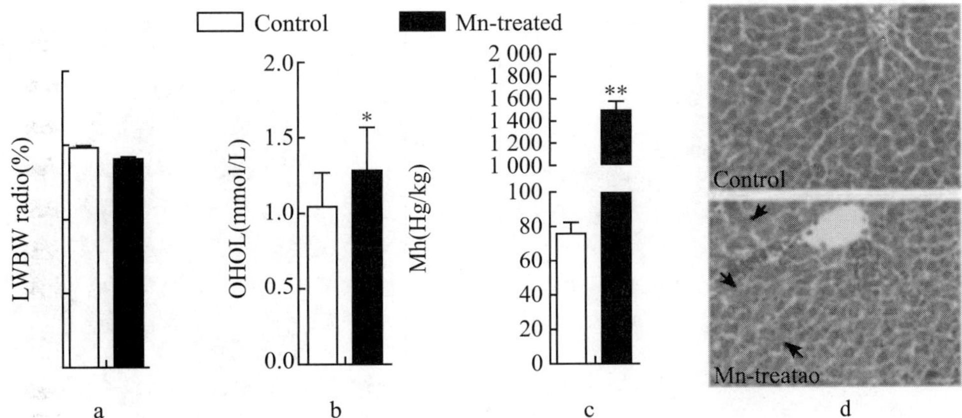

Fig. 2 Hepatic damage in rats following Mn exposure.

Note: (a) The ratios of liver/body weight (LW/BW) were determined for each group, (b) Plasma CHOL levels were measured. (c) Hepatic Mn levels were measured. (d) Liver sections were stained with H&E, and necrotic areas are indicated with arrowheads. The data in (a–c) are shown as the mean±SD. * $P<0.05$ and ** $P<0.01$, as determined by Student's t test.

negative ionization modes as described above. Under experimental conditions, the TIC shared considerable similarity, and the peak shape of each substance was good and the peaks well separated from each other, indicating that the chromatographic and MS conditions were suitable for the measurement of the samples in this study.

3.2 Normalization and multivariate statistical analysis

It is often necessary to normalize metabolomics data before starting any kind of statistical analysis. Normalization can reduce any systematic bias or technical variation, and metabolite concentrations usually span several orders of magnitude, which can lead to misidentification of significant changes.[34] In our study, the "50% rule" was applied to remove the missing values,[35] and the results indicated that the metabolomics data presented a normal distribution after normalization processing (Fig. 4).

To determine whether the global metabolite fingerprints in plasma differed between the control and model rats, we evaluated the separation between the control and Mn-treated rats in both ion modes using unsupervised PCA. PCA is an unsupervised clustering or classification method. PCA showed that 31.9% of the total variance in the data was represented by the first two principal components in the positive mode (Fig. 5a). The total variance in the data represented by the first two principal components in the negative mode was 26% (Fig. 5b). In these plots, two groups showed a slight but not significant separation trend in the 2D-PCA score plots. The plot revealed that the plasma metabolite fingerprints of the Mn-treated rats were closely related to those of the control rats, and were not obviously changed.

To further identify ion peaks that could be used to discriminate between the control and model groups, a supervised PLS-DA model was established that was more focused on the actual class discriminating variation compared to the unsupervised PCA model. PLS-DA is a supervised clustering or

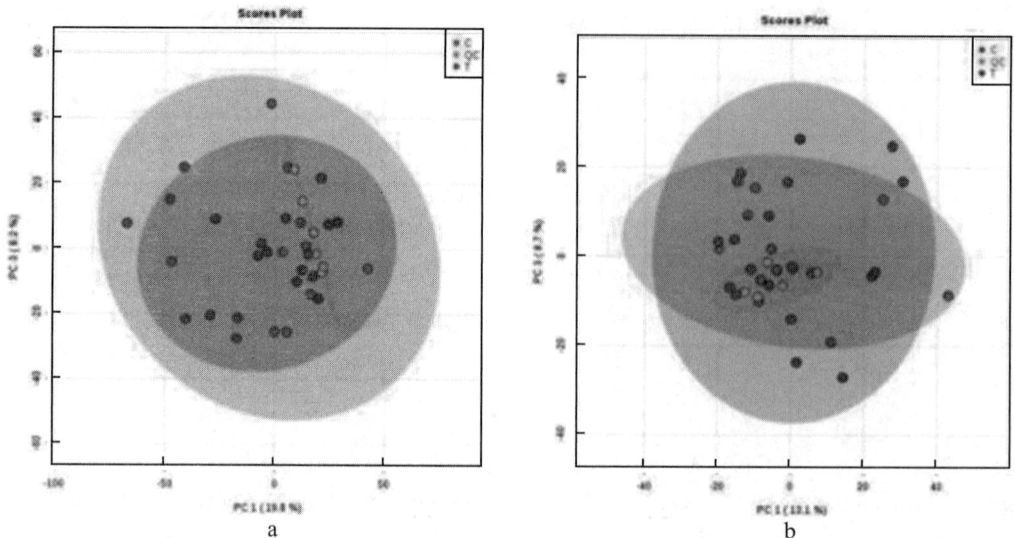

Fig. 3 PCA score plots in ESI positive mode (a) and negative mode (b) based on the UHPLC-Q-TOF/MS data of the plasma samples.

classification method. PLS-DA projects the data into a low-dimensional space that maximizes the separation between different groups in the latent variables. Supervised analysis, PLS-DA, was subsequently performed to maximize the separation and identify the metabolites. A clear separation between the control and Mn-treated groups was observed based on the PLS-DA score plot by the first two components in the positive ion mode ($R^2 = 0.996$, $Q^2 = 0.446$) (Fig. 6a) and negative ion mode ($R^2 = 0.998$, $Q^2 = 0.499$) (Fig. 6b) [indices representing the goodness of the fit (R^2) and the prediction ability of the model (Q^2)[36]]. The models allow for a good classification of the samples into two groups, suggesting the high reliability and predictive power of the model. From these score plots, we found that regardless of whether looking at either the PCA or PLS-DA score plots, the separation trend between the control and Mn-treated groups was good in both positive and negative modes, which indicates that Mn overexpression results in some changes in the levels of a few metabolites.

3.3 Detection and identification of differential metabolites

To identify which variables were responsible for this separation, a study on the variable influence on the projection (VIP) parameter was conducted. VIP values calculated using the PLS-DA model revealed which variables (metabolites) had the greatest influence on the discrimination between the control and Mn-exposed rat metabolic samples. Potential metabolites were selected based on the VIP score (>1). All data in the Mn-treated group were also compared with those in the control groups by Student's t-test analysis. The critical P-value was set to 0.05 for the significantly differential variables in this study. We searched for candidates from the freely accessible databases of HMDB (http://www.hmdb.ca), METLIN (http://metlin.scripps.edu) and KEGG (http://www.kegg.jp) by their masses, then, MS/MS analyses were performed, and due to the possible fragment mechanisms, items without given mass fragment information were removed from the candidate list and only the most

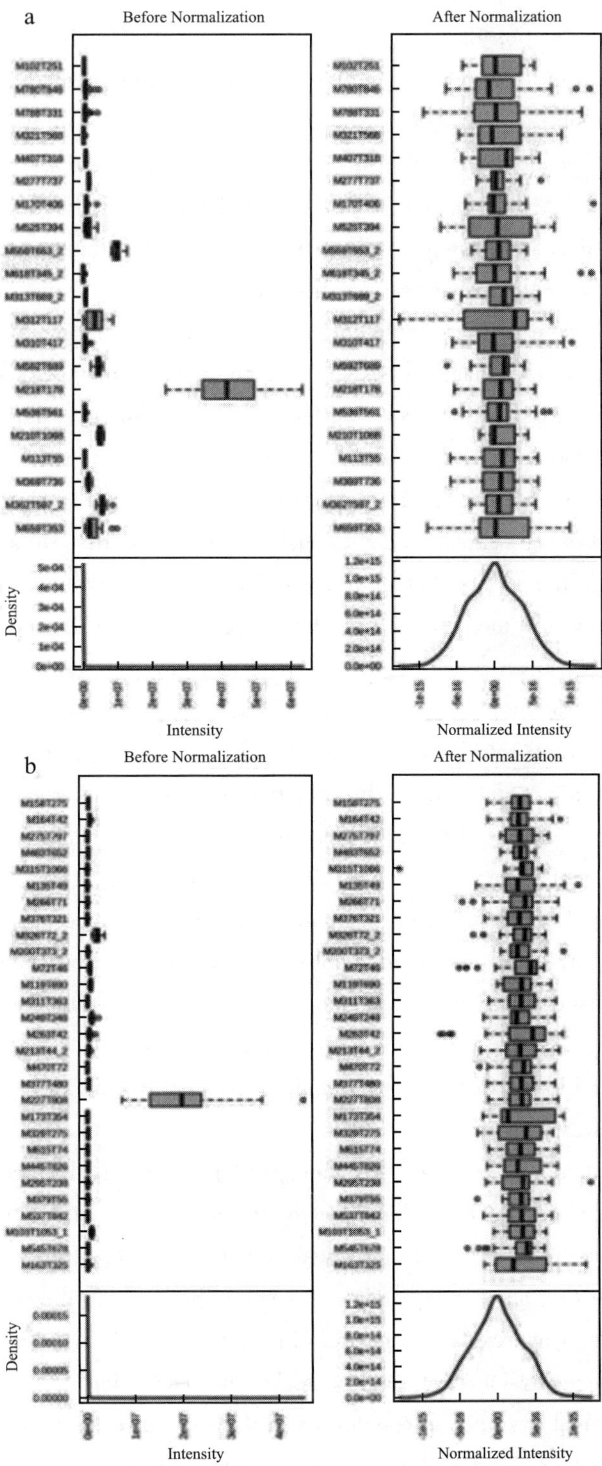

Fig. 4 This graph summarizes the distribution of input data values before and after normalization. (a) Positive mode and (b) negative mode.

probable items were left. By comparing the retention times and mass spectra of the authentic chemicals as well as the standard MS/MS spectra from the above databases, 36 significantly differential plasma metabolites were selected as potential biomarkers related to Mn exposure (Table 2). Table 2 also shows the tentative identification of these metabolites (compound name, molecular formula, adduct and metabolic pathway) and their corresponding concentration fold changes. Among the different changes observed, 12 metabolites showed a remarkable increase in concentration and 24 metabolites reduced significantly in concentration in the Mn-treated group. The increased concentration metabolites of taurodeoxycholic acid, 5-hydroxyindoleacetate, tryptamine and the decreased concentration metabolites of urocanic acid, 3-methoxy-4-hydroxyphenylglycol sulfate, D-proline, beta-hydroxypyruvic acid, stearamide, etc., may be potential biomarkers for manganism based on their VIP scores. The greater VIP score of the metabolite, the greater contribution to the separation of sample classification, and therefore the metabolite may be a potential biomarker.[37]

Hierarchical clustering analysis was performed based on the degree of similarity of the metabolite abundance profiles to show a global overview of all the plasma metabolites that were detected and visualized (Fig.7a and b). This heat map was created using MetaboAnalyst software (the distance was calculated using Pearson and the clustering using Ward). Metabolites with similar abundance patterns were positioned closer together. The heat map and dendrogram indicated the close clustering of the Mn exposure metabolites and their separation from the control groups.

Pathway analysis has been proven to be an invaluable tool for understanding complex relationships among genes and proteins.[38,39] The identified biomarkers responsible for the toxic effect of Mn exposure played an important role in specific metabolic pathways. Therefore, in order to identify possible pathways that are affected by Mn, the metabolites were checked against the KEGG pathway database (http://www.genome.jp/kegg/) for their corresponding metabolic pathways (Table 2). With the power of this metabolomics analysis, the versatile effects of Mn on the different metabolic pathways in plasma could be observed in an untargeted manner. The results showed that the potential biomarkers were responsible for purine metabolism, tryptophan metabolism, tyrosine metabolism, phenylalanine metabolism, taurine metabolism, hypotaurine metabolism, cysteine metabolism, methionine metabolism, arginine metabolism, proline metabolism, etc. (Table 2). The different metabolic pathways were likely due to the dynamic process of Mn exposure and might be closely associated with Mn exposure.

Taurodeoxycholic acid can induce DNA damage depending on the activation of TGR5, CREB and NOX5-S.[40] A significantly increased level of the serotonin metabolite 5-hydroxyindoleacetate was observed in the Mn-treated rats in our research. This paralleled the increased tryptamine in the same group, suggesting that an increased level of tryptamine leads to the increased formation of serotonin, because tryptophan hydroxylase, the rate-limiting enzyme in serotonin synthesis, is not saturated with substrate under physiological conditions.[41] The results indicated that excessive exposure to Mn can induce neurological dysfunction resembling Parkinson's disease (PD), and may be correlated with the higher levels of taurodeoxycholic acid, 5-hydroxyindoleacetate and tryptamine in the plasma.

Urocanic acid has been recognized as an immunosuppressive molecule with anti-inflammatory properties in variousin vitro and in vivo experimental models.[42] 3-Methoxy-4-hydroxyphenyl-glycol levels are decreased in the depressive subjects.[43] Proline catabolism is a component of innate immune signalling, and animals can utilize proline catabolism to promote stress responses and modulate innate

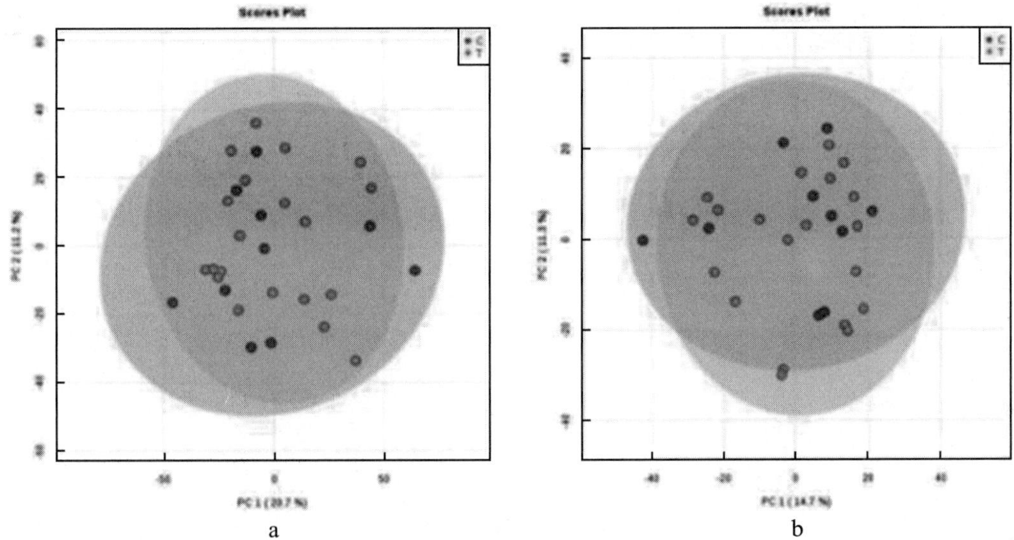

a b

Fig. 5 PCA score map derived from UHPLC-Q-TOF/MS spectra concerning control rats (red dot) and Mn-treated rats (green dot) in the positive mode (a) and negative mode (b).

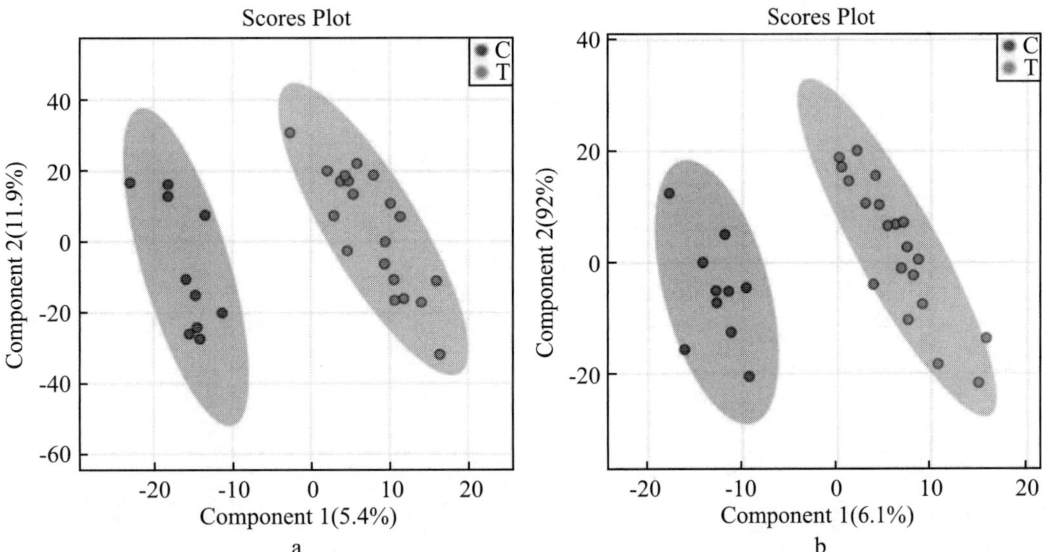

Fig. 6 PLS-DAscore map derived from UHPLC-Q-TOF/MS spectra concerning control (red dot) and Mn-treated (green dot) rats in the positive mode (a) and negative mode (b).

immunity.[44] The decreased levels of urocanic acid, 3-methoxy-4-hydroxyphenylglycol and proline suggest that manganism can inhibit the mental state and immune system.

Mn can preferentially accumulate in the substantia nigra, globus pallidus, and striatum, all of which are dopamine (DA) -rich brain regions.[45] DA is synthesized by tyrosine and its immediate precursor, phenylalanine, and a decreased concentration of phenylalanine would be expected to cause a decreased formation of DA. The key characteristic of PD is that dopamine signalling is reduced in the substantia nigra. From our research, we can speculate that the neurotoxicity of Mn may be connected with the decreased phenylalanine concentration.

Table 2 identification results of candidate biomarkers in rat plasma related to Mn exposure

Ionization mode	Metabolite no.	Adduct	RT (s)	m/z	Formula	Metabolite	Metabolic pathway	VIP	Fold change	p-value
ESI (+)	1	(M+H) +	302. 409	152. 069 8	$C_6H_6N_2O_2$	Urocanic acid	Histidine metabolism	2. 91	0. 60	0. 001 2
ESI (+)	2	$(M+H)^+$	1 153. 065	105. 934 1	$C_{26}H_{44}NO_6S$	Taurodeoxycholic acid	—	2. 60	2. 20	0. 028 6
ESI (+)	3	$(M+H)^+$	1192. 51	105. 9341	$C_{26}H_{45}NO_6S$	Taurochenodeoxycholate	Bile secretion	1. 75	0. 72	0. 036 4
ESI (+)	4	$(M+CH_3CN+H)^+$	802. 701	280. 263	$C_{10}H_{10}N_2O_3$	N-Benzyloxycarbonylglycine	—	1. 87	0. 75	0. 009 4
ESI (+)	5	$(M+H)^+$	398. 565	107. 069 6	$C_{12}H_{13}NO_4$	Nl-Methyl-2-pyridone-5-carboxamide	Nicotinate and nicotinamide metabolism	1. 47	0. 80	0. 028 4
ESI (+)	6	$(M+H)^+$	1 193. 3	141. 958 2	$C_6H_{13}NO_2$	L-Norleucine	—	1. 58	0. 67	0. 053 5
ESI (+)	7	$(M+H)^+$	53. 022	241. 128 9	$C_{11}H_{20}N_2O_3$	L-leucyl-L-proline	—	1. 42	1. 53	0. 077 9
ESI (-)	8	$(M-H)^-$	247. 620 5	249. 087 3	$C_9H_{13}N_3O_5$	Cytidine	Pyrimidine metabolism	1. 499 4	1. 294 938	0. 085 9
ESI (+)	9	$(M+NH_4)^+$	521. 277	225. 608	$C_{11}H_9NO_3$	Indole-3-pyruvic acid	—	1. 39	1. 32	0. 099 7
ESI (+)	10	$(M+H)^+$	60. 181	116. 069 9	$C_9H_9NO_3$	Hippuric acid	—	1. 39	0. 80	0. 098 4
ESI (+)	11	$(M+NH_4)^+$	165. 425	128. 069 7	$C_6H_8O_7$	Citrate	Alanine, aspartate and glutamate metabolism	2. 39	0. 60	0. 097 0
ESI (+)	12	$(M)^+$	897. 205 5	445. 365 7	$C_{29}H_5O_2$	alpha-Tocopherol (vitamin E)	Thiamine, vitamin B6, retinol and biotin metabolism	2. 27	0. 63	0. 086 6
ESI (+)	13	$(M+HCOO+2H)^+$	306. 667 5	161. 106 3	$C_9H_8O_4$	beta-Hydroxypyruvic acid	Glycine, serine and threonine metabolism	3. 10	0. 59	0. 000 2
ESI (+)	14	$(M+H)^+$	1145. 96	130. 006 9	$C_9H_{17}NO_4$	Acetylcarnitine	Fatly acid metabolism	1. 55	0. 39	0. 077 8
ESI (-)	15	$(M-H)^-$	44. 229	111. 007 7	$C_{26}H_{43}NO_6$	Glycocholic acid	Bile secretion	2. 011 2	0. 798 534	0. 003 1
ESI (-)	16	$(M-H_2O-H)^-$	1 056. 81	120. 972 5	$C_9H_{12}O_7S$	3-Methoxy-4-hydroxyphenylglycol sulfate	Norepinephrine metabolism	2. 745 4	0. 122 254	0. 039 1
ESI (+)	17	$(M+H)^+$	251. 646	120. 081	$C_{20}H_{23}N_7O_7$	Folinic acid	—	1. 26	0. 88	0. 048 6
ESI (+)	18	$(M+H)^+$	1 047. 5	115. 052 8	$C_5H_9NO_2$		Arginine and proline metabolism	2. 30	0. 44	0. 053 9
ESI (+)	19	$(M+H)^+$	275. 755	134. 062 5	$C_5H_9NO_2$	D-Proline	Arginine and proline metabolism	2. 73	0. 58	0. 002 7

(continued)

Ionization mode	Metabolite no.	Adduct	RT (s)	m/z	Formula	Metabolite	Metabolic pathway	VIP	Fold change	p-value
ESI (+)	20	$(M+H)^+$	241.064 5	303.070 1	$C_{38}H_{76}N_2O_2$	Stearamide	Cysteine and methionine metabolism	2.85	0.62	0.001 7
ESI (-)	21	$(M-H)^-$	841.205	102.655 3	$C_{11}H_{15}N_5O_3S$	S-Methyl-5′-thioadenosine	Cysteine and methionine metabolism	1.113 2	1.184 661	0.099 9
ESI (+)	22	$(M+H-H_2O)^+$	550.204	100.075 5	$C_{26}H_{44}NNaO_7S$	Sodium taurocholate	Taurine and hypotaurine metabolism	1.19	1.35	0.097 7
ESI (-)	23	$(M-H)^-$	327.713	100.966 9	$C_{26}H_{45}NO_7S$	Taurocholate	Taurine and hypotaurine metabolism	1.644 7	1.494 182	0.052 3
ESI (+)	24	$(M+H)^+$	58.300 5	181.032 3	$C_9H_{11}NO_2$	L-Phenylalanine	Phenylalanine and tyrosine metabolism	1.91	0.72	0.069 3
ESI (+)	25	$(M+H-H_2O)^+$	362.753	158.044 2	$C_8H_8O_2$	Pheny lace tic acid	Phenylalanine metabolism	1.50	1.29	0.049 7
ESI (+)	26	$(M+H-H_2O)^+$	49.576	112.895 1	$C_8H_{11}NO$	Tyramine	Tyrosine metabolism	1.07	1.15	0.042 4
ESI (+)	27	$(M+H)^+$	4.899	112.954 9	$C_{15}H_{11}I_4NO_4$	L-Thyroxine	Tyrosine metabolism	2.30	0.71	0.019 0
ESI (+)	28	$(M+CH_3CN+H)^+$	389.232	256.168 9	$C_{10}H_9NO_3$	5-Hydroxyindoleacetate	Tryptophan metabolism	2.89	2.28	0.031 8
ESI (+)	29	$(M+H)^+$	447.142	178.034 8	$C_{10}H_{12}N_2$	Tryptamine	Tryptophan metabolism	3.20	1.84	0.048 0
ESI (+)	30	$(M+H)^+$	359.695	116.069 7	$C_{10}H_9NO_2$	Indoleacetic acid	Tryptophan metabolism	1.63	0.83	0.013 8
ESI (-)	31	$(M-H)^-$	1 057.07	122.585 8	$C_{10}H_{12}N_5O_7P$	2′-Deoxyguanosine 5′-monophosphate (dGMP)	Purine metabolism	1.794 8	1.283 347	0.090 3
ESI (+)	32	$(M+H)+$	1 135.08	129.957 5	$C_{10}H_{14}N_5O_7P$	Adenosine 3′-monophosphate	Purine metabolism	2.59	0.38	0.062 1
ESI (+)	33	$(M+H-H_2O)^+$	245.698	279.099 9	$C_{10}H_{13}N_5O_4$	Adenosine	Purine metabolism	1.13	0.88	0.066 3
ESI (+)	34	$(M+H)^+$	1 068.62	115.962 5	$C_{10}H_{11}N_4O_8P$	Inosine 5′-monophosphate (IMP)	Purine metabolism	1.87	0.77	0.090 9
ESI (+)	35	$(M+H)^+$	1 174.98	117.934 1	$C_{10}H_{13}N_5O_{11}P$	Guanosine 5′ - monophosphate (GMP)	Purine metabolism	1.41	0.73	0.075 4
ESI (+)	36	$(M+H)^+$	483.785	355.244 3	$C_{10}H_{14}N_5O_7P$	Adenosine monophosphate (AMP)	Purine metabolism	1.35	0.83	0.037 5

Fold change (FC) was calculated as the average metabolite concentration in the Mn-treated rats relative to that of the control.

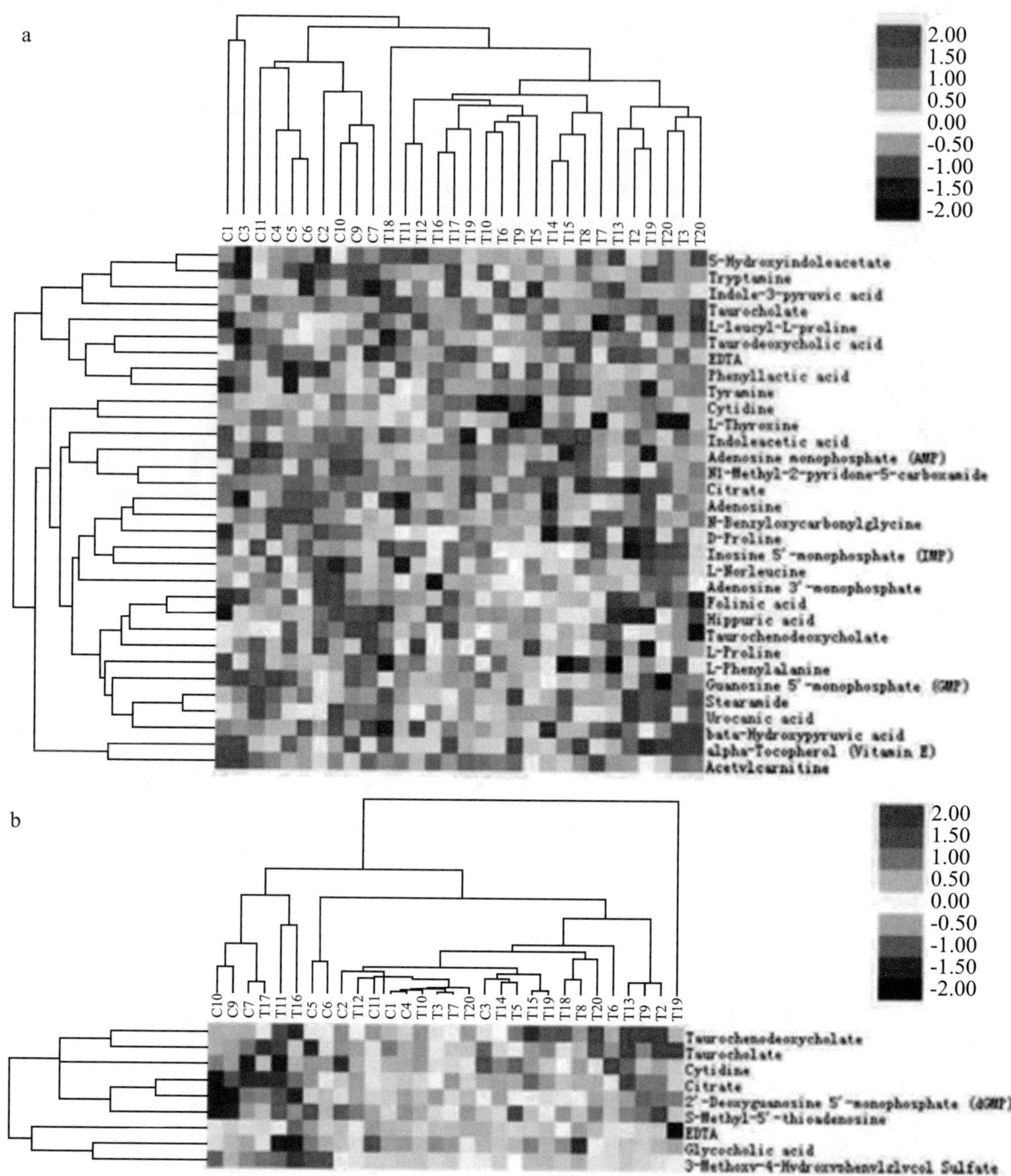

Fig.7 Unsupervised hierarchical clustering heat map of metabolites obtained from the plasma of rats in the positive mode (a) and negative mode (b).

Purine nucleosides are involved in a variety of intracellular neuronal processes and provide the precursors of DNA and RNA.[46] In our studies nearly all detected purines (adenosine 3′-monophosphate, adenosine, IMP, GMP and AMP) were significantly decreased in concentration in rat plasma after Mn exposure, except for dGMP (Table 2). A decrease in the concentration of adenosine and almost all other purines was also observed after chronic Mn treatment in rat brain extracts,[36,47] which is consistent with our results.Their concentration changes are mediated by adenosine receptors which can complex with DA receptors, thus enabling purines to fine-tune the neurotransmission in the basal ganglia.[48] The more purines that interact with adenosine receptors, the less neurotransmitters mentioned are excreted.[7] Adenosine, inosine and guanosine participate in the protection of versatile neurological insults as they are able to reduce the toxic effects of oxygen radicals and oxidative stress.[49,50] The observed reduction in concentration of biochemically important purines in rat plasma due to excessive Mn exposure might prompt the formation of free radicals and lead to an inactivation of these neuroprotective effects.

Furthermore, mineral metabolism can interact with the metabolism of macronutrients such as proteins and fats.[51] The metabolism of amino acids, representing an essential pool of nutrients utilized for cellular energy production, also plays a regulatory role in immune functions.[44] Our stady also demonstrated that Mn exposure is associated with significantly perturbed levels of metabolites in the cysteine, methionine, and arginine metabolism pathways.These results are consistent with the findings described by Bonilla et al.after a nine week lasting daily intraperitoneal injection of 5mg Mn per kg b.w.in mice[52] and Neth et al.after a single low dose $MnCl_2$ injection in rats.[36]

Carnitine and acetylcarnitine, which are involved in the transport of acetyl-CoA into mitochondria,[53] can improve energy and physical function,[54] and play an important role in promoting liver lipid transfer and utilization.[55] Here, a lower level of acetylcarnitine was observed in the Mn-treated rats, which reflected a disorder of fatty acid metabolism.

We also found that Mn-treated rats had lower levels of hippuric acid.Hippuric acid is a loading index of hepatic energy metabolism,[56] and the decreased concentration of hippuric acid indicated that the liver function in the Mn-treated rats was damaged.This finding could be used to further verify the results of histopathology.

4 CONCLUSION

In conclusion, five weeks of oral Mn-exposure led to significantly increased liver Mn concentrations compared to the control, and led to extensive necrosis and dissolved nuclei, suggesting liver damage from hepatic histopathology.Simultaneous changes in the plasma metabolites were also identified using untargeted metabolomics based on the UHPLC-Q-TOF/MS technique, which provided a unique perspective on Mn-induced changes in cellular metabolites.Thirty-six plasma metabolites (12 metabolites showed a remarkable increase in concentration and 24 metabolites reduced significantly in concentration) related to Mn exposure were identified, and were primarily involved in purine metabolism, amino acid metabolism and fatty acid metabolism.Understanding the targets of these metabolites may improve our understanding of cellular and intracellular Mn trafficking and the regulation of Mn homeostasis.

ACKNOWLEDGEMENTS

The financial support from the Gansu Provincial Natural Science Foundation (No. 1606RJYA224), National Key Research and Development Plan (No. 2016YFD0501200) and Central Public-interest Scientific Institution Basal Research Fund (No. 1610322013003) is greatly appreciated.

REFERENCES OMITTED

（发表于《Metallomics》，院选 SCI，IF：3.975）

Synthesis and Antibacterial Activities of Novel Pleuromutilin Derivatives with a Substituted Pyrimidine Moiety

Yunpeng YI[1,2], Ximing XU[3], Yu LIU[1,2], Shuijin XU[4], Xin HUANG[1,2], Jianping LIANG[1,2]*, Ruofeng SHANG[1,2]

(1. Key Laboratory of New Animal Drug Project of Gansu Province, Key Laboratory of Veterinary Pharmaceutical Development, Ministry of Agriculture, China; 2. Lanzhou Institute of Husbandry and Pharmaceutical Sciences of CAAS, 335 Jiangouyan, Lanzhou, 730050, China; 3. Institut Pasteur, Unité de Pathogenèse des infections vasculaires, Département de Biologie Cellulaire et Infection, Paris, 75015, France; 4. Yancheng YouHua Pharmaceutical & Chemical Technology Co., Ltd., Yancheng, 224555, China)

Abstract: The alarming growth of multidrug-resistant bacteria such as methicillin-resistant Staphylococcus aureus (MRSA) and vancomycin-resistant Enterococci (VRE) has become a major global health hazard. Therefore, urgent demand for new antibiotics with a unique mechanism of action is very necessary. The present study reports the design, synthesis, and antibacterial studies of a series of novel pleuromutilin derivatives with substituted 6-amino pyrimidine moieties. Most of the tested compounds exhibited highly potent anti-MRSA or Staphylococcus aureus (S.aureus) activities. 14-O-[(4, 6-Diamino-pyrimidine-2-yl) thioacetyl] mutilin (3) and 14-O-[(2-((3R)-3-Hydroxymethylpiperidine-1-yl)-acetamido-6-aminopyrimidine-2-yl) thioacetyl] mutilin (5h) were the most active compounds and showed higher antibacterial activities. Compound 3 displayed rapid bactericidal activity and affected bacterial growth with the same manner as tiamulin fumarate. Docking experiments for compounds 3 and 5h carried out on the peptidyl transferase center (PTC) of 23S rRNA provided the information about the binding model. In vivo mouse systemic infection experimental results confirmed the therapeutic efficacy of compound 3, with ED_{50} of 4.22mg/kg body weight against MRSA.

Key words: Pleuromutilin derivatives; Synthesis; Antibacterial activity; Molecular docking

---

* Corresponding author. E-mail: liangjpl963@163.com (J. Liang), shangrfl974@163.com (R. Shang).

1 INTRODUCTION

The development of novel antibiotics that can be used effectively for a growing number of multi-drug-resistant bacteria has become extremely challenging[1,2].The emergence of multidrug-resistant bacteria, such as methicillin-resistant Staphylococcus aureus (MRSA) and vancomycin-resistant enterococci (VRE), has led to many available drugs reducing or losing curative effect, and remains a global human threat with the potential for catastrophic consequences in the future[3].To solve the drastic increase of pathogenic bacteria resistance, there is pressing need to develop novel antibiotics with a unique mechanism of action against the dreadful pathogens.

Pleuromutilin (Fig. 1), a diterpene natural product from cultures of two species of basidiomycetes, Pleurotus mutilus and P.passeckerianus, was first discovered and isolated in a crystalline form in 1951[4].The pleuromutilin molecule is constituted of a rather rigid 5-6-8 tricyclic carbon skeleton with eight stereogenic centers and a glycolic acid chain[5-7]. Although, this compound shows modest antibacterial activity[8,9], the chemical modification at C-14 may improve its biological activity, and thus have led to three drugs: tiamulin, valnemulin and retapamulin (Fig. 1)[10-14].While other chemical modifications, including esterification or oxidation of the C-11 hydroxy group, yield inactive products[15].

Pleuromutilin derivatives selectively inhibit bacterial protein synthesis via binding to the 50S subunit of ribosomes at the acceptor (A) and the donor (P) site[16,17].Further studies have demonstrated that the hydrophobic interactions and the hydrogen bonds formed between the nucleotides of domain V and the tricyclic core of the tiamulin are the mainly binding modes[18,19].The cores of pleuromutilin derivatives orient at the A-site and alter the conformation of U2506 to tightly close the binding pocket.

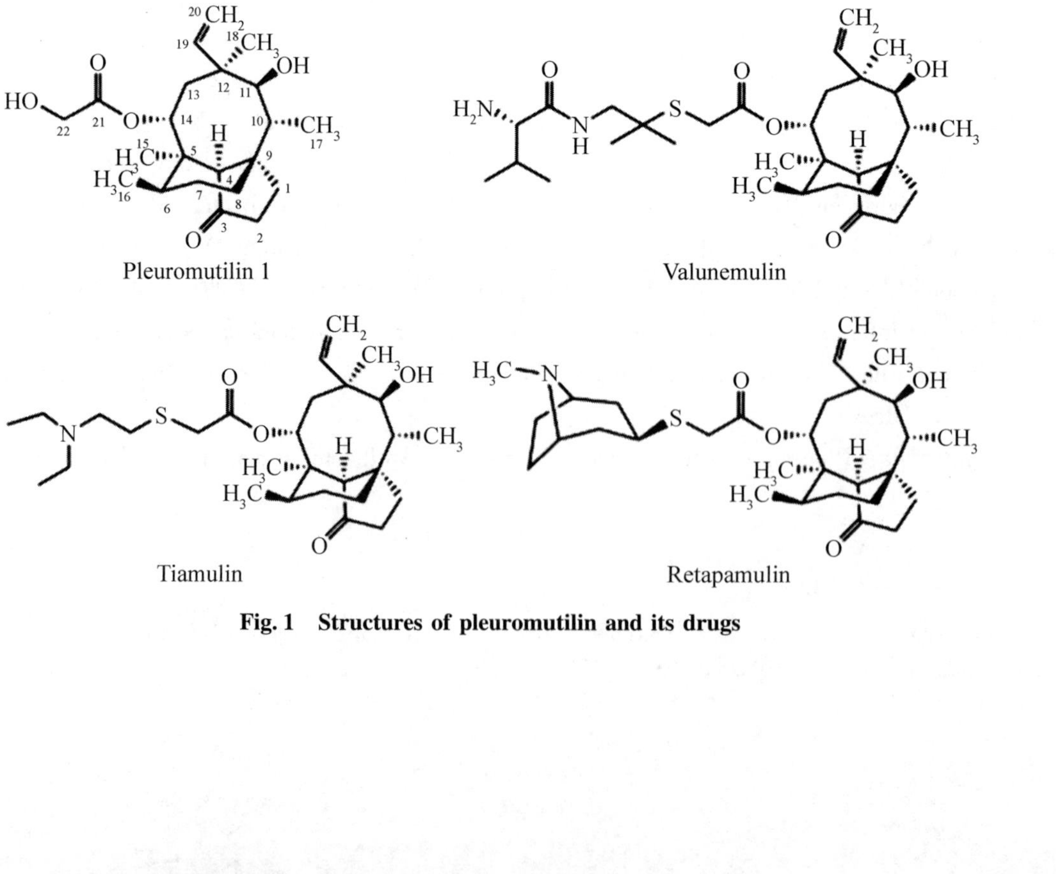

Pleuromutilin 1

Valunemulin

Tiamulin

Retapamulin

Fig. 1 Structures of pleuromutilin and its drugs

While C-14extensions pointtowardthe P-siteandalterlocation of U2585 to preclude tRNA from binding to the P-site[20].

Ling et al.[21] and our previous works[22,23] proposed that heterocyclic rings bearing polar groups at the C-14 side chain of pleuromutilin derivatives raise their antibacterial activity.A series of novel pleuromutilin derivatives with 6-hydroxy pyrimidine moieties were reported, and their antibacterial studies further confirmed that pleuromutilin derivatives with primary amines (polar groups) on the terminal C-14 side chains presented improved activity against resistant Gram-positive bacterial strains[24].In view of the above findings, we decided to revisit extension from the pleuromutilin derivatives with pyrimidine moieties.In this article, we describe the design, synthesis, and antibacterial studies of a series of novel pleuromutilin derivatives with substituted 6-amino pyrimidine moieties for identification of more efficacious drug candidates.

2 RESULTS AND DISCUSSION

2.1 Chemistry

The synthetic approaches for the preparation of target pleuro-mutilin derivatives and their intermediates are illustrated inScheme .The pleuromutilin 1 was converted into the known 22-O-tosyl-pleuromutilin 2, a key intermediate for synthesizing almost all pleuromutilin derivatives. Compound 14-O- [(4, 6-diamino-pyrimi-dine-2-yl) thioacetyl] mutilin (3) was prepared by nucleophilic substitution of 2 with 4, 6-diamino-2-mercapto pyrimidine under basic conditions in 95% yield.The key intermediate, 14-O- [(2-chloroacetamide-6-aminopyrimidine-2-yl) thioacetyl] mutilin 4, was prepared in 72% yield by commercially available chloracetyl chloride and 3 with an aim to construct acetamide linker between the tertiary amine and 6-aminopyrimidine.The target compounds 5a-i were directly obtained in 43%-74% yield from 4 and various secondary amines in the presence of triethylamine.The structures of the synthesized derivatives were characterized by IR, ^{1}H NMR, ^{13}C NMR and HRMS spectra (Supplementary data).For further confirming the structure and conformation, a crystal of compound 3 was obtained as a colorless, block-like from a solution of ethanol and acetone by slow evaporation method at room temperature (Fig. 2)[25].

2.2 Antibacterial activity

The synthesized pleuromutilin derivatives3, 5a-i and tiamulin fumarate used as reference drug were screened for their in vitro antibacterial activity against methicillin-resistant Staphylococcus aureus ATCC 29213 (MRSA-29213), methicillin-resistant Staphylococcus aureus ATCC 33591 (MRSA-33591), Staphylococcus aureus CMCC (B) 26003 (S. aureus), Bacillus subtilis CMCC (B) 63501 (B. subtilis), and Escherichia coli CMCC (B) 44102 (E. coli). The minimum inhibitory concentrations (MICs) of all screened compounds were listed in Table 1. All the tested compounds exhibited excellent antibacterial efficiency against MRSA-29213, MRSA-33591 and S. aureus, with MIC values ranging from 0.0125 to 1, from 0.0625 to 1 and from 0.0625 to 0.25μg/mL, respectively.However, these compounds showed comparatively higher MIC values for B.subtilis and E.coli, with MIC values ranging from 1 to 4 and from 2 to 8μg/mL respectively, than that for MRSA or S.aureus.It can be observed that compound 3 and 5h showed the highest antibacterial activities for all the five strains than the other synthesized compounds and tiamulin fumarate.

The antibacterial activities against the above-mentioned five bacterial strains were evaluated by Oxford cup assays.The zones of inhibition for two concentrations (320 and 160μg/mL) are reported as diameters of growth inhibition (Supplementary data).The results correspond with MICs obtained by agar dilution method as a whole.Compound 3 and 5h showed the best growth inhibition against the pathogens particularly MRSA-29213, MRSA-33591 and S.aureu, the others showed moderate growth of strains except E.coli.

Almost all the synthesized compounds, as well as reference drug, presented slightly higher activity against S.aureus than MRSA in vitro.However, this anti-bacterial difference for the two strains behind the result were not clear.A possible explanation for that may be the membrane permeability alterations of MRSA, which may prohibit influx of certain antimicrobials[26].

Time kill studies were performed to probe the effect of compound 3 and tiamulin fumarate at two different concentrations (1 × MIC and 6 × MIC) against the S.aureu (Fig. 3A) and MRSA-29213 (Fig. 3B).Time-kill kinetics revealed that the two compounds displayed a concentration-dependent effect, and killed bacteria rapidly at 6 × MICs, achieving a >6 log reduction in 4h.However, they were limited to being bacteriostatic (3-4 $\log_{10}$ CFU/mL) at 1 × MICs.Compared to tiamulin fumarate at 6 × MICs, compound 3 displayed more rapid bactericidal kinetics against S.aureu with the same MIC (0.0625μg/mL), but the similar kinetics against MRSA-29213 with only a 2-fold improvement in MIC (0.125μg/mL). These studies suggested that compound 3 affected bacterial growth with the same manner as tiamulin fumarate.

2.3 Molecular docking study

In order to survey the binding model of the synthesized compounds 3, 5a-i within the binding pocket of 50S ribosomal subunit and to understand their structure-activity relationship, we performed molecular docking using Homdock software[27]. The crystal structure (PDB ID: 1XBP)[28] was used for modeling.Tiamulin within 1XBP was used to compare experimental and predicted binding modes and validate our docking protocol.Upon flexible docking into the 50S ribosomal subunit, the 10 compounds present a similar binding mode consistent with that of tiamulin, with a RMSD range of 0.87-1.28 Å within the binding site (Fig. 4 and Table 2).The best pose was selected from the10 poses simulated by the software.The results revealed hydrogen bonding played the most important role in the binding of the compounds to 1XBP, with the binding free energies (ΔGb) in the range of -13.28 to -9.22 kcal/mol (Table 2).

Out of above ten compounds, 3 and 5h exhibited high binding affinity (-12.44 and-13.28 kcal/mol, ΔGb), which were in agreement with they antibacterial activity. For the two docking models (Fig. 5), hydrogen bonds formed between hydroxyl groups (eight-membered ring), C=O (ester) and N (pyrimidine) of compounds 3 and 5h and the residue of G-2484 and G-2044 were the key interactions. In addition, one hydrogen bond was formed between the NH_2 (pyrimidine ring) of compound 3 with C-2046, and two hydrogen bonds were formed between the hydroxyl group (piperidine ring) of the compound 5h with A-2418. However, no π-π or cation-π interactions were found between the pyrimidine rings of the two compounds, as well as the other com-

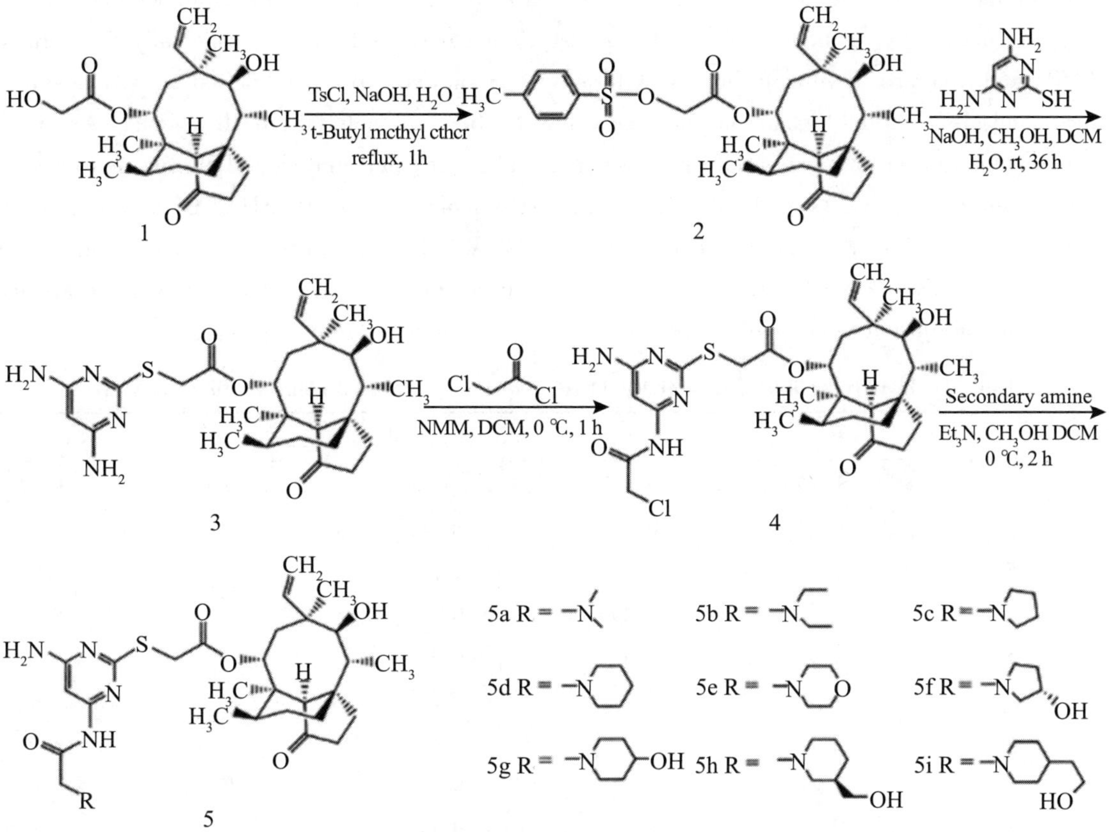

Scheme 1. General synthetic scheme for the pleuromutilin derivatives.

pounds, with the residues.

2.4 In vivo efficacy in mouse model

After being infected with MRSA-29213 (1×10^9 CFU in 0.1mL saline), the mice were intravenously treated with different doses of 3 dissolved in 3.0% vehicle and tiamulin fumarate used as the reference drug. Fig. 6 showed the time related death and survival of the infected mice when treated with different doses of 3 and tiamulin fumarate dissolved in vehicle at a volume of 0.1mL given in a single intravenous injection in the tail vein. The results demonstrated dose dependent effects on the survival against MRSA - 29213 bearing mice, with ED_{50} of 4.22 and 5.94mg/kg body weight, respectively. Thus, compound 3 showed the higher antibacterial activity than that of tiamulin against MRSA-29213 in the mouse systemic model and might act as a potent antibacterial drug.

3 CONCLUSION

In summary, we have synthesized a series of novel pleuro-mutilin derivatives possessing 6-amino pyrimidine moieties. These derivatives were initially evaluated for theirin vitro antibacterial activities against four Gram - positive strains (MRSA - 29213, MRSA - 33591, S. aureus and B. subtilis) and a Gram-negative strain (E. coli). All the screened compounds showed excellent in vitro antibacterial activity against MRSA-29213, MRSA-33591 and S.aureu. Compounds 3 and 5h were the most active antibacterial agents against Gram-positive bacteria, especially the MRSA-33591 and S.aureus in vitro. In time-to-kill assays, 3 showed rapid bactericidal activity against S. aureu and MRSA at the higher concentration, and its effect on bacterial growth was suggested in the same manner as that of tiamulin fumarate. Docking studies for compounds 3 and 5h revealed the binding free energies (ΔGb) of -12.44 and -13.28 kcal/mol, with an RMSD of 0.92 and 1.18 Å, respectively. It is important to note that compound 3 showed higher efficacy than that of tiamulin fumarate against MRSA-29213 when chosen for the further evaluation in vivo activity using systemic infection mode in mice.

Table 1 In vitro antibacterial activity (MIC) of the synthesized pleuromutilin derivatives.

Compounds	MICs (μg/mL)				
	MRSA-29213	MRSA-33591	*S.aureus*	*B.subtilis*	*E.coli*
3	0.125	0.0625	0.0625	2	2
5a	1	0.5	0.25	4	8
5b	0.5	0.25	0.125	4	8
5c	0.5	0.5	0.125	2	4
5d	0.25	0.5	0.25	4	8
5e	0.25	1	0.125	2	4
5f	0.125	0.25	0.125	2	8
5g	0.25	0.25	0.125	2	4
5h	0.125	0.125	0.0625	1	4
5i	1	0.5	0.125	2	8
Tiamulin	0.25	0.125	0.0625	2	4

4 MATERIALS AND METHODS

4.1 Chemistry

The reagents and solvents were obtained commerically and used without furtherpurification. Infrared (IR) spectra were obtained on a Thermo Nicolet NEXUS-670 spectrometer using KBr thin films, and the absorptions are reported in cm^{-1}. ^{1}H NMR spectra were recorded using Bruker 400MHz spectrometers in $CDCl_3$ or DMSO-d_6. The chemical shifts (dδ) were reported in parts per million (ppm) relative to tetramethylsilane. ^{13}C NMR spectra were recorded on 100MHz spectrometers. Purity of the compounds and completion of reactions were monitored by thin layer chromatography (TLC) on silica gel plates (GF254; Qingdao Haiyang Chemical Co., Ltd., Shandong, China), followed by visualization after spraying with a 0.05% $KMnO_4$ aqueous solution or under UV illumination directly. High-resolution mass spectra (HRMS) were obtained with a Bruker Daltonics APEX Ⅱ 47e mass spectrometer equipped with an electrospray ion source.

4.1.1 14-O-[(4, 6-Diamino-pyrimidine-2-yl) thioacetyl] mutilin (3)

To a solution of 4, 6-dihydroxy-2-mercaptopyrimidine (1.44g, 10mmol) in 25mL

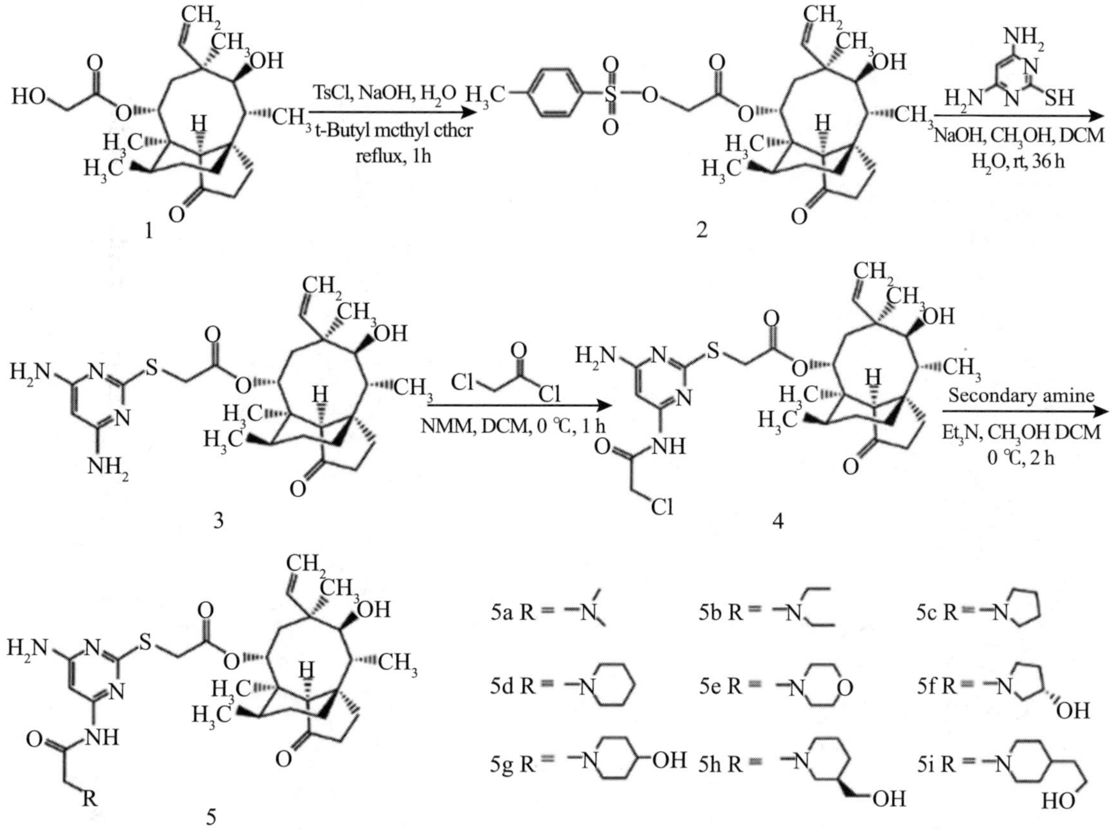

Fig. 2. ORTEP diagram for compound 3 with ellipsoids set at 75% probability.

methanol, 10M NaOH (1.1mL, 11mmol) was added and stirred for 30min. A solution of 22-O-tosylpleuromutilin (5.85, 11mmol) in 30mL of DCM was added dropwise to the reaction mixture. The mixture was stirred for 36h at room temperature. Then the reaction mixture was evaporated under reduced pressure to dryness, and the crude product was extracted with a solution of ethyl acetate (60mL) and water (20mL) and treated with saturated $NaHCO_3$. The target compound 3 was then precipitated and purified by flash silica column chromatography (ethyl acetate : ethanol20 : 1 v/v) toyield 4.77g (95%). IR (KBr): 3 475, 3 377, 2 933, 1 728, 1 619, 1 582, 1 547, 1 465, 1 309, 1 153, 1 117, 1 019 cm^{-1}. ^1H NMR (400MHz, DMSO) δ 6.20-6.11 (m, 1H), 6.06 (d, J = 15.9Hz, 3H), 5.52 (d, J = 7.8Hz, 1H), 5.18 (d, J = 33.9Hz, 1H), 5.04 (dd, J=24.4, 14.5Hz, 2H), 4.50 (d, J=5.8Hz, 1H), 4.03 (dd, J=14.1, 7.1Hz, 1H), 3.80 (d, J=9.7Hz, 1H), 3.42 (s, 1H), 2.39 (s, 1H), 2.17 (dd, J= 25.8, 15.3Hz, 1H), 2.13-1.93 (m, 4H), 1.79-1.53 (m, 2H), 1.51-1.12 (m, 9H), 1.11-0.89 (m, 4H), 0.82 (d, J=6.5Hz, 3H), 0.62 (d, J=6.4Hz, 3H). ^{13}C NMR (101MHz, DMSO) δ 217.6, 168.3, 167.3, 163.8, 141.2, 115.8, 79.6, 73.1, 70.0, 60.2, 57.7, 45.4, 44.5, 44.0, 42.0, 36.8, 34.5, 33.3, 30.6, 29.0, 27.1, 24.9, 21.2, 16.6, 15.0, 12.0. HRMS (ES) calcd $[M+H]^+$ for $C_{26}H_{38}N_{40}N_4S$ 503.2687, found 503.2689.

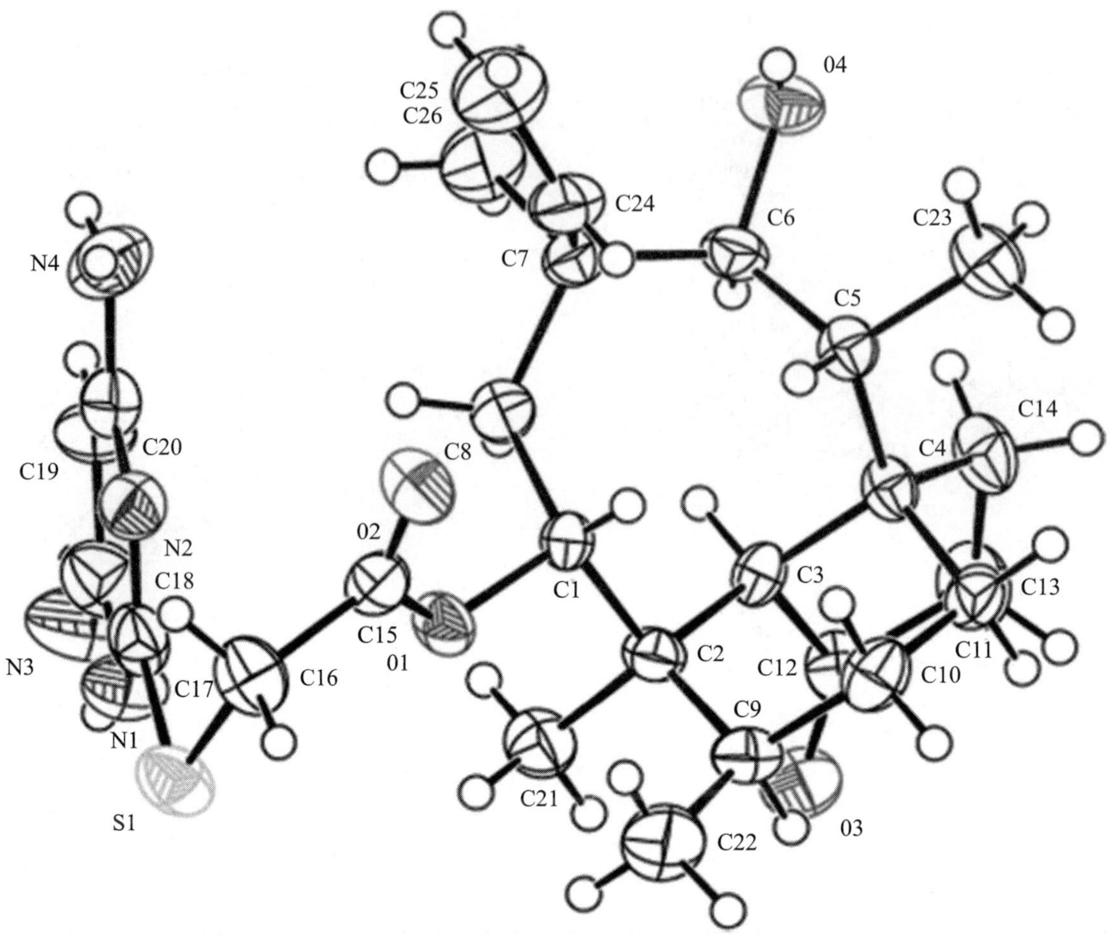

Fig. 3. Time-kill kinetics of compound 3 against S.aureus (A) and MRSA-29213 (B).Mean values of the CFU/mL were obtained from the triplicate measurements.

4. 1. 2 General procedure for the synthesis of compounds 5a-i

To a solution of compound 3 (1. 51g, 3mmol) and 4-methylmorpholine (0. 61g, 6mmol) in 20mL dry DCM, chlor-oacetyl chloride (0. 51g, 4. 5mmol) in 10mL dry DCM was slowly dropped at 0℃.The reaction mixture was stirred for 1h.After the reaction, the solution was washed with ammonium chloride one time, followed by water three times.Then the organic layer was dried with Na_2SO_4 overnight and rotary evaporated to dryness.The crude residue obtained was purified by column chromatography (petroleum ether : ethyl acetate = 1 : 6) to afford the desired compound 4 in 72% yield as a white solid.IR (KBr): 3 388, 2 927, 1 719, 1 625, 1 553, 1 509, 1 459, 1 406, 1 288, 1 153, 1 117cm^{-1}.^{1}H NMR (400MHz, $CDCl_3$) δ 8.57 (s, 1H), 6.94 (s, 1H), 6.43 (dd, J = 17.4, 11.0, 1H), 5.66 (d, J = 8.4, 1H), 5.18 (dd, J = 60.3, 14.3, 2H), 5.06 (s, 1H), 4.07 (s, 2H), 3.77 (d, J = 16.3, 1H), 3.70-3.60 (m, 2H), 3.29 (dd, J=9.7, 6.6, 1H), 2.29-2.20 (m, 1H), 2.19-2.01 (m, 2H), 1.95 (dd, J = 15.9, 8.6, 1H), 1.69 (d, J = 14.2, 1H), 1.52 (ddd, J = 44.3, 27.8, 15.1, 4H), 1.39-1.25 (m, 4H), 1.18 (t, J=7.0, 3H), 1.13-0.96 (m, 4H), 0.79 (d, J=

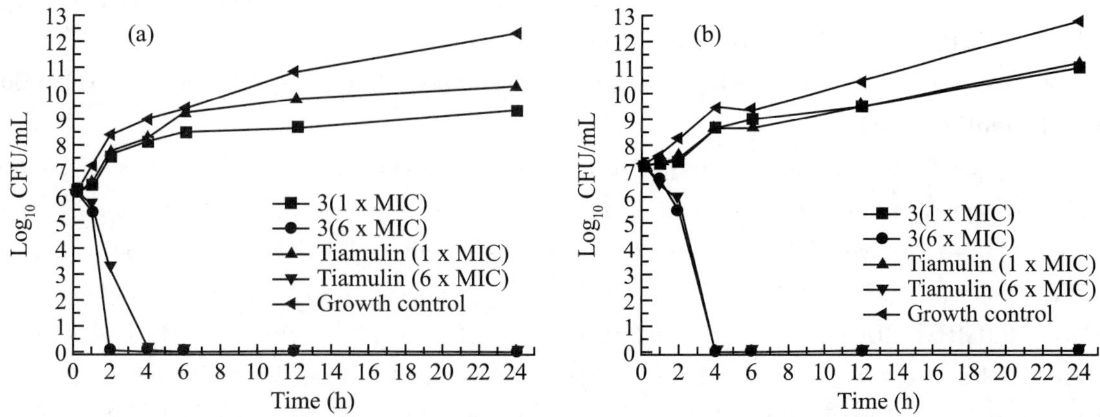

Fig. 4. Superimposition of tiamulin (red) and the best conformations of compounds 3 (green), 5a (blue), 5b (yellow), 5c (magenta), 5d (cyan), 5e (orange), 5f (wheat), 5g (white), 5h (brown) and 5i (pink) docked to the binding pocket of ribosome (1XBP). (For interpretation of the references to colour in this figure legend, the reader is referred to the web version of this article.)

6.9, 3H), 0.67 (d, J = 6.9, 3H). ^{13}C NMR (101MHz, $CDCl_3$) δ 216.12, 168.19, 167.12, 164.23, 163.17, 154.68, 138.27, 116.04, 87.13, 73.57, 68.69, 57.14, 44.44, 43.51, 42.93, 41.70, 40.87, 35.75, 34.98, 33.47, 29.40, 25.89, 25.37, 23.81, 17.41, 15.84, 13.90, 10.44. HRMS (ES) calcd $[M+H]^+$ for $[C_{28}H_{39}ClN_4O_5S]$ 579.2402, found 579.2407.

Secondary amines (4.5mmol) was added to the solution of compound 4 (1.73g, 3mmol) and triethylamine (0.61g, 6mmol) in methanol (60mL) and stirred at 45℃ for 2h. Then the solvent was evaporated in vacuum from the reaction mixture. The residue was added DCM (60mL) and quenched with saturated aqueous NH_4Cl (30mL). The organic layer was separated, washed with water (20mL for three times), dried with anhydrous Na_2SO_4 and rotary evaporated to dryness. The crude residue was purified by silica gel column chromatography (petroleum ether and ethyl acetate) to afford the desired compounds 5a–i.

4.1.3 14-O-[(2-(bis(Methyl)-amino)-acetamido-6-aminopyrimidine-2-yl) thioacetyl] mutilin (5a)

Compound5a was prepared according to the general procedure with a reaction time of 2h. The crude product was purified by column chromatography on silica gel (petroleum ether : ethyl acetate = 1 : 4) to give 43% yield (0.76 g) of 5a as a white solid. IR (KBr): 3 345, 2 935, 1 720, 1 625, 1 593, 1 550, 1 498, 1 457, 1 287cm^{-1}; ^{1}H NMR (400MHz, $CDCl_3$) δ 9.36 (s, 1H), 7.09 (d, J=15.1Hz, 1H), 6.51 (dd, J=17.4, 11.0Hz, 1H), 5.76 (t, J=16.9Hz, 1H), 5.36–5.14 (m, 2H), 5.04–4.82 (m, 2H), 3.93–3.70 (m, 2H), 3.35 (s, 1H), 3.06 (s, 1H), 2.37–2.30 (m, 6H), 2.25–2.13 (m, 2H), 2.02 (dt, J=13.1, 12.4Hz, 3H), 1.79–1.72 (m, 1H), 1.69–1.48 (m, 4H), 1.48–1.42 (m, 4H), 1.38–1.24 (m, 3H), 1.17–1.08 (m, 4H), 0.86 (d, J=7.0Hz, 3H), 0.74 (dd, J=15.9, 5.4Hz, 3H); ^{13}C NMR (101MHz, $CDCl_3$) δ 216.03, 169.57, 167.92, 167.15, 163.04, 155.27, 138.26, 116.05, 86.86, 73.60, 68.57, 62.64, 57.16, 44.96, 44.45, 43.51, 42.95, 40.87, 35.79, 34.99,

33.46, 32.86, 29.42, 25.91, 25.37, 23.82, 15.82, 13.89, 10.42; HRMS (ES) calcd [M + H]$^+$ for [$C_{30}H_{45}N_5O_5S$] 588.3214, found 588.3219.

4.1.4 14-O- [(2- (bis (Ethyl) -amino) acetamido-6-aminopyrimidine-2-yl) thioacetyl] mutilin (5b)

Compound5b was prepared according to the general procedure with a reaction time of 2h. The crude product was purified by column chromatography on silica gel (petroleum ether : ethyl acetate = 1 : 4) to give 52% yield (0.96g) of 5b as a white solid. IR (KBr): 3 348, 2 934, 1 725, 1 625, 1 591, 1 550, 1 500, 1 459, 1 400, 1 287, 1 199, 1 153, 1 117 cm^{-1}; ^{1}H NMR (400MHz, $CDCl_3$) δ 9.53 (d, J = 20.9Hz, 1H), 7.12-7.01 (m, 1H), 6.50 (dt, J = 20.5, 10.3Hz, 1H), 5.74 (d, J = 8.5Hz, 1H), 5.24 (ddd, J = 18.5, 14.2, 3.7Hz, 2H), 4.99 (d, J = 46.1Hz, 2H), 3.95 - 3.64 (m, 2H), 3.35 (dd, J = 9.2, 6.8Hz, 1H), 3.13 (s, 1H), 2.92 (d, J = 29.8Hz, 1H), 2.62 (q, J = 7.0Hz, 3H), 2.37-2.27 (m, 1H), 2.27-2.06 (m, 3H), 2.01 (dd, J = 14.9, 7.5Hz, 1H), 1.73 (dd, J = 17.5, 10.6Hz, 1H), 1.70 - 1.47 (m, 4H), 1.44 (s, 4H), 1.39 - 1.20 (m, 3H), 1.14 (t, J = 6.7Hz, 4H), 1.07 (t, J = 7.1Hz, 5H), 0.86 (d, J = 7.0Hz, 3H), 0.75 (d, J = 6.9Hz, 3H); ^{13}C NMR (100MHz, $CDCl_3$) δ216.03, 171.03, 167.87, 167.17, 163.04, 155.25, 138.25, 116.06, 86.80, 73.60, 68.55, 57.16, 47.77, 44.45, 43.54, 42.96, 40.87, 35.78, 34.99, 33.47, 32.90, 29.43, 25.90, 25.37, 23.82, 15.83, 13.90, 11.04, 10.43; HRMS (ES) calcd [M + H]$^+$ for $C_{32}H_{49}N_5O_5S$ 616.3527, found 616.3521.

Table 2 Bind free energy, noncovalent molecular interaction and RMSD.

Compound	ΔGb (kcal/mol)	Non-covalent molecular interaction				RMSD (A)
		Hydro I interaction	Atom of Compound	Residue	Distance (A)	
3	-12.44	H-bonding	OH (8-membered ring)	G-2484	2.0	0.87
		H-bonding	C=O (ester)	G-2044	2.1	
		H-bonding	C=O (ester)	G-2044	2.3	
		H-bonding	N (pyrimidine ring)	G-2044	2.4	
		H-bonding	NH_2 (pyrimidine ring)	G-2044	2.1	
5a	-12.01	H-bonding	OH (8-membered ring)	G-2484	2.1	0.97
		H-bonding	C=O (ester)	G-2044	2.3	
		H-bonding	C=O (ester)	G-2044	2.2	
		H-bonding	N (pyrimidine ring)	G-2044	1.9	
		H-bonding	NH_2 (pyrimidine ring)	G-2044	2.1	
5b	-10.30	H-bonding	C=O (ester)	G-2044	2.3	1.05
		H-bonding	C=O (ester)	G-2044	2.3	
		H-bonding	N (pyrimidine ring)	G-2044	1.7	
		H-bonding	NH_2 (pyrimidine ring)	G-2044	2.4	

(continued)

Compound	ΔGb (kcal/mol)	Non-covalent molecular interaction				RMSD (A)
		Hydro I interaction	Atom of Compound	Residue	Distance (A)	
5c	-9.22	H-bonding	C=O (ester)	G-2044	2.2	1.02
		H-bonding	C=O (ester)	G-2044	2.3	
		H-bonding	N (pyrimidine ring)	G-2044	1.9	
5d	-9.42	H-bonding	N (pyrimidine ring)	G-2044	2.4	1.28
		H-bonding	NH_2 (pyrimidine ring)	G-2044	2.7	
		H-bonding	NH_2 (pyrimidine ring)	A-2482	2.3	
		H-bonding	NH_2 (pyrimidine ring)	A-2045	2.4	
5e	-10.77	H-bonding	C=O (ester)	G-2044	2.2	0.99
		H-bonding	C=O (ester)	G-2044	2.3	
		H-bonding	N (pyrimidine ring)	G-2044	1.8	
		H-bonding	NH_2 (pyrimidine ring)	G-2044	2.0	
5f	-9.78	H-bonding	NH_2 (pyrimidine ring)	A-2045	1.7	1.08
		H-bonding	N (pyrolidine)	C-2565	2.6	
5g	-9.81	H-bonding	NH_2 (pyrimidine ring)	A-2045	1.9	1.18
		H-bonding	NH_2 (pyrimidine ring)	A-2482	2.2	
5h	-13.28	H-bonding	OH (8-membered ring)	G-2484	2.4	1.19
		H-bonding	C=O (amido bond)	G-2044	2.1	
		H-bonding	C=O (amido bond)	G-2044	2.3	
		H-bonding	N (pyrimidine ring)	G-2044	2.4	
		H-bonding	NH_2 (pyrimidine ring)	G-2044	1.8	
		H-bonding	NH_2 (pyrimidine ring)	G-2044	2.4	
		H-bonding	OH (piperidine)	A-2418	2.5	
		H-bonding	OH (piperidine)	A-2418	2.3	
5i	-11.39	H-bonding	C=O (5-membered ring)	G-2044	2.3	1.25
		H-bonding	C=O (5-membered ring)	G-2044	2.4	
		H-bonding	NH_2 (pyrimidine ring)	G-2044	2.0	
		H-bonding	OH (piperidine)	U-2564	1.8	
Tiamulin	-10.20	H-bonding	OH (8-membered ring)	G-2484	2.1	0.87
		H-bonding	C=O (ester)	G-2044	2.2	
		H-bonding	C=O (ester)	G-2044	2.3	

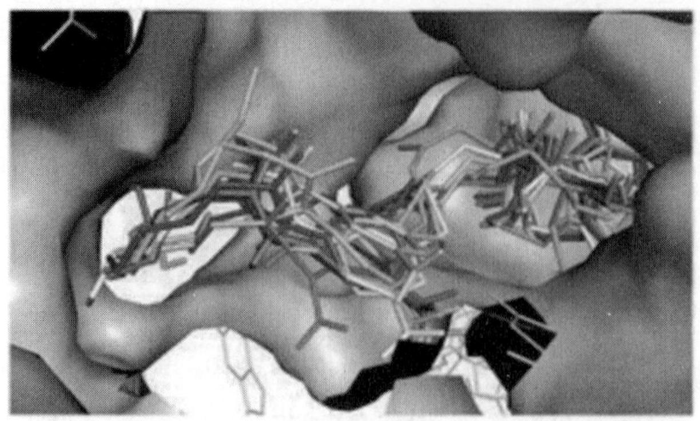

Fig. 5. Docking modes of compounds 3 (A) and 5h (B) into 1XBP.

Note: Important residues are drawn in stick with orange. Hydrogen bonds are showed as dashed red lines. (For interpretation of the references to colour in this figure legend, the reader is referred to the web version of this article.)

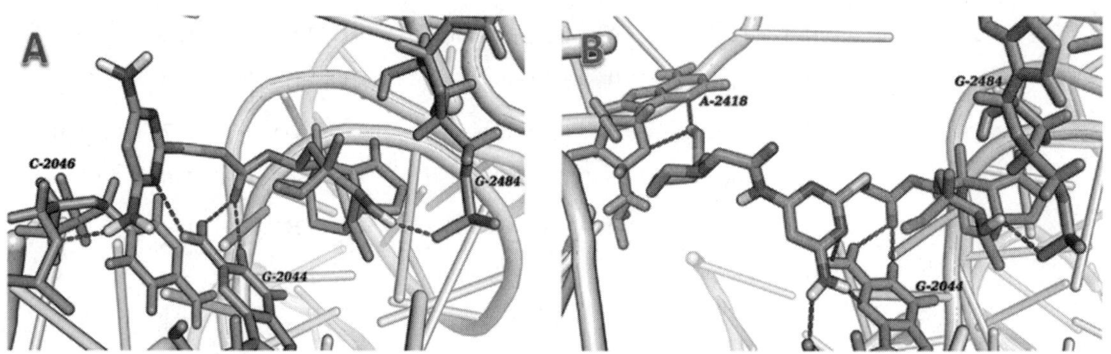

Fig. 6. Efficacy of compound 3 (A) and tiamulin fumarate (B) in mouse systemic infection model.

4.1.5 14-O-[(2-(Pyrrolidine-1-yl)-acetamido-6-aminopyrimidine-2-yl) thioacetyl] mutilin (5c)

Compound 5c was prepared according to the general procedure with a reaction time of 2h. The crude product was purified by column chromatography on silica gel (petroleum ether : ethyl acetate = 1 : 4) to give 49% yield (0.90g) of 5c as a white solid. IR (KBr): 3 357, 2 934, 1 720, 1 624, 1 592, 1 550, 1 500, 1 465, 1 200, 1 018, 939cm^{-1}; 1H NMR (400MHz, $CDCl_3$) δ 9.25 (s, 1H), 7.04 (d, J = 15.2Hz, 1H), 6.52-6.34 (m, 1H), 5.67 (d, J = 8.5Hz, 1H), 5.29-5.07 (m, 2H), 4.90 (d, J = 25.6Hz, 2H), 3.83-3.64 (m, 2H), 3.33-3.25 (m, 1h), 3.19 (s, 1H), 2.59 (s, 4H), 2.28-2.20 (m, 1h), 2.19-2.07 (m, 2H), 2.07-1.90 (m, 3H), 1.78 (s, 4H), 1.69 (dd, J = 14.4, 2.4Hz, 1H), 1.63-1.41 (m, 4H), 1.40-1.33 (m, 4H), 1.23 (dt, J = 14.5, 9.4Hz, 3H), 1.10-1.01 (m, 4H), 0.80 (d, J = 7.0Hz, 3H), 0.66 (t, J = 13.1Hz, 3H); ^{13}CNMR (101MHz, $CDCl_3$) δ 216.02, 169.96, 167.86, 167.13, 163.03, 155.34, 138.27, 116.04, 86.88, 73.60, 68.53, 58.96, 57.15, 53.60, 44.45, 43.50, 42.95, 40.87, 35.79, 34.99, 33.46, 32.89, 29.43, 25.90, 25.37, 23.82, 23.08, 15.81, 13.90,

10.43; HRMS (ES) calcd [M + H]$^+$ for $C_{32}H_{47}N_5O_5S$ 614.3370, found614.3371.

4.1.6 14-O-[(2-(Piperidine-1-yl)-acetamido-6-aminopyrimidine-2-yl) thioacetyl] mutilin (5d)

Compound 5d was prepared according to the general procedure with a reaction time of 2h. The crude product was purified by column chromatography on silica gel (petroleum ether : ethyl acetate = 1 : 4) to give 47% yield (0.87g) of 5d as a white solid. IR (KBr): 3 369, 2 933, 1 734, 1 605, 1 581, 1 549, 1 511, 1 458, 1 401, 1 299, 1 251, 1 160, 1 116, 1 069, 1 024, 846cm^{-1}; ^{1}H NMR (400MHz, $CDCl_3$) d 9.31 (d, J = 15.1Hz, 1H), 7.05-6.96 (m, 1H), 6.54-6.37 (m, 1H), 5.66 (t, J = 8.4Hz, 1H), 5.28-5.08 (m, 2H), 4.96-4.75 (m, 2H), 3.78-3.64 (m, 2H), 3.34-3.24 (m, 1H), 2.97 (s, 1H), 2.42 (s, 4H), 2.24 (dd, J = 13.2, 6.3Hz, 1H), 2.19-2.06 (m, 2H), 1.95 (dt, J = 13.8, 12.6Hz, 3H), 1.68 (d, J = 14.1Hz, 1H), 1.61-1.54 (m, 6H), 1.48-1.40 (m, 3H), 1.38 (d, J = 5.2Hz, 4H), 1.29-1.18 (m, 3H), 1.11-1.03 (m, 4H), 0.80 (d, J = 7.0Hz, 3H), 0.67 (dd, J = 10.8, 4.5Hz, 3H); ^{13}C NMR (101MHz, $CDCl_3$) δ 216.00, 169.81, 167.88, 167.13, 163.04, 155.29, 138.26, 116.03, 86.85, 73.60, 68.53, 62.03, 57.16, 53.96, 44.45, 43.55, 42.95, 40.87, 35.77, 34.99, 33.46, 32.92, 29.42, 25.90, 25.37, 24.94, 23.81, 22.56, 15.82, 13.90, 10.42; HRMS (ES) calcd [M + H]$^+$for [$C_{33}H_{49}N_5O_5S$] 628.3527, found 628.3524.

4.1.7 14-O-[(2-(Morpholine-4-yl)-acetamido-6-aminoprimidine-2-yl) thioacetyl] mutilin (5e)

Compound 5e was prepared according to the general procedure with a reaction time of 2h. The crude product was purified by column chromatography on silica gel (petroleum ether : ethyl acetate = 1 : 3) to give 51% yield (0.96g) of 5e as a white solid. IR (KBr): 3 422, 2 919, 1 719, 1 637, 1 547, 1 500, 1 466, 1 287, 1 118, 1 047cm^{-1}. ^{1}H NMR (400MHz, $CDCl_3$) δ 9.26 (s, 1H), 6.99 (s, 1H), 6.44 (dd, J = 17.4, 11.0Hz, 1H), 5.67 (d, J = 8.4Hz, 1H), 5.18 (dd, J = 58.3, 14.2Hz, 2H), 4.92 (s, 2H), 3.71 (dd, J = 26.5, 17.1Hz, 2H), 3.29 (s, 1H), 3.03 (s, 1H), 2.81-2.65 (m, 2H), 2.31 (dd, J = 20.0, 10.0Hz, 2H), 2.27-2.21 (m, 1H), 2.18-2.00 (m, 3H), 2.00-1.92 (m, 2H), 1.87 (dd, J = 26.1, 14.8Hz, 3H), 1.69 (d, J = 11.9Hz, 1H), 1.67-1.61 (m, 2H), 1.60-1.40 (m, 4H), 1.40-1.30 (m, 4H), 1.20 (dd, J = 18.7, 11.5Hz, 3H), 1.04 (d, J = 17.7Hz, 4H), 0.80 (d, J = 6.8Hz, 3H), 0.67 (d, J = 6.8Hz, 3H); ^{13}C NMR (101MHz, $CDCl_3$) 5 216.00, 168.67, 167.99, 167.07, 163.05, 155.11, 138.28, 116.04, 86.88, 73.59, 68.57, 65.77, 61.70, 57.14, 52.76, 44.45, 43.54, 42.95, 40.87, 35.76, 34.99, 33.46, 32.91, 29.41, 25.90, 25.39, 23.81, 15.84, 13.91, 10.42; HRMS (ES) calcd [M + H]$^+$ for $C_{32}H_{47}N_5O_6S$ 630.3319, found 630.3315.

4.1.8 14-O-[(2-((3S)-3-Hydroxypyrrolidine-1-yl)-acetamido-6-aminopyrimidine-2-yl) thioacetyl] mutilin (5f)

Compound 5f was prepared according to the general procedure with a reaction time of 2h. The crude product was purified by column chromatography on silica gel (petroleum ether : ethyl acetate = 1 : 2) to give 65% yield (1.23g) of 5f as a white solid. IR (KBr): 3 448, 2 930,

2 930, 1 724, 1 629, 1 593, 1 549, 1 500, 1 466, 1 287, 1 150, 1 118 cm^{-1}; ^1H NMR (400MHz, $CDCl_3$) δ 9.34 (s, 1H), 7.08 (d, J = 15.2Hz, 1H), 6.49 (dt, J = 20.0, 10.0Hz, 1H), 5.74 (d, J = 8.5Hz, 1H), 5.24 (ddd, J = 32.5, 14.2, 4.1Hz, 2H), 4.99 (d, J=29.7Hz, 2H), 3.82 (dt, J=30.8, 10.8Hz, 2H), 3.72 (s, 1H), 3.35 (t, J=7.3Hz, 1h), 3.07 (s, 1H), 2.84 (d, J=11.4Hz, 2H), 2.33 (dd, J=13.7, 6.9Hz, 1h), 2.21 (dd, J = 13.9, 10.4Hz, 3H), 2.10 (s, 1H), 2.07–1.97 (m, 2H), 1.73 (dd, J= 22.9, 11.8Hz, 3H), 1.63 (s, 1H), 1.58–1.49 (m, 4H), 1.46–1.40 (m, 4H), 1.36–1.24 (m, 3H), 1.18–1.07 (m, 4H), 0.87 (d, J=7.0Hz, 3H), 0.75 (d, J=6.9Hz, 3H); ^{13}CNMR (101MHz, $CDCl_3$) δ 216.20, 169.56, 167.92, 167.25, 163.09, 155.14, 138.22, 116.06, 86.94, 73.60, 68.63, 63.96, 61.69, 57.16, 53.79, 44.45, 43.52, 42.95, 40.88, 35.76, 35.73, 34.96, 33.48, 32.91, 32.87, 29.42, 25.88, 25.39, 25.07, 23.80, 23.43, 15.87, 15.81, 13.92, 10.42; HRMS (ES) calcd $[M + H]^+$ for $C_{32}H_{49}N_5O_6S$ 630.3320, found 630.3329.

4.1.9 14–O– [(2– (4–Hydroxypiperidine–1–yl) –acetamido–6–aminopyrimidine–2–yl) thioacetyl] mutilin (5g)

Compound 5g was prepared according to the general procedure with a reaction time of 2h. The crude product was purified by column chromatography on silica gel (petroleum ether : ethyl acetate=1 : 2) to give 60% yield (1.16g) of 5g as a white solid. IR (KBr): 3 446, 3 367, 2 928, 2 860, 1 719, 1 685, 1 295, 1 117, 1 012cm^{-1}; ^1H NMR (400MHz, CDCb) δ 9.22 (s, 1H), 7.05 (s, 1H), 6.51 (dd, J=17.4, 11.0Hz, 1H), 5.74 (d, J=8.4Hz, 1H), 5.25 (dd, J = 57.9, 14.2Hz, 2H), 4.95 (s, 2H), 3.88–3.69 (m, 6H), 3.41–3.29 (m, 1H), 3.11 (s, 2h), 2.71–2.47 (m, 4H), 2.42–2.27 (m, 1H), 2.27–2.12 (m, 2H), 2.11–1.97 (m, 2H), 1.84–1.68 (m, 2H), 1.67–1.47 (m, 4H), 1.47–1.39 (m, 4H), 1.39–1.19 (m, 3H), 1.18–1.06 (m, 4H), 0.86 (d, J=7.0Hz, 3H), 0.74 (d, J = 6.9Hz, 3H); ^{13}CNMR (101MHz, $CDCl_3$) δ 216.09, 169.39, 167.93, 167.15, 163.06, 155.19, 138.27, 116.05, 86.88, 73.60, 68.57, 61.19, 59.39, 57.16, 50.38, 44.45, 43.53, 42.95, 40.88, 35.77, 34.98, 33.21, 29.42, 25.39, 23.81, 15.83, 13.93, 13.19, 10.43; HRMS (ES) calcd $[M + H]^+$ for $C_{33}H_{49}N_5O_6S$ 644.3476, found 644.3473.

4.1.10 14–O– [(2– ((3R) –3–Hydroxymethylpiperidine–1–yl) –acetamido–6–aminopyrimidine–2–yl) thioacetyl] mutilin (5h)

Compound 5h was prepared according to the general procedure with a reaction time of 2h. The crude product was purified by column chromatography on silica gel (petroleum ether : ethyl acetate=1 : 2) to give 67% yield (1.32g) of 5h as a white solid. IR (KBr): 3 448, 2 930, 1 720, 1 625, 1 593, 1 550, 1 500, 1 466, 1 400, 1 286, 1 201, 1 153, 1 117 cm^{-1}; ^1H NMR (400MHz, $CDCl_3$) δ 9.33 (d, J = 5.2Hz, 1H), 6.98 (t, J = 8.4Hz, 1H), 6.42 (dd, J=17.3, 11.0Hz, 1H), 5.66 (dd, J = 8.3, 2.3Hz, 1H), 5.33–5.03 (m, 2H), 4.96 (d, J = 18.1Hz, 2H), 3.84–3.63 (m, 2H), 3.58–3.48 (m, 2H), 3.33–3.23 (m, 1H), 3.01 (qd, J = 16.8, 4.6Hz, 2H), 2.67 (dd, J = 30.2, 10.1Hz, 2H), 2.23 (dd, J = 17.0, 9.9Hz, 3H), 2.19–2.06 (m, 3H), 2.05–1.99 (m, 1H), 1.99–1.82

(m, 3H), 1.68 (d, J = 13.7Hz, 2H), 1.61 (d, J = 7.4Hz, 1h), 1.49 (ddd, J = 49.0, 15.8, 7.2Hz, 4H), 1.40-1.34 (m, 4H), 1.31-1.18 (m, 3H), 1.04 (d, J = 17.5Hz, 4H), 0.80 (d, J = 7.0Hz, 3H), 0.71-0.60 (m, 3H); ^{13}C NMR (101MHz, $CDCl_3$) δ 216.10, 169.69, 167.87, 167.14, 163.04, 155.27, 138.24, 116.06, 86.88, 73.60, 68.53, 61.67, 59.39, 57.16, 53.36, 44.45, 43.54, 42.94, 40.88, 38.15, 35.76, 34.98, 33.47, 32.93, 31.29, 30.71, 29.42, 25.89, 25.38, 23.81, 15.84, 13.94, 13.19, 10.43; HRMS (ES) calcd $[M + H]^+$ for $C_{34}H_{51}N_5O_6S$ 658.3628, found 658.3626.

4.1.11 14-O- [(2- (4-Hydroxyethy/piperidine-1-yl) -acetamido-6-aminopyrimidine-2-yl) thioacetyl] mutilin (5i)

Compound 5i was prepared according to the general procedure with a reaction time of 2h. The crude product was purified by column chromatography on silica gel (petroleum ether : ethyl acetate = 1 : 2) to give 74% yield (1.49g) of 5i as a white solid. IR (KBr): 3 448, 2 932, 1 720, 1 625, 1 593, 1 550, 1 466, 1 400, 1 287, 1 201, 1 154, 1 117, 1018cm^{-1}; ^{1}H NMR (400MHz, $CDCl_3$) δ9.32 (s, 1H), 6.99 (s, 1H), 6.43 (dd, J = 17.4, 11.0Hz, 1H), 5.65 (d, J = 8.4Hz, 1H), 5.28-5.07 (m, 2H), 4.95 (s, 2H), 4.39 (dd, J = 6.8, 4.8Hz, 1H), 4.11-3.99 (m, 1H), 3.80-3.61 (m, 2H), 3.28 (s, 1H), 3.24 (d, J = 4.2Hz, 1H), 2.94 (dd, J = 14.7, 8.1Hz, 1H), 2.75 (dt, J = 17.7, 7.5Hz, 2H), 2.50 (dd, J = 14.8, 8.6Hz, 1H), 2.30-2.09 (m, 4H), 2.02 (s, 1H), 1.98 (s, 1H), 1.97-1.86 (m, 1H), 1.79 (dt, J = 13.7, 5.8Hz, 1H), 1.68 (d, J = 13.9Hz, 1H), 1.63-1.42 (m, 4H), 1.42-1.35 (m, 4H), 1.30 (dd, J = 18.2, 15.6Hz, 2H), 1.23-1.12 (m, 4H), 1.09-1.01 (m, 4H), 0.80 (d, J = 7.0Hz, 3H), 0.66 (d, J = 6.9Hz, 3H); ^{13}C NMR (101MHz, $CDCl_3$) δ 216.07, 169.52, 167.83, 167.21, 163.09, 155.19, 138.28, 116.04, 86.91, 73.59, 70.41, 68.63, 62.20, 59.39, 58.55, 57.14, 51.95, 44.45, 42.95, 40.88, 35.76, 34.98, 34.21, 33.47, 29.41, 25.90, 25.39, 23.81, 20.03, 15.79, 13.91, 13.19, 10.42; HRMS (ES) calcd $[M + H]^+$ for $C_{34}H_{53}N_5O_6S$ 672.3789, found 672.3783.

4.2 Biological evaluation

4.2.1 MIC testing

As described previously[22-24], The MIC values were tested using the agar dilution method according to the National Committee for Clinical Laboratory Standards (NCCLS, 4th ed). Briefly, compounds were dissolved in 25% DMSO to a solution with concentration of 1 280μg/mL. Tiamulin fumarate used as a reference drug was directly dissolved in 10mL distilled water. All the solutions were then diluted two-fold with distilled water to provide 11 dilutions (final concentration is 0.625μg/mL). A 2mL volume of the 2-fold serial dilution of each test compound/drug was incorporated into 18mL hot MuellereHinton agar medium, which resulted in the final concentration of each dilutions decreasing ten fold. Inoculums, including four Gram (+), MRSA-29213, MRSA-33591, S.aureus and B.subtilis, and one Gram (-), E.coli were prepared from blood slants and adjusted to approximately 10^5-10^6 CFU/mL with sterile saline (0.90% NaCl). A 10μL amount of bacterial suspension was spotted onto Muellere-Hinton agar plates containing serial dilutions of the compounds/drug. The plates were incubated at 36.5℃ for 24–48h. The same procedure was repeated in triplicate.

4.2.2 Oxford cup assays

Inoculums were prepared in 0.9% saline using McFarland standard and spread uniformly on nutrient agar plates. The 320 and 160μg/mL of all the compounds prepared in the same manner as the MIC test were added individually into the Oxford cups which were placed at equal distances above the agar surfaces. The zone of inhibition for each concentration was measured after a 24–36h cubation at 37℃. The same procedure was repeated in triplicate.

4.2.3 Bactericidal time-kill kinetics

S. aureus and MRSA-29213 were cultured in MuellereHinton broth at 37℃ for 6h with shaking and then diluted to approximately 1.5×10^6 and 1.4×10^7 CFU/mL, respectively. Test compound 3 and tiamulin fumarate having the final concentrations of 1 × MIC and 6 × MIC was inoculated with the aliquots of bacteria resuspended in fresh media. After specified time intervals (0, 1, 2, 4, 6, 12 and 24h), 20μL aliquots were serially diluted 10-fold in 0.9% saline, plated on sterile MuellereHinton agar plates, and incubated at 37℃ for 24h. The viable colonies were counted and represented as $\log_{10}$ (CFU/mL). The same procedure was repeated in triplicate.

4.3 Molecular modeling

Docking studies were performed using Homdock software in the $Chil^2$ package[27]. The starting conformations of compounds were obtained from alignment to tiamulin in the X-ray crystal structure (PDB ID: 1XBP)[28] which was used as a control to evaluate the accuracy of the docking performance. We performed the molecular superposition using the graph-based molecular alignment (GMA) method. Then the optimization of interaction between the receptor and compounds to remove partial overlap using Glamdock, and the binding affinity was evaluated by $Chil^2$ Score. All the compounds were built with Avogadro software[29], including a 5000 steps Steepest Descent and 1000 steps Conjugate Gradients optimization using MMFF94 force field.

4.4 Mouse systemic infection model

The in vivo efficacy of compound3 was determined in 6–8 weeks old Balb/c mice (Laboratory Animal Center of Lanzhou University, Lanzhou, China). All mice were maintained at the central animal facility at the Lanzhou Institute of Husbandry and Pharmaceutical Sciences of CAAS, with a constant 12h light-dark cycle and free access to food and water. After acclimating on site for at least one week, mice were rendered neutropenic upon treatment with 150mg/kg cyclophosphamide intraperitoneally for four days and with 100mg/kg for one day prior to inoculation, respectively. The neutropenic mice (10 per group) were anesthetized with iso-flurane and then received a 0.5mL MRSA-29213 inoculum of 10^9 CFU/mL via intraperitoneal (ip) injection. About 1h after infection, the mice were then intravenously (iv) administered compound 3 dissolved in 0.5mL vehicle (ethyl Oleate : 1, 2-Propanediol : Tween-80 sterile : water = 4 : 10 : 31 : 55) at doses of 2, 3, 4.46, 6.64, and 10mg/kg body weight. Tiamulin fumarate was used as a reference drug in the same manner at the same doses as 3. Mice were euthanized by cervical dislocation under anesthesia with 5% isoflurane and were placed in 70% ethanol for 10min at pre-determined experimental endpoints, including consistent or rapid body weight loss exceeding 20% of the body weight, severe mucous excretion from eyes, severe respiratory distress, ulceration/ infection of the injection site or obvious distress for any reason.The healthy survival of mice at 72h after infection was used as the endpoint. All animals were observed twice daily for symptoms and mortality for the animal procedures.The survival of the mice at 7d after infection was used as the end-point, and the ED50 was calculated by the Bliss method[30].

ETHICS STATEMENT

All animal experiments were performed in accordance with the Ethical Principles in Animal Research and were approved by the Committee for Ethics in Laboratory Animal Center of Lanzhou University (No: SCXK2013-0002).

ACKNOWLEDGMENTS

The authors are grateful for financial support from National Key Technology Support Program (No. 2015BAD11B02) and Agricultural Science and Technology Innovation Program (ASTIP, No. CAAS-ASTIP-2014-LIHPS-04).

APPENDIX A.SUPPLEMENTARY DATA

Supplementary data related to this article can be found at http: // dx.doi.org/10. 1016/j.ejmech. 2016. 11. 054.

REFERENCES OMITTED

（发表于《European Journal of Medicinal Chemistry》，院选 SCI，IF：3. 902）

UPLC-Q-TOF/MS-based Urine and Plasmametabonomics Study on the Cross Mark Ameliorative Effects of Aspirin Eugenol Ester in Hyperlipidemia Rats

Ning MA, Isam KARAM, Xi-Wang LIU, Xiao-Jun KONG, Zhe QIN, Shi-Hong LI, Zeng-Hua JIAO, Peng-Cheng DONG, Ya-Jun YANG*, Jian-Yong LI*

(Key Lab of New Animal Drug Project of Gansu Province, Key Lab of Veterinary Pharmaceutical Development, Ministry of Agriculture, Lanzhou Institute of Husbandry and Pharmaceutical Science of Chinese Academy of Agricultural Sciences, Lanzhou 730050, China**

Abstract: The main objective of this study was to investigate the ameliorative effects of aspirin eugenol ester (AEE) inhy-perlipidemic rat. After five-week oral administration of AEE in high fat diet (HFD) -induced hyperlipidemic rats, the impact of AEE on plasma and urine metabonomics was investigated to explore the underlying mechanism by UPLC-Q-TOF/MS analysis. Blood lipid levels and histopathological changes of liver, stomach and duodenum were also evaluated afterAEE treatment. Without obvious gastrointestinal (GI) side effects, AEE significantly relieved fatty degeneration ofliver and reduced triglyceride (TG), low density lipoprotein (LDL) and total cholesterol (TCH) ($P<0.01$). Clear separations of metabolic profiles were observed among control, model and AEE groups by using principal component analysis (PCA) and orthogonal partial least-squares-discriminate analysis (OPLS-DA). 16 endogenous metabolites in plasma and 18 endogenous metabolites in urine involved in glycerophospholipid metabolism, fatty acid metabolism, fatty acid beta-oxidation, amino acid me-

---

* Corresponding authors, E-mail: yangyue10224@163.com (Y.-J. Yang), lijy1971@163.com, lijianyong@caas.cn (J.-Y. Li).http://dx.doi.org/10.1016/j.taap.2017.07.013 0041-008X/

** Abbreviations: AEE, aspirin eugenol ester; HFD, high fat diet; TG, triglyceride; LDL, low density lipoprotein; TCH, total cholesterol; PCA, principal component analysis; OPLS-DA, orthogonal partial least-squares-discriminate analysis; NSAID, non-steroidal antiinflammatory drug; GI, gastrointestinal; CMC-Na, carboxymethylcellulose sodium; TIC, total ion chromatograms; LV, latentvariables; CV-ANOVA, ANOVA of the cross-validated residuals; LysoPC, lysophosphatidylcholinesi; BCAA, branched chain amino acid; VIP, variance importance for projection; RT, retention time; N1MPC, N1-methyl-2-pyridone-5-carboxamide; 3ICAG, 3-indole carboxylic acid glucuronide; NAD, nicotinamide adenine dinucleotide; AA, arachidonic acid; S1P, sphingosine 1-phosphate.

tabolism, TCA cycle, sphingolipid metabolism, gut microflora and pyrimidine metabolism were considered as potential biomarkers of hyperlipidemia and be regulated by AEE administration.It might be concluded that AEE was a promising drug candidate for hyperlipidemia treatment.These findings could contribute to the understanding of action mechanisms of AEE and provide evidence for further studies.

Key words: Aspirin eugenol ester; Blood lipids; Hyperlipidemia; Metabonomics; Rat

1 INTRODUCTION

Cardiovascular disease is one of the important threats to the health of the population and cause 18 million deaths per year globally.Hyperlipidemia, one of the major risk factors for the development of cardiovascular diseases, such as coronary heart disease, myocardial infarction, cerebral stroke, atherosclerosis and hypertension, is becoming a major health problem in the world (Bowry et al., 2015).Hyperlipidemia is variably defined as a metabolic disorder disease which involves increased serum levels of triglycerides (TG), total cholesterol (TCH), low-density lipoprotein (LDL) as well as decreased levels of high-density lipoprotein (HDL) (Wasan et al., 1997).

Aspirin (acetylsalicylic acid) as a classic non-steroidal anti-inflammatory drug (NSAID) is widely used to treat inflammation and cardiovascular diseases (Nansseu and Noubiap, 2015).In clinical practice, gastrointestinal (GI) side effects are the main constraint in its application, which are mainly caused by the free acidic group and direct contact of drug with gastric mucosa (Kean and Buchanan, 2005).Chemical masking of carboxylic group in aspirin is a simple and efficient method of improv-ing therapeutic efficacy and retarding gastrointestinal side effects (Redasani and Bari, 2012).Eugenol is a natural product and safe essential oil, which is extracted from dry alabastrum of Eugenia caryophyllata Thumb.It has been known for its various therapeutic effects including anti-inflammation, antibacterial, antioxidant, anti-diarrhea, and antiulcer (Capasso et al., 2000; Naidu, 1995; Yogalakshmi et al., 2010).Because of the free hydroxyl group, eugenol is structural instable and vulnerable to oxidation.

Based on the pro-drug principle, aspirin eugenol ester (AEE) was synthesized with the starting precursors of aspirin and eugenol (Li et al., 2012a).There are three advantages in AEE. Firstly, the disappearance of free hydroxyl group increases structural stability of eugenol.Secondly, the disappearance of carboxyl group decreases the acidity and reduces GI side effects caused by aspirin.Finally, eugenol possesses antiulcer activity based on its ability to enhance mucus production which is an important gastroprotective factor (Santin et al., 2011), and thus aspirin and eugenol play complementary roles to reduce GI damages.Therefore, AEE reduce the side effects of its precursors and improve therapeutic effect and stabilization through the disappearance of carboxyl and hydroxyl group.The acute toxicity, subchronic toxicity, teratogenicity, metabolism and pharmacodynamics of AEE had been evaluated in our previous studies, and these studies have indicated that AEE is a promising compound with good druggability (Karam et al., 2015; Li et al., 2012b, 2013b; Ma et al., 2015, 2016; Shen et al., 2015).

With the comprehensive analysis of small molecules, metabonomics provides a powerful approach to discover biomarkers in biological systems (Kell, 2006; Zhao et al., 2014). Metabonomics has been increasingly applied as a versatile tool for evaluating the toxicity and therapeutic effect of numerous compounds (Li et al., 2013a; Liu et al., 2014a). Regulation effects of AEE on blood lipids in hyperlipidemic rats has been confirmed in our previous study, in which 54mg/kg AEE significantly reduced TG, TCH, LDL, and elevated HDL (Karam et al., 2015). However, little information is known concerning the alteration of plasma and urine metabonomics associated with AEE therapeutic effects. With the application ofUPLC-Q-TOF/MS analysis, the objective of this study was to find out more evidences to understand and illustrate the possible underlying mechanism of AEE against hyperlipidemia. Moreover, the blood lipidlowering effects and GI toxicity of AEE were also assessed in this study.

2 MATERIALS AND METHODS

2.1 Reagents and materials

AEE (transparent crystal, purity: 99.5% with RP-HPLC) was prepared in Key Lab of New Animal Drug Project of Gansu Province, Key Lab of Vet-erinary Pharmaceutical Development of Agricultural Ministry, Lanzhou Institute of Husbandry and Pharmaceutical Sciences of CAAS. Carboxy-methylcellulose sodium (CMC-Na) was supplied by Tianjin Chemical Reagent Company (Tianjin, China). MS-grade formic acid was supplied by TCI (Shanghai, China). Deionized water (18MΩ) was prepared with a Di-rect-Q© 3 system (Millipore, USA). MS-grade acetonitrile was purchased from Thermo Fisher Scientific (USA). The TG, TCH, LDL and HDL kits were provided by Ningbo Medical System Biotechnology Co., Ltd. (Ningbo, China). Standard compressed rat feed and high diet feed (HFD) were supplied by Keao Xieli Feed Co., Ltd. (Beijing, China). Standard rat diet consisted of 12.3% lipids, 63.3% carbohydrates, and 24.4% proteins (kcal) and HFD (77.8% standard diet, 10% yolk power, 10% lard, 2% cholesterol and 0.2% bile salts) consisted of 41.5% lipids, 40.2% carbohydrates, and 18.3% proteins (kcal). Erba XL-640 analyzer (German) was used to measure blood lipid levels.

2.2 Animals

Male Sprague-Dawley rats, aged 6 weeks and weighing 165-180g, were purchased from Gansu University of Chinese Medicine (Lanzhou, China). Rats were housed in plastic cages (size: 50cm×35cm×20cm, 10 rats per cage) with stainless steel wire cover and chopped bedding. Rat feed and drinking water were supplied ad libitum. Light/dark regimen was 12/12h and living temperature was 22±2℃ with relative humidity of 55%±10%. Animals were allowed a 2-week quarantine and acclimation period prior to start of the study. Animal welfare and experimental procedures were performed strictly in accordance with the Guidelines for the Care and Use of Laboratory Animals. All of the experimental protocols and procedures were approved by the Institutional Animal Care and Use Committee of Lanzhou Institute of Husbandry and Pharmaceutical Science of Chinese Academy of Agricultural Sciences.

2.3 Drugpreparation

AEE was ground and its suspension was prepared in 0.5% CMC-Na.

2.4 Study design

Fig. 1 showed the study design of this experiment. Rats were randomly separated into two groups. Group I as control group received standard diet (n = 10). Group Ⅱ as model group received high fat diet (HFD) (n = 20). The blood lipid levels were examined after HFD administered for 8 weeks, and the results indicated that the hyperlipidemia disease was established successfully (Table A. 1). After that, Group Ⅱ was divided into two groups including model and AEE groups. Based on individual weekly body weight, AEE was intragastrically administrated at the dosage of 54mg/kg. The rats in control and model groups were received equal volume of 0.5% CMC-Na as AEE group. The administration time of AEE was five weeks and HFD was continuously fed during the experiment period.

2.5 Sample collection

Rats were fasted for 10-12h before blood sampling. At the end of 8th week, rats were euthanatized with 10% chloral hydrate and then the blood samples (1-1.5mL) were withdrawn from the tip of the tail into vacuum tubes to obtain serum samples (4 000×g, 4℃ for 10min). The serum samples were used for blood lipids analysis to evaluate the success of the hyperlipidemia model. At the end of experiment (13th week), all rats were euthanatized and sacrificed. Blood samples were taken from heart to prepare serum and plasma (Na-heparin vacuum tubes). Serum and plasma samples were frozen at -80℃ before biochemical and metabonomic analysis (Barri and Dragsted, 2013). The tissues of liver, stomach and duodenum were removed and fixed in 4% formalin for observations of pathological changes. Individual rats were placed in metabolic cages (1 per cage) to obtain 24-hour urine collections and be stored at-80℃ before analysis.

2.6 Blood lipids and pathological examination

Serum samples were used to measure the levels of HDL, TCH, TG and LDL by using XL-640 automatic analyzer (Erba, Germany). Under the standard protocol, tissues of liver, stomach and duodenum were formalin-fixed and paraffin embedded, sectioned, and stained with hematoxylin and eosin (HE). HE-stained sections were examined using a 13395H2X microscope (Leica, Germany).

2.7 Sample preparation for metabonomics analysis

Plasma samples were thawed at room temperature prior to analysis. 600μL cold acetonitrile (-20℃) was added into 200 of plasma in 2mL centrifuge tube. After vortex-mixing for 30s, the mixture was centrifuged at 12, 000 rpm for 10min to precipitate the proteins. The supernatant was filtered through a 0.22μm nylon filter. An aliquot of 4μL was injected for analysis.

Prior to analysis, urine samples were thawed at room temperature, and then 400μL methanol was added into 400μL of urine, vortex mixed and centrifuged at 12, 000 rpm for 15min to remove solid materials. The supernatant was filtered through a 0.22μm nylon filter and an aliquot of 3μL was injected for analysis.

2.8 Data acquisition and method validation

Chromatographic separation of plasma was performed on an Agilent ZORBAX SB-C18 threaded

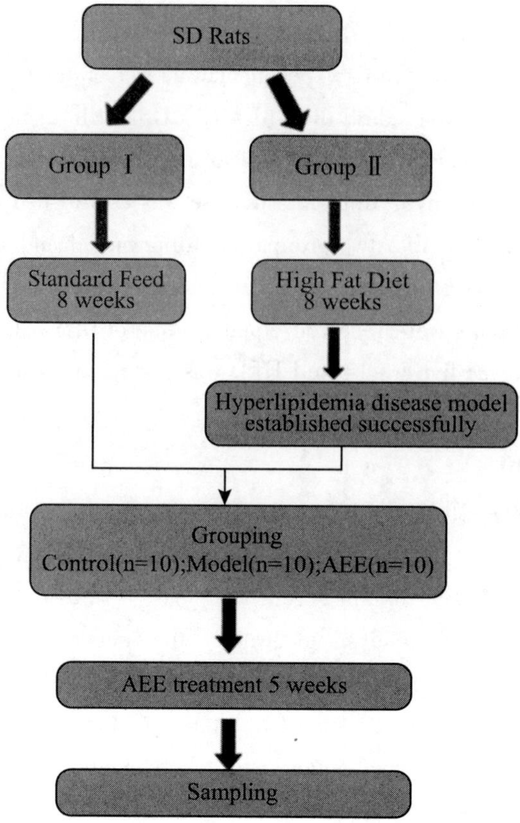

Fig. 1. Study design of the experiment.

column (3. 0mm×100mm, 1. 8μm) using UPLC system consisted of a degasser, thermostat, two binary pumps and autosampler (1290, Agilent Technologies, USA).The column was maintained at 35℃ and eluted at a flowing rate of 0. 3mL/min, using a mobile phase of solvent A-water with 0. 1% formic acid (by volume) and solvent B-acetonitrile with 0. 1% formic acid (by volume). The gradient program was optimized as followed: 0min: 95% A; 0-4min 95% to 60% A; 4-8min 60% to 40% A; 8-18min 40% to 5% A; 18-23min 5% A.The post time was set to 5min for equilibration with 95% A and 5% B.The eluent from the column was directed to the mass spectrometer without split.

Separation of urine was also carried out on Agilent 1290 LC system equipped with ZORBAX SB-C18 threaded column (2. 1 × 100mm, 1. 8μm).Mobile phase A was water with 0. 1% formic acid, while mobile phase B was acetonitrile modified by the addition of0. 1% formic acid.The gradient program was optimized as followed: 0min: 95% A; 02min 95%; 2-19min 95% to 60% A; 19-20min 60% to 5% A; 2021min 5% A; 21-22min 5% to 95% A; 22-23min 95% to 95% A. The flow rate was 0. 35mL/min and the post time was 5min.The analytical column was also maintained at temperature of 35℃.

Agilent 6530 Q-TOF (Agilent Technologies, USA) was used to carry out the mass spectrometry of plasma and urine samples with a dual electrospray ionization source (ESI) operating in positive and negative ion modes.The fragment voltage was set at 135 V in both modes.The capillary voltages were set at 4. 0 kV in positive mode and 3. 5 kV in negative mode, respectively.Used drying gas nitrogen, the desolvation gas rate was set to 10 L/min at 350℃. The scan time was set at 1 spectra/s. Data was collected in centroid mode from 50 to 1 000m/z. The nebulizer pressure was set at 45 psig.

2.9 Multivariate analysis and identification of potential biomarkers

Data processing method previously reported with minor modifications was used in this study (Lu et al., 2016). The raw MS spectra were firstly processed by Mass Hunter Qualitative Analysis software (Agilent technologies, USA) to converted to common data format (.mzData). The program XCMS was used for nonlinear alignment of the data in the time domain and automatic integration and extraction of the peak intensities. The data were filtered by interquantile range and normalized for further multivariate data analysis. The obtained data sets were separately imported into SIMCA-P V13.0 (Umetrics AB, Sweden) to perform unsupervised principal component analysis (PCA) and supervised orthogonal partial least squares discriminant analysis (OPLS-DA). The quality of OPLS-DA models was described by R^2X, R^2Y, and Q^2. R^2X and R^2Y represented the fraction of the sum of squares for the selected component. Q^2 represents the predictive ability of the model. To validate the model, ANOVA of the cross-validated residuals (CV-ANOVA) method were implemented. The results were visualized in the form of score plot to show the group clusters and loading plot to show variables contributing to the classification. The candidate metabolites, with variance importance for projection (VIP) value above 1.0 and P value of ANOVA below 0.05, were considered to be potential biomarkers. Pathway analysis was performed with MetaboAnalyst, which is a web-based tool for visualization of metabonomics.

Identification of the marker metabolites was achieved through a mass-based search followed by manual verification. First, TOF-MS accurate mass value of the molecular ion of interest was searched against the Human Metabolome Database (HMDB) and METLIN Metabolites Database. Then, the putative identifications were verified by comparing the MS/MS fragmentations. Biochemical reactions involved were found through KEGG (Kyoto encyclopedia of genes and genomes) and HMDB. HemI (Heat map Illustrator) was used to generate heatmap and metabolic pathway interpretation was performed with MetaboAnalyst 3.0.

2.10 Statistical analysis

The results of the blood lipids were expressed as mean±standard deviation (SD). The significance of differences between Group I and Group II had been analyzed by Student's t-test. The differences among three experimental groups had been evaluated by one-way ANOVA with Fisher's least significant difference (LSD) test using the Statistical Package for Social Science program (SPSS 16.0, Chicago, IL, USA). The significance threshold was set at $P<0.05$ for the test.

3 RESULTS

3.1 Effect of AEE on blood lipids

The results of blood lipids were shown in Table 1. TCH, TG and LDL levels were significantly higher in the model group than that in the control group ($P<0.01$), whereas the HDL was significantly reduced ($P<0.01$). AEE showed strong effects on reducing TG, TCH and LDL than those in the model group ($P<0.01$). With respect to HDL index, no statistical difference was observed between model and AEE treated groups.

Table 1 Effects of AEE on blood lipid levels in hyperlipidemic rats (n=10).

Variables	Control	Model	AEE
TG	0.44±0.15**	0.67±0.13	0.49±0.16**
HDL	0.52±0.05**	0.35±0.11	0.32±0.09##
LDL	0.25±0.04**	0.46±0.07	0.31±0.09**
TCH	1.24±0.13**	1.63±0.16	1.22±0.2**

AEE: aspirin eugenol ester; TG: triglyceride; HDL: high density lipoprotein; LDL: low density lipoprotein; TCH: total cholesterol. The unit of TG, TCH, HDL and LDL was mmol/L.

** $P<0.01$ significant difference from model group.

$P<0.01$ significant difference from control group.

In comparison with the control group, HDL levels of AEE groups were significantly lower than control group ($P<0.01$). TG, LDL and TCH levels in AEE group did not differ significantly from those in the control group, indicating the improvement of blood lipid levels in hyperlipidemic rats.

3.2 Histological effect of AEE on liver, stomach and duodenum

Pathological results of liver, stomach and duodenum were shown in Fig. 2. Liver histological examinations of control showed normal cell architecture, while significant changes were observed in model group (Fig. 2A). Serious fatty degenerations of liver cells in model group were found and pointed with black arrow (arrow 1). In comparison with the model group, the fatty degenerations in AEE treatment group were significantly decreased and the representative results were shown in Fig. 2A (arrow 2). No histopathological changes in stomach and duodenum were observed from the rats in control and model groups. However, in AEE group, few evidences of hyperemia and edema of lamina propria of stomach and duodenum and the ecclasis of gastric mucosa could be found in rats (Fig. 2B and 2C, arrows 3-5).

3.3 Plasma metabolomics analysis

Representative total ion chromatograms (TICs) of the plasma samples analyzed by Q-TOF in positive and negative modes were shown in Fig. 1. TICs showed the good separation effect and strong sensitivity of the established method. In order to ensure the success of the hyperlipidemia model and understand the global metabolic in plasma, PCA approach was used to distinguish the differences between control and model groups. The PCA score plots showed the samples in model group exhibited a tendency to be away from those in control (Fig. 3a and 3b). The R^2X of positive and negative PCA models were 0.806 and 0.667, indicating the data can be highly elucidated by the two models. Typically, a clear separation of samples from model and control was observed, which indicated remarkable changes of endogenous metabolites in plasma induced by HFD. Thus, the hyperlipemia model induced by HFD was successful, and can be used to explore the effects of AEE.

OPLS-DA model was conducted to determine whether AEE influenced plasma metabolic pattern and the resulting projection model could provide thelatentvariables (LVs) that focused on maximum separation. Score plots of the OPLS-DA model were shown in Fig. 3c and d (ESI +: $R^2X=0.806$, $R^2Y=0.868$, $Q^2=0.843$; ESI-: $R^2X=0.603$, $R^2Y=0.958$, $Q^2=0.902$). The P-values from

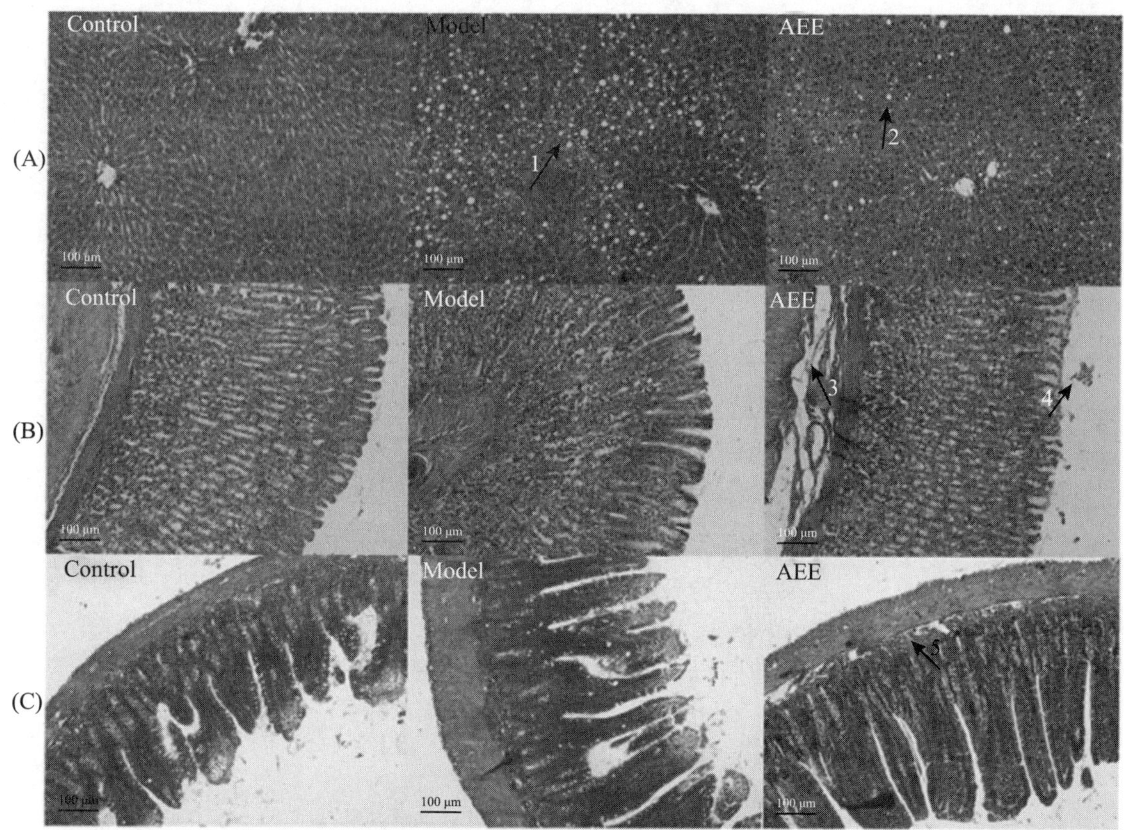

Fig. 2. Histopathological results of liver, stomach and duodenum after a five-week AEE treatment (HE × 100).

Note: A: liver tissue; B: stomach; C: duodenum. In comparison with the model group, fatty degenerations and fat droplets of liver cells in AEE group were significantly decreased (arrows 1-2). Minor pathological changes including hyperemia, edema and ecclasis were found (arrows 3-5).

CV-ANOVA were <0.001, which showed high quality of the OPLS-DA models. In both of positive and negative modes, clear separations were observed among control, model and AEE groups. In positive mode, as shown in Fig. 3c, samples in model group were located at the opposite direction along LV2 compared with those in control group. Compared with the model and control, samples in AEE group were all located between them along LV2. However, samples in AEE groups were located at the opposite direction along LV1 compared with control and model groups. In negative mode, as shown in Fig. 3d, AEE group was distinct from the model group. Samples in AEE group showed a tendency of approaching to the control. Based on the results of score plots, significant difference between model and AEE groups was observed, implying that AEE could restore the metabolic profiling of plasma in hyperlipidemic rat.

Loading plot as a tool was used to reveal the correlation between groups and metabolites. As shown in Fig. 3e and 3f, ions in loading plot clustering close to each group were considered to make great contributions to the classification. Among the ions, with VIP values above 1 and P-values below 0.05, 16 metabolites in plasma including lysophosphatidylcholines (LysoPCs), oleamide, proline, tryptophan, valine, leucine, lysine, carnitine, isoleucine, tyrosine, methionine,

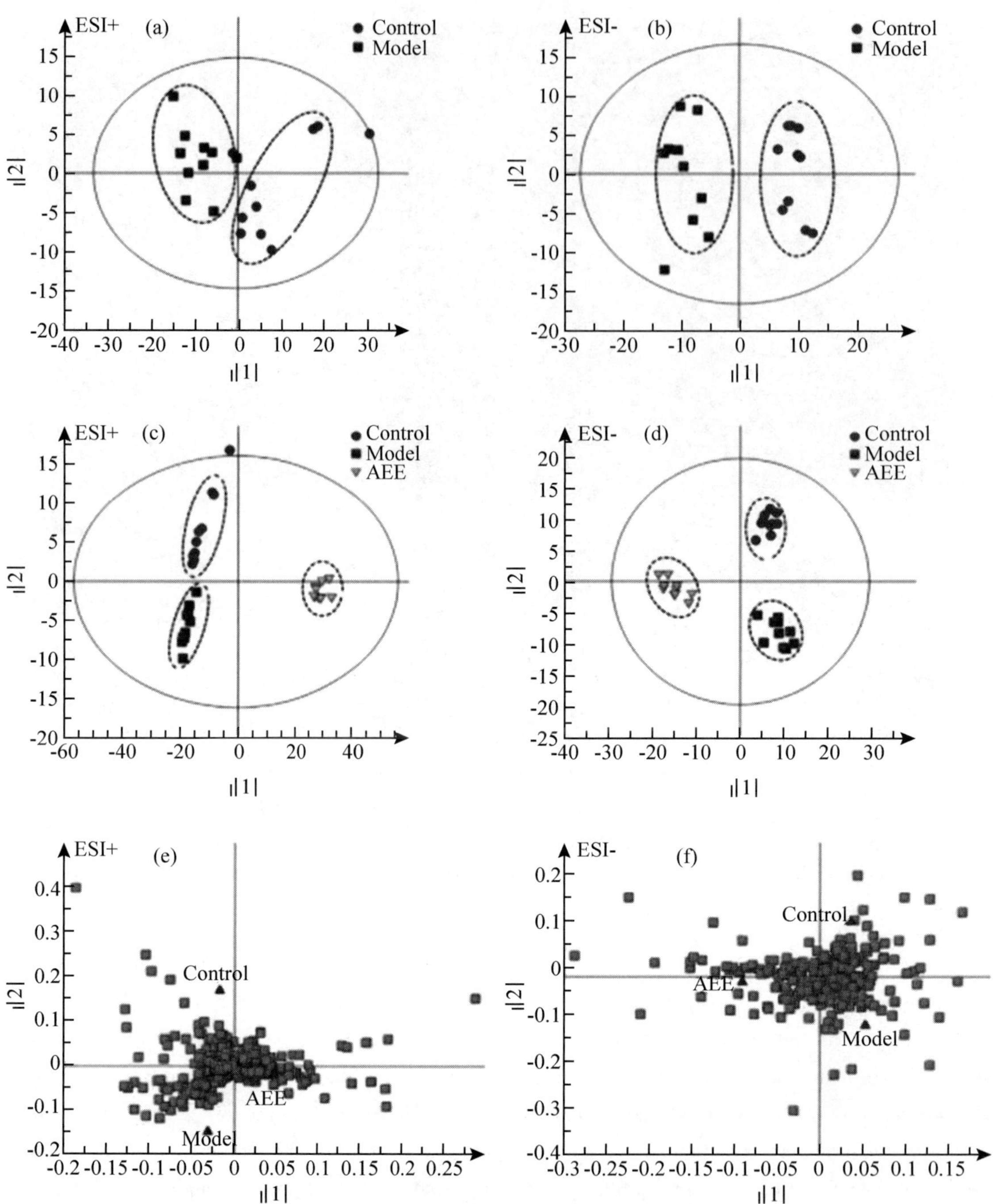

Fig. 3. Score plots and loading plots of plasma in each group based on UPLC-TOF/MS in the positive and negative modes (n=10).

Note: a and b, principal component analysis score plots between control and model groups; c and d, orthogonal partial least-squares-discriminate analysis score plots among control, model and AEE groups; e and f, corresponding loading plots of orthogonal partial least-squares-discriminate analysis models.

arachidonic acid, and sphingosine 1-phosphate (S1P) were selected and identified as potential biomarkers, and fold changes of relative intensity and biological pathways were provided in Table 2.

Table 2 Potential biomarkers in plasma associated with hyperlipidemia based on the UPLC-Q-TOF/MS analysis.

RT	VIP	Formula	Metabolite	Adduction	m/z	Fold change Model/control	AEE/model	KEGG pathway
15. 22	6. 42	$C_{26}H_{54}NO_7P$	LysoPC (18: 0)	[M + H] +	524. 4157	0. 55**	4. 28**	Glycerophospholipid metabolism
13. 13	3. 9	$C_{26}H_{52}NO_7P$	LysoPC (18: 1)	[M + H] +	522. 4005	0. 94	4. 29**	Glycerophospholipid metabolism
20. 40	2. 02	$C_{18}H_{35}NO$	Oleamide	[M + H] +	282. 3128	1. 15**	1. 43**	-
1. 55	1. 99	$C_5H_9NO_2$	Proline	[M + H] +	116. 0931	1. 50**	0. 58**	Arginine and proline metabolism
3. 19	1. 98	$C_{11}H_{12}N_2O_2$	Tryptophan	[M + H] +	205. 1258	1. 65**	0. 51**	Tryptophan metabolism
13. 78	1. 81	$C_{25}H_{52}O_7P$	LysoPC (17: 0)	[M + H] +	510. 4001	0. 25**	5. 27**	Glycerophospholipid metabolism
1. 72	1. 80	$C_5H_{11}NO_2$	Valine	[M + H] +	118. 1090	1. 32**	0. 46**	Amino acid metabolism
1. 88	1. 52	$C_6H_{13}NO_2$	Leucine	[M + H] +	132. 1253	1. 16*	0. 59**	Amino acid metabolism
12. 47	1. 41	$C_{28}H_{52}NO_7P$	LysoPC (20: 3)	[M + H] +	546. 4007	3. 01**	3. 65**	Glycerophospholipid metabolism
1. 35	1. 31	$C_6H_{14}N_2O_2$	Lysine	[M + H] +	147. 1372	1. 49**	0. 24**	Amino acid metabolism
1. 52	1. 27	$C_7H_{15}NO_3$	Carnitine	[M + H] +	162. 1380	0. 82	0. 72	Bile secretion
1. 54	1. 17	$C_6H_{13}NO_2$	Isoleucine	[M + H] +	132. 1004	1. 31	0. 31**	Amino acid metabolism
1. 88	1. 07	$C_9H_{11}NO_3$	Tyrosine	[M + H] +	182. 1081	1. 22**	0. 52**	Amino acid metabolism
1. 86	1. 03	$C_5H_{11}NO_2S$	Methionine	[M + H] +	150. 0830	1. 29**	0. 40**	Amino acid metabolism
18. 60	1. 04	$C_{20}H_{32}O_2$	Arachidonic acid	$[M-H]^-$	303. 2334	0. 77*	2. 08**	Arachidonic acid metabolism
10. 03	1. 03	$C_{18}H_{38}NO_5P$	Sphingosine 1-phosphate	$[M-H]^-$	378. 2424	4. 36**	0. 25**	Sphingolipid metabolism

AEE: aspirineugenolester; KEGG: Kyoto encyclopedia of genes and genomes; VIP: variance impoitance for projection; RT: retention time; -: not available in KEGG database. Statistical analysis was performed by one-way ANOVA followed by Fisher's least significant difference (LSD) test.

* $P<0.05$.

** $P<0.01$.

3.4 Urine metabolomics analysis

Urine metabolomics presented similar results with that of plasma, confirming the ameliorative effects of AEE in hyperlipidemia rats. The representative TIC of urine samples was shown in Fig. A.2. Metabonomic differences between control and model groups were also revealed by PCA model (ESI+, $R^2X=0.581$; ESI-, $R^2X=0.672$). According to score plot (Fig. 4a and 4b), clear separation was observed between model and control groups and the samples of model group clustered away from those of control group, indicating that the hyperlipidemia was successfully induced by the HFD.

To further reveal anti-hyperlipidemia effects of AEE, urine data acquired was also analyzed by OPLS-DA model (ESI+, $R^2X=0.677$, $R^2Y=0.99$, $Q^2=0.929$; ESI-, $R^2X=0.62$, $R^2Y=0.982$, $Q^2=0.956$). The P-values from CV-ANOVA of urine OPLS-DA models were <0.001, which showed that there was no over fitting in the models. As shown in Fig. 4c and 4d, AEE treated group was deviated from model group, and clustered on the same side with the control group along LV1. The distribution indicated that the AEE treatment partially improved the hyperlipidemia state.

Corresponding loading plots of OPLS-DA model were inspected (Fig. 4e and 4f). Metabolites with VIP>1 and P-value<0.05 were selected as potential biomarkers responsible for discrimination. 18 metabolites in urine were identified as creatinine, proline betaine, deoxycytidine, cytosine, N1-methyl-2-pyridone-5-carboxamide (N1MPC), pantothenic acid, xanthurenic acid, hippuric acid, 3-indole carboxylic acid glucuronide (3ICAG), phenylacetylglycine, caprylic acid, indoleacetic acid, equol, citric acid, 4-aetamidobutanoic acid, hydroxyphenyllactic acid, suberic acid and 3-hydroxysebacic acid (Table 3).

3.5 Regulative effect of AEE on potential biomarkers and pathway analysis

To fully and intuitively display the relationships and differences between different samples, the relative concentration of the selected biomarkers in plasma and urine were exhibited with heatmap generated by hierarchical cluster analysis. As shown in Fig. 5, the differences of samples and metabolites were simultaneously hierarchically clustered. The dendrogram of the metabolite differences was shown in the horizontal axis of the figure. Obviously, the relevance of the metabolite differences was displayed in the Fig. 5, that the metabolites in the same or similar metabolic pathways were clustered together first. For example, proline, tryptophan, valine, lysine, tyrosine, methionine, leucine and isoleucine belonging to the amino acid metabolism pathway were first clustered together; LysoPC (18:0), LysoPC (18:1) and LysoPC (20:3) involved in glycerophospholipid metabolism were also clustered together. The vertical axis of the figure showed a dendrogram of the sample differences, which indicated that the samples of the control, model and AEE groups were cluster together. Meanwhile, the relative intensity of the metabolites varied significantly in different groups. For instance, HFD significantly increased the relative intensities on some potential biomarkers such as cytosine, caprylic acid, indoleacetic acid, sphingosine 1-phosphate, deoxycytidine and the biomarkers like hippuric acid, LysoPC (17:0) and arachidonic acid were significantly reduced. Five-week treatment of AEE significantly regulated the trends of potential biomarker and reversed the effects of HFD on these potential biomarkers. These results indicated that the varied tendencies of altered metabolites in the hyperlipidemic rats were partly restored to normal by AEE treatment.

Table 3 Potential biomarkers in urine associated with hyperlipidemia based on the UPLC-Q-TOF/MS analysis.

RT	VIP	Formula	Metabolite	Adduction	m/z	Fold change Model/control	AEE/model	KEGG pathway
0. 78	0. 99	$C_4H_7N_3O$	Creatinine	[M + H] +	114. 0663	1. 30	1. 11	Arginine and proline metabolism
0. 8	1. 49	$C_7H_{13}NO_2$	Proline betaine	[M + H] +	144. 102	0. 32**	2. 75**	-
1. 01	1. 39	$C_9H_{13}N_3O_4$	Deoxycytidine	[M + H] +	228. 0973	2. 55**	0. 5**	Pyrimidine metabolism
1. 02	1. 40	$C_4H_5N_3O$	Cytosine	[M + H] +	112. 0506	2. 65**	0. 67*	Pyrimidine metabolism
1. 51	2. 24	$C_7H_8N_2O_2$	N1MPC	[M + H] +	153. 0647	1. 01	1. 61	Nicotinate and nicotinamide metabolism
2. 35	2. 1	$C_9H_{17}NO_5$	Pantothenic acid	[M + H] +	220. 1167	0. 98	0. 55**	Pantothenate and CoA biosynthesis
4. 56	1. 17	$C_{10}H_7NO_4$	Xanthurenic acid	[M + H] +	206. 0435	1. 35	0. 64*	Tryptophan metabolism
5. 52	4. 48	$C_9H_9NO_3$	Hippuric acid	[M + H] +	180. 0642	0. 54**	1. 31	Phenylalanine metabolism
6. 23	3. 42	$C_{15}H_{15}NO_8$	3ICAG	[M + H] +	338. 0869	0. 32**	0. 64	—
6. 69	3. 26	$C_{10}H_{11}NO_3$	Phenylacetylglycine	[M + H] +	194. 0800	0. 68	1. 50	Phenylalanine metabolism
9. 15	1. 30	$C_8H_{16}O_2$	Caprylic acid	[M + H] +	167. 1053	2. 89**	0. 5	Fatty acid biosynthesis
10. 77	1. 50	$C_{10}H_9NO_2$	Indoleacetic acid	[M + H] +	176. 0693	2. 32**	2. 17*	Tryptophan metabolism
11. 21	1. 08	$C_{15}H_{14}O_3$	Equol	[M + H] +	243. 1009	0. 36**	0. 71	-
0. 93	1. 90	$C_6H_8O_7$	Citric acid	$[M-H]^-$	191. 0191	0. 71	0. 74	-
1. 81	1. 51	$C_6H_{11}NO_3$	4-Acetamidobutanoic acid	$[M-H]^-$	144. 0663	1. 63	2. 35**	Arginine and proline metabolism
3. 49	1. 49	$C_9H_{10}O_4$	Hydroxyphenyllactic acid	$[M-H]^-$	181. 0504	8. 27**	0. 28**	Tyrosine metabolism
8. 52	2. 28	$C_8H_{14}O_4$	Suberic acid	$[M-H]^-$	173. 0817	1. 61**	0. 61**	-
8. 6	1. 07	$C_{10}H_{18}O_5$	3-Hydroxysebacic acid	$[M-H]^-$	217. 1078	2. 67**	0. 75	-

AEE: aspirin eugenol ester; KEGG: Kyoto encyclopedia of genes and genomes; VIP: variance importance for projection; RT: retention time; N1MPC: N1-methyl-2-pyridone-5-carboxamide; 3ICAG, 3-indole carboxylic acid glucuronide; -: not available in KEGG database. Statistical analysis was performed by one-way ANOVA followed by Fisher's least significant difference (LSD).

* $P<0.05$.

** $P<0.01$.

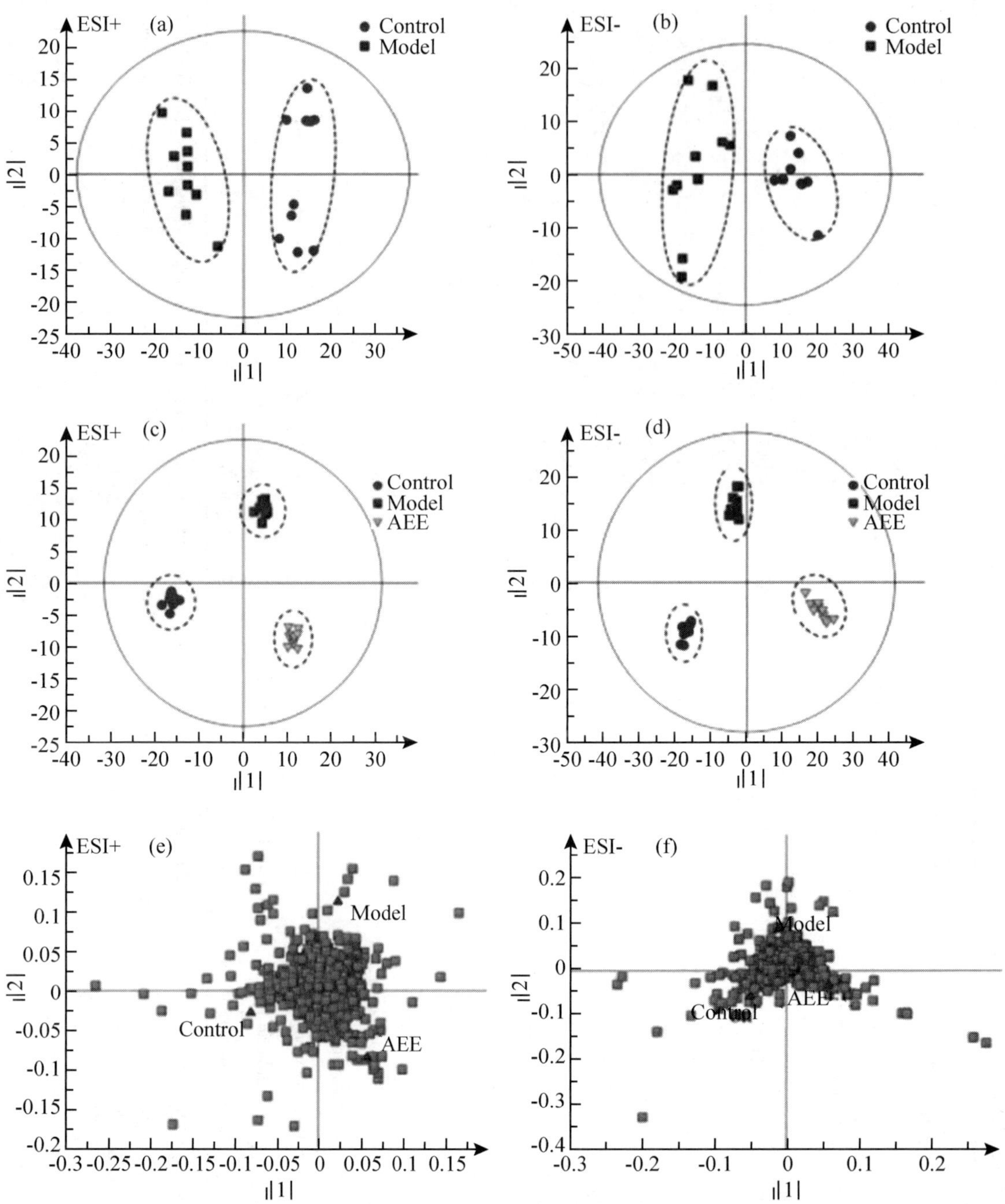

Fig. 4. Score plots and loading plots of urine in each group based on UPLC-TOF/MS in the positive and negative modes (n=10).

Note: a and b, principal component analysis score plots between control and model groups; c and d, orthogonal partial least-squares-discriminate analysis score plots among control, model and AEE groups; e and f, corresponding loading plots of orthogonal partial least-squares-discriminate analysis models.

In order to identify and visualize the affected metabolic pathways inhyperlipidemic rats, metabonomics pathway analysis was performed with MetaboAnalyst 3.0 which was a powerful tool that combined results from pathway enrichment analysis with the pathway topology analysis to identify the most relevant pathways. HMDB IDs of potential biomarkers and Rattus norvegicus (rat, 81 pathways) pathway library were selected for pathway analysis. In present study, the impact-value threshold calculated from pathway topology analysis was set to 0. 10 and the pathway with impact-value above this threshold was filtered out. The results revealed that the disturbed pathways in response to hyperlipidemia and AEE treatment were aminoacyl-tRNA biosynthesis, valine, leucine and isoleucine biosynthesis, pantothenate and CoA biosynthesis, ubiquinone and other terpenoid-quinone biosynthesis, phenylalanine, tyrosine and tryptophan biosynthesis, tryptophan metabolism, glyoxylate and dicarboxylate metabolism, arachidonic acid metabolism and tyrosine metabolism. Fig. 6 and Table 4 showed the summary of pathway analysis. The results suggested that these target pathways displayed marked perturbations in hyperlipidemic rats induced by HFD and could be regulated by AEE treatment to against the development of hyperlipidemia.

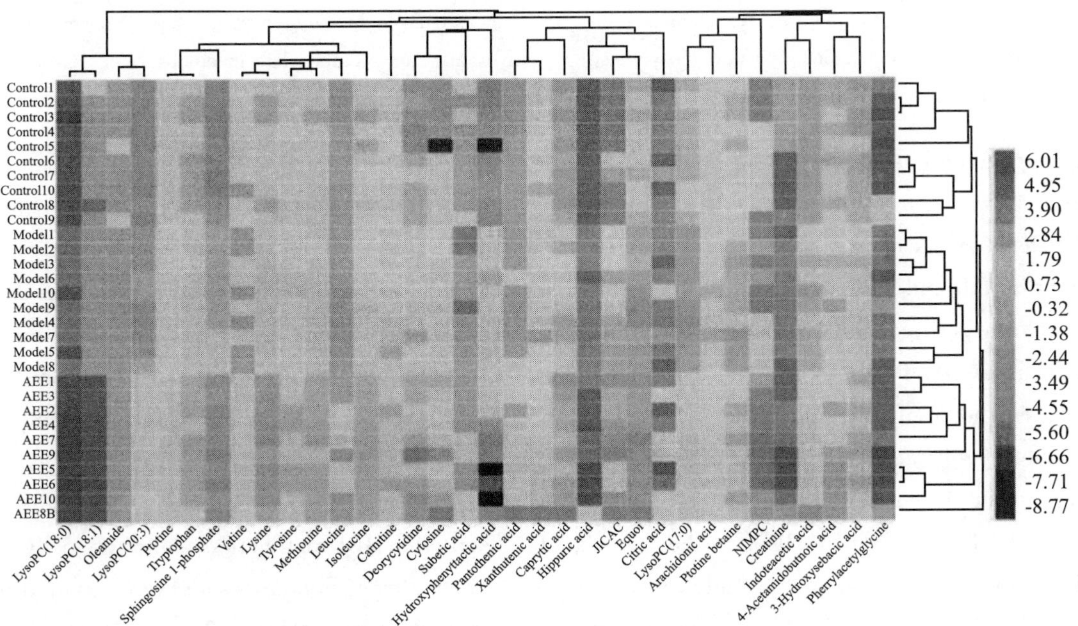

Fig. 5. Heatmap of the 34 plasma and urine metabolites in control, model and AEE groups.

4 DISCUSSION

For hyperlipidemia treatment, therapeutic strategies of many chemical drugs such as statins, fibrates, ezetimibe and nicotinic acid depend on reducing blood lipids. However, most of them are expensive and have undesirable effects such as decline in cognitive function and increased risk of diabetes (Sattar et al., 2010. So there is an obvious need for more efficacious and alternative treatment options for hyperlipidemia. It was reported that aspirin and eugenol could ameliorate hyperlipidemia induced by HFD (Lin et al., 2014; Venkadeswaran et al., 2014, 2016), thus the combination of aspirin and eugenol could enhance the therapeutic efficiency. In AEE 15-day oral toxicity study, blood biochemical results showed AEE could significantly reduce the values ofTG and TCH both in male and female rats with standard diet (Li et al., 2012a, b). In subsequent experiment,

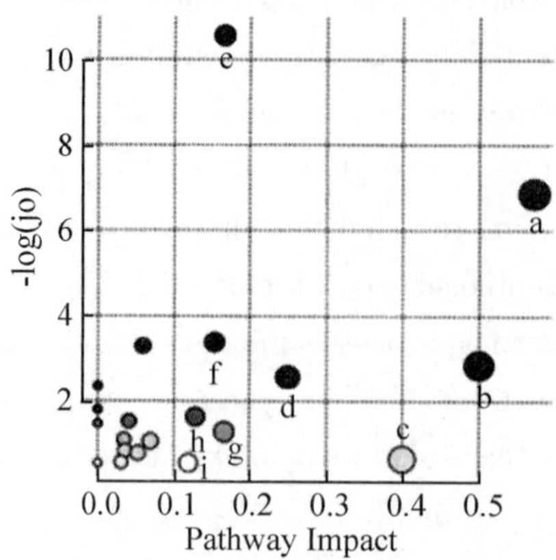

Fig. 6. Disturbed pathways in response to hyperlipidemia and AEE treatment.

Note: a, Valine, leucine and isoleucine biosynthesis; b, ubiquinone and other terpenoid–quinone biosynthesis; c, arachidonic acid metabolism; d, phenylalanine, tyrosine and tryptophan biosynthesis; e, aminoacyl–tRNA biosynthesis; f, pantothenate and CoA biosynthesis; g, glyoxylate and dicarboxylate metabolism; h, tryptophan metabolism; i, tyrosine metabolism.

the regulation effects of AEE on blood lipids in hyperlipidemic Wistar rats had been confirmed and AEE showed more efficacious effects than its precursors (Karam et al., 2015). In present study, increased levels of TCH, TG and LDL and the decreased level of HDL were observed in model group at 8th and 13th week, indicating the blood lipid profiles were significantly changed by HFD and the hyperlipidemia disease was established successfully. Five–week AEE treatment significantly reduced TCH, TG and LDL. These results indicated that this compound might be a new potential drug candidate of anti–hyperlipidemia. HDL synthesized primarily in the liver plays an important role in cholesterol transport and lipid metabolism. In this study, significant difference was observed between control and AEE groups. It may be that the dosage of AEE is insufficient to restore the decreased HDL induced by HFD. On the other hand, disordered HDL production pathways in liver of hyperlipidemic rats may not be fully recovered after AEE treatment. Similar results were reported that HDL level was not significantly affected by drug treatment (Song et al., 2013; Gooda et al., 2016). HDL level is the result of complicating factors and further studies are required to elucidate the effects of AEE on HDL.

Eugenol, an essential constituent found in many plants, is reported to possess gastroprotective activity that eugenol can significantly and dose–dependently reduce gastric ulcers in rat gastric ulcer models (Capasso et al., 2000; Santin et al., 2011). The GI side effects of aspirin, particularly stomach ulceration and bleeding, is the main constraint in its clinical application. In order to overcome these limitations, AEE was designed to get synergistic action through the combination of two pharmacologically active agents, aspirin and eugenol. Metabolism studies proved that AEE was decomposed into salicylic acid (major metabolites of aspirin) and eugenol in vivo, which may play their respective efficient roles and interact in a synergistic manner (Shen et al., 2015). However, in present study, few hyperemia, edema and ecclasis of gastric mucosa could be found in the rats treated with AEE. Pathological injuries in gastric mucosa of AEE group may be attributable to the long duration of the experiments (5 weeks) or the dosage of AEE (54mg/kg) used in the experiment. For the limitation of research objective and content, the comparisons among aspirin, eugenol and AEE are not performed in this study. Further studies are needed to find out the difference between AEE and its precursors, which are helpful to understand the unknown mechanism related with GI side effects of AEE.

Based on the UPLC-Q-TOF/MS analysis method, metabolite profilings of plasma and urine were evaluated. The results indicated that the metabolic patterns of hyperlipidemic rats were obviously changed by AEE treatment. Meanwhile, the biological function and relevant pathway were identified by HMDB, KEGG and MetaboAnalyst platform to elucidate the underlying mechanism. In this study, 34 hyperlipidemia-related endogenous metabolites (e.g. LysoPCs, lysine, malic acid, succinic acid, deoxycholic acid and glycerophosphocholine) were selected as potential biomarkers which were associated with glycerophospholipid metabolism, fatty acid metabolism, fatty acid beta-oxidation, amino acid metabolism, TCA cycle, sphingolipid metabolism, gut microflora and pyrimidine metabolism (Fig.7).

Numerous studies have proved that the disorders of lipid metabolism play a pivotalpathogenetic role in the initiation and progression of hyperlipidemia (Zhao et al., 2015). LysoPCs play essential roles in lipid metabolism and the progress of many diseases including inflammation, atherosclerosis, diabetes, cancer and blood lipid disorders (Jackson et al., 2008; Kabarowski, 2009). In present study, compared with the control group, LysoPC (18 : 0), LysoPC (18 : 1) and LysoPC (17 : 0) of the rats feeding with HFD were significantly decreased and LysoPC (20 : 3) were increased, which was in good agreement with previous reports (Jin et al., 2014; Li et al., 2016). The changes of these LysoPCs in rats fed with HFD may be relevant to the pathogenesis of hyperlipidemia and could be selected as potential biomarkers. In addition, the abnormal changes of LysoPCs in hyperlipidemic rat were intervened by AEE treatment. It has been reported that there are complex interactions between blood lipid levels and LysoPCs (Zhao et al., 2015). So it is speculated that the regulation effects of AEE on LysoPCs might contribute to lowing blood lipid levels, which need further confirmation in future study.

Arachidonic acid (AA), one of the most biologically active n-6 polyunsaturated fatty acids, could be used to produce prostaglandins (PGs) by cyclooxygenase. Many studies have suggested that PGs from AA have multiple beneficial effects against cardiovascular risks such as hyperlipidemia and hypertension (Calder, 2008). AA concentration in HFD-in-duced dyslipidemic rats was significantly lower than the control rats (Gu et al., 2015). Moreover, related evidence proved that patients with cardiovascular disease also had lower AA concentration than normal population (Das, 2008). Therefore, a reduction of AA levels may contribute to increases in cardiovascular risks. In this study, AA was significantly decreased in model group compared with the normal control group and AEE treatment reversed this reduction. One possible explanation might be that the increased level of AA was caused by the inhibition effect of AEE on cyclooxygenase activity, which could enhance cardiovascular protection in hyperlipidemic rat. Oleamide, an amide of the oleic acid, showed significantly increased levels in the model group compared with the control, which was concordant with a previous study (Park et al., 2017). However, oleamide level was also elevated in AEE group.

The abnormal changes of oleamide may indicate the dysfunctions in fatty acid metabolism.

Caprylic acid, involved in fatty acid biosynthesis, is the common name for the eight-carbon saturated fatty acid.Some investigations have reported that caprylic acid can reduce the serum cholesterol level through the excretion of cholesterol in feces and the inhibitation of bile acid absorption in small intestinal (Liu et al., 2017; Xu et al., 2013).In present study, we observed that compared with the control group, HFD could significantly increase caprylic acid level, indicating that HFD consumption may up-regulate the fatty acid biosynthesis to produce more caprylic acid. However, the level of caprylic acid in AEE group was lower than those observed in model group.There are two possible explanations for the decreasing of caprylic acid.First, AEE may restrain the increasing of caprylic acid by inhibitation of fatty acid biosynthesis.Second, caprylic acid may be consumed to reduce serum cholesterol level.

Carnitine cycle is the first step for fatty acid oxidation, in which the fatty acyl Coenzyme A (CoA) enters the mitochondria as fatty acyl carnitines via carnitine transport.It has been reported that, under hyperlipidemia conditions, the process of fatty acid oxidation was promoted to supply the required energy, in which carnitines are consumed to facilitate fatty acyl CoA into mitochondria (Liu et al., 2014b).Suberic acid, also octanedioic acid, is a dicarboxylic acid, which could be increased in the urine of patients with fatty acid oxidation disorders (Hagen et al., 1999).In this study, decreased level of carnitine and increased level of suberic acid were observed in hyperlipidemic rat, suggesting the fatty acid oxidation pathway was promoted. Interestingly, the up-regulation of suberic acid was effectively inhibited by AEE treatment, suggesting AEE could ameliorate the disturbed fatty acid beta-oxidation.

TCA cycle is the main pathway of glucose degradation and the primary energy supplier for universal organisms, and acetyl CoA and citric acid are important intermediary metabolite in TCA cycle.Pantothenic acid is a constituent part of acetyl CoA, thus, variation of pantothenic acid is supposed to exert influence on acetyl CoA metabolism, which consequently affects TCA cycle (Zhu et al., 2013).Compared with the control, citric acid and pantothenic acid in urine were decreased in the model, which indicated the down-regulation of TCA cycle.These results implied that the energy consumption pattern might shift to lipid oxidation in response to hyperlipidemia, which was also observed in mice suffering similar energy metabolism impairment (Mayr et al., 2005).However, pantothenic acid and citric acid levels in AEE group were lower than those in the model, suggesting that the effects of AEE on energy metabolism were limited, and AEE may have other mechanism against hyperlipidemia.

Sphingosine-1-phosphate (S1P), a breakdown product of ceramide metabolism, is involved in the development of metabolic diseases. Recently, many studies have confirmed that HFD could enhance sphingolipid synthesis and elevates S1P levels in plasma (Chen et al., 2016; Kasbi-Chadli et al., 2016).The increased levels of S1P in the plasma samples suggested that the metabolism of sphingolipid was promoted under hyperlipidemia condition, which may decrease the reverse cholesterol transport pathway to lead the increase risk of hyperlipidemia (Liu et al., 2014b).After AEE treatment, the increased levels of S1P were down-regulated, indicating AEE had inhibitation effects on sphingolipid metabolism to reduce hyperlipidemia risk.

Branched chain amino acid (BCAA: leucine, isoleucine and valine) are a group of essential amino acids and important energy source by keto-genesis. Previously, it had been reported elevated circulating BCAA levels are strongly associated with metabolic disorders, such as obesity and type 2 diabetes mellitus (Lynch and Adams, 2014). Moreover, recent studied demonstrated BCAA could cause significant hepatic damage in mice fed with HFD by increasing hepatic oxidative stress, hepatic apoptosis and circulation hepatic enzymes (Zhang et al., 2016a). BCAA levels were increased in model group compared with control, which were consistent with previous reports (Li et al., 2015). AEE treatment showed favorable inhibition on BCAA levels, indicated the depression on BCAA metabolism. It may be inferred that AEE treatment suppressed the BCAA, thus leading to the improvement of liver pathologic changes. As previously reported (Zhang et al., 2016c), elevated levels of proline, tryptophan, lysine, tyrosine and methionine were noticed in the model group. There are several probable explanations for this finding. First, many studies have identified that elevated BCAA are generally attributed to increase muscle protein breakdown and release of amino acids from muscle proteins (Lanza et al., 2010). Second, amino acid metabolism is influenced by many factors such as energy metabolism and liver function. Previous studies found that HFD could cause liver damage and disturbance in glycolysis and the TCA cycle (Jiang et al., 2013), which may lead the disorders ofliver function and energy metabolism to affect plasma amino acid levels.

Table 4 Pathway analysis result with MetaboAnalyst 3. 0.

No.	Pathway name	Total	Expected	Hits	Raw P	-log (P)	Impact
1	Valine, leucine and isoleucine biosynthesis	11	0. 21969	3	0. 0010589	6. 8506	0. 57143
2	Ubiquinone and otherterpenoid-quinone biosynthesis	3	0. 059914	1	0. 058767	2. 8342	0. 5
3	Arachidonic acid metabolism	36	0. 71897	1	0. 52077	0. 65245	0. 4
4	Phenylalanine, tyrosine and tryptophan biosynthesis	4	0. 079886	1	0. 077605	2. 5561	0. 25
5	Glyoxylate and dicarboxylate metabolism	16	0. 31954	1	0. 27714	1. 2832	0. 16667
6	Aminoacyl-tRNA biosynthesis	67	1. 3381	8	2. 59E-05	10. 561	0. 16664
7	Pantothenate and CoA biosynthesis	15	0. 29957	2	0. 034406	3. 3695	0. 15384
8	Tryptophan metabolism	41	0. 81883	2	0. 19619	1. 6287	0. 12821
9	Tyrosine metabolism	42	0. 8388	1	0. 57686	0. 55015	0. 11905

Total: The total number of compounds in the pathways; the hits are the actually matched number from the user upload data; the rawP is the original P value calculated from the enrichment analysis; the impact is the pathway impact value calculated from pathway analysis.

As a tyrosine metabolite, hydroxyphenyllactic acid is in relatively higher concentrations in the urine of patients with phenylketonuria and tyrosinemia (Spaapen et al., 1987). Tyrosine in plasma was significantly increased in model group, which indicated that tyrosine metabolism was up-regulated by HFD under hyperlipidemia condition. Elevated tyrosine may be the reason for the increase of hydroxyphenyllactic acid. AEE showed favorable inhibitation on hydroxyphenyllactic acid. Moreover, the same change trends of hydroxyphenyllactic acid and tyrosine in model and AEE groups were ob-

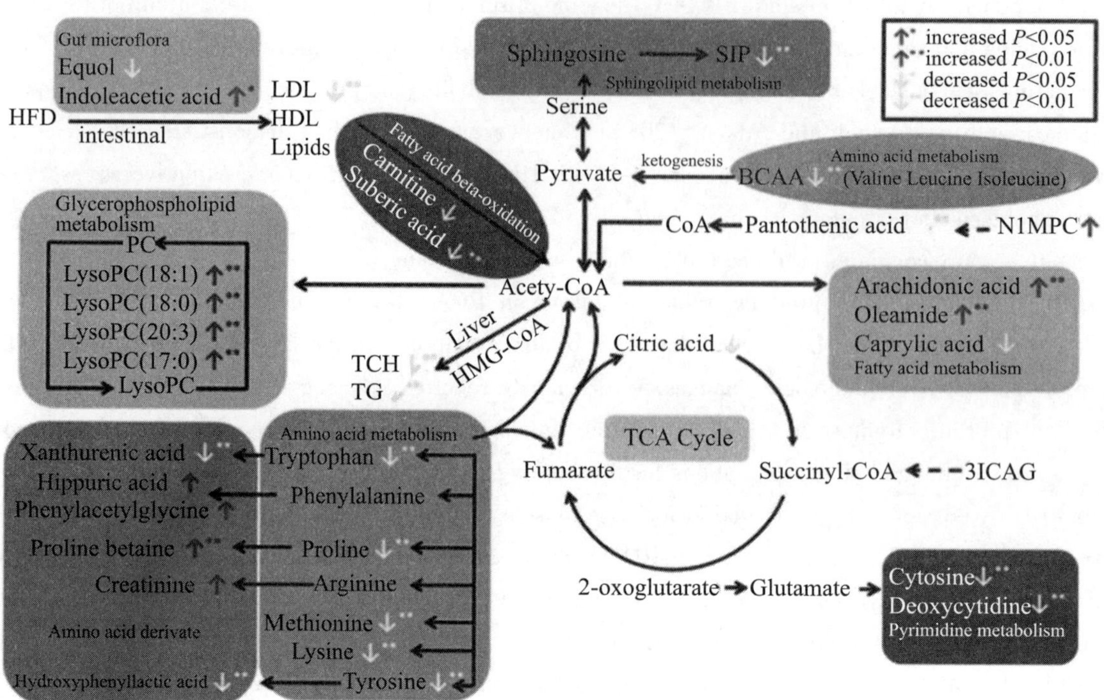

Fig.7. Potential metabolic pathways disturbed in hyperglycemic rat induced by HFD and alterations by AEE treatment.

served.These results indicated that the depression of AEE on tyrosine metabolism might be against the development of hyperlipidemia.

Creatinine, breakdown product of creatine phosphate, is a commonly used indicator of renal function.An increasing number of studies have demonstrated that hyperlipidemia is accompanied with the increase of creatinine (Ben et al., 2017).Proline betaine is an amino acid-derived betaine as osmoprotectant found in urine.Jenna Pekkinen et al.reported that rye barn could increase proline betaine in urine in diet-induced obese mice with the approach of metabonomic study (Pekkinen et al., 2015).In our study, the results showed that compared with the control group, proline betaine decreased in model group, while creatinine increased. Notably, the decreased level of proline betaine was up-regulated after AEE treatment.These results were in accordance with the previous research, and suggested renal damage or dysfunctions caused by hyperlipidemia might be improved by AEE treatment.

Equol is a natural specific metabolic product of daidzein, and gut bacteria play an important role in the conversion from daidzein to equal. Similarly, phenylacetylglycine is the metabolite of phenylalanine by the gut microflora.Hippuric acid is a normal component of urine, which is also involved into phenylalanine metabolism.Equol has various biological benefits such as cardioprotective effects, anticancer effects and reducing blood lipid levels (Zhang et al., 2016b).It has been reported that the abundance and diversity of the gut microbiota in hyperlipidemic rats could be significantly changed by HFD (Zhong et al., 2015).Yue-Tao Liu et al.have reported that hippuric acid was reduced and treated as a urine biomarker related to the HFD-induced atherosclerosis in rabbit

(Liu et al., 2014c). The decreasing levels of equol, hippuric acid and phenylacetylglycine were observed in the model group. Up-regulated levels of hippuric acid and phenylacetylglycine and down-regulated of equol were found in AEE group, which may attribute to the improvement of phenylalanine metabolism and the alteration of gut microflora. Further studies should be conducted to investigate the effects of AEE on gut microbiota and the relationship between metabolites and gut microbiota in hyperlipidemic rat 3-Indole carboxylic acid glucuronide (3ICAG) is a normal urinary in-dolic tryptophan metabolite and has been found to be decreased in HFD-induced atherosclerosis rabbits (Liu et al., 2014c). N1-methyl-2-pyridone-5-carboxamide (N1MPC) is an end product of nicotinamide adenine dinucleotide (NAD) pathways. Increased urine N1MPC concentrations were observed in C57BL/6 mice after HFD feeding, which was positive related with energy overproduction and hepatic fatty acid accumulation (Boulange et al., 2013). Varied changes of 3ICAG and N1MPC were observed in model and AEE groups in this study, suggesting effect of AEE on hyperlipidemia may be associated with tryptophan and NAD metabolism. Deoxycytidine and cytosine are fundamental units of nucleic acids. Elevated levels of cytosine and deoxycytidine in HFD-fed rats in model group might result from RNA degradation. Sharp reductions of cytosine and deoxycytidine were noticed after AEE treatment, which reflected that the perturbations in nucleotide metabolism induced by HFD were inversed by AEE.

Xanthurenic acid and indoleacetic acid are the metabolites from tryptophan catabolism. Alterations of tryptophan metabolism have been reported in diabetes, atherosclerosis and hyperlipidemia (Ragazzi et al., 2006). Levels of xanthurenic acid and indoleacetic acid were increased in hyperlipidemic rats in this study, indicating the up-regulation oftryp-tophan metabolism. AEE was observed to decrease the levels of xanthurenic acid in comparison with the model group. According to these results, the therapeutic effects of AEE may be based on regulating dysfunctional tryptophan metabolism. However, indoleacetic acid was increased in AEE group. There is a possible explanation for the increase of indoleacetic acid. Indoleacetic acid is often produced by the action of bacteria in the mammalian gut, and thus AEE may affect indoleacetic acid level through gut bacteria alteration. 4-Acetamidobutanoic acid is a product of the urea cycle. 3-Hydroxysebacic acid is a normal urinary 3-hydroxydicarboxylic acid metabolite and can be elevated in patients with peroxisomal disorders. Very little is known about their relevance for hyperlipidemia. More studies are needed to discover the functional roles of these metabolites in hyperlipidemia.

5 CONCLUSION

It was demonstrated that AEE treatment ameliorated plasma and urine metabolism in HFD-fed-hyperlipidemic rats that ultimately decreased blood lipids and improved hepatic steatosis. 16 endogenous metabolites in plasma and 18 endogenous metabolites in urine were found as biomarkers to explain the underlying mechanism ofAEE. The effects of AEE on hyperlipidemia might involve in regulating the dysfunction of glycerophospholipid metabolism, fattyacid metabolism, energy metabolism amino acid metabolism, sphingolipid metabolism and pyrimidine metabolism. These findings not only provided new insights into the HFD-induced hyperlipidemia but also revealed the possible action mechanism of AEE for the treatment of hyperlipidemia. This study indicated that AEE might be a promising drug candidate in the application of hyperlipidemia treatment. The selected metabolites might be served as potential drug targets for hyperlipidemia diagnosis or treatment. Future studies are needed to investigate the potential roles of AEE in the regulation of the selected endogenous metabolites associated with hyperlipidemia.

CONFLICT OF INTEREST STATEMENT

The authors declared no conflict of interest.

ACKNOWLEDGEMENTS

The work was supported by the National Natural Science Foundation of China (No. 31402254) and the Natural Science Foundation for Youth of Gansu Province (1506RJYA148).

Appendix A. Supplementary data

Supplementary data to this article can be found online at http://dx.doi.org/10.1016/j.taap.2017.07.013.

REFERENCES OMITTED

（发表于《Toxicology and Applied Pharmacology》，院选 SCI，IF：3.791）

Cardioprotection of Sheng Mai Yin a Classic Formula on Adriamycin Induced Myocardial Injury in Wistar Rats*

Kai ZHANG[1,2], Jingyan ZHANG[1], Xurong WANG[1], Lei WANG[1], Pugliese MICHELA[2], Passantino ANNAMARIA[2], Jianxi LI[1**]

(1. Lanzhou Institute of Husbandry and Pharmaceutical Sciences, Chinese Academy of Agricultural Sciences, Lanzhou 730050, China; 2. Department of Veterinary Sciences, University of Messina, 98168 Messina, Italy)

Abstract: Background: Sheng Mai Yin (SMY), a well-known Chinese herbal medicine, is widely used to treat cardiac diseases characterized by the deficiency of Qi and Yin syndrome in China.SMY-based treatment has been derived from Traditional Chinese Medicine (TCM), officially recorded in the Chinese Pharmacopoeia.

Purpose: We aimed to clarify whether SMY attenuates myocardial injury induced byadriamycin in Wistar rats with chronic heart failure (CHF).

Methods: To quantifyginsenoside Rg1, ophiopogonin D, ophiopogonin D', schisandrin by HPLC.To establish CHF animal model, adriamycin was intraperitoneally injected in Wistar rats for 7 weeks at a dose of 2mg/kg body weight.Overall, 180 rats were randomly assigned to six groups: control, CHF model, captopril (positive control), high dose (HSMY), medium dose (MSMY), and low dose (LSMY).Experimental rats were fed 0.625mg/kg captopril and 90mg/kg, 45mg/kg, and 22.5mg/kg SMY, respectively, over 7 weeks.The inflammatory cytokines TNF-α and IL-6 were measured using ELISA.Matrix metalloproteinases (MMPs) were identified using immunohistochemistry (IHC).Both IHC and RT-PCR were used for quantification of COL-Ⅳ expression levels in the heart tissues.Scanning electron microscopy (SEM) was used for the visualization of myocardium morphology.

Results: The concentration ofginsenoside Rg1, ophiopogonin D, ophiopogonin D' and

* Abbreviations: CA, Captopril administration group; CHF, Chronic heart failure; CHFM, CHF model group; COL-Ⅳ, Collagen-Ⅳ; CON, Control group; ECM, Extracellular matrix; ELISA, Enzyme-linked immunosorbent assay; HF, Heart failure; HSMY, High dose Sheng Mai Yin administration group; IHC, Immunohistochemistry; IL-6, Interleukin-6; LSMY, Low dose Sheng Mai Yin administration group; MI, Myocardial infarction; MMP-2, Matrixmetalloproteinases-2; MMP-9, Matrixmetalloproteinases-9; MMPs, Matrix metalloproteinases; MSMY, Middle dose Sheng Mai Yin administration group; SEM, Scanning electron microscopy; SMY, Sheng Mai Yin; TCM, Traditional Chinese Medicine; TNF-α, Tumor necrosis factor-α

** Corresponding author, E-mail addresses: fyjohn@sina.com (K.Zhang), lzjianxil@163.com (J.Li).

schisandrin in SMY was found to be 25.63±3.42mg, 11.00±1.17mg, 7.02±0.51mg, and 25.31±4.28mg per gram of SMY, respectively. Compared with CHF model group, TNF-α levels were significantly lower ($p<0.01$) in the four drug-administered groups. Moreover, except in the SYM low dose group, IL-6 levels in the other 3 drug-administered groups were also significantly reduced ($p<0.01$). COL-Ⅳ expression was also significantly reduced on treatment with high SYM dose ($p<0.05$). IHC results confirmed that SMY and captopril significantly reduced MMPs expression in the heart.

Conclusion: SMY could control or slow CHF progression by suppressing pathological changes in the myocardium in CHF models. This could be attributed at least partly to the downregulation of IL-6 and TNF-α and inhibition of overexpression of MMPs and COL-Ⅳ, which significantly relieved the cardiac-linked pathologies, decreased the risk of myocardial fibrosis, and inhibited cardiac remodeling. These findings suggested that SMY and captopril have similar efficacy for the treatment of adriamycin-induced myocardial injury. In addition, Chinese herbal preparation SMY may play a role in the treatment of cardiac diseases.

Key words: Chronic heart failure (CHF); Sheng Mai Yin (SMY); Matrixmetalloproteinases (MMPs); Tumor necrosis factor-α (TNF-α); interleukin-6 (IL-6)

1 INTRODUCTION

Natural products possess an enormous structural and chemical diversity, which cannot be matched by synthetic libraries of small molecules and thus continue to inspire novel discoveries in chemistry, biology, and medicine. These natural chemicals have been evolutionarily optimized as drug-like molecules and remain the best known sources of drugs and drug-leads (Newman and Cragg, 2012). In the last decade, de-replication has emerged as a hot research topic, leading to a huge publication boom since 2012. This blending of multiple disciplines in ways that provide important and novel conceptual and/or methodological advances has opened up vast research prospects (Gaudêncio and Pereira, 2015 Leung et al., 2011). This interdisciplinary research has also led to the identification of natural products like 6, 6″-biapigenin, which is only the second inhibitor discovered so far for NEDD8-activating enzyme. Importantly, 6, 6″-biapigenin was found to be enzymatically active in kinetics-and cell-based assays, with a potency in the micromolar range. Fong et al. (2007) showed that the extract of the rhizomes of Alisma orientalis (Sam) Juzep. has synergistic growth inhibitory effect with cancer drugs that are P-glycoprotein substrates including actinomycin D, puromycin, paclitaxel, vinblastine, and doxorubicin. Zhong et al. (2015) firstly reported that natural product-like compound 1 was the first natural product-like inhibitor and only the second inhibitor overall of TLR1-TLR2 heterodimerization, as potential agents for the treatment of inflammatory and autoimmune diseases. Amentoflavone was found to be JAK2 inhibitor by structure-based virtual screening of a natural product library, and its analogues might function as Type Ⅱ inhibitor of JAK2 (Ma et al., 2014). Liu et al. (2014) indicated that natural product-like compound 1inhibited STAT3 DNA-binding activity in vitro and attenuated STATA3-directed transcription in cellulo with selectivity over STAT1 and with comparable potency to the well-known

STAT3 inhibitor S3I－201, and also exhibited selective anti－proliferative activity against cancer cells over normal cells. Pep－tidyl－proline isomerases (PPIases) played a key role in cancer, neuro－degeneration, and psychiatric disorders, however, macrocyclic natural products might create potent and selective inhibitors, such as FK506, rapamycin, and cyclosporin. Manivannan et al. (2017) illustrated that twelve novel silybin analogues had significantly greater efficacy than silybin, and derivative 15k as a novel tubulin inhibitor with significant activity against ovarian cancer cells. Nature has been extremely generous to the mankind in offering life－saving therapies, and the next great drug may be just around the corner: are we ready to seize the opportunity? (Shen, 2015).

Sheng Mai Yin (SMY) is a classical natural and effective formula, which is routinely used in China, and contains Radix Ginseng (Panax ginseng C.A.Mey., Araliaceae), Radix Ophiopogon (Ophiopogon japonicas (Thumb.) Ker－Gawl., Liliaceae) and Fructus Schisandrae (Schisandraechinensis (Turcz.) Bail., Magnoliaceae) (Chen et al., 2007; Wang et al., 2005; Committee of Pharmacopoeia of PR China, 2005)."Sheng" and "Mai" are the Chinese abbreviations for Panax ginseng and Ophiopogon japonicas, which have both been extensively used for centuries in China as effective drugs (Chen et al., 2007; Gillis, 1997). Medicinal plants have been used in patients with congestive heart failure as well as systolic hypertension (Rastogi et al., 2016). Specifically, Panax ginseng, Fructus Schisandrae, and Ophiopogon japonicas, which contain multiple bioactive components, have been shown to be effective against many diseases (Wang et al., 2005; Zhang et al., 2010). Given the excellent activity and safety of its components, SMY is widely used for the treatment of cardiac diseases, which are characterized by deficiency of Qi and Yin syndrome (Mo et al., 2015). Clinically, SMY has been shown to treat shock, coronary heart disease, angina, myocardial infarction (MI), viral myocarditis, pulmonary heart disease, heart failure etc. In addition, SMY can treat myocardial diseases, rheumatoid diseases, systemic lupus erythematosus, as well as epidemic hemorrhagic fever. These diverse curative effects of SMY explain why SMY in combination with other herbs has been used for the treatment of diabetes, nodular lupus erythematosus, mild brain dysfunction syndrome, optic atrophy, recurrent pneumothorax, iron deficiency anemia, severe infectious mononucleosis, and malignant tumor (with a maximum clinical dosage of 0.3g/kg) (Zhang et al., 2010). Interestingly, SMY is also especially prescribed for coronary artery disease (Wang et al., 2002).

On the cellular level, SMY suppresses mitochondrial apoptosis as indicated by reduction in several pro－apoptotic factors (Bax, cytochrome c, and cleaved caspase－3) and up－regulation of the antiapoptosis factor Bcl－2 (Mo et al., 2015). The improvement in cardiac contractile function afforded by SMY treatment is likely mediated by an increase in Ca^{2+} release from SR through L－type Ca^{2+} current－activated RyRs (Zhang et al., 2008). On the other hand, SMY improves the post－ischemic myocardial dysfunction by opening the mitochondrial KTAP channels (Wang et al., 2002)., improving the heart structure and reducing Cx43 expression after MI. SMY also inhibits myocardial fibrosis in rats with diabetic cardiomyopathy, and significantly delays the formation of diabetic cardiomyopathy through multiple signaling pathways (Ni et al., 2011). In view of these considerations, the aim of our study was to investigate the basis of the protective function of SMY on myocardial injury and on the regulation of IL-6 and TNF-α levels. Our study also explores the regulatory role of SMY on MMPs and COL-Ⅳ to achieve myocardial remodeling as well as the protective function of SMY against the pathological changes of the myocardium.

Table 1 Primers (F, forward; R, reverse) used for relative mRNA quantifications by real-time PCR.

Target	Primer sequence (5′ to 3′)	Length (pb)
COL-Ⅳ	F: GAGGGTGCTGGACAAGC	
	R: TAAATGGACTGGCTCGGAATTC	
GAPDH	F: TGTGAGGGAGATGCTCAGTG	
	R: GGCATTGCTCTCAATGACAA	

2 MATERIALS AND METHODS

2.1 Preparation of SMY

SMY was purchased fromChiatai Qingchunbao Pharmaceutical Co. (Zhejiang, China) (Batch No.-1410014). This SMY preparation was a 1 320g mixture of three common Chinese herbal medicines: Ginseng radix, Ophiopogonis Radix and Schisandrae Chinensis Fructus mixed in a ratio of 1 : 2 : 1.

2.2 Reagents and drugs

Adriamycin was purchased from Shenzhen Main Luck Pharmaceuticals Inc. (Shenzhen, China). Captopril was purchased from North China Phar. Co. (Hebei, China). Rat IL-6, TNF-α enzyme-linked immunosorbent assay (ELISA) assay kits were obtained from RayBiotech. Inc. (GA, USA). Histostain-Plus kits was purchased from ZSBIO. Inc. (Beijing, China). Antibodies for type IV collagen, matrixmetalloproteinases-2 and matrixmetalloproteinases-9 were purchased from Boster Co. (Beijing, China). All other agents used in this study were of commercially available grade and purity.

2.3 Animals and CHF model

AdultWistar rats weighing 160-200g, with an equal proportion of males and females, were provided by the Animal Breeding Center of Lanzhou Military Region General Hospital. These animals were housed under controlled conditions at a temperature of 25±2℃, humidity of 40%±5% and on a 12h light-dark cycle. The rats had free access to solid rodent chow and tap water. Animals were allowed a 1 week acclimatization period prior to entry into any experimental protocol. The entire laboratory procedure was carried out under the permission and surveillance of local ethics committee. The experimental procedures were approved by Lanzhou Institute of Husbandry and Pharmaceutical Sciences, CAAS.

Adriamycin was used to establish the animal model of CHF. It was injected intraperitoneally for 7 weeks at a dose of 2mg/kg body weight (Li et al., 2006; Cheng, 2011). The rats were randomly assigned into six groups. Group 1 (CON) was normal controls comprising of 30 healthy rats fed under the same conditions as the experimental group but lacking any treatment. Group 2 (CHF model control, CHFM) were rats, which were not subjected to any treatment after the establishment of CHF in them. Group 3 (CA, positive control) were animals treated with captopril at a dosage of 0.625mg/kg body weight. Group 4 (HSMY) were treated with high dosage of SMY (90mg/kg body weight per day) for 7 weeks after the establishment of CHF model. Group 5 (MSMY) were treated with medium dosage of SMY (45mg/kg body weight per day) for 7 weeks after the establishment of CHF model. Group 6 (LSMY) consisted of animals treated with low dosage of SMY (22.5mg/kg body weight per day) for 7 weeks after the establishment of CHF model. At the end of the protocol, cardiac function was examined physiologically and heart tissues were harvested for performing IHC and biochemical analyses.

Quantification ofginsenoside Rg1, ophiopogonin D, ophiopogonin D', schisandrin by HPLC Precisely weighed samples (ginsenoside Rg1, ophiopogonin D, ophiopogonin D', schisandrin and SMY test sample) were extracted with methanol in an ultrasonic bath and filtered in a 0.45μm filter. An aliquot of 20μl of each sample was injected onto the HPLC column (kromasil100-5C_{18} 250×4.6mm, 5μm particle size) and elution was carried out with acetonitrile : water (ginsenoside Rg1 6 : 4, ophiopogonin D 5.5 : 4.5, ophiopogonin D' 5.5 : 4.5, schisandrin 1 : 1) at a flow-rate of 1mL/min and eluate was monitored at 203nm (ginsenoside Rg1), 205nm (ophiopogonin D, ophiopogonin D') and 250nm (schisandrin) (Committee of Pharmacopoeia of PR China, 2005). The procedure was repeated three times for each sample. Each solution was prepared and injected three times and the curve was plotted using an average of the area. The calculated concentrations of ginsenoside Rg1, ophiopogonin D, and ophiopogonin D' were expressed in terms of mean± standard deviation (SD) (mg/g).

2.4 Determination of TNF-α and IL-6 levels in heart tissues

The inflammatory cytokines TNF-α and IL-6 were quantified using ELISA kits according to the manufacturer's instructions. Myocardial tissues were homogenized in RIPA lysis buffer, and centrifuged at 6 000g for 15min at 4℃. 0.1mL supernatant was removed and mixed with 1ml normal saline. The supernatants were collected for the analyses, and data were expressed as picograms per milligram of protein.

2.5 Real-time PCR (q-PCR)

We quantified the mRNA expression levels of COL-Ⅳ and GAPDH (reference) by One Step STBR© PrimerScript© RT-PCR Kit (TaKaRa, Japan) using the standard curve method (Terova et al., 2011). Primer 5.0 software was used to design the primers for COL-Ⅳ and GAPDH (Zhao et al., 2014). The nucleotide sequences of all primers used in this study are reported in Table 1. Real-time analysis was performed in duplicate for each sample using SYBR© Premix Ex Taq™ Ⅱ (Tli RNaseH Plus, TaKaRa, Japan). The following real-time run conditions were used: 2min at 50℃, 30min at 60℃ and 5min at 95℃, followed by 35 cycles consisting of 20s at 94℃, 30s at 63℃, and 30s at 72℃. We used iQ5 Real-Time PCR system (Bio-Rad, USA) to perform SYBR© Premix Ex Taq™ Ⅱ reactions and collected runs data using the iQ5™ software (Bio-Rad, USA). Cycle threshold (Ct) value obtained by each standard mRNA amplification was used to

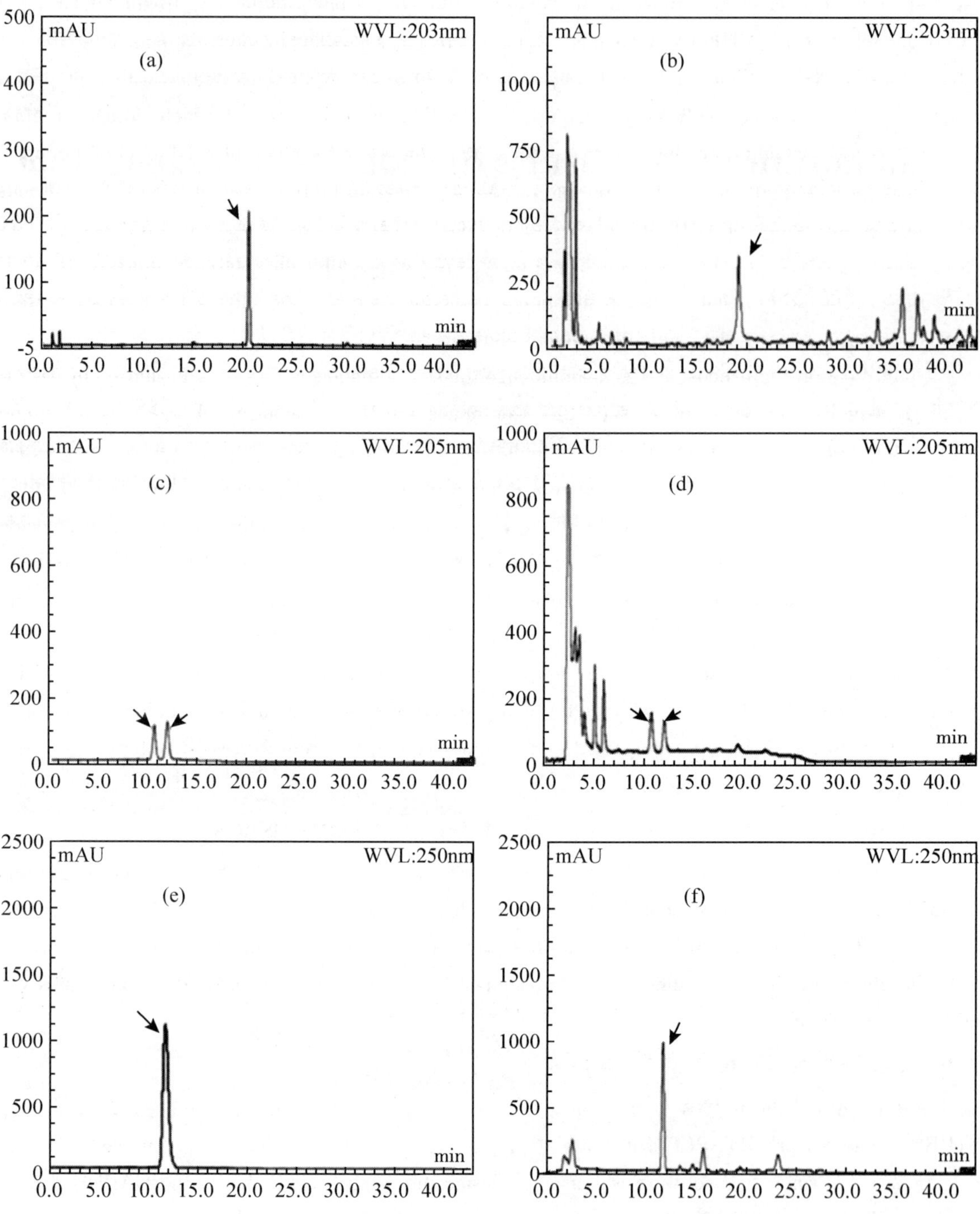

Fig. 1. HPLC chromatogram.

Note: (a) reference substance (ginsenoside Rg1, 203nm), (b) SMY test sample (203nm), (c) reference substance (A-ophiopogonin D, B-ophiopogonin D', 205nm), (d) SMY test sample (205nm), (e) reference substance (schisandrin, 250nm), (f) SMY test sample (250nm).

create a standard curve for each target gene. This curve served as a basis for calculating the unknown mRNA levels of each gene present in the total RNA extracted from each sample.

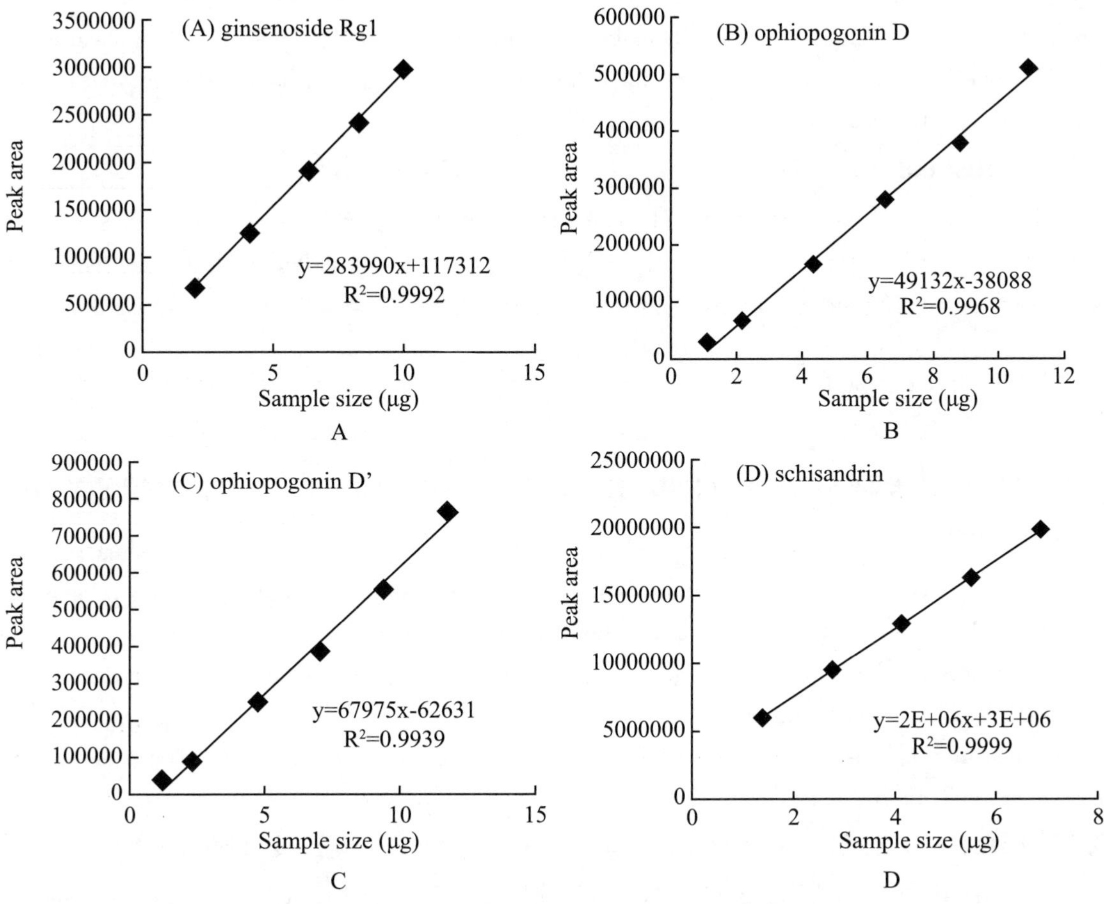

Fig. 2. Calibration curve of standard substances obtained using HPLC.

2.6 Myocardial sample preparation for scanning electron microscopy (SEM)

Using a scanning electron microscope JSM-6510A SEM (JEOL, Japan), the diameter of myocardial fibers was measured using the Smile Shot™ system software. For sample preparation, the following procedure has been proposed (Li, 2008). Cardiac tissue was first washed using Phosphate Buffer Solution (PBS) buffer before being refrigerated in 2%-4% glutaraldehyde at 2-4℃. The treated cardiac tissue was then rinsed with double-distilled water, and washed again with PBS. Ethyl alcohol substitution and dehydration was performed in a series from 40%→50%→60%→70%→80%→90%→95%→100% with each dehydration step being about 15min long. 100% ethyl alcohol substitution was done 3 times before the treated tissue was immersed in tertbutyl alcohol 2h to substitute ethyl alcohol. Finally, vacuum freezing drying and gold coating were performed on the dehydrated and tertbutyl alcohol-substituted samples.

2.7 Immunohistochemistry

IHC staining for type Ⅳ collagen (COL-Ⅳ 1 : 1000, BOSTER, China), matrixmetalloproteinases-2 (MMP-2 1 : 500, BOSTER, China) and matrixmetalloproteinases-9 (MMP-9 1 : 250, BOSTER, China) were performed using the standard streptavidin-biotin-peroxidase on 5μm thick sections of formalin-fixed, paraffin-embedded tissue.The sections were de-paraffinized in xylene, and rehydrated through graded alcohols to distilled water before antigen retrieval by heat method in citrate solution (pH 6.0). An automated detection using Leica ST5010 autostainer XL (Lanzhou, China) was utilized for analysis.

2.8 Statistical analysis

Data were expressed as mean±SD.The effect of treatments on the expression of biotransformation genes were made by one-way variance analysis (ANOVA) test using theBonferroni post-hoc, followed by Student-Newman-Keuls pair-wise test, taking $p = 0.05$ as a significant cut-off.All calculations were performed using GraphPad Prism© 5 software.

3 RESULTS

3.1 Quantitative analysis ofginsenoside Rg1, ophiopogonin D, ophiopogonin D', schisandrin

The chromatogram obtained forginsenoside Rg1, ophiopogonin D, ophiopogonin D', schisandrin (Fig. 1) showed distinct peaks for each of them. The calibration curves (Fig. 2) were prepared using standard substances and were found to be linear in the concentration range used ($R^2 = 0.99$). The concentration of ginsenoside Rg1, ophiopogonin D, ophiopogonin D' and schisandrin in SMY was found to be 25.63±3.42mg, 11.00±1.17mg, 7.02±0.51mg, and 25.31 ±4.28mg per gram of SMY, respectively.

3.2 Effects of SMY on inflammatory cytokine levels in rats with CHF

In comparison to CON (Fig. 3), the levels of both TNF-α and IL-6 in CHF rats were significantly higher ($p<0.01$).There was a significant difference in the levels of TNF-α and IL-6 ($p<0.05$) in CA treated animals compared to controls.IL-6 levels in MSMY and LSMY were also significantly higher ($p<0.05$).Overall, in comparison with CHFM, TNF-α levels in all of the four drug-administered groups were significantly lower ($p<0.01$).Except for LSYM, IL-6 levels for the three other drug-administered groups were significantly lower ($p<0.01$).Among the four drug administration conditions, no significant difference was observed ($p>0.05$), however as the dose of SYM decreased, the levels of TNF-α and IL-6 showed an increasing trend ($p>0.05$).

3.3 Effects of SMY on COL-Ⅳ

In comparison to CON (Fig. 4), all of the other 5 groups significantly increased COL-Ⅳ expression levels in heart tissues ($p<0.01$).In comparison with CHFM, the COL-Ⅳ expression of HSMY was significantly reduced ($p<0.05$) in heart tissues (Fig. 5).Among CA and the 3 SMY-treated groups, there was no significant difference in COL-Ⅳ levels ($p>0.05$) (Fig. 5).The levels of COL-Ⅳ in CHFM were considerably higher than those in the control group, indicating an overproduction of extracellular matrix in model rats (Fig. 5).

3.4 Effects of SMY on MMPs

MMP-2 and MMP-9, two major indicators of ventricular remodeling were also detected by IHC to further evaluate myocardial fibrosis (Figs. 6 and 7).The expression level of MMP-2 and

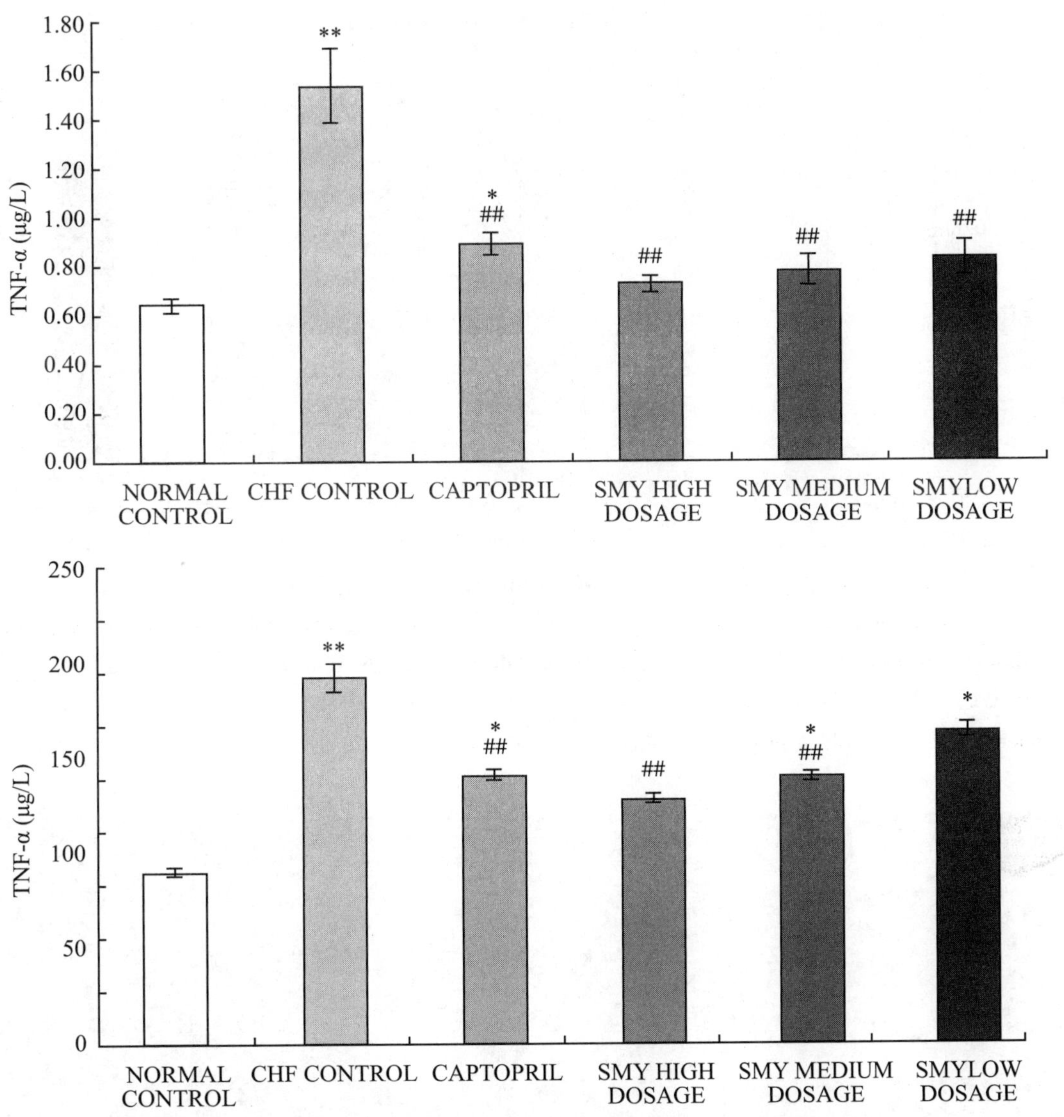

Fig. 3. Effect of SMY on TNF−α and IL−6 among six groups (Mean±SD).

Note: TNF−α and IL−6 of CHFM ($p<0.01$) and CA ($p<0.05$) were significantly increased, IL−6 of MSMY and LSMY was significantly increased ($p<0.05$), respected to CON, * indicates significant differences ($p<0.05$), ** indicates extremely significant differences ($p<0.01$). TNF−α of 4 drug administration groups were extremely significant decreasing ($p<0.01$). Beside of LSMY, IL−6 of the other three drug administration groups were extremely significant decreasing ($p<0.01$), respected to CHFM, # indicates significant differences ($p<0.05$), ## indicates extremely significant differences ($p<0.01$).

MMP−9 was significantly increased in the model CHFM compared to the level in control group. CA and SMY treatment groups reduced their expressions back toward normal level.

3.5 Scanning electron microscopy

In the CON, the myocardium fiber was integrated without any degeneration, necrosis and fibrosis, with a diameter of 8.73−10.15μm (Figs.8 and 9). Drug−treatment in the five other groups

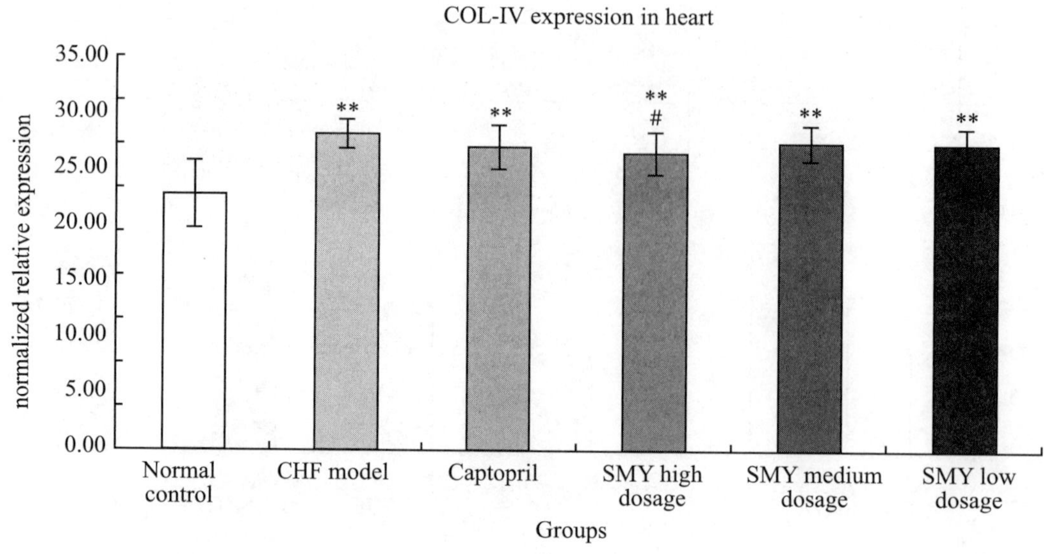

Fig. 4. Expressions of COL-Ⅳ in heart tissues (Mean±SD).

Note: Data were analyzed by one-way ANOVA, with $p<0.05$ indicating statistical significance. All 5 groups significantly increased COL-Ⅳ in heart tissues. CON was the reference group in pairwise comparison. * indicated $p<0.05$, ** indicated $p<0.01$. HSMY significantly decreased COL-Ⅳ in heart tissues, CHFM was the reference group in pairwise comparison. # indicated $p<0.05$, ## indicated $p<0.01$.

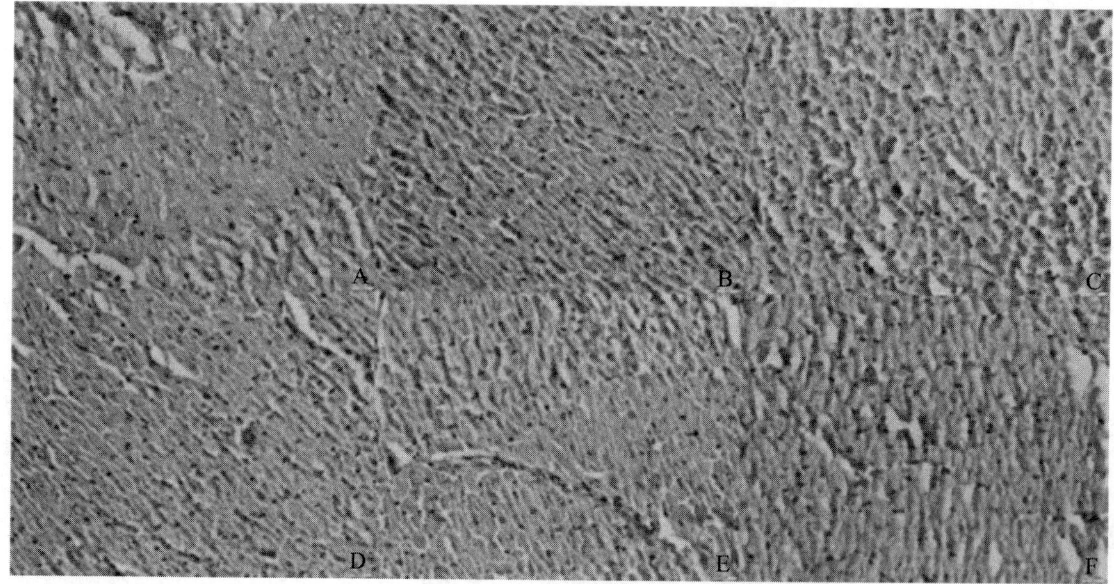

Fig. 5. Results of COL-Ⅳ expression by IHC in heart tissues.

Note: (A) The expression quantity of COL-Ⅳ was below 20%, it could be considered the negative sample (Negative, CON, ×400). (B) CHFM (×400). (C) CA (×400). (D) HSMY (×400). (E) MSMY (×400). (F) LSMY (×400). (For interpretation of the references to color in this figure legend, the reader is referred to the web version of this article.)

presented different degree of degeneration, fibrosis or both. In CHFM, the myocardium fiber was not

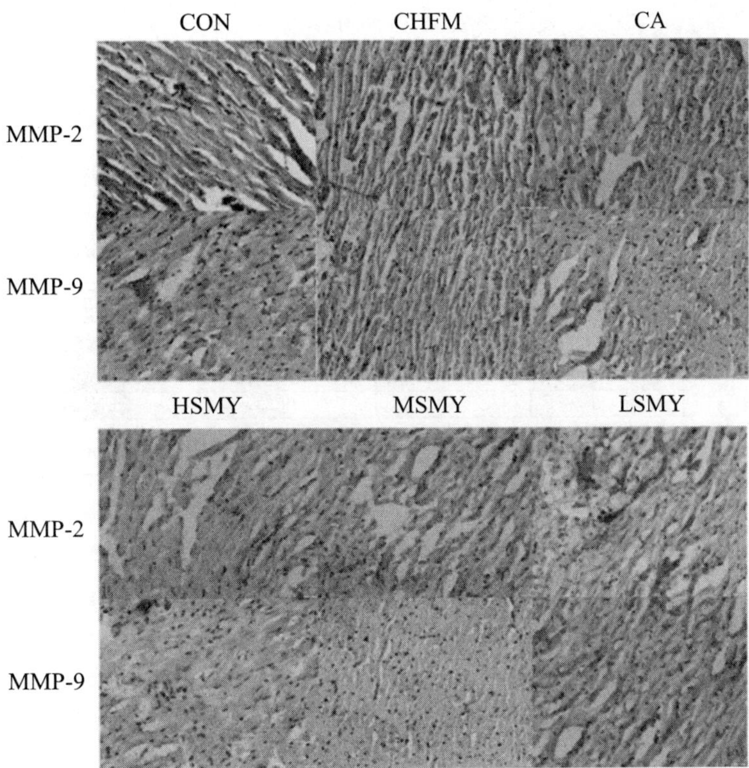

Fig. 6. Results of MMPs expression in heart tissues.

Note: (A) The expression quantity of MMP-2 was below 20%, it could be considered the negative sample (Negative, CON, ×400). (B) CHFM (×400). (C) CA (×400). (D) HSMY (×400). (E) MSMY (×400). (F) LSMY (×400). (For interpretation of the references to color in this figure legend, the reader is referred to the web version of this article.)

integrated, and there was significant degeneration, the myocardial diameter was only 3.52 ± 0.51μm. In CA, fibrosis was not presented, but the myocardium fiber was incomplete with ruptured fragments, and the myocardial diameter was 3.99 ± 0.38μm. In the three SMY-treated groups, HSMY showed an integrated myocardium fiber without any significant degeneration, however slight fibrosis was still observed. Both MSMY and LSMY presented light fibrosis but the completeness of myocardial fiber was higher than in the CHFM without any significantly ruptured fragments. Overall, with a decreasing dosage of SMY, the myocardial diameters showed a decreasing trend.

4 DISCUSSION

Several active components such asschisandrol A, which is one of the key active components of SMY, have been shown to be bioactive in vivo. It has been shown that these natural compounds activate eNOS and expression of Bcl-2, and reduce collagen deposition and expression of Bax and ASK1 in myocardial tissues. They also inhibit caspase-3 activity and mitochondrial permeability in H9c2 cells (Wang et al., 2010; Park et al., 2012; Chen et al., 2013; Chun et al., 2014; Chiu et al., 2008). Schisandrin B has also been shown to prevent doxorubicin-induced cardiac dysfunction by modulation of DNA damage, oxidative stress and inflammation through inhibition of

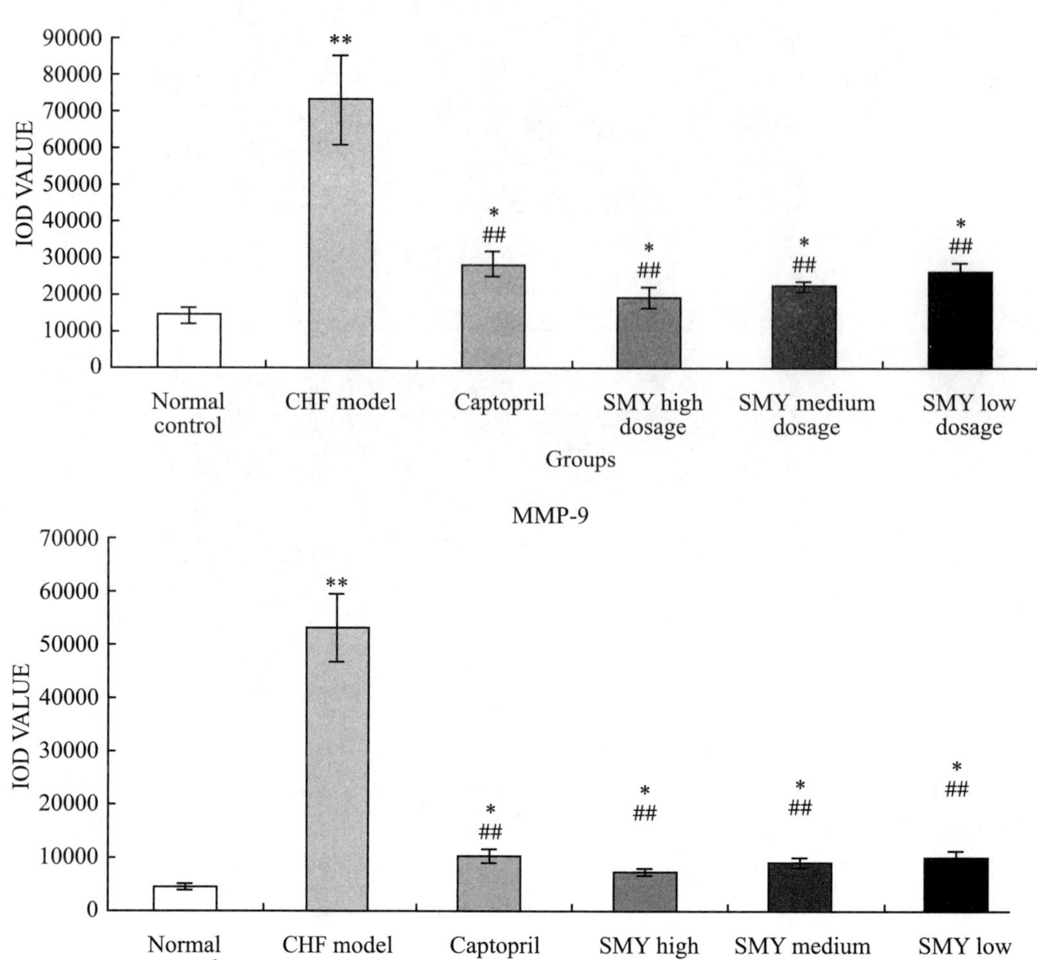

Fig.7. Quantitative analysis of MMPs expression.IOD values represent the Mean±SD.

Note: CON was the reference group in pairwise comparison. * indicated $p < 0.05$, ** indicated $p < 0.01$. CHFM group was the reference group in pairwise comparison.#indicated $p<0.05$, ##indicated $p<0.01$.

MAPK/p53 signaling (Thandavarayan et al., 2015).Eight ginsenoside monomers Rb1, Rg1, Rf, Rh1, Rc, Rb2, Ro, and Rg3 have been reported to act as NF-κB inhibitors, thereby they can downregulate TNF-α, IL-6, IL-8 levels through the inhibition of NF-κB (Xing et al., 2013). The cellular response to IL-6 in the heart has also been well characterized.Cardiac tissues provide revealing examples of how the duration of IL-6 signaling relates to the chronicity of the disease and demonstrates the transition from protective to pathogenic.Cardiac myocytes themselves make IL-6 in response to injury and in addition to an increase in IL-6 signaling, increased IL-6 production is also associated with depressed cardiac function (Yang et al., 2004).Increased IL-6 plays a role in late phase preconditioning that confers cardio protection (Dawn et al., 2004; Smart et al., 2006). However, chronic elevated myocardial production of IL-6 family cytokine, which occurs post-MI

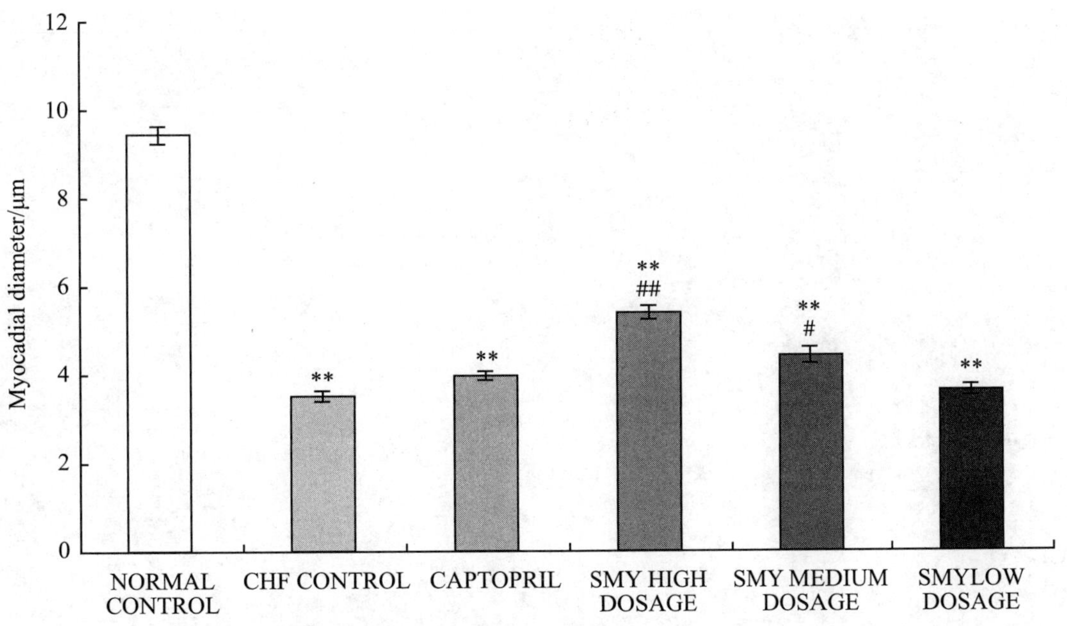

Fig.8. Myocardial diameter (μm) among six groups (Mean±SD).

Note: * indicates significant differences ($p<0.05$) respect to CON, ** indicates extremely significant differences ($p<0.01$) respect to CON, #indicates significant differences ($p<0.05$) respect to CHFM, ##indicates extremely significant differences ($p<0.01$) respect to CHFM.

and in HF, has been associated with worsening of heart outcomes (Terrell et al., 2006; Wollert and Drexler, 2001).SHR-developed cardiac hypertrophy complicated with diastolic heart dysfunction, increased expression of brain natriuretic peptide, downregulation of beta adrenergic receptors and simultaneous up-regulation of IL-6, which indicates active proinflammatory process, at least partly, explain the pathologies during early stage when cardiac hypertrophy associated with diastolic dysfunction occurs (Haugen et al., 2007).According to Chinese Pharmacopoeia, ginsenoside Rg1 was used to control the quantification standards of SMY (Committee of Pharmacopoeia of PR China, 2005).However, SMY contained Ginseng, Ophiopogon, Schisandrae, based on the related literature, ginsenoside Rg1, ophiopogonin D, ophiopogonin D' and schisandrin have various bioactivities on cardiac diseases.For instance, ginsenoside Rg1 showed that protect cardiomyocytes under hypoxic conditions by reducing intracellular Ca^{2+} overload (He et al., 2014).Ophiopogonin D and ophiopogonin D' showed those attenuate doxorubicin induced autophagic cell death by relieving mitochondrial damage in vivo and in vitro (Zhang et al., 2015).Schisandrin showed that enhance glutathione antioxidant response in H9c2 cells (Ko and Chiu, 2005), prevent doxorubicin induced cardiac dysfunction by modulation of DNA damage, oxidative stress and inflammation through inhibition of MAPK/p53 signaling (Thandavarayan et al., 2015), and inhibit NADPH-dependent and CYP450-catalyzed reaction in the myocardial tissues, and enhance glutathione antioxidant response in H9c2 cells (Chen and Ko, 2010).In present study, ginsenoside Rg1, ophiopogonin D, ophiopogonin D', schisandrin in SMY were used to be the quantification standards of SMY, that was not merely quantification standards of SMY, but also a better control of CHF pro-

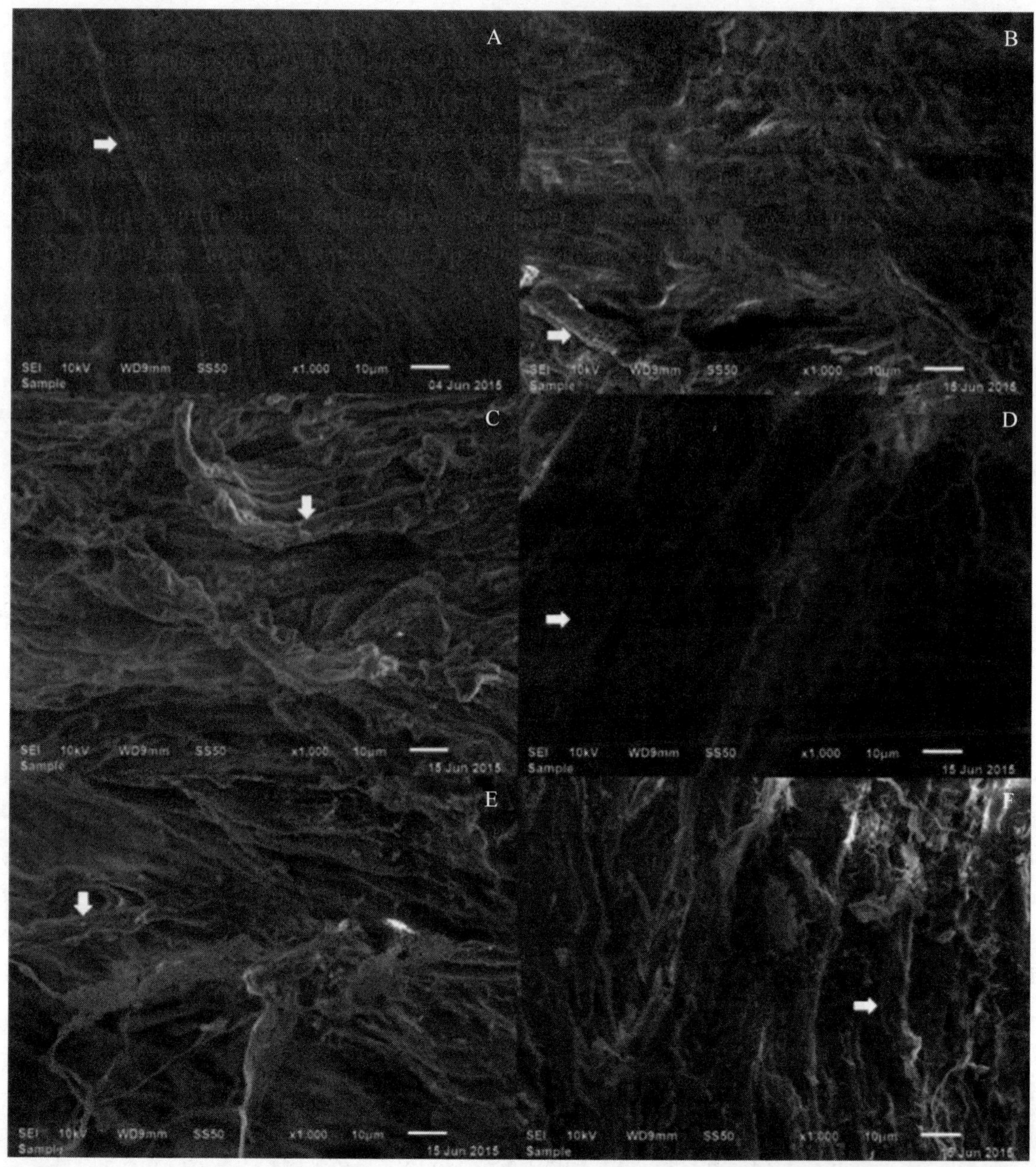

Fig. 9. SEM microphotographs (× 1000) of myocardial fiber without any treatment (CON, A) ; B (CHFM) : Adriamycin-induced was used to establish the rats model with CHF, and without any treatment. C (CA) was treated with Captopril, different dosage of SMY were administrated in rats model with CHF (HSMY, D; MSMY, E and LSMY, F).

gression. Quantification standards of formula would provide the scientific basis for the further research in effects of bioactive components on CHF progression.

In this study, The concentration of ginsenoside Rg1, ophiopogonin D, ophiopogonin D' and schisandrin in SMY was found to be 25. 63±3. 42mg, 11. 00±1. 17mg, 7. 02±0. 51mg, and 25. 31 ±4. 28mg per gram of SMY, respectively. TNF-α of four drug treated groups was significantly lower and IL−6 of CA, HSMY and MSMY dosage group was lower than CHFM.These observations suggest that CA and SMY have protective potential for the treatment of CHF.Thus SMY has similar treatment efficacy as captopril.Among the CA and the 3 SMY groups, no significant difference was observed ($p > 0.05$), however with decreasing dosage of SMY, both IL−6 and TNF−α showed and increasing trend ($p>0.05$).This study suggests that owing to the lack of side effects, SMY can be used as a clinical treatment drug at high doses for the treatment of myocardial injury in patients with chronic heart failure.

Cardiac biomarkers reflecting different aspects of myocardial function, e.g., MR−pro ANP, NT−proBNP−atrial and ventricular wall stress and troponin Ⅰ (TNI) −myocyte injury, are increased upon injury (Gustafsson et al., 2009; Hillege et al., 2000), however it is not certain whether these biomarkers reflect decreased renal clearance or increased cardiac secretion (Bosselmann et al., 2013).Extracellular matrix (ECM) precipitation, with COL−Ⅳ as the major component, is the most characteristic pathological change of CHF (Zhao et al., 2014).Atrophic renal tubular epithelial cells are the main source of COL−Ⅳ expression in kidney (Ogata et al., 2002; Kimura et al., 2005).The results of present study proved that the protein expression and transcription level of COL−Ⅳ are consistent.Although, the increase in COL−Ⅳ expression could play a role in cardiac repair, the overexpression could also affect blood flow, and cause chronic ischemia and anoxic injury (Xie et al., 2009).On the other hand, under conditions of hypoxia and ischemia, epithelial−mesenchymal transdifferentiation would be favored, and lots of COL−Ⅳ would be synthetized by myofibroblasts.The myocardial injury modeling induced by adriamycin, led to muscle lesions, interstitial edema and interstitial fibrosis in the myocardium.IL−6 and TNF−α are important assessment indexes for the myocardial injury, with excessive IL−6 and TNF−α promoting myocardial injury, and myocardial remodeling.Schisandrol A, an active component of SMY, has been reported to activate eNOS, lead to expression of Bcl−2, and reduce collagen deposition in myocardial tissues (Wang et al., 2010; Park et al., 2012).Our study indicated that SMY can reduce the levels of IL−6 and TNF−α ($p<0.05$), and effectively attenuate the pathologic changes of myocardium, including myocardial fibrosis.COL−Ⅳ expression could be used as the main evaluation index for the assessment of degree of fibrosis of the cardiac tissues.COL−Ⅳ expression significantly decreased using high dose of SMY ($p<0.05$).Moreover, with decreasing dosage of SMY, COL−Ⅳ expression showed an increasing trend ($p>0.05$).These results illustrated a positive correlation between inflammatory cytokine (IL−6, TNF−α) levels and myocardial fibrosis index (COL−Ⅳ).

Myocardial fibrosis is a common pathological change in MI−induced heart failure as well as other end−stage cardiovascular diseases (Kong et al., 2014).It can induce detrimental left ventricular remodeling and eventually cardiovascular−related deaths (Eschalier et al., 2013; Nguyen et al., 2014).Myocardial fibrosis is characterized by ECM overproduction of collagen in particular (Wang et al., 2005).MMPs are key enzymes for ECM degradation, which can degrade all the ECM components, except polysaccharides and play an important role in ECM degradation during myocardial interstitial remodeling.The expression of MMPs, ECM and collagen were significantly higher during heart failure, and the changes of MMPs regulate the degradation of collagen and ECM, which causes the cardiac/myocardial remodeling, and play a role in cardiac pathology process (Chen et al., 2004).Therefore, the overexpression of MMPs, ECM and collagen increase the risk of myocardial fibrosis (Huang, 2006).In the heart, MMP-1, MMP-2 and MMP-9 are the mainly expressed isoforms (Ulrich et al., 2004; Visse and Nagase, 2003). MMP-2 and MMP-9 are the limiting enzymes for collagen Ⅰ (COL-Ⅰ) and collagen Ⅲ (COL-Ⅲ) in the heart, that degrade the denatured collagen, gelatin, collagen-Ⅳ (COL-Ⅳ) and collagen Ⅴ (COL-Ⅴ), and regulate the speed of COL-Ⅰ and COL-Ⅲ degradation (Chen et al., 2004; López et al., 2004) Inhibition of MMPs could alleviate MI to some extent by improving the sensitivity of PPAR- [gamma] agonist (Chen et al., 2004).In this study, the results showed that expression of MMPs and collagen were significantly higher in SMY-treated groups as compared to the control group ($p<0.01$), after the establishment of CHF model.This indicates that the changes in levels of MMPs, which are caused by CHF, can be reversed and thus treated using captopril and SMY, which effectively inhibit the expression of MMPs.

5 CONCLUSIONS

In conclusion, our data showed that Chinese herbal preparation SMY decreases the production of cytokine IL-6 and TNF-α to reduce the cardiac or myocardial injury.It also effectively attenuated or suppressed the pathological changes of myocardium on rats with CHF and inhibited the overproduction of MMP-2, MMP-9 and COL-IV, which significantly relieved cardiac pathology process, decreased the risk of myocardial fibrosis and cardiac remodeling. These findings suggest that SMY and captopril have the similar efficacy on adriamycin-induced myocardial injury, in addition, that Chinese herbal preparation SMY holds a vast potential in the treatment of cardiac diseases.

AUTHORS' CONTRIBUTIONS

Z.K.and L.J.designed this study, including search strategy, eligibility criteria.Z.K., W.L.and P.M.selected studies and analyzed the data.P.A.and W.X.checked and repeated the data.Z.K.wrote the manuscript.All authors read and approved the final manuscript.

CONFLICT OF INTEREST

The authors do not have any conflict of interest.

ACKNOWLEDGMENTS

The work of L.J.was supported by the Special Fund for Agro-scientific Research in the Public Interest, "Study and application on some key technologies in 2 biopharmaceuticals manufacture", (Ministry of Agriculture of P.R.China, 201303040).

REFERENCES OMITTED

(发表于《Phytomedicine》, 院选 SCI, IF: 3.526)

iTRAQ-based Proteomic Technology Revealed Protein Perturbations in Intestinal Mucosa from Manganese Exposure in Rat Models

Hui WANG,[iD*1] Shengyi WANG[1], Dongan CUI[1], Shuwei DONG[1], Xin TUO[1], Zhiqi LIU[2], Yongming LIU[*1]

(1. Engineering and Technology Research Center of Traditional Chinese Veterinary Medicine of Gansu Province/Key Lab of New Animal Drug Project of Gansu Province/Key Lab of Veterinary Pharmaceutical Development of Ministry of Agriculture/Lanzhou Institute of Husbandry and Pharmaceutical Sciences of Chinese Academy of Agricultural Sciences, Lanzhou 730050, Gansu, China; 2. Institute of Agro-Products Processing Science and Technology, Chinese Academy of Agricultural Sciences, Beijing 100193, China)

Manganese (Mn) is an essential metal ion as a biological cofactor, but in excess, it is toxic; however, the homeostatic mechanisms of Mn at the cellular level have not been identified. We hypothesized that intracellular Mn can be modulated by proteins that could provide an insight into Mn regulatory mechanisms at the cellular level. Our experiment demonstrated that rats exposed to Mn ($200mg \cdot L^{-1}$ in water for 5 weeks) showed a significant Mn-induced decrease in total distance traveled, but daily body weight gain and feed intake were not significantly changed. The major organs showed a normal orientation and well defined histological structures without any signs of histopathological changes. We performed a qualitative and quantitative analysis of 3012 proteins for modifiers of cellular Mn content in intestinal mucosa cells by isobaric tags for relative and absolute quantitation (iTRAQ). Following stringent validation assays and bioinformatics, a total of 175 intestinal mucosa proteins were found that may regulate Mn levels under biologically relevant Mn exposures. These proteins are related to pancreatic secretion, protein digestion and absorption, fat digestion and absorption, biosynthesis of amino acids, glycerolipid metabolism, Alzheimer's and Parkinson's disease, mineral absorption associated protein, etc. The identified proteins and biological targets may have potential applications in vivo for modulating cellular Mn content, and will help us understand the homeostatic mechanisms of Mn at the cellular level.

* Corresponding authors, E-mail: wanghui01@caas.cn; myslym@sina.com

1 INTRODUCTION

Metal ions have been demonstrated to play important roles in biological systems. These roles include the structural components of biomolecules, as signaling molecules, as catalytic cofactors in hydrolytic reactions and oxidation-reduction, and in the structural rearrangements of electron transfer chemistry and organic molecules.[1] However, at high intracellular concentrations, metal ions are toxic because they can perturb cellular redox potential and produce highly reactive hydroxyl radicals.[2] The study of metal ion homeostasis and their importance for cell viability through protein-coding and regulatory machinery has been extensive.[3,4] A battery of regulatory proteins in all cells mediate metal homeostasis by regulating the expression of genes that encode intracellular chelators, metal transporters or other detoxification enzymes.[5] The homeostasis mechanisms of metal ions are maintained by a concerted effect including metallochaperone proteins, metal transporters, transcription factors and metal responsive signaling pathways.[6]

Manganese (Mn) is an essential metal ion in some important physiological processes such as immune function, development, energy metabolism, reproduction, and antioxidant defenses.[7] Incorporation of Mn into metalloproteins has beneficial effects, as these metalloproteins are involved with transferases, oxidoreductases, hydrolases, etc.[8] But in excess, Mn brings neurotoxicity, and as a result, the body will experience a variety of psychiatric and motor disturbances due to basal ganglia dysfunction.[9,10] In our previous publication, we found that Mn-induced alterations are associated with metabolic responses to manganese poisoning, or manganism. Major alterations were observed in purine metabolism, amino acid metabolism and fatty acid metabolism.[11] However, little is known about the complex regulatory biology involving the intracellular handling of Mn.

The critical function of the gastrointestinal tract is absorbing nutrients and water.[12] Gastrointestinal changes could be detrimental to the performance, health and welfare of mammals. Mn uptake is regulated by the gastrointestinal tract by responding to dietary Mn levels; however, the molecular mechanisms are not fully understood. The action of the transporters involved in regulating uptake and efflux of Mn into cells is not well established and signaling pathways responsive to physiological Mn concentrations have not been described.[13]

Proteomics is a mechanism-based high throughput screening approach used to identify proteins expressed that may lead to pathology or disease outcomes in tissue types or certain cells at a given moment.[14] Isobaric tags for relative and absolute quantitation (iTRAQ) is an isotope labeling technique, characterized by high throughput, high sensitivity and high accuracy-based proteomic technology.[15] iTRAQ coupled with liquid chromatography-tandem mass spectrometry (LC-MS) can produce a full set of information through simultaneously isolating and identifying hundreds of proteins.[16,17] Proteomics technologies have the potential ability to solve ecotoxicological issues and elucidate the underlying molecular mechanisms through screening critical proteins based on the protein expression signature (PES).[18,19] The PES is defined as the expression signature or expression pattern of a set of proteins as biomarkers.[20] Mn-specific signaling pathways or binding proteins responsive to physiological Mn concentrations in vivo have not been systematically described. To identify the differentially expressed proteins that modify cellular Mn content, the proteomics of iTRAQ was performed to identify the modifiers of intracellular and cellular regulation of Mn homeostasis and trafficking. Understanding the targets of PES could provide an insight into the regulatory mechanisms of Mn at the cellular level.

2 MATERIALS AND METHODS

2.1 Ethics statement

All experiments were carried out with respect and humanely for alleviation of suffering following the protocols approved by the Institutional Animal Care and Use Committee of Lanzhou Institute of Husbandry and Pharmaceutics Sciences of the Chinese Academy of Agricultural Sciences (animal use permit: SCXK20008-0013). We received approval for the specific procedures used in this study. Animals were maintained and experiments were conducted following the recommendations in the Guide for the Use and Care of Laboratory Animals. The weight, health status and body temperature of the rats were monitored on a daily basis. They were anesthetized using sodium pentobarbital ($50mg \cdot kg^{-1}$) by intraperitoneal injection at the end point of the study, and then mounted in a stereotaxic device with heads held in a horizontal plane. All efforts were made to minimize suffering using analgesia, anesthesia, and post-injury care and monitoring. We confirm that the rats did not sustain any injury in our study.

2.2 Animals and experimental design

The experiment was performed on male Sprague-Dawley rats. Sixteen rats (148±2g) were randomly allocated to either the control or Mn-treated group (eight rats per treatment group) and were housed in plastic cages (5 days adaptation and 5 weeks of feeding with experimental diets). All of the rats were housed under pathogen-free conditions on a 12h light-dark cycle at 20±3℃ and 40-60% relative humidity. During the first 5 days, all groups were fed the basal diet (AIN-93G). This was prepared according to guidelines described elsewhere (35, 10, and $6mg \cdot kg^{-1}$ of Fe, Mn, and Cu, respectively).[21] The rats in the control group received the AIN-93G diet and non-ionic water, and those in the Mn-treated group were fed the AIN-93G diet and Mn-rich water ($200mg \cdot L^{-1}$) for 5 weeks. According to WHO guidelines, the risk concentration for Mn in drinking water is $0.5mg \cdot L^{-1}$.[22] All animals had free access to their diets and water. The body weight (BW) of all rats in each group was measured weekly.

2.3 Behavior analysis

An open-field measurement system was used to measure rats' activity at the end of the experiment; this system can automatically record all activities of the animal. Rats were transported (within their home cage) to acclimatize to the testing room for 1 hour before testing started. Control and Mn-treated rats were put in the separated testing open field chamber (60cm±60cm±60cm), and 6min locomotor activity was automatically recorded by video surveillance. Cameras were mounted on tripods and placed parallel to the shoebox cages. The experiments were repeated three times in an hour period. End points for analysis were the total distance travelled.

2.4 Tissue collection

At the end point of the study, rats were anesthetized using sodium pentobarbital. The small intestines were obtained immediately and thoroughly rinsed using normal saline then cut into 1cm length segments and fixed in 10% neutral formalin used for morphological analysis. Mucosa was obtained from the remaining segment as described previously and snap frozen using liquid nitrogen then stored at−80℃ until biochemical and molecular analyses.[23]

The intestines, lung, spleen, kidney and testicles were harvested and rinsed using normal saline, then immersed immediately in 10% neutral buffered formalin for 24h. A gradient of alcohols was used for dehydration, and then the tissues were embedded in paraffin using conventional methods.[11,24] The tissues were sectioned into 4mm thick slices; slices were collected and mounted on microscope slides. For observation of morphological changes, hematoxylin−eosin staining was performed and a light microscope (Olympus x51 model, Tokyo, Japan) equipped with a camera (Olympus 20, Tokyo, Japan) was used.

2.5 Intestinal mucosa protein extraction

Total intestinal mucosa proteins were extracted according to Wang et al. 23 with slight modifications. In brief, small samples of mucosal scrapings from the intestine were crushed using a mortar with liquid nitrogen. Approximately 100mg of each powdered sample was transferred to sterile tubes with 600μL of lysis buffer (SDT; containing 4% SDS (w/v), 100mM Tris−HCl pH 7.6, 0.1M DTT).

Each mixture was thensonicated in a boiling water bath using an ultrasonicator (Sonics & Materials, Newtown, CT, USA) with 100W power output. They were sonicated 10 times with 10s on and 15s off cycles. After that, each lysed cell suspension was sonicated in a boiling water bath for 15min to solubilize the proteins, then the suspension was centrifuged at 13 400 rpm for 25min. The supernatant protein was collected and the protein concentration measured using the BCA (bicinchoninic acid) protein assay. The protein concentration was $23.64 \pm 4.07 mg \cdot mL^{-1}$.

2.6 Protein digestion and peptidequantitation

Firstly, DTT was added to 40mL of protein from each sample (until the DTT concentration was 100mM) in a boiling water bath for 5min, then it was cooled to room temperature. Nextly, 200μL of UA buffer (150mM Tris−HCl pH 8.0, 8M urea) was used to remove the DTT and detergent components by ultrafiltration (Microcon 10 kDa centrifugal filter units), followed by centrifugation at 14 000g for 15min. Subsequently, 100μL of 0.05M iodoacetamide (IAA) in UA buffer was added to the concentrate followed by centrifugation at 14 000g for 10min. Then 100μL UA buffer was added to the resulting concentrate with centrifugation at 14 000g for 10min and repeating this step a further two times. Afterwards, 100μL of Dissolution buffer was added, followed centrifugation at 14 000g for 10min, and this step was repeated two times. Finally, the protein suspensions were digested overnight at 37℃ with 40μL trypsin buffer (3mg trypsin in 40μL dissolution buffer). The peptides were quantified using UV light at 280nm.[25]

2.7 iTRAQ labeling

TheiTRAQ Reagent 8−plex Multiplex Kit (Applied Biosystems, Foster City, CA, USA) was used to label each sample according to the manufacturer's instructions. The controls of normal small intestine tissue samples were labeled as (Sample 1) −113, (Sample 2) −114, (Sample 3) −

115 and (Sample 4) -116, while Mn-treated samples were labeled as (Sample 5) -117, (Sample 6) -118, (Sample 7) -119 and (Sample 8) -121.

2.8 Peptide fractionation and LC-MS/MS analysis

Strongcation exchange (SCX) chromatography was used to fractionate iTRAQ labeled mixed peptides on a 20AD HPLC system (Shimadzu; Kyoto, Japan) using a polysulfoethyl column (5mm, 200 Å, 2.1 × 100mm, The Nest Group, MA, USA).

Then the samples were analyzed by QExactive mass spectrometer (Thermo Finnigan).The samples were added to a $^{C}18$-reversed phase column in buffer A, then separated with buffer B.Peptides were separated using a flow rate of 300nL · min^{-1}.The gradient profile was: 0-50% B in 0min to 220min, 50%-100% B in 220min to 228min and 100% B in 228 to 240min.

Tandem mass spectrometry was used to identify the different components on the top 15 ions.Full MS scans were applied at 70k resolution with a 3e6 AGC target and a scan range of 300-1 800m/z. The 10 fragment patterns were collected after each full scan. MS2 scans were monitored with a 2.0m/z isolation window, 17.5k resolution with a 1e5 AGC target, 60ms maximum injection time, and a normalized collisional energy (NCE) of 30 eV.

2.9 Data analysis

MS/MS spectra were embedded in the software of Proteome Discoverer 1.4 (thermo) against the uniprot_ rat_ 33675_ 20141101.fasta (33 675 sequences, downloaded on 11 November 2014) and decoy databases and analyzed using Mascot 2.2 (Matrix Science).The following options were used for protein identification: peptide mass tolerance = 20 ppm, enzyme = trypsin, missed cleavage = 2, and fragment MS/MS tolerance = 0.1 Da; carbamidomethyl (C), iTRAQ8plex (K) and iTRAQ8plex (N-term) were set asfixed modifications; oxidation (M), iTRAQ8plex (Y), and false discovery rate (FDR) $\leq$ 0.01 were set as variable modifications.[26] The peak intensity was extracted using Proteome Discoverer 1.4 from each fragmentation spectrum.

The peptides identified that expressed changes in comparison with the control were calculated based on the iTRAQ reporter ion intensities.Twofold changed cut-offs were selected as significantly changed to categorize proteins based on statistical analysis and relative quantification, that is, the proteins were considered to be up-regulated if iTRAQ ratios >1.2, whereas they were considered to be down-regulated if iTRAQ ratios<0.83.

2.10 Bioinformatics

Proteins were further analyzed for biological and functional relevance based on statistically significant differences.These proteins were classified using their gene function and the freely available Gene Ontology (GO) by biological pathways.NCBI BLAST+ client software was used to identify homologous proteins based on the functional annotation to transfer to the targeted proteins.The top 10 blast hits for each query protein with an E-value less than 1×10^{-3} were retrieved and loaded into Blast2GO (Version 2.7.2) for GO pathway annotation and mapping using the Kyoto Encyclopedia of Genes and Genomes (KEGG) Automatic Annotation Server (http://www.genome.jp/tools/kaas/).In addition, a functional and differentially expressed protein association network was built using Search Tool of the Retrieval of Interacting Genes/Proteins (STRING; http://string.

embl.de/).

2.11 Statistical analysis

Statistical analyses were conducted using SPSS version 19.0 software (SPSS, Inc., Chicago, IL, USA) and a difference at $P<0.05$ was considered to be statistically significant. Data were expressed as mean±S.D.

3 RESULTS AND DISCUSSION

3.1 Mn exposure significantly decreased locomotion

The openfield test is a widely used procedure for examining the behavioral effects of drugs.[27] Because altered locomotion is associated with Mn exposure, we conducted behavioral analysis during the fifth week of Mn exposure using video surveillance to monitor Mn-induced changes in activity. Depicting behaviors revealed a significant Mn-induced decrease in total activity in the Mn-exposed group ($P<0.01$), measured as total distance traveled at the end of the experiment (Fig. 1). Excess Mn in the brain is neurotoxic and known as manganism.[28] Previous studies have reported that excess Mn will produce a variety of psychiatric and motor disturbances including decreased motor skills, memory deficits and psychotic behavior resembling Parkinson's disease due to basal ganglia dysfunction.[9,10,28] The results indicated that a model of Mn poisoning was established and the behavioral patterns displayed significant changes in the body produced by motor disturbances due to excess Mn which could lead to the accumulation of neurofibrillary tangles (striatum, globus pallidus and substantia nigra),[29] a hallmark of Parkinson's disease.

3.2 Effect of manganese exposures on body weight of rats

Body weights (BW) of rats during each week of the experiment are shown in Fig. 2. The BW increased gradually, but the doses and period of Mn exposures had no significant effect on the BW of Mn-treated rats compared with the control group in this study ($P>0.05$). Our experiment demonstrated that longer Mn exposure time influenced the behavioral pattern of animals, but feed intake and daily BW gain were not significantly different.

3.3 Effect of manganese exposure on histopathology

The histopathology of the major organs was studied. No histopathological changes were observed in the intestine, lung, spleen, kidney and testicle of the rats after exposure to Mn at $200mg \cdot L^{-1}$ for 5 weeks (Fig. 3). The small intestine of mammalians provides an extensive contact surface for ingested compounds. Absorption and metabolism of xenobiotics, nutrients and drugs are enabled by the unique morphological and functional properties of the small intestine.[30] Horning et al. reported that the nervous system is the primary target for excessive Mn, and that Mn exposure has not been linked with damage to the kidney, liver, skin, blood, or stomach.[7] But in our previous study, we found extensive necrosis and dissolved nuclei in hepatic tissue in the Mn-treated group of rats.[11] The reason for this phenomenon is because the liver plays a major role in Mn excretion, and impairment of liver function results in excessive retention of Mn.[31] Our results indicated that Mn exposure did not damage every organ in the body.

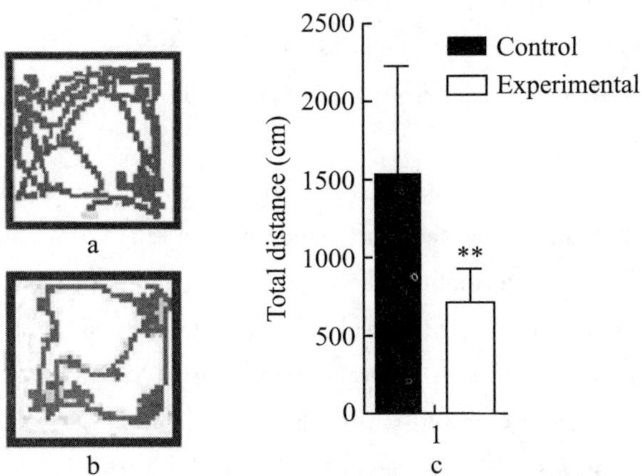

Fig. 1 Decreasedlocomotor behavior in Mn-treated rats.

Note: An openfield test was performed to test the locomotor activity of the Mn-treated rats. Motion trails of (a) controls and (b) Mn-treated rats, and distance traveled (c) were recorded in the open-field test. It showed a significant decrease in locomotor behavior in the Mn-treated rats.

** $P<0.01$.

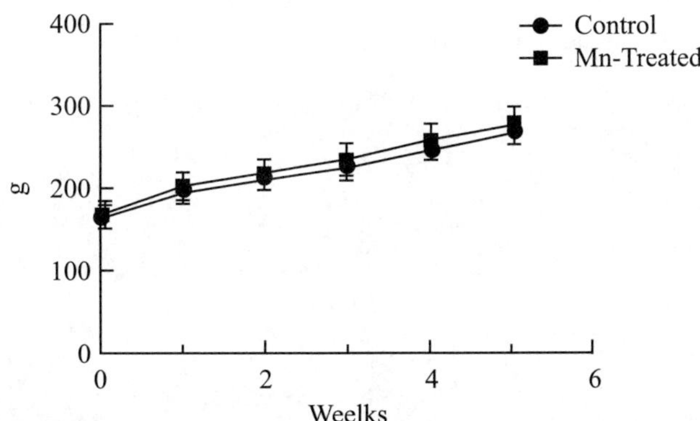

Fig. 2 Body weight of rats during the experimental period.

Note: No statistically significant difference was observed in the body weight of rats between Mn-treated and control groups ($P>0.05$).

3.4 Proteomic alterations of small intestine in response to Mn-treated

The homeostasis of transition metal ions in all cells are mediated by a list of regulatory proteins through modulating the expression of genes that encode intracellular chelators, metal transporters and/or other detoxification enzymes.[5] Small intestinal enterocytes contain a large variety of regulatory proteins, plasma membrane transporters, intracellular binding proteins, metabolic enzymes, and others within the basolateral membrane.[30,32] The enterocytes have a rapid turnover, which allows quick cellular adaptation and quick replacement of cells in response to diets or exposure to xenobiotics.[30] The advantages of proteomics technology are high throughput, high sensitivity, rapid testing of small molecules, and discovery of new proteins that are related to disease.[33] iTRAQ along with LC-MS is a robust protein discovery technique, allowing the determination of which proteins are affected in intestinal mucosa cells under biologically relevant Mn exposures. Our application of iTRAQ-based proteomic technology targeted specific enzymatic or protein processes with rapid testing of small molecules. This approach would be expected to lead to many potential protein or biological targets.

In the current study, rats were treated withMn for 5 weeks, and at the end of the experiment, the small intestine mucosa were collected and used for protein extraction, protein digestion and iTRAQ labeling to provide more information about proteomes on exposure to Mn. A total of 3012 proteins in intestinal mucosa cells were quantitatively and qualitatively analysed by iTRAQ in this study. The statistically significant proteins were detected using the ANOVA test ($P<0.05$), and if the iTRAQ ratio was less than 1.2, the proteins were discarded. A total of 175 intestinal mucosa proteins passed the criteria, and may modulate intracellular Mn levels based on their statistically significant abundance ($P<0.05$) (Table 1). Table 1 shows that the expression of 108 different proteins in rat small intestinal mucosa were down-regulated, whereas 67 were up-regulated. The 175 proteins were assigned to seven functional groups, i.e. metabolism (amino acid, lipid and sugar metabolism), binding proteins, disease (Alzheimer's disease, Parkinson's disease, amyotrophic lateral sclerosis, etc.), cellular component and process, oxidative stress, signaling pathway, and enzymatic activity based on Gene Ontology (GO) classification (Table 1). This small group of proteins will help us understand the homeostatic processes mediating normal cellular Mn content.

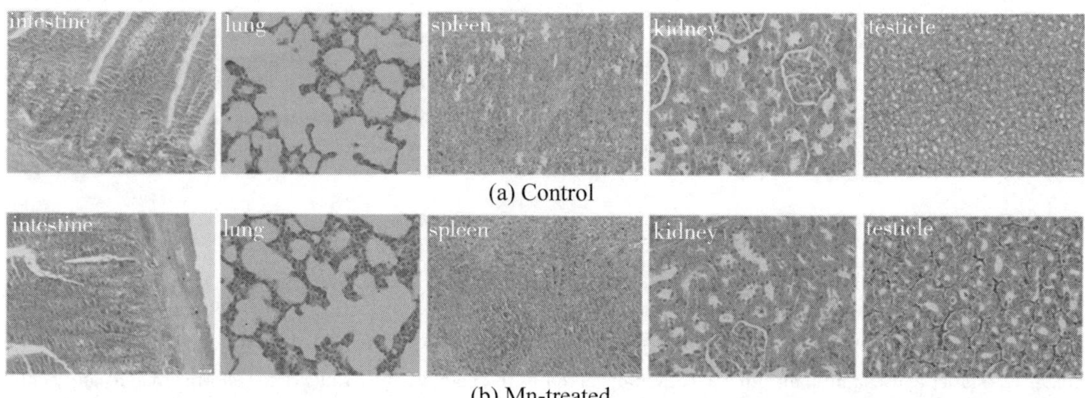

Fig. 3 Hematoxylin and eosin staining of the tissue sections collected from the different groups of rats after the experiment. The major organs showed normal orientation and well-defined histological structures without any signs of histopathological changes.

Among these proteins, fatty acid-binding protein (P02692) was down-regulated after Mn exposure. Fatty acid-binding protein is an important cytosolic protein in small intestine epithelial cells, and participates in the uptake, intracellular metabolism and transport of long chain fatty acids.[34] The down-regulation of fatty acid-binding protein in our study suggested that Mn exposure might have interrupted the lipid metabolism and fatty acid transport of the intestine. Apart from this protein, chymotrypsin (P00774, F1MA56, P00773 and G3V8J3) and trypsin (G3V7Q8, P00762 and P00763), two important proteolytic enzymes involved in the digestive systems of mammals,[35] were significantly down-regulated after Mn exposure. Hence, we speculated that a high concentration of

Mn ion might destroy the absorptive functions of the intestinal epithelium through inhibiting the activities of digestive enzymes.

The oxidation states of Mn^{2+} and Mn^{3+} are the two common species in the human body and can cycle through redox.[36] The Mn redox cycle provides a "double-sword" effect on nutritional metabolism and toxic effects on biological function. Mn serves as a cofactor for Mn superoxide dismutase (MnSOD) that catalyzes superoxide (O_2^-.) to hydrogen peroxide (H_2O_2) and thereby detoxifies free radicals in the mitochondria to prevent oxidative stress through the Mn^{2+}/Mn^{3+} cycle.[28] But the Mn^{2+}/Mn^{3+} cycle can also accelerate dopamine auto-oxidation to form cytotoxic quinones and produce reactive oxygen species (ROS).[37] Oxidative stress and dysregulation of neurological signaling seem to be the mechanisms involved in the neurodegeneration caused by Mn, and they have been found as a consequence of Mn toxicity in the brain.[38-40] Our proteomic analysis found that 10 proteins related to oxidative stress were down-regulated significantly in Mn-treated rats. Peroxiredoxin (Q9Z0V5) is a highly expressed antioxidant enzyme that catalyzes and scavenges hydrogen peroxide and other ROS.[35] The present study also demonstrated that Mn exposure decreases the activities of 17β-hydroxysteroid dehydrogenase (Q5M875). Interactions between the antioxidant and steroidogenic enzyme systems are physiologically relevant.[41] The data on antioxidant status in Mn-treated rats obtained in the present study are in agreement with the observed changes in the activities of steroidogenic enzymes indicating an imbalance in pro-oxidant and antioxidant systems, leading to oxidative stress in manganese-exposed rats. Heat shock proteins are highly regulated proteins that are involved in normal cellular activity, but when the cell is exposed to heat or excess ROS production, they will be up-regulated.[35,42] Our results found that the level of heat shock protein (G3V913) increased after Mn exposure. This result indicated that Mn exposure might induce excess ROS production in the intestine, which enhances the expression of heat shock protein and depresses the expression of antioxidative enzymes, and subsequently destroys the antioxidant defense barrier of the intestines.

The altered expression of components of protein complexes that are involved in RNA processing and protein translation was also observed in our study. Although changed protein concentrations cannot be directly correlated with changed functions, the resulting changes strongly indicate altered digestion, absorption, and metabolism as well as oxidative stress within the small intestine.

Table 1 The list of differentially expressed proteins in intestinal mucosa cells in rats induced byMn exposure by iTRAQ identification.

Accession	Description	Coverage	Unique peptides	MW [kDa]	Calc.pI	Fold change	P-Value
Metabolism							
P00774	Chymotrypsin-like elastase family member 2A	21. 03	5	28. 9	8. 60	0. 421	0. 00
D3ZAM3	Carboxypeptidase B	43. 37	13	47. 5	5. 58	0. 475	0. 00
G3V976	Carboxypeptidase A2	35. 49	12	46. 9	5. 33	0. 480	0. 00
G3V7G2	Bile salt-activated lipase	24. 02	11	67. 1	5. 57	0. 481	0. 00

(continued)

Accession	Description	Coverage	Unique peptides	MW [kDa]	Calc.pI	Fold change	P-Value
G3V7Q8	Cationictrypsinogen	35. 22	7	26. 3	7. 49	0. 558	0. 00
P00762	Anionic trypsin-1	21. 14	3	25. 9	4. 89	0. 599	0. 00
P00731	Carboxypeptidase A1	31. 74	10	47. 2	5. 76	0. 621	0. 00
P00763	Anionic trypsin-2	10. 16	2	26. 2	4. 97	0. 629	0. 00
G3V844	Alpha-amylase	57. 48	17	57. 1	7. 80	0. 372	0. 00
Q5I0L0	Alpha-amylase	13. 11	2	57. 5	6. 42	0. 362	0. 00
G3V8A7	Pancreatic lipase	36. 56	1	51. 4	6. 73	0. 486	0. 00
P27657	Pancreatictriacylglycerol lipase	36. 56	1	51. 4	6. 79	0. 492	0. 00
G3V8J3	Chymotrypsin-like	23. 48	5	28. 1	8. 19	0. 678	0. 00
D3ZFG3	Elastase 3B	29. 74	5	28. 9	5. 53	0. 719	0. 00
D3ZQV0	Protein LOC100365995	8. 13	1	26. 2	5. 33	0. 806	0. 01
B1WC02	CTPsynthase	2. 71	1	66. 6	6. 58	1. 296	0. 00
B1WBV8	Pld4 protein	3. 18	1	55. 9	6. 96	1. 326	0. 00
G3V8A9	Inactive pancreatic lipase-related protein 1	30. 87	10	52. 3	6. 11	0. 620	0. 00
P50442	Glycine amidinotransferase	18. 68	6	48. 2	7. 49	0. 573	0. 00
P54318	Pancreatic lipase-related protein 2	6. 41	2	52. 5	6. 34	0. 543	0. 00
P18757	Cystathionine gamma-lyase	20. 10	7	43. 6	7. 96	0. 678	0. 00
P49088	Asparagine synthetase	6. 42	3	64. 2	6. 46	0. 750	0. 00
P02692	Fatty acid-binding protein	73. 23	8	14. 3	8. 07	0. 798	0. 01
D4A746	GDP-mannosepyrophosphorylase B	5. 56	2	39. 9	6. 74	0. 819	0. 02
P43427	Solute carrier family 2	2. 19	1	55. 5	6. 60	0. 824	0. 02
Q6P784	Branched-chain-amino-acidamin-otransferase	17. 27	5	43. 6	8. 31	0. 820	0. 02
F1LNA9	Colipase	34. 82	3	12. 2	7. 80	0. 572	0. 00
E9PSV5	Phosphoserine aminotransferase	5. 68	2	40. 5	7. 97	0. 761	0. 00
D3ZXI0	Pyrroline-5-carboxylatereductase	4. 21	1	32. 2	6. 84	0. 769	0. 00
Q562C3	Glycine C-acetyltransferase	28. 85	7	45. 2	7. 65	0. 692	0. 00
G3V960	Guanidinoacetate N-methyltrans-ferase	2. 97	1	26. 4	6. 27	0. 726	0. 00
Q7M0C4	Cytochrome P450 2B4	9. 54	3	41. 4	7. 09	0. 736	0. 00
F1LSR9	Alcoholdehydrogenase 1	24. 00	8	39. 7	8. 56	0. 772	0. 00

(continued)

Accession	Description	Coverage	Unique peptides	MW [kDa]	Calc.pI	Fold change	P-Value
Q3B7D0	Oxygen - dependentcoproporphyrinogen-Ⅲ oxidase	4.97	2	49.2	8.53	0.782	0.00
P04797	Glyceraldehyde - 3 - phosphatedehydrogenase	64.56	1	35.8	8.03	0.787	0.00
P24329	Thiosulfate sulfurtransferase	10.44	2	33.4	7.85	0.790	0.00
Q5BJY6	Putative N-acetylglucosamine-6-phosphatedeacetylase	3.42	1	43.5	6.39	0.799	0.01
P12336	Solute carrier family 2	6.32	4	57.0	6.54	0.806	0.01
P52873	Pyruvate carboxylase	3.23	3	129.7	6.81	0.812	0.01
Q91XT9	Neutralceramidase	1.97	1	83.4	7.01	0.817	0.01
P51538	Cytochrome P450 3A9	15.51	6	57.8	8.82	0.825	0.02
Q5HZE4	Methylthioribose - 1 - phosphateisomerase	5.42	2	39.6	6.05	0.828	0.02
D4A7D7	Hexose - 6 - phosphatedehydrogenase	3.76	2	89.8	7.23	1.221	0.01
P15337	Cyclic AMP-responsive element-binding protein 1	4.11	1	36.6	5.57	1.275	0.00
P29975	Aquaporin-1	7.43	1	28.8	7.83	1.230	0.01
B2GV54	Neutral cholesterol esterhydrolase 1	6.13	1	45.8	7.05	1.241	0.00
M0R451	Glyceraldehyde - 3 - phosphatedehydrogenase	34.34	1	35.7	8.19	1.208	0.01
Q921A4	Cytoglobin	5.79	1	21.5	6.80	1.212	0.01
Binding proteins							
Q4KLG2	GTPase IMAP family member 8	2.18	1	77.0	8.19	0.565	0.00
E9PU29	Protein Rnf31	2.35	1	119.2	6.62	0.600	0.00
P02634	Protein S100-G	48.10	3	9.0	5.01	0.612	0.00
A0A068F1Y2	Beta-actin	53.23	1	20.4	5.91	0.652	0.00
G3V8I8	Alkalinephosphatase	22.67	7	59.4	6.28	0.692	0.00
D3ZXR2	Protein Wdr52	0.33	1	209.4	5.44	0.694	0.00
Q2XTB0	Metallothionein 1	44.44	1	1.7	7.62	0.749	0.00
D4AA47	ATPase	1.03	1	144.7	7.15	0.766	0.00
P60905	DnaJ homolog subfamily C member 5	12.63	1	22.1	5.07	1.204	0.01
Q63862	Myosin-11	57.72	1	152.4	6.18	1.205	0.01

(continued)

Accession	Description	Coverage	Unique peptides	MW [kDa]	Calc.pI	Fold change	P-Value
G3V9F3	Myosinphosphatase Rho-interacting protein	1.65	1	116.9	6.21	1.205	0.01
Q9EPF2	Cell surface glycoprotein MUC18	10.19	6	71.3	6.09	1.210	0.01
G3V9E3	Caldesmon 1	39.17	20	60.6	6.58	1.210	0.01
Q99N97	H-Caldesmon	27.75	6	26.6	4.98	1.212	0.01
P47875	Cysteine and glycine-rich protein 1	67.36	11	20.6	8.57	1.213	0.01
F1M265	Palladin	22.63	12	79.0	7.17	1.220	0.01
Q62896	BET1 homolog	15.25	1	13.2	9.07	0.798	0.01
F1LP76	Elongator complex protein 1	0.83	1	149.1	6.34	0.801	0.01
Q8VIL3	ZW10interactor	9.02	1	30.1	6.10	0.802	0.01
D4A929	WD repeat-containing protein 81	1.03	1	212.1	5.81	0.799	0.01
Q9QYF3	Unconventional myosin-Va	0.66	1	211.6	8.69	0.805	0.01
G3V8M5	Serine/threonine-protein phosphatase	8.79	1	35.1	5.06	0.808	0.01
F1MAD9	Structural maintenance of chromosomes protein	1.01	1	146.7	6.96	1.232	0.01
A2NB82	Productively rearranged V-lambda-2	9.91	1	12.2	6.27	1.233	0.01
Q62951	Dihydropyrimidinase-related protein 4	2.84	1	61.0	6.77	1.237	0.00
D3ZHB7	Protein Ube3c	1.57	1	124.3	6.64	1.249	0.00
Q5U2Z5	Cap-specific mRNA-methyltransferase 1	1.79	1	95.6	7.02	1.258	0.00
B2GUV0	Cyth1 protein	5.29	1	46.2	5.63	1.260	0.00
P0C0R5	Phosphoinositide 3-kinase regulatory subunit 4	1.18	1	152.4	7.18	1.269	0.00
F1LRJ0	Protein Kcmf1	4.52	1	41.1	5.66	1.282	0.00
Q5U214	ProteinSnrpa	2.85	1	31.2	9.83	1.337	0.00
P68136	Actin	71.88	2	42.0	5.39	1.357	0.00
B2RYP0	ProteinRhoc	38.34	1	22.0	6.58	1.598	0.00
P62747	Rho-related GTP-binding protein RhoB	20.92	1	22.1	5.24	1.666	0.00
Q6IFV3	Keratin	11.19	1	48.8	4.86	2.109	0.00
P00684	Ribonuclease pancreatic beta-type	15.13	2	16.8	8.29	0.589	0.00
F1M9X2	Glycoprotein GP2	3.58	2	58.8	5.07	0.667	0.00

(continued)

Accession	Description	Coverage	Unique peptides	MW [kDa]	Calc.pI	Fold change	P-Value
P15865	Histone H1. 4	49. 32	5	22. 0	11. 11	1. 212	0. 01
A8QIC3	DnaJ (Hsp40) homolog	11. 97	2	30. 0	7. 33	1. 212	0. 01
F1LSL1	Transcription factorPur-beta	8. 52	1	33. 5	5. 33	1. 356	0. 00
M0R6T4	Protein Mms19l	2. 23	1	112. 9	6. 24	1. 784	0. 00
M0R5F9	Protein LOC100910864	6. 38	1	33. 7	9. 82	1. 432	0. 00
F1M1U0	Protein Mxra7	9. 59	1	15. 9	4. 20	1. 321	0. 00
D3ZA93	Protein Acot13	8. 57	1	15. 3	8. 84	0. 806	0. 01
D4A9L9	Protein Erp27	8. 46	2	30. 7	5. 03	0. 555	0. 00
D4A970	Protein Ovca2	6. 61	1	24. 3	5. 99	0. 724	0. 00
D3ZHM7	Protein Dnttip2	1. 71	1	84. 7	5. 81	0. 755	0. 00
D4A8G0	Protein Lsm12	4. 10	1	21. 7	7. 74	0. 826	0. 02
B2RYE6	Uncharacterized protein	13. 51	1	20. 7	9. 52	1. 256	0. 00
Q38PG1	AHNAK 1	17. 13	1	58. 7	5. 45	1. 320	0. 00
P63055	Purkinje cell protein 4	41. 94	2	6. 8	6. 71	1. 281	0. 00
Q499T8	Intelectin 1	4. 47	1	34. 5	5. 78	1. 582	0. 00
D4A0X4	Elongation factor 1-alpha	27. 37	1	48. 7	8. 72	0. 697	0. 00
M0RE01	Uncharacterized protein	17. 65	1	36. 9	6. 34	0. 750	0. 00
Q569A6	ER lumen protein - retaining receptor 1	3. 77	1	24. 5	8. 62	0. 766	0. 00
Q4FZS2	Budding uninhibited bybenzimidazoles 3 homolog	10. 74	3	36. 9	6. 84	0. 796	0. 01
D4A5L9	Protein LOC679794	54. 29	6	11. 6	9. 58	0. 811	0. 01
Q06645	ATPsynthase F (0) complex subunit C1	5. 15	1	14. 2	9. 91	0. 827	0. 02
D4A817	Histone H2B	67. 46	1	13. 9	10. 32	1. 205	0. 01
Q4QQV6	Lymphocyte specific 1	17. 22	5	36. 5	4. 67	1. 210	0. 01
Q9R066	Coxsackievirus and adenovirus receptor homolog	6. 58	1	39. 9	7. 27	1. 214	0. 01
Q6P725	Desmin	73. 77	27	53. 4	5. 27	1. 229	0. 01
G3V913	Heat shock 27kDa protein 1	50. 49	7	22. 8	6. 21	1. 241	0. 00
D3ZJ08	Histone H3	60. 29	2	15. 4	11. 27	1. 356	0. 00
Q5FVG5	Similar totropomyosin 1	69. 37	10	32. 9	4. 67	1. 279	0. 00
Q5U1W8	High - mobility groupnucleosome binding domain 1	8. 33	1	10. 1	9. 76	1. 264	0. 00

(continued)

Accession	Description	Coverage	Unique peptides	MW [kDa]	Calc.pI	Fold change	P-Value
Q91XN7	Tropomyosin alpha isoform	50.00	2	28.5	4.74	1.313	0.00
D3ZBP2	Protein Sin3a	2.13	1	127.7	6.61	1.458	0.00
P49793	Nuclear pore complex protein Nup98-Nup96	0.83	1	197.2	6.23	1.761	0.00
Cellular component and process							
Q5XI41	Translocating chain-associated membrane protein 1	2.94	1	43.0	9.67	0.747	0.00
B2RZ37	Receptor expression - enhancing protein 5	13.76	2	21.4	8.12	0.743	0.00
D4AC92	Sulfotransferase	9.66	2	33.5	6.95	0.751	0.00
Q5U3Y7	Transmembrane protein 97	13.07	2	20.9	9.29	0.752	0.00
D3ZE21	Protein Tmed11	7.94	1	24.6	5.57	0.771	0.00
Q5U3Z3	Isochorismatase domain-containing protein 2	7.62	1	23.1	7.83	0.772	0.00
P19944	60S acidic ribosomal protein P1	14.04	1	11.5	4.32	0.786	0.00
B5DEQ0	Translocon-associated protein subunit beta	3.83	1	20.0	8.35	0.792	0.01
F1M8X9	Protein Gbf1	0.97	1	206.4	5.80	0.797	0.01
D3ZPJ2	Protein RGD1564095	10.53	1	11.6	4.64	0.801	0.01
F1LNE5	Protein MEMO1	6.12	1	31.6	6.95	0.803	0.01
Q5XIG4	OCIA domain-containing protein 1	7.29	1	27.6	7.42	0.813	0.01
Q6IG04	Keratin	3.65	1	56.8	7.36	0.814	0.01
P0C5H9	Mesencephalic astrocyte	31.84	5	20.4	8.25	0.823	0.02
D3Z952	Microfibrillar-associated protein 2	5.41	1	20.8	4.89	0.824	0.02
P62250	40S ribosomal protein S16	45.21	8	16.4	10.21	0.827	0.02
P15693	Intestinal-type alkalinephosphatase 1	33.15	11	58.4	6.00	0.832	0.02
Q9JHW0	Proteasome subunit beta type-7	3.25	1	29.9	7.97	1.204	0.01
Q62713	Neutrophil antibiotic peptide NP-3A	19.54	1	9.3	7.65	1.211	0.01
B1WC16	BCL2 - associated transcription factor 1	3.47	3	106.0	9.99	1.211	0.01
F1LLV6	Protein Ces1f	5.35	1	62.4	7.12	1.215	0.01
Q5BJS4	FUN14 domain - containing protein 1	10.97	1	17.1	8.63	1.229	0.01

(continued)

Accession	Description	Coverage	Unique peptides	MW [kDa]	Calc.pI	Fold change	P-Value
Q6AYC8	SH2 domain-containing protein 4A	2. 84	1	48. 5	9. 04	1. 238	0. 00
M0R4Z0	Protein Cbx1	41. 33	2	17. 2	5. 03	1. 254	0. 00
P24051	40S ribosomal protein S27-like	28. 57	1	9. 5	9. 45	1. 254	0. 00
Q6IM78	DNAtopoisomerase Ⅰ	4. 38	1	69. 0	9. 22	1. 268	0. 00
Q6EIX2	Mitochondrial import inner membranetranslocase	10. 48	1	13. 7	6. 80	1. 469	0. 00
D3Z9K7	Protein Pdia2	41. 18	17	58. 4	4. 82	0. 524	0. 00
P00773	Chymotrypsin-like elastase family member 1	12. 78	2	29. 0	8. 47	0. 565	0. 00
G3V7A3	Protein Serpini2	5. 68	1	45. 9	6. 15	0. 652	0. 00
Q5XFW5	Peptidyl-prolyl cis-trans isomerase	10. 42	1	10. 4	4. 70	0. 764	0. 00
E9PTI2	Putativelysozyme C-2	5. 98	1	13. 1	6. 84	1. 209	0. 01
D4A305	Coiled-coil domain containing 58	5. 56	1	16. 7	8. 16	0. 819	0. 02
Oxidative stress							
Q5M875	17-Beta-hydroxysteroiddehydrogenase 13	23. 00	6	33. 5	9. 38	0. 582	0. 00
D3ZTP0	Protein Aldh1l2	2. 26	1	102. 9	6. 39	0. 618	0. 00
Q9Z0V5	Peroxiredoxin-4	28. 94	4	31. 0	6. 65	0. 632	0. 00
Q6AYT0	Quinone oxidoreductase	3. 65	1	35. 0	8. 22	0. 735	0. 00
D3ZEN2	Glyceraldehyde-3-phosphatedehydrogenase	13. 03	1	35. 9	8. 12	0. 770	0. 00
Q5PPL3	Sterol-4-alpha-carboxylate 3-dehydrogenase	1. 93	1	40. 4	8. 85	0. 771	0. 00
D3Z7Y1	Uncharacterized protein	5. 47	2	37. 5	6. 96	0. 770	0. 00
Q6AYS8	Estradiol 17-beta-dehydrogenase 11	20. 81	4	32. 9	8. 63	0. 799	0. 01
F1M201	Dehydrogenase/reductase SDR family member 7B	4. 63	1	35. 2	9. 55	0. 806	0. 01
Q5XIB4	Ufm1-specific protease 2	3. 25	1	52. 3	6. 54	0. 753	0. 00
Signaling pathway							
Q7TP21	Cb1-727	1. 47	1	75. 0	8. 10	1. 217	0. 01
Q6AYK1	RNA-binding protein with serine-rich domain 1	4. 92	1	34. 2	11. 84	1. 217	0. 01
G3V9A3	Protein LOC679816	19. 76	1	27. 7	4. 78	0. 775	0. 00

(continued)

Accession	Description	Coverage	Unique peptides	MW [kDa]	Calc.pI	Fold change	P-Value
Q63450	Calcium/calmodulin-dependent protein kinase type 1	4. 81	1	41. 6	5. 35	0. 822	0. 02
Q5XI64	Monoacylglycerol lipase ABHD6	13. 95	4	38. 3	8. 62	0. 830	0. 02
Q06486	Caseinkinase I isoform delta	3. 37	1	47. 3	9. 74	1. 304	0. 00
Enzymatic activity							
Q6AYS4	Plasma alpha-L-fucosidase	2. 40	1	53. 2	6. 48	0. 736	0. 00
D3ZD11	Protein Spcs2	16. 37	2	25. 0	8. 57	0. 771	0. 00
D3ZP98	Histocompatibility 13	12. 09	4	40. 2	6. 71	0. 797	0. 01
D4A6W4	Solute carrier family 52	2. 96	1	60. 3	8. 59	0. 798	0. 01
D3ZL24	Calpastatin	6. 95	3	71. 7	5. 08	1. 244	0. 00
Q6MG88	G-Protein-signaling modulator 3	11. 39	1	17. 5	5. 85	1. 257	0. 00
Q5GAM3	Ribonuclease 16	15. 38	2	17. 7	9. 22	1. 493	0. 00
D3Z8M2	Serine/threonine-protein kinase WNK4	2. 13	2	173. 0	6. 33	0. 762	0. 00

3. 5 Gene Ontology annotation

Gene Ontology (GO) functional annotation and enrichment analysis are useful for the analysis of large genomic and proteomic datasets.[43] The biological events behind the data can be speculated through examining significantly over-represented GO terms and provide a primary overview of the intestinal mucosa proteome. Blast2GO (Version 2. 8. 0) uses NCBI BLAST tofind homologous sequences for each obtained hit. Functional classification of the identified proteins to the query sequence according to GO resulted in an assessment of the cellular compartments, molecular function, and biological process represented. Enrichment analysis of GO category was conducted using 175 differentially expressed proteins that represent the overall trends of the specific functional categories enriched in intestinal mucosa cells in order to gain more insight into the biological significance of the differentially expressed proteins. Here, 138 proteins (79. 77%) from intestinal mucosa cells with 5079 annotation terms were involved in the GO categories molecular function, biological process, and cellular component (Fig. 4). In the category of molecular function, the majority of differential proteins were found in binding (45. 7%), catalytic activity (41. 7%), transporter activity (3. 5%) and structural molecule activity (3. 5%) (Fig. 4a). The binding functions included nucleic acid and nucleotide binding, ion binding and protein binding that were mainly at the intracellular level, while catalytic activities included hydrolase and transferase activities.[44] The category cellular component involved locations, subcellular structures, and macromolecular complexes, such as telomere, nucleus, and origin recognition complex. The differentially abundant proteins in our study were related to cellular components in the intracellular space (38. 7%), organelle

(26.9%), membrane (12.5%), macromolecular complex (11.8%) and extracellular region (5.4%) (Fig. 4b).The final category, biological process included broad biological goals, such as purine metabolism or mitosis. The majority of proteins in our study were found in cellular process (17.4%), metabolic process (16.1%), single organism process (15.6%), response to stimulus (10.2%), biological regulation (9.8%) and multicellular organismal process (8.5%) (Fig. 4c). The functional annotation will help us to effectively analyze the differentially expressed unigenes caused by Mn exposure.

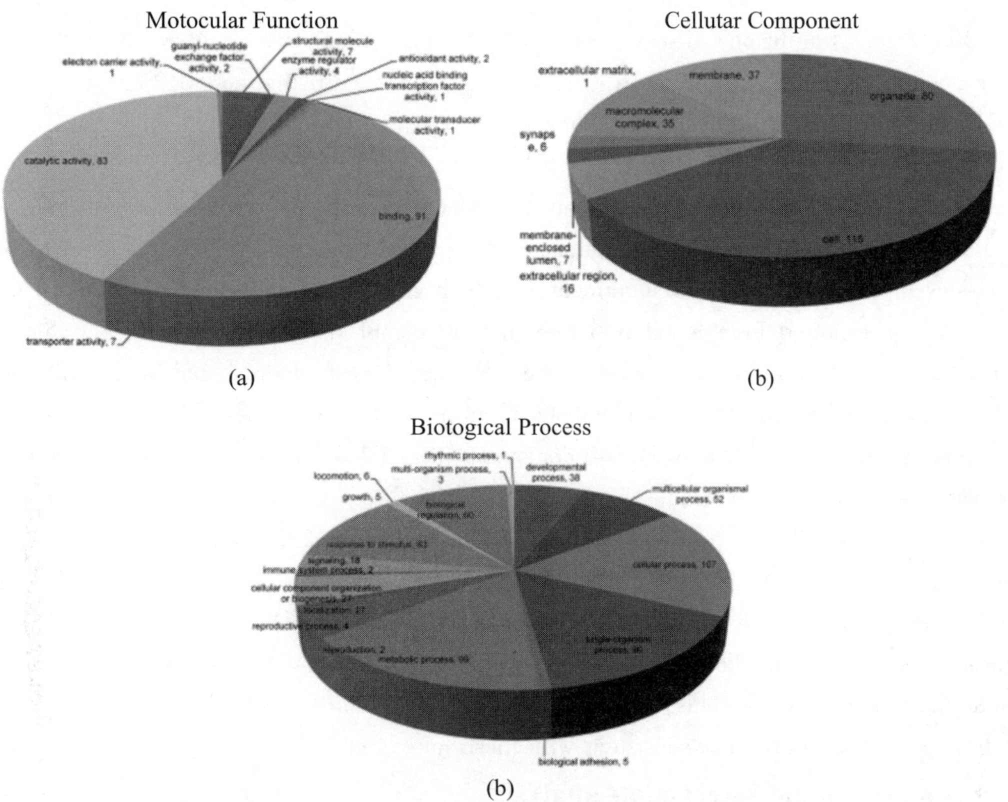

Fig. 4 Distribution of GO terms in the categories molecular function (a), cellular component (b), and biological process (c).

Note: In (a), binding and catalytic activity were the most represented biological processes. In (b), the most represented categories were cell, followed by organelle. In (c), the most represented categories were metabolic process and cellular process.

3.6 KEGG automatic annotation server (KAAS)

KEGG pathway enrichment can reveal protein functional information in the metabolic pathway through analyzing the differentially expressed proteins.[45] The KEGG pathway is a collection of manually drawn pathway maps graphically displaying the distribution of differentially expressed genes among the biochemical pathways. In our study, intestinal mucosa proteins in Mn-treated rats were annotated to biochemical pathways from the KO number of the homologous/similar proteins through aligning with rat protein sequences in the KEGG GENES database using KAAS.The signal/metabolic pathway associations for 93 differential proteins with 140 unique KEGG orthologues were established. The most represented KEGG maps were protein processing in pancreatic secretion (15 members, path: ko04792), bile secretion (2 members, path: ko04976); protein digestion and absorption (10 members, path: ko04974), fat digestion and absorption (7 members, path: ko04975), biosynthesis of amino acids (6 members, path: ko01230), glycerolipid metabolism (5 members, path: ko00561), glycine, serine and threonine metabolism (5 members, path: ko00260), arginine and praline metabolism (3 members, path: ko00330), cysteine and methionine metabolism (3 members, path: ko00270), carbohydrate digestion and absorption (3 members, path: ko04973); Huntington's disease (4 members, path: ko05016), Alzheimer's disease (3 members, path: ko05010), Parkinson's disease (2 members, path: ko05012), etc. (Fig. 5). Upon Mn exposure, profound metabolic changes and activation/deactivation occur in a vast number of intracellular pathways.

Mn^{2+} is taken up by glucose-activated b-cells, resulting in a robust signal increase in glucose-stimulated rodent b-cell lines and in islets.[46] Mn^{2+} can be used to characterize isolated islet potency in vitro and assess the functionality of both grafted and endogenous pancreatic islets in vivo.[47] Mn^{2+} is excreted from blood into bile to form complexes with bile acids.[28] Our results showed that high concentrations of Mn^{2+} disturbed the secretion of pancreas and bile, which is consistent with the above theories. Mineral metabolism can interact with the metabolism of macronutrients such as proteins and fats.[48] Our study also demonstrated that Mn exposure is associated with significantly perturbed amino acid metabolism, fatty acid metabolism and carbohydrate metabolism. These results are consistent with our previous finding that Mn exposure significantly altered the concentrations of 36 metabolites in the plasma of rats, which were associated with amino acid metabolism and fatty acid metabolism.[11] Exposure to high Mn levels causes neurodegenerative diseases including Parkinson's disease, Huntington's disease and Alzheimer's disease,[49] caused by Mn preferentially accumulating in the globus pallidus, striatum, substantia nigra and subthalamic nucleus.[50] The results observed in this study were consistent with these important findings.

3.7 Protein-protein interaction analysis

The elementary part of protein complexes is protein function in living cells. However, protein function does not act independently. In addition to KEGG, the protein-protein interactions (PPI) network was also constructed using STRING Database version 10.5. Fig. 6 shows the results that were in accordance with the Pathway and GO analysis highlighting different groups of PPI; 16 proteins have various relationships in PPI and were recognized as key nodes: anionic trypsin-1 (Prss1), chymotrypsin-like elastase family member 2A (Cela2a), chymotrypsin-like protease CTRL-1 precursor (Ctrl), chymotrypsin-like elastase family member 3B precursor (Cela3b), glyceraldehyde-3-phosphate dehydrogenase (Gapdh), peroxiredoxin-4 (Prdx4), chymotrypsin-like elastase family member 1 (Cela1), Rho-related GTP-binding protein RhoB (Rhob), carboxypeptidase A1 (Cpa1), electron carrier protein (LOC690675), histone H3 (Hist2h3c2), GDH/6PGL endoplasmic bifunctional protein precursor (H6pb), calcium/calmodulin-dependent protein kinase type 1 (Camk1), protein LOC298795 (Sfn), actin (Acta1) and solute carrier family 2 (Slc2a5). A number of proteins were associated with pancreatic secretion, protein digestion and absorption: Cela2a, Cela3b, Cpa1, Ctrl, Pnlip, Prss1, Try4; with metabolic pathways: Alpi, Asah2, Asns, Atp5g1, Cpox, Cth, Ctps, Gatm, Gmppb, H6pd, LOC690675, Mri1, Nsdhl, Pc, Pnlip, Psat1, Pycr1, Tst; with proteasome: psma1, psma2, psma3, psma4, psmb7; and with structural molecule activity: Rplp1, Rps27l, RGD1564095, Rps3a, Rps11, Rps15, Rps16, Csnk1d; which were differentially expressed through Mn exposure. Mn is a cofactor in numerous enzymatic processes, so it is essential for brain biology and physiology. But excessive levels of Mn in two neurodegenerative syndromes (Huntington and Parkinson's disease) result in neuronal loss in the basal ganglia,[51,52] pathological changes in the striatum,[53] as well as having effects on mitochondrial dysfunction and disruption of cellular energy metabolism.[7,54] Specifically, we found that four proteins were involved in Huntington's disease, Alzheimer's disease and Parkinson's disease: Gapdh, Atp5g1, OC690675, and Creb1. S100 proteins are calciumbinding proteins that are involved in controlling diverse extracellular and intracellular processes such as differentiation, cell growth, and antimicrobial function.[55,56] S100 protein family members exhibit variable affinities for oligo-merization properties, post-translational modifications, divalent metal ions, and unique spatial/temporal expression patterns.[57] In our study, the mineral absorption associated protein (S100G) was also differentially expressed. Results of the current study indicated that Mn exposure altered the small intestinal proteomes in rats.

4 CONCLUSIONS

All organisms require mechanisms for sensing smallfluctuations to maintain a controlled balance of efflux, uptake, and sequestration in metal levels and to ensure that metal availability is physiologically controlled.[2] In conclusion, chronic exposure to a high concentration of Mn significantly changed the behavioral patterns in the rat producing motor disturbances and altering the protein expression profile in rats' intestinal mucosa cells. A total of 175 intestinal mucosa proteins that may regulate Mn levels were identified using iTRAQ-based proteomic technology under biologically relevant Mn exposures. These proteins are involved in pancreatic secretion, protein digestion and absorption, fat digestion and absorption, biosynthesis of amino acids, Alzheimer's and Parkinson's disease, etc. The targets of these proteins related to Mn exposure may improve our understanding of the regulation of Mn homeostasis and intracellular and cellular Mn trafficking.

CONFLICT OF INTEREST

The authors declare no competingfinancial interest.

ACKNOWLEDGEMENTS

Financial support from theGansu Provincial Natural Science Foundation (No. 1606RJYA224), the National Key Research and Development Plan (No. 2016YFD0501200) and the Central Public-interest Scientific Institution Basal Research Fund (No. 1610322013003) is greatly appreciated. We would also like to thank Shanghai Applied Protein Technology Co. Ltd. for their technical assistance.

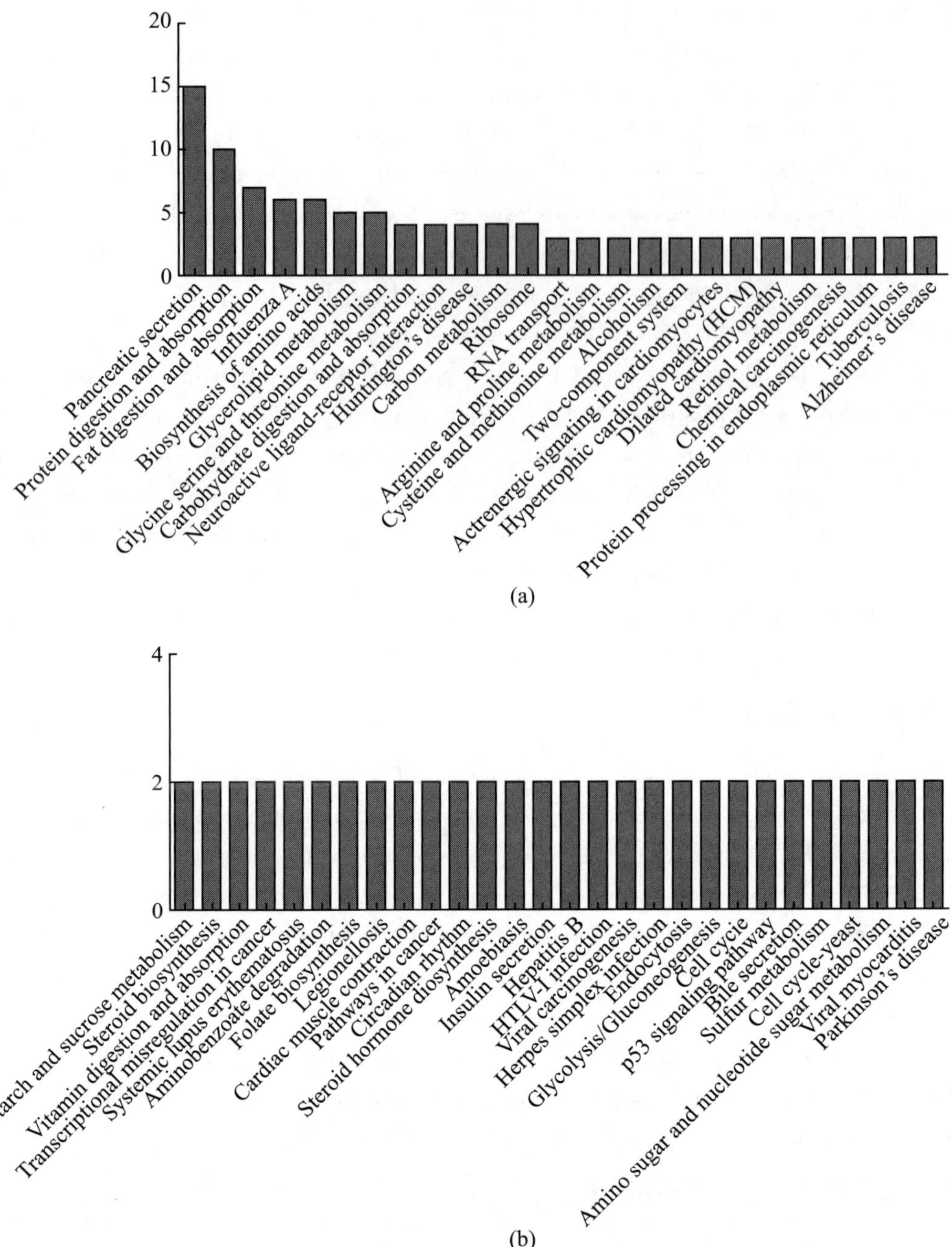

Fig. 5 The most represented KEGG maps (a and b).

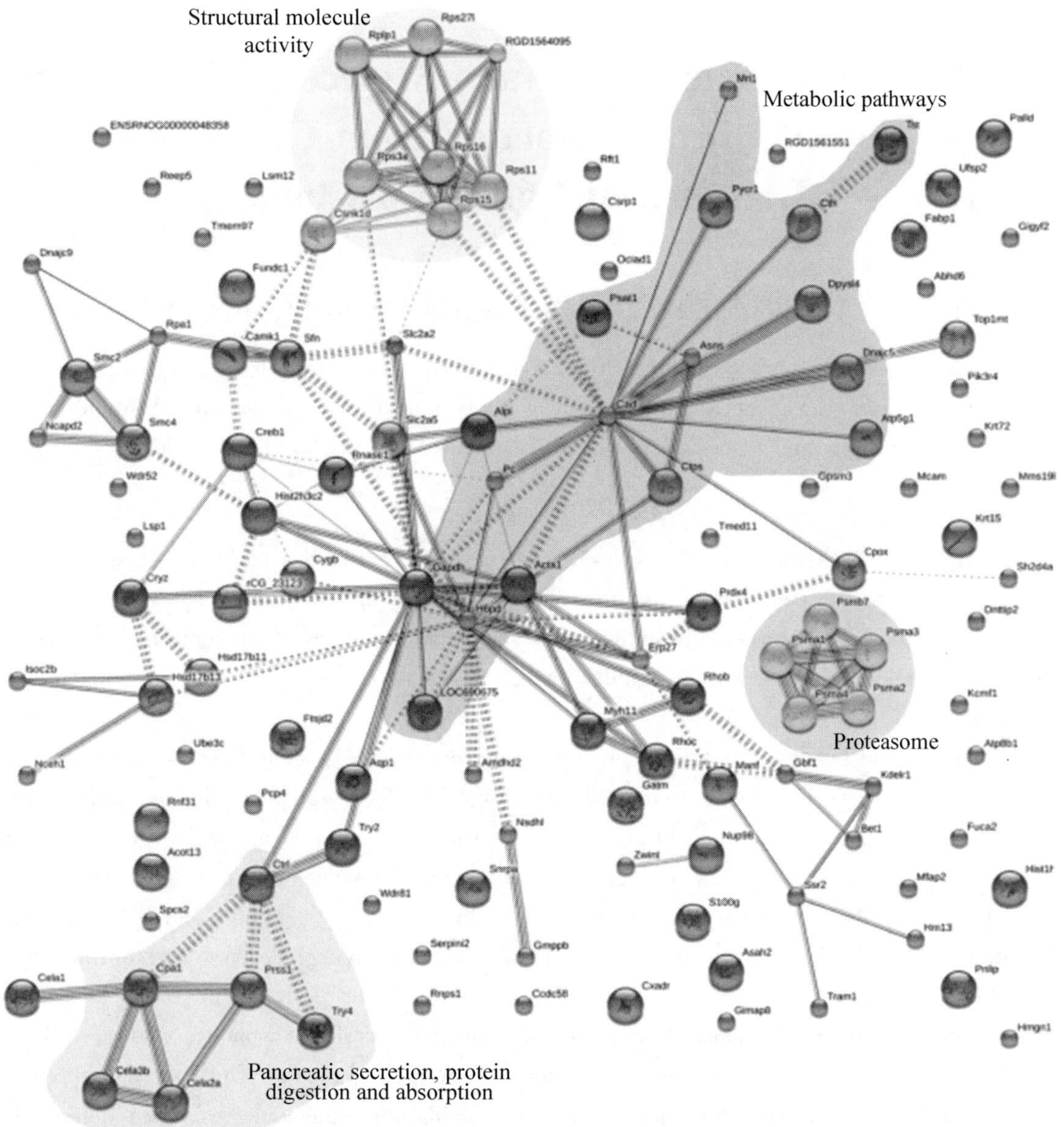

Fig. 6 Protein–protein interaction analysis of cluster 4 highlighting protein networks involved in pancreatic secretion, protein digestion and absorption (orange), metabolic pathways (purple), proteasome (green) and structural molecule activity (yellow). In the network, proteins are represented as nodes. Colors of the lines connecting the nodes represent different evidence types for protein linkage.

REFERENCES OMITTED

（发表于《RSC Advances》，院选 SCI，IF：3.108）

Epidemic Characterization and Molecular Genotyping of *Shigella flexneri* isolated from Calves with Diarrhea in Northwest China

Zhen ZHU[+], Mingze CAO[+], Xuzheng ZHOU, Bing LI, Jiyu ZHANG[*]

(Key Laboratory of New Animal Drug Project of Gansu Province/Key Laboratory of Veterinary Pharmaceutical Development of Ministry of Agriculture/Lanzhou Institute of Husbandry and Pharmaceutical Sciences of CAAS, 730050 Lanzhou, China)

Abstract: [Background]: The widespread presence of antibiotics resistance genes in pathogenic bacteria can cause enormous problems. Food animals are one of the main reservoirs of intestinal pathogens that pose a potential risk to human. Analyzing the epidemiological characteristics and resistance patterns of Shigella flexneri in calves is necessary for animal and human health.

[Methods and results]: A total of54 *Shigella flexneri* isolates, including six serotypes (1a, 2a, 2b, 4a, 6 and Xv), were collected from 837 fecal samples obtained from 2014to 2016. We performed pulsed-field gel electrophoresis (PFGE) and applied the restriction enzyme NotI to analyze the genetic relatedness among the 54 isolates and to categorize them into 31 reproducible and unique PFGE patterns. According to the results of antimicrobial susceptibility tests, all 26 *Shigella flexneri* 2a serotypes were resistant to cephalosporin and/or fluoroquinolones. The genes bla_{TEM-1}, bla_{OXA-1}, and $bla_{CTX-M-14}$ were detected in 19cephalosporin-resistant S.flexneri 2a isolates. Among 14 fluoroquinolone-resistant isolates, the aac (6′)-lb-cr gene was largely present in each strain, followed by qnrS (5). Only one ciprofloxacin-resistant isolate harbored the qepA gene. Sequencing the quinolone resistance determining regions (QRDRs) of the fluoroquinolone-resistant isolates revealed two point mutations in gyrA (S83 L, D87N/Y) and a single point mutation in parC (S80I). Interestingly, two gyrA (D87N/Y) strains were resistant to ciprofloxacin.

---

* Corresponding author, E-mail: infzjy@sina.com

[+]Equal: contributors

[Conclusions]: The current study enhances our knowledge of Shigella in cattle, although continual surveillance is necessary for the control ofshigellosis. The high level of cephalosporin and/or fluoroquinolone resistance in Shigella warns us of a potential risk to human and animal health.

Key words: *Shigella flexneri*; Antimicrobial susceptibility; Resistant

1 BACKGROUND

The majority of Enterobactericeae family bacteria, including *Salmonella*, *E.coli* and *Shigella* spp., the major etiological agent of diarrheal disease, are a global public health burden, particularly in low-income countries[1-3]. Shigella is phylogenetically distinct from several independent E. coli strains and has evolved through convergent evolution[4]. The genus Shigella consists of four subgroups differentiated according to their biochemical and serological properties: A (*S.dysenteriae*), B (S.flexneri), C (S.boydii), and D (S.sonnei). All four species of *Shigella* cause shigellosis, but *S.flexneri* is the predominant subgroup found in developing countries, whereas S.sonnei is found in industrialized countries[5]. The first *Shigella* species identified was *S.dysenteriae*, followed by S. flexneri at the end of the 19th century. Shigellosis became a notorious and widespread epidemic during World War 1 with the transmission of *S.flexneri* strain NCTC1, a 2a lineage[6,7]. Based on the differing structural characteristics of the antigenic determinants of the O antigen, *S.flexneri* is divided into no fewer than 20 serotypes: 1a, 1b, 1c, 1d, 2a, 2b, 2v, 3a, 3b, 4a, 4av, 4b, 4c, 5a, 5b, X, Xv, Y, Yv, 6, and 7b[8,9].

Given that shigellosis is a global public health burden, previous studies have focused on the human gastrointestinal pathogens but have ignored animal groups. *Shigella* spp. infect and also cause corresponding clinical symptoms in monkeys, cows, pigs, chickens and other animals[10-13]. Indeed, animals that live in environments characterized by poor sanitary hygiene, restricted access to clean drinking water and long-term exposure to contaminated food are prone to dysentery[14,15]. Many antibiotics are used to control disease and promote growth during the breeding process, leading to the widespread dissemination of antibiotic resistance genes (ARGs). The spread of drug resistance among pathogenic bacteria in humans and animals may be disastrous.

The present study investigated the *Shigella* epidemic in cows in the northwest region of China. *S. flexneri* 2a was first isolated from a yak with diarrhea in Tibet in 2014. In this study, we used pulsed-field gel electrophoresis (PFGE) to analyze the relationships among *S.flexneri* isolates and tested for antimicrobial susceptibility patterns. Our results will help prevent diarrhea in calves and will assist in the selection of effective antibiotics against Shigella.

2 METHODS

2.1 Bacterial isolation and identification

Fresh stool samples were isolated from 2014 to 2016 in Northwest China (Gansun, Shanxi, Qinghai, Xinjiang and Tibet) from calves (3 to 20 days) with diarrhea. Samples were stored in transport medium, cultured directly on *Salmonella-Shigella* (SS) agar and incubated at 37℃ for

24h to select for Shigella.Resultant colonies (colorless, semitransparent, smooth, and moist circular)[16] were picked and grown at 37℃ for 24h on MacConkey (MAC) Agar to verify identity.Colonies were selected and cultured in brain heart infusion broth at 37℃ for 5h with shaking at 250 rpm. All isolates were confirmed using API20E test strips (bioMer-ieux Vitek, Marcy-l' Etoile, France) according to the manufacturer's recommendations.*Shigella* was serotyped using a commercially available kit (Denka Seiken, Tokyo, Japan) and confirmed by PCR[17].

2.2 Antimicrobial susceptibility testing

The antimicrobial susceptibility of *S.flexneri* isolates was determined via the Kirby-Bauer disc-diffusion method in accordance with the guidelines of the Clinical and Laboratory Standards Institute (CLSI)[18].

The antibiotic discs (OXOID, UK) included penicillin G (P, 10μg), ampicillin (AMP, 10μg), amoxicillin/clavulanic acid (AMC, 30μg), cephalothin (KF, 30μg), cephazolin (KZ, 30μg), cefamandole (MA, 30μg), cefoxitin (FOX, 30μg), ceftriaxone (CRO, 30μg), cefotaxime (CTX, 30μg), cefoperazone (CFP, 75μg), cefepime (FEP, 30μg), meropenem (MEM, 10 pg), imipenem (IPM, 10μg), norfloxacin (NOR, 10μg), enrofloxacin (ENR, 5μg), levoflox-acin (LEV, 5 pg), ciprofloxacin (CIP, 5 pg), erythromycin (E, 15 pg), chloramphenicol (C, 30 pg), tetracycline (TE, 30μg), streptomycin (S, 10μg), gentamicin (CN, 10μg), and amikacin (AK, 30μg).E. coli strain ATCC25922 was used as a quality control strain in each test batch.

2.3 PCR amplification of ARGs

We performed PCR assays that targeted 24 different ARGs using the primers described in Table 1. To determine the underlying resistance mechanism of β-lactam antibiotics, we amplified extended-spectrum β-lactamase (ESBL) genes, specifically bla_{CTX-M}, bla_{SHV}, bla_{TEM}, and bla_{OXA}, as well as ampC genes, specifically bla_{MOX}, bla_{FOX}, bla_{DHA}, bla_{CIT}, bla_{ACC}, and bla_{MIR}[19-21].Plasmid-mediated quinolone resistance (PMQR) determinant genes, including qnrA, qnrB, qnrD, qnrS, qepA and aac (6') -Ib-cr and four quinolone resistance determining region (QRDR) genes as well as DNA gyrase (gyrA, gyrB) and topoisomerase IV (par C, par E) were amplified to determine the underlying mechanism of quinolone resistance[16,21-23].The PCR fragments were sequenced after purification and compared to sequences in GenBank.

2.4 PFGE

Genotypes and transmission patterns were determined by performing PFGE according to the method described in a previous study[19].S.flexneri isolates were digested with the restriction enzyme NotI (TaKaRa, Japan) at 37℃ for 3h to generate a DNA fingerprinting profile.*Salmonella enterica* serotype Braenderup strain H9812 was digested with xbaI (TaKaRa, Japan) and used as a molecular size standard.Electrophoresis was performed on the CHEF Mapper XA system (Bio-Rad) with a 1% agarose SeaKem Gold gel (Lonza, USA).Electrophoretic parameters were determined by performing multiple screening runs and included switching times of 2.16 to 54.17 s, a voltage of 6v/cm, a 120° angle and a run time of 21h.PFGE images were obtained using a Universal Hood II (Bio-RAD, USA) and analyzed using BioNumerics software version 7.1 (Applied Maths, Sint-

Martens – Latem, Belgium). A clustering tree that indicated relative genetic similarity was constructed using UPGMA (Unweighted Pair Group Method with Arithmetic Mean) and the Dice-predicted similarity value with a 1.0% pattern optimization and 1.5% band position tolerance.

3 RESULTS

3.1 Bacterial isolation and identification

During our epidemiological survey of *Shigella*, we collected 873 fecal samples from calves with diarrhea and obtained 54 S. flexneri isolates from five provinces in northwest China from 2014 to 2016. Isolate information is shown in detail in Table 2. Among the 54 *S.flexneri* isolates, there were six serotypes: five (9.26%) isolates were 1a, twenty-six (48.15%) isolates were 2a, four (7.41%) isolates were 2b, six (11.11%) isolates were 4a, eight (14.81%) isolates were 6, and five (9.26%) isolates were Xv (Fig. 1). Our surveillance of the Gansu isolates identified all of the serotypes, except 4a. All 4a serotypes were isolated from Shanxi, while all Xv and 1a serotypes were from Gansu. Additionally, serotype 2a was widely isolated from each province, with the exception of Xinjiang, and serotype 6 was found only in yaks. Interestingly, *Shigella* was primarily isolated in the first quarter and fourth quarter, accounting for 54% (29/54) and 30% (16/54), respectively (Fig. 2).

Table 1 Primers for the detection of antibiotic resistance genes.

Target	Primer sequence (5′ to 3′)	Amplicon size (bp)	Reference
β-lactamase $bla_{CTX-M-1}$	F: GGTTAAAAAATCACTGCGTC	873	Cui et al., 2015[16]
$bla_{CTX-M-2}$	R: TTACAAACCGTCGGTGACGA F: CGACGCTACCCCTGCTATT	552	Zong et al., 2008[17]
$bla_{CTX-M-8}$	R: CCAGCGTCAGATTTTTCAGG F: TCGCGTTAAGCGGATGATGC	689	Zong et al., 2008[17]
$bla_{CTX-M-9}$	R: AACCCACGATGTGGGTAGC F: AGAGTGCAACGGATGATG	868	Cui et al., 2015[16]
$bla_{CTX-M-25}$	R: CCAGTTACAGCCCTTCGG F: TTGTTGAGTCAGCGGGTTGA	497	Liu et al., 2015[18]
bla_{SHV}	R: GCGCGACCTTCCGGCCAAAT F: CGCCGGGTTATTCTTATTTGTCGC	1015	Zong et al., 2008[17]
bla_{TEM}	R: TCTTTCCGATGCCGCCGCCAGTCA F: ATGAGTATTCAACTTTCCG	876	This study
bla_{OXA}	R: CCAATGCTTAATCAGTGAG F: ATTAAGCCCTTTACCAAACCA	890	Cui et al., 2015[16]
bla_{MOX}	R: AAGGGTTGGGCGATTTTGCCA F: GCTGCTCAAGGAGCACAGGAT	520	Cui et al., 2015[16]
bla_{FOX}	R: CACATTGACATAGGTGTGGTGC F: AACATGGGGTATCAGGGAGATG	190	Cui et al., 2015[16]

(continued)

Target	Primer sequence (5′ to 3′)	Amplicon size (bp)	Reference
blaDHA	R: CAAAGCGCGTAACCGGATTGG F: AACTTTCACAGGTGTGCTGGGT	405	Cui et al., 2015[16]
bla_{CIT}	R: CCGTACGCATACTGGCTTTGC F: TGGCCAGAACTGACAGGCAAA	462	Cui et al., 2015[16]
bla_{ACC}	R: TTTCTCCTGAACGTGGCTGGC F: AACAGCCTCAGCAGCCGGTTA	346	Cui et al., 2015[16]
bla_{MIR}	R: TTCGCCGCAATCATCCCTAGC F: TCGGTAAAGCCGATGTTGCGG	302	Cui et al., 2015[16]
PMQRs qnrA	R: CTTCCACTGCGGCTGCCAGTT F: ATTTCTCACGCCAGGATTTG	516	Colobatiu et al., 2015[19]
qnrB	R: GATCGGCAAAGGTTAGGTCA F: GATCGTGAAAGCCAGAAAGG	476	Colobatiu et al., 2015[19]
qnrD	R: ACGATGCCTGGTAGTTGTCC F: CGAGATCAATTTACGGGGAATA	656	Cui et al., 2015[13]
qnrS	R: AACAAGCTGAAGCGCCTG F: ACGACATTCGTCAACTGCAA	417	Colobatiu et al., 2015[19]
aac (6′) -Ib-cr	R: TAAATTGGCACCCTGTAGGC F: CCCGCTTTCTCGTAGCA	544	Colobatiu et al., 2015[19]
qepA	R: TTAGGCATCACTGCGTCTTC F: CGTGTTGCTGGAGTTCTTC	403	Colobatiu et al., 2015[19]
QRDR gyrA	R: CTGCAGGTACTGCGTCATG F: TACACCGGTCAACATTGAGG	648	Hu et al., 2007[20]
	R: TTAATGATTGCCGCCGTCGG		
gyrB	F: TGAAATGACCCGCCGTAAAGG R: CTGTGATAACGCAGTTTGTCCGGG	309	Hu et al., 2007[20]
parC	F: GTACGTGATCATGGACCGTG R: TTCGGCTGGTCGATTAATGC	531	Hu et al., 2007[20]
parE	F: ATGCGTGCGGCTAAAAAAGTG R: TCGTCGCTGTCAGGATCGATAC	290	Hu et al., 2007[20]

3.2 Antimicrobial susceptibility testing

The 54 *S.flexneri* isolates were examined for susceptibility to 23 antibiotics. More than 50% of isolates were resistant to 8 antibiotics. Among them, resistance to P (54/54, 100%), AMP (51/54, 94.44%) and TE (49/54, 90.74) was most common, followed by E (46/54, 85.19%), S (38/54, 70.37%), KZ (34/54, 62.96%), KF (29/54, 53.70%) and CN (29/54, 53.70%). None of the isolates were resistant to IMP, MEM and the fourth-generation cephalosporin FEP. In addition, although a certain number of isolates were resistant to second - and third - generation cephalosporins (MA, FOX, CRO, CTX and CFP) and fluoroquinolones (CIP NOR,

ENR and LEV), these comprised no more than 30% of the total number of isolates, and the resistance rate was lower than those of other antibiotics (Table 3, Fig. 3).

Remarkably, all 26 *S.flexneri* 2a isolates demonstrated varying degrees of resistance to cephalosporins and/or fluoroquinolones and exhibited multidrug resistance (MDR). The *S. flexneri* 2a isolates were resistant to 14 diverse cephalosporins/fluoroquinolones. Among them, 73.06% (19/26) of isolates were resistant to cephalosporin, 53.85% (14/26) of isolates were resistant to fluoroquinolones, and 26.92% (7/26) of isolates were resistant to both cephalosporin and fluoroquinolones. Furthermore, isolate GBSF1512433 was resistant to all cephalosporins (with the exception of FEP) and fluoroquinolones (with the exception of CIP). Compared with the S.flexneri 2a isolates collected from beef calves, the 4 yak isolates were sensitive to most cephalosporins and fluoroquinolones but resistant to KF, KZ and MA (Table 4).

3.3 ARGs analysis of cephalosporin-and/or fluoroquinolone-resistant *S. flexneri* 2a isolates

In this study, only three β-lactamase gene types (bla_{OXA-1}, bla_{TEM-1} and $bla_{CTX-M-14}$) were identified among the 19 cephalosporin-resistant S. flexneri 2a isolates (Table 5). All isolates harbored bla_{TEM-1}type ARGs (100%), 15 isolates harbored bla_{OXA-1} (15/19, 78.95%), and 14 harbored $bla_{CTX-M-14}$ (14/19, 73.68%). In total, 63.16% (12/19) of isolates harbored three β-lactamase gene types. All *S.flexneri* 2a isolates from yaks were negative for bla_{CTX-M} type ARGs.

Both PMQR genes and SNPs in QRDRs were identified for 14 quinolone-resistant isolates (Table6). According to the PCR results, all quinolone-resistant isolates were positive for aac (6′)-Ib-cr but negative for qepA, except strain GBSF1602098. Only five (5/14, 35.71%) strains isolated from Gansu harbored qnrS, and no isolate harbored all three ARGs simultaneously. The point mutations in the QRDR genes play important roles in determining quin-olone and/or fluoroquinolone resistance[24]. In the present study, we successfully amplified all four QRDR genes and compared them to reference sequences. We found two point mutations in gyrA and one point mutation each in gyrA and parC (Table 6). All quinolone-resistant strains carried mutations that altered the amino acid sequences of gyrA (S83 L) and parc (S80I). In addition, each strain carried the mutation 87 (D→N or Y) in gyrA, with the exception of GBSF1510390. Interestingly, GBSF1505314 and GBSF1602098 harbored the gyrA D87Y mutation, which confers resistance to ciprofloxacin.

3.4 PFGE pattern analysis

PFGE was performed to determine the genetic relatedness among the isolates and to study the molecular epidemiology in specific geographical regions[25]. The PFGE patterns of the 54 NotI-digested *S.flexneri* isolates were heterogeneous, and multiple PFGE patterns were present among the strains. Thus, diverse factors such as geography and environment may affect PFGE patterns. At an 80% similarity level, *S.flexneri* isolates generated 31 reproducible and unique PFGE patterns, including 11 common types (CT) and 20 single types (ST) (Fig. 3).

Table 2 **Strain information of *S.flexneri* isolates from diarrheal calves, 2014 to 2016.**

Strain name	Serotype	Isolation year	Origin	Province
TYSF1412001	2a	2014	Yak	Tibet
GBSF1412056	2a	2014	Beef cattle	Gansu
GBSF1501026	2a	2015	Beef cattle	Gansu
GBSF1501071	Xv	2015	Beef cattle	Gansu
GYSF1501076	6	2015	Yak	Gansu
QYSF1501088	6	2015	Yak	Qinghai
XBSF1501093	2b	2015	Beef cattle	Xinjiang
GBSF1501105	2a	2015	Beef cattle	Gansu
SBSF1501123	4a	2015	Beef cattle	Shanxi
QYSF1502130	6	2015	Yak	Qinghai
GBSF1502176	2a	2015	Beef cattle	Gansu
GYSF1502197	6	2015	Yak	Gansu
SBSF1502219	4a	2015	Beef cattle	Shanxi
XBSF1502236	2b	2015	Beef cattle	Xinjiang
GBSF1503241	2a	2015	Beef cattle	Gansu
GYSF1503270	1a	2015	Yak	Gansu
GBSF1503288	1a	2015	Beef cattle	Gansu
GBSF1505314	2a	2015	Beef cattle	Gansu
SBSF1505331	2a	2015	Beef cattle	Shanxi
GBSF1506340	Xv	2015	Beef cattle	Gansu
GBSF1507358	1a	2015	Beef cattle	Gansu
GBSF1509369	2a	2015	Beef cattle	Gansu
GBSF1510375	2a	2015	Beef cattle	Gansu
GBSF1510390	2a	2015	Beef cattle	Gansu
QYSF1511395	2a	2015	Yak	Qinghai
GBSF1511401	2a	2015	Beef cattle	Gansu
GYSF1511409	2a	2015	Yak	Gansu
SBSF1512413	4a	2015	Beef cattle	Shanxi
GBSF1512419	2b	2015	Beef cattle	Gansu
GBSF1512425	2a	2015	Beef cattle	Gansu
GBSF1512433	2a	2015	Beef cattle	Gansu
GBSF1601015	Xv	2016	Beef cattle	Gansu

(continued)

Strain name	Serotype	Isolation year	Origin	Province
GBSF1601024	Xv	2016	Beef cattle	Gansu
TYSF1601031	2b	2016	Yak	Tibet
GBSF1601064	2a	2016	Beef cattle	Gansu
GYSF1601073	6	2016	Yak	Gansu
GBSF1602082	2a	2016	Beef cattle	Gansu
QYSF1602094	6	2016	Yak	Qinghai
GBSF1602098	2a	2016	Beef cattle	Gansu
GBSF1602103	2a	2016	Beef cattle	Gansu
SBSF1603115	4a	2016	Beef cattle	Shanxi
SBSF1603121	4a	2016	Beef cattle	Shanxi
GBSF1603138	2a	2016	Beef cattle	Gansu
GBSF1603149	2a	2016	Beef cattle	Gansu
QYSF1603158	6	2016	Yak	Qinghai
SBSF1604173	2a	2016	Beef cattle	Shanxi
SBSF1604195	4a	2016	Beef cattle	Shanxi
GBSF1605203	Xv	2016	Beef cattle	Gansu
GBSF1608241	2a	2016	Beef cattle	Gansu
GYSF1610256	2a	2016	Yak	Gansu
GYSF1610266	6	2016	Yak	Gansu
GBSF1610275	1a	2016	Beef cattle	Gansu
GBSF1611283	1a	2016	Beef cattle	Gansu
GBSF1611290	2a	2016	Beef cattle	Gansu

Among all isolates, the majority of S.flexneri 2a (26/54, 48.15%) isolates were classified into 11 PFGE patterns (4 CT and 7 ST).These PFGE patterns were closely related to each other, except the Tibet (TYSF1412001) and Qing-hai (QYSF1511395) isolates, suggesting the strains isolated from different geographical locations exhibit diverse PFGE patterns and a capricious genetic diversity.

4 DISCUSSION

ARGs are widespread and cause problems when present in pathogens[26].Over the past decade, MDR *Shigella* has been reported in many countries[27].However, only a few studies have described the prevalence of *Shigella* in animals worldwide.In the present study, we investigated the epidemiology of *S.flexneri* in cows in northwest China.During a 2-year survey, 54 *S.flexneri* isolates were ob-

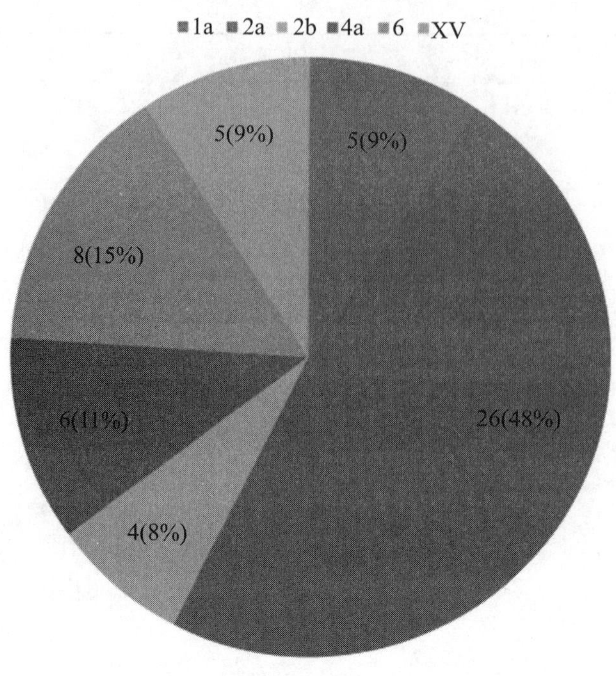

Fig. 1

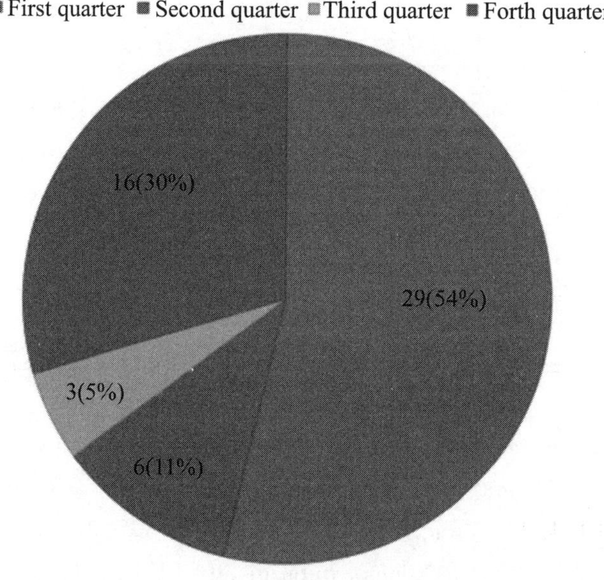

Fig. 2

tained. Unfortunately, 16S rRNA gene sequence analysis does not effectively distinguish between closely related strains in a superfamily, such as *Shigella* and *E. coli*[28], and conventional biochemical and serological techniques are also insufficient. Therefore, PFGE was utilized to analyze the molecular characteristics of these isolates, to determine the relatedness among isolates and to study the molecular epidemiology in specific geographical regions. The clustering results allowed us to analyze the epidemiological trends of *S. flexneri*. Characterization of these isolates will be helpful for clinical diagnosis, treatment, prevention and the control of shigellosis[15].

Antimicrobial resistance has emerged as a serious problem[29], particularly for conventional, older-generation antibiotics such as P, AMP, TE, and E. According to the results of our antimicrobial susceptibility tests, cephalosporin and fluoroquinolone resistance rates in our isolates were higher than those in human isolates[19,26]. Notably, the predominant S. flexneri 2a isolates were all resistant to cephalosporins, fluoroquinolones and multiple antibiotics. Two isolates (GBSF1505314 and GBSF1602098) were also resistant to ciprofloxacin, which is the first-line antibiotic treatment for shigellosis. The universal emergence of resistant and MDR strains in animals may be attributable to the unrestricted and excessive use of antibiotics in veterinary clinics. The widespread presence of MDR strains has reduced the selectivity of clinical medications to treat shigellosis[30]. Notably, our PFGE dendrogram showed various genetic patterns for S. flexneri, and there were diverse resistance profiles associated with each pattern. Based on these results, *S. flexneri* has the ability to adapt to the selective pressures of different antibiotics.

Table 3 Statistical analysis of the results of antimicrobial susceptibility to 23 antibiotics for 54 *S. flexneri*

Antibiotic	Antimicrobial resistance rate No. (%)					
	Total (n=54)	Gansu (n=37)	Shanxi (n=8)	Xinjiang (n=2)	Qinghai (n=5)	Tibet (n=2)
Penicillin G (P)	54 (100%)	37 (100%)	8 (100%)	2 (100%)	5 (100%)	2 (100%)
Ampicillin (AMP)	51 (94.44%)	37 (100%)	8 (100%)	2 (100%)	3 (60%)	1 (50%)
Amoxycillin/Clavulanic acid (AMC)	5 (9.62%)	3 (8.11%)	1 (12.5%)	1 (50%)	0	0
Cephalothin (KF)	29 (53.70%)	19 (51.35%)	5 (62.5%)	2 (100%)	2 (40%)	1 (50%)
Cephazolin (KZ)	34 (62.96%)	21 (56.76%)	7 (87.5%)	2 (100%)	3 (60%)	1 (50%)
Cefamandole (MA)	16 (29.63%)	12 (32.43%)	2 (25%)	1 (50%)	1 (20%)	0
Cefoxitin (FOX)	3 (5.56%)	2 (5.41%)	1 (12.5%)	0	0	0
Ceftriaxone (CRO)	12 (22.22%)	9 (24.32%)	2 (25%)	1 (50%)	0	0
Cefotaxime (CTX)	14 (25.93%)	10 (27.03%)	2 (25%)	1 (50%)	1 (20%)	0
Cefoperazone (CFP)	6 (11.11%)	6 (16.22%)	0	0	0	0
Cefepime (FEP)	0	0	0	0	0	0
Meropenem (MEM)	0	0	0	0	0	0
Imipenem (IPM)	0	0	0	0	0	0
Norfloxacin (NOR)	16 (29.63%)	12 (32.43%)	3 (37.5%)	1 (50%)	0	0
Enrofloxacin (ENR)	13 (24.07%)	11 (29.73%)	2 (25%)	0	0	0
Levofloxacin (LEV)	14 (25.93%)	11 (29.73%)	2 (25%)	1 (50%)	0	0
Ciprofloxacin (CIP)	2 (3.70%)	2 (5.41%)	0	0	0	0
Erythromycin (E)	46 (85.19%)	35 (94.59%)	6 (75%)	2 (100%)	3 (60%)	0

(continued)

Antibiotic	Antimicrobial resistance rate No. (%)					
	Total (n=54)	Gansu (n=37)	Shanxi (n=8)	Xinjiang (n=2)	Qinghai (n=5)	Tibet (n=2)
Tetracycline (TE)	49 (90.74%)	35 (94.59%)	8 (100%)	2 (100%)	3 (60%)	1 (50%)
Chloramphenicol (C)	17 (31.48%)	10 (27.03%)	6 (75%)	1 (50%)	0	0
Streptomycin (S)	38 (70.37%)	30 (81.08%)	4 (50%)	2 (100%)	2 (40%)	0
Gentamicin (CN)	29 (53.70%)	23 (62.16%)	4 (50%)	2 (100%)	0	0
Amikacin (AK)	3 (5.56%)	3 (8.11%)	0	0	0	0

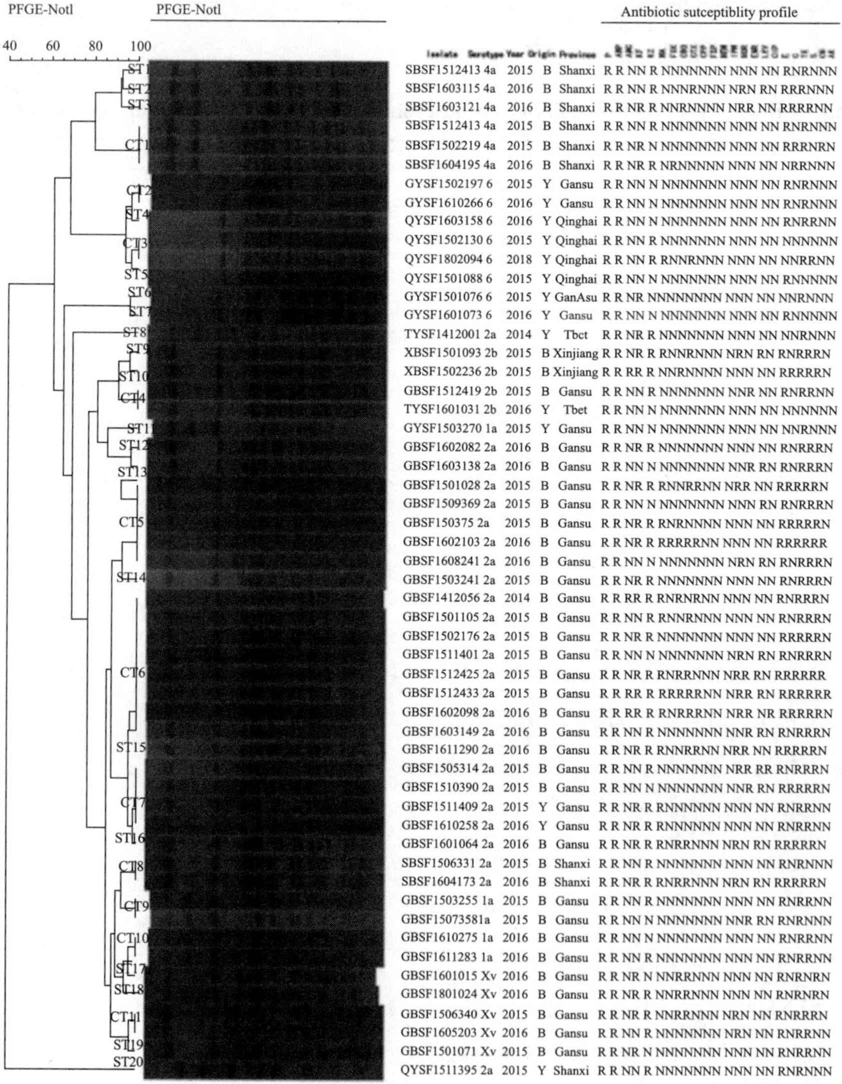

Fig. 3

The high levels of resistance of S.flexneri 2a to ceph-alosporin/fluorquinolones, which are the most effective treatments for severe gastrointestinal infections caused by pathogenic bacteria, prompted us to study potential molecular resistance mechanisms. The emergence of ESBL-producing Shigella spp.has been observed in many countries[31].In the current study, only 3 ARG genotypes (bla_{OXA-1}, bla_{TEM-1} and $bla_{CTX-M-14}$) were detected.Among them, the bla_{TEM-1} gene was detected in all 19 cephalosporin-resistant isolates.In total, 174 bla_{TEM} variants resistant to penicillin and other β-lacta-mase antibiotics have been recorded.TEM-1 confers resistance to ampicillin and cephalothin[32].bla_{OXA}-type ARGs are class D β-lactamases, which were named for their ability to hydrolyze oxacillin[32].Initially, bla_{OXA}-beta-lactamases were reported in P. aeruginosa, although now thebla_{OXA} gene has been detected in plasmids and integrons in many Gramnegative organisms[32,33]. According to some studies, the probable host preference for bla_{OXA}-type β-lactamase is S.flexneri[34]. In the present study, 15/19 (78.95%) isolates harbored bla_{OXA}-type genes, and sequencing results indicated that all the bla_{OXA} genes were bla_{OXA-1}.

Table 4 Statistical analysis of the cephalosporin and/or fluoroquinolone suscentibilitv for 26 *S.flexneri* 2a

Cephalosporin and/orFluorquinolones resistance spectrum	Cephalosporin and/orFluorquinolones resistance rate No. (%)				
	Total (n=26)	Gansu (n=22[a])	Shanxi (n=2)	Qinghai (n=1[a])	Tibet (n=1[a])
KF/KZ	5 (19.23%)	3 (13.64%)	0	1[*] (100%)	1[*] (100%)
KF/KZ/MA	3 (11.54%)	2[a] (9.09%)	1 (50%)	0	0
KF/KZ/MA/CRO	1 (3.85%)	1 (4.55%)	0	0	0
KF/KZ/MA/CTX	1 (3.85%)	1 (4.55%)	0	0	0
KF/KZ/MA/CRO/CFP	1 (3.85%)	1 (4.55%)	0	0	0
KF/KZ/MA/FOX/CRO/CTX/CFP	1 (3.85%)	1 (4.55%)	0	0	0
NOR/LEV	3 (11.54%)	3 (13.64%)	0	0	0
ENR/LEV	3 (11.54%)	3 (13.64%)	0	0	0
NOR/ENR/LEV/CIP	1 (3.85%)	1 (4.55%)	0	0	0
KF/KZ/MA/CRO/CTX/NOR/LEV	2 (7.69%)	1 (4.55%)	1 (50%)	0	0
KF/KZ/MA/CRO/CTX/NOR/ENR/LEV	1 (3.85%)	1 (4.55%)	0	0	0
KF/KZ/MA/CTX/CFP/CIP/NOR/ENR	2 (7.69%)	2 (9.09%)	0	0	0
KF/KZ/MA/CRO/CTX/CFP/NOR/ENR/CIP	1 (3.85%)	1 (4.55%)	0	0	0
KF/KZ/MA/FOX/CRO/CTX/CFP/NOR/ENR/LEV	1 (3.85%)	1 (4.55%)	0	0	0

[*] a yak origin S.flexneri 2a isolate

Additionally, bla_{CTX-M} has become one of the most prevalent extended-spectrum-β-lactamases (ESBLs)[35].This gene was widely harbored by *S.flexneri* 2a isolated from beef cattle.Interestingly, all *S.flexneri* 2a isolated from yaks were negative for bla_{CTX-M} type ARGs.

Table 5 Antimicrobial spectrum and ARGs analysis of S.flexneri 2a with resistance to cephalosporin.

Strain name	Antimicrobial spectrum	ARGs in plasmid TEM	OXA	CTX-M-9
TYSF1412001	KF/KZ	TEM-1	OXA-1	——
QYSF1511395	KF/KZ	TEM-1	——	——
GBSF1503241	KF/KZ	TEM-1	OXA-1	CTX-M-14
GBSF1502176	KF/KZ	TEM-1	OXA-1	CTX-M-14
GBSF1602082	KF/KZ	TEM-1	——	CTX-M-14
SBSF1505331	KF/KZ/MA	TEM-1	OXA-1	CTX-M-14
GYSF1511409	KF/KZ/MA	TEM-1	——	——
GYSF1610256	KF/KZ/MA	TEM-1	OXA-1	——
GBSF1510375	KF/KZ/MA/CRO	TEM-1	OXA-1	CTX-M-14
GBSF1501105	KF/KZ/MA/CTX	TEM-1	OXA-1	CTX-M-14
GBSF1412056	KF/KZ/MA/CRO/CFP	TEM-1	OXA-1	CTX-M-14
GBSF1602103	KF/KZ/MA/FOX/CRO/CTX/CFP	TEM-1	OXA-1	CTX-M-14
GBSF1601064	KF/KZ/MA/CRO/CTX/NOR/LEV	TEM-1	——	CTX-M-14
SBSF1604173	KF/KZ/MA/CRO/CTX/NOR/LEV	TEM-1	OXA-1	CTX-M-14
GBSF1611290	KF/KZ/MA/CTX/CFP/NOR/ENR	TEM-1	OXA-1	CTX-M-14
GBSF1501026	KF/KZ/MA/CTX/CFP/NOR/ENR	TEM-1	OXA-1	CTX-M-14
GBSF1512425	KF/KZ/MA/CRO/CTX/NOR/ENR/LEV	TEM-1	OXA-1	——
GBSF1602098	KF/KZ/MA/CRO/CTX/CFP/NOR/ENR/CIP	TEM-1	OXA-1	CTX-M-14
GBSF1512433	KF/KZ/MA/FOX/CRO/CTX/ CFP/NOR/ENR/LEV	TEM-1	OXA-1	CTX-M-14

Table 6 Antimicrobial spectrum and amino acid types in QRDR and PMQRs genes analysis of S.fLexneri 2a with resistance to fluoroquinolones.

Strain name	Antimicrobial spectrum	QRDR			ARGs in plasmid		
		gyrA 83	87	par C 80	aac (6′) -ib-cr	qnrS	qepA
GBSF1509369	NOR/LEV	S83L	D87N	S80I	+	-	-
GBSF1511401	NOR/LEV	S83 L	D87N	S80I	+	-	-
GBSF1608241	NOR/LEV	S83 L	D87N	S80I	+	-	-
GBSF1510390	ENR/LEV	S83 L	D87D	S80I	+	+	-
GBSF1603138	ENR/LEV	S83 L	D87N	S80I	+	-	-

(continued)

Strain name	Antimicrobial spectrum	QRDR			ARGs in plasmid		
		gyrA 83	87	par C 80	aac (6′) -ib-cr	qnrS	qepA
GBSF1603149	ENR/LEV	S83 L	D87N	S80I	+	−	−
GBSF1505314	NOR/ENR/LEV/CIP	S83 L	D87Y	S80I	+	+	−
GBSF1601064	KF/KZ/MA/CRO/CTX/ NOR/LEV	S83 L	D87N	S80I	+	−	−
SBSF1604173	KF/KZ/MA/CRO/CTX/ NOR/LEV	S83 L	D87N	S80I	+	−	−
GBSF1611290	KF/KZ/MA/CTX/CFP/ NOR/ENR	S83 L	D87N	S80I	+	−	−
GBSF1501026	KF/KZ/MA/CTX/CFP/ NOR/ENR	S83 L	D87N	S80I	+	+	−
GBSF1512425	KF/KZ/MA/CRO/CTX/ NOR/ENR/LEV	S83 L	D87N	S80I	+	+	−
GBSF1602098	KF/KZ/MA/CRO/CTX/ CFP/NOR/ENR/CIP	S83 L	D87Y	S80I	+	−	+
GBSF1512433	KF/KZ/MA/FOX/CRO/ CTX/CFP/NOR/ENR/LEV	S83 L	D87N	S80I	+	+	−

+: Presence corresponding genes

−: Absence corresponding genes

Fluoroquinolones are highly effective for the treatment of shigellosis worldwide[36]. The primary mechanism of quinolone resistance involves the accumulation of sequential mutations in QRDRs that encode DNA gyrase and topoisomerase IV[37]. The most prevalent mutations in Shigella spp. are the point mutations in gyrA codons 83, 87 and 211, and parC codon 80[38,39]. Novel mutations in QRDRs are also being discovered[39]. In the present study, three mutations in gyrA codon 83 (S → L) and/or 87 (D → N or Y) and parC codon 80 (S → I) were detected in each fluoroquinolone-resistant isolate. All substitutions are responsible for reduced affinity. In addition, the amino acid diversity at the same position may lead to different levels of quinolone resistance[40,41]. GyrA D87Y mutations were detected in only two ciprofloxacin-resistant isolates. However, the role of this mutation in ciprofloxacin resistance is unclear and requires further investigation.

Over the past few years, PMQR determinants have been deemed the most common ARGs in Enterobacteri-aceae worldwide[42]. PMQR determinants mediate only low-level quinolone resistance. However, these resistance genes are usually associated with mobile or transposable elements that allow for dissemination among Enterobac-teriaceae. In addition, the presence of PMQR genes may facilitate the selection of QRDR mutations that result in higher levels of quinolone resistance[37,43,44]. The aac (6′) -Ib-cr gene encodes an acetyltransferase that is known to reduce quinolone activity.

In the present study, all 14 isolates resistant to fluoroquinolones were positive for aac (6′) -Ib-cr, indicating the aac (6′) -Ib-cr gene is widespread in S. flexneri 2a. Compared with the aac (6′) -Ib-cr gene, the transmembrane segment efflux pump qepA gene was scarcely detected in Shigella, and we found only one ciprofloxacin-resistant isolate that was qepA-positive. The qnr family (which includes the first PMQR genes) contains a variety of subtypes, including qnrA, qnrB, qnrC, qnrD and qnrS and several qnr family genes that have been reported in Shigella[39,45]. The Qnr proteins protect DNA gyrase against quinolones and facilitate the selection of QRDR mutations that improve resistance to these antimicrobials.

5 CONCLUSIONS

In conclusion, cephalosporin and/or fluoroquinolone resistance in *Shigella* has been widely reported. To increase our understanding of Shigella in cattle, we investigated *Shigella* in calves with diarrhea and analyzed the genetic relatedness, antimicrobial susceptibility, QRDR mutations, and prevalence of PMQR and β-lactamase in *S. flexneri* 2a isolates from five provinces in northwest China. However, this study also had limitations, including the lack of a systematic surveillance system to prospectively or retrospectively detect and analyze shigellosis in veterinary clinics. Furthermore, we are unable to effectively monitor and control antibiotic abuse and the resulting spread of ARGs. Therefore, it is essential to continually monitor rates of shigellosis and the development of resistance patterns.

ACKNOWLEDGEMENTS

The authors thank all the farms for providing the samples in this study. Funding

This work was supported by funds from the National Natural Science Foundation of China (31, 272, 603, 31, 101, 836).

AVAILABILITY OF DATA AND MATERIALS

The data supporting the findings of this study are contained within the manuscript.

AUTHORS' CONTRIBUTIONS

Conceived and designed the experiments: JYZ and ZZ. Performed the experiments: ZZ and MZC. Analyzed the data: ZZ, BL, and XZZ. Contributed reagents/materials/analysis tools: MZC and FSC. Wrote the paper: ZZ. All authors read and approved the final manuscript.

ETHICS APPROVAL AND CONSENT TO PARTICIPATE

Permission to work in specific locations, information regarding the number of samples harvested, and an associated permit number for calves were not required, and no endangered or protected species were involved or harmed during this study.

COMPETING INTEREST

The authors declare that they have no competing interests.

REFERENCES OMITTED

（发表于《Antimicrobial Resistance and Infection Control》，院选 SCI，IF：2.989）

iTRAQ-based Quantitative Proteomic Analysis Reveals Bai-Hu-Tang Enhances Phagocytosis and Cross-Presentation against LPS Fever in Rabbit

Shidong ZHANG[1,2], Dongsheng WANG[1,2], Shuwei DONG[1,3], Zhiqiang YANG[1,2,3*], Zuoting YAN[1,3*]

(1. Lanzhou Institute of Husbandry and Pharmaceutical Science of Chinese Academy of Agricultural Sciences, Lanzhou 730050, China; 2. Key Laboratory of Veterinary Pharmaceutics Discovery, Ministry of Agriculture, Lanzhou 730050, China; 3. Key Laboratory of New Animal Drug Project of Gansu Province, Lanzhou 730050, China)

Abstract: [Ethnopharmacological relevance]: Bai-Hu-Tang (BHT), a classical anti-febrile Chinese formula comprising of liquorice, anemarrhena rhizome, gypsum and rice, has been traditionally used to anti-febrile treatment and promote the production of body fluid to relieve thirst. In this paper, we aim to explore anti-febrile mechanism of BHT at protein level through analyzing alteration of differentially expressed proteins (DEPs) both lipopoly-saccharide (LPS) fever syndrome and that was treated with BHT in rabbits.

[Materials and methods]: Febrile model was induced by LPS injection (i.v.) in rabbits, and BHT (750mg dry extract/kg body weight) was gavaged to another group of LPS fever rabbits. After sacrifice of animals, total protein of liver tissue was isolated, and two-dimensional liquid chromatography (LC) -tandem mass spectrometry (MS) coupled with isobaric tags for relative and absolute quantification (iTRAQ) labeling analysis was employed to quantitatively identify differentially expressed proteins in two group animals, which were compared with control group. Then bioinformatic analysis of DEPs was conducted through hierarchical Clustering, Venn analysis, gene ontology (GO) annotation enrichment, and kyoto encyclopedia of genes and genomes (KEGG) pathways enrichment.

[Result]: The results demonstrated there were 63 and 109DEPs in LPS fever group and BHT-treated group, respectively. Enrichment analysis of GO annotations indicated that BHT mainly regulated expression of some extracellular structural proteins for response to stimulus and stress. KEGG analysis showed that ribosome and phagosome were the most significant pathways. Thereinto, several proteins in phagosome pathway were significantly up-regulated by BHT, including F-actin, coronin, Rac, and major histocompatibility complex class Ⅰ (MHC Ⅰ), which work in phagocytosis and cross-presentation.

---

* Corresponding authors, E-mail: zhiqyang2006@163.com (Z.Yang), yanzuoting@caas.cn (Z.Yan).

[Conclusion]: BHT may contribute topyrogen clearance by boosting antigenic phagocytosis, degradation, and cross presentation in the liver.
Key words: Bai-Hu-Tang; LPS fever; Differentially expressed proteins; Phagocytosis; Cross presentation

1 INTRODUCTION

Bai-Hu-Tang (BHT) is a classical traditional Chinese prescription that has important research value.It is firstly recorded in a medicine classics of Treatise on Cold Damage and Miscellaneous Diseases (in Chinese Shanghan Zabing Lun) compiled by Zhong-Jing Zhang about 2000 years ago in China.It is formulated with four herbs of liquorice, anemarrhena rhizome, gypsum and rice (Fig. 1), and is traditionally used to clear excessive heat and promote the secretion of saliva or body fluids (Charles, 2000; Zhu, 2007).In recent years, its new usage refers to many clinical diseases including diabetes mellitus (Chen et al., 2008), eczema, pruritus, some anxiety and emotional disorders (Fred and Bob, 2004).In Western countries, the efficacy of Chinese herbal medicine has been recognized (Tu, 2011), and traditional Chinese medicine (TCM) is becoming frequently used in Western countries (Cheung, 2011).BHT is widespreadly concerned, and also practiced in European countries (European Herbal and Traditional Medicine Practitioners Association, 2007).

It reported that BHT significantly decreased rectum temperature that was increased by lipopolysaccharide (LPS) in the rabbit (Zhang et al., 2011; Wang et al., 2010), and BHT was a valuableantifebrile and anti-inflammatory natural drug in LPS fever symptom (Jia et al., 2013). Because LPS can be used to imitate bacterial infection and caused a typical fever in animal by intravenous or intraperitoneal injection (Shibata et al., 2005; Steiner et al., 2006; Ravanelli et al., 2007), researchers often used LPS to induce reproducible febrile responses in animals (Roth et al., 2002).Likewise, many TCM researchers believed that LPS fever rabbit shows a representative febrile syndrome of Chinese medicine pattern, and the febrile syndrome can be used to study antipyretic traditional herb medicine and ethnopharmacological research (Yu et al., 2010; Ai et al., 2011; Ni et al., 2012).Actually, according to the principle of defensive-qi-nutrient-blood syndrome differentiation (in Chinese Wei-Qi-Ying-Xue Bianzheng) in TCM, LPS fever is considered Qifen syndrome with the main symptom of unknown high fever (Yang, 2002; Ni et al., 2012).Based on the theory of prescription corresponding to syndrome (in Chinese Fangzheng Xianguan), BHT is the classical formula corresponding to Qifen syndrome, because it can significantly eliminate or improve febrile symptom (Li et al., 2007).However, the mechanism of BHT remains unclear.

As an essential part of TCM theory, TCM syndromes might be related to diverse and dynamic proteomics change, which is often along with alteration of different syndrome behaviors (Su et al., 2012).Proteomics can be used to elucidate the differential expressions of many proteins from body fluids, cells, tissues, and blood (Liu and Guo, 2011), and it allows the systematic quantitative and qualitative mapping of the whole proteome during disease (Hussain and Huygens, 2012).By

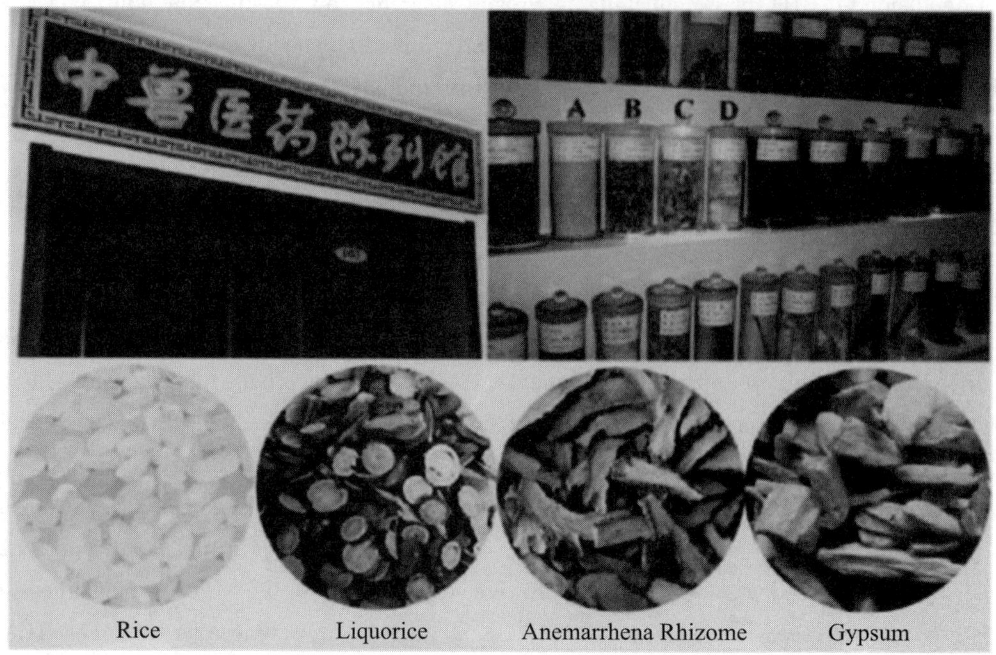

Fig. 1. Four herb voucher specimens are deposited in the institute herbarium.

Note: The specimen bottles were (A) Rice, (B) Liquorice, (C) Anemarrhena Rhizome, and (D) Gypsum.The pie charts were macro photography of herb specimens from the specimen bottles of A, B, C, and D.

analyzing these protein changes in tissue before and after TCM treatment, proteomics is powerful to study the mechanism of action of TCM remedies (Cho, 2007), and understanding of the characteristic changes in proteomics associated with a specific syndrome will facilitate syndrome identification and novel diagnostic approaches (Su et al., 2012), it also greatly promotes the quality evaluation and standardization, and the modernization and internationalization of TCM (Lu et al., 2010). Furthermore, many researchers have paid more attention to proteomics studies based on treatment-TCM syndrome animal models (Ji et al., 2015).

Against the above background, the study herein attempted to determine hepatic proteomic profiles using isobaric tags for relative and absolute quantification (iTRAQ) combined with two-dimensional liquid chromatography and tandem mass spectra analysis in LPS fever rabbits and that treated with BHT. Differently expressed proteins (DEPs) were analyzed using bioinformatics to explore the mechanism and effect of BHT, and it will support BHT traditional use at protein levels.

2 MATERIALS AND METHODS

2.1 Herbal materials and BHT extraction

The herbs of Liquorice (sliced root of *Glycyrrhiza glabra* L. in Leguminosae) and Anemarrhena Rhizome (sliced root of Anemarrhena asphodeloides Bunge in Liliaceae) and Gypsum (crystal of calcium sulfate) and Rice (nonglutinous rice, polished seed of *Oryza sativa* L. in Gramineae) were purchased from Antaitang Pharmaceutical Co., Ltd., China. All of the herbal pieces are in line with the standards of Chinese pharmacopoeia (National Pharmacopoeia Committee, 2015), which were identified by TCM expert Dr. Zuoting Yan. As showed in Fig. 1, the herb samples were deposited in a TCM Specimen Room with voucher numbers: NO. 100720 for Liquorice, NO. 101208 for Anemarrhena Rhizome, NO. 100906 for Gypsum, and NO. 101202 for Rice.

Liquorice 7.2g, Anemarrhena Rhizome 21.8g, Gypsum 60.2g and Rice 10.8g were extracted together by refluxing boiling water (1 : 10, w/v) for 2h, and water (1 : 5, w/v) for another 1h, respectively. The two-times decoction was blended together, then filtered with three-layer gauzes, and concentrated to 100ml. The experimental decoction was autoclaved at 121℃ for 15min, and kept in airtight containers at 4℃ until used. In order to calculate the concentration of the decoction, 10ml of the experimental juice was completely dried at 65℃ for 24h in an electric heating air-blowing drier, and yielded 1.08g crude powder. Thus, the concentration of the experimental decoction was 108mg/ml.

2.2 Animals and treatment

Adult New Zealand white rabbits weighting between 2.0 and 2.5kg was used in the experiments, and housed individually in rabbit stocks under controlled room humidity of (50±5)% and a temperature of (25±1)℃ with a 12h lights and 12h darkness cycles. Animals were fed commercial stock diet and water ad libitum, and allowed to stabilize for at least 3d in new surroundings before any experiments. The rabbits were randomly divided into three groups (6 animals/group). The group Ⅰ was served as a control, and received only physiological saline; the group Ⅱ was fever model, which received only LPS (15μg/kg body weight) by intravenous injection (i.v.) into the ear vein; the group Ⅲ was gavaged with BHT (a representative dose of 750mg dry extract/kg body weight) at the same time of LPS intravenous injection. Rabbits then received standard diet and water ad libitum. After LPS injection and BHT gavage for 6h, the rabbits were anesthetized with an intraperitoneal injection of sodium pentobarbital (60mg/kg) for 5min, and sacrificed immediately. A piece of liver tissue was cut from the lower edge of the right lobe of the liver, flushed with saline (0.9% NaCl), and froze in liquid nitrogen immediately until used. The animal study and bioethical allowance were approved by the Animal Care and Bioethical Committee (LZMY2015-7) of the Lanzhou Institute of Husbandry and Pharmaceutical Science of Chinese Academy of Agricultural Sciences.

2.3 Protein isolation, digestion and labeling withi TRAQ reagent

The liver tissue was ground into powder in liquid nitrogen, extracted with Lysis buffer (7M urea (Sangon, UB0148), 2M Thiourea (Sangon, TB1943), 4% CHAPS (Sigma, 19899), 40mM Tris-HCl [(BBI, TB0194), pH 8.5] containing complete protease inhibitor. The cell was lysed by sonication at 200W for 15min and then centrifuged at 4℃, 30 000g for 15min. The supernatant was mixed well with 5×volume of chilled acetone (Sangon, A6854) containing 10% (v/v) trichloroacetic acid (Sangon, TB0968) and incubated at -20℃ overnight, then centrifuged at 4℃, 30 000g, retained precipitate and washed it with chilled acetone three times. The pellet was air-dried and dissolved in Lysis buffer [7M urea (Sangon, SB0485), 2M thiourea, 4% NP40 (Fluka, 74385), 20mM Tris-HCl, pH 8.0-8.5]. The suspension was sonicated and centrifuged at 4℃, 30 000g for 15min, transferred to another tube. The disulfide bonds were reduced by 10mM dithiothreitol, and cysteines was blocked by 55mM iodacetamide (Sigma, I6125).Then the supernatant was washed by acetone, and centrifuged with the former condition.Took the pellet, air-dried it, and dissolved in 500μL 0.5M tetraethylammonium bromide (Fluka, T0761), sonicated one more time.Finally, samples were centrifuged, and supernatant was transferred to a new tube and quantified using the Bradford method (Sangon, C503041).The proteins in the supernatant were stored at-80℃ for further analysis.

Before the protein processing, each 3 individual protein samples were mixed equally into 1 specimen.As a result of the strategy, every group contained 2 sample pools, and these sample pools were enrolled to be conducted in subsequent experiments.Total protein (100μg) was took out of each sample pool solution and then the protein was digested with Trypsin Gold (Promega, PRV5280) with the ratio of protein: trypsin=30 : 1 at 37℃ for 16h.After trypsin digestion, peptides were dried by vacuum centrifugation.Peptides were reconstituted in 0.5M tetraethylammonium bromide and processed according to the manufacturer's protocol for 8-plex iTRAQ reagent (Applied Biosystems Sciex, 4390812).Briefly, one unit of iTRAQ reagent was thawed and reconstituted in 24μL isopropanol.Samples were labeled respectively with different isobaric tags as control samples labeled 113 and 114, LPS fever samples labeled 115 and 116, and BHT-treated samples labeled 117 and 118. Then labeled samples were incubated at room temperature for 2h.The labeled peptide mixtures were then pooled and dried by vacuum centrifugation.

2.4 LC-MS/MS analysis

TheiTRAQ labeled peptide mixtures were fractionated by strong cation exchange chromatography, which was conducted on a LC-20AB HPLC Pump system (Shimadzu, Kyoto, Japan) using a 4.6mm×250mm Ultremex SCX column containing 5μm particles (Phenomenex).The peptides were eluted at a flow rate of 1 ml/min with a gradient of buffer A [25mM NaH_2PO_4 in 25% acetonitrile (Fisher, A998-4), pH 2.7] for 10min, 5%-60% buffer B (25mM Na_2PO_4, 1M KCl in 25% acetonitrile, pH 2.7) for 27min, 60%-100% buffer B for 1min.The system was then maintained at 100% buffer B for 1min before equilibrating with buffer A for 10min prior to the next injection.Under a monitoring absorbance of 214nm, a total of 20 fractions were collected, then desalted with a Strata X C18 column (Phenomenex) and vacuum-dried.

Each fraction was dissolved in buffer liquid (2% acetonitrile, 0.1% formic acid (Dikma, 50144)) and centrifuged at 20 000g for 10min, then 10μL supernatant was loaded on a LC-20AD nanoHPLC (Shimadzu, Kyoto, Japan) by the autosampler onto a 2cm C18 trap column (inner diameter 75μm).The peptides were subjected to nanoelectrospray ionization followed by tandem mass spectrometry (MS/MS) in a QEXACTIVE (Thermo Fisher Scientific, San Jose, CA) coupled online to the HPLC.Intact peptides were detected in the Orbitrap at a resolution of 70 000. Peptides were selected for MS/MS using high-energy collision dissociation operating mode with a normalized collision energy setting of 27.0; ion fragments were detected in the Orbitrap at a resolution of 17 500. A data-dependent procedure that alternated between one MS scan followed by 15 MS/MS scans was applied for the 15 most abundant precursor ions above a threshold ion count of 20 000 in the MS survey scan with a following Dynamic Exclusion duration of 15s.The electrospray voltage applied was 1.6 kV.Automatic gain control (AGC) was used to optimize the spectra generated by the orbitrap.The AGC target for full MS was 3E6 and 1E5 for MS2. For MS scans, the m/z scan range was 350–2 000 Da.For MS2 scans, the m/z scan range was 100–1 800.

2.5 Data analysis

MS raw data was converted into MGF using Proteome Discoverer 1.2 (PD 1.2, Thermo), and the MGF data file were searched by using the Mascot search engine (Matrix Science, London, UK; version 2.3.02) to identify proteins.Each confident protein identification involves at least one unique peptide.For protein quantification, it was required that a protein contains at least two unique spectra.The quantitative protein ratios were weighted and normalized by the median ratio in Mascot. We only used ratios with $P<0.05$, and only fold changes of>1.5 or<0.5 for samples were considered as significant.

2.6 Bioinformatic analysis

Gene ontology annotation of proteins was performed using Blast2GO program against the non-redundant protein database (NR, NCBI).Gene Ontology (GO) is an international standardization of gene function classification system.It has 3 Ontologies which can describe the molecular function, cellular component, and biological process, respectively. The main types of annotations were obtained from the gene ontology consortium website (http://www.geneontology.org).Then DEPs were mapped to different biological pathways according to functional categories of the kyoto encyclopedia of genes and genomes (KEGG) pathway database (http://www.genome.ad.jp/kegg/pathway.html).The significantly altered GO terms and KEGG pathways were identified based on the criteria of having two DEPs and a hypergeometric test.DEPs identified by GO and KEGG analysis were applied to obtain more biological information by cluster analysis.Cluster analysis was carried out by the Cluster 3.0, and the results were visualized and browsed using the TreeView program.Venn diagram was drawn at online Venny 2.0 (http://bioinfogp.cnb.csic.es/tools/venny/index.html).P value used in statistical analysis was calculated by a hyper geometric distribution probability formula, and it reflected whether the statistical result was true or false.In this analysis, $P<0.05$ was considered statistically significant.

3 RESULTS

3.1 Clustering analysis

All differentially expressed proteins were clustered through hierarchical clustering.Proteins expression levels were indicated by a color scale in two group rabbits.Corresponding to their individual members, it also showed a complex expression profile in different groups.Visual inspection of liver proteins expression profiles (Fig. 2) displayed that individual group proteins expression profiles may predict the LPS fever group and BHT-treated group. Among two groups of samples, 148 proteins were clustered, most of them displayed in a specific manner.In general, the subfamilies of A, C, and H were associated with LPS fever, whereas A to F, and H were nearly all related to BHT treatment.

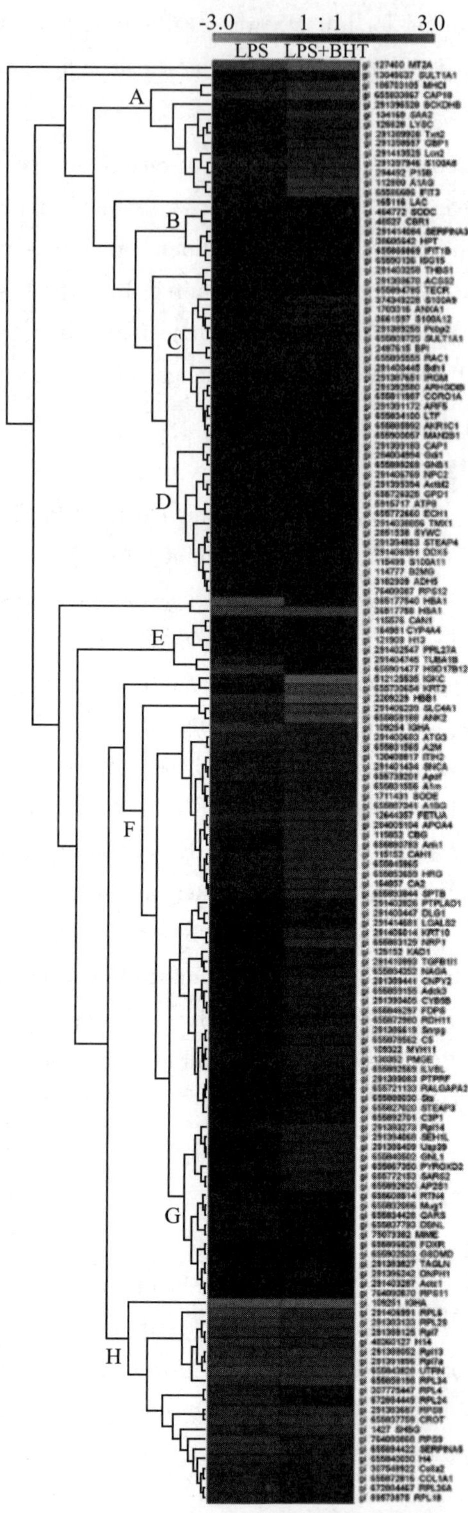

Fig. 2. Hierarchical clustering of differentially expressed proteins in rabbit liver.

Note: The log2-transformed ratios are shown in red-green color scale where red represents a higher expression and green represents a lower expression. Gene identities and symbols corresponding to proteins were listed on the right side of each row. A-H are different clustering subfamilies.

3.2 Numbers and distribution of DEPs

A total of 2798 proteins were determined byiTRAQ-coupled 2D LC-MS/MS analysis in rabbit liver tissue. Among them, proteins that displayed significantly altered expression levels with fold cut-off of>1.5 or<0.5 comparing with the control group were considered as differentially expressed proteins ($P<0.05$). There were 63 DEPs in LPS fever group, and 109 proteins in BHT-treated group. Regarding to DEPs from LPS fever group, 26 proteins were up-regulated, and 37 proteins were down-regulated. However, there were 49 proteins up-regulated and 60 proteins down-regulated in BHT-treated group. The Venn analysis further displayed that LPS fever group and BHT-treated group shared 21 up-regulated proteins and 21 down-regulated proteins, respectively (Fig. 3).

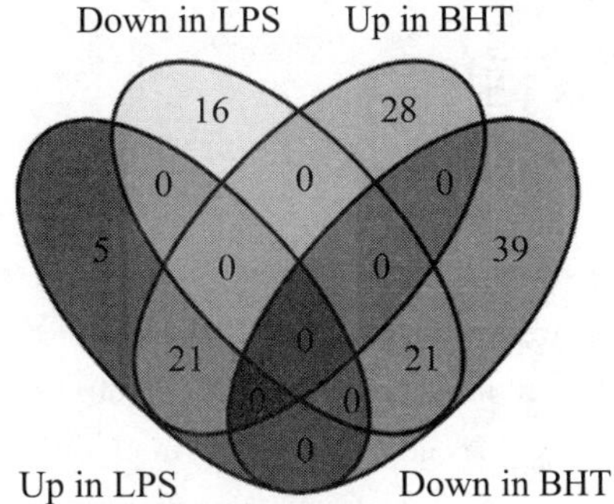

Fig. 3. A Venn diagrams showing the distribution of shared DEPs compared with control in LPS fever group and BHT treated group, respectively.

Note: Blue part was up-regulated proteins in the LPS fever group (Up in LPS), yellow part was down-regulated proteins in the LPS fever group (Down in LPS); while green part was up-regulated proteins in BHT-treated group (Up in BHT), red part was down-regulated proteins in BHT-treated group (Down in BHT). The overlap sections were the numbers of common proteins in different groups.

3.3 Functional classification of DEPs

Functional annotation ofDEPs was performed using Blast2GO program. Three main types of annotations were obtained from the gene ontology consortium website: cellular component, metabolic functions, and biological processes. Sixty three proteins were significantly altered in LPS fever animals, and enrichment analysis of GO showed that DEPs mainly were structural constitutes of macromolecular complex, and took part in processes of transport and localization (Fig. 4A). Enrichment analysis for BHT-treated group indicated that DEPs primarily involved in processes of response to stimulus and stress, and the proteins were structural molecule located at the extracellular region (Fig. 4B).

3.4 Kyoto encyclopedia of genes and genomes pathways

Differentially expressed proteins was selected to identify enriched pathways in dataset, and pathways analysis on DEPs data were conducted based on KEGG. These DEPs of LPS fever group were found to be enriched in ribosome (20%), phagosome (10.91%), african trypanosomiasis (9.09%), tuberculosis (9.09%), and malaria (7.27%). The DEPs from BHT-treated group

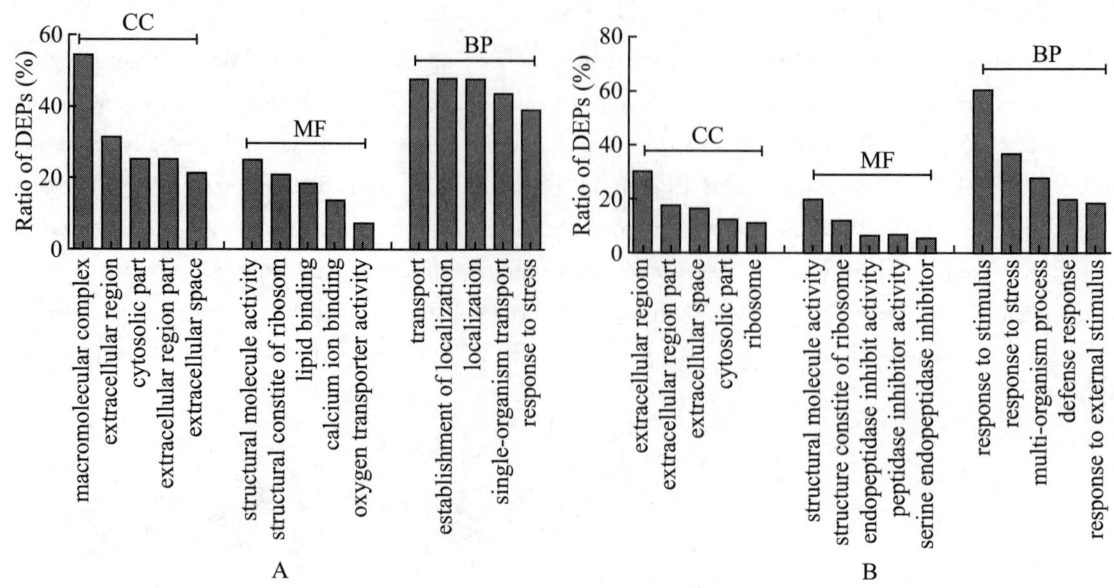

Fig. 4. Classification of DEPs based on their functional annotations using Gene Ontology analysis.

Note: A is LPS fever group, B is BHT-treated group. BP means biological processes, CC is cellular components, and MF is molecular functions. Vertical coordinates are different items of functional annotations, and abscissas are the percentage of proteins referred to the corresponding annotation in all DEPs.

were found to been enriched in ribosome (9.09%), phagosome (7.07%), systemic lupus erythematosus (6.06%), viral myocarditis (6.06%), and complement and coagulation cascades (6.06%) (Table 1).

Table 1 Pathway enrichment analysis of DEPs.

Pathway	Ratio of DEPs (%)	P-value	Pathway ID
LPS			
Ribosome	20	4.62E-07	ko03010
Phagosome	10.91	0.000101566	ko05144
VAfricantrypanosomiasis	9.09	0.000121674	ko05143
Tuberculosis	9.09	0.004113382	ko04966
Malaria	7.27	0.005244643	ko05152
Amoebiasis	7.27	0.007030718	ko05320
Viralmyocarditis	7.27	0.007030718	ko05330
Collecting duct acid secretion	5.45	0.009468653	ko04145
Autoimmune thyroid disease	5.45	0.01387688	ko05146
Allograft rejection	5.45	0.01408804	ko05202
LPS+BHT			
Ribosome	9.09	0.002866728	ko03010

(continued)

Pathway	Ratio of DEPs (%)	P-value	Pathway ID
Phagosome	7.07	0.004683464	ko04145
Systemic lupuserythematosus	6.06	0.00624934	ko05322
Viralmyocarditis	6.06	0.00702792	ko05416
Complement and coagulation cascades	6.06	0.009785233	ko04610
Tuberculosis	6.06	0.01597254	ko05152
Staphylococcusaureus infection	5.05	0.01565514	ko05150
Dilatedcardiomyopathy	5.05	0.04004298	ko05414
Mineral absorption	4.04	0.001891274	ko04978
Africantrypanosomiasis	4.04	0.01262041	ko05143

3.5 Phagosome pathway involved in the response to LPS fever and BHT action

According to the result of KEGG pathways, phagosome pathway has important roles in response to LPS fever and BHT treatment. As showed in Fig. 5, some proteins involved in phagocytosis were up-regulated significantly, including F-actin, coronin, and Rac, particularly in BHT-treated group. Additionally, as important antigenic presentation molecule, MHC Ⅰ was up-regulated in LPS fever group, and further enhanced in BHT-treated group, but MHC Ⅱ molecule was not changed in two groups. This finding indicated that phagocytosis and cross presentation were boosted by BHT treatment to LPS fever.

4 DISCUSSION

Liver, as the largest internal organ in mammals, plays an important role in maintenance of normal physiological functions of tissues and organs (Chen et al., 2011). The modern medicine indicates that liver mainly function in biotransformation and detoxification of xenobiotics (Wu et al., 2013). In traditional Chinese medicine, liver sites at lower energizer (in Chinese Xia Jiao), and governs blood storing and free flow of Qi according to TCM theories of triple-energizer syndrome differentiation (in Chinese Sanjiao Bianzheng) and defensive-qi-nutrient-blood syndrome differentiation (in Chinese Wei-Qi-Ying-Xue Bianzheng). It is representative to select liver for elucidating dysfunction of viscera when febrile syndrome transmits and changes. Although differentially expressed genes in liver was detected in febrile syndrome rabbits by microarray analysis (Zhang et al., 2015), the changes in mRNA expression may poorly correlate with the change in protein expression (Lao et al., 2014). It is better to understand the essence of TCM using proteomics combined with other omics research (Altelaar et al., 2013). Therefore, proteomics research was set out with the aim of identifying alteration of proteins expression profile in liver tissue of LPS fever rabbits and that also treated with BHT.

Based on the data from iTRAQ analysis, different ontological classes of DEPs were analyzed to explore anti-febrile mechanism of BHT. The top five altered GO items showed that proportion of extracellular proteins were increased by BHT (Fig. 4B), which participated in response to stimulus. At the same time, components of macromolecular complex decreased (Fig. 4).This alteration demonstrated that the predominant DEPs mainly contributed to intracellular processes of transport and location in LPS fever syndrome, whereas after BHT treatment mostDEPs were extracellular structural molecules for response to stimulus and stress.Because extracellular structural molecules often provide structural support and contextual information to cells within tissues and organs (Cichon and Radisky, 2014).Thus, BHT may influence cellular structure, biomechanical signals recognition and response in the liver.

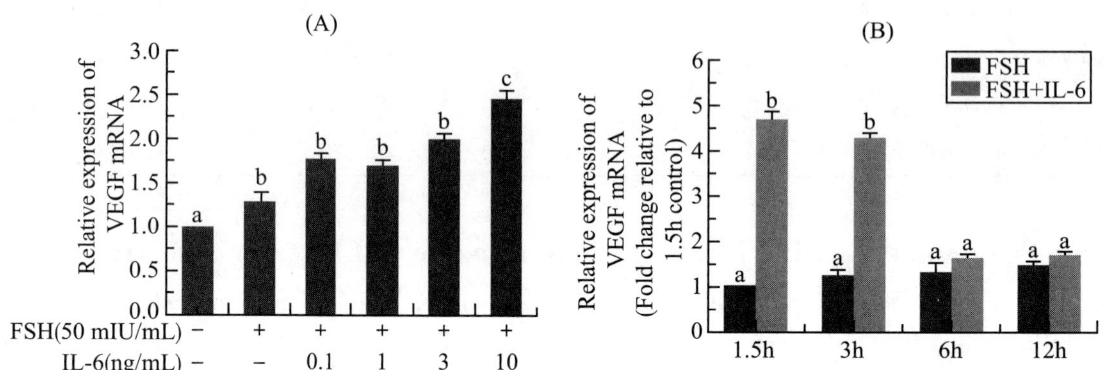

Fig. 5. Phagocytosis and exogenous antigen presentation pathway mediated by major histocompatibility complex class Ⅰ (MHC Ⅰ) and class Ⅱ (MHC Ⅱ).

Note: MHC Ⅰ involves an important pathway of cross presentation.Exogenous antigens is internalized through phagosome, degraded at reactive oxygen species (ROS) response, then antigen fragments is loaded on MHC Ⅰ in endoplasmic reticulum (ER) or translocated back into the phagosome for local MHC Ⅰ peptide loading.The exogenous antigen fragment can also be loaded on MHC Ⅱ in endosome.Proteins in Red box are DEPs.Red diamonds + numbers mean fold changes of DEPs in LPS group, and red triangles are fold changes of DEPs in BHT-treated group.

Another important analysis of DEPs was biological pathways comparison.When categorized by KEGG enrichment, phagosome pathways was highlighted, which accounted for 10.91% and 7.07% of the total altered proteins in LPS fever animals and that treated with BHT, respectively ($P<0.01$, Table 1).It reported that phagosome accumulation often increases nonspecific uptake and retention in the liver (Choi, 2014), and BHT promoted LPS clearance by increasing genes expression of phagosome (Zhang et al., 2015).Actually, phagosome pathway depends particularly on a process of phagocytosis to engulf antigen and sequester it into a phagosome (Kim et al., 2014). Phagocytosis plays a major role in ingesting and destructing large particles including cellular debris, dead cells, foreign objects, and pathogens.It also serves as a basis for antigen processing in the immune response and tissue maintenance.Phagocytosis is professionally conducted by phagocytes with a vast and complex machinery of regulatory proteins.As depicted in Fig. 5, after phagocytes recognize the particles or pathogens, these non-self antigenic materials are internalized into a sealed phagosome through formation of early phagocytic cups.Then the phagosome is matured through fusion of successive endosomes until form an acidic and degradative environment in phagolysosome that is critical for pathogen destruction, presentation, and disposal (Davies et al., 2013).The initiation and coordination of early stages of phagocytosis involve F-actins that was rapidly polymerized by actin around the phagocytic cup (Yokoyama et al., 2011) and normally disassembled upon closure of the early phagosome (Swanson, 2008). Additionally, nicotinamide adenine dinucleotide phosphate (NADPH) oxidase complex executes reactive oxygen species response for $O_{2.-}$ production to degrade antigen. Rac is a critical part of the NADPH and a key component of ROS response (Bylund et al., 2010), especially Rac1 and Rac2 are critical components of the NADPH oxidase activation in macrophages and neutrophils, respectively (Panday et al., 2015). After treatment with BHT, F-actin, coronin and Rac were up-regulated significantly, which were not altered in LPS fever group (Fig. 5). These findings suggested that BHT mediated the phagocytosis of LPS by stimulating both phagocytic cup formation and NADPH oxidase activation.

Antigen presentation is important to immune recognition and response. Two important molecules of major histocompatibility complex class Ⅰ (MHC Ⅰ) and class Ⅱ work in different pathways of antigen presentation. All nucleated cells express MHC Ⅰ molecules that present MHC I-peptide complexes to CD^{8+} T cells. But only a restricted set of cells express MHC Ⅱ molecules that present MHC Ⅱ-peptide complexes to CD^{4+} T cells. Usually, MHC Ⅰ presents endogenous antigens, and MHC Ⅱ presents exogenous antigens (Rock et al., 2016). However, an additional antigen presentation pathway is cross-presentation that allows exogenous antigens to access the processing and presentation machinery so that exogenous antigenic peptides are displayed on MHC Ⅰ for immunity recognition and consequently priming of CD^{8+} T cell responses (Platzer et al., 2014; Segura et al., 2015). This pathway operates in dendritic cells and other phagocytes and plays a central role in immune surveillance (Kurts et al., 2010). In this study, the expression of MHC Ⅰ was up-regulated in LPS fever group, and also was furtherly boosted by BHT. But, there were not any expression changes of MHC Ⅱ (Fig. 5). Furthermore, it proved that BHT has no effect on peripheral CD^{4+} T cells number in LPS fever rabbit, but significantly increased the quantity of CD^{8+} T cells (Zhang et al., 2013). Therefore, BHT can enhance the cross presentation ability to clear pyrogenic LPS in rabbit liver.

5 CONCLUSIONS

This study identified 63 and 109DEPs in the liver of LPS fever animals and that treated with BHT. Bioinformatics analysis demonstrated that BHT mainly affect extracellular proteins for response to stimulus and stress for LPS fever treatment. Further analysis showed that BHT not only mediated antigen capture and internalization but also enhanced antigenic degradation in a Rac-dependent manner in phagosome pathway. Additionally, BHT boosted antigen cross presentation through enhancing the expression of MHC Ⅰ molecule. Therefore, it may be an important mechanism that BHT contributed to clear pyrogen by enhancing antigen phagotosis, degradation, and cross-presentation in liver. Further experimental investigations are needed to explore how BHT enhance the expression of these key proteins.

Abbreviations: TCM, traditional Chinese medicine; BHT, Bai-Hu-Tang; LPS, lipopolysaccharide; iTRAQ, isobaric tags for relative and absolute quantification; LC-MS/MS, liquid chromatography-tandem mass spectrometry; DEPs, differently expressed proteins; GO, Gene Ontology; KEGG, kyoto encyclopedia of genes and genomes; MHC Ⅰ, major histocompatibility complex class Ⅰ; MHC Ⅱ, major histocompatibility complex class Ⅱ; NADPH, nicotinamide adenine dinucleotide phosphate; ROS, reactive oxygen species

AUTHORS' CONTRIBUTIONS

SZ, ZY and ZTY conceived the study. SZ designed and performed all of experiments, and analysed the data and wrote the manuscript. DW and SD contributed to animals treatment and tissue samples collection. All authors read and approved the final version of the manuscript.

ACKNOWLEDGEMENTS

This study was supported by the National Natural Science Foundation of China (31402244) and the Central Scientific Research Institutes for Basic Research Fund of China (1610322015012). Gratitude is also sent to the anonymous and editor whose suggestions greatly improved the manuscript.

REFERENCES OMITTED

(发表于《Journal of Ethnopharmacology》, 院选 SCI, IF: 2.981)

Synthesis and Biological Activity Evaluation of Novel Heterocyclic Pleuromutilin Derivatives

Yunpeng YI, Yunxing FU, Pengcheng DONG, Wenwen QIN, Yu LIU, Jiangping LIANG, Ruofeng SHANG*

(Key Laboratory of New Animal Drug Project of Gansu Province/Key Laboratory of Veterinary Pharmaceutical Development, Ministry of Agriculture/Lanzhou Institute of Husbandry and Pharmaceutical Sciences of CAAS, Lanzhou 730050, China)

Abstract: A series ofpleuromutilin derivatives were synthesized by two synthetic procedures under mild reaction conditions and characterized by Nuclear Magnetic Resonance (NMR), Infrared *Spectroscopy* (IR), and High Resolution Mass *Spectrometer* (HRMS). Most of the derivatives with heterocyclic groups at the C-14 side of pleuromutilin exhibited excellent in vitro antibacterial activities against *Staphylococcus aureus*, methicillin-resistant *Staphylococcus aureus* (MRSA), methicillin-resistant *Staphylococcus epidermidis* (MRSE), and vancomycin-resistant *Enterococcus* (VRE) in vitro antibacterial activity. The synthesized derivatives which contained pyrimidine rings, 3a, 3b, and 3f, displayed modest antibacterial activities. Compound 3a, the most active antibacterial agent, displayed rapid bactericidal activity and affected bacterial growth in the same manner as that of tiamulin fumarate. Moreover, molecular docking studies of 3a and lefamulin provided similar information about the interactions between the compounds and 50S ribosomal subunit. The results of the study show that pyrimidine rings should be considered in the drug design of pleuromutilin derivatives.

Key words: Pleuromutilin derivatives; Antibacterial activity; Synthesis; Molecular docking

1 INTRODUCTION

Antibiotics have been necessary life-saving drugs since the advent of penicillin in 1928[1]. However, antibiotic-treatment faces total defeat as a result from drug resistance. Meanwhile, the discovery of antibiotics has gotten stuck in bottleneck. Although there have been more than 31 lead classes of antibiotics found and 160 kinds of antibiotics approved from the 1890s-1980s, few lead

* Corresponding author, E-mail: shangruofeng@ caas.cn

class antibiotics have been found in the past thirty years[1].

Many antibiotic treatment failures were associated with the spread of bacterial drug resistance, such as methicillin-resistant Staphylococcus aureus (MRSA), methicillin-resistant Staphylococcus epidermidis (MRSE), and vancomycin-resistant Enterococcus (VRE)[2]. Consequently, the study for clinically-available potent antibiotics that are effective against multidrug-resistant pathogens is becoming exceedingly urgent[3].

Pleuromutilin was first discovered and isolated from basidiomycetes, Pleurotus mutilus, and P. passeckerianus in 1951[4]. Tiamulin (Figure 1) was the first derivative approved for veterinary use in 1979, following which valnemulin (Figure 1) was approved in 1999. It was not until 2007 that retapamulin (Figure 1) was used in human medicine[4]. The C-14 side of pleuromutilin is the modest modification position for designing a high antibacterial agent and thus improves biological activity and enhances water-solubility[5-8].

The antibacterial mechanisms ofpleuromutilin derivatives are to inhibit protein synthesis by blocking the peptidyl transferase center (PTC). Their tricyclic mutilin core of pleuromutlin derivative intervenes the A-site and the C-14 side chain extends to the P-site, which disturbs the translation of mRNA[4,9-12].

Many authors realized thatthioether's presence in C-14 could improve its antibacterial activities[11,12]. Recently, three new potential drugs, BC-3205, BC-7013, and lefamulin, which contain heterocyclic groups, have been approved in clinical trials. Nabriva's lead product, lefamulin, has entered phase Ⅲ of clinical trials on community-acquired bacterial pneumonia[12] and has received U.S. Food and Drug Administrationfast-track status[13].

In the present project, we decided to screen potential derivatives on thepleuromutilin skeleton, as well as to increase our knowledge of the binding site by molecular docking.

2 RESULTS AND DISCUSSION

2.1 Chemistry

Mutilin 14-tosyloxyacrtate (2) was prepared by pleuromutilin and p-toluene sulfonyl chloride under basic conditions in 78% yield. The pleuromutilin derivatives 3a-h were formed, as shown in Scheme 1. Mercaptan, used as an nucleophilic agent, easily attacked the C21-position of 14-tosyloxyacrtate under basic conditions. The synthetic route for compound 3i is depicted in Scheme 2. Compound 4 was prepared by reacting mutilin 14-tosyloxyacrtate (2) with sulfocarbamide under alkaline conditions. Compound 3i was obtained directly from compound 4 and 5 with a one pot reaction.

All synthesis compounds were characterized by Nuclear Magnetic Resonance (NMR), Infrared Spectroscopy (IR) (Thermo Nicolet Corporation, Waltham, MA, USA.), and High Resolution Mass Spectrometer (HRMS) (Bruker Corporation, Billerica, MA, USA.), A colorless crystal of compound 3a, block-like, was obtained by slow evaporation of a chloroform solution (Figure 2).

2.2 Antibacterial activity

The synthesized derivatives 3a-i were evaluated for their antibacterial activity against several

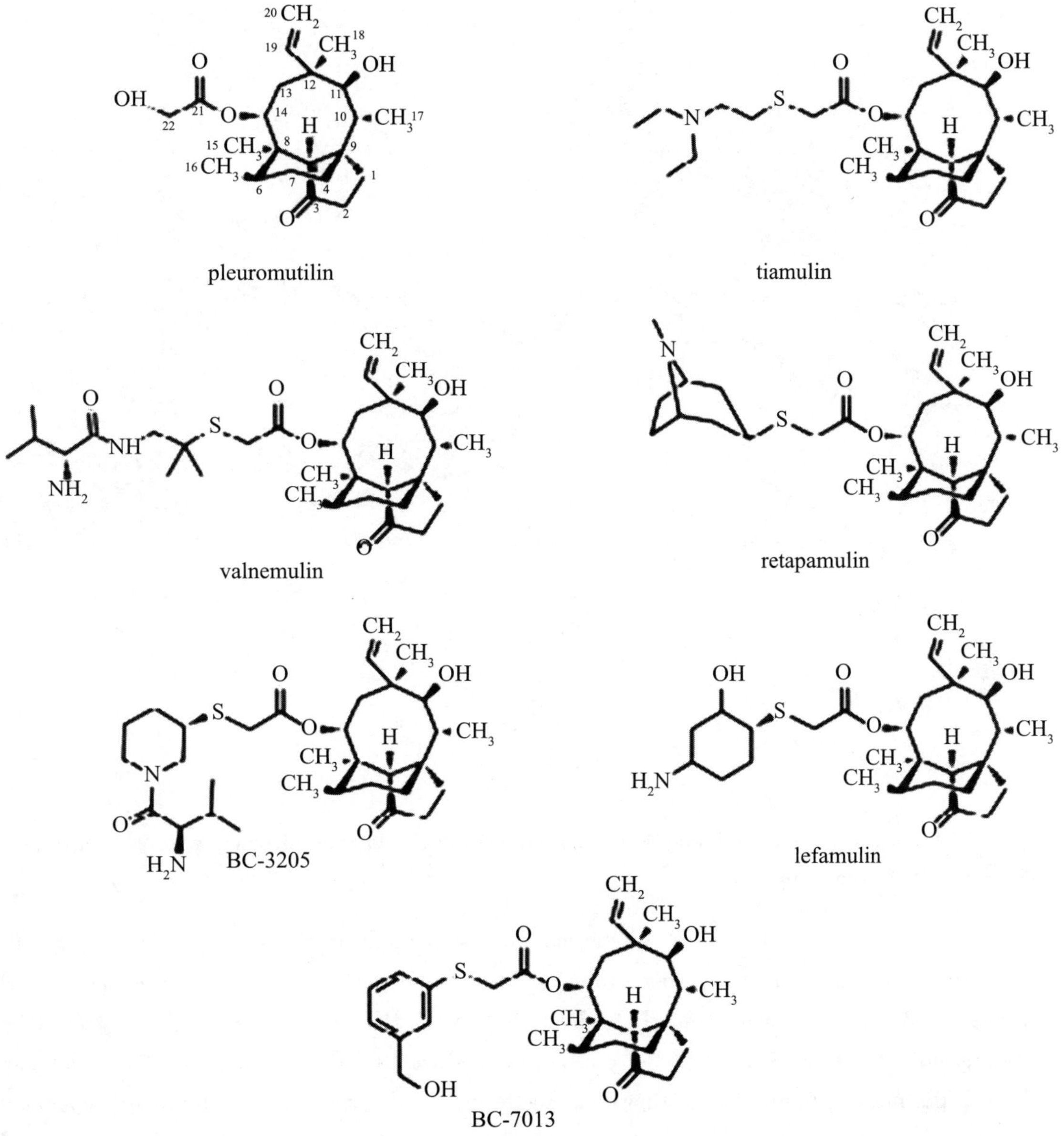

Figure 1. Structural formulas of pleuromutilin and derivation thereof.

representative Gram-positive strains, including Staphylococcus aureus, methicillin-resistant Staphylococcus aureus (MRSA), methicillin-resistant Staphylococcus epidermidis (MRSE), and vancomycin-resistant Enterococcus (VRE), and a Gram-negative bacterium, E. coli. Tiamulin fumarate served as a control. The results of minimum inhibitory concentrations (MICs) and lipophilicities (Clogp), which were predicted by ACD/Labs (Toronto, Ontario, ON, Canada), are shown in Table 1. The synthesized pleuromutilin derivatives exhibited modest to excellent antibacterial activity against S.aureus, MRSA, MRSE, and VRE (MIC, 0.0625-2μg/mL), respectively, while they showed poor activity against E.coli. Compounds 3a, 3b, and 3f with a pyrimidine group showed higher antimicrobial activity than the other derivatives.

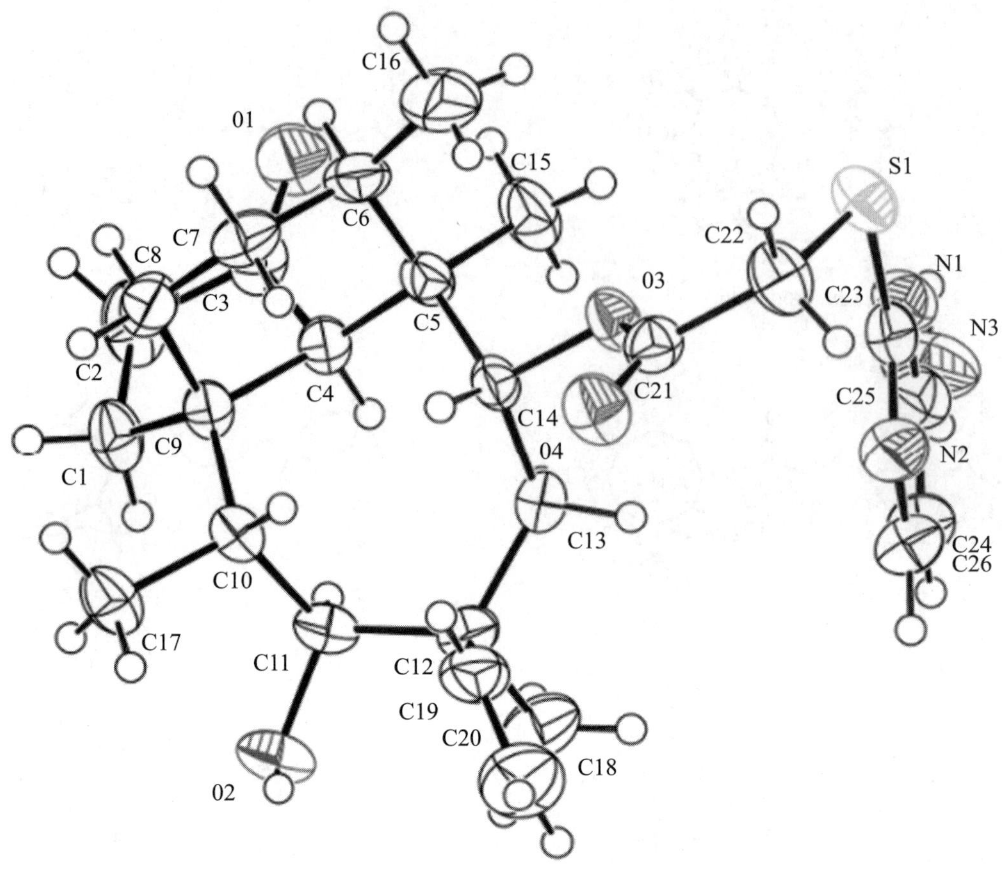

Figure 2. Oak Ridge Thermal-Ellipsoid Plot Program (ORTEP) diagram for compound 3a with ellipsoids set at 75% probability.

The most potent compound, 3a, displayed promising antibacterial activity, the lowest MIC value in 3a-i and was therefore further evaluated for in vitro time-kill assay[14]. The bactericidal properties of 3a were compared at 1 × MIC and 6 × MIC against S. aureus (3a MIC = 0.0625μg/mL, tiamulin MIC = 0.0625μg/mL), and MRSA (3a MIC = 0.125μg/mL, tiamulin MIC = 0.25μg/mL) (Figure 2). As shown in Figure 3, 3a displayed a concentration-dependent effect, with faster killing kinetics at higher concentrations. Although 1 × MIC of compound 6a and tiamulin slowed bacterial propagation, their 6 × MIC achieved a 3 - $\log_{10}$ reduction in 4 - 6h. Compared to tiamulin, 3a showed more rapid bactericidal kinetics against S.aureus and MRSA with the same concentration (1 × MIC). Notably, 3a at 6 × MIC concentration reduced the viable count by approximately 6.8-$\log_{10}$ and achieved complete killing after 6h of incubation.

Further comparison of the physicochemical parameters of compounds 3a-i, having a relatively hydrophilic C-14 side chain, shows that at good solubility of all of these derivatives. The lipophilicity of compound 3a (ClogP = 3.23) is close to that of tiamulin, which contributes to its antibacterial activity comparable to that of tiamulin[15]. In summary, 3a exhibited signigicantly improved bactericidal activity compared to that of tiamulin.

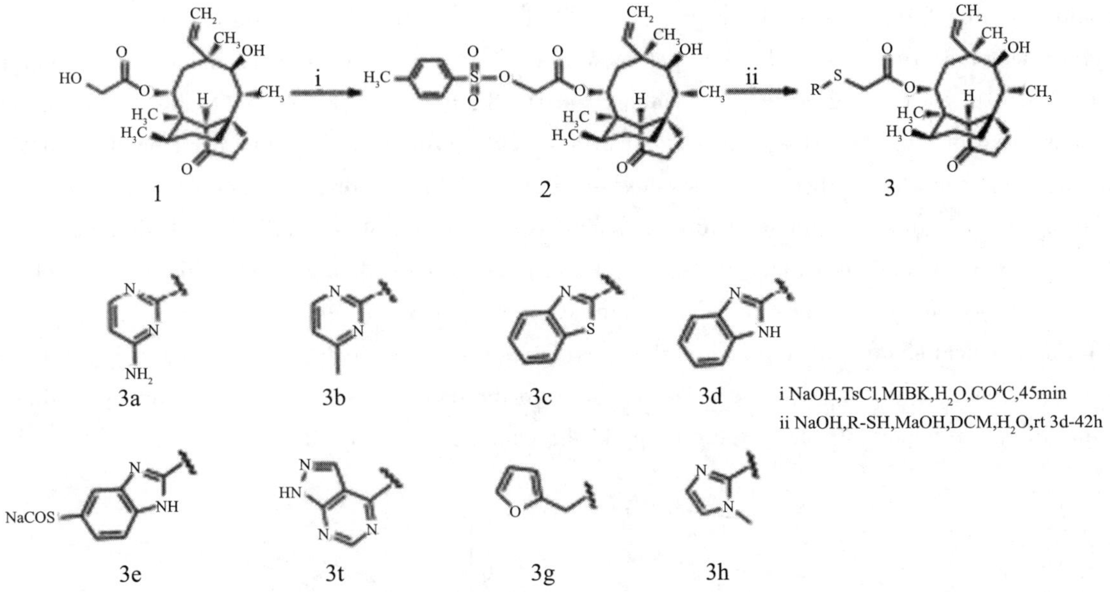

3a 3b 3c 3d

i NaOH,TsCl,MIBK,H_2O,CO℃,45min
ii NaOH,R-SH,MaOH,DCM,H_2O,rt 3d-42h

3e 3t 3g 3h

Scheme 1. Synthesis of compounds 3a–3h.

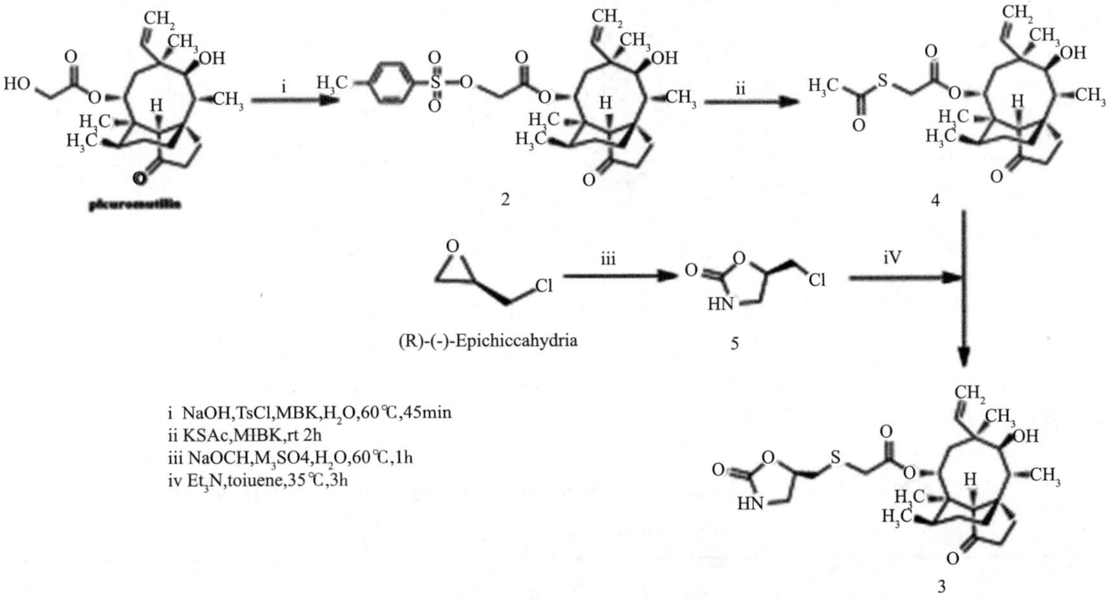

(R)-(-)-Epichiccahydria

i NaOH,TsCl,MBK,H_2O,60℃,45min
ii KSAc,MIBK,rt 2h
iii NaOCH,M_3SO4,H_2O,60℃,1h
iv Et_3N,toiuene,35℃,3h

Scheme 2. Synthesis of compound 3i.

2.3 Molecular docking studies

Molecular docking was performed with the aim of revealing the relations of the derivatives with their antibacterial activity at the atomic level. The suggestion that all pleuromutilin derivatives have been demonstrated to be effective might owe to their binding of the bacterial ribosome to PTC. To understand this phenomenon, we performed molecular docking to determine the reliability of this proposal. On applying Homdock software[16], the redocking of lefamulin into 5HI7[12] placed the compound in the PTC as that X-ray crystallography structure. The crystal structures was a typical complex of the 50s ribosome about S. aureus and lefamulin. The docking results showed that the eight-membered ring can bind to the active site of the ribosome in the same manner as lefamulin with RMSD (Root-mean-square Deviation) at 0.88-1.12 Å (Figure 4, Table 2). Hydroxyl groups in all of the derivatives were located in a suitable position to form hydrogen bonds with G-2532, and some ester carbonyl groups were bound to C-2088 as hydrogen bonds. Although the tricyclic mutilin core did not form any hydrogen bonds with the PTC, it is stabilized by hydrophobic and Van der Waals interactions. Compounds 3b and 3d formed three hydrogen bonds with similar docking modes. Compound 3a, the most promising of the candidates (ΔGb = -8.855 kcal/mol), bonded to PTC as a typical characteristic based on the docking results. Interestingly, 3a, 3b, and 3f exhibited generally low energy and MIC values. The results were probably caused by its amine in pyrimidine, which plays an important role in the binding of the complex (Figure 4B).

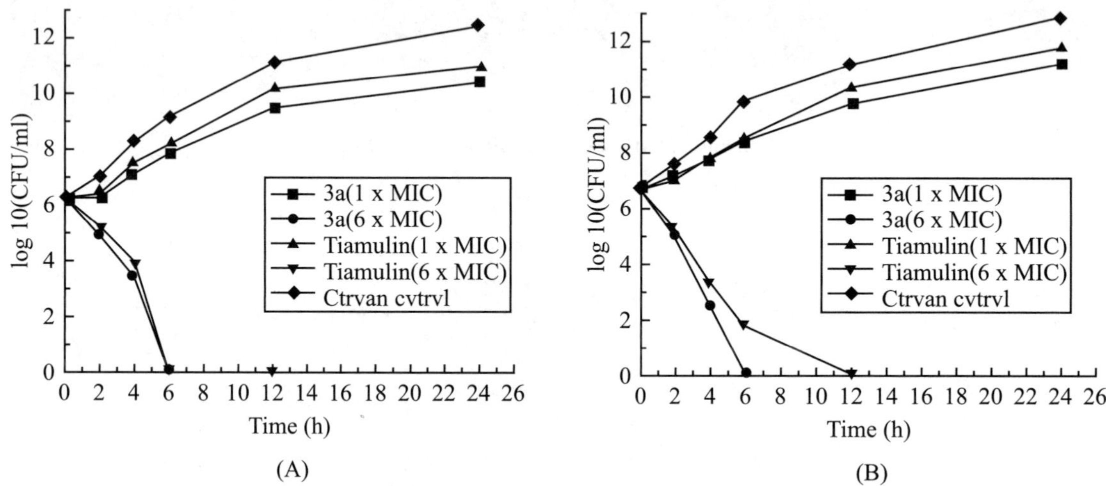

Figure 3. Time-kill kinetics of compound 3a against *S.aureus* (ATCC 25923) (A) and MRSA (ATCC 43300) (B).

Note: Mean values of the CFU/mL (colony forming units per milliliter) were obtained from measurements taken in triplicate.

Table 1. Antibacterial activity (Minimum Inhibitory Concentration) of the synthesized pleuromutilin derivatives.

Compound	MIC (fxg/mL)					ClogP
	E.coli (ATCC25922)	MRSA (ATCC43300)	S.aureus (ATCC25923)	MRSE (ATCC51625)	VRE (ATCC51559)	
3a	4	0.125	0.0625	0.0625	0.0625	3.23
3b	4	0.5	0.0625	0.5	0.25	4.96
3c	8	0.25	0.5	0.5	1	4.99
3d	8	0.25	0.125	0.25	0.25	4.46
3e	16	1	0.5	1	2	2.49
3f	2	0.125	0.0625	0.25	2	2.95
3g	8	0.5	0.5	1	1	3.83
3h	16	0.5	1	1	0.125	2.89

(continued)

Compound	MIC (fxg/mL)					ClogP
	E.coli (ATCC25922)	MRSA (ATCC43300)	S.aureus (ATCC25923)	MRSE (ATCC51625)	VRE (ATCC51559)	
3i	8	0. 25	0. 0625	0. 25	0. 5	1. 85
Tiamulin	8	0. 25	0. 0625	0. 5	0. 5	3. 63

Despite their similarities in H-binding, the activity of 3a was better than 3d. When we modeled the structural overlaps of compounds 3a and 3d, their side chain appeared markedly different. The pocket of PTC showed hydrophilicity (blue in Figure 4C) at the A-2478 position, which could make 3a more stable than 3d.

The binding sites of thepleuromutilin derivatives were different from that of other antibiotics, like macrocyclic and clindamycin. The macrocyclic ring was bound at aminoacyl-tRNAs at the bacterial ribosome (A-site)[17], which is the same side of the ribosomal tunnel as its side chain. In the crystal structure (PDB ID: 1JZX), clindamycin also binds to the A-site of their PTC of bacterial ribosomes[18].

Table 2. Binding RMSD (Root-mean-square Deviation), number of noncovalent molecular interactions and free energy.

Compound	RMSD[a]	Residue	Atom of Compound	Hydro I Interaction	Distance (Å)[b]	Angles (°)	ΔGb (kcal/mol)
3a	0. 98	G2532	OH (eightmembered ring)	H-bonding	2. 06	168. 5	-8. 855
		G2088	C=O ester	H-bonding	2. 2	128. 1	
		A2478	NH_2	H-bonding	1. 67	170. 6	
3b	0. 91	G2532	OH (eightmembered ring)	H-bonding	2. 08	175. 1	-8. 583
		G2088	C=O ester	H-bonding	2. 21	132. 6	
		C2090	NH	H-bonding	1. 97	143. 9	
3c	0. 94	G2532	OH (eightmembered ring)	H-bonding	1. 67	175. 8	-7. 816
		G2088	C=O ester	H-bonding	2. 22	133. 1	
3d	0. 92	G2532	OH (eightmembered ring)	H-bonding	1. 67	175. 1	-8. 065
		G2088	C=O ester	H-bonding	2. 257	131. 5	
		C2090	imidazole NH	H-bonding	1. 87	147. 5	
3e	1. 12	G2532	OH (eightmembered ring)	H-bondingπ-π	1. 86	170. 3	-7. 523
		A2089	benzene ring	interaction	4. 37		
3f	0. 91	G2532	OH (eightmembered ring)	H-bondingπ-π	1. 75	163. 3	-8. 218
		A2089	pyrimidine ring	interaction	4. 34		

(continued)

Compound	RMSD[a]	Residue	Atom of Compound	Hydro I Interaction	Distance (Å)[b]	Angles (°)	ΔGb (kcal/mol)
3g	0. 92	G2532	OH (eightmembered ring)	H-bonding	1. 67	177. 0	-7. 501
		G2088	C=O ester	H-bnnding	2. 15	131. 9	
3h	0. 91	G2532	OH (eightmembered ring)	H-bonding	1. 78	170. 7	-8. 483
		G2088	C=O ester	H-bonding	2. 04	128. 1	
3i	0. 92	G2532	OH (eightmembered ring)	H-bonding	2. 02	170. 6	-8. 413
		G2088	C=O ester	H-bonding	2. 20	140. 0	
			C24 C=O		2. 7	123. 6	
			C24 C=O	H-bonding		105. 9	
		A2089	NH		1. 812. 34	137. 6	

[a] predicted pose was considered successful if the RMSD between the predicted pose and the native ligand was less than 2. 0 Å.[b] The H-binding and $\pi-\pi$ interaction of distance. These bonding distance could evaluate their strong or weak interaction.

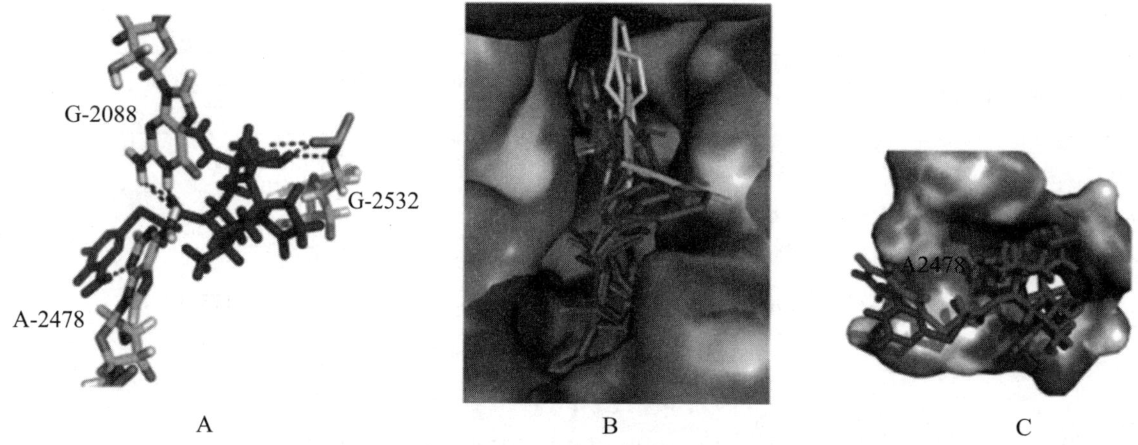

Figure 4. (A) Docking modes of the synthesized compounds into 5HL7. Key amino acid residues and ligand-active site interactions are shown. 3a (blue) to 5HL7; Important residues are drawn in sticks and different color. Hydrogen bonds are showed as dashed red lines; (B) The best pose of the compounds obtained from the docking study in the active site of the peptidyl transferase center. (3a blue, 3b cyan, 3c yellow, 3d magentas, 3e oranges, 3f wheat, 3g gray, 3h green, 3i salmon, lefamulin red); (C) 3a (blue) and 3d (magentas) in 5HL7.

3 EXPERIMENTAL SECTION

3.1 Synthesis

All chemical reagents were purchased from J and K Chemical or Sigma-Aldrich Chemistry Co. (Darmstadt, Germany). Unless otherwise noted, all reactions were conducted under atmosphere.

Thin-layer chromatography (TLC) analysis (Qingdao Haiyang Chemical Co., Ltd., Shandong, Qingdao, China) was used to monitor the reaction process.Column chromatography was carried out on silica gel (200-300 mesh).The products were eluted in an appropriate solvent mixture under air pressure.Concentration and evaporation of the solvent after reaction or extraction was carried out on a rotary evaporator.IR spectra were obtained on a Thermo Nicolet NEXUS-670 spectrometer (Thermo Nicolet Corporationcompany, Waltham, MA, USA) and recorded as KBr thin films and absorptions are reported in cm^{-1}.HRMS were obtained with a Bruker Daltonics APEX II 47e mass spectrometer. NMR spectra (Bruker Corporation, Billerica, MA, USA) were recorded on a Bruker - 400MHz spectrometer (Bruker Corporation, Billerica, MA, USA) in appropriate solvents.Chemical shifts (δ) were expressed in parts per million (ppm) relative to the tetramethylsilane.Multiplicities of NMR signals are designated as s (singlet), d (doublet), t (triplet), q (quartet), m (multiplet), br (broad), etc.^{13}C-NMR spectra were recorded on 100MHz spectrometers.The single-crystal structure of the title compound was determined on a Bruker SMART APEX II X-diffractometer (Bruker Corporation, Billerica, MA, USA).All NMR, IR and HRMS datum were been added in Supplementary Materials.

14-O- (p-Toluenesulfonyloxyacetyl) mutilin (2).A 5mL of NaOH aqueous solution (2g, 50mmol) was added dropwise to a mixture of pleuromutilin (7.57g, 20mmol) and p-toluenesulfonyl chloride (4.2g, 22mmol) in methyl isobutyl ketone (10mL) and water (5mL). The mixture was vigorously stirred for 45min at 60℃, then the reaction mixture was cooled to 10℃ and separated.The organic layer was washed with 5mL water and 5mL saturated sodium carbonate solution.The organic phase was dried overnight with anhydrous sodium sulfate.After filteration, the solvent was concentrated in vacuo to give 10.56g of yellow oil.It was used in the next step without further purification.Yield: 93%.IR (KBr): 3 446 (OH), 2 924 (CH_2), 2 863 (CH_2), 1 732 (C=O), 1 633 (C-C), 1 597 (C=C), 1 456 (C=C), 1 371 (CH_3), 1 297 (C-O-C), 1 233 (CH), 1 117 [C- (C=O) -C], 1 035 (C-O-C), 832 (CH), 664 (CH_2=), 560 (CH_2=) cm^{-1}.^1H-NMR (400MHz, $CDCl_3$) δ 7.74 (d, J=8.3Hz, 2H), 7.26 (t, J=13.3Hz, 2H), 6.34 (dd, J=17.4, 11.0Hz, 1H), 5.70 (d, J=8.5Hz, 1H), 5.19 (dd, J=55.1, 14.2Hz, 2H), 4.47-4.33 (m, 2H), 3.28 (s, 1H), 2.38 (s, 3H), 2.25-2.09 (m, 3H), 2.01 (s, 1H), 1.99-1.88 (m, 1H), 1.71-1.62 (m, 1H), 1.61-1.51 (m, 2H), 1.46-1.38 (m, 2H), 1.35 (d, J=7.8Hz, 3H), 1.27 (d, J=11.5Hz, 1H), 1.18 (dd, J=11.6, 4.5Hz, 2H), 1.12-1.00 (m, 4H), 0.80 (d, J=7.0Hz, 3H), 0.55 (d, J=7.0Hz, 3H).^{13}C-NMR (100MHz, $CDCl_3$) δ 215.71 (C=O), 163.87 (C=O), 144.29 (benzene-C), 137.70 (CH=), 131.63 (benzene-C), 128.91 (benzene-C), 127.09 (benzene-C), 116.38 (CH_2=), 73.54 (CH), 69.29 (CH), 64.03 (CH), 57.02 (CH), 44.39 (C), 43.51 (CH_2), 42.97 (C), 40.84 (C), 35.54 (CH), 35.03 (CH), 33.40 (CH_2), 29.34 (CH_3), 25.77 (CH_2), 25.39 (CH_2), 23.81 (CH_2), 20.68 (CH_3), 15.53 (CH_3), 13.76 (CH_3), 10.47 (CH_3).HRMS (ESI) calcd. $[M+H]^+$ for $C_{29}H_{40}O_7S$ 533.250, found 533.2507.

14-O- (Aceticacidthioacetyl) mutilin (4).Compound 3 was prepared by stirring a mixing of compound 2 (10mmol), potassium thioglycolate (20mmol), and methyl isobutyl ketone (30mL) in room temperature, and the mixture was stirred for 2h. The mixture was extracted with water (10mL). The organic phase were combined, dried over Na_2SO_4, and concentrated to give compound 3. Yield: 80%. IR (KBr): 3 448 (OH), 2 967 (CH_2), 2 924 (CH_3), 2 865 (CH_2), 1 731 (C=O), 1 702 (C=O), 1 453 (C-C), 1 419, 1 384 (CH_3), 1 295 (C-O-C), 1 183 (CH), 1 154(C-O),1 115 [C- (C=O) -C], 1 017 (C-O-C) cm^{-1}.^{1}H-NMR (400MHz, $CDCl_3$) δ 6.58-6.30 (m, 1H), 5.72 (d, J=8.4Hz, 1H), 5.43-5.14 (m, 2H), 3.63 (s, 2H), 3.36 (d, J=6.4Hz, 1H), 2.36 (dd, J=13.5, 3.5Hz, 3H), 2.35-2.27 (m, 1H), 2.21 (dd, J=17.4, 7.9Hz, 1H), 2.13 (d, J=14.9Hz, 1H), 2.04 (dd, J=26.1, 17.5Hz, 1H), 1.73 (dd, J=31.8, 10.6Hz, 2H), 1.68-1.62 (m, 2H), 1.62-1.41 (m, 6H), 1.41-1.24 (m, 2H), 1.24-0.99 (m, 4H), 0.89 (t, J= 8.4Hz, 3H), 0.78-0.61 (m, 3H).^{13}C-NMR (101MHz, $CDCl_3$) δ 216.90 (C=O), 193.45 (C=O), 167.32 (C=O), 138.92 (CH=), 117.19 (CH_2=), 74.60 (CH), 70.09 (CH), 58.13 (CH), 45.45 (C), 44.71 (CH_2), 44.01 (C), 41.89 (C), 36.74 (CH), 36.00 (CH_3), 34.45 (CH_2), 32.20 (CH_2), 30.42 (CH_2), 30.06 (CH_2), 26.84 (CH_2), 26.41 (CH_3), 24.82 (CH_2), 16.74 (CH_3), 14.81 (CH_3), 11.43 (CH_3).HRMS (ESI) calcd. $[M+H]^+$ for $C_{24}H_{36}O_5S$ 437.2356, found 437.2339.

(R) -5-ChloroMethyl-2-oxazolidinone (5).The title compound was prepared by stirring a mixing of magnesium sulphate (10mmol), sodium cyanate (10mmol) and water (50mL) in room temperature. (R) - (-) -Epichlorohydrin to the solution (5mmol) was added dropwise to the mixture.The result mixture was stirred under 60℃ for 1h.The reaction mixture was concentrated in vacuo and extracted by ethyl acetate.The two layers were separated, and the organic layer was dried over Na_2SO_4, filtered, and concentrated to dryness.Yield: 57%.IR (KBr): 3 365 (NH), 1 744 (C=O), 1 429 (C-N), 1 240 [C-C (=O) -O], 736 (C-Cl) cm^{-1}.^{1}H-NMR (400MHz, DMSO) δ 7.60 (s, 1H), 4.84 (dd, J=9.6, 4.7Hz, 1H), 4.04-3.73 (m, 2H), 3.59 (t, J=9.0Hz, 1H), 3.25 (dd, J=9.0, 6.2Hz, 1H).^{13}C-NMR (101MHz, d6-DMSO) δ 158.68 (C=O), 74.35 (CH), 46.66 (CH_2), 43.03 (CH_2).HRMS (ESI) calcd. $[M+H]^+$ for C_4H_6ClNO 136.0159, found 136.0150.

General Procedure for Synthesis of Compounds 3a-3g.A mixing of thiols (1mmol), sodium hydroxide (1.1mmol), water (0.5mL) and methanol (3mL) were stirred in room temperature. After 30min, compound 2 (1.1mmol) in 5mL CH_2Cl_2 was added dropwise to the mixture for 36h-42h.The mixture was concentrated in vacuo.The residue was dissolved by CH_2Cl_2.The solution was extracted three times with water. The organic phase was dried overnight with anhydrous sodium sulfate.The solvent was concentrated in vacuo to give crude products.The crude product was purified by silica gel column chromatography.

14-O- [(4-Amino-pyrimidinone-2-yl) thioacetyl] mutilin (3a).Compound 3a was prepared according to the general procedure from 14-O- (p-toluene sulfonyloxyacetyl) mutilinmutilin (2) and 4-amino-2-Pyrimidinone.The crude product was purified over silica gel column chromatography to give 4.09g.Yield: 84%.IR (KBr): 3 448 (OH), 2 933 (CH_2), 1 730 (C=O), 1 629 (C-C), 1 583 (C=N), 1 543 (C=C), 1 467 (C=C), 1 372 (CH_3), 1 249 [C-C (C=O) -O], 1 153 (C-O), 1 117 (C- (C=O) -C), 1 018 (C-O-C) cm^{-1}.^{1}H-NMR

(400MHz, $CDCl_3$) δ 7.89 (d, J=5.1Hz, 1H), 6.42 (dd, J=17.1, 11.2Hz, 1H), 6.04 (d, J = 5.2Hz, 1H), 5.68 (d, J = 7.9Hz, 1H), 5.17 (dd, J = 51.9, 14.0Hz, 2H), 4.94 (s, 2H), 3.72 (dd, J=32.8, 16.1Hz, 2H), 3.28 (s, 1H), 2.24 (d, J=6.5Hz, 1H), 2.14 (dd, J = 14.8, 9.1Hz, 2H), 2.02 (s, 1H), 1.94 (dd, J = 15.6, 8.5Hz, 1H), 1.69 (d, J = 13.7Hz, 1H), 1.61 – 1.52 (m, 2H), 1.48 (d, J = 12.0Hz, 2H), 1.37 (s, 4H), 1.32–1.22 (m, 2H), 1.07 (s, 4H), 0.79 (d, J=6.2Hz, 3H), 0.68 (d, J=6.2Hz, 3H). ^{13}C-NMR (101MHz, $CDCl_3$) δ 216.11 (C=O), 168.72 (pyrimidine-C), 167.20 (C=O), 161.35 (pyrimidine-C), 154.88 (pyrimidine-C), 138.26 (CH=), 116.02 (CH_2=), 100.18 (pyrimidine-C), 73.60 (CH), 68.51 (CH), 57.36 (CH), 57.21 (CH_2), 44.46 (CH_2), 43.47 (C), 42.93 (C), 40.87 (CH), 35.81 (CH), 35.02 (CH), 33.48 (CH_2), 33.06 (CH_2), 29.45 (CH_2), 25.89 (CH_3), 23.84 (CH_2), 15.74 (CH_3), 13.92 (CH_3), 10.44 (CH_3). HRMS (ESI) calcd. $[M+H]^+$ for $C_{26}H_{37}N_3O_4S$ 488.2578, found 488.2570.

14-O-[(4-Methylpyrimidine-2-yl) thioacetyl] mutilin (3b). Compound 3b was prepared according to the general procedure from 14-O-(p-toluene sulfonyloxyacetyl) mutilin (2) and 4-methy-2-pyrimidinone. The crude product was purified over silica gel column chromatography to give 3.55g. Yield: 73%. IR (KBr): 3 439 (OH), 2 935 (CH_2), 1 733 (C=O), 1 658 (C-C), 1 580 (C=N), 1 535 (C=C), 1 458 (C=C), 1 396 (CH_3), 1 285 (C-O-C), 1 118 [C-(C=O)-C] cm^{-1}. ^1H-NMR (400MHz, $CDCl_3$) δ 6.37 (dt, J=33.7, 16.8Hz, 1H), 5.99 (s, 1H), 5.69 (d, J = 8.4Hz, 1H), 5.19 (dd, J = 56.5, 14.2Hz, 2H), 3.88–3.73 (m, 2H), 3.29 (d, J=5.6Hz, 1H), 2.27–2.15 (m, 2H), 2.11 (d, J = 13.1Hz, 3H), 2.02 (d, J = 6.4Hz, 1H), 1.96 (d, J = 10.9Hz, 1H), 1.69 (d, J = 14.2Hz, 1H), 1.58 (dd, J = 21.0, 10.8Hz, 2H), 1.49 (dd, J = 26.7, 13.3Hz, 2H), 1.43–1.26 (m, 6H), 1.20 (dd, J = 17.5, 11.2Hz, 2H), 1.05 (d, J = 20.9Hz, 4H), 0.80 (d, J = 6.9Hz, 3H), 0.67 (d, J = 6.9Hz, 3H). ^{13}C-NMR (101MHz, $CDCl_3$) δ 215.94 (C=O), 165.80 (C=O), 164.69 (pyrimidine-C), 164.11 (pyrimidine-C), 157.85 (pyrimidine-C), 137.90 (CH=), 116.27 (CH_2=), 107.63 (pyrimidine-C), 73.55 (CH), 69.14 (CH), 57.08 (CH), 44.43 (C), 43.49 (CH_2), 42.94 (C), 40.87 (C), 35.70 (CH), 35.00 (CH_3), 33.44 (CH_2), 32.23 (CH_2), 29.39 (CH_2), 25.84 (CH_3), 25.36 (CH_2), 23.82 (CH_2), 23.09 (CH_2), 15.84 (CH_3), 13.84 (CH_3), 10.46 (CH_3). HRMS (ESI) calcd. $[M+H]^+$ for $C_{27}H_{38}N_2O_4S$ 487.2625, found 487.2623.

14-O-[(Benzimidazole-2-yl) thioacetyl] mutilin (3c). Compound 3c was prepared according to the general procedure from 14-O-(p-toluene sulfonyloxyacetyl) mutilin (2) and 2-mercaptobenzothiazole. The crude product was purified over silica gel column chromatography to give 3.54g. Yield: 67%. IR (KBr): 3 442 (OH), 2 929 (CH_2), 1 731 (C=O), 1 458 (C=C), 1 429 (C-C), 1 274 (C-O-C), 1 153 (C-O), 1 117 [C-(C=O)-C] cm^{-1}. ^1H-NMR (400MHz, $CDCl_3$) δ 7.77 (dd, J=19.9, 8.0Hz, 2H), 7.39 (t, J=7.7Hz, 1H), 7.29 (t, J=7.6Hz, 1H), 6.42 (dd, J=17.4, 11.0Hz, 1H), 5.76 (d, J=8.5Hz, 1H), 5.19 (dd, J=57.6, 14.2Hz, 2H), 4.08 (dd, J=41.4, 16.2Hz, 2H), 3.31 (d, J=6.4Hz,

1H), 2.29 (dd, J = 14.1, 7.2Hz, 1H), 2.20 (dd, J = 11.6, 6.3Hz, 1H), 2.06 (s, 1H), 1.98 (dd, J = 16.0, 8.5Hz, 1H), 1.82 - 1.64 (m, 2H), 1.64 - 1.51 (m, 2H), 1.50 - 1.33 (m, 6H), 1.24 (dd, J = 14.3, 7.4Hz, 2H), 1.13 - 0.97 (m, 4H), 0.85 (d, J = 7.0Hz, 3H), 0.77 (d, J = 6.9Hz, 3H). ^{13}C-NMR (101MHz, $CDCl_3$) δ 215.94 (C=O), 165.84 (C=O), 163.54 (benzothiazole-C), 151.74 (benzothiazole-C), 137.76 (CH=), 134.49 (benzothiazole-C), 125.02 (benzothiazole-C), 123.46 (benzothiazole-C), 120.68 (benzothiazole-C), 120.05 (benzothiazole-C), 116.21 (CH_2=), 73.57 (CH), 69.20 (CH), 57.10 (CH), 44.42 (C), 43.41 (CH_2), 42.90 (C), 40.86 (C), 35.74 (CH), 34.98 (CH), 34.67 (CH_2), 33.43 (CH_2), 29.40 (CH_2), 25.84 (CH_2), 25.28 (CH_3), 23.81 (CH_2), 15.80 (CH_3), 13.81 (CH_3), 10.44 (CH_3). HRMS (ESI) calcd. $[M+H]^+$ for $C_{29}H_{37}NO_4S$ 528.2237, found 528.2234.

14-O-[(Benzothiazole-2-yl) thioacetyl] utilin (3d). Compound 3d was prepared according to the general procedure from 14-O-(p-toluene sulfonyloxyacetyl) mutilin (2) and 2-mercaptobenzimidazole. The crude product was purified over silica gel column chromatography to give 3.68g. Yield 72%. IR (KBr): 3 423 (OH), 2 925 (CH_2), 1 726 (C=O), 1 458 (C=C), 1 439(C-C), 1 271 (C-O-C), 1 152 (C-O), 1 117 [C-(C=O)-C] cm^{-1}. ^{1}H-NMR (400MHz, $CDCl_3$) δ 7.50 (dd, J = 5.8, 3.1Hz, 2H), 7.20 (dd, J = 6.0, 3.2Hz, 2H), 6.41 (dd, J = 17.4, 11.0Hz, 1H), 5.79 (d, J = 8.4Hz, 1H), 5.18 (dd, J = 43.7, 14.2Hz, 2H), 3.91 (s, 2H), 3.73 (q, J = 7.0Hz, 1H), 3.35 (d, J = 6.4Hz, 1H), 2.35 - 2.28 (m, 1H), 2.27 - 2.17 (m, 1H), 2.08 (s, 1H), 2.03 (dd, J = 16.1, 8.6Hz, 1H), 1.82 - 1.65 (m, 2H), 1.64 - 1.52 (m, 2H), 1.49 - 1.32 (m, 6H), 1.24 (t, J = 7.0Hz, 2H), 1.16 - 1.05 (m, 4H), 0.88 (d, J = 6.9Hz, 3H), 0.72 (d, J = 7.0Hz, 3H). ^{13}C-NMR (101MHz, $CDCl_3$) δ 215.89 (C=O), 167.78 (C=O), 147.20 (benzimidazole-C), 137.75 (CH=), 121.68 (benzimidazole-C), 116.28 (CH_2=), 73.61 (CH), 69.82 (CH), 57.43 (CH), 57.09 (benzimidazole-C), 44.42 (C), 43.52 (CH_2), 42.97 (C), 40.84 (C), 35.68 (CH), 35.04 (CH_2), 34.16 (CH_2), 33.42, 29.38 (CH_2), 25.84 (CH_2), 25.44 (CH_3), 23.83 (CH_2), 17.42 (CH_2), 15.80 (CH_3), 13.80 (CH_3), 10.50 (CH_3). HRMS (ESI) calcd. $[M+H]^+$ for $C_{29}H_{38}N_2O_4S$ 511.2526, found 511.2531.

14-O-[(5-Benzimidazolesulfonate-2-yl) thioacetyl] utilin (3e). Compound 3e was prepared according to the general procedure from 14-O-(p-toluene sulfonyloxyacetyl) mutilin (2) and sodium 2-mercapto-5-benzimidazolesulfonate dihydrate. The crude product was purified over silica gel column chromatography to give 3.2g. Yield: 54.2%. IR (KBr): 3448 (OH), 2940 (CH_2), 1732 (C=O) 1298 (C-O-C), 1190 (S=O), 1178 (S=O) cm^{-1}. ^{1}H-NMR (400MHz, DMSO) δ 7.80 (s, 1H), 7.69 (d, J = 8.5Hz, 1H), 7.60 (d, J = 8.5Hz, 1H), 6.04 (dd, J = 17.8, 11.2Hz, 1H), 5.50 (d, J = 8.1Hz, 1H), 4.96 (dd, J = 38.2, 14.5Hz, 2H), 4.38 (d, J = 4.4Hz, 1H), 3.36 (d, J = 5.5Hz, 1H), 2.51 (s, 1H), 2.35 (s, 1H), 2.13 (dd, J = 21.0, 10.7Hz, 1H), 2.08 - 1.96 (m, 2H), 1.90 (dd, J = 16.0, 8.1Hz, 1H), 1.60 (s, 2H), 1.44 (d, J = 6.9Hz, 1H), 1.33 (d, J = 7.6Hz, 2H), 1.27 - 1.12 (m, 6H), 1.10 - 1.02 (m, 1H), 1.02 - 0.90 (m, 4H), 0.78 (d, J=

6. 8Hz, 3H), 0. 58 (t, J=9. 3Hz, 3H). ^{13}C-NMR (101MHz, DMSO) δ 217. 48 (C=O), 166. 34 (C=O), 150. 52 (benzimidazole-C), 145. 67 (benzimidazole-C), 141. 16 (benzimidazole-C), 133. 41 (CH =), 128. 59 (benzimidazole - C), 125. 97 (benzimidazole - C), 123. 22 (benzimidazole-C), 115. 68 (CH_2=), 113. 22 (CH_2), 110. 71 (benzimidazole-C), 72. 92 (CH), 71. 44 (CH), 57. 42 (CH), 45. 35 (C), 44. 58 (CH_2), 41. 95 (C), 36. 89 (CH), 36. 64 (CH), 35. 03 (CH_2), 34. 40 (CH_2), 30. 53 (CH_2), 29. 06 (CH_2), 26. 99 (CH_2), 24. 85 (CH_2), 16. 47 (CH_3), 14. 64 (CH_3), 11. 96 (CH_3). HRMS (ESI) calcd. $[M + H]^+$ for $C_{29}H_{38}N_2O_7S_2$ 591. 2193, found 591. 2192.

14-O- [(Pyryrazolo [3, 4d] pyrimidine-4-yl) thioacetyl] utilin (3f). Compound 3f was prepared according to the general procedure from 14-O- (p-toluene sulfonyloxyacetyl) mutilin (2) and 4-mercaptopyrazolo [3, -d] yrimidine. The crude product was purified over silica gel column chromatography to give 4. 30g. Yield: 84%. IR (KBr): 3 431 (OH), 2 934 (CH_2), 1 732 (C=O), 1 567 (C=C), 1 456 (C=C), 1 406 (C-C), 1 271 (C-O-C), 1 152 (C-O), 1 117 [C- (C=O) -C], 981 (CH) cm^{-1}.

^{1}H-NMR (400MHz, $CDCl_3$) δ 8. 65 (d, J = 37. 7Hz, 1H), 8. 29 - 8. 08 (m, 1H), 6. 43 (dt, J = 21. 8, 10. 9Hz, 1H), 5. 80 (d, J = 8. 4Hz, 1H), 5. 38 - 5. 29 (m, 1H), 5. 24 (dd, J = 33. 0, 14. 5Hz, 2H), 4. 18 - 4. 02 (m, 2H), 3. 38 (d, J = 6. 3Hz, 1H), 2. 31 (dd, J = 13. 9, 7. 1Hz, 1H), 2. 22 (dd, J = 13. 0, 7. 5Hz, 1H), 2. 11 (t, J = 8. 4Hz, 1H), 2. 04 (d, J = 7. 9Hz, 1H), 1. 73 (dd, J = 31. 9, 9. 2Hz, 2H), 1. 65 (dd, J=17. 1, 7. 1Hz, 2H), 1. 58 - 1. 37 (m, 6H), 1. 36 - 1. 22 (m, 2H), 1. 19 - 1. 06 (m, 4H), 0. 88 (t, J=11. 3Hz, 3H), 0. 79 (t, J=12. 6Hz, 3H). ^{13}C-NMR (101MHz, $CDCl_3$) δ 216. 06 (C = O), 166. 18 (C = O), 162. 56 (pyrimidine - C), 153. 05 (pyrimidine - C), 151. 46 (pyrimidine-C), 137. 88 (CH=), 131. 82 (pyrazolo-C), 116. 28 (CH_2=), 110. 65 (pyrimidine-C), 73. 59 (CH), 69. 22 (CH), 57. 14 (CH), 44. 46 (C), 43. 57 (CH_2), 42. 92 (C), 40. 88 (C), 35. 75 (CH), 35. 01 (CH_2), 33. 46 (CH_2), 31. 11 (CH_2), 29. 41 (CH_2), 25. 44 (CH_3), 23. 83 (CH_2), 17. 40 (CH_2), 15. 74 (CH_3), 13. 86 (CH_3), 10. 47 (CH_3). HRMS (ESI) calcd. $[M + H]^+$ for $C_{27}H_{36}N_4O_4S$ 513. 2530, found 513. 2527.

14-O- [(Furfuryl-2-yl) thioacetyl] mutilin (3g). Compound 3g was prepared according to the general procedure from 14-O- (p-toluene sulfonyloxyacetyl) mutilin (2) and furfuryl mercaptan. The crude product was purified over silica gel column chromatography to give 3. 53g. Yield: 74%. IR (KBr): 3 547 (OH), 2 933 (CH_2), 2 882 (CH_2), 1 731 (C=O), 1 455 (C=C), 1 281 (C-O-C), 1 150 (C-O), 1 115 [C- (C=O) -C] cm^{-1}. ^{1}H-NMR (400MHz, $CDCl_3$) δ 7. 29 (s, 1H), 6. 42 (dd, J = 17. 4, 11. 0Hz, 1H), 6. 31 - 6. 06 (m, 2H), 5. 71 (d, J=8. 4Hz, 1H), 5. 23 (dd, J=57. 0, 14. 2Hz, 2H), 3. 75 (s, 2H), 3. 30 (s, 1H), 3. 02 (s, 2H), 2. 29 (dd, J = 13. 7, 6. 8Hz, 1H), 2. 15 (dt, J = 19. 6, 8. 8Hz, 2H), 2. 09- 1. 98 (m, 2H), 1. 75 - 1. 66 (m, 1H), 1. 64 - 1. 55 (m, 2H), 1. 53 - 1. 35 (m, 6H), 1. 30 (t, J=14. 9Hz, 2H), 1. 17- 1. 05 (m, 4H), 0. 82 (d, J=7. 0Hz, 3H), 0. 68 (d, J=6. 8Hz, 3H). ^{13}C-NMR (101MHz, $CDCl_3$) δ 215. 99 (C=O), 167. 67 (C=O), 149. 27 (furan-C), 141. 51 (furan-C), 138. 12 (CH=), 116. 19 (CH_2=), 109. 37

(furan-C), 107.39 (furan-C), 73.65 (CH), 68.31 (CH), 57.21 (CH), 44.46 (C), 43.85 (CH_2), 42.94 (C), 40.77 (C), 35.78 (CH), 35.04 (CH), 33.45 (CH_2), 32.11 (CH_2), 29.44 (CH_2), 27.36 (CH_2), 25.85 (CH_3), 25.42 (CH_2), 23.86 (CH_2), 15.81 (CH_3), 13.92 (CH_3), 10.49 (CH_3).HRMS (ESI) calcd. $[M+H]^+$ for $C_{27}H_{38}O_5S$ 475.2513, found 475.2522.

14-O- [(1-Methylimidazole-2-yl) thioacetyl] mutilin (3h).Compound 3h was prepared according to the general procedure from 14-O- (p-toluene sulfonyloxyacetyl) mutilin (2) and 2-mercapto-1-methylimidazole.The crude product was purified over silica gel column chromatography to give 3.46g.Yield: 73%.IR (KBr): 3 423 (OH), 2 924 (CH_2), 2 863 (CH_2), 1 717 (C=O), 1 455 (C=C), 1 410 (C-C), 1 280 (C-O-C), 1 145 (C-O), 1 117 [C-(C=O) -C] cm^{-1}.^{1}H-NMR (400MHz, $CDCl_3$) δ 6.88 (d, J=49.1Hz, 2H), 6.35 (dd, J=17.4, 11.0Hz, 1H), 5.63 (d, J=8.5Hz, 1H), 5.32-5.01 (m, 2H), 3.92-3.62 (m, 2H), 3.56 (d, J=11.0Hz, 3H), 3.27 (s, 1H), 2.22 (dt, J=13.8, 7.0Hz, 1H), 2.19-2.07 (m, 2H), 2.03 (d, J=21.6Hz, 1H), 1.93 (dd, J=16.0, 8.6Hz, 1H), 1.77-1.61 (m, 1H), 1.58-1.46 (m, 3H), 1.45-1.35 (m, 2H), 1.31 (d, J=14.3Hz, 3H), 1.26-1.13 (m, 2H), 1.12-0.97 (m, 4H), 0.80 (d, J=7.0Hz, 3H), 0.59 (d, J=6.9Hz, 3H).^{13}C-NMR (101MHz, $CDCl_3$) δ 215.93 (C=O), 166.75 (C=O), 139.06 (imidazole-C), 138.01 (CH=), 128.55 (imidazole-C), 121.39 (imidazole-C), 116.07 (CH_2=), 73.57 (CH), 68.77 (CH), 57.12 (CH), 44.42 (C), 43.48 (CH_2), 42.94 (C), 40.77 (C), 36.08 (CH), 35.70 (CH_2), 35.00 (CH_3), 33.44 (CH_2), 32.35 (CH_2), 29.40 (CH_2), 25.83 (CH_2), 25.41 (CH_3), 23.82 (CH_2), 15.61 (CH_3), 13.79 (CH_3), 10.44 (CH_3).HRMS (ESI) calcd. $[M+H]^+$ for $C_{26}H_{38}N_2O_4S$ 475.2625, found 475.2630.

14-O- [2-oxazolidinone, 5- (methyl) -] (thioacetyl) mutilin (3i).A mixing of (R) -5-chloroMethyl-2-oxazolidinone (5) (1mmol), sodiumiodide (0.1mmol), and acetone (10mL) were stirred in room temperature.After 30min, the reaction solution was filtered and concentrated.Compound 4 (1.1mmol) and triethylamine (20mL) was added under N 2. The solvent was stirred in 40℃ for 10h.The mixture was extracted with water (10ml) and HCl (2N, 10mL). The organic layers were concentrated in vacuo to give crude products.The crude product was purified by silica gel column chromatography.Yield: 62%.IR (KBr): 3 422 (O), 2 933 (CH_2), 1 735 (C=O), 1 686 (C=O), 1 458 (C=C), 1 420 (C-C), 1 284 (C-O-C), 1 151 Molecules 2017, 22, 996 11 of 13 (C-O), 1 117 [C- (C=O) -C] cm^{-1}.^{1}H-NMR (400MHz, $CDCl_3$) δ 6.63-6.35 (m, 1H), 6.17 (d, J=8.2Hz, 1H), 5.75 (d, J=8.2Hz, 1H), 5.27 (dd, J=51.9, 14.2Hz, 2H), 4.92-4.67 (m, 1H), 3.82-3.62 (m, 1H), 3.45-3.32 (m, 2H), 3.30-3.07 (m, 2H), 2.99 (ddd, J=9.9, 8.9, 4.4Hz, 1H), 2.87 (dt, J=12.8, 5.2Hz, 1H), 2.34 (s, 1H), 2.28-2.17 (m, 2H), 2.11 (s, 1H), 2.08 (d, J=8.6Hz, 1H), 1.77 (d, J=14.3Hz, 1H), 1.65 (d, J=10.4Hz, 2H), 1.52 (dd, J=25.3, 6.7Hz, 2H), 1.44 (d, J=1.0Hz, 4H), 1.39 (s, 1H), 1.35-1.26 (m, 1H), 1.16 (d, J=14.9Hz, 4H), 0.89 (d, J=6.8Hz, 3H), 0.73 (d, J=6.8Hz, 3H).^{13}C-NMR (101MHz, $CDCl_3$) δ 216.99 (C=O), 168.63 (C=O), 159.41 (C=O), 139.16

(CH=), 117.14 (CH_2=), 75.66 (CH), 74.61 (C), 69.70 (CH), 58.17 (CH), 45.46 (C), 45.06 (CH_2), 44.88 (CH_2), 43.95 (C), 41.78 (C), 36.75 (CH), 36.03 (CH), 35.99, 34.81 (CH), 34.45 (CH_2), 30.42 (CH_2), 26.86 (CH_3), 26.41 (CH_2), 24.84 (CH_2), 16.82 (CH_3), 14.88 (CH_3), 11.48 (CH_3).HRMS (ESI) calcd. $[M + Na]^+$ for $C_{26}H_{39}NO_6S$ 516.2395, found 516.2394.

3.2 Biological evaluation

3.2.1 MIC testing

The MIC values of 3a-i and tiamulin fumarate against bacteria were determined using the broth dilution method[13]. Stock solutions of compounds were prepared in DMSO. The compounds were added to the test tube and serially diluted in Mueller-Hinton broth (the final concentration is 0.0625μg/mL). Five bacteria, including S. aureus, MRSA, MRSE, VRE, and one Gram-negative bacterium, E.coli, were cultivated and added to the tube.The initial concentration of bacteria cannot be lower than 10^5CFU/mL.The broth was incubated at 36.7℃ for 18-24h.MICs were read when the change of clarity in the broth was observed in the control test tube.

3.2.2 Bactericidal time-kill kinetics

The twobacterias were prepared in Muellere Hinton broth at 37℃ for 6h with shaking.The solution of compound 3a and tiamulin fumarate, 1 × MIC and 6 × MIC, were added to the bacterial suspension so that the final concentrations were 10^6 ~ 10^7CFU/mL, respectively.After specified time intervals (0, 1, 2, 4, 6, 12, and 24h), 20mL aliquots were serially diluted in 0.9% saline, plated on sterile Muellere Hinton agar plates, and incubated at 37℃ for 24h.The viable colonies were counted and represented as $\log_{10}$ (CFU/mL).The same procedure was repeated in triplicate.

3.3 Molecular modeling studies

The 50s ribosomal of S.aureus in complex with lefamulin (PDB ID: 5HL7)[12] was simulated useHomdock software in the Chil2 package (University of Pittsburgh, Pittsburgh, PA, U.S., version0.99).The package contains a Graph-based molecular alignment (GMA) tool and a Monte-Carlo/Simulated Annealing (MC/SA) algorithm-based docking (GlamDock) (University of Pittsburghcompany, Pittsburgh, PA, USA, version 0.99) tool. Lefamulin was the template for flexible molecular alignment, and the interaction was optimized by GlamDock according to the ChillScore scoring function based on ChemScore with a smooth, improved potential. All the compounds were prepared with Avogadro software[19], including a 5 000 steps Steepest Descent and 1 000 steps Conjugate Gradients geometry optimization based on the MMFF94 force field.The docking site was set to base lefamulin.All compounds have compared to original conformation of 5HL7, which was kept for binding affinity comparison. Compounds and receptors were estimated by ChillScore. Hydrogen bonds and other interactions were detected and generated by PyMol 1.5.03[20,21].

4 CONCLUSIONS

In summary, a series of novelpleuromutilin derivatives bearing heterocyclic ring at the C-14 side chain were synthesized.Our results show that a heterocyclic substituent bearing a amine group leads to excellent in vitro antibacterial activity against S.aureus, MRSA, MRSE, and VRE.Compounds that contain pyrimidine rings prepared in this thesis showed moderate to excellent biological activities, and the hydrophilicity of the side chain showed some correlations to its activity. Compound 3a, the most effective compound, showed rapid bactericidal activity against S.aureus and MRSA in time-kill assay. Molecular docking studies also revealed that 3a displayed lower Gibbs free energy. Thus, compound 3a has been selected for further evaluation as a promising candidate for treating bacterial infection.

Supplementary Materials: Samples of Compounds 2, 4, 5, 3a, 3b, 3c, 3d, 3e, 3f, 3g, 3h and 3i are available from Supplementary Materials.

Acknowledgments: This work was financed by Central Agricultural Scientific Research Institutions (No. 1610322016007), National Key Technology Support Program (No. 2015BAD11B02) and Agricultural Science and Technology Innovation Program (ASTIP, No.CAASASTIP-2014-LI-HPS-04).

Author Contributions: Supervision of the whole work was conducted byYunpeng Yi; Study conception and experiments were designed by Pengchegn Dong, Yunpeng Yi, Yunxing Fu and Ruofeng Shang; Experiments were conducted by Wenwen Qin; Experimental data were analyzed and interpreted by Yu Liu and Jiangping Liang; Manuscript was prepared by Yunpeng Yi and Ruofeng Shang.

Conflicts of Interest: The authors declare no conflict of interest.

REFERENCES OMITTED

（发表于《molecules》，院选 SCI，IF：2.861）

The Response of Gene Expression Associated with Lipid Metabolism, Fat Deposition and Fatty Acid Profile in the Longissimus Dorsi Muscle of Gannan Yaks to Different Energy Levels of Diets

Chao YANG[1,2,3], Jianbin LIU[1], Xiaoyun WU[1], Pengjia BAO[1], Ruijun LONG[3,4*], Xian GUO[1*], Xuezhi DING[1*], Ping YAN[1*]

(1. Key Laboratory of Yak Breeding Engineering, Lanzhou Institute of Husbandry and Pharmaceutical Sciences, Chinese Academy of Agricultural Sciences, Lanzhou, China; 2. State Key Laboratory of Pastoral Agricultural Ecosystem, College of Pastoral Agriculture Science and Technology, Lanzhou University, Lanzhou, China; 3. International Centre for Tibetan Plateau Ecosystem Management, Lanzhou University, Lanzhou, China; 4. School of Life Sciences, Lanzhou University, Lanzhou, China)

Abstract: The energy available from the diet, which affects fat deposition in vivo, is a major factor in the expression of genes regulating fat deposition in the longissimus dorsi muscle. Providing high-energy diets to yaks might increase intramuscular fat deposition and fatty acid concentrations under a traditional grazing system in cold seasons. A total of fifteen adult castrated male yaks with an initial body weight 274.3±3.14kg were analyzed for intramuscular adipose deposition and fatty acid composition. The animals were divided into three groups and fed low-energy (LE: 5.5MJ/kg), medium-energy (ME: 6.2MJ/kg) and high-energy (HE: 6.9MJ/kg) diets, respectively. All animals were fed ad libitum twice daily at 08: 00-09: 00 and 17: 00-18: 00 and with free access to water for 74 days, including a 14d period to adapt to the diets and the environment. Intramuscular fat (IMF) content, fatty acid profile and mRNA levels of genes involved in fatty acid synthesis were determined. The energy levels of the diets significantly ($P<0.05$) affected the content of IMF, total SFA, total MUFA and total PUFA. C16: 0, C18: 0 and C18: 1n9c account for a large proportion of total fatty acids. Relative expression of acetyl-CoA carboxylase (ACACA), fatty acid synthase (FASN), stearoyl-CoA desaturase (SCD), sterol regulatory element-binding protein-1c (SREBP-1c), peroxisome proliferator-activated receptor γ (PPARr) and fatty acid-binding protein 4 (FABP4) was greater in HE than in LE yaks ($P<0.05$). Moreover, ME yaks had higher ($P<0.05$) mRNA expression levels of PPARγ, ACACA, FASN, SCD and FABP4 than did

* Corresponding authors, E-mail: longrj@lzu.edu.cn (RL); guoxian@caas.cn (XG); dingxuezhi@caas.cn (XD); pingyanlz@163.com (PY)

the LE yaks.The results demonstrate that the higher energy level of the diets increased IMF deposition and fatty acid content as well as increased intramuscular lipogenic gene expression during the experimental period.

1 INTRODUCTION

Yaks (Bos grunniens), one of the world's most extraordinary herbivores, are a multifunctional principal livestock species unique to the Qinghai-Tibetan Plateau, where altitudes range from 3, 000 to 5, 000m and which is characterized by hypoxia, low annual average temperatures, a short forage growing season (approximately 90-120 days) and a shortage of forage nutrients[1,2].There are approximately 15 million yaks on the Qinghai-Tibetan Plateau, accounting for more than 90% of total yak population in the world.These yaks provide many major resources (e.g., meat, milk, hair, hides and dung), and yak meat is the major economic resource for local herders[3,4].

As quality life improves worldwide, people increasingly expect that the meat they consume should be healthy and high quality (low in fat; high in protein, vitamins and minerals).The intramuscular fat (IMF) and intramuscular fatty acid content play an important role in meat quality[5,6]. However, yak meat has a low IMF content because intramuscular adipose deposition is difficult under the conditions of long-term malnutrition in a traditional grazing system in cold seasons[7].Meat quality, including palatability, tenderness and juiciness, could be improved by increasing the IMF content[8], and increasing the IMF content and fatty acid profile may improve the quality of yak meat.Many studies have demonstrated that the IMF content and fatty acid levels are influenced by age[9], breed[10] and diets[11-15].In vivo, lipogenesis, lipolysis and fatty acid transport lead to IMF deposition.A higher-energy content diet could contribute to lipogenesis[16].The concentration of intramuscular fatty acids is mainly regulated by inducing and inhibiting genes encoding specific metabolic enzymes associated with lipid metabolism or transcription factors[17].Sterol regulatory element-binding protein-1c (SREBP-1c) and peroxisome proliferator-activated receptor γ (PPARγ) are the most important genes involved in lipid metabolism in muscle tissue[18,19].Recently, effects of different dietary energy levels on IMF deposition and fatty-acid composition in beef cattle[20], double-muscled Belgian Blue bulls[21] and Angus × Chinese Xiangxi yellow cattle[22] have been reported, as being affected by diets with different protein levels on intramuscular fat deposition in yaks[23]. However, little information is available regarding the influence of different energy diets on IMF deposition and fatty acid profiles in yaks.

The effects of different dietary energy content on IMF deposition and fatty acid composition in yaks are unclear. Therefore, the objective of this study was to analyze the relationship between dietary energy and fat deposition and fatty-acid profile, as well as expression of the regulatory genes PPARγ, SREBP-1c, stearoyl-CoA desaturase (SCD), acetyl CoA carboxylase (ACACA), lipoprotein lipase (LPL), fatty acid-binding protein 4 (FABP4) and fatty acid synthase (FASN) in the muscle of adult yaks.

2 MATERIALS AND METHODS

2.1 Ethics statement

This feeding experiment was conducted between February and May 2016 at Hongtu Yak Breeding Cooperatives (located in Qinghai-Tibetan Plateau, at 35°08′38″N, 102°99′36″E and with average altitude 3, 230m) of Tibetan Autonomous Prefecture of Gannan, Gansu Province, China. Before the experiment, all animal studies and the barn environment were examined and approved by the Institutional Animal Care and Use Committee of Lanzhou Institute of Husbandry and Pharmaceutical Sciences, China. The yaks were provided for use as experimental animals by the owner, and all animals were cared for according to the Guide for the Care and Use of Laboratory Animals, Lanzhou Institute of Husbandry and Pharmaceutical Sciences, China.

2.2 Animals, diets and management

A total of fifteen adult castrated male yaks with similar body conditions (body weight 274.3 ± 3.14kg) were included in a single-factor completely randomized experimental design. The animals were randomly divided into three treatment groups with five replicates each. Three diets with different levels of energy and a concentrate-to-forage ratio of 30 : 70 (DM basis) were formulated to be isonitrogenous, containing similar roughage mixtures (40% oats silage, 40% microbial corn stalk silage and 20% highland barley hay) and different energy concentrates: low energy level (LE: 5.5MJ/kg), medium energy level (ME: 6.2MJ/kg) and high energy level (HE: 6.9MJ/kg). The ingredients and nutrient composition of the three diets are shown in Table 1.

The experiment lasted for 60d, and the animals were given a 14d adaptation period to familiarize them with the diets, facilities and staff before the experiment. All animals were weighed, labeled before feeding in the morning and then individually housed in tie-stalls; each animal had 9 m^2 for normal activities with bedding, and the barn was cleaned every day. The yaks were fed twice a day at 08: 00-09: 00 and 17: 00-18: 00 with a total of 2.45kg concentrates and 5.75kg roughage mixtures; they were given free access to water and mineral blocks. The final weights of the animals were determined by weighing them after fasting them

Table 1 Ingredients and nutrient composition of the diets during the experiment.

Item	LE	ME	HE
Ingredient (%)			
Corn	29.00	44.80	56.00
Corn germ	30.00	20.00	12.00
Wheat bran	4.00	4.00	—
DDGS	15.00	7.00	6.30
Prickly ash seed	10.00	2.00	4.00
Cottonseed meal	6.00	12.00	16.00
Soybean meal	—	5.00	—
Salt	0.80	0.80	0.80
White stone powder	2.00	2.00	2.00

(continued)

Item	LE	ME	HE
Dicalcium phosphate	0. 60	0. 60	0. 60
Urea	0. 80	—	0. 50
Sodium bicarbonate	1. 00	1. 00	1. 00
Premix[a]	0. 80	0. 80	0. 80
Nutrient composition, % of DM			
Crude protein	16. 53	16. 74	17. 21
Crude fat	3. 73	4. 18	5. 57
NEg[b] (MJ/kg)	5. 5	6. 2	6. 9
Neutral detergent fiber	15. 93	13. 15	12. 32
Acid detergent fiber	4. 54	4. 14	3. 72
Calcium	0. 64	0. 84	0. 75
Phosphorus	0. 31	0. 34	0. 36

[a] Premix was provided per kilogram of total diet DM, and the composition was as follows: 22, 520 IU of vitamin A, 1, 920 IU of vitamin D_3, 18 IU of vitamin E, 0. 36 IU of vitamin K_3, 5. 28mg of vitamin B_2, 0. 008mg of vitamin B_{12}, 21. 2mg of D-calcium pantothenate, 9mg of Cu, 132. 8mg of Zn, 240mg of Fe and 8mg of Mn, 0. 28mg of Co.

[b] NEg, net energy for gain; DDGS, dried distillers grains with solubles; LE, low energy level; ME, medium energy level; HE, high energy level.

https: //doi.org/10. 1371/journal.pone. 0187604. t001 for 12h. They were transferred to a slaughter house and all yaks received a jugular vein injection of sierra oxazine hydrochloride injection (Shengda Animal Pharmaceutical Co. Ltd., Dun-hua, Jilin, China) with 0. 2mL/kg liveweight to ease their pain before slaughter, and exsanguination via the carotid artery was executed after animals completely under anesthesia.

2. 3 Sample collection

After slaughter, thecarcasses were immediately washed and divided into halves. Subsequently, two samples of the longissimus dorsi muscle between 12th and 13th ribs were collected from each animal. All instruments used to collect the tissue were sterilized in advance. The muscle samples were washed in 0. 9% NaCl solution, and the first samples were stored in Ziploc bags at −20℃ for proximate nutrient and fatty acid analysis. The second were placed in 2mL cryogenic vials (Corning Incorporated, New York, USA), transported in liquid nitrogen and stored at −80℃ for total RNA extraction.

2. 4 Intramuscular fat content analysis

A 5g meat sample was used to measure intramuscular fat content in each animal. Sea sand was added to evaporate moisture in a water bath. Lipids were extracted using the petroleum ether of Soxhlet apparatus (AOAC method) and expressed as grams per 100g of fresh muscle tissue.

2. 5 Fatty acid profile analysis

Samples of yak muscles (100mg) were trimmed of connective tissue and finely chopped. Intramuscular fat was extracted in 3mL chloroform-methanol 1 : 1[24] and methylated in 2mL 4% methanol solution in HCl[25] containing 100μL C19 : 0 (methyl non - adecanoate) as an internal standard, and then the mixtures were heated in a water bath at 85℃ for 1h. Subsequently, 1mL n-hexane was added to the mixture after it reached room temperature and shaking extraction was performed for 2min. The mixture was allowed to stand for 1h until layering occurred; following this, 100μL supernatant was transferred to a new centrifuge tube and diluted with n-hexane to 1mL. The extract was then filtered through filter membrane with an aperture of 0.45μm. Fatty acid profiles were determined by gas chromatograph - mass spectrometer (ThermoFisher Trace 1310, Thermo Scientific, MA, USA) with a flame ionization detector and a 30m-long capillary column 0.25mm in internal diameter and 0.25μm thick (TG-5MS, Thermo Scientific, MA, USA). Nitrogen was used as a carrier gas at a flow rate of 1.2mL/min. The chromatographic conditions were as follows: the capillary column was incubated at 80℃ for 1min, and the temperature was increased by 10℃/min until it reached 200℃. Subsequently, the temperature was increased by 4℃/min until it reached 250℃ and then increased at 2℃/min to 270℃. Following this, the temperature was held for three minutes at 270℃. During analysis, the temperature of injector was kept at 290℃ and splitless injection was performed with a 1min opening valve time. Next, the capillary column was directly subjected to mass spectrum analysis under the following conditions: the interface temperature set at 290℃ and the ion source maintained at 280℃. Otherwise, 70 eV ionization energy of impact ionization (EI) was used to fragment the eluents from the capillary gas chromatograph. The scanned area was from 30 to 400 amu (atomic mass units). Individual fatty acid contents were determined through comparison to the mixed fatty acid standard product and the internal standard, and the fatty-acid levels were calculated by following formula:

$$X_i = m \times A_{si} / \sum si$$

where X_i is the level of each fatty acid, expressed asmg/kg muscle tissue; m is the intramuscular fat content of the sample; A_{si} represents the peak area of all types of fatty acids in the sample; $A\sum si$ is the sum of the peak area of all fatty acids in the sample.

2.6 RNA extraction and quantitative RT-PCR

The target and reference gene primers were designed using integrated mRNA sequences based on sequences published by the National Center for Biotechnology Information (NCBI) (www.ncbi.nlm.nih.gov) primers were designed using Premier Primer 5 software (PREMIER Biosoft International, CA, USA), and sequences of each are shown in Table 2.

The total RNA of thelongissimus dorsi muscle was extracted by RNAiso Plus (TaKaRa, Dalian, China) according to the manufacturer's instructions. The extracted RNA was dissolved in 30μL diethylpyrocarbonate (DEPC) - treated water (Solarbio LIFE SCIENCES, Beijing, China), and the concentration was measured by spectrophotometer (NanoDrop 2000, Thermo Scientific, MA, USA). Purity and integrity were checked using agarose-gel electrophoresis with ethidium bromide. Total RNA was reverse transcribed into five copies using a PrimeScript™.

RT reagent kit (TaKaRa, Dalian, China) in accordance with the manufacturer's instructions, and the RT-PCR products were stored at-20℃ for quantitative real-time PCR.

Quantitative real-time PCR was performed in triplicate to determine mRNA relative expression using a SYBR © Premix Ex Taq™ II (TaRaKa, Dalian, China). Each 25-μL real-time reaction contained 12.5μL SYBR Premix Ex Taq II (2x), 1μL each of 10μM primers, 2.0μL cDNA and

8. 5μL RNase Free dH_2O.Reactions were run on a fluorescence thermal cycler (CFX96, Bio-Rad, CA, USA), and the program was as follows: 95℃ for 30s, 40 cycles of 95℃ for 5s, annealing at 60℃ for 30s and a melting curve with a temperature increase of 0. 5℃ every 5s starting at 65℃. The RT-PCR analyses for each studied gene were performed using cDNA from five biological replicates with three technical replicates per biological replicate.

The threshold cycle (C_T) resulting from quantitative RT-PCR was analyzed using the $2^{-\Delta\Delta Ct}$ method, and all data were normalized with the reference gene beta-actin (β-actin) gene[26].

Table 2 Gene names, primer sequences, accession, product size of the used genes.

Gene name*		Primer sequence (5′→3′)	Accession	Product size
p-actin	F	ACCATCGGCAATGAGCG	XM_ 005887322. 2	150bp
	R	CACCGTGTTGGCGTAGAG		
LPL	F	TCCTGGAGTGACCGAATC	XM_ 005902304. 2	125bp
	R	AGGCAGCCACGAGTTTT		
ACACA	F	AAGCAATGGATGAACCTTCTTC	XM_ 005888165. 2	197bp
	R	GATGCCCAAGTCAGAGAGC		
PPARy	F	CATTTCCACTCCGCACTA	XM_ 005902845. 2	122bp
	R	GGGATACAGGCTCCACTT		
FASN	F	GACGGTCGCATCATCTTCC	XM_ 005905364. 2	156bp
	R	GAGCACAATCCCTGTCTTCG		
FABP4	F	TGAGATTTCCTTCAAATTGGG	XM_ 014478668. 1	101bp
	R	CTTGTACCAGAGCACCTTCATC		
SCD	F	TACTGCGGTCCAAGTCGTT	NM_ 173959. 4	165bp
	R	CAGCCTTGTCTGGAGTCATC		
SREBP-1c	F	AGCTCAAGGACCTGGTGGTG	XM_ 014477492. 1	140bp
	R	GACAGCAGTGCGCAGACTCA		

* LPL, lipoprotein lipase; ACACA, acetyl-CoA carboxylase; PPARγ, peroxisome proliferator-activated receptors gamma; FASN, fatty acid synthase; FABP4, adipocyte fatty acid binding protein 4; SCD, stearoyl-CoA desaturase; SREBP-1c, sterol regulatory element-binding protein-1c.

https: //doi.org/10. 1371/journal.pone. 0187604. t002

2. 7 Statistical analysis

The fatty acid content and mRNA abundance in the longissimus dorsi muscle were processed by one-way ANOVA analysis using the LSD procedure to perform multiple comparisons in SPSS 19. 0 (SPSS Inc., Chicago, USA).The Pearson correlation analyses were performed to calculate correlation coefficients among gene expression, as well as correlation coefficients between gene expression and intramuscular fat content.Furthermore, the correlations between gene expression and fatty acid concentrations were also determined.Values at $P<0.05$ were considered significant difference.

3 RESULTS

3. 1 Intramuscular fatty acid composition

The effect of diets with different energy contents on the fatty acid profile of yaks is shown in Table 3. Intramuscular fat content was highly affected (P=0. 001) by dietary energy levels. The fatty acids content varied significantly among diets and was positively associated with higher-energy diets. Significant (*P*<0. 001, P = 0. 011 and P = 0. 001) differences were found between HE and LE yaks regarding saturated fatty acids (SFA), monounsaturated fatty acids (MUFA) and polyunsaturated fatty acids (PUFA), and there were also significant (P = 0. 037 and P = 0. 005) differences between HE and ME yaks in terms of SFA and PUFA content. However, ME yaks were not significantly different from LE and HE yaks (P = 0. 227 and P = 0. 111) in terms of MUFA, and the level of PUFA was similar (P=0. 363) between the ME and LE yaks.

Table 3 IMF content and fatty acid composition in longissimus dorsi of yaks fed diets supplying different energy levels.

Item *	LE	ME	HE	SEM	P-value
IMF (g/100g)	0. 56[c]	0. 92[b]	1. 34[a]	0. 102	0. 001
Fatty acid (mg/kg)					
C14: 0 Myristic acid	15. 35[c]	21. 55[b]	25. 11[a]	1. 138	<0. 001
C14: 1 Myristoleic acid	3. 33[c]	4. 14[b]	5. 73[a]	0. 291	<0. 001
C15: 0 Pentadecanoic acid	2. 08[b]	3. 01[ab]	4. 03[a]	0. 308	0. 020
C16: 0 Palmitic acid	1951. 06[c]	2748. 31[b]	3367. 25[a]	174. 237	<0. 001
C16: 1 Palmitoleic acid	69. 80[c]	81. 91[b]	118. 14[a]	5. 60	<0. 001
C17: 0 Margaricacid	6. 31[b]	7. 81[b]	10. 26[a]	0. 595	0. 010
C17: 1 Heptadecanoic acid	5. 20[b]	6. 02[ab]	8. 44[a]	0. 566	0. 038
C18: 0 Stearic acid	1044. 74[c]	1242. 88[b]	1588. 32[a]	63. 775	<0. 001
C18: 1n9c Oleic acid	1952. 68[b]	2441. 40[ab]	3059. 18[a]	186. 659	0. 038
C18: 1n9ttrans oleic acid	37. 37[b]	44. 45[b]	75. 05[a]	5. 866	0. 008
C18: 2n6c Linolenic acid	64. 49[b]	66. 39[b]	80. 41[a]	2. 846	0. 029
C20: 1 Eicosenoic acid	1. 22[b]	1. 82[a]	2. 11[a]	0. 136	0. 012
C20: 3n3 Eicosatrienoic acid	0. 71[b]	1. 14[a]	1. 31[a]	0. 085	0. 002
C20: 4n6 Arachidonic acid	8. 61[c]	11. 81[b]	18. 81[a]	1. 185	<0. 001
C20: 5n3 EPA	1. 84[b]	2. 22[ab]	3. 08[a]	0. 212	0. 036
C22: 6n3 DHA	1. 26[b]	1. 54[ab]	2. 17[a]	0. 157	0. 038
ΣSFA	3019. 55[c]	4023. 57[b]	4994. 97[a]	232. 906	<0. 001
ΣMUFA	2069. 60[b]	2579. 74[ab]	3268. 65[a]	195. 509	0. 027
ΣPUFA	76. 91[b]	83. 09[b]	105. 78[a]	4. 134	0. 002

[a,b,c] Means in a row with different small letter superscripts differ significantly (*P*<0. 05).

[*] IMF, intramuscular fat; ΣSFA, saturated fatty acid (without any double bonds, C14: 0-C18: 0); ΣMUFA, monounsaturated fatty acid, all fatty acids with a single double bond (C14: 1-C20: 1); Σ PUFA, polyunsaturated fatty acid, all fatty acids with 2 or more double bonds (including C18: 2n6c, C20: 3n3, C20: 4n6 and C20: 5n3); EPA, eicosapentaenoic acid; DHA, docosahexaenoic acid.

LE, low energy level; ME, medium energy level; HE, high energy level.

https: //doi.org/10. 1371/journal.pone. 0187604. t003

The fatty acids C16: 0, C18: 0 and C18: 1n9c stand out, accounting for a large proportion of total fatty acids. HE yaks had the highest levels of C14: 0, C14: 1, C16: 0, C16: 1, C17: 0, C18: 0, C18: 1n9t, C18: 2n6c and C20: 4n6, and HE yaks had significantly higher (*P*<

0.001) concentrations of C14: 0, C14: 1, C15: 0, C16: 0, C16: 1, 17: 0, C17: 1, C18: 0, C18: 1n9c, C18: 1n9t, C18: 2n6c, C20: 1, C20: 3n3, C20: 4n6, C20: 5n3 and C22: 6n3 than did the LE yaks. Otherwise, significant ($P<0.001$, P = 0.022, P = 0.003, P = 0.001, P = 0.005, P = 0.034, P = 0.008 and P = 0.003) differences were found between ME and LE yaks in the concentrations of C14: 0, C14: 1, C16: 0, C16: 1, C18: 0, C20: 1, C20: 3n3 and C20: 4n6. mRNA abundance of longissimus dorsimuscle.

The mRNA abundance of the candidate genes in each diet is shown in Fig 1. The mRNA levels of ACACA, FASN, SCD, SREBP - 1c, PPARγ and FABP4 increased ($P < 0.05$) with increasing dietary energy; HE and ME yaks had higher ($P<0.05$) mRNA abundance of above genes (except for SREBP-1c) than did the LE yaks. In addition, significant (P = 0.025, P = 0.01, P = 0.041 and $P<0.001$) differences were found in gene expression between HE and ME yaks for PPARγ, ACACA, FASN and SCD; ME yaks had higher ($P<0.001$, $P<0.001$, $P<0.001$, $P<0.001$ and P = 0.016) expression levels of PPARγ, ACACA, FASN, SCD and FABP4 than did the LE yaks. Furthermore, no significant (P = 0.508) differences were found of LPL among three treatments, and HE and ME yaks had similar (P>0.05) expression of SREBP-1c and FABP4.

Relationships among the expression levels of genes involved in lipid metabolism in longissimus dorsi muscle.

The expression of PPARγ was positively correlated with those of ACACA ($P<0.01$), FABP4 ($P<0.01$), SCD ($P<0.01$), SREBP-1c ($P<0.05$) and FASN ($P<0.01$). The expression of SREBP-lc was also positively correlated with those of ACACA ($P<0.01$), SCD ($P<0.05$) and FASN ($P<0.01$). In addition, FASN expression was positively correlated with those of ACACA ($P<0.01$), FABP4 ($P<0.05$) and SCD ($P<0.01$). There was a significant ($P<0.01$) correlation between SCD and ACACA and a positive correlation ($P<0.05$) between SCD and FABP4 (Table 4).

Relationships among gene expression levels, intramuscular fat content and fatty acid content in longissimus dorsi muscle.

IMF content was positively correlated with the expression levels of PPARγ ($P<0.05$), ACACA ($P<0.01$), FASN ($P<0.01$), SCD ($P<0.01$) and SREBP-1c ($P<0.01$), but no significant correlation was found between IMF content and the expression of other genes (Table 5).

Considering the correlations between gene expression and fatty acid content, FASN and SCD are notable (Table 5). These genes were positively correlated with levels of C14: 0 ($P<0.01$), C14: 1 ($P<0.01$), C16: 0 ($P<0.01$), C16: 1 ($P<0.01$), C18: 0 ($P<0.01$), C18: 1n9c ($P<0.01$), C18: 1n9t ($P<0.01$), C18: 2n6c ($P<0.05$ and $P<0.01$), C20: 1 ($P<0.01$), C20: 3n3 ($P<0.01$), C20: 4n6 ($P<0.01$), C20: 5n3 ($P<0.01$), C22: 6n3 ($P<0.01$), SFA ($P<0.01$), MUFA ($P<0.01$) and PUFA ($P<0.01$). The expression levels of ACACA and PPARγ were positively correlated with C14: 0 ($P<0.01$), C14: 1 ($P<0.01$), C16: 0 ($P<0.01$), C16: 1 ($P<0.01$), C18: 0 ($P<0.01$), C20: 4n6 ($P<0.01$) and SFA ($P<0.05$ and $P<0.01$), and ACACA was positively correlated with C18: 1n9c ($P<0.05$), C18: 1n9t ($P<0.05$) and MUFA ($P<0.05$). PPARγ had positive correlation with C18: 2n6c ($P<0.05$), C20: 3n3 ($P<0.01$), C20: 5n3 ($P<0.05$), C22: 6n3 ($P<0.05$)

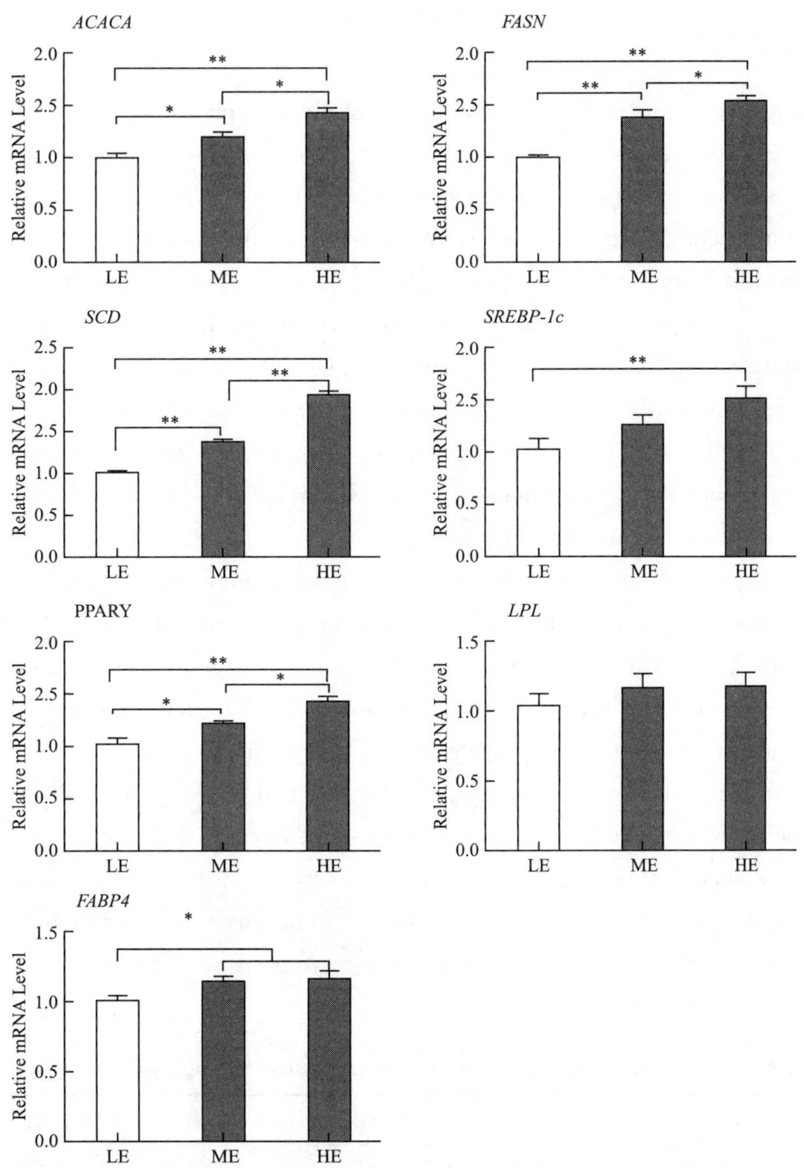

Fig 1. Relative expression levels of seven candidate genes in longissimus dorsi muscle of yaks fed diets supplying different energy levels.

Note: ACACA: acetyl – CoA carboxylase, FASN: fatty acid synthase, SCD: stearoyl – CoA desaturase, SREBP–1c: sterol regulatory element–binding protein–lc, PPARγ: peroxisome proliferator – activated receptors gamma, LPL: lipoprotein lipase, FABP4: fatty acid binding protein 4. Values are represented as the Mean±SEM. * and ** indicate $P<0.05$ and $P<0.01$, respectively, between two groups.

https://doi.org/10.1371/journal.pone.0187604.g001 and PUFA ($P<0.01$). In addition, SREBP–1c was positively correlated with C14: 0 ($P<0.01$), C14: 1 ($P<0.01$), C16: 0 ($P<0.01$), C16: 1 ($P<0.01$), C18: 0 ($P<0.01$), C20: 4n6 ($P<0.05$) and SFA ($P<0.05$). Significant correlations were found between the expression level of FABP4 and the concentrations of C14: 0, C16: 0, C18: 0, C20: 4n6, SFA ($P<0.05$) and C20: 3n3 ($P<0.01$).

4 DISCUSSION

This paper describes for the first time the effects of diets supplying different energy levels on fat deposition and the fatty acid profile of the longissimus dorsi muscle in the domesticated yak and the role of genes involved in lipid metabolism in changing the fatty acid composition across three treatments.Non-structural carbohydrate (starch) was priority to support skeletal and muscle growth resulting in low rate of fat deposition during the "growing phase" in beef cattle.Conversely, the intake of starch, first and foremost, contributed to fat deposition during the "finishing phase" of beef cattle[27].Consistently, several studies have demonstrated that IMF content increases with increased energy content in finishing cattle[20] and buffalo cattle[28], which was also found in current study.Bovine intramuscular fat is characterized by its abundant saturated fatty acids (SFA), especially palmitic acid and stearic acid[29], as we found for the yaks.Moreover, previous research showed that the level of SFA depends on the degree of fat deposition[30,31] and that SFA content increased with increasing IMF content.It is, therefore, possible that higher energy content regulates SFA levels by enhancing IMF content.Consistent with our data, Smet et al[21] reported that the levels of C14 : 0, C15 : 0, C16 : 0, C16 : 1 and C18: 1 increased with higher-energy diets in Belgian Blue bulls. By contrast, the level of C18 : 0 decreased in previous studies[32-34] but increased in the current research when higher-energy diets were provided.Pentadecanoic acid (C15 : 0) and heptadecanoic acid (17 : 0) belong to odd-chain fatty acids (OCFAs) and only have a small proportion of total saturated fatty acids in ruminant meat.De novo synthesis of linear OCFAs is different from palmitic acid which used propionyl-CoA as primer, instead of acetyl-CoA and a large portion of OCFAs in ruminants was synthesized by rumen bacteria, and it also originates from diets including maize silage, grass silage and other plants[35,36].Therefore, the synthesis of OCFAs may either not occur in adipose tissue or regulate by lipogenic genes.

Table 4 Pearson correlation coefficients among gene expression in longissimus dorsi muscle of yaks.

Gene[a]	ACACA	LPL	FABP4	SCD	SREBP-1c	FASN
PPARγ	0.682**	0.075	0.730**	0.812**	0.534*	0.663**
FASN	0.684**	0.246	0.571*	0.850**	0.765**	
SREBP-1c	0.741**	0.220	0.267	0.631*		
SCD	0.748**	0.286	0.621*			
FABP4	0.351	-0.213				
LPL	0.407					

Pearson correlation coefficients are across all treatments. Number of observations = 15. * $P < 0.05$; ** $P < 0.01$; *** $P<0.001$.

[a] LPL, lipoprotein lipase; ACACA, acetyl-CoA carboxylase; PPARγ, peroxisome proliferator-activated receptors gamma; FASN, fatty acid synthase; FABP4, adipocyte fatty acid binding protein 4; SCD, stearoyl-CoA desaturase; SREBP-1c, sterol regulatory element-binding protein-1c.

https: //doi.org/10.1371/journal.pone.0187604.t004

Table 5 **Pearson correlation coefficients among gene expression, IMF and fatty acid content inlongissimus dorsi muscle of yaks.**

Item[a]	PPARγ	FASN	ACACA	LPL	SREBP-1c	SCD	FABP4
IMF(g/100g)	0.549*	0.791**	0.743**	0.339	0.785**	0.786**	0.216
Fatty acid[b](mg/kg)							
C14:0 Myristic acid	0.735**	0.959**	0.753**	0.356	0.736**	0.887**	0.566*
C14:1 Myristoleic acid	0.693**	0.819**	0.788**	0.212	0.743**	0.889**	0.422
C16:0 Palmitic acid	0.805**	0.911**	0.710**	0.129	0.668**	0.851**	0.588*
C16:1 Palmitoleic acid	0.724**	0.812**	0.825**	0.244	0.730**	0.920**	0.481
C18:0 Stearic acid	0.762**	0.838**	0.830**	0.175	0.754**	0.886**	0.584*
C18:1n9c Oleic acid	0.446	0.896**	0.524*	0.177	0.224	0.651**	0.313
C18:1n9t trans oleic acid	0.489	0.727**	0.600*	-0.055	0.500	0.666**	0.267
C18:2n6c Linolenic acid	0.586*	0.672**	0.313	0.036	0.400	0.670**	0.309
C20:1 Eicosenoic acid	0.434	0.869**	0.447	0.404	0.360	0.750**	0.373
C20:3n3 Eicosatrienoic acid	0.776**	0.780**	0.466	0.187	0.449	0.853**	0.673**
C20:4n6 Arachidonic acid	0.822**	0.778**	0.744**	0.282	0.589*	0.961**	0.581*
C20:5n3 EPA	0.624*	0.779**	0.382	0.337	0.396	0.727**	0.291
C22:6n3 DHA	0.629*	0.776**	0.379	0.358	0.378	0.723**	0.289
ΣSFA	0.696**	0.913**	0.641*	0.096	0.544*	0.785**	0.524*
ΣMUFA	0.443	0.891**	0.532*	0.165	0.248	0.652**	0.304
ΣPUFA	0.711**	0.757**	0.472	0.140	0.488	0.819**	0.419

Pearson correlation coefficients are across all treatments. Number of observations = 15.

* $P<0.05$; ** $P<0.01$; *** $P<0.001$.

[a] ACACA, acetyl-CoA carboxylase; FASN, fatty acid synthase; SCD, stearoyl-CoAdesaturase; SREBP-1c, sterol regulatory element-binding protein-1c; PPARγ: peroxisome proliferator-activated receptors gamma; LPL, lipoprotein lipase; FABP4, adipocyte fatty acid binding protein 4.

[b] IMF, intramuscular fat; ΣSFA, total saturated fatty acid (without any double bonds, C14: 0-C18: 0); ΣMUFA, total monounsaturated fatty acid, all fatty acids with a single double bond (C14: 1-C20: 1); ΣPUFA, total polyunsaturated fatty acid, all fatty acids with 2 or more double bonds (including C18: 2n6c, C20: 3n3, C20: 4n6 and C20: 5n3); EPA, eicosapentaenoic acid; DHA, docosahexaenoic acid.

https://doi.org/10.1371/journal.pone.0187604.t005

Fatty acid composition of the longissimus dorsi muscle is affected by diets, including protein content, energy content[37] and fatty acid profile[17,38,39]. Abundant MUFA and PUFA not only improve meat characteristics, such as flavor, cholesterol content and nutritional benefits, but are also beneficial for human health. Contrary to our expectations, dietary MUFA and PUFA can't be directly deposited in muscle due to the biohydrogenation of the rumen, and a portion of unsaturated fatty acids (UFA) become saturated[40]. The concentrations of MUFA and PUFA were greater ($P<$

0.05) in the HE group than in the LE group, which might be influenced by two factors: On one hand, the diets contained different amounts of dried distillers grains with solubles (DDGS), and a previous study found that high-fat DDGS is a good source of UFA and to some extent, protects the UFA from ruminal biohydrogenation[41].On the other hand, high dietary energy level may affect the normal function of rumen microbes associated with biohydrogenation, allowing more UFA to reach the small intestine, where they can be absorbed by the intestinal epithelium and deposited in muscle[42].There were small amounts of eicosapentaenoic acid (EPA) and docosahexaenoic acid (DHA) in yak muscle, as described by Qin et al.; however, EPA and DHA were not detected in finishing cattle[22,43].

To some degree, yak meat is more nutritious than cattle, as EPA and DHA contribute to reduced serum levels of fat and cholesterol and improved brain function[44].In the HE condition, the SFA content was higher than those of MUFA and PUFA, and the meat therefore does not meet human nutritional needs.Therefore, future research should focus on developing yak meat with higher MUFA and PUFA levels and lower SFA levels.

IMFdeposition is the result of a balance between dietary energy content and animals maintenance requirements. The net energy for gain (NEg) provided in the three diets was 5.5MJ/kg, 6.2MJ/kg and 6.9MJ/kg.As a result, the high-energy diet had the highest IMF content.It is possible that more available substrate (glucose or acetate) involved in fat synthesis resulted in a higher degree of intramuscular fat deposition[45].Additionally, the fatty acids profile of bovines is not strongly affected by diet because most dietary unsaturated fatty acids are saturated by microbial biohydrogenation in the rumen[46,47].Our results differed because higher-energy diets were provided.Fatty acid composition may be regulated by genes involved in lipid synthesis and fatty acid metabolism.Nevertheless, no studies have evaluated in detail the molecular mechanism underlying changes in fatty acid composition in finishing yaks.We found that expression of lipogenic and fatty acid metabolic genes affect fatty acid concentrations.Therefore, more attention should be paid to transcription factors and other key proteins related to lipid metabolism in the longissimus dorsi muscle.

The process of IMF deposition can be regulated by diets (including different energy or protein levels) through the potentially complex regulation mechanism. PPARγ, which belongs to the nuclear receptor superfamily, plays a vital role in regulating the expression of some genes encoding proteins involved in fat accumulation and adipocyte differentiation in adipose and muscle tissue.It has been reported that PPARγ was identified as an important candidate gene which has positive role in the several lipogenesis-related pathways of intramuscular fat using protein-protein interaction networks[48].In detail, PPARγ promotes the expression of some adipocyte proteins or enzymes, such as fatty acid binding protein (FABP4), fatty acid synthase (FASN) and lipoprotein lipase (LPL)[49].Few studies examined the effects of dietary energy levels on PPARγ gene expression in longissimus dorsi muscle of yak.We observed that the mRNA expression levels of PPARγ, FABP4 and FASN were increased with increasing IMF content in the HE diets as compared to LE diet, but no expression changes occurred for the LPL gene.The result was in line with a previous study where the expressions of PPARγ, FABP4 and FASN were activated with an increase in IMF content in

Angus×Simmental cattle fed high-starch or low-starch diets[50], and in vitro study also found that bovine perimuscular pre-adipocytes cultured with insulin and dexamethasone to stimulate the differentiation had greater mRNA expression of PPARγ, FABP4 and FASN compared with control[51]. It was expected that the expression of PPARγ was positively (P<0.05) correlated with the expression of FABP4 and FASN as well as IMF content, however, had no correlation with LPL in our study. Interestingly, the expression of LPL was not significantly (P>0.05) influenced by dietary energy content in this study, indicating that the rate of lipogenesis was greater than that of lipolysis due to adequate energy intake and that little triglyceride was hydrolyzed into non-esterified fatty acids in the muscle to provide energy for the peripheral tissues. In brief, our results indicated that the IMF deposition in yaks fed high dietary energy was attributed to increased expression of PPARγ, FABP4 and FASN which accelerate the rate of lipogenesis and low expression level of LPL during the finishing stage.

Peroxisome proliferator-activated receptors (PPARs), which belong to the nuclear receptor superfamily, are responsible for nutrient metabolism and energy homeostasis[52,53]. All isoforms of PPARs can be activated or inhibited by the agonists (long-chain fatty acids or chemically associated derivatives) binding in the area of nutrition stems, and C16: 0, C18: 0, C18: 1, C18: 2 and C20: 5n3 can indirectly activate PPARγ[54]. Long-chain fatty acids (LCFAs) such as C18: 1 and C18: 2, provided ligand to activate PPARγ and increased its mRNA expression level, can stimulate porcine adipocyte differentiation[55]. The significant positive correlations were found between the mRNA expression of PPARγ and the concentrations of C16: 0, C18: 2, C20: 3n3, C20: 4n6, C20: 5n3 and C22: 6n3 in our study. We speculated that these LCFAs activate the PPARγ to up-regulate its downstream genes and expedite the process of lipogenesis in intramuscular fat. FABP4 is a carrier protein binding intracellular LCFAs into nucleus to activate peroxisome proliferator-activated receptors. The mRNA expression level of FABP4 positively correlated with C16: 0, C20: 3n3 and C20: 4n6 may be due to its greater affinity for these LCFAs.

It is well known that FASN produces a multipurpose enzyme that catalyzes the entire pathway of palmitate synthesis from malonyl-CoA[56]. In agreement with our results, Saburi et al[57] observed that FASN affected the synthesis of C14: 0, C16: 0, C16: 1, C18: 0, C18: 1 and C18: 2 as well as the ratio of monounsaturated to saturated fatty acids in Japanese black cattle and its TW haplotype enhanced C18: 0, C18: 1 content and decreased C14: 0, C14: 1, C16: 0 and C16: 1 content. The 'AA' genotype of the FASN SNP was dramatically related to higher concentrations of C14: 0, C14: 1, C16: 1 and C18: 2, but lower concentrations of 18: 1n9c and C20: 3n6 in Canadian commercial cross-bred beef steers[58]. However, we found the mRNA expression of FASN was positively (P < 0.05) correlated with the content of large proportion of fatty acids which detected in current study. It suggested that the mechanism of FASN to regulate fatty acid synthesis may be influenced by different diets and genotypes or by the regulation of transcription factors (PPARγ and SREBP-1c).

In IMF accumulation, SREBP-1c is a critical transcription factor that regulates the expression of lipogenic genes (ACACA, FASN and SCD), and positive (P<0.05) correlations were found among these genes in current study (Table 4). The expression of FASN, ACACA, SREBP-1c and

SCD was positively ($P < 0.05$) correlated with IMF content, in agreement with Jeong and Kwon[59], who observed that expression of ACACA and FASN were significantly positively correlated ($P<0.05$) with IMF content and that higher IMF content up-regulated the expression of FASN, ACACA, SREBP-1c and SCD[37]. By contrast, no clear correlation between these genes and the IMF content of beef cattle fed soybeans or rumen-protected fat[17] may be due to higher dietary energy up-regulating the expression of genes related to lipid metabolism to enhance the rate of fat synthesis in our study. Acetyl CoA carboxylase (ACACA), a crucial rate-limiting enzyme in the mammalian cytosol, catalyzes the first step in de novo fatty-acid synthesis in the short term[60], resulting in the biosynthesis of long-chain fatty acids, principally palmitic acid. In this study, C16: 0, C18: 1n9t and C20: 4n6 were positively associated with ACACA, but ACACA was not positively associated with C16: 0 or other long-chain fatty acids in beef cattle fed high-or low-silage diets[61]. Two closely synonymous mutations in the ACACA gene significantly affected the polyunsaturated/saturated fatty acid ratio in the meat[62]. Therefore, the effect of ACACA on fatty acid composition may have been not evident because of single-nucleotide polymorphisms. Stearoyl-CoA desaturase (SCD) is the rate-limiting enzyme and catalyzes biosynthesis of MUFA from SFA, as well as conjugated linoleic acid from vaccenic acid. SCD increased the concentrations of C16: 1 and C18: 1 via C16: 0 and C18: 0 as substrates[63], which is similar to our results. However, SCD was negatively correlated with a-linolenic and arachi-donic acid (C20: 4n6) concentrations in beef cattle, and feeding linolenic acid (C18: 2n6c), eico-sapentaenoic acid (C20: 5n3), and docosahexaenoic acid (C22: 6n3) reduced SCD expression in rats by 50%[17,64]. The results of this study, in which the expression of SCD was positively correlated with levels of C18: 2n6c, C20: 4n6 and C20: 1, indicate that gene expression and fatty acids content might be primarily affected by dietary energy. SREBP-1c affects fatty acid composition through binding the sterol-regulatory-element sequences in the SCD promoter region[65], but the positive correlation between the expression of SREBP-1c and C20: 4n6 levels may also be attributed to the diet. This suggested that the increased IMF content by high dietary energy level probably is due to increased expression of SREBP-1c, FASN, ACACA and SCD, and the above genes may further regulate the biosynthesis of fatty acids.

5 CONCLUSIONS

The present study investigated the impact of dietary energy on the IMF deposition, partial fatty acid content and the mRNA expression of lipogenic genes in the longissimus dorsi muscle during the experimental period in detail. The results indicated that the HE diets promoted the deposition and partial fatty acid content of longissimus dorsi muscle mainly by up-regulation of mRNA expression of ACACA, SCD, FASN, SREBP-1c, PPARγ and FABP4. The data might be used to manipulate or provide useful information to improve IMF deposition and fatty acids accumulation in muscle.

SUPPORTING INFORMATION

S1 File. Experimental animals use permission.

(PDF)

S1 Table.Sequences of PCR product for each gene.
(DOCX)

ACKNOWLEDGMENTS

The authors thank Animal Husbandry Sciences Institute of Tibetan Autonomous Prefecture of Gannan for their help, and we are also grateful to the staff of Hongtu Specialized Yak Breeding Cooperative for their assistance in sample collection.

AUTHOR CONTRIBUTIONS

Conceptualization: Chao Yang, Ruijun Long, Xuezhi Ding, Ping Yan.

Data curation: Chao Yang, Jianbin Liu, Xiaoyun Wu.

Formal analysis: Chao Yang, Jianbin Liu, Xian Guo.

Funding acquisition: Xuezhi Ding.

Investigation: Chao Yang, Jianbin Liu, Pengjia Bao, Xuezhi Ding.

Methodology: Chao Yang, Pengjia Bao, Xuezhi Ding.

Project administration: Xuezhi Ding, Ping Yan.

Resources: Xuezhi Ding, Ping Yan.

Software: Chao Yang, Xiaoyun Wu.

Supervision: Ruijun Long, Xuezhi Ding.

Validation: Ruijun Long, Xuezhi Ding.

Visualization: Chao Yang, Xuezhi Ding.

Writing-original draft: Chao Yang.

Writing-review & editing: Chao Yang, Jianbin Liu, Xiaoyun Wu, Pengjia Bao, Ruijun Long, Xian Guo, Xuezhi Ding, Ping Yan.

REFERENCES OMITTED

(发表于《PLOSONE》，院选 SCI，IF：2. 806)

Palmatine Inhibits TRIF-dependent NF-κB Pathway Against Inflammation Induced by LPS in Goat Endometrial Epithelial Cells

Baoqi YAN[1], Dongsheng WANG[1], Shuwei DONG[1], Zhangrui CHENG[3], Lidong NA[2], Mengqi SANG[1], Hongzao YANG[2], Zhiqiang YANG[1], Shidong ZHANG[1*], Zuoting YAN[1*]

(1. Lanzhou Institute of Animal Husbandry and Pharmaceutical Sciences, Chinese Academy of Agricultural Sciences, Lanzhou 730050, China; 2. College of Veterinary Medicine, Gansu Agricultural University, Lanzhou 730070, China; 3. Department of Production and Population Health, Royal Veterinary College, North Mymms, Hertfordshire, United Kingdom)

Abstract: Palmatine, a natural pharmaceutical drug, possesses many biological activities. But its clinical application is rarely reported in the veterinary medicine. The aim of this study was to investigate the anti-inflammatory effects of palmatine on lipopolysaccharide (LPS) -induced inflammation in goat endometrial epithelial cells (gEECs), and the possible molecular mechanisms. Palmatine cell toxicity was determined by MTT assay, and the production of inflammatory cytokine in the cultured medium was measured with ELISA, qRT-PCR and Western blotting. Our results showed that palmatine treatment inhibited the release of tumor necrosis factor (TNF) -α, interleukin (IL) -1β, IL-6, nitric oxide (NO), matrix metalloproteinase (MMP) -9 and MMP-2. Furthermore, palmatine enhanced the secretion of prostaglandins E_2 (PGE_2) and IL-10. Palmatine significantly down-regulated the expression of Toll-like receptor 4 (TLR4), cluster of differentiation 14 (CD14), Toll/interleukin1 receptor (TIR) -domain-containing adaptor protein inducing interferon-β (TICAM, TRIF) and nuclear factor-κB (NF-κB) in LPS stimulated gEECs, but did not alter the production of MyD88. In conclusion, palmatine inhibits TRIF-dependent NF-κB pathway to reduce LPS-induced inflammatory responses in goat endometrial epithelial cells.

Key words: Palmatine; LPS; anti-inflammatory mechanism; TLR_4; NF-κB pathway

1 INTRODUCTION

Reproductive performance is a major concern in the livestock industry because of its important economic impact. Endometritis often leads to infertility, prolonged calving interval, reduced milk production, and increased treatment costs, which caused huge economic losses to the dairy farming industry[1]. In the past decades, the understandings of some important clinical disorders, including metritis, endometritis, and subclinical endometritis (SE), have advanced. It is reported that overall prevalence of cytologicalendometritis was 36. 2%[2,3]. Endometritis is considered a chronic inflammatory disease because mammals fail to completely clear postpartum bacterial contaminants in time[4].

The uterine immune response is generated not only by professional immune cells but also by endometrial epithelial and stromal cells. Endometrial cells express receptors that recognize microbe-associated molecular patterns (MAMPs), previously known as pathogen-associated molecular patterns (PAMPs). Lipopolysaccharide (LPS) is an important component of the outer membrane of Gram-negative bacteria and has been reported to be an important virulence factor. Bacteria which overwhelm the early mucosal defenses in the genital tract activate an innate immune response by signaling through receptors such as Toll-like receptor (TLR) 4 which recognizes bacterial LPS. This in turn triggers an inflammatory response involving an influx of inflammatory cells, a cascade of up-regulated inflammatory mediators (cytokines, chemokines, growth factors and lipid mediators) and tissue damage. In addition to immune cells, endometrial epithelial and stromal cells play crucial roles in uterine inflammation and immunity. The endometrium can regulate inflammatory response by releasing cytokines and chemokines[5]. Endometrial cells can respond to LPS through the Toll-like receptor (TLR) 4/CD14/MD2 complex signaling pathway. They activate nuclear factor-κB (NF-κB) to regulate pro-inflammatory genes expression and initiate inflammation[6,7]. Once activated, these receptors trigger an inflammatory response aiming to eliminate the invading bacteria[8,9].

Palmatine, a natural pharmaceutical drug, was listed in "Pharmacopoeia of the People's Republic of China" in 1977[10]. It is a kind of quaternary amine isoquinoline alkaloid, with various biological activities. Clinically, palmatine is used as a remedy for abdominal pain, enteritis, gastritis, chronic endometritis, and pelvic inflammation[11,12]. Also it has protective effects on acute and chronic inflammation in experimental animal models, such as ear edema, acetic acid induced capillary permeability test, and cotton pellet-induced inflammation[13,14].

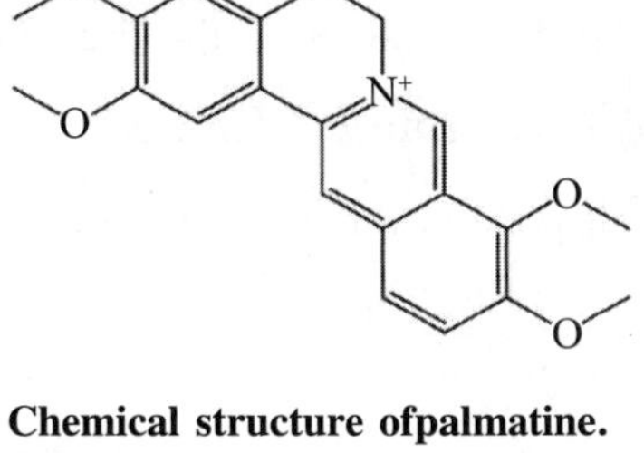

Fig. 1. Chemical structure ofpalmatine.

Meanwhile, Wu et al. demonstrated that palmatinecan induce cell apoptosis to kill breast cancer cells MCF-7[15]. Previous studies have reported its anti-inflammatory mechanisms. It reduced levels of inflammation in patients with acute episode of bronchial asthma by inhibiting the NF-κB activation, and relieved osteoarthritis via regulating the Wnt/β-catenin and Hedgehog signaling pathways[16,17].

Compared with the above research, clinical application of Palmatine is rarely reported in veterinary. In this study, we assessed the anti-inflammatory effect of Palmatine on LPS-stimulated goat endometrium epithelial cells (gEECS), and investigated the possible signaling pathways of Palmatine involved endometritis.

2 MATERIALS AND METHODS

2.1 Reagents and antibodies

Palmatine (purity≥98%, Fig. 1) was purchased from Shanghai source leaf Biological Technology Co., Ltd. (Shanghai, China). LPS (Escherichia coli 0111: B4), 3- (4, 5-dimethylthiazol-2-yl) -2, 5-diphenyl tetrazolium bromide (MTT) was purchased from Sigma Chemical Co. (St. Louis, MO, USA). DMEM-F12 and fetal bovine serum (FBS) were supplied by Gibco BRL (Invitrogen Corporation, Carlsbad, CA, USA). The enzyme-linked immunosorbent assay (ELISA) kits for goat of IL-1β, PGE_2 and IL-6 were purchased from Shanghai enzyme-linked Biological Technology Co., Ltd. (Shanghai, China). TNF-α, MMP-9, MMP-2 and IL-10 kits were purchased from Trust Specialty Zeal biological Trade Co. Ltd. (Boston, MA, USA). Nitric oxide (NO) test kit was bought from the Jiancheng Bioengineering Institute (Nanjing, Jiangsu, China). Anti-NF-κB p65 (ab131109, Abcam, USA) antibody was purchased from Abcam Biological Technology Co. (Cambridge, UK).

2.2 Cell culture andcytotoxicity

Goatendometrium epithelial cells (gEECS) which are immortalized cells[18] were cultured at 37℃ and humidified with 5% CO_2. They were cultured with the medium Dulbecco's modified eagle's medium-F12, DMEM-F12, containing 10% fetal bovine serum (FBS), 100U/mL penicillin and 100U/mL streptomycin. GEECS were seeded into 96-well plate ata densityof 1×10^4 cells/well with0. 2mL of the above culture medium for 24h (control group), then exposed to different concentrations of Palmatine for 48h. Viability of cells was estimated with an MTT assay[19,20], then the inhibitory rate of cell proliferation was calculated as (A2-A0) / (A1-A0) ×100%[21], where A_0 was the optical density of cells before the drug treatment; A_1 was the optical density of cells in control group and A_2 was the optical density of cells after the drug treatment for 48h.

2.3 Cytokines assays in LPS-stimulated EECS

GEECS were seeded into 6-well plate at a density of 2×10^5 cells/well with the culture medium for 24h (control group), exposed to LPS (5μg/mL) for 12h (LPS group), then treated with Palmatine at various concentrations (80, 40, 20μg/mL) for additional 8h. The levels of TNF-α, IL-1β, IL-6, PGE 2, MMP-9, MMP-2 and IL-10 in the supernatant were quantified using ELISA kits according to the manufacturer's instructions.

2.4 NO assays in LPS-stimulated EECS

The release of NO into supernatant of gEECS was spectrophotometrically determined by assaying the nitrite, a stable product of NO. Nitric oxide was detected using a Griess reaction. Absorbance was measured under 550nm with a microplate reader, and sodium nitrite was used as a standard curve (Jiancheng Bioengineering Institute, Nanjing).

2.5 The mRNA expression determined by reverse transcription polymerase chain reaction

Total RNA was extracted withRNAiso Plus reagent (Takara, Dalian, China), then measured using 260/280 UV spectrophotometry. Equal quantities of RNA were reverse transcribed into cDNA using the Prime Script Reagent Kit with gDNA Eraser (TaKaRa, Dalian, China) according to the manufacturer's protocol. The primers were synthesized by Takara Biotechnology Co. (Dalian, China), and the primers sequences are given in Table 1. PCR reactions were set up with the SYBR © Premix Ex TaqTMII (Takara, Dalian, China) in BioRad Light cycler. After denature at 95℃ for 30s, PCR was performed with 40 cycles as follows: 95℃ for 5s, 60℃ for 30s and 55℃ for 20s. All PCR experiments were performed in triplicates, including non-template controls. The relative levels of mRNA were calculated using the $2^{(-\Delta\Delta Ct)}$ method by comparing to GADPH.

2.6 The protein expression of NF-κBp65 by Western blot

Collected gEECS after them were washed in phosphate-buffered saline (PBS, pH 7.4). Total protein was extracted with a total protein extraction kit according to the manufacturer's instructions (Bestbio, Shanghai, china). The protein concentrations were determined by BCA protein assay kit (Takara, Tanapan). An equal amount of protein was loaded onto 10% sodium dodecyl sulfate-polyacrylamide gel, then transferred to cellulose nitrate membrane. The membrane was blocked with 5% skim milk in Phosphate-buffered Saline containing 0.1% Tween 20 (PBS-T) for 2h at room temperature, then incubated with the primary anti-body against NF-κBp65 (ab131109, Abcam, USA, 1: 1000) for 2h. After this, the membrane was washed with PBS-T at 5min per time for five times. The membrane was incubated with horseradish peroxidase-conjugated goat anti-rabbit antibody (dilution 1: 15, 000), and washed 8 times with PBST. The immunoactive proteins were detected with an enhanced chemi-luminescence Western blotting detection kit (Takara, Dalian, China). Blots were normalized against β-actin to correct for differences in protein loading[22]. The result was analyzed using Image J gray level analysis software (Broken Symmetry software, UK).

Table 1 Primers used in real-time quantitative RT-PCR assays.

Gene name	Primers Sequence (5′-3′)	Product size	Gene ID
TNF-α	ACGGCGTGGAGCTGAAACTGATGGTGTGGGTGAGGAA	127bp	NM-001286442.1
IL-1β	TGAAGGCTCTCCACCTCCTCTTCTTGTTGTCTCTTTCCTCTCCTTG	90bp	D63351.1
TLR4	CTTGCGTCCAGGTTGTTCCCTCGGTTGATACGGGGATGT	89bp	NM_ 001285574.1
CD14	ACGACGATTTCCGCTGTGTATACTGCTTCGGGTCGGTGT	155bp	XM_ 005683042.1
(RelA) NF-κB (p65)	CAGCTCACAGATCGGGAAAAGCGGTGCTGTCTGGAAGGAA	115bp	JQ342.0881
MyD88	GGGACTAAGCGGAAGACCAGCTTCACCATTTCCCACGA	138bp	JQ308783.1

(continued)

Gene name	Primers Sequence (5′-3′)	Product size	Gene ID
TICAM2	TTTCCTCTCAGTTTTCCTGGTTCTCGTTTTCTCCTGTGTGTCCTTT	150bp	JQ923482. 1
GAPDH	AAGTTCCACGGCACAGTCAAACCACATACTCAGCACCAGC	125bp	XM_ 005680968

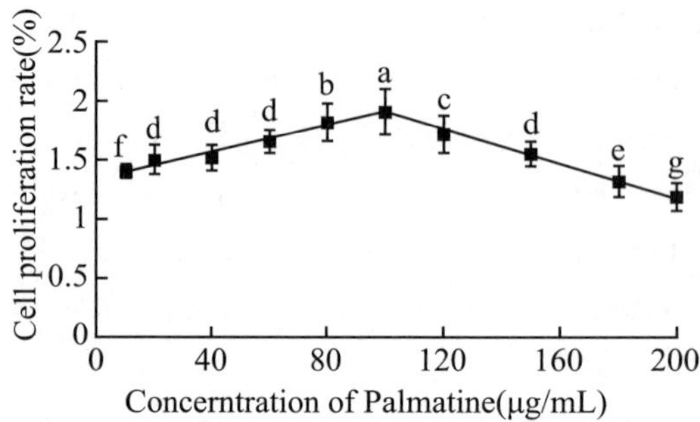

Fig. 2. Effect of palmatine on the viability of gEECS.Cells were treated with palmatine for 48h, and cytotoxicity was determined by the MTT assays (mean±SD, n=8).

2. 7 Statistical analysis

Statistical data analysis was carried out using SAS 9. 1 software (SAS Institute, NC, USA). Differences between groups were determined with one - way analysis of variance (ANOVA), followed by Duncan's multiple range test. Statistical differences between groups represented with (**) P<0. 01 vs.control group, (##) P<0. 01 vs.LPS group and (#) P<0. 05 vs.LPS group.

3 RESULTS

3. 1 Effect ofpalmatine on gEECS viability

Our results showed thatpalmatine at 10-227μg/mL promoted the proliferation of gEECs.The effect of promoting proliferation was dosedependent with a peak effect at 100μg/mL.However, when the concentration of palmatine in 100-227μg/mL, the effect of promoting proliferation decreased with the increase of drug concentrations (Fig. 2).

3. 2 Palmatine modulated inflammatory cytokines in LPS-stimulated EECS

The effects ofpalmatine on pro-inflammatory cytokines TNF-a, IL-1β, IL-6, and anti-inflammatory cytokine IL-10 were investigated in LPS-stimulated EECS.As depicted in Fig. 3, LPS significantly increased the TNF-α, IL-1β, IL-6 and IL-10 levels in LPS group compared with the control group (P<0. 05).Palmatine treatment significantly inhibited the production of TNF-α, IL-6 and IL-1β in a dose-dependent manner compared with the LPS group (P<0. 05).It presented a different regularity about anti-inflammatory cytokine IL-10 from pro-inflammatory.With the 80μg/mL palmatine treatment, the level of IL-10 decreased significantly, whereas palmatine at 40 and 20μg/mL stimulated the secretion of IL-10 evidently (P<0. 05).

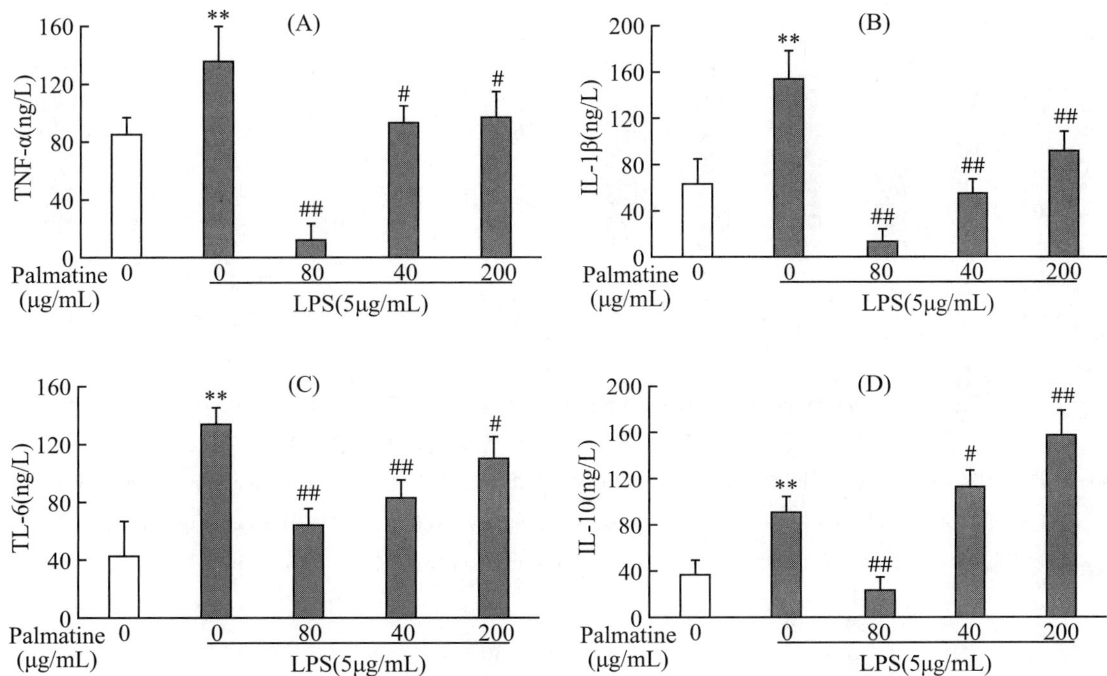

Fig. 3. Modulatory effects of the palmatine (20, 40 and 80μg/mL) on the level of TNF-a (A), IL-1β (B), IL-6 (C) and IL-10 (D) in LPS-stimulated gEECs.

Note: The cells were treated with the palmatine for 8h after LPS administration.The control group was treated without palmatine and LPS.The concentrations of pro-inflammatory cytokines were determined with ELISA.Statistical differences between groups represented with (**) $P<0.01$ vs.control group, (##) $P<0.01$ vs.LPS group and (#) $P<0.05$ vs.LPS group, values are expressed as mean ± SD (n=3).

3.3 Palmatine modulated the production of PGE_2 and NO in LPS-stimulated EECS

As shown in Fig. 4, LPS caused significant increase of NO and PGE_2 production compared with the control group ($P<0.05$).Palmatine significantly inhibited the NO production in a dose-dependent manner.In contrast, palmatine significantly promoted the PGE_2 secretion.

3.4 Palmatine inhibited MMP-2 and MMP-9 in LPS-stimulated EECS

The result of Fig. 5 indicated that LPS significantly promoted the production of MMP-2 and MMP-9 compared with the control group ($P<0.05$) and palmatine treatment significantly inhibit the secretion of MMP-2 and MMP-9.

3.5 Effects ofpalmatine on gene expression in LPS-stimulated EECS

As shown in Fig. 6 LPS significantly up-regulated gene expression of IL-1β, TLR4, CD14 and TICAM compared with control group ($P<0.05$).When LPS stimulated-gEECs were exposed to palmatine, the gene expressions of IL-1β, TLR4, CD14 and TICAM were down-regulated intensively.Noted that there was no distinct changes of MyD88 in either LPS group or palmatine treatment groups ($P<0.05$).

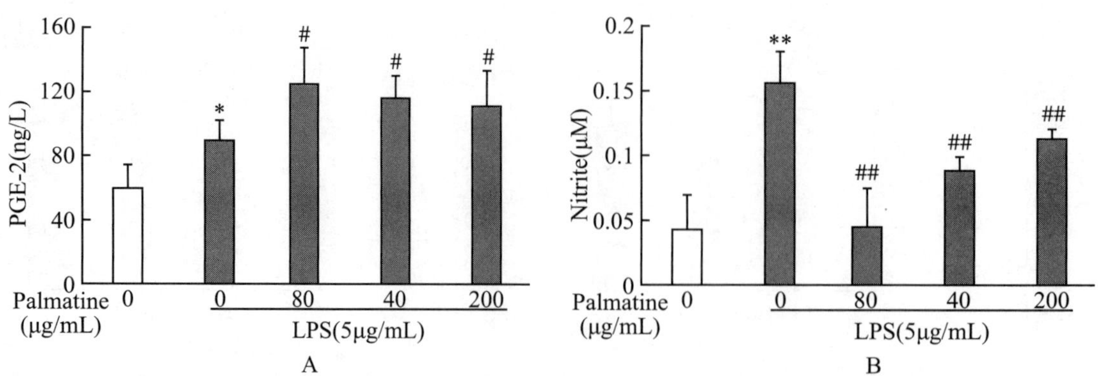

Fig. 4. Modulatory effects of the palmatine (20, 40 and 80μg/mL) on the level of NO (B) and PGE 2 (A) in LPS-stimulated gEECs.

Note: The cells were treated with the palmatine for 8h after LPS administration.The control group treated without palmatine and LPS.Statistical differences between groups represented with (**) $P<0.01$ vs.control group, (##) $P<0.01$ vs.LPS group and (#) $P<0.05$ vs.LPS group, values are expressed as mean±SD (n=3).

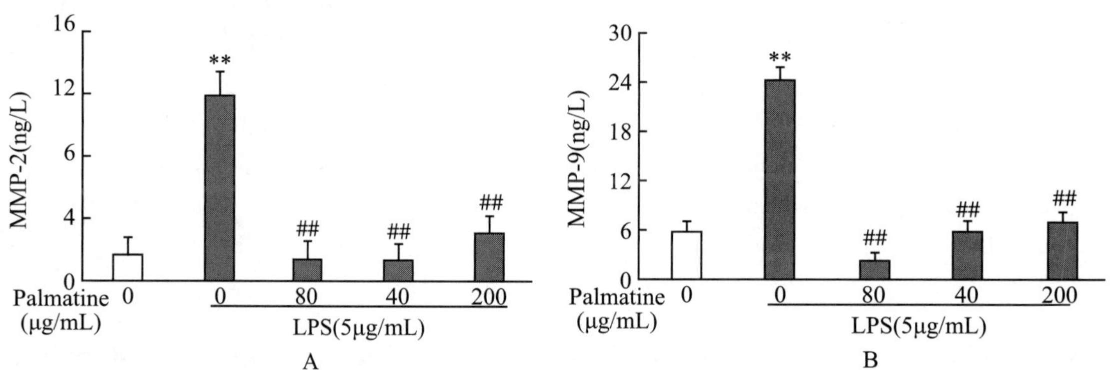

Fig. 5. Modulatory effects of the palmatine (20, 40 and 80μg/mL) on the level of MMP-9 (B) and MMP-2 (A) in LPS-stimulated gEECs.

Note: The cells were treated with the palmatine for 8h after LPS administration.The control group treated without palmatine and LPS.Statistical differences between groups represented with (**) $P<0.01$ vs.control group, (##) $P<0.01$ vs.LPS group and (#) $P<0.05$ vs.LPS group, values are expressed as mean±SD (n=3).

3.6 Palmatine inhibited protein and mRNA expression of NF-κB

The Fig.7 demonstrated that the relative mRNA level and protein expression of NF-κBp65 increased markedly with the stimulation of LPS compared with the control group ($P<0.05$).Palmatine at 40 and 20μg/mL did not affect the mRNA expression of NF-κBp65. But palmatine at 80μg/mL significantly inhibited the mRNA level of NF-κBp65 with the LPS group ($P<0.05$).It showed that when palmatine reached a certain concentration it will significantly inhibit the gene expression of NF-κBp65. The relative protein level of NF-κBp65 was calculated as NF-κBp65/β-actin. The result declined that palmatine inhibit the protein expression of phosphorylated p65. Palmatine at 80μg/mL exhibited the most obvious inhibitory effect.In general, palmatine suppressed the gene expression and the phosphorylation of NF-κBp65.

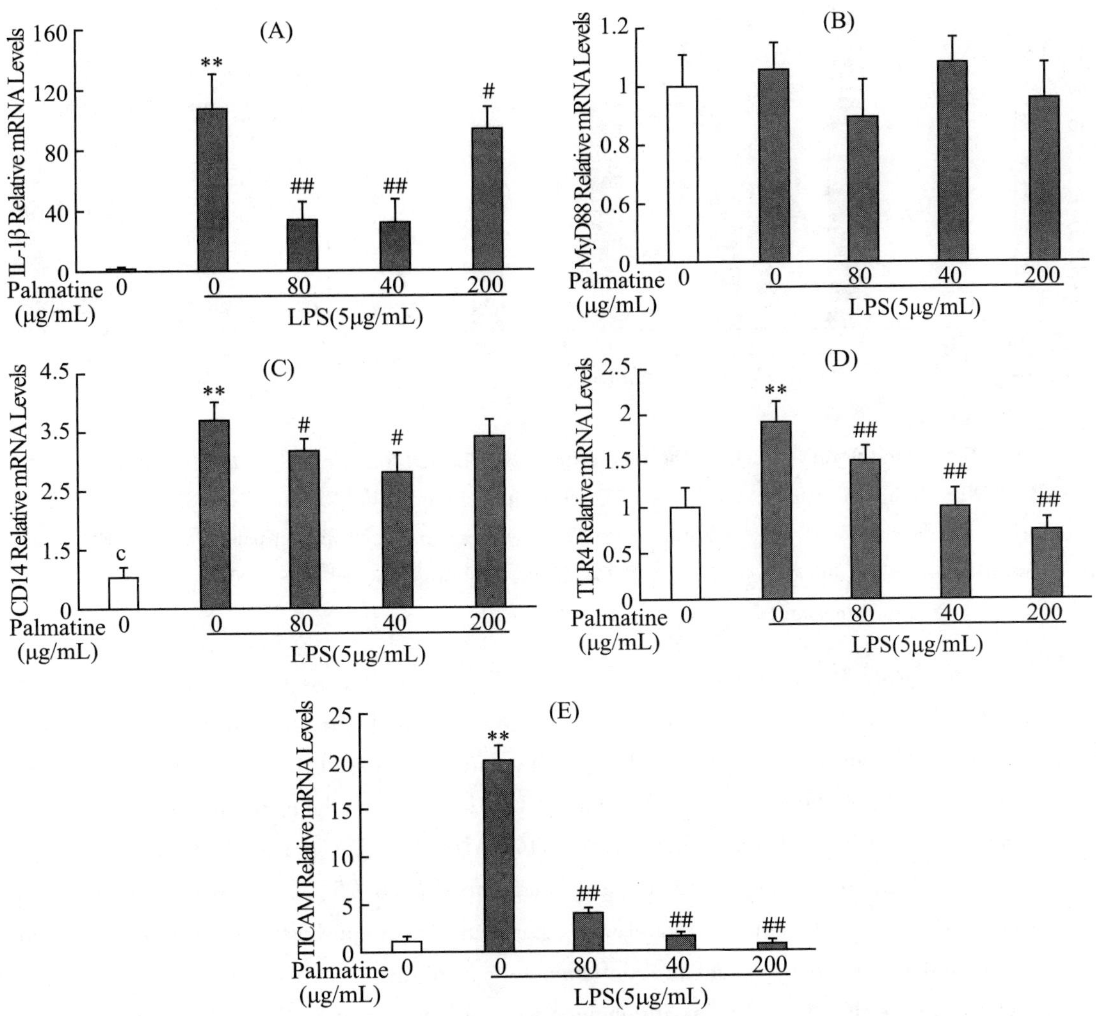

Fig. 6. Modulatory effects of the palmatine (20, 40 and 80μg/mL) on the gene expression in the gEECS induced by LPS, IL-1β (A), MyD88 (B), CD14 (C), TLR 4 (D), TICAM (E).

Note: The cells were treated with the palmatine for 8h after LPS administration. The control group treated without palmatine and LPS. Statistical differences between groups represented with (**) $P<0.01$ vs. control group, (##) $P<0.01$ vs. LPS group and (#) $P<0.05$ vs. LPS group, values are expressed as mean±SD (n=3).

4 DISCUSSION

Endometritis is one of the top problems veterinarians face in clinical practice. It has been reported that 14%-53% of cows as well as 10%-20% of mares will develop into some types of endometritis[23,24]. In recent years, with the multiplication of drug resistant bacteria and the excessive residues of antibiotics, the study of Chinese herbal medicine for the treatment of endometritis became a research hotspot. Palmatine, an important traditional medicinal compound, has various biological activities usually used to treat jaundice, hypertension, dysentery, inflammation, and liver-related diseases[17]. However there are few reports about the anti-inflammatory effects of palmatine on endometritis in veterinary. EECs have been used as an in vitro model system in endometrial disease re-

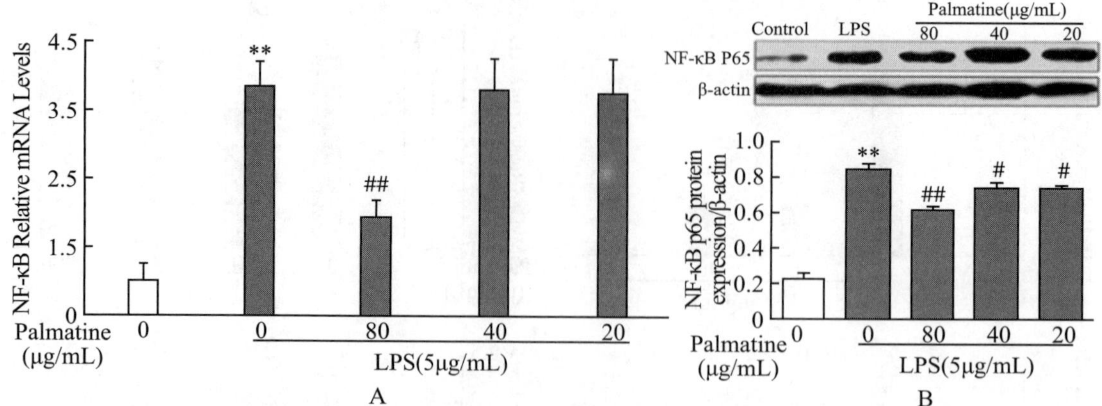

Fig.7. Modulatory effects of the palmatine (20, 40 and 80μg/mL) on the NF-κB.

Note: (A) NF-κB mRNA expression and (B) NF-κBp65 protein levels in LPS-stimulated gEECs. The cells were treated with the palmatine for 8h after LPS stimulation. The control group treated without palmatine and LPS. Statistical differences between groups represented with (**) $P<0.01$ vs. control group, (##) $P<0.01$ vs. LPS group and (#) $P<0.05$ vs. LPS group, values are expressed as mean±SD (n=3).

search for many years. It was used to study the first line of defense against invading organisms, known as innate immunity[25]. In the present study, we used an LPS-induced gEECs inflammation model to evaluate the anti-inflammatory activity of palmatine. We found that 5μg/mL LPS could successfully induce inflammation after its treatment for 12h in gEECs (data not shown). The results of Fig. 3 showed that the level of pro-inflammatory factors TNF-α, IL-1β and IL-6 increased significantly in the LPS group compared with the control group ($P<0.05$). These results indicated that LPS-stimulated gEECS inflammation model was successful. And it can be used to further experiment.

TNF-α and IL-6 are important pro-inflammatory cytokines[26]. Previous studies reported that animals with endometritis have higher inflammatory cytokines concentrations in uterine flush than healthy ones[27]. TNF-α is able to activate other inflammatory cytokines like IL-1β and induce the expression of adhesion molecules[28]. IL-6 is responsible for the coordination of the acute phase response and plays an important role in the local inflammatory reaction by amplifying leukocyte accumulation[29]. IL-1β mobilizes neutrophils followed by phagocytosis of invading pathogens within the uterine lumen[30]. NO, an inflammatory mediator, its high concentration can cause vasodilatation and increase inflammation-based permeability and amount of mononuclear cells at the site of inflammation[31]. In present study, palmatine had showed an anti-inflammatory effect by significantly inhibiting the secretion of the above pro-inflammatory factors in a dose-dependent manner (Figs. 3, 4).

IL-10 acts as a potent anti-inflammatory cytokine that functions to resolve inflammatory response by limiting the expression of pro-inflammatory cytokines and chemokines. It protects the host from excessive inflammation[32,33]. As reported previously, endometritis had a concomitant increase of anti-inflammatory cytokines, such as IL-10, it may be desirable to administer immune stimulatory preparations[34]. In our study, palmatine at 80μg/mL significantly decreased the level of IL-10, but its lower concentrations (40 and 20μg/mL) significantly raised IL-10 production

(Fig. 3).The reason of this was that anti-inflammatory effect of the palmatine at 80μg/mL had already reached a certain level, even below normal level (vs. control group b 0.05). Hardly there were any inflammatory effects in model, so the secretion of anti-inflammatory cytokine IL-10 was inhibited. While, after the treatment of palmatine at 40 and 20μg/mL, inflammation still existed in LPS-induced EECS. So the secretion of IL-10 was up-regulated by palmatine to inhibit pro-inflammatory cytokines.

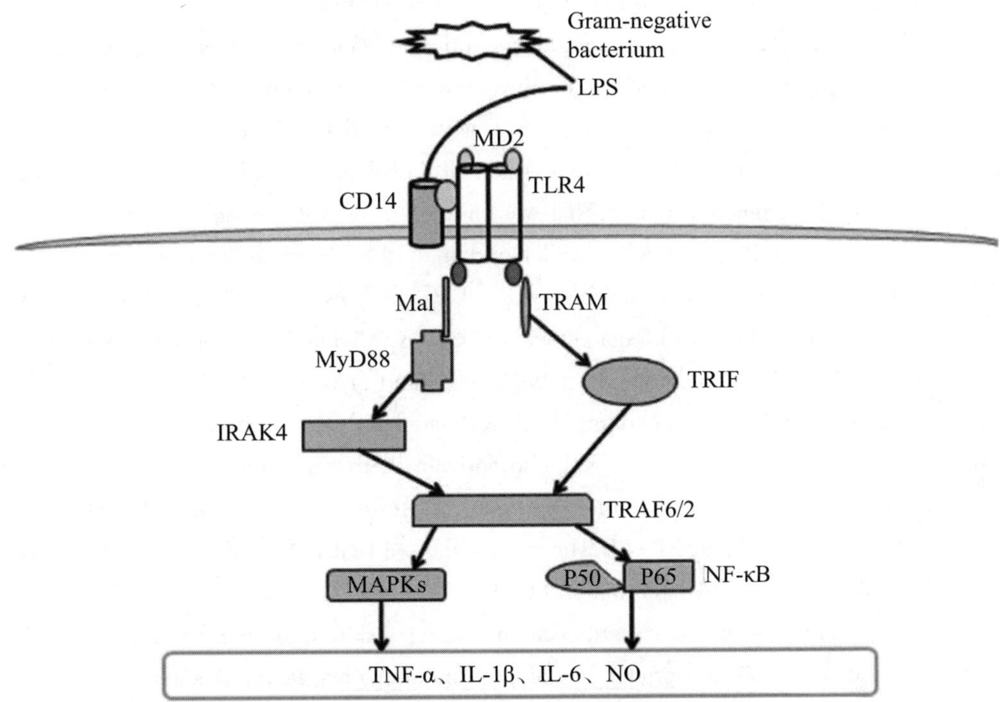

Fig.8. The signal pathways of the anti-inflammatory mechanism of palmatine.

Note: LPS significantly increased the gene expressions of TLR4, CD14, TRIF and NF-κB ($P<0.05$). Palmatine can significantly decrease the levels of them. Otherwise the level of MyD88 had no changes ($P<0.05$).

In previous research, the synthesis of PGE_2 was stimulated by various factors involved in inflammation. It modulates the inflammatory response, including vasodilatation and the production of various cytokines. Previous studies reported the higher concentration of PGE_2 in blood plasma or uterine fluid during endometritis[35,36]. Our results demonstrated that palmatine promoted the release of PGE_2. It suggested that palmatine may enhance PGE_2 to inhibit inflammation as PGE_2 exerts both pro-and anti-inflammatory effects combining with changes of the cytokines[37,38].

Matrix metalloproteinase (MMP) is a key enzyme in collagen metabolism. It can occur in fibroblasts and epithelial cells as an important regulator of inflammation[39]. MMP-2 and MMP-9 are important enzymes in the degradation of the extracellular matrix (EMC) and with a destructive effect on the integrity of the basement membrane[40]. When Endometritis occurred early, MMPs produced massively. Inhibition of its production would protect endometrium from structure degradation (Fig. 5).

Previous studies have indicated that LPS is recognized by TLR4, which leads to the activation of two different signaling pathways: MyD88-dependent and TRIF-dependent. TLR4-and MyD88-dependent signaling pathways are essential for the response to LPS by the epithelial and stromal cells in the bovine endometrium and activated NF-κB up-regulates the transcription of pro-inflammatory genes such as TNF-α, IL-6[41-43]. In MyD88 knock-out mice macrophage, TRIF is recruited to TLR4 and activates an alternative pathway that triggers the activation of NF-κB, MAPKs, and IRF3. These signaling cascades lead to the production of pro-inflammatory cytokines, type Ⅰ interferons (IFNs), chemokines, and antimicrobial peptides to remove the invading pathogens[44,45]. It was reported that the TLR4 mRNA and protein levels significantly increased in LPS-induced mammary gland tissues[46]. According to these theories, we detected the genes expression of TLR4 and CD14. Our results revealed that the gene expression of TLR4 and CD14 increased significantly in LPS-induced gEECs inflammation model. That was in agreement with previous studies. We examined the mRNA expression of TRIF (TICAM) and found that LPS significantly up-regulated the transcription of TRIF. Palmatine treatment down-regulated the gene levels of TLR4, CD14 and TRIF in TLR4 signaling pathways, but it did not affect MyD88 expression. It is well known that IL-1β, TNF-α and IL-6 expression is modulated by the NF-κB and MAPK pathways, which are signaling pathways down-stream of TRIF-dependent TLR4. Activated NF-κB subunits translocate to the nucleus and bind to the promoters of target genes to activate their transcription[42,47]. To further assess the mechanism of Palmatine anti-inflammatory effect, we tested the gene expression of NF-κB and the protein expression of activated NF-κB. The results showed that Palmatine down-regulated the expression of NF-κB and suppressed the action of NF-κB.

In summary, Palmatine possesses anti-inflammatory bioactivities in LPS-induced gEECs inflammation model. It decreases the production of pro-inflammatory factors, such as TNF-α, IL-1β, IL-6 and NO and increases the production of anti-inflammatory IL-10 and pro-resolution mediator and PGE 2 . In addition, palmatine protect gEECS from inflammatory injury by inhibiting the production of MMP2. As summarized in Fig.8, our results suggest that the anti-inflammatory mechanism of palmatine was related to the inhibition of TRIF-dependent and NF-κB signaling pathways at the gene level.

ACKNOWLEDGMENTS

This work was supported byLanzhou Institute of Husbandry and Pharmaceutical Sciences of CAAS and guided by Zuo-ting Yan research and doctor Shi-dong Zhang. This dissertation was funded jointly by three projects, which are the project of National Science and Technology Supporting Plan of China (2012BAD12B03), and the project of National Science and Technology Supporting Plan of Gansu Province (1506RJYA146) and the Agricultural Science and Technology Innovation Program (CAAS-ASTIP-2014-LIHPS-03).

REFERENCES OMITTED

（发表于《International Immunopharmacology》，院选 SCI，IF：2.551）

Differential Proteomic Profiling of Endometrium and Plasma Indicate the Importance of Hydrolysis in Bovine Endometritis

Shi-Dong ZHANG[1,2], Shu-Wei DONG[2], Dong-Sheng WANG[2], Chike F.OGUEJIOFOR[3], Ali A.FOULADI-NASHTA[4], Zhi-Qiang YANG[1,2], Zuo-Ting YAN[1,2]

(1. Lanzhou Institute of Husbandry and Pharmaceutical Sciences of Chinese Academy of Agricultural Sciences, Lanzhou 730050, China; 2. Key Laboratory of Veterinary Pharmaceutics Discovery, Ministry of Agriculture, Lanzhou 730050, China; 3. Department of Veterinary Obstetrics and Reproductive Diseases, University of Nigeria, Nsukka 410001, Nigeria; 4. Reproduction Group, Royal Veterinary College, Hawkshead Campus, Hatfield AL97TA, United Kingdom)

Abstract: Endometritis is an important disease of dairy cows that leads to significant economic losses in the dairy cattle industry.To investigate the alteration of proteins associated with endometritis in the dairy cow, the isobaric tags for relative and absolute quantification (iTRAQ) technique was applied to quantitatively identify differentially expressed proteins (DEP) in the endometrium and peripheral plasma of Chinese Holstein cows with endometritis.Compared with the normal (control) group, 159 DEP in the endometrium and 137 DEP in the plasma were identified in cows with endometritis.Gene ontology analysis demonstrated that the predominant endometrial DEP were primarily involved in responses to stimulus and stress processes and mainly played a role in hydrolysis in the extracellular region.The predominant plasma DEP were mainly components of the cytosol and non-membrane-bound organelles, and they were involved in the response to stress and regulation of enzyme activity.Protein-protein interaction of tissue DEP revealed that some core seed proteins, such as RAC2, ITGB2, and CDH1 in the same network as CD14, MMP3, and MMP9, had important functions in the cross-talk of pathways related to extracellular proteolysis. In summary, significant enzymatic hydrolase activity in the extracellular region is proposed as a molecular mechanism by which altered proteins may promote inflammation and hence endometritis.

Key words: Isobaric tags for relative and absolute quantification (iTRAQ), Bovine endometritis; Differentially expressed proteins; Hydrolysis

1 INTRODUCTION

Uterine infection andendometritis are important causes of infertility and reproductive losses in dairy cattle worldwide (Gilbert, 2011; LeBlanc, 2014). The protective physical barriers to contamination provided by the uterine luminal epithelium, cervix, vagina, and vulva are all disrupted during normal parturition (Sheldon and Dobson, 2004). This disruption and the large volumes of fluid and tissue debris in the uterine lumen both allow bacteria from the environment or the animal's feces to contaminate the uterus postpartum (Bondurant, 1999; Sheldon and Dobson, 2004). Several studies showed the presence of bacteria in the uterus of more than 90% of cows within the first 2 wk after calving (Dohmen et al., 1995; Sheldon et al., 2008). Although most bacterial contaminants are gradually cleared from the uterus via uterine involution and innate immune response (Azawi, 2008; Singh et al., 2008), up to 20% of cows have clinical metritis and 20% to over 50% of cows subsequently develop subclinical inflammation of the uterus (endometritis) beyond 3 wk postpartum (Sheldon et al., 2009a; LeBlanc, 2014). Endometritis causes uterine tissue damage, embryonic death and early abortion, delayed onset of ovarian cyclicity, extended luteal phases, and reduced conception rates in affected cattle (Sheldon et al., 2009b; Gilbert, 2011). These problems lead to infertility and substantial financial losses in the dairy industry (Sheldon et al., 2009a).

Theendometrium has crucial roles in female reproduction including the regulation of the estrous cycle, sperm transit, nourishment of the early embryo, and formation of the placenta. Furthermore, the endometrial luminal epithelial cells and the underlying stromal cells express pathogen recognition receptors and mount an innate immune response to microbes or microbial ligands (Hickey et al., 2011; Oguejiofor et al., 2015a, b). The major bacteria isolated from cows with endometritis include Escherichia coli and Trueperella pyogenes, followed by other anaerobic species, such as Fusobacte-rium and Bacteroides (Dohmen et al., 2000; Williams et al., 2005; Bicalho et al., 2012). Although the effects of different types of bacteria on uterine disease mechanisms are not completely understood (Westermann et al., 2010), unregulated inflammation is known to lead to disease (Maybin et al., 2011). This outcome can impair reproductive performance not only during the infection but also after the clinical signs of endometritis have resolved (Plöntzke et al., 2010; LeBlanc, 2014). Interestingly, some evidence suggests that certain viruses such as bovine viral diarrhea virus (Grooms, 2004; Oguejiofor et al., 2015a; Cheng et al., 2016) and bovine herpes virus-4 (Donofrio et al., 2009) contribute to the etiology of uterine disease.

Endometritis is described histopathologically as a superficial inflammation of the endometrium, with histological evidence of inflammation (Sheldon et al., 2006). Endometritis has been defined based on the presence of abnormally increased PMN in uterine luminal fluid (Kasimanickam et al., 2004; Gilbert et al., 2005) or the biopsy of endometrial tissue (Galvão et al., 2011). In addition, increased endometrial expression of inflammatory genes may be present, including genes coding for granulocyte chemotactic protein 2, IL-1β and IL-8, and tumor necrosis factor (TNF) (Fischer et al., 2010; Kasimanickam et al., 2014). Elevated levels of inflammatory cytokines (TNF-α, IL-1β, IL-6, IL-10) in serum have also been used to identify the development of bovine endometritis (Islam et al., 2013; Kasimanickam et al., 2013). Previously, genome-wide transcriptomic profiling of bovine endometrium using mRNA-Seq showed that immune activation and inflammation preceded tissue proliferation and repair in the healthy postpartum endometrium (Foley et al., 2012). This finding highlighted the importance of a balanced inflammatory immune response as key to sufficient bacterial clearance and restoration of an endometrial environment capable of supporting a new pregnancy (Jabbour et al., 2009). Limited studies exist on the bovine uterine proteome. One study identified 9 proteins using 2-dimensional (2D) electrophoresis to compare the proteins present in the uterine luminal fluid between pregnant and nonpregnant animals (Ledgard et al., 2009). Another study used 2D fluorescence difference in-gel electrophoresis to compare the proteome of the pregnant and non-pregnant endometrium (Berendt et al., 2005). However, scant information is available on the proteomic profile of cows with endometritis. A previous study identified 14 endometritis-associated proteins using 2D electrophoresis (Choe et al., 2010).

Recent developments in the field of proteomics have led to a renewed interest in animal disease diagnosis and treatment. Isobaric tags for relative and absolute quantification (iTRAQ) is the most popular technique, and combined with multidimensional liquid chromatography (LC) and tandem MS, it is used to study differentially expressed proteins (DEP). The aim of this study was to use the iTRAQ technique to characterize the proteomic changes in endometrial tissue and plasma from Chinese Holstein dairy cows with endometritis. The bioinformatics analysis of the differential proteome may enable a new understanding of the main proteins or pathways associated with bovine endometritis and provide a foundation for future diagnosis and treatment of diseased animals.

2 MATERIALS AND METHODS

2.1 Animals and Collection of Samples

Sample collection was carried out under license in accordance with national guidelines (Ministry of Agriculture of China 2015/No. 18). Rectal palpation and cervico-vaginal mucus observation were conducted to identify normal animals or cows with endometritis. Endometrial biopsies and blood specimens were collected from Chinese Holstein cows, aged from 2 to 5 yr, at 21 to 35d postpartum. Ten candidate cows had obvious signs of clinical endometritis (i.e., presence of purulent or mucopurulent vaginal discharge), but they did not show any symptoms of other local or systemic diseases beyond the focus of this study. Another 10 cows without signs of endometritis were categorized as the normal control animals. Briefly, 5mL of blood was collected from the jugular vein into commercial vacuum tubes and thoroughly mixed with the anticoagulant EDTA K_2. This samples were centrifuged at 2, 500×g for 15min, and 0.5mL of plasma was recovered and quickly frozen with liquid nitrogen. Endometrial biopsies were quickly frozen in liquid nitrogen and fixed in 4% formalin, respectively. After formalin fixation, tissue specimens were embedded in paraffin blocks using routine procedures, followed by hematoxylin and eosin staining and histopathological examination under the microscope to identify pathological state.

2.2 Protein Isolation, Digestion, and Labeling with iTRAQ Reagent

Protein extraction was performed on 4 nonendometritis samples and 4 endometritis samples that were selected according to histopathological results. Briefly, the frozen tissue was ground into powder and then dissolved in lysis buffer Ⅰ (7M urea, 2M thiourea, 4% pro-panesulfonic acid, 40mM Tris-HCl, pH 8.5) containing complete protease inhibitor. The cells were lysed by sonication at 200W for 15min and then centrifuged at 4℃, 30, 000×g for 15min. The supernatant was mixed with 5×volume of chilled acetone containing 10% (vol/vol) trichloroacetic acid, incubated at-20℃ overnight, and then centrifuged as described. The precipitate was washed 3 times in chilled acetone. The recovered pellet was air-dried and dissolved in lysis buffer Ⅱ (7M urea, 2M thiourea, 4% Nonidet P40, 20mM Tris-HCl, pH 8.0-8.5). The suspension was sonicated, centrifuged as described, and transferred to another tube. The disulfide bonds were reduced by treatment with 10mM dithiothreitol, whereas cysteines were blocked by treatment with 55mM iodoacetamide. The supernatant was then washed in acetone and centrifuged as previously described. Recovered pellets were air-dried, dissolved in 500μL of 0.5M tetraethylammonium bromide (TEAB; Applied Biosystems, Milan, Italy), and sonicated once more. Finally, samples were centrifuged, and the supernatant was transferred to a new tube and quantified using the Bradford method. The proteins in the supernatant were kept at-80℃ for further analysis. Correspondingly, the plasma samples (matching the endometrial tissues) were cleared of highly abundant proteins using ProteoMiner kit (Bio-Rad, Hercules, CA), and other processes, including washing, precipitation, centrifugation, sonication, and the reduction of disulfide bonds, were performed as already described.

Total protein (100μg) was measured from each sample solution and the protein was digested at 37℃ for 16h with Trypsin Gold (Promega, Madison, WI) with a protein-to-trypsin ratio of 30: 1. After trypsin digestion, peptides were dried by vacuum centrifugation. Peptides were reconstituted in 0.5M TEAB and processed according to the manufacture's protocol for 8-plex iTRAQ reagent (Applied Biosystems). Briefly, 1 unit of iTRAQ reagent was thawed and reconstituted in 24μL of isopropanol. Samples were labeled respectively with different isobaric tags and incubated at room temperature for 2h. The labeled peptide mixtures were then pooled and dried by vacuum centrifugation.

Liquid Chromatography-Electrospray Ionization Tandem Mass Spectrometry Analysis Based on Q Exactive

TheiTRAQ-labeled peptide mixtures were fractionated by strong cation exchange chromatography, which was conducted on a LC-20AB HPLC Pump system (Shimadzu, Kyoto, Japan) using a 4.6×250mm Ultremex SCX column containing 5-μm particles (Phenomenex, Torrance, CA). The peptides were eluted at a flow rate of 1mL/min with a gradient of buffers as follows: buffer A [25mM NaH_2PO_4 in 25% acetonitrile (ACN), pH 2.7] for 10min, 5 to 60% buffer B (25mM NaH_2PO_4, 1M KCl in 25% ACN, pH 2.7) for 27min, and 60 to 100% buffer B for 1min. The system was then maintained at 100% buffer B for 1min before equilibrating with buffer A for 10min before the next injection. Under a monitoring absorbance of 214nm, 20 fractions were collected, desalted with a Strata X C18 column (Phenomenex), and vacuum-dried.

Each fraction was dissolved in buffer liquid (2% ACN, 0.1% formic acid) and centrifuged at 20, 000 g for 10min; the final concentration of peptide was about 0.5μg/μL on average. Ten microliters of supernatant was loaded on a LC-20AD nanoHPLC (Shimadzu, Kyoto, Japan) by the autosampler onto a 2-cm C18 trap column (i.d.75μm).The peptides were subjected to nanoelectrospray ionization followed by tandem mass spectrometry (MS/MS) in a Q Exactive (Thermo Fisher Scientific, San Jose, CA) coupled online to the HPLC.Intact peptides were detected in the Q Exactive orbitrap at a resolution of 70 000. Peptides were selected for MS/MS using high-energy collision dissociation operating mode with a normalized collision energy setting of 27.0; ion fragments were detected in the Orbitrap at a resolution of 17 500. A data-dependent procedure that alternated between 1 MS scan followed by 15 MS/MS scans was applied for the 15 most abundant precursor ions above a threshold ion count of 20 000 in the MS survey scan with a following a dynamic exclusion duration of 15 s.The electrospray voltage applied was 1.6 kV.Automatic gain control was used to optimize the spectra generated by the orbitrap.The automatic gain control target for full MS and MS2 was 3E6 ion counts and 1E5 ion counts, respectively.The m/z scan range was 350 to 2 000 Da.For MS2 scans, the m/z scan range was 100 to 1 800 Da.

2.3 Data Analysis

Raw MS data were converted into Mascot generic file (MGF) using Proteome Discoverer 1.2 (PD 1.2, Thermo Fisher Scientific), and the MGF data file were searched by using Mascot search engine (version 2.3.02, Matrix Science, London, UK) to identify proteins. Each confident protein identification involved at least 1 unique peptide.For protein quantification, a protein had to contain at least 2 unique spectra.The quantitative protein ratios were weighted and normalized by the median ratio in Mascot (http://www.matrixscience.com/help/quant_ statistics_ help.html).We only used ratios with P-values <0.05, and we set fold changes of >1.5 or <0.67 for endometrial tissue and >1.2 or <0.67 for plasma as the criteria to determine up-or down-regulated proteins, respectively.

2.4 Bioinformatics Analyses

Functional annotations of proteins were performed using Blast2GO program against the nonredundant protein database (NR, NCBI; http://www.ncbi.nlm.nih.gov/protein/? term=txid9913 [Organism: noexp]).Gene Ontology (GO) is an international standardization of gene function classification system.It has 3 ontologies that can describe molecular function, cellular component, and biological processes, respectively.The main types of annotations were obtained from the GO consortium website (http://david.abcc.ncifcrf.gov/).A Venn diagram was drawn using online Venny 2.0 (http://bioinfogp.cnb.csic.es/tools/venny/index.html). The protein-protein interaction (PPI) network was analyzed using STRING (Search Tool for the Retrieval of Interacting Genes/Proteins) software (http://string.embl-heidelberg.de/). The DEP were compared by a hierarchial cluster analysis using Cluster 3.0 (http://bonsai.hgc.jp/~mdehoon/software/cluster/software.htm). The data were displayed using Tree View software (https://sourceforge.net/projects/jtreeview/).

2.5 Western Blot

The same endometrial tissues that were used for iTRAQ analysis were divided into non-endometritis and endometritis groups.Total proteins were extracted from each sample using a commercial kit (BestBiotech, Shanghai, China). Protein concentration was determined using the BCA Protein

Assay Kit (Biomiga, Shanghai, China). Briefly, equivalent amounts of protein (20μg) were loaded on 12% SDS-PAGE gels and then transferred onto 0. 45μm nitrocellulose membrane (Millipore Corp., Billerica, MA). After blotting, the membrane was blocked with 5% skim milk and incubated with primary antibody and second antibody at 37℃ for 60min, respectively. The incubated membrane was then washed in 0. 05% PBS Tween 3 times. Detection was performed using chemiluminescence luminal reagents (CWbiotech, Beijing, China) and recorded by film exposure. The polyclonal antibodies to serum amyloid A (sc-20275), myeloperoxidase (sc-16128-R), matrix metalloproteinase 3 (sc-6839), fibrinogen (sc33916), integrin β6 (sc-6632), and integrin β2 (sc-6623) were purchased from Santa Cruz Biotechnology (Santa Cruz, CA). Mouse monoclonal antibody to β actin (ab8226) and goat polyclonal antibody to lactoferrin (ab112968) were obtained from Abcam (Cambridge, UK). The horseradish peroxidase-conjugated goat anti-mouse IgG and rabbit anti-goat IgG were obtained from Proteintech (Wuhan, China).

3 RESULTS

3.1 Histopathology of endometrial tissue

A representative section showing typical histopathologic changes observed in the endometrial tissue specimens from cows with endometritis are shown in Figure 1. The normal endometrium comprises a single layer of columnar luminal epithelial cells. The endometrial gland lies deeper in the stromal layer and comprises the glandular epithelia, which release their secretions into the lumen of the gland. The lamina propria contained few red blood cells or migrant granulocytes and lymphocytes (Figure 1A). During inflammation in the uterus or endometritis (Figure 1B), degenerative necrosis was apparent in the luminal epithelial cells, with nuclear condensation and damage of the mucous layer and basement membrane. Capillaries were expanded in the lamina propria, with significant infiltration of red blood cells and leukocytes, resulting in a general increase in cell density. This pattern was typical of hemorrhagic endometritis (Figure 1B).

3.2 Protein profile by iTRAQ-Coupled 2D LC-MS/MS analysis

A total of 3, 493 proteins were determined byiTRAQ-coupled 2D LC-MS/MS analysis in the endometrial tissue. Among them, 159 proteins displayed significantly altered levels with fold cut-off of ≥1.5. As shown in Supplemental Table S1 (https://doi.org/10.3168/jds.2016-12365), these significantly altered proteins include 109 increased and 50 decreased ones. Analysis of the plasma identified 137 DEP with fold cut-off of ≥1.2, out of which 49 proteins were increased and 88 were decreased (Supplemental Table S1).

3.3 Functional classifications of DEP

Functional annotation of the DEP was initially performed using Blast2GO program. Three main types of annotations were obtained from the GO consortium website: cellular component, metabolic functions, and biological processes. One hundred fifty-nine proteins were significantly altered in the endometritic uterus, and enrichment analysis of biological processes showed that endometritis primarily affected response to stimulus and stress (Figure 2A). Enrichment analysis also identified hydrolase activity as a predominant event for stress and stimulus during inflammation (Figure 2C). Furthermore, cellular component-based enrichment analysis identified the significantly altered proteins as being located in the extracellular region (Figure 2B). The same GO analysis was conducted on 137 proteins identified in the plasma. Biological processes associated with DEP mainly involved in stress response (Figure 3A) and molecular function regulation of enzyme activities (Figure 3C). Some plasma DEP were involved in components of non-membrane-bound organelle and cytoskeletal processes (Figure 2B), such as PDZ and LIM domain 2 (PDLI2), coronin-1A (COR1A), annexin A1 (ANXA1), and ribosomal protein L14 (RL14; Supplemental Table S2; https://doi.org/10.3168/jds.2016-12365).

3.4 Protein distribution between endometrium and plasma

The endometrial tissue contained 159 DEP, and the plasma had 137 DEP. However, only 9 proteins occurred in both the endometrial tissue and plasma. Of these proteins, 3 were similarly increased and 2 were similarly decreased in the 2 specimens; 3 proteins were increased in the tissue, but decreased in the plasma; and 1 was increased in the plasma, but decreased in the tissue. The distribution of these proteins is shown in a Venn diagram (Figure 4). Cluster analysis was performed for the 9 proteins as shown in the dendrogram (Figure 4). These include hemoglobin subunit β (HBB), resistin (RETN), serum amyloid A protein (SAA), inter-α-trypsin inhibitor heavy-chain H2 precursor, IgA heavy-chain constant region, immunoglobulin gamma 1 heavy-chain constant region, CD14, matrix metalloproteinase 3 (MMP3), and serpin peptidase inhibitor, clade A (α-1 antiproteinase, antitrypsin), member 10.

3.5 Validation of changes in protein level by western blot

To validate the DEP identified by the iTRAQ analysis, 7 increased proteins were selected for Western blot analysis, including SAA, lactoferrin, myeloperoxidase, MMP3, fibrinogen, integrin β6, and integrin β2. Equal amounts of proteins from endometrial tissue were detected with antibodies as shown in Figure 5. The results showed that the ratios of the selective proteins were consistent with those obtained from the iTRAQ analysis.

3.6 Analysis of protein-protein interaction

The interactions between proteins are complex and interrelated during disease development. Studying the interaction network is of strategic importance to researchers for understanding how disease-related proteins interact with each other and how they affect cell function. To investigate these issues, the STRING database was searched for protein interactions with the 159 DEP in endometrial tissue (Figure 6). As observed from the network, RAC2, ITGB2, and CDH1 were the core seed proteins in the interaction network, and they had important functions in several signal transduction pathways including MAPK signaling pathway, VEGF endometritis (B). ul = uterine lumen; eg = endometrial gland; le = luminal epithelium; ge = glandular epithelium; lp = lamina propria; g = granulocytes; l = lymphocytes; r = red blood cells. signaling pathway, Wnt signaling pathway, focal adhesion, adherens junction, cell adhesion molecules, leukocyte transendothelial migration, and regulation of actin cytoskeleton.

4 DISCUSSION

Endometritis is prevalent in dairy herds worldwide, causing significant economic losses

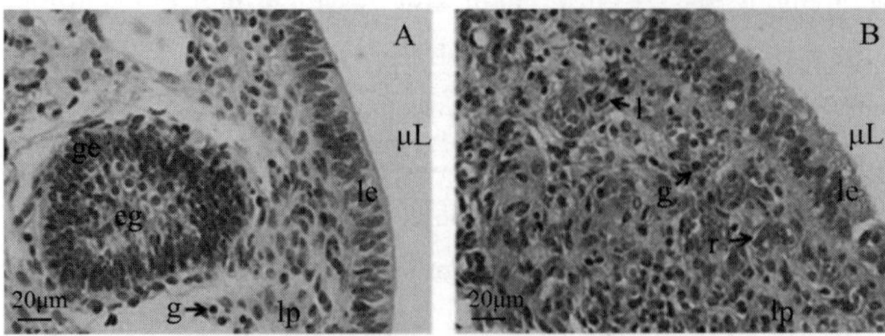

Figure 1. Histopathologic change in endometrial tissue without endometritis (A) and with endometritis (B) investigated following hematoxylin and eosin staining (400×).

Note: Micrographs of the bovine endometrium showing normal tissue (A) and histopathologic changes observed during

resulting from infertility (Sheldon et al., 2009a). Nevertheless, research on endometritis in dairy cows is limited by high costs, a long gestation period, and lack of uniform genetic backgrounds. With the development of omics research, proteomics is emerging as a useful tool for studying endometritis in cows. Understanding the molecular mechanisms of endometritis at the protein level is likely to contribute to the design of new methods in disease control and fertility enhancement.

Diagnosis ofendometritis in cows is based on the presence of pathologic vaginal discharge. In the past 10 yr, standardized definitions and characterizations of uterine disease were suggested by several authors (Williams et al., 2005; Barlund et al., 2008; LeBlanc et al., 2011). Different physiological conditions well understood to possibly influence a cow's endometrium include DIM, estrus cycle, and age. However, correcting for all variables is difficult when working with farm animals such as cattle as opposed to working with controlled laboratory animals and conditions. We were unable to determine the exact days of the estrous cycle and DIM of the animals, and therefore we cannot account for any influence due to ovarian hormones. In the absence of the accurate information of cow's physiology, we herein conducted histopathologic study using hematoxylin and eosin staining to characterize endometritis (Figure 1), and only tissues with the same type of hemorrhagic endometritis were selected for proteomic analysis.

According to the results of iTRAQ analysis, approximately 3 500 bovine proteins were identified, and with a P-value cut-off of <0. 05, 159 and 137 proteins were found to expressed differentially in the endometrial tissue and plasma, respectively (Supplemental Tables S1 and S2; https: // doi .org/ 10 . 3168/ jds . 2016-12365). As shown in Supplemental Table S1, DEP in the endometrial tissue of cows with endometritis included many specific and nonspecific immunity-associated proteins, providing evidence of on-going inflammatory activity. Five cathelicidin proteins (cathelicidins 2, 3, 5, 6, and 7), a family of host-defense peptides, were significantly increased. These small amphipathic peptides exhibit rapid and potent activity against many pathogens, including bacteria, viruses, and fungi (Braff et al., 2005).

The complement proteins, complement component 1, r subcomponent (C1R) and complement component 1, s subcomponent (C1S), were also increased in the endometrium. Complement proteins are important in innate and adaptive immunity by opsonisation of pathogens, recruitment of phagocytes to the site of infection, and direct killing of pathogens through the formation

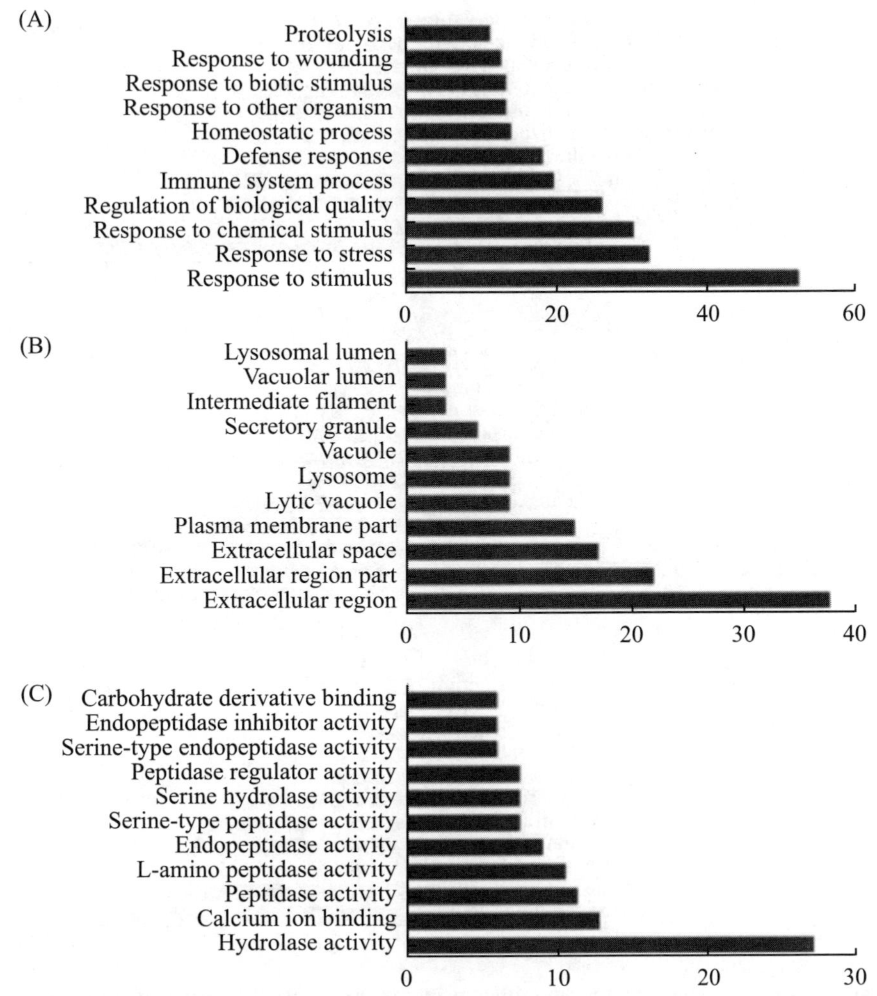

Figure 2. Top 11 classifications of identified proteins differentially expressed in theendometrium based on their functional annotations using the Gene Ontology enrichment analysis: (A) biological processes, (B) cellular components, and (C) molecular functions.Color version available online.

of the membrane-attack complex (Medzhitov, 2007).Here, several complement proteins were also increased in the plasma of cows with endometritis, including complement component-7 (C7) and-9 (C9); complement factor-H (CFH), -D (CFD), and-properdin (CFP); and complement C4-A precursor (C4A13).The presence and activity of serum complement proteins in the bovine uterus during endometritis are not well known.However, physiological hemorrhage from the endometrium and the increased vascular permeability due to mucosal inflammation are likely to facilitate the leakage of cellular and serum components, including complement, to the uterine lumen (Singh et al., 2008).

The protein S100A12 is a member of the S100A calcium-binding proteins that possess inflammatory and innate immune properties against different pathogens (Foell et al., 2013; Zackular et al., 2015). Antimicrobial peptides (AMP), including defensins, have innate and adaptive immune properties and are secreted at the mucosal surface of the female reproductive tract (Hickey

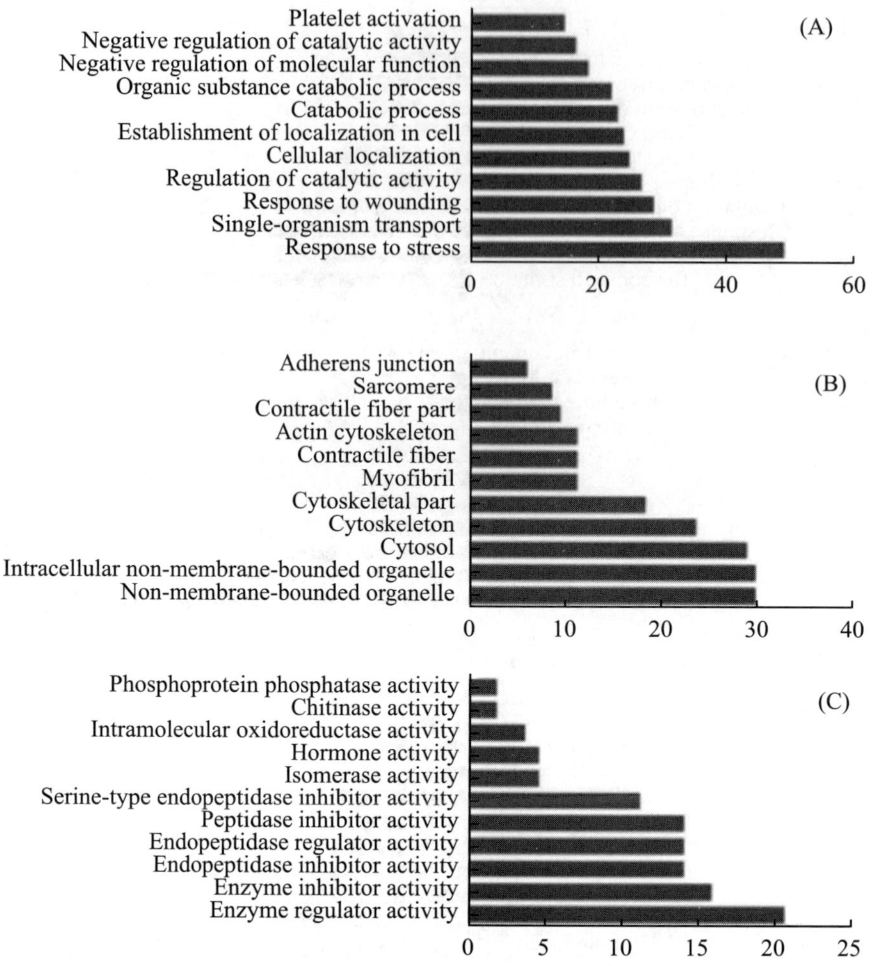

Figure 3. Top 11 classifications of the identified proteins differentially occurring in plasma based on their functional annotations using the Gene Ontology enrichment analysis: (A) biological processes, (B) cellular components, and (C) molecular functions.Color version available online.

et al., 2011; Oppenheim et al., 2003).We observed increased protein levels of S100A12 and β-defensin in the endometrial tissue during endometritis, which reflects previous marked expression of S100 genes and several AMP genes in cows with severe endometrial inflammation (Wathes et al., 2009).We also observed increased levels of lactoferrin protein, which is known to have important roles in chemotaxis, activation of immune cells, antigen processing, and adaptive immune response (Puddu et al., 2009).

Acute phase proteins (APP) are blood proteins primarily synthesized by hepatocytes as part of the acute phase response, which is in turn part of the early-defense or innate immune response to different stimuli including infection and inflammation (Cray et al., 2009).Here, we observed increased levels of the APP, SAA, and haptoglobin (HP) in the tissues with endometritis (Supplemental).Interestingly, these proteins were not significantly elevated in the plasma of the same cows with endometritis.Moreover, SAA-1, SAA-4, and C-reactive protein showed a tendency for decreased levels in the plasma of these cows.Previously, the plasma concentrations of the APP,

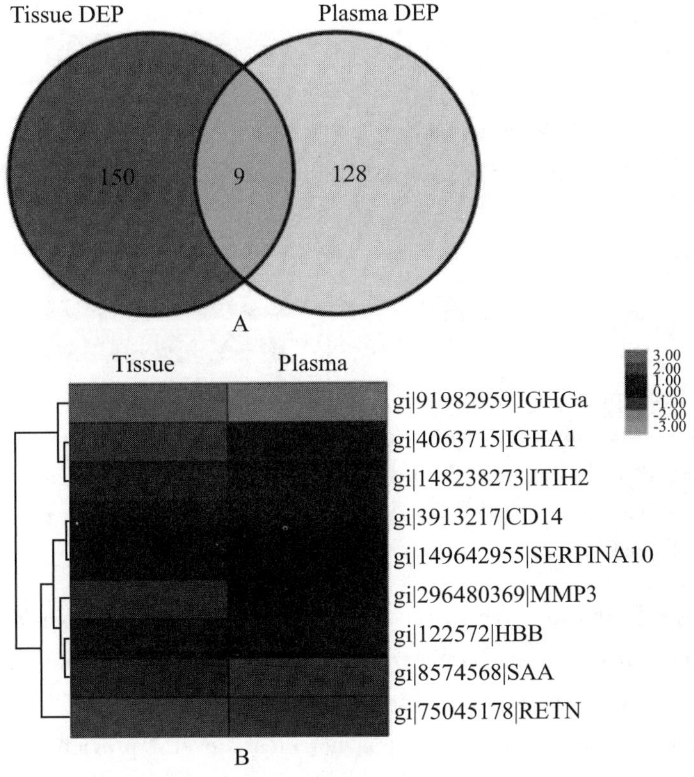

Figure 4. Venn diagram showing the distribution of proteins between endometrial tissue and plasma (A), and cluster analysis of 9 shared differentially expressed proteins (DEP) in tissue and plasma (B).

Note: IGHGa = immunoglobulin gamma 1 heavy chain constant region; IGHA1 = IgA heavy chain constant region; ITIH2 = inter-α-trypsin inhibitor heavy chain H2 precursor; CD14 = cluster of differentiation 14; SERPINA10 = serpin peptidase inhibitor, clade A (alpha-1 antiproteinase, antitrypsin), member 10; MMP3 = matrix metalloproteinase 3; HBB = hemoglobin subunit β; SAA = serum amyloid A protein; RETN = resistin. Color version available online.

α1-acid glycoprotein, ceruloplasmin, and HP, were increased in postpartum cows with uterine bacterial contamination, but these proteins returned to basal levels within 14 to 21d of uterine involution (Sheldon et al., 2001). Although the plasma concentrations of HP and SAA were higher in cows with postpartum metritis (Huzzey et al., 2009; Chan et al., 2010), the levels of these proteins in prolonged uterine disease (subclinical or clinical endometritis) are largely unknown. The gene transcripts for APP (HP and SAA3) were increased in the endometrium or endometrial cells in response to bacterial presence or inflammation (Gabler et al., 2010; Chapwanya et al., 2012) but not in cows with subclinical or clinical endometritis (Fischer et al., 2010). However, it is not clear whether APP are derived entirely from the liver, or if the endometrial cells can secrete significant levels of these proteins locally in response to infection or inflammation; our findings suggest the latter. However, endometrial cells infected with E. coli or exposed to bacterial LPS increased their gene expression of HP and SAA3 (Chapwanya et al., 2013; Oguejiofor et al., 2015b), but the secreted protein concentrations of HP and SAA were not altered (Davies et al., 2008).

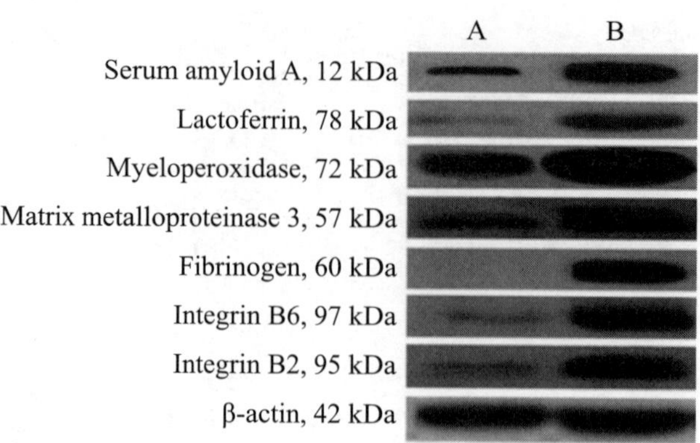

Figure 5. Confirmation of 7 differentially expressed proteins by Western blot analysis in (A) thenonendometritis group, and (B) the endometritis group.

Note: β-Actin was used as an internal reference to normalize the quantitative data. The figure is a representative result, which was tested in each animal one time, with a similar result in 4 samples. Color version available online.

Furthermore, we did not observe marked changes in the concentrations of inflammatory cytokines in either the endometrial tissue or the plasma from cows with endometritis. The expression of these cytokines differs depending on the time postpartum and the health condition of the uterus, but their roles in the development of endometritis is not clear. Several previous studies associated the differences in endometrial mRNA expression of inflammatory cytokine genes (e.g., IL1, IL6, IL8, and TNF) with the development of clinical or subclinical endometritis (Gabler et al., 2009; Fischer et al., 2010; Kasimanickam et al., 2014). The protein concentration of inflammatory cytokines in the endometrial tissue during endometritis is not known. Their concentrations in uterine flush were observed to differ in cows with clinical or subclinical endometritis (Kim et al., 2014). Moreover, the serum concentrations of proinflammatory cytokines were increased in the cows with clinical and subclinical endometritis in one study (Kasimanickam et al., 2013), but another study showed no differences (Kim et al., 2014).

The results of GO enrichment analysis of endometrial DEP showed that the most predominant proteins were proteolytic enzymes distributed at the extracellular region, which were activated to respond to stress and stimulus during endometritis (Figure 2). Matrix metalloproteinases (MMP) are a family of mainly extracellular zinc-dependent endoproteinases whose prominent functions include the proteolytic degradation of the extracellular matrix (ECM) proteins and tissue membranes as well as the remodeling of biologically active proteins during morphogenesis, tissue development, wound healing, and reproduction (Page-McCaw et al., 2007; Klein and Bischoff, 2011). Matrix metalloproteinases also regulate inflammation and innate immunity by modulating cytokine and chemokine activity by recruiting leucocytes to the site of inflammation (Parks et al., 2004; Van Lint and Libert, 2007). Uterine infection may alter the levels of MMP and therefore interfere with the mechanism of endometrial repair in the postpartum uterus. Bovine endometrial cells respond to bacterial LPS with increased gene expression of MMP1, MMP3, and MMP13 (Oguejiofor et al., 2015b). The increased endometrial expression of these MMPs, in addition to MMP9, was also associated with de-

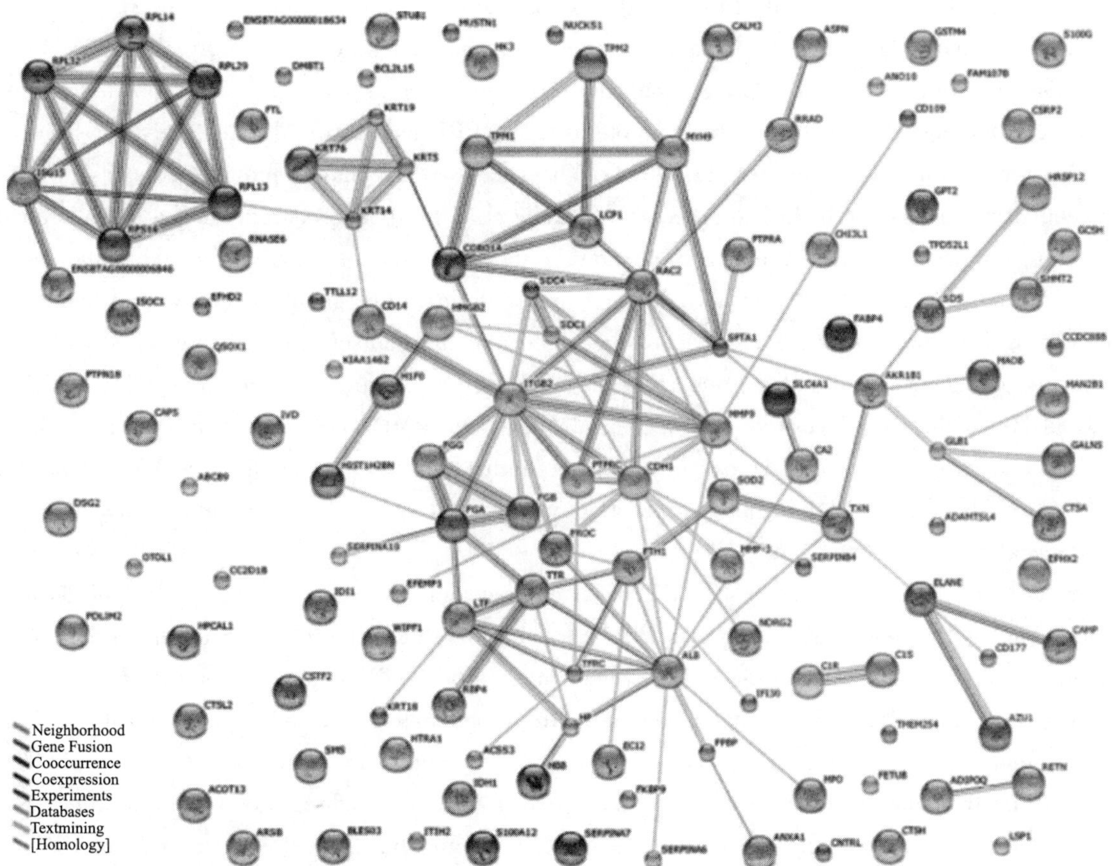

Figure 6. The protein-protein interaction network of differentially expressed proteins from endometrial tissue following analysis by STRING software (https://string-db.org/).

Note: An edge was drawn with 8 color lines as shown (see legend), which represent the existence of the 8 types of evidence used in predicting the associations. Color version available online.

layed uterine remodeling in postpartum cows (Wathes et al., 2011). We observed significant increases in both MMP3 and MMP9 proteins in the endometrial tissue during endometritis (Figure 5, Supplemental Table S1; https://doi.org/10.3168/jds.2016-12365). Matrix metalloproteinase-9 is a gelatinase that can digest several other ECM molecules, while MMP3 is a stromelysin that can also activate other MMP (Nagase et al., 2006). We hold that increased MMP promoted inflammation and played a role in the ECM degradation and cell or basement membrane damage in endometrial tissue during endometritis (Figure 1B). An imbalance in the proteolytic and antiproteolytic activity of MMP is known to be able to lead to several conditions ranging from tissue destruction in chronic inflammation to cancer metastasis (Klein and Bischoff, 2011). Although ECM remodeling is critical to endometrial receptivity for the embryo (Bauersachs et al., 2006), escalated MMP proteolytic activities are likely to cause chronic inflammation and a hostile uterine environment for reproduction.

In addition to MMP, many other proteins are also known to have extracellular proteolytic activities. Moreover, proteins rarely carry out their functions alone, but often by physically interacting with other proteins (Spirin and Mirny, 2003). In this study, PPI was analyzed using STRING to reveal functional links among the endometrial tissue proteins that were significantly altered in endometritis (Figure 6). The PPI data are commonly shown as networks, which include a node (a protein) and an edge (interaction between proteins). The PPI data are used extensively to assign or predict protein function (Xiong et al., 2013; Yu et al., 2015) and to provide more insight into molecular mechanisms of biological processes (Huang et al., 2015). According to the neighborhood counting approach of protein function prediction, interacting proteins may have some similar functions (Chua et al., 2006; Ng et al., 2010). Our results identified significant enzymatic hydrolase activity in the extracellular region as a possible molecular mechanism through which the DEP may promote inflammation and endometritis. Therefore, some core seed proteins such as RAC2, ITGB2, and CDH1 in the same network with CD14, MMP3, and MMP9 may be related to extracellular hydrolase activity and proteolysis (Figure 6). The definite roles of the identified core seed proteins RAC2, ITGB2, and CDH1 in this hydrolase activity are not clear, but they have critical roles in cell signaling, migration, and adhesion (Ridley, 2006; Sarantos et al., 2008; West and Harris, 2016). Their interactions with CD14, a co-receptor involved in bacterial recognition and innate immune response (Kumar et al., 2011), and MMP highlights the complex network of recognition molecules, adhesion molecules, proteolytic factors, and immune cell migration in the inflammatory process and endometritis. Essentially, the functional roles of adhesion molecules in cell proliferation (George and Dwivedi, 2004) and leucocyte migration (Madri and Graesser, 2000) are facilitated by the cleavage and proteinase activity of MMPs. Therefore, the protein network complemented GO annotation and provided some detailed information about the interactions of DEP in endometritis. Some of these important seed proteins were determined by Western blot, which validated the fold changes from iTRAQ analysis in endometrial tissue (Figure 5).

Additionally, DEP in the plasma were mainly components of the cytosol and non-membrane-bound organelles involved in response to stress and regulation of enzyme activity (Figure 3). A recent study showed that the bovine uterine proteome is dynamic, with the concentrations of proteins including enzymes, antioxidants, and immune molecules in the uterine fluid differing from those of plasma (Faulkner et al., 2012). Interestingly, 9 DEP (Figure 4) overlapped in the endometrial tissue and plasma. Endometrial proteome may have included some blood-derived proteins. Thus, some increased proteins, such as HBB, SAA, and RETN, may have been transferred from blood to the uterus (Figure 4). However, many plasma proteins also differed significantly from the endometrium (Supplemental Table S2, Figure 4). Although plasma and endometrial tissue came from the same cows, the blood cell damage was not confirmed to arise from the inflammatory lesion of the endometrial tissue. The predominant DEP in endometrial tissue were extracellular (Figure 2), whereas the main plasma DEP were intracellular (Figure 3). This pattern indicated that cell damage may have also taken place in the peripheral vessels instead of the uterine endometrium, which is actively regulated by local factors. After all, the blood system is sensitive to any other disease or dysfunction. Therefore, endometrial proteome profiling is thought to be more significant in understanding the endometritis mechanism as well as contributing to the exploration of disease treatment.

5 CONCLUSIONS

We used iTRAQ proteomic analysis to determine protein alterations in endometrial tissue and plasma of Chinese Holstein dairy cows with endometritis.The analysis identified 159 and 137 DEP in endometrial tissue and plasma, respectively.Profiling of the endometrial DEP demonstrated that the host defense was activated to protect the endometrium from infection and inflammation. However, bioinformatics analysis of DEP indicated that extracellular hydrolase activity was significantly increased in uterine tissue.Therefore, we propose that inflammatory damage induced by hydrolysis may contribute significantly to the aggravation of endometritis.These findings can facilitate further studies to better understand the molecular mechanisms through which altered proteins may promote inflammation and hence endometritis.

ACKNOWLEDGMENTS

This work was supported by the Increased Fund of Project Budget of Chinese Academic of Agricultural Science (2014ZL012), the Fund of Innovative Project in Science and Technology of CAAS (CAAS-ASTIP-2014-LIHPS-03), the National Key Science and Technology Support Project in the 12th Five-year Plan of China (2012BADB12B03), and the Central Scientific Research Institutes for Basic Research Fund of China (1610322014001).The authors declare no conflict of interest.

REFERENCES OMITTED

(发表于《Journal of Dairy Science》，院选 SCI，IF：2.474)

The Toxicity and Theacaricidal Mechanism against *Psoroptes cuniculi* of the Methanol Extract of Adonis Coerulea Maxim*

Xiaofei SHANG[Δ], Xiao GUO[Δ], Feng YANG, Bing LI, Hu PAN, Xiaolou MIAO**, Jiyu ZHANG**

(Key Laboratory of New Animal Drug Project of Gansu Province/Key Laboratory of Veterinary Pharmaceutical Development of Ministry of Agriculture/Lanzhou Institute of Husbandry Pharmaceutical Sciences, Chinese Academy of Agricultural Sciences Lanzhou 730050, China)

Abstract: [Scope]: Adoniscoerulea Maxim. is a perennial herbaceous plant that grows in scrub, grassy slope areas, and as traditional medicine it has been used to treat animal acariasis for thousands of years. In this paper, we aimed to study the acute toxicity and cytotoxicity of the methanol extract of A. coerulea (MEAC) in vivo and in vitro for supporting the clinic uses. The acaricidal activity and the mechanism of action against Psoroptes cuniculi were investigated.

[Results]: The results showed thatisoorientin, luteolin and apigenin were the primary compounds in MEAC. The toxicity test showed that median lethal dose (LD_{50}) and the 50% inhibitory concentration (IC_{50}) of MEAC were estimated to be more than 5 000mg/kg in mice in vivo and more than 50mg/ml against RAW 264. 7 and GM00637 cells in the 3-(4, 5-dimethylthiazol-2-yl)-2, 5-diphenyltetrazolium bromide (MTT) test. After culturing with MEAC, the activities of superoxide dismutase (SOD), catalase (CAT), malonyldialdehyde (MDA), glutathione-S-transferase (GST), acetylcholinesterase (AChE) and Na^+-K^+-ATPase of mites were evaluated. Compared with the control group, SOD activity of MEAC-treated group of mites was inhibited, and CAT activity was activated at the preliminary phase but was gradually inhibited over the period of incubation. MDA content reached a peak at 6h and

* Abbreviations: AChE, Acetylcholinesterase; CAT, Catalase; CLL model, The complementary log-log model; GST, Glutathione-S-transferase; HPLC, High performance liquid chromatography; IC_{50},: 50% inhibitory concentration; LD_{50}, Median lethal dose; MDA, Malonyldialdehyde; MEAC, Methanol extract of Adonis coerulea Maxim; MTT, 3-(4, 5-dimethylthiazol-2-yl)-2, 5-diphenyltetrazolium bromide; SOD, Superoxide dismutase; UV, Ultraviolet spectrophotometer

** Corresponding authors, E-mail address: gx_ 139417@163.com (J.Zhang).

[Δ] First co-author.

then gradually decreased. However, GST activity in the mites was activated in a dose-and time-dependent manner. AChE and Na^+-K^+-ATPase activities related to neural conduction, vital functions and the transmembrane ion gradient of the mites were inhibited.

[Conclusion]: MEAC is safe in the given doses in both the in vitro and the in vivo tests, can be applied in the clinic and it had good acaricidal activity. The extension of the incubation time in the mites led to dynamic disequilibrium between the production and clearing of superoxide anions, a disruption of the energy metabolism and the transmembrane ion gradient, and the inhibition of motor function. These factors may have resulted in mite death.

Key words: Adoniscoerulea Maxim; Toxicity; *Psoroptes cuniculi*; Acaricidal activity; Mechanism of action

1 INTRODUCTION

Acariasis is an important ectoparasitic disease that is caused by mites of the groups *Sarcoptidae*, *Psoroptidae* and *Demodicidae*. This disease could reduce the productivity and quality of afflicted animals (Dagleish et al., 2007; O'Brien, 1999). Globally the increasing number of pets (cats, dogs, birds, domestic animals) cohabitating with humans induced the spread and increase of a wide spectrum of zoonotic diseases, especially acariasis (Moskvina and Zheleznova, 2015). In order to treat and control sarcoptic mange, chemical drugs have been widely used in the clinic, and these drugs exhibit relatively satisfactory treatment effectiveness. Concurrently, the acaricide resistance has become a major threat in the control of this disease. During the last 20 years, after the first report of arthropod resistance (Melander, 1914), an increase in new cases of resistance has been reported for various parasite species (FAO, 2004). In order to avoid the resistance and enhance the effects of chemical drugs, the identification of alternative medicines, particularly natural products for mite control, has attracted the interest of many people (Qin and Zhang, 2013). Integrated mite management and bio-acaricidal agents have become an important trend in China and other countries (Zhong et al., 2016). Knowledge of acaricidal cytotoxicology and an understanding of the resistance mechanism will provide a fresh framework for the development of new plant-based acaricides.

Adoniscoerulea Maxim. is a perennial herbaceous plant that grows in scrub, grassy slope areas. This plant is distributed in northeast Tibet and the Sichuan, Qinghai and Gansu Provinces in China at altitudes of 2 300- 5 000 m (Chinese materia editorial committee, 2002). In our field investigation of folk veterinary medicine in the Ruoergai region in 2011, we found that the aerial parts of this plant has been widely used to treat animal acariasis by local people (Shang et al., 2013). In our previous studies, *A.coerulea* extract presented marked acaricidal activity against P.cuniculi with a median lethal time (LT_{50}) of 3. 137h at a concentration of 250mg/ml in vitro, and it could cure rabbit acariasis after three treatments in the clinic (Shang et al., 2012). To date, the safety of this plant in humans and animals to support clinic uses, and the acaricidal mechanism of action haven't been evaluated.

In this paper, high performance liquid chromatography (HPLC) was used to analyze the main compounds of the methanol extract of *A.coerulea* (MEAC), the cytoxicity in three cell lines and the acute toxicity in mice were evaluated in vitro and in vivo.Then, the enzyme's activities of mites related to the oxidant stress reactions and the metabolic pathway, which also corresponded to the resistance mechainsm of chemical drugs, such as ATP binding cassette transporters, acetylcho-linesterase (AChE) and glutathione - S - transferases (GST), were studied for understanding the possible mechanism.

2 MATERIALS AND METHODS

2.1 Plant material and extraction

Adoniscoerulea Maxim.was collected from Ruoergai County in Sichuan Province, southwest China in Sep. 2011. The raw material was identified by Prof.Zhigang Ma, Pharmacy College of Lanzhou University, China.A voucher specimen with accession number ZSY112 was submitted to the Herbarium of Lanzhou Institute of Animal and Veterinary Pharmaceutics Sciences, Chinese Academy of Agricultural Science (Lanzhou, China).

The methanol extracts of *A.coerulea* (MEAC) were prepared as follows.Firstly, 100g powder of herbal *Adonis coerulea* Maxim. (aerial parts) was soaked in 1000 ml methanol for 12h and then decocted in 80℃ for 5h two times. After filtering with filter, concentrating and evaporating with rotary evaporator, MEAC was dried in vacuum dryer at 60℃ to a constant weight. Finally, the methanol extract (8.78g) was obtained, and the yield was 8.78%.

2.2 Analysis of the constituents via UV and HPLC

MEAC was dissolved in methanol and filtered with a 2μm filter membrane.The extract was then scanned with an ultraviolet spectrophotometer at between 200nm and 400nm to determine peak characterization.

An RP-HPLC analysis of the MEAC was performed on a Waters apparatus (two solvent delivery systems, model 600, and a Photodiode Array detector, model 996) using a gradient solvent system comprised of water (A) and CH_3CN (B).The gradient profile was as follows: 0-20min (90%-85% A), 20-35min (85%-70% A), 35-70min (70%-40% A) and 70-85min (40%-10% A) at 0.5 ml/min.On-line UV spectra were recorded from 200 to 400nm.Data acquisition and quantification were performed using Millennium software version 2.10 (Waters). A SunFire C18 column (250mm * 4.6mm, 5μm, Waters, Ireland) was maintained at ambient temperature (30.0℃).The mobile phase was filtered through a Millipore 0.45μm filter and degassed prior to use.The peaks were detected, and isoorientin, luteolin and apigenin were detected via comparison with chemical standards.

2.3 Toxicological evaluation

2.3.1 In vivo acute toxicity test

The up-and-down or staircase method for acute toxicity testing was carried out as previously described (Bruce, 1985; Shang et al., 2015). 500-5 000mg/kg MEAC was administered orally to mice with a gradual increase in dose.The behavioral changes and mortality of animals were observed continuously for the first 4h and 7 days after the drug administration.The LD_{50} was calculated from the graph of percentage (%) of mortality (converted to probit) against log-dose of the extract (Taïwe et al., 2011).

2.3.2 In vitrocytotoxicity test

The experiments were carried out according to (Xu et al., 2016; Huang et al., 2011). Three cell lines, normal human hepatocyte L02 cells, murine macroghage RAW264.7 cells and human fibroblasts GM00637 cells were purchased from the Shanghai Institute of Biochemistry and Cell Biology, Chinese Academy of Sciences. The cells were grown in DMEM medium supplemented with 10% (v/v) heat inactivated fetal bovine serum, penicillin (100 kU/L), streptomycin (100mg/L) and 2mM of glutamine. The cell cultures were maintained at 37℃ in a humidified CO_2 (5%) incubator. Cells were cultured at a density of 5000 cells/well into a 96-well plate, and treated with MEAC and isoorientin, luteolin and apigenin at various concentrations for 48h. Then, a fresh solution of 3- (4, 5-dimethylthiazol-2-yl) -2, 5-diphenyltetrazolium bromide (MTT) (0.5mg/ml) was added to each well and incubated for an additional 4h. After that, the supernatants were discarded followed by addition of 150μl DMSO and vibration for 10min. The absorbance was measured at 570nm using BIO-RAD model 680 multiwell plate reader. Three replicates were performed for each group.

2.4 Acaricidal mechanism of action

2.4.1 Mite collection

P.cuniculi was isolated from the ear cerumen of naturally infested rabbits. The ear cerumen were collected and placed in Petri dishes and then incubated at 35℃ for 30min in an incubator. The adult mites were then collected for testing under a stereomicroscope (Walton and Currie, 2007), and the rabbits were treated immediately following material collection. The experiments complied with the rulings of Gansu Experimental Animal Center (Gansu, China) and were approved by the Ministry of Health, P.R.China in accordance with NIH guidelines.

2.4.2 Acaricidal activity in vitro

The experiments were carried out according to (Bruce, 1985; Shang et al., 2016). 0.1 ml (250 and 125mg/ml) of MEAC and isoorientin, luteolin and apigenin (5mg/ml, purchased from Shanghai R& D Centre for Standardization of Chinese Medicines, LC ≥ 98%) were separately added into 24-cell culture plates, and the liquid excess was absorbed with filter paper. As the positive drug, ivermectin (10mg/ml) was added into a 24-cell culture plate, and 10% glycerin was added in an untreated group. Then, 10 adult mites were collected with a needle from the ear cerumen of naturally infested rabbits and placed in each well. All plates were incubated at 25±1℃ under 75% relative humidity. For the next 24h, they were checked every 3h. The viability of the mites was checked regularly via stimulation with a needle, and the mites were recorded as dead if body and appendages did not move under a microscope. Five replicates were performed for each extract concentration.

2.4.3 Determination of enzyme activities

The experiments were carried out according to a previously described method (Chen et al., 2014; Xiong et al., 2013). The 0.2 ml MEAC (250mg/ml and 125mg/ml diluted using 10% glycerin and filtered using a 0.45μm filter membrane) was added into plastic Petri dishes, and 10% glycerin was added in a control group. After absorbing the liquid excess via the filter papers,

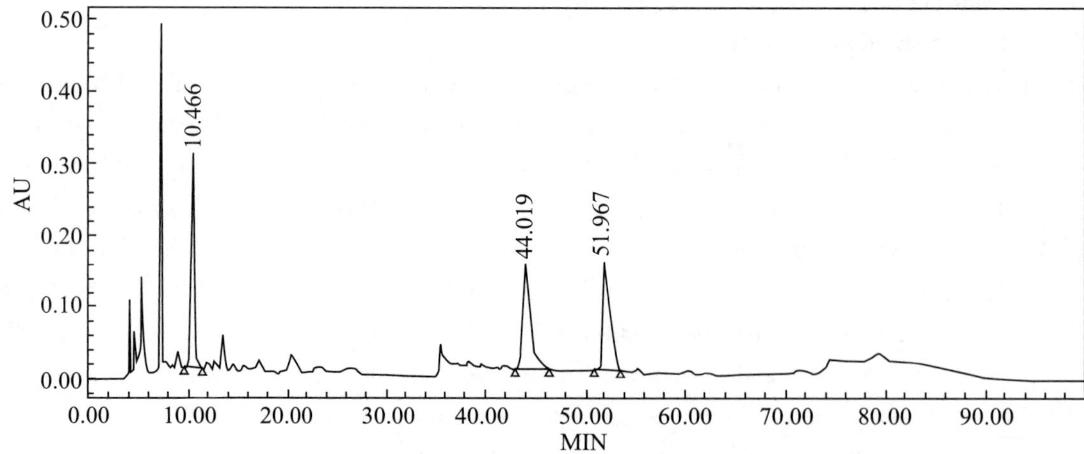

Fig 1. HPLC chromatogram of Adoniscoerulea Maxim.methanol extract.

the mites were placed in each dish (200 mites/group). All plates were incubated at 25±1℃ and 75% relative humidity. For the next 24h, they were checked every 3h. The status of mites was observed under a microscope, and viability of the mites was checked regularly via stimulation with a needle. When body and appendages did not move, the mites were recorded as dead.

After culturing for 3, 6, 9, 12, 18 and 24h, the mites were collected immediately and placed in a homogenizer. They were homogenized for 5min with 400μl physiological saline in ice water. Then, the homogenates of the mites were transferred to a centrifuge tube and centrifuged at 2 500×g at 4℃ for 10min. The supernatant was collected and used to assay the activities of superoxide dismutase (SOD), catalase (CAT), malonyldialdehyde (MDA), glutathione-S-transferase (GST), acetylcholinesterase (AChE) and Na^+-K^+-ATPase according to the manual of the Nanjing Jiancheng assay kit (Nanjing Jiancheng Bioengineering Institute, Nanjing, China). Three replicates were performed for each group.

Specific, SOD activity (U/mgprot) was measured using nitro blue tetrazolium as a substrate after suitable dilution, and the absorbance was detected with a microplate reader (ELISA) at 450nm (Multiskan MK3, Thermo Scientific, USA). CAT activity (U/gHb) was estimated using H_2O_2 as a substrate, and the absorbance was detected at 240nm using an ultraviolet spectrophotometer (Evolution 300 UV-VIS, Thermo Scientific, USA). Because ammonium molybdate can pause H_2O_2 decomposing reaction catalyzed by CAT immediately, and residual H_2O_2 can react with ammonium molybdate to produce a yellowish complex.

Lipidhydroperoxide decomposition products can condensate with thibabituric acid (TBA) to produce red compounds which has absorption peak at 532nm. Therefore, MDA levels (nmol/mg prot) were assessed by determining the degree of lipid peroxidation in mites, and absorbance was determined at 532nm. Meanwhile, glutathione peroxidase can catalyze H_2O_2 and reduced glutathione (GSH) to produce H_2O and oxidized glutathione. GST activity (U/mg prot) was determined by assaying the concentration of GSH, and the absorbance was detected at 412nm via ultraviolet spectrophotometer.

AChE hydrolizes acetylcholine to ethanoic acid and choline. The addition of a hydrosulfide color- developing reagent produces symtrinitrobenzene with a yellow color. According to this principle, AChE activity (U/mg prot) was determined, and the absorbance was determined at 412nm via ultraviolet spectrophotometer; Finally, ATPase catalyzes the hydrolysis of ATP with the product of ADP and phosphate, and the amount of phosphate can be determined and thus measure the activity of ATPase.The changes in the Na^{+}-K^{+}-ATPase activity (U/mg prot) of Psoroptes cuniculi treated using MEAC were determined, and the absorbance was detected at 636nm via ultraviolet spectrophotometer.

2. 5 Statistical analysis

The data obtained were analyzed using SPSS software version 14. 0 and expressed as the means ±SD.The data were analyzed using oneway ANOVA, followed by Student's two-tailed t-test for comparisons between the test and control and Tukey's test when the data involved three or more groups.P-values of less than 0. 05 ($P<0.05$) were considered significant.The median lethal time value (LT 50) was calculated using the complementary log-log (CLL) model.

3 RESULTS

3. 1 Analysis of the constituents of MEAC by UV and HPLC

In UV scans between 200nm and 400nm, the methanol extract of A.coerulea had three absorption peaks at 203nm, 270nm and 328nm.In these, the peak at 203nm may have represented the end absorption of ultraviolet light. In the HPLC analysis, isoorientin, luteolin and apigenin were found in MEAC (Fig. 1).

3. 2 In vivo acute toxicity test

In this test, administration of MEAC (500-5 000mg/kg) to mice did not cause death or acute behavior changes during the observation periods, and we also did not notice any pathology changes in mice.The LD 50 was estimated to more than 5 000mg/kg, and MEAC was safe at the given dose in mice.

3. 3 In vitrocytotoxicity test

In this test, L02, RAW 264. 7 and GM00637 cell lines had been adopted to evaluate the cytotoxicity of MEAC and three compounds.The results showed that MEAC was inactive or had no cytotoxicity against RAW 264. 7 and GM00637 cells due to IC_{50} (50% inhibitory concentration) values being more than 50mg/mL.It had moderate activity against L02 cell.Further studies showed that apigenin and luteolin had moderate activity against L02, RAW 264. 7 and GM00637 cells with IC_{50} values ranging from 13. 67 to 19. 50mg/mL.Isoorientin was considered non-cytotoxic on all three cell lines, and IC_{50} values were more than 50mg/mL (Table 1).

3. 4 Acaricidal activity in vitro

The results showed that the percentages of mortalities of MEAC (250mg/mL) against P.cuniculi were 52. 00%, 64. 00%, 68. 00%, 86. 00%, 96. 00% and 100% at 3h, 6h, 9h, 12h, 18h and 24h, respectively.The LT 50 of 250mg/mL and 125mg/mL were 3. 368h and 9. 792h, respectively.The positive drug ivermectin presented the best acaricidal activity, and it could kill all mites within 9h.The mortalities of control group were 0%, 0%, 0%, 2%, 4% and 8% at the various times, respectively (Tables 2 and 3).

Then, theacaricidal activities of three main compounds (apigenin, luteolin and isoorientin) were evaluated.Compared with the control group, apigenin demonstrated moderate activity with LT_{50} at 24. 902h; however, the acaricidal activity of other two compounds was very weak with the mortalities at 24h of only 18. 00% (Tables 2 and 3).

Table 1 Thecytotoxicity (IC 50mg/mL) of the methanol extract of Adonis coerulea Maxim.and three compounds against three cell lines.

Group	IC50 (mg/mL)		
	L02 cell	RAW264. 7 cell	GM00637 cell
MEAC	16. 16	>50	>50
apigenin	15. 04	19. 50	15. 91
luteolin	13. 67	16. 06	19. 17
isoorientin	>50	>50	>50

3. 5 Superoxide dismutase (SOD) activity

In this test, the activity of SOD in mites was inhibited by MEAC compared with the control group.At the same time points, the high-dose group exhibited a larger decrease than the low-dose group; additionally, the higher the drug concentration, the more the enzyme activity was inhibited. SOD values of the MEAC group (250mg/mL) were 28. 114 ($P<0. 05$), 32. 034 ($P<0. 01$), 28. 187 ($P<0. 05$), 30. 227 ($P<0. 05$), 30. 490 ($P<0. 01$) and 29. 642U/mg prot ($P<0. 01$) compared with the control group levels of 34. 977, 41. 094, 38. 423, 37. 728, 40. 255 and 40. 936U/mg prot at 3h, 6h, 9h, 12h, 18h, and 24h, respectively (Fig. 2A).

3. 6 Catalase (CAT) activity

Compared with the control group, the CAT activity of the MEAC-treated group was activated at the preliminary phase (3h, 0. 1745U/gHb).Then, the CAT activity was returned to basal levels as the control group at 6h (0. 1387U/gHb).After that it was gradually inhibited by MEAC as the incubation time increased (9h, 0. 1449U/gHb; 12h, 0. 1377U/gHb; 18h, 0. 1218U/gHb).Finally, the CAT activity of the MEAC-treated group (0. 0951U/gHb) was below that of the control group (0. 1157U/gHb) at 24h ($P<0. 05$) (Fig. 2B).

3. 7 Malonyldialdehyde (MDA) value

The changes in the MDA content of the mites treated using MEAC are shown in Fig. 2C.The MDA contents in the high-dose group increased: 1. 0093, 3. 4608 ($P<0. 01$), 2. 6054 ($P<0. 01$), 2. 3899 ($P<0. 05$), 2. 3457 ($P<0. 05$) and 2. 3057nmol/mg prot ($P<0. 05$) at 3h, 6h, 9h, 12h, 18h, and 24h, respectively.Meanwhile, the MDA levels of the control group were 1. 1868, 2. 4804, 2. 0295, 1. 9520, 1. 9255 and 1. 9147nmol/mg prot at 3h, 6h, 9h, 12h, 18h, and 24h. As the duration of incubation with MEAC increased, the MDA content also increased, peaking at 6h and then gradually decreasing.

3.8 Glutathione-S-transferase (GST) activity

The results showed that MEAC markedly activated the GST activity in the mites in a dose-and time-dependent manner. Compared with the control group (51.3872, 53.1992, 55.7274, 56.6968, 57.9249 and 60.4244U/mg prot), the GST activities of the 250mg/mL group were 64.6206, 69.6317, 80.1180, 89.6422, 92.6964 and 96.5159U/mg prot at culturing durations of 3h, 6h, 9h, 12h, 18h and 24h (all $P<0.01$), respectively (Fig. 2D).

3.9 Acetylcholinesterase (AChE) activity

In the assay, following treatment with MEAC, AChE activity was activated by MEAC at 3h and was then inhibited and gradually decreased with incubation time. In the test, AChE values of the MEAC group were 1.4423 ($P<0.01$), 0.8774, 0.7184, 0.6739 ($P<0.01$), 0.6727 ($P<0.01$) and 0.3135U/mg prot ($P<0.01$) compared with those of the control group (0.9536, 0.8616, 0.8959, 0.8957, 0.9658 and 0.9482U/mg prot) at culture durations of 3h, 6h, 9h, 12h, 18h and 24h, respectively (Fig. 2E).

3.10 Na^+-K^+-ATPase activity

In the assay, MEAC markedly and gradually inhibited the Na^+-K^+-ATPase activity in mites in a dose-and time-dependent manner. The inhibition rates of Na^+-K^+-ATPase activity were 52.710% and 45.055% compared with the control group at the concentrations of 250 and 125mg/mL, respectively, at 24h ($P<0.01$). Compared with the control group (1.6640, 1.7888, 1.7751, 1.7266, 1.7081 and 1.7048U/mg prot), the Na^+-K^+-ATPase activities of the 250mg/mL group were 1.1133, 0.7455, 0.8452, 0.8012, 0.7567 and 0.8062U/mg prot at culturing durations of 3h, 6h, 9h, 12h, 18h and 24h (all $P<0.01$), respectively (Fig. 2F).

4 DISCUSSION

Now, psoroptic acariasis is a global disease that causes substantial losses in the United States, Turkey, South Korea, India, China and other countries (Fichi et al., 2007; Dua et al., 2013). Many plant-based acaricidal agents have been developed and used to treat and control psoroptic and sarcoptic mange in veterinary clinics (Qin and Zhang, 2013). As a folk medicine, A.coerulea has been used to treat animal acariasis for thousands of years in China (Chinese materia editorial committee, 2002). But until now, no one has used modern technology to study the acaricidal effects and the toxicity.

In this paper, firstly, in order to the explain the main compounds of MEAC, a UV method was adopted to investigate absorbance peaks and wavelengths. In addition to the wavelengths and end absorption of 203nm, 270nm and 328nm were found. HPLC analysis indicated that flavonoids were the primary components, and isoorientin, luteolin and apigenin were found to be present in the extract.

Subsequently, the toxicity in vitro and in vivo was evaluated in order to support the clinic uses of this plant. The results showed that LD_{50} and IC_{50} of the extract were estimated to be more than 5 000mg/kg in mice and more than 50mg/mL against RAW 264.7 and GM00637 cells; we thought

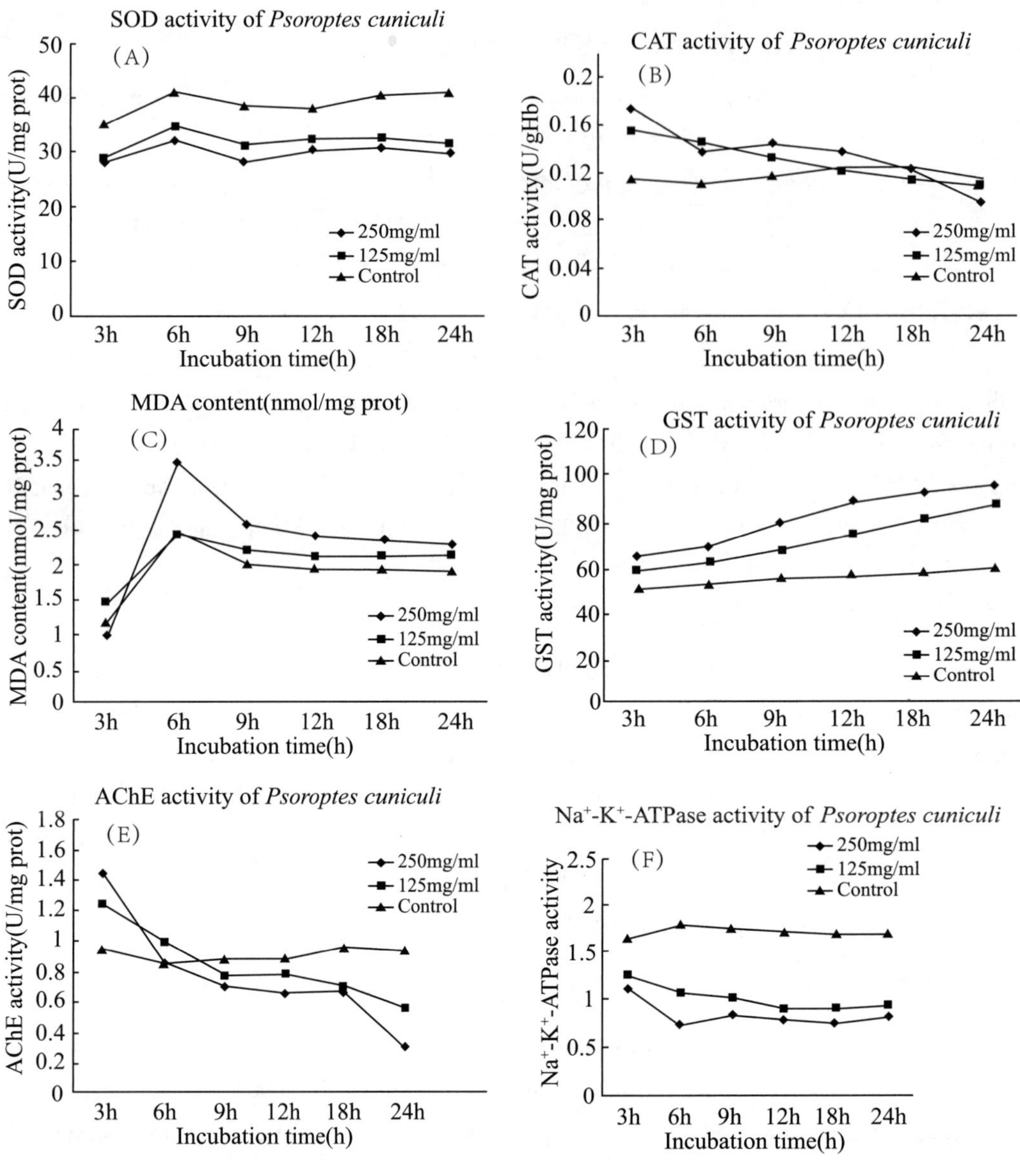

Fig. 2. Changes in the SOD activity (A), CAT activity (B), MDA content (C), GST activity (D), AChE activity (E) and Na^+-K^+-ATPase activity (F) of *Psoroptes cuniculi* treated with the methanol extract of Adonis coerulea Maxim.

that MEAC was safe at the given dose. And it could be widely used in clinic to treat animal acariasis. Of course, the sub-acute toxicity of MEAC should be studied further.

Meanwhile, apigenin and luteolin presented the moderate cytotoxicity in vitro against L02, RAW 264. 7 and GM00637 cells with IC_{50} values ranging from 13. 67 to 19. 50mg/mL. But the subsequent studies showed that the acaricidal activities of three compounds were weak against P.cuniculi; we speculated that these compounds may not be active, or they had a synergistic or additive effect. The active compound should be identified and isolated in a further study.

Table 2　Acaricidal activities of the methanol of *Adonis coerulea* Maxim.and three compounds against *Psoroptes cuniculi* in vitro test.

Group		Time (h)					
Drug (s)	Concentration (s)	3h Mean mortality*	6h Mean mortality*	9h Mean mortality*	12h Mean mortality*	18h Mean mortality*	24h Mean mortality*
MEAC	250mg/mL	52. 00± 13. 30A	64. 00± 11. 40 A	68. 00± 8. 37A	86. 00± 11. 40Aa	98. 00± 5. 48 A	100. 00± 0. 00A
	125mg/mL	22. 00± 8. 37 B	36. 00± 8. 94 B	42. 00± 8. 37B	56. 00± 8. 94 B	68. 00± 13. 04 B	76. 00± 11. 40 B
Apigenin	5mg/mL	0. 00± 0. 00c	4. 00± 5. 48C	8. 00± 4. 47C	32. 00± 8. 37C	36. 00± 8. 94C	42. 00± 13. 04C
Luteolin	5mg/mL	0. 00± 0. 00c	4. 00± 5. 48C	4. 00± 5. 48C	10. 00± 7. 07 D	14. 00± 5. 48 D	18. 00± 8. 37 D
Isoorientin	5mg/mL	0. 00± 0. 00C	4. 00± 5. 48C	4. 00± 5. 48C	8. 00± 4. 47 D	16. 00± 8. 94Da	18. 00± 8. 37 D
Ivermectin	10mg/mL	76. 00± 5. 48 D	98. 00± 5. 48 D	100. 00± 0. 00D	100. 00± 0. 00Ab	100. 00± 0. 00A	100. 00± 0. 00A
Control	–	0. 00± 0. 00C	0. 00± 0. 00C	0. 00± 0. 00C	2. 00± 4. 70 D	4. 00± 5. 48 Db	8. 00± 4. 47 D

The difference between data with the different capital letter within a column is very significant ($P<0.01$), and the difference between data with the different small letters within a column is significant ($P<0.05$).
Mean mortality* represent Mean mortality (%) ±SD.

Table 3　The LT_{50} values of the methanol extract of *Adonis coerulea* Maxim. and three compounds against *Psoroptes cuniculi* in vitro by CLL model.

Group (s)		Regression line	LT_{50} (h) (95%FL)	Pearson Chi-square
Drug (s)	Concentration (s)			
MEAC	250mg/mL	Y = 2. 203x−1. 236	3. 368 (0. 781−5. 734)	10. 883
	125mg/mL	Y = 1. 661x−1. 646	9. 792 (7. 880−12. 140)	0. 913
Apigenin	5mg/mL	Y = 2. 656x−3. 708	24. 902 (20. 459−34. 442)	6. 559
Luteolin	5mg/mL	Y = 1. 755x−3. 290	74. 902 (39. 717−580. 539)	1. 126
Isoorientin	5mg/mL	Y = 1. 837x−3. 386	69. 710 (38. 463−439. 892)	1. 142

Because the various active components or compounds extracted during plant medicine development may affect the metabolic pathways or targets, the study of the acaricidal mechanism of plant-based medicines to provide new approaches or methods for developing better acaricidal agents has been attracted the tremendous attention of researchers. In our works, the acaricidal mechanisms of the extract were investigated by determining the enzyme activities related to the oxidant stress reactions of mites and their metabolic pathways.

Under normal conditions, the antioxidant protective enzymes of mites are composed of SOD and CAT, among others (Jiang et al., 2003), which are in dynamic equilibrium with the production and clearing of superoxide anions in insects.These enzymes also play a role in protecting the structure and function of cell membranes.However, influences of environments or drugs can destroy this equilibrium and harm mites.After treatment with MEAC, the SOD activity of the mites was gradually inhibited, leading to the production of excess free radicals and the prevention of the catalytic decomposition of H_2O_2 in the body (Chen et al., 2014).The increase in the H_2O_2 content may be related to the increase in CAT activity in mites.When applying treatments with MEAC, the mites first increased their CAT activity to catalyze the breakdown of abundant H_2O_2; however, the mites were gradually killed, and CAT activity was decreased as incubation time increased.The degree of lipid peroxidation in the mites increased, as did their MDA content.All of the above suggested that MEAC destroys the dynamic equilibrium and induces the accumulation of superoxide anion to kill mites.Cell energy metabolism was also blocked in this process (Chou and Chen, 2007; Du et al., 2008).

Glutathione S-transferases (GST) are a group of enzymes that catalyze the conjugation between glutathione (GSH) and several molecules and play a central role in the detoxification of xenobiotic and endogenous compounds in mites (Mounsey et al., 2010; Prapanthadara et al., 2000).In this paper, MEAC markedly increased GST activity in mites compared with the control group in a dose-and time-dependent manner, and some endogenous compounds and hydrogen peroxides may have been released and accumulated by mites, leading to mite death.

AChE could hydrolize ACh to terminate the signal transmission mediated by ACh, and the activity of AChE is related to the vital movements of insects.Thus, AChE plays an important role in insect neural conduction.In the test, after 3h of activation, AChE activity was inhibited by MEAC and gradually decreased with incubation duration.These results suggest that the neural conduction and the vital movement of mites were destroyed and blocked, resulting in mite tissue damage and death (Jeyaprakash and Hoy, 2007).This may be one of the reasons for the observed mite death. Additionally, Na^+-K^+-ATPase is an important enzyme present in the plasma membrane of insect cells and plays a very important role in protecting the transmembrane ion gradient in mites (Eo and Oh-Deog, 2010).MEAC gradually and markedly inhibited the Na^+-K^+-ATPase activity in mites in a dose-and time-dependent manner.These results indicated that when cells are affected by noxious factors, the transmembrane ion gradient of mites may be destroyed and the energy metabolism may be inhibited (Jiang et al., 2003).

In the test, we found that theacaricidal activity of ivermectin is better than MEAC in vitro.MEAC is, however, effortless degraded in the environment, dose not remain in livestock, is not as prone to resistance, is relatively safe for humans, animals, and the environment and also has other advantageous features (Samish and Rehacek, 1999).MEAC is worthy to study and develop as an alternative medicine.Meanwhile, the inhibition of the glutathione transferase system, AChE and Na^+-K^+-ATPase activities, relate to the resistance mechanism against chemical drugs resistance mechanism, were proved.Thus, MEAC could be used to overcome the resistance of mites against chemical drugs or increase the acaricidal activity of chemical drugs against mites.

5 CONCLUSIONS

The methanol extract of A.coerulea had good acaricidal activity and safety in both the in vitro and in vivo tests, and could be applied in the clinic. MEAC inhibits the protective enzyme system and the glutathione transferase system, which play roles in clearing away superoxide anions and toxins in mites. AChE and Na^+-K^+-ATPase activities were also inhibited, which were related to mite neural conduction, vital movements, and the transmembrane ion gradient. Thus, we speculated that the mechanism of death in P.cuniculi involved the inhibition of the dynamic equilibrium between the production and clearing of super-oxide anions, which destroyed motor function. These studies would provide the evidence for the traditional uses of this plant. Meanwhile, the exact mechanism should be determined in light of the proteomics and genomics of mites to determine the various proteins or genes affected by the drug in attempts to determine the pathway or mechanism by which mites are affected by this plant.

COMPETING INTERESTS

All authors declare that they have no competing interests.

AUTHORS' CONTRIBUTIONS

XS and XG conceived the study, FY, XM and XG determined the index, XS, JZ and BL wrote the manuscript, and HP performed the statistical analyses. All authors read and approved the final version of the manuscript.

ACKNOWLEDGEMENTS

This work was financed by the National Natural Science Foundation of China (31302136), Key Technology R& D Program of Gansu Province (2016GS10130), Youth Science Foundation of Gansu Province (1506RJYA144) and the Special Fund of the Chinese Central Government for Basic Scientific Research Operations in Commonwealth Research Institutes (No. 1610322016012).

REFERENCES OMITTED

（发表于《Veterinary Parasitology》，院选 SCI，IF：2. 356）

Genomic Analysis and Resistance Mechanisms in *Shigella flexneri* 2a Strain 301

Zhen ZHU, Xuzheng ZHOU, Bing LI, Sihan WANG,
Fusheng CHENG, Jiyu ZHANG

(Key Laboratory of New Animal Drug Project of Gansu Province/Key Laboratory of Veterinary Pharmaceutical Development of Ministry of Agriculture/Lanzhou Institute of Husbandry and Pharmaceutical Sciences, CAAS, Lanzhou, China)

Abstract: *Shigella flexneri* is one of the most prominent pathogenic bacteria in developing countries. In the battle against shigellosis and other bacterial diseases, antibiotic resistance has become an increasing global public health threat. Although the serious phenomenon of multidrug resistance (MDR) has been identified as one of the top three burdens on human health, resistance mechanisms are still poorly understood at the molecular level. In this study, we analyzed genomic data and the evolution of resistance in Shigella flexneri under sequential selection stress from three separate antibiotics: ciprofloxacin (CIP), ceftriaxone (CRO), and tetracycline. Through whole-genome sequencing, 82 chromosomal antibiotic resistance genes were identified. Re-sequencing of the evolved populations identified single nucleotide polymorphisms (SNPs) that contributed to MDR and SNPs that were specific to a single drug. A total of 40 SNPs in 8 genes and 3 intergenic regions, including mutations in metG (L582R) and 1538924, 1538924, and 1538924, appeared under each antibiotic. Several nonsynonymous mutations in gyrB (S464Y), ydgA (E378A), rob (R156H), and narX (K75E) were observed under selective pressure from CIP or CRO. Based on a bioinformatic analysis and previous reports, we discuss the contribution of these mutated genes to resistance. Therefore, more circumspect selection and use of antimicrobial drugs for treating shigellosis is necessary.

Key words: Pathogenic bacteria; Resistance; Plasmid

1 INTRODUCTION

SHIGELLA SPP., The major etiologic agent of human shigellosis, is a global public health burden, especially in low-income countries. Based on biochemical and serological properties, the genus *Shigella* consists of four species: *Shigella dysenteriae* (*S. dysenteriae*), *Shigella flexneri* (*S. flexneri*), *Shigella boyddi* (*S. boydii*), and *Shigella sonnei* (*S. sonnei*). All four species of *Shigella* can cause shigellosis; however, *S. flexneri* is the predominant subgroup in developing countries, and *S. sonnei* is the predominant subgroup in industrialized countries.[1,2]

The first identified *Shigella* species was *S.dysenteriae*, followed rapidly by the identification of *S.flexneri* at the end of the 19th century.*Shigellosis* became a notorious, widespread epidemic during World War 1 with the transmission of strain NCTC1, a 2a lineage of S.flexneri.[3,4] S.flexneri has been identified in all areas of the world and has developed unprecedented diversity, with no fewer than 20 serotypes: 1a, 1b, 1c, 1d, 2a, 2b, 2v, 3a, 3b, 4a, 4av, 4b, 4c, 5a, 5b, X, Xv, Y, Yv, 6 and 7b,[5,6] although serotype 2a is still highly prevalent.

To counter the blight caused by shigellosis, a large number of antibiotics have been used in clinical therapy, ultimately leading to widespread antibiotic-resistant bacteria.The systematic appearance of antibiotic resistance in pathogenic bacteria remains a global problem.[7,8] One of the main reasons for the rapid accumulation of resistance in Shigella is likely the inappropriate and excessive use of antibiotics among outpatients in China.[9]

Bacteria can achieve resistance via the sequential accumulation of multiple spontaneous mutations, the horizontal transfer of resistance genes, or changes in the expression of chromosomal genes.[10-12] The evolution of resistance through access to single spontaneous mutations is particularly relevant for several types of antibiotics, such as quinolones, for which high levels of resistance can result from a single-locus mutation in genes in the quinolone resistance-determining region (QRDR) (gyr A, gyr B, par C, and par E).[13,14] However, for most antibiotics, developing a high level of resistance requires multiple mutations.[15]

Despite decades of research, our knowledge of the mechanisms underlying bacterial drug resistance is still incomplete. Therefore, extensive investigations are required to expand on and supplement our knowledge of antibiotic resistance mechanisms.

In this study, we not only analyzed the antibiotic resistance genes (ARGs) in the S.flexneri 2a 301 chromosome but also explored the undiscovered mechanisms underlying resistance under the continuous action of antibiotics.Through this multi-level study, we have assessed a considerable amount of genomic information on S.flexneri 301 and many other Shigella isolates reported over a long period, which will contribute toward understanding antibiotic resistance and provide guidance for addressing shigellosis.

2 MATERIALS AND METHODS

2.1 Storage and antimicrobial susceptibility testing

There are many existing records of the collection of S.flexneri 2a, strain 301 because it was identified by the General Hospital of the People's Liberation Army (PLAGH) in 1984. In this study, S.flexneri 2a 301 isolated from bacillary dysentery patients decades ago and stored by the Chinese Center of Disease Control and Prevention (CDC) was used.The bacterium was grown in brain heart infusion broth (BD) for 24 hr under aerobic conditions and stored in 25% glycerol at -80℃.

We tested S.flexneri 301 for resistance against 25 modern antimicrobial drugs by using the microdilution method in 96-well plates, as recommended by the Clinical and Laboratory Standards Institute (CLSI).[16] The antibiotics (obtained from Sigma) included ampicillin (AMP), piperacillin (PIP), amoxicillin (AMX), cefazolin (CFZ), cefuroxime (CFX), cefoxitin

(FOX), ceftriaxone (CRO), ceftazidime (CAZ), cefotaxime (CTX), cefepime (FEP), cefoperazone (CFP), imipenem (IPM), aztreonam (ATM), chloramphenicol (CHL), tetracycline (TC), erythromycin (E), rifampin (RIF), streptomycin (STR), gentamicin (GM), amikacin (AK), cotrimoxazole (COT), trimethoprim (TMP), norfloxacin (NOR), ciprofloxacin (CIP), and levofloxacin (LVX).The Escherichia coli ATCC25922 strain was used for quality control.

2.2 De novo sequencing and genome annotation

Genomic DNA from Shigella flexneri 301 strain (SF) for de novo sequencing was extracted with a DNA Isolation Kit (OMEGA) and sequenced on the HiSeq 2000 technology platform (Illumina, Inc., San Diego, CA) by using a pairedend strategy at Beijing Genomics Institute (BGI), Shenzhen, China. Automatic generation of qualified clusters and sequencing were performed on paired-end 2 × 90 bp reads.Clean data were acquired by filtering out adapters, duplications, and low-quality reads from the total data.

The assembly of short reads into genome sequences used NC_ 004337 (Sf. 2a 301) as a reference genome and was performed with the SOAPdenovo[17,18] assembler, with the key parameter Kmer value set to 71 for optimal results.The assembly results were then locally assembled and optimized according to paired-end and overlap relationships via mapping reads to contigs.

Functional annotation was accomplished by an analysis of protein sequences.We aligned genes with databases to obtain their corresponding annotations.[19-26] In particular, we identified protein orthologs of ARGs by using the Antibiotic Resistance Genes Database (ARDB).[27] A gene was annotated as an ARG if the BLAST value was beyond the set threshold of amino acid sequence identity over 90%. To ensure biological meaning, the highest-quality result was chosen as the gene annotation and was completed by performing BLAST searches of the genes against each database.

2.3 Bioinformatics analysis

We compared the SF genome data assembled in this study with the reference genomes (complete and draft) of *Shigella* and *E.coli* from the National Center for Biotechnology Information (NCBI).A comparative genomics analysis was performed to explore the diversity of the relationships and the evolution of some *Shigella* and *E.coli* reference genomes.We used annotations to define core and pan genomes[28] and to establish a linear synteny analysis of the SF and reference genomes at the amino acid level to study the differences between each reference genome.[29] To perform a phylogenetic tree analysis, we generated a multiple sequence alignment (54 S. flexneri, 2 S. boydii, 2 S.sonnei, 1 S.dysenteriae, and 1 E.coli) by using TreeBeST.[30]

2.4 Laboratory evolution experiment and re-sequencing

After determining minimum inhibitory concentrations (MICs), we conducted experiments to establish a series of drug-resistant strains.*S.flexneri* 301 was grown in Luria-Bertani (LB) medium (BD) that was supplemented with CIP, CRO, or TC at 0.25 × MIC at 37℃ with shaking at 250 rpm until the culture reached an OD_{600} of-0.5. At this OD value, the bacteria had adapted to this fixed antibiotic concentration, and a higher drug concentration was necessary to maintain selection pressure to ensure that the population achieved a higher resistance level.Hence, the strain was successively cultured in medium containing two fold – increased concentrations to obtain 10 drug – resistant generations, in which the MICs had significantly improved. Under the same conditions, we also cultivated S.flexneri 301 in LB medium with no antibiotic as the control group.

The re–sequencing of nine antibiotic–induced strains was also performed on the HiSeq 2 000, and the sequencing conditions were similar to those for the de novo sequencing, except that different key parameter Kmer values ranging from 23 to 81 were used for optimal results when they were assembled. The nine re – sequenced genomes were compared with SF to identify resistance – related single nucleotide polymorphisms (SNPs) in mutated genes. All the SNPs in mutated genes were tested in the control group by Sanger sequencing.

3 RESULTS

3.1 Antibiotic–resistant phenotype

After being revived, the *S. flexneri* 301 isolate was tested to determine MICs by using 25 modern antimicrobial drugs. The results of the susceptibility testing suggested that the strain was sensitive to most of the antibiotics, but its resistance to several common drugs was still quite serious.

The resistance to TMP (64μg/mL) was the highest, followed by AMP (32μg/mL), COT (32μg/mL), CHL (32μg/mL), ATM (16μg/mL), RIF (16μg/mL), and E (8μg/mL). The 301 strain exhibited resistance to PIP, AMX, and STR, with MICs ranging from 2 to 4μg/mL. Fortunately, the 301 strain did not show resistance to any of the cephalosporin (CFZ, CFX, FOX, CRO, CAZ, CTX, FEP, CFP) or fluoroquinolone (CIP, NOR, LVX) antibiotics, with CRO (0.0625mg/mL), CAZ (0.125μg/mL), CTX (0.125μg/mL), CIP (0.02μg/mL), and NOR (0.16μg/mL) all exhibiting MICs under 0.2μg/mL. In addition, the strain was susceptible to the aminoglycosides STR, GM, and AK at a concentration of 1μg/mL. Although the MIC for TC was 1μg/mL, this is within the scope of inhibiting bacteria. The results showed that this strain possessed slight multidrug resistance (MDR).

3.2 Overview of de novo sequencing data

We sequenced the most epidemic *S. flexneri* strain (301) by using next–generation sequencing technology and constructed a high–quality complete genome (SF) (Supple–mentary Fig. S1; Supplementary Data are available online at www.liebertpub.com/mdr).

Using the IlluminaHiSeq 2,000 sequencing platform, 732 MB of data were produced for the sampled SF. Based on the assembly results for the sampled SF, we present a chromosome sequence of 4,660,005bp in length, with an average guanine–cytosine (GC) content of 50.87%; the number of scaffolds was 28, and the number of contigs was 95. Based on the genomic analysis of SF, we found that the genome contained 4,832 genes, and the total length of genes was 4,009,839bp, accounting for 86.05% of the genome. The number of tandem repeat sequences was 78, and the total length of tandem repeat sequences was 11,024bp, constituting 0.2366% of the genome.

3.3 Bioinformatic analysis

After capturing genomic information and genetic variation through Illumina sequencing, we compared the SF genome with some available Shigella and E.coli genomes.

A phylogenetic analysis of gene families was used to identify the evolutionary relationship of SF with other *Shigella* and *E.coli* strains that have diverged in different countries and regions. SF showed the closest evolutionary relationship with NC_ 004337, despite the presence of 51 InDels and 317 SNPs in SF. We used other *Shigella* species and *E.coli* genomes along with the completed or draft genomes of 53 S.flexneri serotyping strains available at NCBI to establish the phylogenetic relationships among a broader set of isolates (Fig. 1). Analysis of this phylogenetic tree showed that S.flexneri serotypes 1-5 and X, Xv, and Y were virtually monophyletic within Shigella and Shigella spp. originating in the Escherichia genus, as previously reported.[31,32] However, we also found that the evolution of S.flexneri was characterized by national independence.

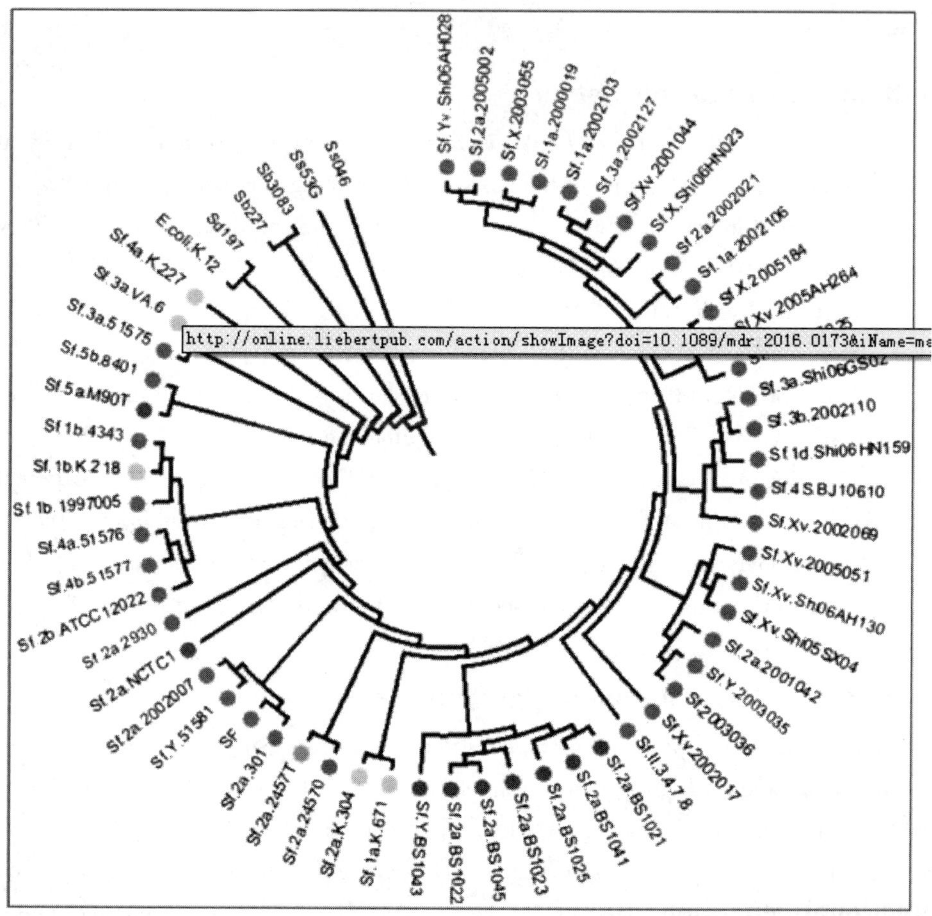

Fig. 1. Phylogenetic analysis of the Shigella flexneri 301 strain genome and reference genomes.

Note: Our phylogenetic tree shows the evolutionary relationships among reference genomes available at NCBI. These strains were isolated in different countries and regions. Red circle, China; purple circle, France; yellow circle, Bangladesh; blue circle, the United States; green circle, Japan. SF, Shigella flexneri 301 strain.

To investigate similarities with other Shigella and E.coli K12 genomes, we compared their basic information and genome features with those of SF. Although the years and areas of these strains were different, they had a similar GC%, ranging from 50.60% to 51.10% (SF 50.87%). However, there was a large difference in SNPs and coverage between SF with the S.flexneri genome and SF with other Shigella and E.coli K12 genomes (Table 1).

Table 1 genome characteristics of reference compared with shigella flexneri

Reference genome	Serotype	Isolation year	Separate area	Length (bp)	GC (%)	Core genes (total genes)	SNPs	Coverage (%)
NCTC1	B (2a)	1915	France	4,526,576	51.00	4,455 (4,567)	459	99.68
2457T	B (2a)	1954	Japan	4,599,354	50.99	4,123 (4,617)	809	99.41
2002017	B (Xv)	2002	China	4,650,856	50.66	4,293 (4,706)	725	97.02
Ⅱ: 3(4), 7 (8)	B	2008	China	4,866,173	50.60	4,312 (4,497)	729	97.24
Sd197	A	1950	China	4,560,911	50.92	3,852 (4,505)	65,674	87.18
Sb227	C	1950	China	4,646,520	51.10	4,044 (4,287)	47,651	88.51
Ss046	D	1950	China	5,055,316	50.76	4,109 (4,475)	47,596	84.42
Escherichia coli K12	—	—	—	4,527,247	50.80	4,091 (4,140)	57,950	82.58

Coverage: similarity between SF and the reference genome.

GC%, guanine-cytosine content; SF, Shigella flexneri 301 strain; SNP, single nucleotide polymorphism.

The genome of SF was similar in amino acid linearsynteny to that of S.flexneri, especially NCTC1 (only an inversion), but there were great differences compared with other Shigella strains and E.coli K12 (Fig. 2 and Supplementary Fig.2).They shared the most (4 455) core genes and a high degree (99.68%) of coverage, but with the least (459) SNPs compared with the S.flexneri 2a NCTC1 (Table 1). However, although there was a large number (3 852 to 4 109) of core genes and a high similarity (84.58% to 87.18%) of coverage, the number of SNPs was tremendous compared with SF with Sd197 (65 674), Sb227 (47 651), and Ss046 (47 596). Unexpectedly, although there were 57 950 SNPs in E.coli K12 compared with SF, they shared 4 091 core genes, 82.58% genome similarity, and good amino acid linear synteny (Table 1 and Supplementary Fig.S2).

3.4 Functional gene annotation

After completion of the analyses described earlier, we annotated the 4 832 genes by using 10 databases that are commonly employed in bioinformatics. Functional annotation was completed through BLAST searches of the genes against different databases, and we ultimately obtained abundant gene annotation information (Table 2).According to an analysis of the annotation results, we identified 82 ARGs, 52 of which encoded MDR efflux pumps (Table 3).This high number of efflux pump genes was organized into five different families: the major facilitator superfamily (MFS), the multidrug and toxic compound extrusion family (MATE), the resistance nodulation cell division family (RND), the small multidrug resistance family (SMR), and the ATP-binding cassette

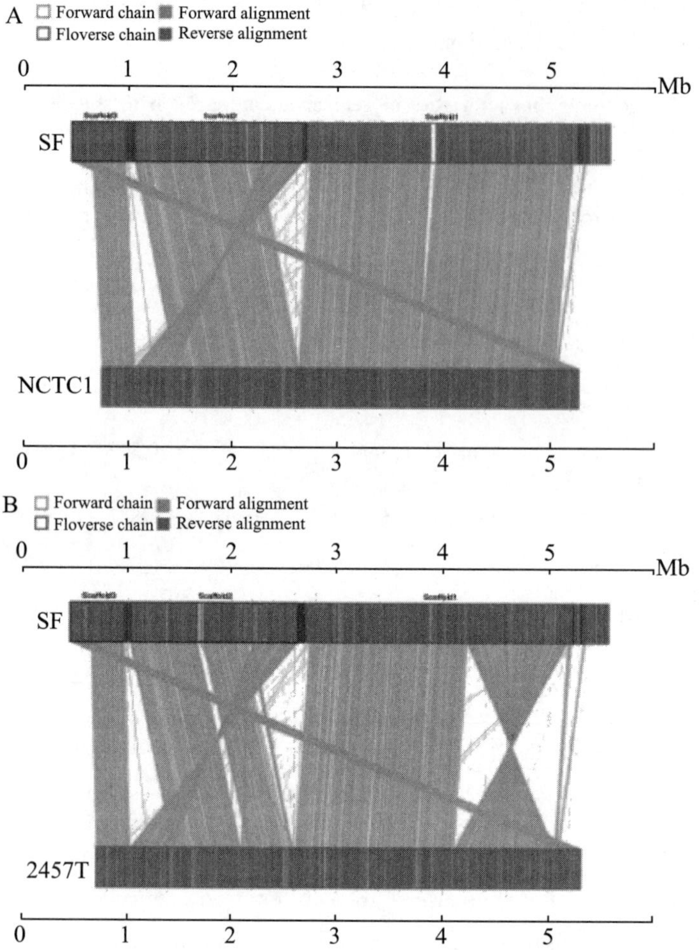

Fig. 2. Linearsynteny analysis of SF and *S.flexneri* 2a reference genomes at the amino acid level.

Note: From top to bottom; the first axis is the first reference sequence; the second and third axes are the reference sequences; and the target sequence is located between the axes. The *yellow* box represents the forward chain, and the *blue* box represents the reverse chain within the axis region. In the box on the axes, the *yellow* region represents the amino acid sequence on the forward chain of this genome sequence, and the *blue* region represents the amino acid sequence on the reverse chain of this genome sequence. (a) Linear synteny analysis of SF with Sf_ 2a_ NCTC1; (b) Linear synteny analysis of SF with Sf_ 2a_ 2457T.

family (ABC).[33,34]

Table 2 results of annotation by each database

Database	Function group number	Gene number
GO		
Biological process	21	6406
Cellular component	8	2902
Molecular function	9	4084

(continued)

Database	Function group number	Gene number
KEGG	33	3265
COG	22	3620
NR	—	4775
Swiss-Prot	—	3939
ARDB	—	20
PHI	—	30
VFDB	—	167
CAZY	—	188
T3SS	—	0

Table 3 complete list of antibiotic resistance genes in shigella flexneri detected by multiple databases

Gene ID	Gene name	Family	Annotation database	System	Function
SFGL000107	emrE	SMR	ARDB/NR/KEGG/COG/Swiss-Prot	Inner/multi-pass membrane protein	MDR efflux pump
SFGL000280	mdtA	RND	NR/KEGG/COG/Swiss-Prot	Inner membrane fusion protein	MDR efflux pump
SFGL000281	mdtB	RND	NR/KEGG/COG/Swiss-Prot/GO	Inner/multi-pass membrane protein	MDR efflux pump
SFGL000283	mdtC	RND	NR/KEGG/COG/Swiss-Prot/GO	Inner/multi-pass membrane protein	MDR efflux pump
SFGL000284	mdtD	MFS	NR/COG/Swiss-Prot/GO	Inner/multi-pass membrane protein	MDR efflux pump
SFGL000362	pbpG	PBP[a]	NR/KEGG/COG/Swiss-Prot/GO	Periplasm	β-Lactam resistance
SFGL000368	mdtQ	RND	NR/COG/Swiss-Prot/GO	Outer-membrane protein	MDR efflux pump
SFGL000420	bcr	MFS	ARDB/NR/KEGG/COG/Swiss-Prot/GO	Inner/multi-pass membrane protein	MDR efflux pump
SFGL000449	yoji	ABC	NR/KEGG/COG/Swiss-Prot	Inner-membrane (ATP-binding) protein	MDR efflux pump
SFGL000470	gyrA	Enzyme	NR/KEGG/COG/Swiss-Prot/GO	DNA gyrase subunit A	QRDR
SFGL000495	arnA	Enzyme	ARDB/KEGG/COG/Swiss-Prot	Formyltransferase	Polymyxin resistance protein

(continued)

Gene ID	Gene name	Family	Annotation database	System	Function
SFGL000599	emrY	MFS	NR/KEGG/COG/ Swiss-Prot	Inner/single-pass membrane protein	MDR efflux pump
SFGL000600	emrK	RND	NR/KEGG/COG/ Swiss-Prot	Inner/single-pass membrane protein	MDR efflux pump
SFGL000692	acrD	RND	NR/COG/Swiss-Prot	Inner/multi-pass membrane protein	Aminoglycoside efflux pump
SFGL000751	pbpC	PBP[a]	NR/KEGG/COG/ CAZy/Swiss-Prot/GO	Inner/single-pass type Ⅱ membrane protein	β-Lactam resistance
SFGL000909	emrA	RND	NR/KEGG/COG/ Swiss-Prot/GO	Inner/single-pass membrane protein	MDR efflux pump
SFGL000910	emrB	MFS	NR/KEGG/COG/ Swiss-Prot/GO	Inner/multi-pass membrane protein	MDR efflux pump
SFGL000964	hosA	HTH[b]	GO	Transcriptional regulator	Multiple antibiotic resistance
SFGL001284	parC	Enzyme	NR/KEGG/COG/ Swiss-Prot/GO	DNA topoisomerase Ⅳ subunit A	QRDR
SFGL001291	parE	Enzyme	NR/KEGG/COG/ Swiss-Prot/GO	DNA topoisomerase Ⅳ subunit B	QRDR
SFGL001296	tolC	RND	ARDB/NR/KEGG/ COG/Swiss-Prot/GO	Outer/multi-pass membrane protein	MDR efflux pump
SFGL001318	bacA	Enzyme	ARDB/COG	Inner/multi-pass membrane protein	Pyrophosphate phosphatase
SFGL001452	dacB	PBP[a]	NR/KEGG/COG/ Swiss-Prot/GO	Periplasm	β-Lactam resistance
SFGL001454	yhbE	DMT[c]	COG/Swiss-Prot/GO	Inner/multi-pass membrane protein	MDR efflux transporter
SFGL001512	aaeA	RND	NR/COG/Swiss-Prot	Inner/single-pass membrane protein	MDR efflux pump
SFGL001642	mrcA	PBP[a]	NR/KEGG/COG/ CAZy/Swiss-Prot/GO	Inner/single-pass type Ⅱ membrane protein	Penicillin resistance
SFGL001688	yhgN	MarC[d]	NR/KEGG/COG/ Swiss-Prot/GO	Inner/multi-pass membrane protein	Multiple antibiotic resistance protein
SFGL001734	yhhJ	ABC	NR/KEGG/COG/ Swiss-Prot/GO	Inner/multi-pass membrane protein	MDR efflux pump
SFGL001735	yhiH	ABC	NR/COG/Swiss-Prot	Inner-membrane (ATP-binding) protein	MDR efflux pump
SFGL001785	mdtE	RND	ARDB/NR/KEGG/ COG/Swiss-Prot/GO	Inner-membrane fusion protein	MDR efflux pump

(continued)

Gene ID	Gene name	Family	Annotation database	System	Function
SFGL001846	mdtF	RND	ARDB/NR/KEGG/COG/Swiss-Prot/GO	Inner/multi-pass membrane protein	MDR efflux pump
SFGL001884	yibH	RND	COG/Swiss-Prot	Inner/multi-pass membrane protein	MDR efflux pump
SFGL002015	mdtL	MFS	ARDB/KEGG/COG/Swiss-Prot	Inner/multi-pass membrane protein	MDR efflux pump
SFGL002027	gyrB	Enzyme	NR/KEGG/COG/Swiss-Prot/GO	DNA gyrase subunit B	QRDR
SFGL002057	emrD	MFS	ARDB/NR/KEGG/COG/Swiss-Prot/GO	Inner/multi-pass membrane protein	MDR efflux pump
SFGL002104	hsrA	MFS	NR/COG/Swiss-Prot/GO	Inner/multi-pass membrane protein	Drug resistance transporter
SFGL002168	rarD	DMT[C]	NR/KEGG/COG/GO	Inner/multi-pass membrane protein	Chloramphenical resistance permease
SFGL002458	mdtP	RND	ARDB/NR/COG/Swiss-Prot/GO	Outer-membrane protein	MDR efflux pump
SFGL002459	mdtO	RND	ARDB/NR/COG/Swiss-Prot	Inner/multi-pass membrane protein	MDR efflux pump
SFGL002460	mdtN	RND	ARDB/NR/COG/Swiss-Prot	Inner-membrane fusion protein	MDR efflux pump
SFGL002480	dinF	MATE	KEGG/COG/GO	Inner/multi-pass membrane protein	MDR efflux pump
SFGL002610	ampG	MFS	NR/KEGG/COG/Swiss-Prot/GO	Inner/multi-pass membrane protein	β-Lactam resistance
SFGL002626	mdlA	ABC	NR/KEGG/COG/Swiss-Prot	Inner-membrane (ATP-binding) protein	MDR efflux pump
SFGL002627	mdlB	ABC	NR/KEGG/COG/Swiss-Prot	Inner-membrane (ATP-binding) protein	MDR efflux pump
SFGL002641	acrB	RND	ARDB/NR/COG/Swiss-Prot/GO	Inner/multi-pass membrane protein	MDR efflux pump
SFGL002642	acrA	RND	ARDB/NR/Swiss-Prot	Inner-membrane fusion protein	MDR efflux pump
SFGL002657	rosB	MFS	ARDB/NR/KEGG/COG/Swiss-Prot/GO	Inner/multi-pass membrane protein	MDR efflux pump
SFGL002658	rosA	MFS	ARDB/KEGG/COG/Swiss-Prot/GO	Inner/multi-pass membrane protein	Fosmidomycin resistance protein
SFGL002747	fepB	ABC	NR/KEGG/COG/Swiss-Prot/GO	Ferrienterobactin-binding periplasmic protein	MDR efflux pump

(continued)

Gene ID	Gene name	Family	Annotation database	System	Function
SFGL002906	mrdA	PBP[a]	NR/KEGG/COG/ Swiss-Prot/GO	Inner/single-pass membrane protein	β-Lactam resistance
SFGL002909	dacA	PBP[a]	NR/KEGG/COG/ Swiss-Prot/GO	Inner/peripheral membrane protein	β-Lactam resistance
SFGL002918	pagP	Enzyme	NR	Lipid Apalmitoyltrans-ferase	Antimicrobial peptide resistance
SFGL003198	mdtK	MATE	ARDB/NR/KEGG/ COG/Swiss-Prot	Inner/multi-pass membrane protein	MDR efflux pump
SFGL003204	ydhC	MFS	NR/COG/Swiss-Prot/GO	Inner/multi-pass membrane protein	MDR efflux pump
SFGL003226	ydhJ	RND	COG	Single - pass mem-brane fusion protein	MDR efflux pump
SFGL003229	slyA	HTH[b]	GO	Transcriptional regu-lator	Multiple antibiotic re-sistance
SFGL003276	mdtJ	SMR	NR/KEGG/COG/ Swiss-Prot/GO	Inner/multi-pass membrane protein	MDR efflux pump
SFGL003277	mdtl	SMR	NR/KEGG/COG/ Swiss-Prot/GO	Inner/multi-pass membrane protein	MDR efflux pump
SFGL003334	marC	MarC[d]	NR/KEGG/COG/ Swiss-Prot/GO	Inner/multi-pass membrane protein	Multiple antibiotic re-sistance protein
SFGL003335	marR	HTH[b]	NR/KEGG/COG/ Swiss-Prot/GO	Transcriptional repressor	Multiple antibiotic re-sistance protein
SFGL003336	marA	HTH[e]	NR/KEGG/COG/ Swiss-Prot/GO	Transcriptional regu-lator	Multiple antibiotic re-sistance protein
SFGL003520	yebQ	MFS	NR/KEGG/COG/ Swiss-Prot/GO	Inner/multi-pass membrane protein	MDR efflux pump
SFGL003522	prc	PBP[a]	NR/KEGG/GO	Tail-specific protease	β-Lactam resistance
SFGL003686	ychE	MarC[d]	KEGG/COG/Swiss-Prot/GO	Multi-pass membrane protein	Multiple antibiotic re-sistance protein
SFGL003877	mdtH	MFS	ARDB/KEGG/COG/ Swiss-Prot	Inner/multi-pass membrane protein	MDR efflux pump
SFGL004045	msbA	ABC	NR/KEGG/COG/ Swiss-Prot/GO	ATP-binding/ permease protein	MDR efflux pump
SFGL004134	macB	ABC	ARDB/NR/KEGG/ COG/Swiss-Prot	ATP-binding/ permease protein	Macrolide - specific efflux pump
SFGL004135	macA	ABC	NR/KEGG/COG/ Swiss-Prot/GO	Inner/single-pass membrane protein	Macrolide - specific efflux pump
SFGL004177	mdfA	MFS	ARDB/NR/KEGG/ COG/Swiss-Prot	Inner/multi-pass membrane protein	MDR efflux pump

(continued)

Gene ID	Gene name	Family	Annotation database	System	Function
SFGL004183	dacC	PBP[a]	NR/KEGG/COG/ Swiss-Prot/GO	Inner/peripheral membrane protein	β-Lactam resistance
SFGL004232	ybhF	ABC	NR/KEGG/COG/ Swiss-Prot	ATP-binding protein	MDR efflux pump
SFGL004233	ybhS	ABC	NR/KEGG/COG/ Swiss-Prot	Inner/multi-pass membrane protein	MDR efflux pump
SFGL004234	ybhR	ABC	NR/KEGG/COG/ Swiss-Prot/GO	Inner/multi-pass membrane protein	MDR efflux pump
SFGL004337	ampH	Enzyme	NR/COG/Swiss-Prot/GO	Inner-membrane protein	β-Lactam resistance
SFGL004441	mrcB	PBP[a]	NR/KEGG/COG/ CAZy/Swiss-Prot/GO	Inner/single-pass type Ⅱ membrane protein	β-Lactam resistance
SFGL004458	yadH	ABC	NR/KEGG/COG/ Swiss-Prot	Inner/multi-pass membrane protein	MDR efflux pump
SFGL004459	yadG	ABC	NR/KEGG/COG/ Swiss-Prot	ATP-binding protein	MDR efflux pump
SFGL004503	ftsl	PBP[a]	NR/KEGG/COG/ Swiss-Prot/GO	Inner/single-pass membrane protein	β-Lactam resistance
SFGL004539	ksgA	Enzyme	ARDB	rRNA small subunit methyltransferase A	Kasugamycin resistance
SFGL004719	Bll_ ec	Enzyme	ARDB/NR/KEGG/ COG/Swiss-Prot/GO	β-Lactamase	β-Lactam resistance
SFGL004721	sugE	SMR	ARDB/KEGG/COG/ Swiss-Prot/GO	Inner/multi-pass membrane protein	MDR efflux pump
SFGL004750	ytfp	DMT[c]	COG/Swiss-Prot/GO	Inner/multi-pass membrane protein	MDR efflux transporter

[a]PBP penicillin-binding protein.

[b]HTH (helix-turn-helix) marR-type DNA-binding domain.

[c]drug/metabolite transporter (dmt) superfamily.

[d]MarC multiple antibiotic resistance protein.

[e]HTH (helix-turn-helix) araC-type DNA-binding domain.

ATP, Adenosine triphosphate; MDR, multidrug resistance; MFS, major facilitator superfamily; QRDR, quinolone resistance-determining region.

Genes encoding various types of enzymes related to drug resistance, such as the Bl1_ ec gene, encoding the enzymeampC, which can hydrolyze beta (β) -lactam antibiotics, and QRDE genes (DNA gyrase: gyrA, gyrB and DNA topoi-somerase: parC, parE) were present in SF, as well as other enzymes mediating bacterial drug resistance in different indirect ways. We also found 10 penicillin-binding proteins, which contribute to β-lactam resistance.

Among the examined databases, ARDB is a special database containing a vast array of ARGs.

Using this database, we identified 20 ARGs encoding enzymes and MDR efflux pumps (Fig. 3). Among these pumps, three efflux pump systems have been reported to interact with tolC, an outer-membrane channel protein, to form tripartite transperiplasmic complexes: macAB-tolC, mdtEF-tolC, and acrAB-tolC, which are responsible for the direct extrusion of aminoglycosides, glycylcyclines, macrolides, β-lactams, fluoroquinolones, and acriflavine from the cell.

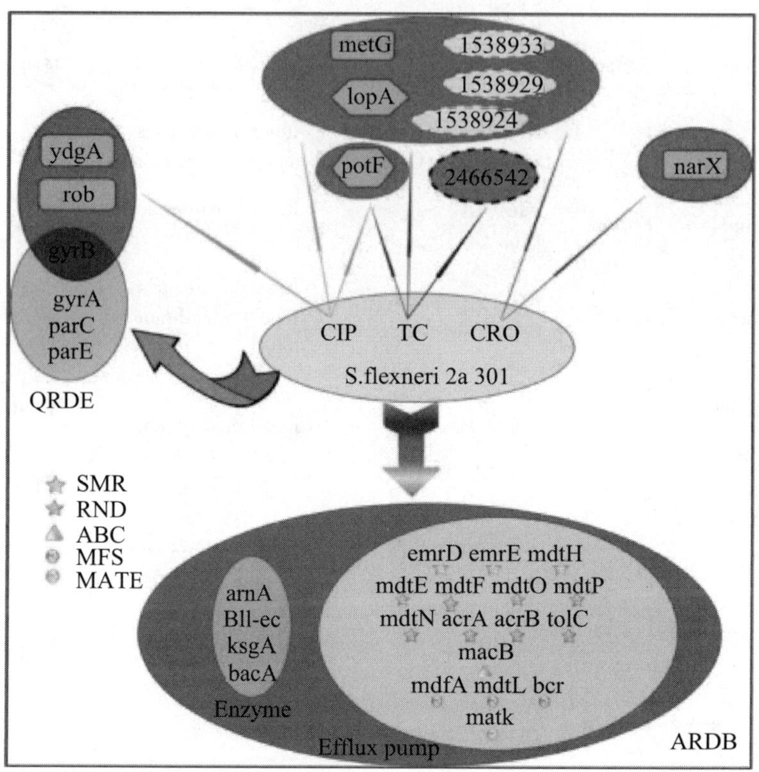

Fig. 3. SNPs and ARGs identified in this study.

Note: This figure mainly includes three groups: (1) The red region at the top represents the SNPs identified by re-sequencing. The rectangles and polygons represent nonsynonymous and synonymous mutations, respectively, and the dotted line represents mutated loci in noncoding regions of genes. Different arrows indicate the mutation caused by each antibiotic. (2) QRDE consists of four resistance genes. DNA gyrase subunit B (gyrB) also exhibits a non-synonymous mutation under the pressure of ciprofloxacin. (3) The chromosomal ARGs are annotated in the ARDB database. ARGs, antibiotic resistance genes; SNP, single nucleotide polymorphism.

Next, we established a statistical map for the ARGs of all reference genomes (Fig. 4) and compared the ARGs that were consistent between SF and other phenotypes. Interestingly, the efflux pump genes acrAB-tolC, metH, mdtK, and mdtL existed in all the reference strains, and all the *S.flexneri* lineages exhibited the bcr gene (except for M90T) but lacked mdtG and mdtM. In addition, there was a significant β-lactamase (bl1_ ec) conferring resistance to penicillin and cephalosporin. This enzyme can break open the β-lactam antibiotic ring and deactivate the molecule's antibacterial properties, and it is widely conserved throughout most of *Shigella*. Some special genes (dfra5, sul1, ctx-m, qnrS, dfra14, tem1, and tetA) detected in only one or two strains are capable of conferring resistance to specific antibiotics.

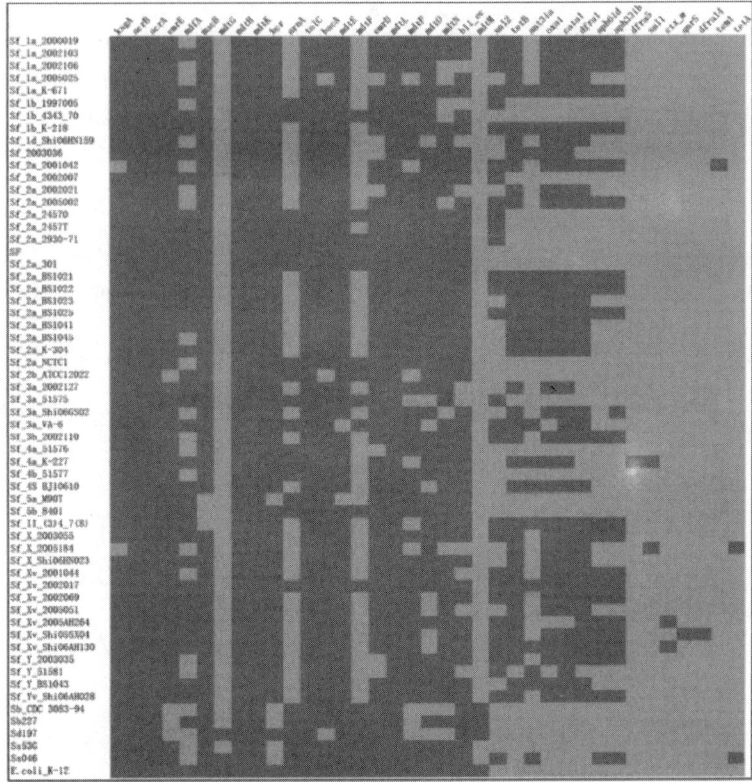

Fig. 4 Statistical chart of ARGorthologs in reference sequences from Shigella and Escherichia coli from the ARDB database.

Note: Red and green represent the presence or absence of the gene in the corresponding genome, respectively.

The chromosomal ARG complements of these strains identified by using ARDB were similar, whereas the ARGs present on plasmids showed large differences.

3.5 Evolution and re-sequencing of induced drug-resistant strains

According to the results of susceptibility testing, we selected three different types of antibiotics (CIP, CRO, and TC) to select directional resistance in S.flexneri 301. The laboratory evolution experiments accumulated ample data related to the genetic changes underlying the multiple resistance phenotypes. Over time, the resistance level increased dramatically, and 30 strains acquired various resistance levels, ranging from sensitive to highly resistant. At the end of the evolution experiments, the MIC values for CIP, CRO, and TC had become 64, 64, and 128mg/mL, respectively (increases of 3,200-, 1,024-, and 128-fold, respectively).

A comparison of the patterns of evolution revealed that resistance to the three antibiotics increased gradually and consistently in the first five generations, then increased in a stepwise fashion in the generations of CIP5 - 7, CIP8 - 9, CRO8 - 10, and TC6 - 7, 8 - 9. Following the characteristics of the evolving dose-response curves, we selected CIP-and TC-induced strain generations 5, 7, and 10 and CRO-induced generations 5, 9, and 10 for re-sequencing (Fig. 5).

Whole-genome re-sequencing of the nine evolved populations identified mutations that were either specific for resistance to a particular drug or shared in resistance to multiple drugs. Through Illumina whole-genome re-sequencing, we obtained a vast amount of high-quality genetic information, with the only exception being CRO-9, which showed a low average sequencing depth (Supplementary Table 1). All sequences have been submitted to NCBI Sequence Read Archive (SRA), and the accession number is SRP106446.

Subsequently, we identified a list of SNPs by comparing the re-sequenced genomes with SF and verified them through Sanger sequencing, with most of the SNPs being confirmed (Fig. 3, Supplementary Fig. 1, and Table 4). Each sequenced strain exhibited two or more SNPs, but we reliably identified only one SNP in each gene.

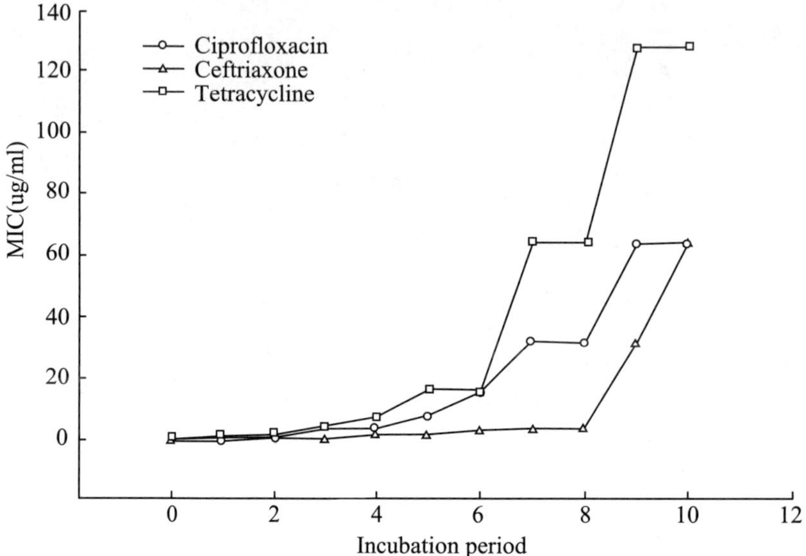

Fig. 5. Gradient antibiotic stress induces high levels of drug resistance in spontaneous adaptive trajectories.

Note: The relationship between the incubation period and MICs is shown. The horizontal coordinates were the number of generations under the action of antibiotics. Strains from the successive evolution experiment were grown in 10 gradient concentrations of the antibiotics CIP, CRO, and TC, and the corresponding MICs were measured. Based on the characteristics of this line graph, the generations indicated by filled points were selected for re-sequencing. CIP, ciprofloxacin; CRO, ceftriaxone; MICs, minimum inhibitory concentrations; TC, tetracycline.

Table 4 shows all 40 verified SNPs, located in eight genes and three intergenic regions. The majority of these gene mutations resulted in nonsynonymous amino acid substitutions (15 of 40) in five genes and synonymous substitutions (7 of 40) in two genes; the other SNPs were located in the ppiA gene promoter (18 of 40). Among these mutated genes, only gyrB of CIP-5, 7, and 10 was directly related to ciprofloxacin resistance. Notably, the same mutations in metG and the promoter of ppiA were found to have occurred in each antibiotic-induced population. The mutated genes rob (CIP-10) and narX (CRO-10) are components of the mar-sox-rob regulon activator and the narX-narL two-component regulatory system, respectively.

Table 4 complete list of single nucleotide polymorphisms and their predicted effect

Populations	Induced drug	Site	SNP	Mutation	Gene name	Gene ID	Ref ID	Function
CIP-5, 7, 10	Ciprofloxacin	284497	T→G	L582R	metG	SFGL000324	Scaffold1	Methionine-tRNA ligase
CRO-5, 10	Ceftriaxone	284497	T→G	L582R	metG	SFGL000324	Scaffold1	Methionine-tRNA ligase
TC-5, 7, 10	Tetracycline	284497	T→G	L582R	metG	SFGL000324	Scaffold1	Methionine-tRNA ligase
CIP-7, 10	Ciprofloxacin	1365696	C→A	A298A	lpoA	SFGL001416	Scaffold1	Penicillin-binding protein activator
CRO-5, 10	Ceftriaxone	1365696	C→A	A298A	lpoA	SFGL001416	Scaffold1	Penicillin-binding protein activator
TC-10	Tetracycline	1365696	C→A	A298A	lpoA	SFGL001416	Scaffold1	Penicillin-binding protein activator
TC-5, 10	Tetracycline	1538924	G→C	Promoter	ppiA promoter	SFGL001608 promoter	Scaffold1	Peptidy1-proly1 cis-trans isomerase A
CIP-10	Ciprofloxacin	1538924	G→C	Promoter	ppiA promoter	SFGL001608 promoter	Scaffold1	Peptidy1-proly1 cis-trans isomerase A
CRO-10	Ciprofloxacin	1538924	G→C	promoter	ppiA promoter	SFGL001608 promoter	Scaffold1	Peptidy1-proly1 cis-trans isomerase A
CIP-5, 10	Ciprofloxacin	1538929	G→A	Promoter	ppiA promoter	SFGL001608 promoter	Scaffold1	Peptidy1-proly1 cis-trans isomerase A
CRO-5, 10	Ceftriaxone	1538929	G→A	Promoter	ppiA promoter	SFGL001608 promoter	Scaffold1	Peptidy1-proly1 cis-trans isomerase A
TC-5, 10	Tetracycline	1538929	G→A	Promoter	ppiA promoter	SFGL001608 promoter	Scaffold1	Peptidy1-proly1 cis-trans isomerase A
CIP-5, 7, 10	Ciprofloxacin	1538933	A→T	Promoter	ppiA promoter	SFGL001608 promoter	Scaffold1	Peptidy1-proly1 cis-trans isomerase A
CRO-5, 10	Ceftriaxone	1538933	A→T	Promoter	ppiA promoter	SFGL001608 promoter	Scaffold1	Peptidy1-proly1 cis-trans isomerase A
TC-5, 7, 10	Tetracycline	1538933	A→T	Promoter	ppiA promoter	SFGL001608 promoter	Scaffold1	Peptidy1-proly1 cis-trans isomerase A
CIP-5, 7, 10	Ciprofloxacin	1953446	C→A	S464Y	gyrB	SFGL002027	Scaffold1	DNA gyrase
CIP-7, 10	Ciprofloxacin	670796	A→C	E378A	ydgA	SFGL003261	Scaffold2	Conserved hypothetical protein
CRO-10	Ceftriaxone	1072783	A→G	K75E	narX	SFGL003701	Scaffold2	Nitrate/nitrite sensor protein
TC-5	Tetracycline	1507145	G→T	G297G	potF	SFGL004164	Scaffold2	Putrescine-binding periplasmic protein
CIP-10	Ciprofloxacin	1507145	G→T	G297G	potF	SFGL004164	Scaffold2	Putrescine-binding periplasmic protein
CIP-10	Ciprofloxacin	335049	G→A	R156H	rob	SFGL004594	Scaffold3	DNA-binding transcriptional activator

4 DISCUSSION

Shigella is one of the major etiologic causes of shigellosis worldwide.There are currently no specific therapies available for this disease, and good hygiene is the best precautionary measure against bacillary dysentery.[35] Despite improvements in public sanitation, outbreaks of shigellosis are still quite serious.[36,37]

Due to the lack of an available vaccine, antimicrobial drugs have been extensively used for the treatment of shigellosis.However, the MDR of Shigella means that treatment is not straight forward, although some regional variation in resistance exists.[35] Mather confirmed that resistance existed in S.flexneri before the appearance and widespread use of antibiotics, and S.flexneri is already a well-adapted pathogen and continues to respond to selection pressures.[38] Although shigellosis is still intractable due to its epidemiological distribution and wide-ranging antibiotic resistance, whole-genome sequencing might provide information that could help researchers develop an effective treatment scheme.

Through abioinformatic analysis, we studied S.flexneri strains ranging from the earliest isolate, NCTC1, to the newly isolated X and Y serotypes, in addition to analyzing *E.coli* and other species of Shigella.Our phylogenetic analysis and linear synteny showed that the genome of Shigella presents high homology and has been relatively stable during the process of evolution.The 301 chromosome is very similar to those of other S.flexneri strains and shares more than 90% of its core genes with reference genomes.In addition, our study suggested that the 301 strain may have evolved directly from NCTC1 and that these strains share a common ancestor in *E.coli*, as previously reported.[35]

Intrinsic ARGs are also highly conserved and could be responsible for the persistent inheritance of antimicrobial resistance in Shigella.[4] Most currently used antibiotics are produced from environmental microorganisms or their derivatives.[39] Therefore, the original resistance genes in environmental bacteria had different ecological purposes, such as detoxification, signal trafficking, or metabolic functions.[40,41]

The intrinsic ARGs in the 301 chromosome are similar to other chromosomal ARGs of *Shigella*, most of which existed before the wi : despread clinical use of antibiotics and the development of MDR.However, the ARGs present on plasmids show clear differences in each reference strain and usually mediate resistance to a single drug or class of drugs. Compared with chromosomal ARGs, plasmid ARGs may cause greater harm because of their transferability and rapid spread from one strain to another, leading tothe dissemination of resistance.[42]

In this study, a panel of chromosomes were discovered in the 301 strain.These ARGs encode enzymes, efflux pumps, and transcriptional regulation factors, conferring the potential ability to resist most types of antimicrobial drugs.Although the 301 strain is sensitive to most modern antibiotics, resistance is likely to speed the development of adaptation to evolutionary pressure.

Accumulation of mutations in genes encoding transporters, targets, and proteins could activate conferred resistance.[41] Our evolution experiments generated important mutations related to drug resistance that allowed us to clarify potential resistance mechanisms.Through a comparison of the mutation patterns associated with the three antibiotics, we found that the bacteria exhibited different measures to cope with sequentially increasing selective pressure. For each of the populations, the L582R mutation in metG, which is located in the tRNA-binding region, was ubiquitous.

MetG encodes methionyl-tRNA synthetase (MetRS), an enzyme essential for viability that functions to covalently link methionine amino acids to their cognate tRNA molecules.[43] MetRS is required not only for elongation in protein synthesis but also for the initiation of all mRNA translation through initiator tRNA (fMet) aminoacylation. A mutation in methionyl-tRNA synthetase (MetG) might result in reduced growth rates relative to the wild type, which would affect sensitivity and susceptibility to antibiotics.[43,44] The bacterial metG gene, or MetRS, is an unsurpassed target that presents unlimited potential for the development of antibacterial agents. Topical research is mainly concentrated on MetRS inhibitors as late-model, effective antibiotics.

Regarding selection pressure from the first-line drug for treating shigellosis, CIP, resistance occurs mainly through mutations in the QRDRs of gyrA and gyrB, encoding DNA gyrase, andparC and parE, encoding topoisomerase. Previous studies have shown that most quinolone resistance in Shigella is mediated by mutations in gyrA (Ser83Leu, As-p87Gly/Asn, or His211Tyr) and parC (Ser80Ile).[45,46] Although the Ser464 mutation in gyrB has been described as being associated with quinolone resistance,[47,48] its prevalence in clinical isolates is very low, and it was discovered in Shigella for the first time.

With the increasing use of CIP, the new mutation R156H in rob (CIP-10) has developed to cope with the pressure to survive. A previous study on the evolution of resistance byToprak et al.[15] revealed another mutation in rob, E262K, that caused a high level of resistance to CHL. The right origin-binding protein rob, which belongs to the AraC/XlyS family, is a component of the so-called marbox (MarA-SoxS-rob) global transcription regulator of MDR.[49] Many bacteria possess a rob gene that responds to drugs by activating the expression of genes encoding proteins that are involved in antibiotic efflux and detoxification through interacting as a monomer with the highly degenerate sequence recognized by MarA and SoxS.

Many multidrug efflux systems, such as the AcrAB-TolC complex, play an important role in the intrinsic resistance of bacteria against antimicrobial compounds and are regulated by marbox.[50] A great deal of evidence shows that rob is involved in the regulation of genes conferring an MDR phenotype in various species of the Enterobacteriaceae family and promotes resistance to antibiotics through up-regulation of the multidrug efflux pump AcrAB-TolC.[50,51] Another principle role of rob is to mediate MDR by up-regulating the expression of micF antisense mRNA, a small inhibitory RNA that downregulates the outer-membrane porin OmpF to limit the uptake of antibiotics such as fluoroquinolones and β-lactams.[52-54]

Similarly, two nonsynonymous mutations found in populations that evolved under CRO inhibition emerged in metG and another regulatory system gene, narX. However, regarding the mechanism underlying the inhibition of bacterial cell wall synthesis, β-lactam antibiotics are bactericidal drugs. The reduction in the growth rate caused by metG mutations may lead to increased function under the threat of CRO. Therefore, the MICs increase slowly and continuously before CRO-8.

Under the pressure of high concentrations of CRO, one member of the two-component system (TCS) of signal transduction, narX, developed a mutation (K75E). Bacterial species have a wide range of TCSs to control numerous physiological behaviors, such as virulence, environment stress responses, biofilm formation, central metabolism, and antibiotic resistance.[55,56]

TCSs play essential roles in prokaryotic organisms to monitor and rapidly deploy physical responses to meet changing environmental conditions. As a typical TCS, thenarX−narL regulator pair consists of two signal transducers. The sensor kinase narX senses certain environmental changes and accordingly modulates the phosphorylation state of the response regulator. The response regulator narL controls the level of gene expression to perform related physiological behaviors.[56,57] Although the mechanism of resistance mediated by narX is unclear, previous research showed that this TCS can confer increased β−lactam antibiotic resistance through the overexpression of the response regulator narL.[58]

Under pressure from TC, only one nonsynonymous mutation appeared. However, it definitely did not represent a separate mechanism of resistance. Intrinsic MDR ARGs may play an important role in resistance.

We cannot ignore the indispensable gene peptidyl − prolyl cis − trans isomerases (ppiA) identified in association with each of the tested drugs. This enzyme has mainly been studied in relation to cancer but also shows crucial functions in bacteria. It facilitates proper protein folding by increasing the rate of transition of proline residues between the cis and trans states, which is a rate−limiting step in protein folding.[59] Although ppiA is not as essential as surA for viability, it influences the growth rate and has substantial effects on antibiotic sensitivity.[60]

In this study, in addition to identifying major resistance potential through whole−genome sequencing, we analyzed the coping mechanisms under each tested antibiotic selective pressure, as previously reported. Regardless of whatever effect is conferred by mutations, these genes contribute to the tolerance of strains to antibiotic threats in various manners. However, this is just the tip of the iceberg regarding bacterial drug resistance mechanisms. With their enormous ability to select resistance to antimicrobial drug stress, bacteria become resistant to all types of antibiotics. Therefore, we can turn our attention to the development of functional gene inhibitors for decreasing the accumulation of resistance ability.

Overall, multiple antibiotic resistance in bacteria has become an inevitable problem that is still poorly understood at the molecular level. Widespread drug resistance is a complex process, involving various types of resistance genes, regulatory mechanisms, and quorum − sensing synergies, resulting in high−level resistance ability. Continued efforts to study the underlying resistance mechanisms, achieve the appropriate use of antibiotics, and develop novel antimicrobials or inhibitors of bacterial survival and metabolic activity are needed to improve treatment strategies and to ameliorate worldwide resistance against frequently used antibiotics.

ACKNOWLEDGMENTS

This work was supported by funds from the National Natural Science Foundation of China (31272603, 31101836). The authors are grateful to theChinese Center for Disease Control and Prevention for providing the bacteria for this study.

AUTHORS' CONTRIBUTIONS

Conceived and designed the experiments: J.Y.Z.and Z.Z.Performed the experiments: Z.Z.Analyzed the data: Z.Z., B.L., and X.Z.Z.Contributed reagents/materials/analysis tools: F.S.C.and J.R.N.Wrote the article: Z.Z.

DISCLOSURE STATEMENT

No competing financial interests exist.

REFERENCES OMITTED

（发表于《MICROBIAL DRUG RESISTANCE》，院选 SCI，IF：2.306）

Acaricidal Activities of the Essential Oil from *Rhododendron nivale* Hook.f.and its main Compund, δ-cadinene against *Psoroptes cuniculi*

Xiao GUO**, Xiaofei SHANG**, Bing LI, Xuzheng ZHOU, Hao WEN, Jiyu ZHANG*

(Key Laboratory of New Animal Drug Project of Gansu Province/Key Laboratory of Veterinary Pharmaceutical Development of Ministry of Agriculture/Lanzhou Institute of Husbandry and Pharmaceutical Sciences, Chinese Academy of Agricultural Sciences, Lanzhou 730050, China)

Abstract: In this paper, the acaricidal activities of *Rhododendron nivale* Hook.f.and its main compound, δ-cadinene were investigated, and the chemical composition of the essential oil was analyzed. The results showed that among aqueous, 70% ethanols, acetic ether, chloroform, petroleum ether and essential oil extracts from the shoots and leaves, the essential oil showed the best in vitro acaricidal activity against adult *P. cuniculi*, which occurred in a concentration-and time-dependent manner.The median lethal time (LT_{50}) values of four concentrations (33. 33-4. 17mg/mL) of the essential oil ranged from 1. 476 to 25. 900h, respectively.After the treatment of P.cuniculi with the essential oil and ivermectin, infected rabbits were free of scabs or secretions in the ear canal by day 20. Then, the percent yield of essential oil from the leaves and shoots was 2. 45% (w/w), which includes 50 compounds.The primary component identified was terpenes, and among of compounds identified from the essential oil of R.nivale the highest relative content was δ-cadinene, which also presented the marked acaricidal activity against *Psoroptes cuniculi* in vitro.These findings provide evidence for the use of acaricides as a traditional medicine and indicate that the essential oil and δ-cadinene could be used to control mites in livestock.

Key words: Rhododendronnivale; Psoroptes cuniculi; Acaricidal activity; Essential oil; δ-Cadinene;

1 INTRODUCTION

Animalacariasis is a parasitic infection that occurs on the body surface or epidermis of animals

* Corresponding author.E-mail: gx_ 139417@ 163. com; infzjy@ sina.com (J.Zhang).

** First co-author.

and is caused by mites of *Sarcoptidae*, *Psoroptidae* and *Demodicidae*.This infection can reduce the production and the quality of animal products and is often fatal (Dagleish et al., 2007).As one pathogen of animal acariasis, *Psoroptes* is a parasite that can be found on sheep, horses, rabbits, goats, cattle, and buffaloes and can induce serious scabs and exudates on an animal's skin when there is a high parasite load.The afflicted animals may stop eating, become severely emaciated and eventually die (Bates, 1999; Walton and Currie, 2007; Qin and Zhang, 2013).

For a long time, organophosphates, organochlorine, pyrethrins, macrolides, and other chemical drugs were widely used to treat and control psoroptic and sarcoptic mange in veterinary clinics (ÓBrien, 1999).Although these drugs led to relatively good treatment outcomes, they have also resulted in increased resistance of the target species, toxicity, environmental hazards and other side effects (Currie et al., 2004; Halley et al., 1993).Now, due to the acaricidal effects in vitro and/or in vivo, many plant-based acaricides those are highly toxic to *Psoroptes* cuniculi and *Sarcoptes scabiei* have been found and developed from *Adonis coerulea*, *Eugenia caryophyllata* and *Eupatorium adenophorum* (Shang et al., 2013; Fichi et al., 2007; Nong et al., 2013).

To control the parasitic diseases of yaks, Tibetan sheep and other animals, a number of plants have been used in veterinary clinics throughout history by local people in the Tibetan region of China.Rhododendron nivale Hook.f.is a perennial evergreen undershrub with a height of 30-120cm, and it is distributed in the northeastern and southeastern areas of the Tibet Autonomous Region of China, Nepal, India, Bhutan and Sikkim (Chinese materia editorial committee, 2002). In China, the shoots and leaves of the plant are used to prevent cough and expel phlegm and to treat some parasitic diseases by people.

In this paper, we evaluated the in vitroacaricidal activity against *P.cuniculi* of aqueous, 70% ethanol extract, acetic ether extract, chloroform extract, petroleum ether extract and the essential oil.Then, the in vivo activity of the essential oil was further studied.In additional, the chemical compounds in the essential oil of R.nivale were analyzed by GC-MS, and the acaricidal activity of δ-cadinene was investigated.

2 MATERIALS AND METHODS

2.1 Plant material

Rhododendronnivale was collected from the Linzhi Region of the Tibet Autonomous Region of China in July 2014, the plant was identified by the School of Resource and Environment, Agricultural and Animal Husbandry College, Tibet University, China.A voucher specimen with accession number ZSY411 was submitted to the Herbarium of the Lanzhou Institute of Animal and Veterinary Pharmaceutics Sciences, Chinese Academy of Agricultural Science (Lanzhou, China).

In the test, the dry shoots and leaves were used to extract the different extracts, and the fresh shoots and leaves were used to extract the essential oil.For preparation of the dry material, shoots and leaves collected from the plant were placed in the drying oven at 30℃ for 5 days.When the shoots and leaves were dried, they were smashed to pieces and stored in refrigerator for 4℃.

2.2 Extraction

The extracts of R.nivale were prepared as follows.Firstly, 100g of dry shoots and leaves were soaked in 1 000mL of 70% ethanol for 12h and extracted at 85℃ for 5h. After filtering with filter, concentrating and evaporating, the extract was dried in vacuum dryer at 60℃. The 70% ethanol extract of R. nivale was obtained, and the percent yield was 45. 259% (The percent yield = [the weight of obtained ethanol extract/the weight of dry material] × 100%). Secondly, the different extracts were obtained after partitioning in separation funnel from the 70% ethanol extract. Detailed, 2 times volume of petroleum ether was added to the ethanol extract solution, and a separatory funnel was used to isolate the petroleum ether extract for 3 times. After comprehensive demixing of the two solvents, the petroleum ether portion was separated. Then, this portion was filtered, concentrated and evaporated, and the petroleum ether extract was obtained. The chloroform extract, the ethyl acetate extract and the aqueous extract of R. nivale were obtained gradually by the same method. The percent yield of the petroleum ether, chloroform, ethyl acetate and aqueous extracts was 5. 78%, 10. 03%, 7. 60% and 15. 53%, respectively.

Finally, the essential oil of the fresh shoots and leaves from R. nivale was extracted. After soaking for 12h with 1 500mL distilled water, the essential oil was obtained from 100g of fresh shoots and leaves of R. nivale by steam distillation using a volatile oil extractor, for 12h. The oil was taken up in diethyl ether, dried using sodium sulfate and stored in vials at 4℃.

2. 3 Analysis of the essential oil

The oil was analyzed by a gas chromatograph−mass spectrometer (Agilent 7890−5975C−GC−MS, USA) with electron impact ionization (70 eV) using a chromatograph equipped with a fused silica capillary column (30m × 0. 25mm, 0. 25μm film thickness). The flow of the carrier gas (helium) was 1mL/min, and the split ratio was 50: 1. The analysis was performed using the following temperature program: oven isotherm at 35℃ for 2min, rising to 80℃ (3min) at a rate of 5℃/min, then rising to 130℃ (2min) at a rate of 8℃/min and to 140℃ (6min) at a rate of 0. 5℃/min, finally rising to 150℃ at a rate of 2℃/min for 2min. The injection volume was 2μL, and the injector temperature was 260℃. The mass range of the MS analysis was 30−550m/z. The identification of compounds was based on a comparison of retention indices and mass spectra to those of standard samples, data from the NIST11 GC−MS library and data reported in the literature. The percentages of compounds were calculated by the area normalization method.

2. 4 Collection of mites

P. cuniculi was isolated from the infested ear cerumen of naturally infected rabbits. First, infested ear cerumen was collected and placed in petri dishes. Then, petri dishes were incubated at 35℃ for 30min. Finally, under a stereomicroscope the adult mites (P. cuniculi) that had eight legs and were apparently healthy with the good movement ability were counted and collected for testing using a needle. After collection of materials, rabbits were immediately treated with ivermectin tablets (Pharmaceutical factory of the Traditional Chinese Veterinary Medicine Institute of the Chinese Academy of Agricultural Science, 2012).

2. 5 Acaricidal activity in vitro

To study and compare theacaricidal activity of the different extracts of R. nivale in vitro, all extracts were diluted to 25mg/mL in 10% glycerin. A 10% glycerin solution was used for the control. In additional, the acaricidal activity of essential oil of R.nivale and the main compound, δ-cadinene (BOC Sciences, USA) also were investigated, they were diluted to 33.33, 16.67, 8.33 and 4.17mg/mL, and 6.25, 3.13 and 1.56μL/mL, respectively.Experiments were conducted according to Shang et al. (2013) and Du et al. (2008).Each extract (100μL) was separately placed into a 24-well cell culture plate.After collecting and placing 10 adult mites into each well, all plates were incubated at 25℃ under 75% relative humidity in an incubator.The number of dead mites was counted every 3h, and the total incubation period was 24h.The status of mites was observed under a microscope, and viability of the mites was checked regularly via stimulation with a needle.Five replicates were performed for each extract.

2.6 Acaricidal activity in vivo

The middle dose (16.67mg/mL) was selected and used to evaluate theacaricidal activity of essential oil in vivo by treating naturally infested New Zealand White rabbits.The experimental procedures were performed as previously described by Guillot and Wright (1981), Fichi et al. (2007) and Nong et al. (2013).Fifteen naturally infested rabbits with the similar ages (80-100 days of ages), weights (2.0kg±0.4kg) and clinical score were examined and randomly divided into three groups.The extent of the scabbing in the external ear canal and the presence or absence of mites was clinically evaluated.There were no significant differences of the clinical score in the right ear between the groups of the rabbits.Prior to treatment, none of the rabbits had been treated with any anti-acariasis drugs or other agents, and no other complicating diseases were observed in the rabbits.

All the treatments were topical.The right ear of the rabbits of one group was treated with 2mL of 16.67mg/mL essential oil diluted by 50% with glycerin, and those of another group were treated with 2mL of 1% ivermectin.The animals of the control group were treated with 2mL of 50% glycerin. These drugs were applied three times at 0, 5 and 10 days.Subsequently, on days 0, 8, 15 and 20 after the beginning of treatment, the right ears of all the rabbits were examined using a stereomicroscope to evaluate the presence of scabs and mites.The scoring system used to evaluate the degree of infestation is presented according to the description of Fichi et al. (2007).Detailed: 0 was absence of scabs and or mites; 0.5 was irritation in ear canal but no mites observed; 1 was small number of scabs in the ear canal, mites present; 2 was external ear canal filled with scabs, mites presents; 3 was scabs in ear canal and proximal 1/4 of pinna, mites present; 4 was 1/2 pinna filled with scabs, mites present; 5 was 3/4 of the pinna filled with scabs, mites present; 6 was all internal surface of the pinna full of scabs, mites present.

Table 1 In vitroacaricidal activity of the five extracts and the essential oil of *Rhododendron nivale* against *Psoroptes cuniculi*.

Group (s)	Concentrations (mg/mL)	Mean mortality* (%) ±SD					
		3h	6h	9h	12h	18h	24h
Aqueous extract	25mg/mL	8.00± 8.37CDb	20.00± 7.07D	22.00± 8.37C	28.00± 8.37C	52.00± 8.37C	68.00± 8.37B

(continued)

Group (s)	Concentrations (mg/mL)	Mean mortality* (%) ±SD					
		3h	6h	9h	12h	18h	24h
75% Ethanol extract	25mg/mL	22.00± 4.47BCa	48.00± 8.37C	56.00± 11.40B	62.00± 8.37B	70.00± 7.07B	72.00± 8.37B
Ethyl acetate extract	25mg/mL	10.00± 7.07CD	50.00± 7.07C	58.00± 8.37B	68.00± 8.37B	76.00± 5.48B	94.00± 5.48A
Chloroform extract	25mg/mL	28.00± 8.37B	78.00± 8.37B	84.00± 8.94Ab	90.00± 7.07A	98.00± 4.47A	100.0± 0.00A
Petroleum ether extract	25mg/mL	78.00± 4.47A	88.00± 8.37AB	98.00± 4.47A	100.0± 0.00A	100.0± 0.00A	100.0± 0.00A
Essential oil	25mg/mL	80.00± 7.07A	98.00± 4.47A	100.0± 0.00Aa	100.0± 0.00A	100.0± 0.00A	100.0± 0.00A
Control	25mg/mL	0.00± 0.00D	0.00± 0.00E	0.00± 0.00D	2.00± 4.47D	0.00± 0.00D	6.00± 5.48C

The difference between data with the different capital letter within a column is $P<0.01$, the difference between data with the different small letter within a column is $P<0.05$. Mean mortality * represent Mean mortality (%) ±SD.

Table 2 Probit regression analysis of toxicity values of the essential oil of *Rhododendron nivale* and δ-cadinene using a CLL model.

Concentration	Regression line	LT_{50} (h) (95%FL*)	Pearson Chi-square
The essential oil of *Rhododendron nivale*			
33.33mg/mL	Y=7.834X-1.325	1.476 (1.289-1.804)	0.431
16.67mg/mL	Y=3.303X-2.493	5.685 (3.753-7.647)	16.716
8.33mg/mL	Y=3.148X-2.873	8.181 (7.198-9.230)	7.679
4.17mg/mL	Y=1.873X-2.647	25.900 (19.981-39.688)	1.102
δ-Cadinene			
6.25μl/mL	Y=4.224X-1.115	1.836 (0.240-4.204)	17.856
3.13μl/mL	Y=1.868X-0.870	2.924 (1.400-4.480)	12.885
1.56μl/mL	Y=1.753X-1.474	6.940 (4.831-9.657)	8.222

95% FL*: 95% Confidence limits for times.

Table 3 In vivoacaricidal activity of the essential oil of *Rhododendron nivale*.

Groups	Day (s)			
	0 (Score±SD)	8 (Score±SD)	15 (Score±SD)	20 (Score±SD)
Essential oil	3.00±0.71	1.60±0.55A	0.50±0.35A	0.00±0.00
1%Ivermectin	3.20±0.84	1.00±0.71 A	0.00±0.00A	0.00±0.00

(continued)

Groups	Day (s)			
	0 (Score±SD)	8 (Score±SD)	15 (Score±SD)	20 (Score±SD)
Control[a]	3. 20±0. 45	3. 20±0. 45B	3. 40±0. 55B	-

The difference between data with the different capital letter within a column is $P<0.01$.

[a] After the observation at day 15, rabbits of control were treated.

2. 7 Statistical analysis

The data obtained were analyzed using SPSS software version 18. 0 and were expressed as the mean±SD.The datas of in vitro test (mean mortality percentage±SD) and in vivo test (mean score ±SD) were analyzed by one-way ANOVA followed by Student's two-tailed t-test for the comparison between test and control groups and by Tukey's test when the data involved three or more groups.P-values less than 0. 05 ($P<0.05$) were considered significant.Regression tests followed probit analysis was used to calculate the median lethal time value (LT_{50}) with the complementary log-log (CLL) model.

3 RESULTS

3. 1 Acaricidal activity in vitro

In this experiment, all extracts showed significant effects against P.cuniculi at all test times and concentrations compared to the control.Of these, essential oil had the best acaricidal activity, and at 9h all mites were killed.The mean mortality at 24h of the aqueous, 70% ethanol, ethyl acetate, chloroform, and petroleum ether extracts and of the essential oil were 68. 00%, 72. 00%, 94. 00%, 100%, 100% and 100%, respectively (Table 1).Meanwhile, the essential oil showed significant acaricidal activity against P.cuniculi at all concentrations.The mean mortalities at 24h of the four concentrations (33. 33, 16. 67, 8. 33 and 4. 17mg/mL), and the LT_{50} values were 1. 476h, 5. 685h, 8. 181h and 25. 900h, respectively (Table 2).

Then, as the main compound, δ-cadinene also presented significant acaricidal activity against P.cuniculi.The mean mortalities at 24h of the three concentrations (6. 25, 3. 13 and 1. 56μL/mL) were 100%, 98% and 88. 00%, respectively. The LT 50 values were 1. 836h, 2. 924h and 6. 940h, respectively (Table 2).

3. 2 Acaricidal activity in vivo

After studying theacaricidal activity of the essential oil in vitro, the topical treatment effects in vivo on infested rabbits was evaluated.The results showed that before treatment, no difference was observed between the three selected groups.After the second treatments with the essential oil or ivermectin, at day 8 the numbers of scabs on the right ears were substantially decreased with significant differences between the treatment groups and the control group.After the third treatments, at day 15 the right ear of the rabbits that were treated with the essential oil had small scabs or secretions in the ear canals, and the group of ivermectin treated rabbits were free of scab.Finally, the rabbits in all of the treatment groups showed a positive mental and physical status with the normal movement ability, good appetite for food and rabbit fur or skin at day 20, but the status of rabbits in the untreated control group declined and they became emaciated (Table 3).

3.3 Analysis of the essential oil

In this study, 2.45% (w/w) essential oil was obtained from the leaves and shoots of R. nivale. In total, 50 compounds, comprising 54 X. Guo et al./ Veterinary Parasitology 236 (2017) 51–54, 87.28% of the oil, were identified by GC–MS. The major compound identified was δ–cadinene, which was 17.17% of the oil, followed by T–muurolol (7.85%), δ–eudesmol (7.79%), δ–eudesmol (6.15%) and T–cadinol (5.20%).

4 DISCUSSION

Currently, screening and finding newacaricidal medicine from natural plants has attracted the interest of many people. Many papers report that certain plants or their extracts have marked acaricidal activity, such as Azadirachta indica, Eugenia caryophyllata and Eupatorium adenphorum (Du et al., 2008; Fichi et al., 2007; Nong et al., 2013). In our field of investigation of folk veterinary medicine, we found that the aromatic leaves of R. nivale were used to treat some parasitic diseases in regions of the Tibet Autonomous Region of China. In this paper, the in vitro and in vivo acaricidal activities against P.cuniculi were tested.

First, to screen the active compositions of R.nivale in vitro, the acaricidal activities of the five extracts and the essential oil from the plant were studied and compared. The results demonstrated that among all extracts, the essential oil presented the best acaricidal activity, and all mites were killed within 9h. This result indicated that the active ingredients of R.nivale could be in low polarity compositions, particularly the essential oil.

Then, the in vitroacaricidal activity of the essential oil from *R. nivale* against *P. cuniculi* was evaluated. The results showed that it exhibited significant acaricidal activity against *P. cuniculi* in a dose–and time–dependent manner (Table 2). To further study the acaricidal activity of the essential oil in vivo, we investigated the effects of topical treatment on infested rabbits. After the third treatments, the rabbits in all of the treatment groups at day 20 were free of scabs and mites and showed a favorable mental and physical status, compared to the control group rabbits (Table 3). Thus, we concluded that the essential oil has significant acaricidal activity both in vitro and in vivo and could be widely applied as a potential acaricidal agent for the treatment of acariasis in animals.

Finally, the primary component identified in GC–MS wasterpenes, and among of compounds identified from the essential oil of R.nivale the highest relative content was δ–cadinene. The result of the further test showed that δ–cadinene has the significant acaricidal activity against *P.cuniculi* (Table 2). So, the active ingredient of the essential oil may be terpenes and that the bioactivity of δ–cadinene and other predominant compounds should be investigated further. And the synergistic action and the structure–activity relationships associated with acaricidal activity should be studied further.

5 CONCLUSION

In short, these preliminary tests demonstrated that the essential oil of R. nivale and its main compound, δ–cadinene have significant acaricidal activity against *P.cuniculi*. As an alternative medicine for commercial acaricidal agents, they could be used to control animal acariasis, especially for infestations that are ivermectin-resistant.Further studies should be carried out to isolate more active acaricidal compounds from the essential oil and to develop a new potential medication to control mites in livestock.In addition, the acaricidal mechanism is still unclear and should be studied further.

COMPETING INTERESTS

All authors declare that they have no competing interests.

ACKNOWLEDGEMENTS

This work was financed by the National Science and Technology Infrastructure Program of China (2015BAD11B01) and China Agricultural Research System (CARS-38).The authors would also like to express their gratitude to Dr.Allan Grey at Lanzhou University for editing this paper for English language use.

APPENDIX A.SUPPLEMENTARY DATA

Supplementary data associated with this article can be found, in the online version, at http://dx.doi.org/10.1016/j.vetpar.2017.01.028.

REFERENCES OMITTED

(发表于《Veterinary Parasitology》，院选 SCI，IF：2.242)

Effects of Biotic Andabiotic Factors on Soil Organic Carbon in Semi-arid Grassland

Fu-Ping TIAN[1], Zhi-Nan ZHANG[2], Xiao-Feng CHANG[1,2], Lei SUN[3], Xue-Hong WEI[3], Gao-Lin WU[2,3] *

(1. Lanzhou Scientific Observation and Experiment Field Station of Ministry of Agriculture for Ecological System in the Loess Plateau Area/Lanzhou Institute of Animal and Veterinary Pharmaceutics Sciences, Chinese Academy of Agricultural Sciences, Lanzhou, Gansu 730050, China; 2. State Key Laboratory of Soil Erosion and Dryland Farming on the Loess Plateau, Institute of Soil and Water Conservation, Chinese Academy of Sciences and Ministry of Water Resources, Yangling, Shaanxi 712100, China; 3. Agricultural and Animal Husbandry College of Tibet University, Nyingchi 860000, China)

Abstract: Carbon sequestration in grassland soil has been paid considerable attention in recent decades. However, the changes of soil organic carbon (SOC) still need clarification under the effect of environmental factors in semi-arid area. Here, twenty sampling sites were selected to study the effects of plant community (plant cover, biomass, litter, composition and diversity) and environmental factors (i. e. mean annual precipitation, mean annual temperature) on SOC sequestration in the semi-arid grasslands. The results showed that SOC was significant positively related to mean annual precipitation, soil water content and soil pH. The higher above-and below-ground biomass, species evenness and diversity presented the higher SOC. Specially, the species richness and proportion of gramineous species functional group significantly increased SOC. Below-ground biomass affected SOC mainly in the top 30cm soil. Our results suggest that higher plant species richness and gramineous species proportion play a positive role in increasing the potential of soil carbon sequestration in semi-arid grassland.

1 INTRODUCTION

Soil carbon (C) is the largest C pool in terrestrial biosphere, which stored more C than that contained in plants and the atmosphere (Jobbógy and Jackson, 2000). Even a slight change in the amount of soil C may dramatically influence atmospheric CO_2 concentration. Thus, soil carbon pays a

* Corresponding author, E-mail: gaolinwu@ gmail.com

critical role in global C cycling (Shi et al., 2012; Song et al., 2012).The climate change could profoundly affect vegetation, soil and the carbon cycle of terrestrial ecosystem (Jobbágy and Jackson, 2000).

As the important form of human managing and utilizing grassland, cultivation and abandonment significantly affected soil organic carbon cycles (Shang et al., 2012; 2014).So it is necessary to determine more effective strategies for land use and vegetation management to ameliorate the rising level of atmospheric CO_2 (Yang et al., 2015).

Changes in environmental factors can greatly affect C and nitrogen (N) cycling in terrestrial ecosystems (Allison and Treseder, 2008; Liu et al., 2009).Previous studies have documented that soil organic carbon (SOC) content would increase or stay constant with precipitation change (Song et al., 2012; Zhou et al., 2009).Grassland represents an important global C reservoir, and stores as much as 20% of global soil C (Jobbágy and Jackson, 2000).Thus, the changes of soil C in grassland will be a significant component of the global soil C feedback to plant and climate change (Jobbágy and Jackson, 2000; Song et al., 2012).However, there were few field experimental evidences on the responses of SOC to community composition and abiotic factors in the natural grassland (Fornara and Tilman, 2008).

Plant-soil feedbacks play an important role in the maintenance of both plant community structures and soil properties (Bezemer et al., 2006). Changes in plant composition can alter soil organic matter, soil nutrient availability, and the composition of soil microbial communities (Casper and Castelli, 2007; Fornara and Tilman, 2008; Shang et al., 2014).These influences on soil chemical properties result in either net positive or negative effects, associated with plant performance and plant community composition (Kardol et al., 2006; Wu et al., 2016).Additionally, some previous studies found that climate change influenced plant growth and soil microbial activities (Bai et al., 2010; Song et al., 2012).However, these studies only focus on one or two aspects among climate change, plant and soil property.As the best of our knowledge, the relationships between plant community and soil property under the condition of climate change are still poorly understood.So it is significant for us to understand the relationship among the three factors, which contribute to the ecological systems.

The Chinese government has implemented the most ambitious ecological program titled "Grain-for-Green" project (converting degraded, marginal land and cropland into grassland, shrubland and forest).The large scale of project can enhance C sequestration capacity in China, especially in arid and semi-arid areas.Our aims are offering a new perspective in the study of SOC response to plant-soil interface microhabitats (biotic and abiotic factors) in semi-arid grassland.

The specific objectives of this study are that (1) to find the role of (biotic and abiotic factors on SOC in the plant-soil interface microhabitats; (2) understanding the relationship between plant community composition and abiotic factors and SOC in typical semi-arid temperate grasslands

2 MATERIALS AND METHODS

2.1 Study area

This study was conducted at Gansu Province, China.The region is characterized by semi-arid climate, mean annual temperature (MAT) ranges from 6. 73 to 9. 57℃ and mean annual precipitation (MAP) ranges from 183. 81 to 387. 77mm.75% of the annual rainfall in this region is concentrated in July-September. We selected 20 sites (longitude 103°24′-54′E, latitude 35°15′-22′ N, and altitude from 1 434 to 2 523m) (Figure 1) for soil and plant community sampling (Wu et al., 2014), and the soil pH value ranges from 7. 70 to 9. 06. The main species in study sites include Labiatae (Thymus mongolicus), Gramineae (Stipa bungeana, Stipa grandis and Agropyron cristatum), Compositae (Heteropappus altaicus, Artemisia gmelinii, Artemisia ordosica, Artemisia frigida and Peganum multisectum).

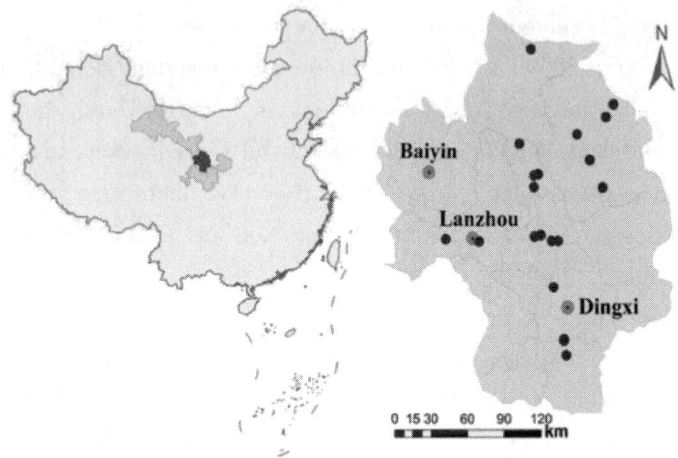

Figure 1. Location of sampling sites used in this study.

2. 2 Experimental design

At each site, one plot (20 m × 20 m) was set up randomly.Tenquadrats (1 m × 1 m) were emplaced in diagonal line for community survey.Five plots were chosen to harvest all plants [above-ground biomass (AGB) and litter], the other five plots were collected plant species (plant cover, name, height, number). Three samples were taken with a 9cm diameter corer to sample belowground biomass at a soil depth of 0-10, 10-20, 20-30, 30-50, 50-70 and 70-100cm in each quadrat.The samples in same layers were mixed together to make one sample.The majority of the roots were found in the soil samples thus obtained and then isolated using a 2mm sieve.The remaining fine roots taken from the soil samples were isolated by spreading the samples in shallow trays, overfilling the trays with water and allowing the outflow from the trays to pass through a 0. 5mm mesh sieve.All the isolated roots were oven-dried at 65℃ and weighed to within 0. 01g.Five replicate (four corners and center of the quadrat) soil samples were taken with a 6cm diameter soil corer, and the depths were consistent with belowground biomass.All soil samples were air-dried and then passed through a 2mm sieve.The soil bulk density ($g \cdot cm^{-3}$) was measured using a soil bulk sampler with a 5cm diameter and 5cm high stainless steel cutting ring.The original volume of each soil core and its dry mass after oven-drying at 105℃ were measured.

Species richness (SR) represents the numbers of species in eachquadrat.All the species in the quadrats were divided into two functional groups (Allen et al., 2011): grass (grass family) and forbs (any herbaceous, dicotyledonous broad-leaved plant). The composition functional groups represent the ecological structure which contributed to the predictions of species assemblages at a more practicable and more general level, compared with the individual species. Shannon-Wiener index (H) and Evenness index (E) were calculated as:

Richness index (R): R=S;

$$H = -\sum_{i=1}^{..} (Pi \ln Pi)$$

Shannon-Wiener diversity index (H):

Evenness index (E):

$$E = \frac{H}{\ln S}$$

where S is the total species numbers of grassland community, H is the Shannon-Wiener diversity index and Pi is the density proportion of i species.

2.3 Determination and relative calculation

Soil water content was measured gravimetrically and expressed as a percentage of soil water to dry soil weight. Soil bulk density was calculated depending on the inner diameter of the core sampler, sampling depth and the oven-dried weight of the composite soil samples. Soil pH was measured with a glass electrode. When measuring the soil pH, soil samples were diluted with water (the ratio of soil to water was 1 : 2.5). SOC was measured by the $K_2Cr_2O_7$ method.

Calculation of SOC:

$$SOC = \frac{BD \times SC \times D}{10}$$

SOC is SOC storages ($Mg \cdot ha^{-1}$); BD is soil bulk density ($g \cdot cm^{-3}$); SC is soil organic carbon concentration ($g \cdot kg^{-1}$); and D is soil thickness (cm).

Climate data were obtained based on the 1957-2005 climate data (average) of 4 weather stations in the sampling regions. All the meteorological data were obtained from the China Meteorological Data Sharing Service system.

2.4 Data analysis

Correlation analyses were used to determine the relationship between the meteorological features (i.e. mean annual precipitation, mean annual temperature) and plant community (cover, above-ground biomass, litter, proportion of gramineous, Shannon-Wiener index and species richness), as well as the relationship between below-ground biomass and SOC. Simple linear regression was used to account for the effects of environmental factors and plant community on below-ground biomass in 0-30cm soil-depth. Principal Component Analysis (PCA) was applied to eliminate any redundancy among the environmental factors, biological factors and soil property variables. A variable was significantly correlated with a PC when the correlation (its loading value) with this component was at least 0.70. A PC was considered significant when its eigenvalue was at least 1. This stage allows us to determine the important factors in each component which affect the SOC.

To determine the specific variable impacting SOC, we conducted multiple linear stepwise regression analysis at the last stage. All statistical analysis was conducted by SPSS, ver. 18.00 (SPSS Inc., Chicago, 2009).

3 RESULTS

3.1 Effects of environmental factors on plant community

Correlation results showed that MAP positively and significantly affected plant community (Table 1, $P<0.01$); MAT negatively and significantly affected Cover, AGB, POG, H and SR ($P<0.05$) except the litter ($P>0.05$). SR is significant correlation with MAP and MAT (Table 1, $P<0.01$).

Linear regressionof BGB versus nine possible predictive variables showed a possible relationship with MAP, MAT, Cover, AGB, H and SR and no apparent relationships with Litter, POG and E (Figure 2). In addition, MAP, Cover, AGB, H and SR were significantly and positively correlated with BGB (Figure 2a, c, d, e, g, i, respectively). In contrast, BGB decreased with the increased MAT which can explain 18% of the variation of BGB (Figure 2b, $R^2=0.18$). BGB was mainly affected by MAP and SR ($R^2=0.30$, 0.20, respectively).

Multiple regression analysis, using as predictor variables in the environmental factors and plant community but except Litter, POG and E, showed that the MAP and H had significantly positive effects on BGB ($F=24.62$, $R^2=0.34$, $P<0.01$). The average of SOC (1 m soil depth) was (7.09±0.49) (mean±SE) kg · m^{-2} and plant BGB was (1.20±0.093) (mean±SE) kg · m^{-2} in the twenty sites (Figure 3). SOC showed a positive and significant relationship with BGB, especially in 0–30cm soil depth ($R=0.56$, $P<0.01$). There was no relationship between BGB and SOC in the 30–100cm soil depth ($R=0.073$, $P>0.05$).

Table 1. The relationship of meteorological features (mean annual precipitation (MAP, mm), mean annual temperature (MAT ℃,) and plant community (cover, %), above-ground biomass (AGB, g · m^{-2}), litter (g · m^{-2}) the proportion of gramineous species (POG, %), Shannon-wiener index (H), and species richness (SR)).

Abiotic factors	Characteristic of plant community					
	Cover	AGB	Litter	POG	H	SR
MAP	0.53**	0.34**	0.32**	0.33**	0.37**	0.58**
MAT	−0.21*	−0.27**	−0.056	−0.37**	−0.42**	−0.60**
SR	0.54**	0.65**	0.27	0.22	0.62**	1

3.2 Effect of environmental factors on SOC

The PCA results showed that the first two PCs (Comp. 1 and Comp. 2) explained 37.34% and 14.41% of the variance in the twenty sites data (Table 2).

As shown in Figure 4, SOC significantly correlated with environmental factors and biological factors ($P<0.05$). MAT, R/S ratio, BD and pH were negatively correlated with SOC. Cover, AGB, BGB, Litter, POG, H, E, SR, SWC, MAP were positively correlated with SOC, and

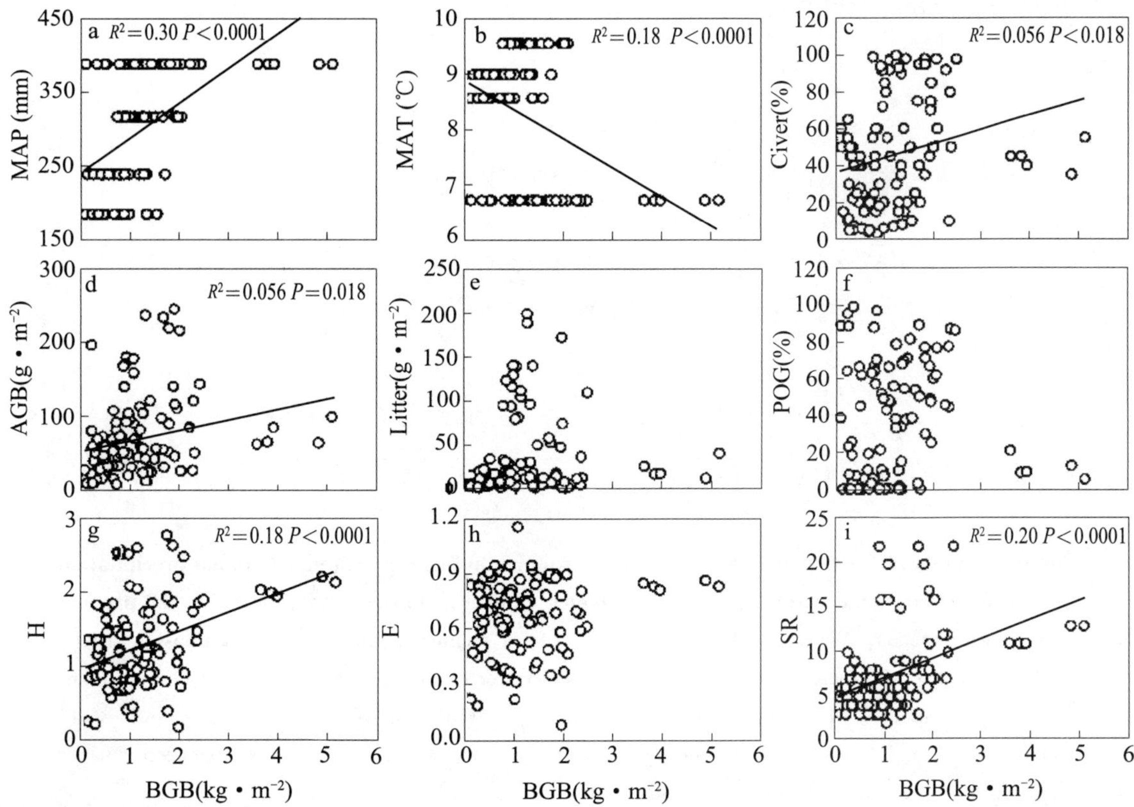

Figure 2. Linear relationship of mean annual precipitation (MAP, a), mean annual temperature (MAT, b), Cover (c), above-ground biomass (AGB, d), Litter (e), the proportion of gramineous (POG, f), Shannon-Wiener index (H, g), Evenness index (E, f) and species richness (SR, i) with below-ground biomass (BGB).

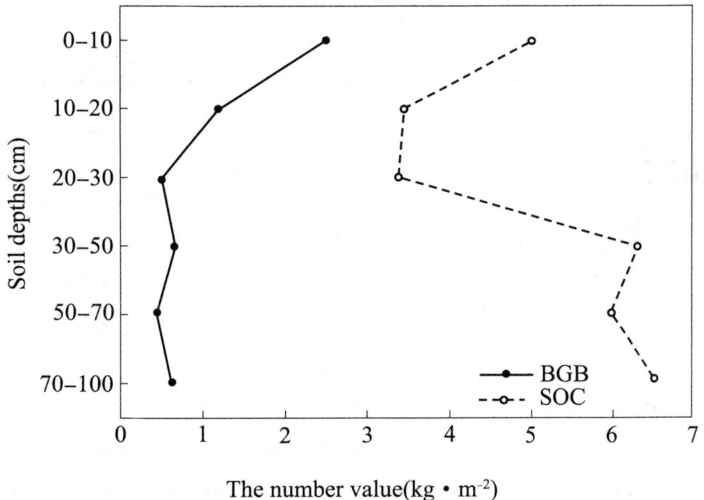

Figure 3. Changes between soil organic carbon (SOC, kg · m^{-2}) and plant below-ground biomass (BGB, kg · m^{-2}) at different soil depths in the semi-arid grassland.

SOC showed stronger positively with MAP, SR and SWC.

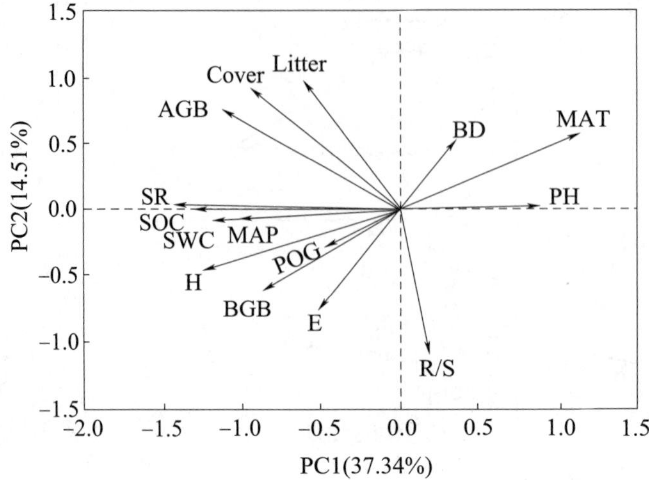

Figure 4. Biplot of the first two PCA axes of environmental factors (e.g.mean annual precipitation and mean annual temperature), biological factors (Cover, above-ground biomass, the proportion of gramineous, Shannon-Wiener index, species richness, belowground biomass, Litter, R/S ratio) and soil factors (soil pH value and soil water content). Abbreviations: MAT: mean annual temperature, MAP: mean annual precipitation, AGB: above - ground biomass, BGB: belowground biomass, POG: the proportion of gramineous, E: Evenness index, H: Shannon-Wiener index, SWC: soil water content, SOC: soil organic carbon.

Table 2. Eigenvalues and contribution on of principal component analysis (PCA).

Component	Eigenvalues	Contribution rate (%)	Cumulative contribution rate (%)
1	5. 60	37. 34	37. 34
2	2. 16	14. 41	51. 74
3	1. 53	10. 18	61. 93
4	1. 45	9. 68	71. 61

4 DISCUSSION

SOC plays a critical role in controlling soil fertility and cropping system productivity and sustainability (Liu et al., 2016a).The changes in SOC were influenced by both biotic and abiotic factors whether in the agriculture, forest or grassland (Song et al., 2012; Zhou et al., 2013; Merino et al., 2015).In our study, the main abiotic factors influencing SOC were MAP and SWC, and the main biotic factors were BGB and SR. MAP and SWC changed the environment of plant growing, but the BGB and SR changed the competition of plant species in community.Greater plant species number was associated with soil C accumulation in N - limited grassland communities (Tilman et al., 2006). Moreover, higher species diversity occurs in more fertile habitats without fertilizer disturbance in an alpine natural grassland community (Wu et al., 2014).

Our results showed that plant species richness had a positive relationship with soil organic carbon in the semi-arid grassland. High species richness enhanced primary productivity (Wu et al., 2011), and above-ground plant resources will be returned to the soil through decomposing in litter layers (Bardgett and Wardle, 2003). The amount of plant residuals returned to the soil affected the SOC. Below-ground biomass increased with plant species richness. Our results showed that the SOC was positively associated with root biomass in the 0-30cm soil layer. The vertical distribution of SOC had a slightly stronger association with vegetation (Jobbágy and Jackson, 2000), and the root distributions affect the vertical placement of C in the soil. Soil C and N storage tended to increase at higher species diversity due to greater root biomass accumulation in the soil top 20cm (Fornara and Tilman, 2008). The higher species richness promote greater root biomass accumulation. The increase in the C input from fine root enhanced organic matter protection, which also promotes SOC accumulation in the surface soil (Liu et al., 2016b). Plants regulate SOC by controlling, assimilating and accumulating C in the plant root system and then release from soil to atmosphere through respiration and leaching. Below-ground biomass coupling with a simultaneous change in the rooting pattern would be expected to alter rhizosphere oxygen concentrations and anaerobic microbial processes (Bouchard et al., 2007). However, in the 30-100cm soil layer, with the decrease of root biomass, there is no significant correlation was showed between BGB and SOC. There are significantly positive correlations between BGB, SWC, BGB and SWC in the soil of 0-10 and 10-20cm. So higher soil water promoted more root accumulation in the top soil. BGB was not associated with SWC in the soil layer of 20-100cm. In this case, it is not good for root growth, and BGB was not good for SOC accumulation in the deeper soil (30-100cm).

In addition, the composition of plant functional group (i.e. POG) plays an important role in affecting SOC. Meanwhile, Wu et al. (2011) has documented positive association of gramineous grass with soil nutrient properties in alpine meadow grassland. Differences in plant functional groups can create different soil biotic and abiotic microhabitats by affecting soil organic matter, soil nutrient availability, and soil microbial communities (Shang et al., 2012, 2014; Casper and Castelli, 2007). Gramineous species had higher productivity, which can result in greater amounts of nutrients fixed within their tissues (Harris et al., 2007), and a relatively higher proportion of belowground biomass (Wu et al., 2016). Gramineous species had higher belowground biomass, which is related to higher litter decomposition rate and soil nitrogen availability than that in the forbs species. Furthermore, fine roots of C 4 and C 3 grasses (graminoids) can contribute to an increase in the soil organic matter pool (Fornara et al., 2009).

There was a significant negative correlation between soil pH and SOC. It is well known that acid deposition is a major problem to influence the ecological system (Guo et al., 2010). Acid rain causes acidification of the soil surface, and reducing soil pH value. Many studies showed that the soil acidification affected on the cropland and forest (Guo et al., 2010), but few studies reported the effects of soil acidification in grassland (Shi et al., 2012). Our study found SOC increased with decreased in soil pH. This is inconsistent with the report that acidification led to a decrease in SIC, carbon stock in the topsoil would not necessarily decline with soil acidification (Guo et al., 2010). It may be partly because of that (1) acidification inhibits soil microbial activities and thus reduces the SOC decomposition rate; (2) N deposition, a major cause of acidification, will lead to an increase of SOC inputs with increasing vegetation productivity (Neff et al., 2002), which induce a decrease in microbial biomass and oxidase activity (Zak et al., 2008; Dalmonech et al., 2010). Therefore, low soil pH value may promote SOC accumulation in the semi-arid grassland.

5 CONCLUSIONS

We examined the effects of biotic andabiotic factors on SOC in semi-arid grassland.The results showed that MAT negatively affected on SOC and MAP positively affected on SOC.SR and the composition of plant functional group significantly affected SOC.Soil pH was negatively associated with SOC.Below-ground biomass affected SOC mainly in the soil top 30cm.Our findings indicated that the main biotic factors affecting SOC are SR and belowground biomass, and the main abiotic factors affecting SOC are MAT, MAP and soil pH in the semi - arid grassland. We suggested that establishment of grasses-dominated artificial grassland could improve the potential sequestration of SOC in semi-arid grassland ecosystem.

ACKNOWLEDGEMENTS

This research was funded by the National Natural Science Foundation of China (31372368, 41371282), the Youth Innovation Promotion Association CAS (2011288), the "Light of West China" Program of CAS (XAB2015A04), Lanzhou Institute of Animal and Veterinary Pharmaceutics Sciences of Chinese Academy of Agricultural Sciences (CAAS-ASTIP-2014-LIHPS-08).

The authors have declared no conflict of interest.

REFERENCES OMITTED

（发表于《Journal of Soil Science and Plant Nutrition》，院选 SCI，IF：1.600）

Trace Elements may be Responsible for Medicinal Effects of Saussurea Laniceps, Saussurea Involucrate, Lycium barbarum and Lycium Ruthenicum

Hui WANG, Shengyi WANG, Dongan CUI, Xia LI, Xin TUO, Yongming LIU *

(Engineering and Technology Research Center of Traditional Chinese Veterinary Medicine of Gansu Province/Key Lab of New Animal Drug Project of Gansu Province/Key Lab of Veterinary Pharmaceutical Development of Ministry of Agriculture/Lanzhou Institute of Husbandry and Pharmaceutical Sciences of Chinese Academy of Agricultural Sciences, Lanzhou 730050, Gansu, China)

Abstract: [Background]: The pharmacodynamics of *Saussurea laniceps*, *Saussurea involucrata*, *Lycium barbarum* and *Lycium ruthenicum* have been researched, and trace elements have been considered as the essential elements, but little attention has been paid to the trace elements of the herbal medicine. We would like to report on the content of copper (Cu), manganese (Mn), iron (Fe), zinc (Zn) and selenium (Se) levels in the four herbal medicines.

[Materials and Methods]: A total of 20 whole plant materials were collected of each species inChina. The content of Cu, Mn, Fe and Zn in the dried aerial parts was estimated by the standard atomic absorption spectrophotometry. The level of Se was detected using hydride generation atomic fluorescence spectrometry.

Results: The mean concentrations of Cu, Mn, Fe, Zn and Se in S. laniceps were (7.758±0.924) μg/g, (201.3±16.24) μg/g, (222.7±35.10) μg/g, (18.48±2.913) μg/g and (1.42±0.16) μg/g, respectively; S. involucrata were (19.56±2.20) μg/g, (88.75±8.53) μg/g, (812.7±126.9) μg/g, (34.85±3.81) μg/g and (1.04±0.05) μg/g, respectively; L. barbarum were (10.83±0.26) μg/g, (9.598±0.32) μg/g, (55.65±3.83) μg/g, (11.92±0.27) μg/g and (11.84±0.59) μg/kg, respectively; L. ruthenicum were (12.67±0.39) μg/g, (13.78±1.13) μg/g, (98.04±5.03) μg/g, (14.46±1.27) μg/g and (35.12±2.34) μg/kg, respectively.

Conclusion: This study provided the trace elements content of Cu, Mn, Fe, Zn and Se in the four herbal medicines. The trace elements are maybe other functional compounds for medicinal effects. Deep relationship between pharmacological and trace elements contents, especially its

* Corresponding author, E-mail: myslym@ sina.com.

mechanism of action should be future research.

Key words: *Saussurea laniceps*; *Saussurea involucrata*; *Lycium barbarum*; *Lycium ruthenicum*; trace elements content; traditional Chinese Medicine

1 INTRODUCTION

Related medicinal plants from the same family or genus have been and are being used for similar therapeutic purposes in various Chinese Medicines (Yi et al., 2010).— Snow lotus || flowers are famous Chinese Herbal Medicine (CHM) for the treatment of rheumatoid arthritis, stomachache and dysmenorrhea in traditional Chinese Medicine (TCM), Tibetan medicine, Uyghur medicine, Mongolian medicine and Kazakhstan medicine (Chinese Pharmacopoeia Commission, 2010; Yi et al., 2010, Chik et al., 2015; Chen et al., 2016).*Saussurea involucrata* and *Saussurea laniceps* are two species of "snow lotus".*S. involucrata* is Composite family and grows in the mountains at heights of 4 000–5 000m in Xinjiang areas in China.S.laniceps mainly distributed in the Qinghai–Tibet plateau at heights of 3 500–5 300m.The two snow lotuses have different macroscopic and microscopic features (Chen et al., 2014).The main compositions of *S.involucrata* were phenolic acids, flavonoids, and lignanoids; but the main compositions of *S.laniceps* were phenylpropanoids, lignans, flavonoids, coumarins, sesquiterpenes, steroids, ceramides, and polysaccharides (Yi et al., 2009a and b; Wang et al., 2010, Chik et al., 2015; Chen et al., 2016). Although *S.laniceps* are primarily harvested as medicinal plants, they also have become popular souvenir items with tourists because they are strange-looking, rare, and grow in exotic locations (Yang et al., 2003).Lycium barbarum, known as goji berry, belonging to the Solanaceae family, has been used as a TCM to nourish the liver and kidney, to protective against chronic diseases, and to improve vision (Bo et al., 2016).It has long been favorited in Southeast Asia and China, and is increasingly becoming popular as a so-called "superfruit" in western diets due to its potential health–promoting constituents (Hempel et al., 2017).*Lycium ruthenicum* is also *Solanaceae* family and widely distributes in salinized desert of Qinghai–Tibet plateau, which is a unique nutritional food, and has widely been used for treatment of abnormal menstruation, heart disease and menopause (Zheng et al., 2011).Active constituents of *L.ruthenicum* are reported to be having a variety of biological activities, including anti-aging, immunoregulation, lowering blood-fat and blood-sugar levels and anti–fatigue.Because of overexploitation and deterioration of its habitat, *L.ruthenicum* is now considered to be threatened in China (Liu et al., 2012).

Whereas the studies of pharmacological properties and chemical composition have been mainly researched, little attention has been paid to the inorganic components of the rare medicine.Unlike synthetic drugs, herbal medicine is a synergistic system comprising multiple and complicated components (Yi et al., 2014 and 2016).Trace metal levels are important for human health.Trace elements play an important role in the plant metabolism and biosynthesis as cofactors for enzymes, and also play an important role in the formation of the active chemical constituents present in medicinal plants (Tokalıoğlu, 2012).Copper (Cu), manganese (Mn), iron (Fe), zinc (Zn) and selenium (Se) are essential for correct growth and development of all animals and can be considered as trace minerals with a central role in many metabolic processes throughout the body (Rahman et al., 2006; Stef and Gergen, 2012). Cu is required for iron metabolism, enzyme systems, immune systems, connective tissue metabolism and mobilization (Wang et al., 2014b). Mn participates in a wide range of adsorptive and redox reactions (Tebo et al., 2004). Fe is needed for normal growth of animals, and synthesis of Fe-requiring enzymes (Hostetler et al., 2003). Zn deficiency causes increased susceptibility to infection and impaired immune function (Hotz and Brown, 2001). Selenoenzymes have a major role as antioxidant that inhibit proinflammatory cell metabolisms and protect cell components (Rayman, 2000). Trace elements have been added to the list of elements considered as essential components of herbal medicine (Rajasekaran et al., 2005). In order to better clarify the pharmacodynamics, the objective of the present study was to estimate the trace elements content in *S. laniceps*, *S. involucrate*, *L. barbarum* and *L. ruthenicum.*

Materials and Methods

S. laniceps, *S. involucrate*, *L. barbarum* and *L. ruthenicum* (Figure 1) were collected in August, 2016 in Shannan district (east longitude 92°14′-93°22′, north latitude 28°08′-29°07′, altitude 3800-5200m) of Tibet Autonomous Region, Tianshan Mountains (east longitude 88°05′-88°54′, north latitude 42°05′-43°00′, altitude 4 000-4 500m) of Xinjiang Autonomous Region, Zhongning county (east longitude 105°30′-105°44′, north latitude 37°15′-37°28′, altitude 1 200-1 500m) of Ningxia Autonomous Region, and Yumen district (east longitude 96°20′-96°36′, north latitude 39°54′-40°05′, altitude 1 400-1 700m) of Gansu province, People's Republic of China, respectively. A total of 20 samples were collected for each species. The four herbal medicines were identified by professor Chaoying Luo, Lanzhou Institute of Husbandry and Pharmaceutical Sciences of CAAS. Voucher specimens with accession numbers 20160805, 20160814, 20160816 and 20150827 were submitted to the Herbarium of the Traditional Chinese Veterinary Medicine (Lanzhou Institute of Husbandry and Pharmaceutical Sciences of CAAS).

2 PROCESSING AND DIGESTION OF SAMPLES

All chemical reagents were guaranteed reagent grade. Standard stock solutions were prepared from Chinese Certified Referenced Material (100μg/mL) and were diluted to the corresponding metal solution. The working solution was freshly prepared by diluting an appropriate aliquot of the stock solutions. Ultra pure water was used for all dilutions.

The four herbal medicines samples were dried in an oven at 70℃ for 48h, crushed, and sieved through 0. 25mm sieve. The powder (1. 0g) were placed in PTFE digestion tubes and 10mL diacid mixture ($HNO_3 : H_2O_2 = 8 : 2$) was added to each digestion tubes for overnight. After that the samples were digested using microwave digestion system (MARS5, CEM Company, America). The optimal digestion procedure was described in our previous publication (Wang et al., 2014a). The digested samples were cooled to room temperature, transferred to glass tube, and removed the diacid mixture until 1mL was left using intelligent temperature controller in 130℃. The digested solution was transferred to volumetric flask, and volume made up to 50mL with ultra pure water. A blank digest was carried out in the same way.

Four elements (Cu, Mn, Fe, Zn) were measured using flame atomic absorption

Figure 1　Photos of *S. laniceps* (a), *S. involucrata* (b), *L. barbarum* (c) and *L. ruthenicum* (d) plants and their medicinal materials.

spectrometry (FAAS) (ZEEnit 700, Analytik Jena, Germany) with a deuterium background corrector. The hydride generation atomic fluorescence spectrometry was used for Se determination (Lu et al., 2004). The optimal operating condition and procedure was developed as described in our previous publication (Wang et al., 2014a) with slightly updated for higher sensitivity and lower detection limits in Table 1.

Table 1　Working conditions of the atomic absorption spectrometer and atomic fluorescence spectrometry

Element	Absorption line (nm)	Slit (nm)	Lamp current (mA)	Gas consumption ($L \cdot h^{-1}$)	Burner height (mm)	Flame type	Input dosage ($mL \cdot min^{-1}$)	Linear equations
Cu	324.8	1.2	3.0	50	6	C_2H_2-air	5.0	A=0.1325C−0.0023 R^2=0.9999
Mn	279.5	0.8	6.5	60	6	C_2H_2-air	5.0	A=0.1212C+0.0021 R^2= 0.9998
Fe	248.3	0.2	6.0	65	6	C_2H_2-air	5.0	A=0.0209C+0.0011 R^2=0.9999
Zn	213.9	0.5	4.0	60	6	C_2H_2-air	5.0	A=0.1542C+0.0009 R^2=0.9990
Se	217.6	-	60.0	24	-	Ar	0.5	A=0.1059C−0.0014 R^2=0.9989

Where A denotes absorbance value, C denotes concentration (μg/g or μg/kg)

3 RESULTS AND DISCUSSION

Medicinal herbs have been particularly recognized as being readily available, valuable and affordable resource for health care, and as an integral part of traditional medical practices for thousands of years (Tripathy et al., 2015). Potential healthcare applications of herbal medicines are very much related to their chemical composition and bioactive constituents, such as total phenolics, flavonoids, polysaccharides, anthocyanins, proanthocyanidins, alkaloids, as well as minerals (macro-, micro-and trace elements) (Szentmihalyi et al., 2006; Zengin et al., 2008). Trace elements are essential elements required to all cells for a number of biochemical functions, acting as structural components of tissues, constituents of the body fluids, and enzymes in major metabolic pathways (Lozak et al., 2002; Wang et al., 2016). Determination of the element composition of herbal medicines is essential for understanding their health effects and nutritive.

As show in Figure 2, the mean concentrations of Cu, Mn, Fe, Zn and Se in S.laniceps were (7.758±0.924) μg/g, (201.3±16.24) μg/g, (222.7±35.10) μg/g, (18.48±2.913) μg/g and (1.42±0.16) μg/g, respectively; in S.involucrata were (19.56±2.20) μg/g, (88.75±8.53) μg/g, (812.7±126.9) μg/g, (34.85±3.81) μg/g and (1.04±0.05) μg/g, respectively. And the trace elements content has significant differences between S.laniceps and *S. involucrate* ($P<0.01$ or $P<0.05$). The mean value of the total trace elements contents in "snow lotus" followed a descending order as: Fe>Mn>Zn>Cu>Se. In our study, we found that the "snow lotus" is rich in Mn, Fe and Zn, especially abundant in Mn compared with other plants (Niu et al., 2009; Zhao et al., 2009). Mn participates in a wide range of redox and adsorptive reactions, and plays a significant role in the bioavailability and geochemical cycling of many essential or toxic elements (Tebo et al., 2004). The therapeutic usage of herbal medicine in curing of various diseases due to presence of trace elements is already a widespread approach in medicinal purposes (Rajan et al., 2014). Most of the herbal medicines are rich in trace elements, thereby providing a possible link to the therapeutic action of the medicine (Kolasani et al., 2011). The trace elements present in "snow lotus" may play a direct or indirect role in the formation of the active chemical constituents and pharmacological nature especially Mn. Therefore, trace elements can be responsible for medicinal properties. The further research in the relationship of pharmacodynamics and trace elements should be needed.

The TCM plants *L. barbarum* and *L. ruthenicum* are valued for the abundance of bioactive carotenoids and anthocyanins in their fruits, respectively (Liu et al., 2014). However, the trace elements contents remain poorly understood. As show in Figure 3, the mean concentrations of Cu, Mn, Fe, Zn and Se in *L. barbarum* were (10.83 ± 0.26) μg/g, (9.598 ± 0.32) μg/g, (55.65±3.83) μg/g, (11.92±0.27) μg/g and (11.84±0.59) μg/kg, respectively. But the contents of Cu, Mn, Fe, Zn and Se in *L. ruthenicum* were (12.67±0.39) μg/g, (13.78±1.13) μg/g, (98.04±5.03) μg/g, (14.46±1.27) μg/g and (35.12±2.34) μg/kg, respectively; which were significantly higher than *L. barbarum* ($p<0.01$). L. barbarum is one kind of the famous food plants and traditional medicine (Gong et al., 2016), have been gradually regarded as a health functional food in Asia, Europe and United States based on it is beneficial in immunity improvement, antioxidation, enhancing hemopoiesis, antiradiation, anti-age and anti-cancer

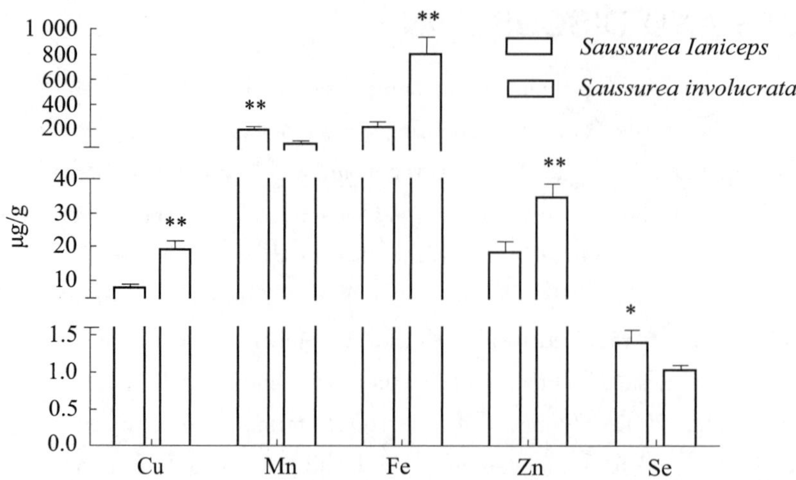

Figure 2 Mean concentrations of the trace elements analyzed in *S.laniceps* and *S.involucrate*.
**** means $P<0.01$, * means $P<0.05$.**

(Dong et al., 2009).*L.ruthenicum* processed abundant anthocyanins, and with wide range of physiological and biological activities such as antitumor effect and antioxidant activity.Adding certain amount of L.ruthenicum to daily diet will be beneficial for people's health (Liu et al., 2012).The biochemical mechanisms of the medicinal effects are primarily attributed to the anthocyanins, flavonoids, polysaccharides, essential oils and carotenoids (Yao et al., 2011).Researchers reported that for enhancing the immune system and curing of skin diseases, body required small quantity of trace elements, which recover quickly from serious infection and defend from pathogen (Rajan et al., 2014).From our research the contents and composition of trace elements in *L.barbarum* and *L. ruthenicum* is maybe another functional compound. And *L. ruthenicum* can be regarded as an additional supplementary source of many important trace elements in human daily diet and nutrition.

4 CONCLUSIONS

S.laniceps, *S.involucrata*, *L.barbarum* and *L.ruthenicum* are worth additional attention because of wide uses, extensive biological activities, and reliable clinical efficacy in various ethno-medical systems.This study provided the trace elements content of Cu, Mn, Fe, Zn and Se in the four herbal medicines.The results suggest that the four herbs may provide a beneficial source of some elemental micronutrients. The trace elements are maybe other functional compounds for medicinal effects.The therapeutic activity of the four herbs is maybe associated with the content of trace elements to some extent.Deep relationship between pharmacological and trace elements contents, especially its mechanism of action should be future research.

ACKNOWLEDGMENTS

The financial support from the CentralPublicinterest Scientific Institution Basal Research Fund (No. 1610322013003) and Gansu Provincial Natural Science Foundation (No. 1606RJYA224) is greatly appreciated.

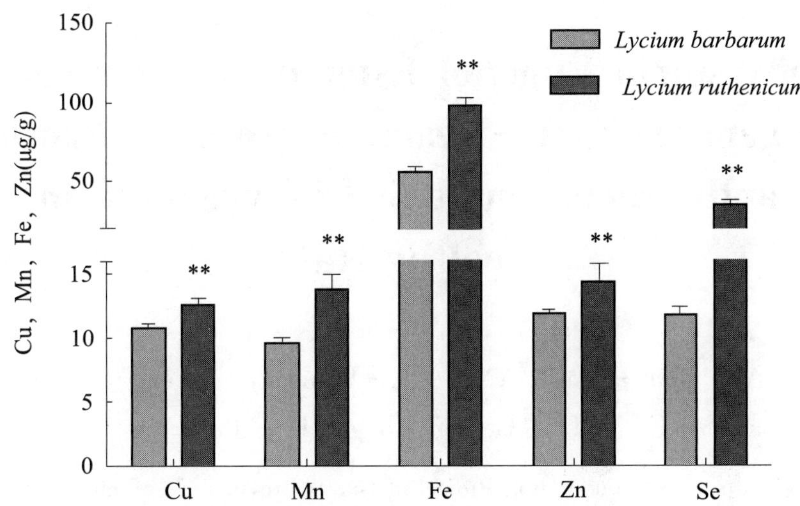

Figure 3: Mean concentrations of the trace elements analyzed in L.barbarum and L.ruthenicum. ** means $p<0.01$, * means $p<0.05$.

Conflict of Interest: Authors declare that there is no conflict of interest

REFERENCES OMITTED

(发表于《Afr J Tradit Complement Altern Med》，院选 SCI，IF：1.553)

Impact of Aspirin Eugenol Ester on Cyclooxygenase-1, Cyclooxygenase-2, C-Reactive Protein, Prothrombin and Arachidonate 5-Lipoxygenase in Healthy Rats

Ning MA[1], Guan-Zhou YANG[2], Xi-Wang LIU[1], Ya-Jun YANG[1], Isam MOHAMED[1], Guang-Rong LIU[1], Jian-Yong LI[1] *

(1. Key Lab of New Animal Drug Project of Gansu Province/Key Lab of Veterinary Pharmaceutical Development, Ministry of Agriculture/Lanzhou Institute of Husbandry and Pharmaceutical Science of CAAS, Lanzhou730050, China; 2. School of Pharmacy Lanzhou University, Lanzhou 730020, China)

Abstract: Aspirineugenol ester (AEE) is a promising drug candidate which is used for the treatment of inflammation, pain, fever, and the prevention of cardiovascular diseases. This study focuses on the effect of AEE on five proteins which are related to inflammation and thrombosis, including cyclooxygenase-1 (COX-1), cyclooxygenase-2 (COX-2), C-reactive protein (CRP), prothrombin (F Ⅱ) and arachidonate 5-lipoxygenase (ALOX5). Meanwhile, the study was administrated to compare the drug effect between AEE and its precursor from the view of chemical-protein interactions. Healthy rats were given AEE, aspirin, eugenol and integration of aspirin and eugenol. Carboxyl methyl cellulose sodium (CMC-Na) was used as control. After drugs were administered intragastrically for seven days, the blood samples were collected to measure the proteins concentration by enzyme linked immuno-sorbent assay (ELISA). The results showed that the concentrations of key endogenic bioactive enzymes were significantly reduced in AEE groups when compared with CMC-Na and aspirin groups ($P<0.01$). Drug effects of AEE on five proteins were stronger than aspirin and eugenol. From the view of chemical-protein interactions, AEE had positive effects on anti-inflammation and anti-thrombosis and showed stronger effects than aspirin and eugenol.

Key words: Aspirineugenol ester (AEE); Chemical-protein interactions; Thrombosis; Inflammation; Rats.

1 INTRODUCTION

As a classical drug, aspirin has been widely used for treatment of inflammation, headache,

* Corresponding author, E-mail: lijy1971@163.com

fever and cardiovascular diseases more than one century[1-4]. The biochemical mechanism of aspirin has been described previously[5-7]. Aspirin produces its therapeutic and side effect via inhibitation of cyclooxygenase (COX), which is a key enzyme to catalyze prostaglandin formation[6]. Later on, aspirin was successfully extended to prevent and treat of cardiovascular diseases since the inhibition of thromboxane A2 (TXA2) which is crucial factor for blood clotting[2]. Recently, the mechanism of aspirin in anti-cancer and anti-aging also has been confirmed[8-10].

Eugenol is the main component of volatile oil extracted from dry alabastrum of Eugenia.

Table 1 The basic information of keyendogenic bioactive enzymes.

Substance (Abbreviation)	Full Name	PDB Number	Inflammation	Thrombosis
ALOX5	Arachidonate 5-lipoxygenase	P09917	+	-
COX-1	Prostaglandin-endoperoxide synthase 1	P23219	+	-
COX-2	Prostaglandin-endoperoxide synthase 2	P35354	+	-
FⅡ	Coagulation Factor Ⅱ	P00734	-	+
CRP	C-reaction protein	P02741	+	-

Notes: The keyendogenic bioactive enzymes including five proteins were found out in PDB. Based on the different effects in inflammation and thrombosis, the key endogenic bioactive enzymes were divided into two categories: one category has a close relationship with inflammation; another has a close relationship with thrombosis. +: close relation; -: loose relationship.

Caryophyllata Thumb and used as a fragrance ingredient in many dietary products. Several pharmacological activities of eugenol have been demonstrated, including antivirus, antibacterial, antipyretic, analgesia, anti-inflammation, anti-platelet aggregation, anticoagulation, antioxidation, anti-diarrhea, anti-hypoxia and, antiulcer, and inhibition of intestinal movement and arachidonic acid metabolism[11-19].

However, the adverse reactions of aspirin such as gastrointestinal damage, have limited the long-term use of this old drug[20]. Eugenol which contains phenolic hydroxyl group, is irritative and can be easily oxidized. In order to reduce these side effects and improve the efficacy, an aspirin eugenol ester (AEE) was synthesized recently[21]. AEE is a colorless transparent crystal. The acute toxicity, pharmacodynamics, stability, teratogenicity and mutagenicity of AEE have been investigated. The LD_{50} of AEE is 10. 39g/kg, which is 0. 02 times of aspirin and 0. 27 times of eugenol[22]. A 15-day oral dose toxicity study as well as a comprehensive evaluation of genotoxicity by Ames test and the mouse bone marrow micronucleus assay were carried out. The results showed that its toxicity was not significant[22,23]. Its anti-inflammatory, analgesic, and antipyretic effects were similar to aspirin and eugenol, but lasted for a longer period[24,25]. Recently, metabolites of AEE in beagle dog also have been confirmed by HPLC-MS/MS[26].

STITCH (Search Tool for Interactions of Chemicals), is a well-known database containing both of the known and predicted interactions of chemicals and proteins (http://stitch.embl.de/)[27]. A general way to evaluate candidate drugs is to analyze the relationship between the candidate drugs and the targeted substances which are related to disease[28-30].Thus, the information detailing chemical - protein interactions of aspirin and eugenol was employed and retrieved from STITCH (Figure 1).Several proteins which are related to inflammation and thrombosis were chosen in the study, including COX-1, COX-2, CRP, F Ⅱ, and ALOX5 (Table 1).The main idea behind this method is based on the fact that the compounds and their precursor often share similar functions.Therefore, this study focused on investigating the influence of AEE on these proteins and comparing effects between AEE and its precursor.This study will help to partly clarify the effects of AEE from protein-chemical interactions, which can provide guidance for the further study of AEE.

2 EXPERIMENTAL

2.1 Chemicals and reagents

AEE; transparent crystal (purity: 99.5% with RE-HPLC), was prepared in key lab of new animal drug project of Gansu Province, key lab of veterinary pharmaceutical development of Agricultural Ministry, Lanzhou institute of Husbandry, and pharmaceutical sciences of CAAS.CMC-Na (carboxyl methyl cellulose sodium) was supplied by Tianjin Chemical Reagent Company (Tianjin, China). Aspirin and Tween-80 were supplied by Aladdin Industrial Corporation (Shanghai China).Eugenol was supplied bySinopharm Chemical Reagent Co., Ltd. (Shanghai China).ELISA kits of COX-1, COX-2, CRP, FⅡ and ALOX5 were purchased from USA TSZ biological Trade Co., Ltd. (Lexington, USA). Multiskan Go 1510 as spectrophotometer was supplied by Thermo Fisher Scientific (USA).

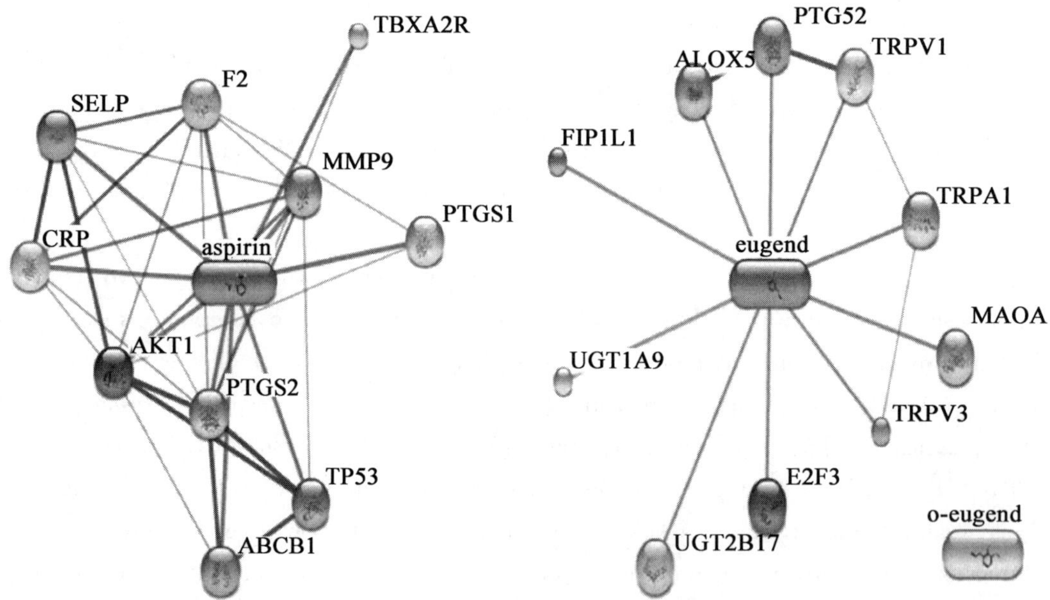

Figure 1. The confidence view of aspirin and eugenol from chemical-protein interactions.

Note: Through searching the database of STICH, proteins linked to aspirin and eugenol were found out.Stronger associations are represented by thicker lines.Protein-protein interactions are shown in blue, chemical-protein interactions in green.Through the network, the key substances from inflammation and thrombosis were selected. A: Aspirin as the center, B: Eugenol as the center

2.2 Animals

SeventyWistar male rats with clean grade, aged 7 weeks and weighing 150~160g, were purchased from the animal breeding facilities of Lanzhou Army General Hospital (Lanzhou, China). They were housed in plastic cages of appropriate size with stainless steel wire cover and chopped bedding. Light/dark regime was 12/12h and living temperature is (22±2)℃ with relative humidity of (55±10)%. Standard compressed rat feed from Beijing Keao Xieli Co., Ltd. (Beijing China) and drinking water were supplied ad libitum. The study was performed in compliance with the Guidelines for the care and use of laboratory animals as described in the US national institutes of healt and approved by institutional animal care and use committee of Lanzhou institute of husbandry and pharmaceutical science of CAAS. Animals were allowed a 2-week quarantine and acclimation period prior to start of the study.

2.3 Drug preparation

AEE and aspirin suspension liquid were prepared in 0.5% of CMC-Na. Eugenol and Tween-80 at the mass ratio of 1 : 2 were mixed with the distilled water to prepare eugenol emulsion.

2.4 Dose and sample collection

Animal disease models have been used to evaluate the effects of AEE on anti-inflammatory, analgesic, and antipyretic. Doses of 1.1 and 0.56mmol/kg were chosen as high and low dosage in xylene-induced ears swelling model to investigate the anti-inflammatory effects of AEE. The results showed that xylene-induced ear swelling degree was significantly inhibited by AEE administration for three days at 0.56mmol/kg (182.56mg/kg)[21]. When the factors including the using of normal rats, administration time and dose were taken into consideration in this experiment, the high-, medium-, and low-doses of AEE were selected as 200, 100, and 50mg/kg, and finally, seven days as administration time were designed in the experiment. For the comparability of the experimental results, the mole of the doses in medium-dose of AEE, aspirin andeugenol groups are the same as 0.3067mmol. Based on this experimental design, the doses of aspirin and eugenol were selected as 55.2mg/kg and 50.3mg/kg, respectively. In integration of aspirin and eugenol group, aspirin and eugenol at the molar ratio 1 : 1 (0.3067 mole) were also designed in order to compare AEE and its precursor. The 0.5% of CMC-Na at the dosage of 50mg/kg was as the negative control and the volume of CMC-Na was close to equal in comparison with AEE and aspirin groups (seen in Table 2). The drugs were intragastrically administered in each rat based on individual daily body weights for seven days. After seven days drug administration, the rats were anesthetized with 10% chloral hydrate and blood was collected from the abdominal aorta. Blood samples were collected into vacuum tubes without anticoagulant and then centrifuged at 1 000×g for 20min at 4℃. Serum was stored at -80℃ until the day of analysis.

Table 2. The experimental design in the study.

Groups	Drug	Dosage (mg/kg)	Average volume per rat (mL)
A	CMC-Na	50	1.74
B	Eugenol	50.3	1.63
C	Aspirin	55.2	1.73
D	AEE	50	1.65
E	AEE	100	1.72
F	AEE	200	1.76
G	Aspirin + eugenol	55.2 + 50.3	1.68

Note: A: CMC-Na group, B: model group, C: aspirin group, D: AEE low dose, E: AEE medium dose, F: AEE high dose, G: Ratio 1 : 1, integration of aspirin and eugenol, molar ratio 1 : 1, 0.3067mmol. In order to compare the difference between AEE and its precursors, the mole of the doses in medium-dose of AEE, aspirin and eugenol groups are the same as 0.3067mmol. The averaged intragastrically volume in different groups were approximate equal.

2.5 Study design

Based onprodrug principal, aspirin and eugenol were combined into AEE. After absorption, AEE was decomposed into salicylic acid and eugenol which are the major metabolites of AEE (26). Therefore, the proteins interacted with aspirin and eugenol were found out by the SMILE string of aspirin and eugenol by searching STITCH (seen in Figure 1). Chemicals are linked to proteins by evidence derived from experiments, databases, and the literature. In these linked proteins, the substances COX - 1, COX - 2, CRP, F Ⅱ and ALOX5, which have close relationship with inflammation and thrombosis, were selected as the research objects. COX-1, COX-2, CRP, and ALOX5 are related to inflammation, while F Ⅱ is responsible for the formation of thrombosis (seen in Table 1).

2.6 Statistical analysis

In the experiment, the OD value was tested four times at last; averaged, then the value into equation for sample concentration analyzed by ANOVA for unpaired groups was tested too. All data are expressed as mean±standard deviation (SD), using SPSS 19 (Statistical Product and Service Solutions). P-value below 0.05 was considered as statistical significance.

3 RESULTS

The OD_{450} value and the concentration In the experiment, the serum was diluted 5-fold by sample dilution. In order to reduce the experimental error, the sample was tested four times with spectrophotometer under 450nm. After getting the first-hand data, the average value of sample wells was calculated. Based on the straight line regression equation of the standard curve, the enzyme concentrations were got with the average sample OD_{450} value in the equation, and as multiplied by the dilution factor, the results were the actual concentrations (seen in Table 3).

Table 3 The actual concentrations of keyendogenic bioactive enzymes in each group with different drugs.

Group	ALOX5 (pg/mL)	CRP (μg/L)	F Ⅱ (IU/L)	COX-1 (ng/mL)	COX-2 (ng/mL)
CMC-Na	78. 8±6. 1	9093±652	626±51	8. 46±0. 63	6. 4±0. 35
Eugenol	72. 8±6. 8	8921±503	634±34	7. 64±0. 41**	6. 05±0. 27*
Aspirin	65. 4±3. 7**	8731±480	550±66*	6. 83±0. 36**	5. 15±0. 28**
Low-dose AEE	59. 8±3. 1**	7585±360**	540±54**	5. 87±0. 18**	4. 7±0. 21**
Medium-dose AEE	57±4. 2**	7592±815**	470±35**	5. 72±0. 34**	4. 6±0. 26**
High-dose AEE	55. 6±4**	7935±347**	438±37**	5. 36±0. 09**	4. 43±0. 24**
Ratio 1 ∶ 1	55	7142±510**	435±55**	5. 34±0. 3**	4. 36±0. 28**

Notes: After getting the average value of OD_{450} and regression equation, the diluted sample concentrations were got firstly, and then the actual concentrations were got by multiplying dilution factor.The results were expressed as mean± SD.ANOVA in SPSS was used in data analysis. * $P<0.05$, ** $P<0.01$. Significant difference from CMC-Na group.Ratio 1 ∶ 1, integration of aspirin and eugenol, molar ratio 1 ∶ 1, at 0. 3067mmol.

The effects of AEE: When compared with CMC-Na group, the concentration of COX-1 and COX-2 were reduced significantly in eugenol group.There were also significant differences between aspirin and CMC-Na groups.The values of ALOX5, F Ⅱ, COX-1 and COX-2 were reduced significantly.When compared aspirin and eugenol groups, the values of COX-1, COX-2 and F Ⅱ in aspirin group significantly showed difference from the values in eugenol group (Figure 2).The results indicated that both eugenol and aspirin had an influence on COX-1 and COX-2. In addition, the influence of aspirin on COX-1, COX-2 and F Ⅱ is stronger than eugenol.However, aspirin and eugenol had no effect on CRP index.

All rats in low -, medium - and high - dose AEE groups showed a significant difference compared with the CMC-Na group ($P<0.01$, Table 3).AEE was designed at different dose in the experiment.The concentration changes of the target proteins had a certain relationship with the AEE dosage.With the increase of AEE dose, the mean concentration values of F Ⅱ, ALOX5, COX-1, and COX-2 were in decline besides CRP value.The CRP concentration in AEE high-dose group was higher than the values in low and medium groups. The results in different groups were shown in Figure 2. When compared with the values in low-dose AEE group, the values in high-dose AEE group possessed significant difference.High-dose AEE produced stronger effect on the concentration of the target proteins than medium-dose AEE group.However, the difference between medium-and high-dose AEE groups was not statistically significant.

From the results, when compared with aspirin and eugenol groups, medium-dose AEE can significantly reduce the concentration values of all selected key endogenic bioactive enzymes ($P<0.01$).These data indicated that AEE had greater effects than aspirin and eugenol at the same molar quantity and also demonstrated that there was a significant difference between AEE and its precursor. Meanwhile, there was no statistical difference between medium-dose AEE group and Ration 1 ∶ 1 group (integration of aspirin and eugenol group, molar ratio 1 ∶ 1, equal molar quantity with medium-dose AEE).

4 DISCUSSION

Felix Hoffman synthesized acetylsalicylic acid in 1897, which was the first nonsteroidal anti-inflammatory drug (NSAID)[31]. Nowadays, NSAIDs are among the most commonly used drugs worldwide and their analgesic, anti-inflammatory and anti-pyretic therapeutic properties are thoroughly accepted. More than 30 million people use NSAIDs every day, and they account for 60% of the US over-the-counter analgesic market[32]. AEE is being developed for anti-inflammatory, analgesic, antipyretic, anti-atherosclerosis and anti-thrombosis pharmaceutical. Metabolites of AEE in-vivo and in-vitro have been confirmed by HPLC-MS/MS[26]. Five metabolites of AEE were detected in the experimental condition, including salicylic acid, salicylic acid glucuronide, salicylic acid glycine, eugenol glucuronide and eugenol sulfate. Previous studies showed that the effect of AEE is similar to aspirin and eugenol but last for long time, which indicated that AEE could be a promising NSAID candidate[21]. Due to the significant efficacy against inflammation, the action mechanism of AEE is necessary to be understood. In this study, STITCH database and ELISA method were used to measure the concentration changes of five proteins which are related to inflammation and thrombosis processes.

The primary anti-inflammatory effects of NSAIDs is competitive inhibition of cyclooxygenases (COX) 1 and 2 which catalyze the rate-limiting conversion of arachidonic acid to the pro-and anti-inflammatory prostaglandins and thromboxanes. COX-1 and COX-2 convert arachidonic acid to prostaglandin H2 (PGH2), a committed step in prostanoid synthesis. COX-2 is responsible for production of inflammatory prostaglandins, which could be induced by inflammatory stimuli[33]. COX-1 plays a major role in the generation of prostaglandins in gastric epithelial cells, such as prostaglandin E2 (PGE2), which plays an important role in cytoprotection[34]. CRP displays several functions associated with host defense and be treated as a maker of inflammation[35]. F Ⅱ converts fibrinogen to fibrin and activates factors Ⅴ, Ⅶ, Ⅷ, Ⅹ Ⅲ and functions in blood homeostasis, inflammation and wound healing[36]. ALOX5 catalyzes the first step in leukotriene biosynthesis, and thereby plays a role in inflammatory processes.

After strictly biological and toxicological research and test, pure CMC had been approved as food additive by (WHO) and (FAO). CMC has found wide applications in the pharmaceutical industry as a reliable carrier of drug[37,38]. In this study, the effect of CMC-Na was eliminated by administrating CMC-Na to control group. Therefore, it manifests that the effect of AEE is not related to CMC-Na. Tween-80 is widely used as emulsifier in food industry and drug production. In general, the body has a great tolerance to Tween-80[39] but the biological potency of Tween-80 on target protein in this experiment is not investigated.

Compared with AEE groups, the ALOX5 values of integration of aspirin and eugenol group did not show statistical differences. This indicated that AEE may have similar effect to integration of aspirin and eugenol at molar ratio 1 : 1. The COX-1 and COX-2 values in AEE groups decreased with the increase of AEE dose. This change trend may result from the dose changes of salicylic acid and eugenol from enzymolysis of AEE which is crucial for inhibition of COX-1 and COX-2[26]. The

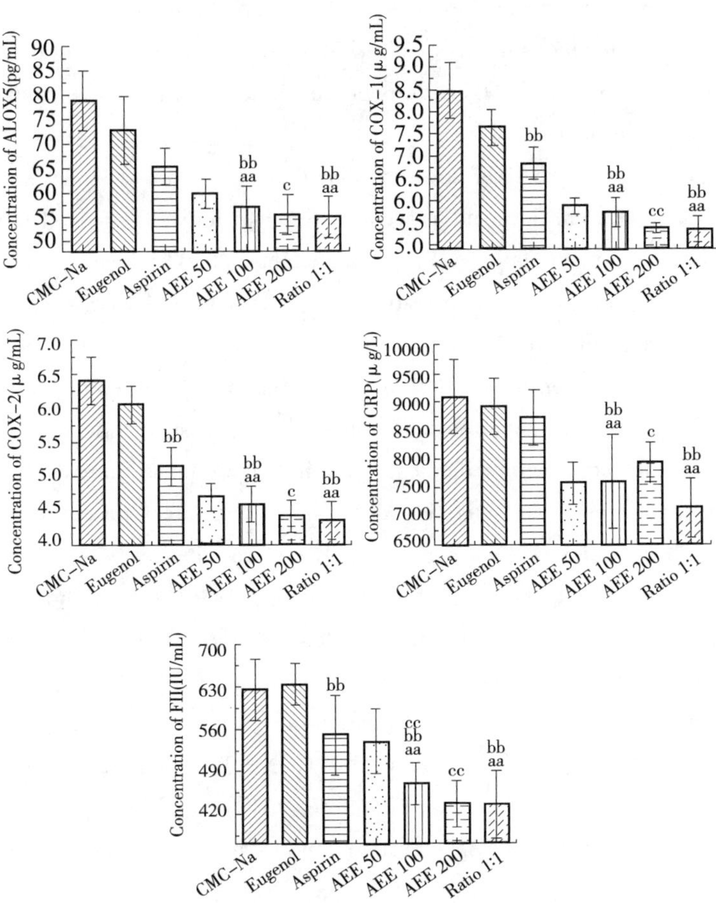

Figure 2. Effects of different drugs on keyendogenic bioactive enzymes after drugs administration for seven days (n=10).

Note: The concentrations of enzymes in serum were measured by ELISA method. Values were presented as mean±SD. Multiple comparisons were carried out to find out the difference between AEE and its precursor. [aa]$P<0.01$ compared with aspirin group. [bb]$P<0.01$ compared with eugenol group. [c]$P<0.05$, [cc]$P<0.01$ compared with low-dose of AEE group. Ratio 1 : 1, integration of aspirin and eugenol, molar ratio 1 : 1, at 0.3067mmol.

COX-1 and COX-2 values in eugenol group were significantly lower than the values in CMC-Na group, which indicated that eugenol had influence in COX-1 and COX-2 and also showed positive an impact on inflammation. However, the effect of aspirin on COX-1 and COX-2 was stronger than eugenol.

The CRP concentration of rats in low-, medium-and high-dose groups did not have great differences. However, its concentration was decreased compared to CMC-Na, eugenol, and aspirin groups. The reduction of CRP in AEE groups may be caused by synergistic effect of aspirin and eugenol. In addition, CRP concentration in the high-dose AEE group was higher than low-and medium-group, which may be caused by the side effects of salicylic acid from hydrolysis of AEE[26]. The concentration change of FⅡ in AEE groups appeared dose-dependent relationship and showed significant differences in comparison with CMC-Na and eugenol groups. The results in CMC-Na and eugenol groups were very close, which may indicate that eugenol have little or no effect on FⅡ.There were statistical differences among aspirin, medium-dose AEE and integration of aspirin and eugenol three groups (same molar quantity).The differences in three groups may be from eugenol which increased the effect of aspirin on FⅡ.

Like many other drugs, NSAIDs are associated with a broad spectrum of side effects, including gastrointestinal and cardiovascular events, renal toxicity, increased blood pressure, and deterioration of congestive heart failure. The side effects of AEE were not evaluated in this experiment.More studies are necessary to investigate on drug action of AEE such as side effects, aspirin-eugenol interactions, and clinical use.

From the results of the experiment, the concentrations of all key endogenic bioactive enzymes were reduced in AEE groups, which indicated that AEE had positive effects on anti-inflammation and anti-thrombosis.From the view of chemical-protein interactions, the results showed that AEE had stronger effects than aspirin and eugenol. Meanwhile, the effect difference between medium-dose and high-dose AEE groups was not significant.Therefore, it may be suggested that the clinical use dose of AEE is 100mg/kg.In the present study, it is possible to conclude that AEE is a good candidate for the development of new anti-inflammatory NSAIDs.At the same time, these findings lay the groundwork for further pharmacological and clinical studies.

The study provides the first evidence that AEE is an effective COX inhibitor.AEE can significantly reduce the concentrations of CRP, ALOX5, COX-1, COX-2, and FⅡ, which show better effects than that of aspirin and eugenol.Therefore, AEE could be a potential candidate for the treatment of inflammation as a new NSAID.Further studies of its clinical use and drug effects are necessary.

ACKNOWLEDGEMENTS

The work was supported by science-technology innovation engineering of CAAS (CAAS-ASTIP-2014-LIHPS) and the National Natural Science Foundation of China (No. 31572573).

REFERENCES OMITTED

(发表于《Iranian Journal of Pharmaceutical Research》，院选 SCI，IF：1.352)

Identification of Optimal Reference Genes for Examination of Gene Expression in Different Tissues of Fetal Yaks

Mingna LI, Xiaoyun WU, Xian GUO, Pengjia BAO, Xuezhi DING, Min CHU, Chunnian LIANG, Ping YAN*

(Key Laboratory for Yak Breeding Engineering of Gansu Province, Lanzhou Institute of Husbandry and Pharmaceutical Sciences, Chinese Academy of Agricultural Sciences, Lanzhou, China)

Abstract: Li M., Wu X., Guo X., Bao P., Ding X., Chu M., Liang C., Yan P. (2017): Identification of optimal reference genes for examination of gene expression in different tissues of fetal yaks.Czech J.Anim.Sci., 62, 426-434.

Reverse transcription quantitative real-time PCR (RT-qPCR) is widely used to study the relative abundance of mRNA transcripts because of its sensitivity and reliable quantification.However, the reliability of the interpretation of expression data is influenced by several complex factors, including RNA quality, transcription activity, and PCR efficiency, among others.To avoid experimental errors arising from potential variation, the selection of appropriate reference genes to normalize gene expression is essential. In this study, 10 commonly used reference genes-ACTB, B2M, HPRT1, GAPDH, 18SrRNA, 28SrRNA, PPIA, UBE2D2, SDHA, and TBP - were selected as candidate reference genes for six fetal tissues (heart, liver, spleen, lung, kidney, and forehead skin) of yak (Bos grunniens).The transcription stability of the candidate reference genes was evaluated using geNorm, NormFinder, and BestKeeper. The results showed that the combination of TBP and ACTB provided high-quality data for further study.In contrast, the commonly used reference genes 28SrRNA, SDHA, GAPDH, and B2M should not be used for endogenous controls because of their unstable expression in this study.The reference genes that could be used in future gene expression studies in yaks were indentified.

Key words: Transcription stability; RT-qPCR; TBP gene; ACTB gene

Reverse transcription quantitative real-time PCR (RT-qPCR) is a rapid and powerful method

* Supported by the Innovation Project of Chinese Academy of Agricultural Sciences (CAAS-ASTIP-2014-LI-HPS - 01), the Key Technologies R&D Program of China during the 12th Five - Year Plan Period (2012BAD13B05), and the Program of National Beef Cattle and Yak Industrial Technology System (CARS-37).

** Corresponding author, E-mail: pingyanlz@ 163. com

for the examination of the expression levels of transcripts.Relative quantification by RT−Qpcr can determine if the transcript level changes in given samples relative to control samples (Chen et al., 2015).The accuracy of RT−qPCR largely depends on the stability of the reference gene for normalization, which allows for the elimination of potential variants in RNA quality (Huggett et al., 2005), transcription activity (Vandesompele et al., 2002), PCR efficiency (Rekawiecki et al., 2012), and run−to−run variation during multistage experimental processes (Zeng et al., 2016).

The most frequently used reference genes include glyceraldehyde 3 phosphate dehydrogenase (GAPDH), βactin (ACTB), hypoxanthine guanine phosphoribosyl transferase 1 (HPRT1), β2 microglobulin (B2M), peptidylprolyl isomerase A (PPIA), TATA box binding protein (TBP), 18S ribosomal RNA (18SrRNA), 28S ribosomal RNA (28SrRNA), ubiquitin conjugating enzyme E2 D2 (UBE2D2), and succinate dehydrogenase complex subunit A (SDHA).It is generally accepted that the expression level of reference genes is constant across all cells, tissues, and environmental conditions (Svingen et al., 2015).However, there has been an increase in the number of studies that have evaluated the suitability of classic reference genes because of differences in age (Touchberry et al., 2006), tissue types (Sakai et al., 2014), thermal stress (Purohit et al., 2015), hormones (Das et al., 2013), and other treatment protocols (Young et al., 2006; Mihi et al., 2011).Studies in different tissue types have revealed that the most commonly used reference genes vary with the tissues studied (Zhang et al., 2013; Zeng et al., 2016).The application of unproven reference genes could lead to erroneous conclusions. For many important target genes, which are typically expressed at a low level, minor fluctuations in the reference gene could result in erroneous findings (Das et al., 2013).Taken together, it is extremely important to validate one or more suitable reference genes for normalizing the RT−qPCR data under every experimental condition across a range of samples or tissues.

Yaks (Bos grunniens) are important livestock animals, which are mainly distributed in the Qinghai−Tibetan Plateau and the adjacent alpine regions. Yaks provide milk, meat, fuel, service labour, and other daily necessities for local pastoralists.A number of studies on growth (Hu et al., 2016), cloned embryo development (Pan et al., 2015), and hypoxia (Wu et al., 2015) in yak have used expression analysis with RT−qPCR.In these studies, and in fact in most studies using RT− qPCR methodologies, normalization of expression data was conducted using a single, traditional reference gene as an internal standard.To the best of our knowledge, the reliability of reference genes was only evaluated for yak mammary tissue (Jiang et al., 2016), and no research has been conducted on other yak tissues. Therefore it is essential to identify the optimal reference genes for normalizing the RT−qPCR data.

In this study, we examined the stability of 10 commonly used reference genes (ACTB, B2M, HPRT1, GAPDH, 18SrRNA, 28SrRNA, PPIA, UBE2D2, SDHA, and TBP) in six different tissues of fetal yaks.This work will certainly facilitate future research on gene expression in yaks.

1 MATERIAL AND METHODS

Sample collection and preservation.Six fetuses of Datong yaks were collected at a slaughterhouse in the Qinghai Tibetan Plateau of Datong County.The fetuses were selected at 80−90 days (as estimated based on the crown-rump length of fetus) (Liu et al., 2010). Heart, liver, spleen, lung, kidney, and forehead skin were collected and frozen in liquid nitrogen until analysis. The entire study was conducted in strict accordance with the recommendations in the Guide for the Care and Use of Laboratory Animals of the National Institutes of Health, China.

RNA extraction and reverse transcription. Total RNA was extracted from the collected tissues using the TRIzol reagent (Invitrogen, USA). RNA purity and quantity were detected by NanoDrop 2000 (Thermo Scientific, USA). RNA integrity was verified by 1% agarose gel electrophoresis. The complementary DNA (cDNA) was synthesized from 500 ng total RNA using the PrimeScript™ RT Reagent Kit with gDNA Eraser (TaKaRa, China) following the manufacturer's protocol.

Quantitative real-time PCR. Ten reference genes (Table 1) commonly used as reference genes in RT-qPCR assays were selected for evaluation. The primers were designed using Primer Premier 5.0 software, and synthesized by Takara Biotechnology Co.Ltd. (Dalian, China). The qPCR reaction was performed using a CFX-96 Touch™ Real-Time PCR Detection System (Bio-Rad, USA). Each reaction (20μl) consisted of 1μl cDNA, 10μl SYBR© Premix Ex Taq™ Ⅱ (TaKaRa), 0.8μl of each forward and reverse primers (10μmol/l), and 7.4μl ddH_2O. Standard amplification conditions were 95℃ for 30s, 40 cycles of 95℃ for 5s, and 55℃ for 30s. Next, melting curve analysis was conducted to determine the specificity of PCR products. Three PCR reactions were performed for each sample and then averaged. No template controls (NTC) were included for each primer. Plate controls were conducted on each plate to normalize the Ct value from multiple plates into a single study dataset. PCR efficiency of reference genes was derived from a standard curve generated from serial dilution of pooled cDNA. The mean Ct value of each ten-fold dilution was plotted against the logarithm of the cDNA dilution factor. RT-qPCR efficiency was determined for each gene using slope analysis with a linear regression model. An estimate of PCR efficiency was derived from the formula $E = (10^{-1/slope} - 1) \times 100\%$ (Kubista et al., 2006).

SYNJ1 encodessynaptojanin 1, which is a key neural protein highly expressed in nerve terminals with an essential role in the regulation of synaptic vesicles. The SYNJ1 gene was used as a target gene to evaluate the performance of candidate reference genes. The primers for the SYNJ1 gene were designed using Primer Premier 5.0 software for RT-qPCR (Table 1), and the expression profile was assessed in the heart, liver, spleen, lung, kidney, and forehead skin tissue of six fetuses. Two different factors were used for selecting the reference: (1) a single reference gene with stable transcription levels, and (2) the geometric mean of the combination of stable reference genes. The reaction system and program was the same as that of the qPCR experiment. The PCR reaction was performed using a CFX-96 Touch™ Real-Time PCR Detection System (Bio-Rad).

Data analysis. The stability of the 10 reference genes was evaluated using the softwares geNorm (Vandesompele et al., 2002), NormFinder (Andersen et al., 2004), and BestKeeper (Pfaffl et al., 2004). The geNorm analysis calculated the gene stability measure (M value), which was arbitrarily required to be lower than 1.5 (with lower values indicating increased gene stability across samples), and pairwise variation (V value) for a single gene compared to all other reference gene candidates to determine the benefit of adding an extra reference gene to the normalization process. The arbitrary cut-off V value of 0.15 indicated acceptable stability of the reference gene combination. The NormFinder program was used to introduce raw data into a MS Excel spreadsheet following the manufacturer's instructions. The NormFinder algorithm used the least variation (intragroup and intergroup) to estimate expression stability between subgroups of the sample set, and lower values indicated increased stability in gene tran-scription. The BestKeeper software determined optimal reference genes by employing a pair-wise correlation analysis of all pairs of candidate genes and calculating the geometric mean of the best-suited genes. The raw Ct values were introduced directly into the MS Excel spreadsheet using a macro. The results were immediately calculated by the algorithm. The $2^{-\Delta\Delta Ct}$ method (Schmittgen and Livak 2008) was used to analyze the relative expression of mRNA of each gene by quantitative fluorescence.

Table 1 The primer list of genes used in the study.

Gene	Primers	Accession number	Amplicon length (bp)	PCR efficiency (%)
TBP-TATA box binding protein	F: GTCCAATGATGCCTTACGG R: TGCTGCTCCTCCAGAATAGA	NW_ 005395834. 1	82	96. 8
ACTB-βactin	F: ATTGCCGATGGTGATGAC R: ACGGAGCGTGGCTACAG	NW_ 005392900. 1	177	97. 6
UBE2D2-ubiquitin conjugating enzyme E2 D2	F: TCATTTCCCAACAGATTACC R: AGTTAGTGCTGGAGACC	XM_ 005900181. 1	133	95. 5
PPIA-peptidylprolyl isomerase A	F: TTTTGAAGCATACAGGTCC R: CCACTCAGTCTTGGCAGT	XM_ 005891872. 2	98	100. 2
18SrRNA-18S ribosomal RNA	F: GACAGGATTGACAGATTGAT R: CCCAGAGTCTCGTTCGTTAT	NR_ 036642. 1	117	95. 3
HPRT1-hypoxanthine guanine phosphoribosyl transferase 1	F: GTGATGAAGGAGATGGG R: ACAGGTCGGCAAAGAAC	NW_ 005397637. 1	79	106. 8
B2M-β2 microglobulin	F: CTGAGGAATGGGGAGAAG R: TGGGACAGCAGGTAGAAA	NW_ 005398298. 1	80	108. 9
GAPDH-glyceraldehyde 3 phosphate dehydrogenase	F: TCACCAGGGCTGCTTTTA R: CTGTGCCGTTGAACTTGC	EU195062. 1	126	103. 7
28SrRNA-28S ribosomal RNA	F: TCTTCCTGGAGTTGGGTTGC R: GGTTTCACGCCCTCTTGC	NR_ 036644. 1	145	101. 2
SDHA-succinate dehydrogenase complex subunit A	F: GGGAACATGGAGGAGGACA R: CCAAAGGCACGCTGGTAGA	XM_ 005894659. 2	188	97. 1
SYNJ1-synaptojanin 1	F: TGGTACGGTGTTGGTCTCAA R: TTCAAGGCAGAGCTTCCCT	XM_ 005893341. 2	171	103. 3

PCR = polymerase chain reaction

Table 2 Mean RT-qPCR threshold values (means±standard deviation) of 10 reference genes in yak tissues

Gene	Heart	Liver	Spleen	Lung	Kidney	Forehead skin
TBP	25. 27±0. 34	24. 36±0. 33	24. 63±0. 56	24. 49±0. 78	24. 12±0. 51	2477±0. 36
ACTE	18. 19±0. 32	17. 15±0. 26	17. 53±0. 85	17. 46±0. 90	17. 28±0. 60	17. 74±0. 32
UBE2D2	22. 14±0. 27	22. 30±0. 32	22. 24±0. 58	22. 11±1. 01	21. 48±0. 76	22. 29±0. 40
PPIA	19. 79±0. 44	18. 82±0. 35	18. 93±0. 28	19. 44±1. 10	18. 71±0. 58	20. 22±0. 33
18SrRNA	5. 23±0. 49	4. 72±0. 42	5. 28±0. 08	5. 14±0. 41	5. 26±0. 52	5. 58±0. 48
HPRT1	22. 60±0. 47	22. 97±0. 39	23. 47±0. 46	21. 90±0. 68	23. 47±0. 66	24. 58±0. 45
B2M	21. 80±0. 32	20. 99±0. 29	19. 80±0. 62	21. 56±0. 83	21. 54±0. 71	20. 97±0. 77
GAPDH	18. 97±0. 11	20. 77±0. 30	20. 27±0. 35	20. 39±0. 68	20. 28±0. 72	22. 64±0. 12
28SrRNA	26. 44±0. 81	26. 01±0. 70	26. 65±0. 43	26. 62±0. 66	25. 66±0. 78	28. 85±0. 51
SDHA	21. 37±0. 51	21. 38±0. 33	23. 874±0. 50	24. 34±1. 14	21. 62±0. 89	24. 44±0. 39

2 RESULTS

Transcript levels of candidate reference genes.Ten reference genes were amplified from different tissues of six yaks.Melting curves for each primer showed a single peak, confirming a single product from the RT-qPCR.Standard curves were generated using a ten-fold serial dilution of a cDNA pool, and these employed a linear correlation coefficient (R^2) of >0. 95. Based on the slopes of the standard curves, the estimated PCR amplification efficiencies ranged from 95. 3 to 108. 9% (Table 1).The mean Ct values are shown in Table 2. There was low inter-assay variation for the heart, liver, spleen, and forehead skin tissue, with most standard deviation (SD) values<0. 6 and the highest 0. 85. For lung and kidney tissue, there was larger variation and all SD values were >0. 41, with those of UBE2D2, PPIA, and SDHA for lung tissue exceeding 1. 0.

Transcript levels were used to establish three arbitrary categories: highly expressed 18SrRNA (mean Ct values = 5. 17), moderately transcribed (mean Ct values = 17. 53-23. 07), including the ACTB, B2M, 28SrRNA, HPRT1, SDHA, UBE2D2, and PPIA gene, and low transcript levels (mean Ct values >24), including GAPDH and TBP.

Table 3 Results of stability among 10 candidate genes computed by 3 algorithms on all yak tissues

Gene	geNorm		NormFinder		BestKeeper	
	M value	rank	S value	rank	SD	CV
TBP	0. 822	1	0. 314	4	0. 48	1. 97
ACTB	0. 841	2	0. 336	5	0. 55	3. 12
UBE2D2	0. 841	2	0. 293	2	0. 47	2. 12

(continued)

Gene	geNorm		NormFinder		BestKeeper	
	M value	rank	S value	rank	SD	CV
PPIA	0. 847	3	0. 287	1	0. 64	3. 33
18SrRNA	0. 863	4	0. 300	3	0. 36	6. 95
HPRT1	1. 089	5	0. 571	6	0. 75	3. 27
B2M	1. 117	6	0. 632	9	0. 68	3. 21
GAPDH	1. 134	7	0. 603	7	0. 86	4. 17
28SrRNA	1. 148	8	0. 617	8	0. 88	3. 30
SDHA	1. 454	9	0. 902	10	1. 37	6. 05

M value = gene stability measure, S value = stability value, SD = standard deviation, CV = coefficient of variation

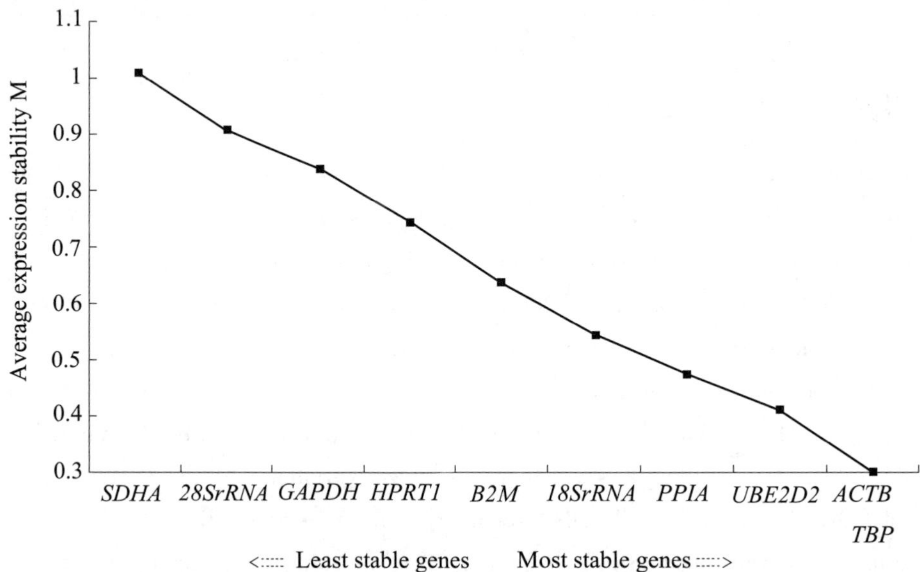

Figure 1 Stability of reference gene expression (M) calculated using *thegeNorm* software. Low M values indicate the best reference genes and high M values the worst reference genes

Stability of reference gene expression. The stability ranking of the 10 reference genes was evaluated based on the entire dataset (Table 3). The TBP gene had the lowest M value. The geNorm algorithm also eliminated the least stable expressed gene and recalculated new M values for the remaining genes. Figure 1 illustrates the average expression stability values (M); TBP and ACTB were the most stable genes. Next, the geNorm algorithms calculated the normalization factor (NF) and used V values to determine the minimal number of genes mandatory for normalization. Additional genes were included when V exceeded the cutoff value 0. 15. Figure 2 shows the first V value < 0. 15 (0. 148) that emerged at V2/3, suggesting that the two reference genes (ACTB and TBP) were sufficient for reliable normalization.

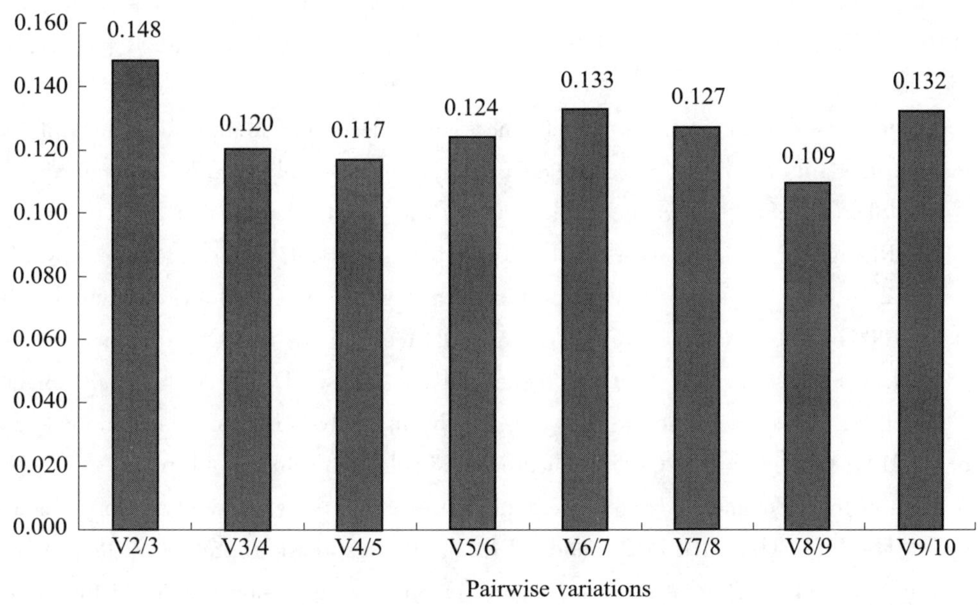

Figure 2. Determination of the optimal number of reference genes for gene expression normalization.

Note: To determine the number of stable reference genes needed for normalization, the pairwise variation Vn/n + 1 was calculated between the normalization factors NFn and NFn + 1 by the geNorm software. A threshold of pairwise variation V<0. 15 was suggested for the valid normalization

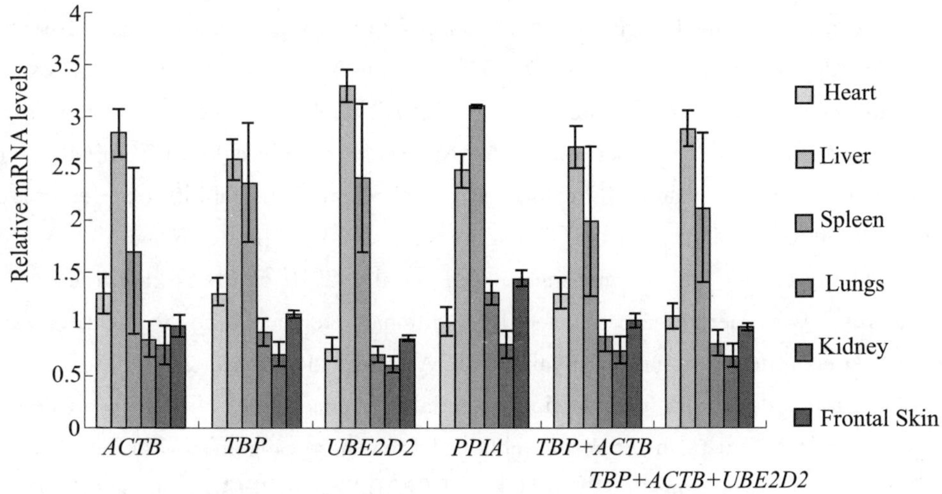

Figure 3. Validation of the recommended reference genes.

Note: Expression profiles of SYNJ1 gene were investigated using different reference genes. The expression of SYNJ1 was normalized using the different groups: ACTB, TBP, UBE2D2, PPIA, the combination of ACTB and TBP, the combination of ACTB, TBP, and UBE2D2. Bars represent the means and standard error of six biological replications

NormFinder calculated the stability value (S value) based on a variance estimation approach. An upper S value of 0. 5 indicated genes were relatively stable, and greater stability of gene expression was indicated by lower S values. The overall ranking of the genes from the most to the least stable was PPIA, UBE2D2, 18SrRNA, TBP, ACTB, HPRT1, GAPDH, 28SrRNA, B2M, and SDHA (Table 3).

BestKeeper measured the stability of reference genes according to the coefficient of variance (CV) and the standard deviation (SD) of the Ct values. The most stable reference gene exhibited the lowest CV and SD (CV±SD) (Radonic et al., 2004). The results showed that the TBP, UBE2D2, ACTB, and PPIA genes had the most stable expression across all tissues, followed by HPRT1, GAPDH, and 28SrRNA. The expression of most reference genes exhibited Ct variation less than 1 (Pfaffl et al., 2004) ($0.36<SD<0.88$), whereas SDHA exhibited high Ct variation (SD = 1.37 Ct) and was deemed unstable. The 18SrRNA gene had the lowest SD (0.36) and the highest CV (6.95); therefore, we could not define its stability with this analysis. B2M displayed a moderate CV±SD value (3.21±0.68), but P = 0.051 indicating it should be excluded (Table 3).

Validation of the recommended reference genes. To examine the validity of the top ranked reference genes (TBP, ACTB, UBE2D2, and PPIA), the expression profile of the target gene SYNJ1 was investigated in different tissues (Figure 3). Similar expression profiles of the target gene were observed using either two (TBP and ACTB) or three (TBP, ACTB, and UBE2D2) stable reference genes. The expression of the SYNJ1 gene in spleen was numerically higher than that in liver with single PPIA gene normalization. This indicated that the PPIA gene was not suitable as a reference gene in this study.

3 DISCUSSION

Reference genes are defined as genes that are stably expressed across all tissues/cell types; however, no single gene is constitutively expressed in all tissues/cell types and under all experimental conditions (Andersen et al., 2004; Yang et al., 2015). Therefore, selection of suitable reference genes is a crucial precondition for a successful gene expression study based on RT-qPCR (Liu et al., 2014). In addition, numerous studies have focused on the expression stability of reference genes in cattle (Lisowski et al., 2008), goats (Zhang et al., 2013), pigs (Nygard et al., 2007), humans (Touchberry et al., 2006), and mice (Zeng et al., 2016). Little is known about the reference genes in yak tissues. Herein, we selected 10 traditional reference genes to identify the superior reference gene for six different tissues from yak fetuses. Although the geNorm, NormFinder, and BestKeeper softwares are widely used for assessing potential reference genes, it has been demonstrated that discrepancy typically exists among results obtained from these programs (Purohit et al., 2015). The top four ranked reference genes (TBP, ACTB, UBE2D2, and PPIA) obtained through geNorm were consistent with those of BestKeeper, although the ranking order of the second and third gene was different. Based on NormFinder, PPIA was identified as the most stable reference gene, followed by UBE2D2 and 18SrRNA, whereas in BestKeeper, 18SrRNA was deemed unsuitable as a reference gene. The different software programs are based on different algorithms (Liu et al., 2014). Thus, the use of more than one algorithm for ranking is necessary.

Based on the aforementioned discrepancies, there was no scientific evidence to allow for determination of a single reference gene in this study. The V values calculated by geNorm software were used to determine the optional number of reference genes, and we must balance the trade-off between accuracy and practicality by using this value (Van-desompele et al., 2002). The V value of V2/3 was 0.148 (<0.15), which indicated that two reference genes are needed for reliable normalization, and the addition of one more reference gene would not significantly improve reliability. Furthermore, the target gene validated this result in that the expression profile with the geometric mean of two reference genes was similar to that of three genes. This showed that the combination of TBP and ACTB would provide high quality data. In contrast, the results suggested that 28SrRNA, SDHA, GAPDH, and B2M should be used for endogenous controls because of their unstable expression.

It has been demonstrated that utilization of different functional reference genes reduces coregulation effects, which may affect the accuracy of pairwise comparison results (Andersen et al., 2004). The candidate reference genes ACTB and TBP were responsible for cell locomotion and trans-cription, respectively (Mihi et al., 2011). ACTB is an important actin isomer and cytoskeleton actin (Zeng et al., 2016), the basal level transcription of which is essential for all cellular physiological conditions. It has been widely used as an internal control for experimental testing (Mihi et al., 2011). TBP is an indispensable basal transcription factor, and the RNA polymerase Ⅱ binds to the TBP-DNA complex to initiate transcription (Ponomarenko et al., 2016). A knockout (Martianov et al., 2002) or knockdown (Muller et al., 2001) of the TBP gene is lethal, and TBP expression might not be subjected to significant regulation. TBP was determined to be the optimal reference gene in low-abundance transcripts of goat (Zhang et al., 2013) and pig (Nygard et al., 2007) tissue expression studies. Appropriate reference genes were suggested to have the same transcript levels as the target gene to enhance the uniformity of the analysis (Spinsanti et al., 2006). In this study, TBP gene was classified in the low transcript level (mean Ct values >24), and the transcript level of the ACTB gene (mean Ct values = 17.15-18.19) was higher than that of the TBP gene. Based upon these concepts, the ACTB would be the logical reference gene for studying high tran script-level target genes, and the combination of ACTB and TBP genes would be more appropriate for modest transcript-level studies. Thus, we recommend the geometric averaging of the ACTB and TBP genes to normalize the relative gene expression levels for further study.

4 CONCLUSION

This study firstly validated the suitable reference genes for data normalization for gene expression in yak splanchnic and skin tissues. The combination of the ACTB and TBP genes was optimal for determination of gene expression in this study. In conclusion, the result may provide methodological support for further candidate gene identification and gene-expression pattern studies. It also clearly advocates for a reference gene assessment prior to performing a target gene expression-level analysis.

REFERENCES OMITTED

(发表于《Czech journal of animal science》，院选 SCI，IF：0.741)

Penicillin-resistant Characterization of *Staphylococcus aureus* Isolated from Bovine Mastitis in Gansu, China

Feng YANG*, Long-hai LIU*, Ling WANG, Xu-rong WANG, Xin-pu LI, Jin-yin LUO, Zhe ZHANG, Shi-dong ZHANG, Zuo-ting YAN, Hong-sheng LI*

(Lanzhou Institute of Husbandry and Pharmaceutical Sciences, Chinese Academy of Agricultural Sciences, Lanzhou 730050, China)

Abstract: Bovine mastitis caused by Staphylococcusaureus is difficult to treat because of increasing resistance against antibiotics, especially penicillin. β-Lactamase and biofilm are responsible for penicillin resistance of S. aureus. The aim of this study was to investigate the β-lactamase activity and biofilm formation capacity of 37 penicillin-resistant S.aureus strains (35 were blaZ positive and 2 were blaZ negative) from bovine mastitis in Gansu Province, China, as well as to measure the intercellular adhesion genes icaA and icaD of these strains. β-Lactamase test kit was used to determine the β-lactamase activity, biofilm formation was tested by semi-quantitative adherence assay method. Moreover, the presence of icaA and icaD were measured by PCR. A total of 32 penicillin-resistant S.aureus strains, including the two blaZ-negative strains, were identified as β-lactamase producers. All tested S. aureus isolates produced biofilm in the microtiter plate assay. Meanwhile, all these strains were PCR-positive for the ica locus, icaA and icaD. The study indicated high prevalence of β-lactamase activity, biofilm-forming capacity, and the ica genes among the penicillin-resistant S.aureus isolates, and implied that S.aureus resistant to penicillin was attributed to multiple mechanisms.

Key words: Staphylococcusaureus; Penicillin-resistant; β-lactamase; Biofilm; Ica

1 INTRODUCTION

Staphylococcusaureus is one of the most prevalent contagious pathogens causing bovine mastitis (Pereyra et al., 2016). Penicillin is used widely for treating the infection caused by this bacterium (Guérin-Faublée et al., 2003). However, the therapeutic effect has been disappointing due to the increasing resistant strains (Pereira et al., 2011). S.aureus resistant to antimicrobial agents mainly

* Received 21 June, 2016 Accepted 18 November, 2016

Corresponding author, E-mail: yangfeng@caas.cn (Yang F); lihsheng@sina.com (LiHS)

* These authors contributed equally to this study.

 Published by Elsevier Ltd. doi: 10.1016/S2095-3119(16)61531-9

depends on various antibiotic resistant genes carried by the pathogen (Peacock et al., 2002). It is generally appreciated that staphylococcal resistance to penicillin is mainly attributed to a structural gene blaZ. β-Lactamase encoded by this gene can hydrolyze the β-lactam ring, causing the β-lactam inactive (Lowy 2003). Besides antibiotic resistant genes, in recent years, biofilm formation is believed to play an important role in staphylococcal resistance (Mootz et al., 2015). Bacteria enclosed in a self-produced extracellular polysaccharide matrix exhibited high level of antibiotics toler-ance and resistance to host defense. These characteristics facilitate the adherence and colonization of S.aureus on the mammary gland epithelium, which often leads to persistent infections (Oliveira et al., 2006). A major component of staphylococcal biofilm matrix is the polymeric N-acetylglu-cosamine, which is related to intercellular adhesion and synthesized by proteins encoded by the intercellular adhesion (ica) locus (icaABCD). Among the ica genes, icaA and icaD have been reported to play a significant role in biofilm production in S.aureus (O'Gara 2007).

The detection of β-lactamase activity and biofilm formation capacity as well as ica locus is therefore necessary for exposition of penicillin-resistant characterization of S.aureus strains. Although the β-lactamase activity (Wang et al., 2011) and the phenotypic and genotypic basis for biofilm formation (Li et al., 2011; He et al., 2014) in S.aureus isolates associated with bovine mastitis has been reported in different regions in China, little is known about β-lactamase and biofilm formation and ica genes in penicillin-resistant S.aureus isolates from bovine mastitis. The present study was carried out to investigate β-lactamase activity and biofilm formation capacity of penicillin-resistant S.aureus strains from bovine mastitis cases in Gansu Province, China, as well as to detect the intercellular adhesion genes icaA and icaD of these strains.

2 MATERIALS AND METHODS

2.1 Bacterial strains

The 37 penicillin-resistant S.aureus isolates (35 were blaZ positive and 2 were blaZ negative) used here were the same as those in our early study (Yang et al., 2015).

2.2 β-Lactamase activity test

The β-lactamase activity of S. aureus strains was tested by β-Lactamase Test kit (Sigma-Aldrich, Lyon, France) according to the manufacturer's recommendation. S. aureus ATCC 25923 was used as the control isolate.

2.3 Biofilm assays

The ability of S.aureus strains to produce biofilm in microtiter plates was determined as previously described (Stepanović et al., 2007), with minor modification. Briefly, S.aureus strains were incubated in trypticase soy broth (TSB) without shaking at 37℃ for 18h, and thereafter diluted 1:100 in TSB supplemented with 1% glucose (w/v). A total of 200μL of the cell suspension was transferred into each well of sterile flat-bottomed 96-well microtiter plate. The plates were incubated aerobically for at 37℃ 24h. After incubation, each well was washed three times with 200μL of sterile phosphate buffered saline (PBS, pH 7.2), dried at room temperature and fixed with 150μL of methanol for 20min. After washing and drying, each well was stained with 150μL of crystal violet used for Gram−staining for 15min.Then the microtiter plate was three times with distilled water and subsequent drying, 150μL of ethanol was transferred into each to resolubilize the dye bound to the strains for 30min at room temperature.Then the optical destiny (OD) of each well was measured at 570nm by microplate reader (Model 550, Bio−Rad, Hercules, CA, USA).Each strain was tested in triplicate.Uninoculated wells containing TSB with glucose served as negative control.The ATCC29213 human isolate was used in each assay as a positive control, as described by Prenafeta et al. (2010).Biofilm formation was interpreted as no biofilm producer ($OD_{570} \leqslant OD_c$), weak biofilm producer ($OD_c < OD_{570} \leqslant 2 \times OD_c$), moderate biofilm producer ($2 \times OD_c < OD_{570} \leqslant 4 \times OD_c$) and strong biofilm producer ($4 \times OD_c < OD_{570}$).

2.4 Detection oficaA and icaD genes

The chromosomal DNA of S.aureus strains was extracted by Bacterial DNA Kit (Omega Bio−Tek, USA) according to the manufacturer's recommendation.The concentration of purified DNA was adjusted at 50ng · μL^{-1} using spectroscopy (Ultraspec 2100 Pro; Amersham Biosciences Europe GmbH, France).The presence of icaA with primers ICAAF (TCTCTTGCAGGAGCAATCAA) and ICAAR (TCAGGCACTAACATCCAGCA) and icaD with primers ICADF (ATGGTCAAGCCCAGA-CAGAG) and ICADR (CGTGTTTTCAACATTTAATGCAA) were screened by PCR as Arciola et al. (2001) described.The PCR mixture (50μL) contained 25μL reaction mixtures (Premix Ex Taq^{TM} ver.2.0, TaKaRa, China), 1μL primer ICAAF and 1μL primer ICAAR (or 1μL primer ICADF and 1μL primer ICADR), 3μL genomic DNA and 20μL ddH_2O.Thermal cycling conditions for both of icaA and icaD were 5min of denaturation at 94℃; 30 cycles of 30s of denaturation at 94℃, 30s of annealing at 94℃, and 30s of elongation at 72℃, and a final elongation step at 72℃ for 7min.PCR products (5μL) were analysed on 2% (w/v) agarose gel stained with ethidium bromide (0.5μg · μL^{-1}), and visualized under ultraviolet transillumination and photographed using Gel Doc XR apparatus (Bio−Rad, USA).

3 RESULTS

The results of β−lactamase activity, biofilm formation and the presence of icaA and icaD in S. aureus isolates are summarized in Table 1. Among the blaZ−positive isolates, 85.7% (30/35) of penicillin−resistant S.aureus isolates were positive for β−lactamase activity.For the two bla−negative S.aureus isolates, both of them were β−lactamase−positive too.

The culture of theblaZ penicillin−resistant S.aureus strains in microtiter plates revealed that all of the 37 isolates were biofilm producers, 5 isolates were weak biofilm producers, 19 isolates were moderate biofilm producers, and the remaining 13 strains were identified as strong biofilm producers.In addition, both of icaA and icaD genes were present in all of the 37 penicillin−resistant S.aureus strains.

Table 1 **β-Lactamase activity, biofilm formation and icaA and icaD genes presence among bovine penicillin-resistant *Staphylococcus aureus* isolates[1)]**

Isolates	blaZ	p-Lactamase activity	OD_{570} (SD)	Biofilm formation[2)]	icaA/icaD
CQ1	+	-	2. 100 (0. 568)	++	+/+
CQ4	+	+	0. 156 (0. 041)	++	+/+
CQ5	+	+	0. 281 (0. 057)	++	+/+
CQ6	+	+	0. 289 (0. 057)	+++	+/+
CQ8	+	+	0. 281 (0. 062)	++	+/+
CQ12	+	+	0. 120 (0. 017)	+	+/+
CQ18	+	+	0. 137 (0. 079)	+	+/+
DN1	+	+	0. 261 (0. 007)	++	+/+
DN5	+	-	2. 872 (0. 257)	+++	+/+
DN6	+	+	0. 338 (0. 088)	+++	+/+
DN7	+	+	0. 217 (0. 666)	++	+/+
DN8	+	+	2. 008 (0. 502)	+++	+/+
DN9	+	+	0. 266 (0. 114)	++	+/+
DN13	+	+	0. 109 (0. 014)	+	+/+
DN15	+	+	0. 179 (0. 032)	++	+/+
DN16	+	+	0. 291 (0. 028)	+++	+/+
DN18	+	-	0. 188 (0. 074)	++	+/+
DN19	+	+	0. 330 (0. 051)	+++	+/+
DN22	+	+	0. 210 (0. 055)	++	+/+
DN23	+	+	0. 167 (0. 428)	++	+/+
DN24	+	+	0. 108 (0. 013)	+	+/+
HZ3	+	+	0. 169 (0. 042)	++	+/+
HZ4	+	+	0. 238 (0. 028)	++	+/+
HZ5	+	+	0. 386 (0. 744)	+++	+/+
HZ8	+	-	0. 229 (0. 158)	++	+/+
HZ12	+	+	0. 938 (0. 499)	+++	+/+
HZ15	+	+	0. 332 (0. 173)	+++	+/+
HZ16	+	+	0. 163 (0. 047)	++	+/+
HZ19	+	+	0. 253 (0. 098)	++	+/+
HZ22	+	+	0. 298 (0. 046)	+++	+/+
HZ23	+	+	0. 708 (0. 102)	+++	+/+

(continued)

Isolates	blaZ	p-Lactamase activity	OD_{570} (SD)	Biofilm formation[2)]	icaA/icaD
HZ25	+	+	0.216 (0.058)	++	+/+
HZ29	+	−	0.110 (0.105)	+	+/+
HZ34	+	+	0.340 (0.051)	+++	+/+
HZ38	+	+	0.307 (0.168)	+++	+/+
HZ39	−	+	0.189 (0.076)	++	+/+
HZ42	−	+	0.264 (0.056)	++	+/+

[1)] In the column of blaZ, β-lactamase activity and icaA/icaD, the "+" and "−" represented positive and negative for each characteristic, respectively.

[2)] Biofilm formation is indicated by the mean optical density (OD_{570}) in the microtiter plates and the standard deviation (SD) of the mean. The isolates were classified as weak biofilm producer (+), moderate biofilm producer (++), and strong biofilm producer (+++).

4 DISCUSSION

S. aureus is usually resistant to most of the antibiotics used for the treatment of bovine mastitis, especially penicillin. In China, Shi et al. (2010) reported that 87.30% *S. aureus* isolates, from cases of mastitis in cow herds, were resistant to penicillin G. Data from other countries also have exhibited high prevalence of penicillin-resistant S. aureus ranged from 32.4% to 71.4% (Güler et al., 2005). Resistance to β-lactam antibiotics is most often caused by β-lactamases production (Fluit et al., 2001). Watts and Salmon (1997) determined that penicillin was the antimicrobial agents most affected by β-lactamase activity in *S. aureus* isolates from bovine mastitis cases. In this study, the two blaZ-negative penicillin-resistant strains exhibited unexpectedly β-lacta-mase activity. The phenotypic resistance may be caused by point mutations rather than gene acquisition (Gao et al., 2012). Among the blaZ-positive strains, both of β-lactamase production and penicillin resistance were detected in the majority of the blaZ strains. The β-lactamase detection rate was much higher than Wang et al. (2011) reported that 50% *S. aureus* isolates from bovine mastitis cases were positive for β-lactamase in Ningxia Hui Autonomous Region in China. Only 14.3% penicillin-resistant strains in our study were found to be negative in the β-lactamase activity test. This can be attributed to the lack of expression of the penicillin-resistant gene blaZ, which is inactivated (Hammad et al., 2014). Additionally, other factors such as biofilm formation may be the principal factor influence the penicillin resistance (Pantosti et al., 2007).

Biofilm formation of mastitis isolates *S. aureus* is associated with a reduced susceptibility to antibiotics (Amorena et al., 1999), bacteria exist in biofilm became 10-1 000 times more tolerant to antibiotics than equivalent plank-tonic cultures (Mah and O'Toole 2001). The results of our study indicated that the biofilm-forming ability was present in all of the tested *S. aureus* strains, including the 5 β-lactamase-negative penicillin-resistant strains. These results highlighted that, except for β-lactamase, biofilm formation may also play a significant role in penicillin resistance of these

S. aureus strains.For the β-lactamase-negative strains, biofilm formation may be the primary cause for penicillin resistance.The biofilm-forming rate in this study was much higher than that of Li et al. (2011) and He et al. (2014) observed in *S. aureus* isolates from Shandong Province (87.6%) and 9 different regions (48.0%) in China, respectively.This discrepancy could be due to the S. aureus samples used by Li et al. (2011) and He et al. (2014) to determine biofilm formation ability in a microtiter assay, collected from bovine subclinical mastitis cases without special screening; whereas, in our study, all tested S.aureus were penicillin-resistant strains.Nevertheless, our results agree with earlier study that reported high prevalence of the biofilm formation among penicillin-resistant S.aureus mastitis isolates (Bardiau et al., 2013).

Theica locus produce polymeric N-acetylglucosamine (PNAG) that mediates intercellular adherence involved in biofilm formation.This adhesion polysaccharide protects the bacteria from opsonophagocytosis (McKenney et al., 2000).The ica locus could be detected in majority of the mastitis S.aureus isolates (Dhanawade et al., 2010).In this study, the icaA and icaD genes were detected in all penicillin-resistant isolates analyzed.This result was in agreement with another report which found that all S. aureus isolates examined to be icaADBC positive (Fowler et al., 2001). However, it is higher than other researchers in China reported that the prevalence of icaAD ranged from 31.4 to 88.2% (Li et al., 2011; He et al., 2014).Although ica-independent biofilm mechanisms formation have been demonstrated in S. aureus isolates (O'Neill et al., 2007), in our study, all S.aureus strains with biofilm-forming ability were positive for both icaA and icaD genes, which was in accordance with earlier study reported that icaA and icaD were detected only in biofilm-producing S.aureus strains (Kouidhi et al., 2010).This result also confirmed the important role of ica genes in biofilm formation of S.aureus.

5 CONCLUSION

The present study indicates high-level prevalence of β-lactamase, formation biofilm and the intercellular adhesion genes icaA and icaD in penicillin-resistant S. aureus strains in Gansu Province, China, which could promote the chronicity of bovine mastitis, with the consequence of persistent infections and increased spread from infected cases.In addition, the results also confirmed that the S.aureus resistant to penicillin is based on β-lactamase alone or in combination with biofilm formation.

ACKNOWLEDGEMENTS

This study was supported by the Central Public-Interest Scientific Institution Basal Research Fund, China (1610322015007), the Key Technologies R&D Program of China during the 12th Five-Year Plan period (2012BAD12B03) and the Natural Science Foundation of Gansu Province, China (145RJYA311).

REFERENCES OMITTED

(发表于《Journal of Integrative Agriculture》, 院选 SCI, IF: 0.724)

An Efficient Novel Synthesis of 14-O-[(4-Amino-6-hydroxy-pyrimidine-2-yl) thioacetyl] Mutilin and the Antibacterial Evaluation*

Yun-Peng YI[1], Wen-Jun XU[1,2], Xin HUANG[1],
Jian-Ping LIANG[1], Ruo-Feng SHANG[1]*

(1. Key Laboratory of New Animal Drug Project of Gansu Province/
Key Laboratory of Veterinary Pharmaceutical Development, Ministry of Agriculture/
Institute of Husbandry and Pharmaceutical Sciences of CAAS, Lanzhou 130050, China;
2. College of Life Science and Engineering, Lanzhou University of
Technology, Lanzhou 730050, China)

Abstract: The title structure of 14-O-[(4-amino-6-hydroxy-pyrimidine-2-yl) thioacetyl] muti-lin, $C_{26}H_{47}N_3O_5S$, has been synthesized using 22-O-tosyl pleuromutilin and 4-amine-6-hydroxy-2-mercatopyrimidine monohydrate, and its structure was characterized by IR, NMR, H RMS and single-crystal X-ray diffraction.This compound has a 5-6-8 tricyclic carbon skeleton and a pyrimidine ring.It crystallizes in orthorhombic, space group $P2_12_12_1$ with a= 10.494 (3), b = 16.997 (5), c = 16.997Å, Z = 4, D_c = 1.275mg · m^{-3}, μ = 0.220mm^{-1}, F (000) = 1248, wR (F^2) = 0.1159 and R = 0.0381. The preliminary biological test showed that the title compound has more potent inhibitions to Staphylococcus aureus, MRSA and MRSE than that of tiamulin fumarate in vitro.

Key words: Pleuromutilin; Single-crystal structure; Antibacterial activities

1 INTRODUCTION

The discovery of pleuromutilin in 1950 started the studies of its derivatives[1].The activation mechanism of pleuromutilin derivatives was different from those of common clinically used antimicrobial agents[2].The function of inhibiting protein synthesis was identified as its most important antibacterial activity.The chemical modifications of pleuromutilin on C (14) side could cause higher

* Supported by Basic Scientific Research Funds in Central Agricultural Scientific Research Institutions (No. 1610322016007), National Key Technology Support Program (No.2015BAD11B02) and Agricultural Science and Technology Innovation Program (ASTIP, No.CAASASTIP-2014-LIHPS-04)

Corresponding author, E-mail: shangrf1974@ 163. com

DOI: 10. 14102/j.cnki.0254-5861. 2011-1612

activity than those on other sites[3], and thus led to three drugs: tiamulin, valnemulin and retapamulin.Tiamulin became the first pleuromutilin drug approved for veterinary, and retapamulin was approved by United States Food Drug Administration[4].Pleuromutilin derivatives selectively inhibit bacterial protein synthesis via binding to the peptidyl transferase center at 50S subunit of ribosomes[5].

The title compound is a novel semi-synthetic pleu-romutilin derivative that inhibits the colonization of gram - positive bacterium, for example methicillin - resistant Staphylococcus aureus (MRSA) and methicillin-resistant Staphylococcus epidermidis (MRSE)[2].In the study, the title compound was synthesized and characterized by IR, NMR, H RMS and single-crystal X-ray diffraction.Furthermore, its in vitro antibacterial activity along with tiamulin used as a reference drug was also investigated.The synthesis route is depicted in Scheme 1.

In the present project, we disclose the synthesis and antibacterial activity of a novel pleuromutilin derivative that features a pyrimidine ring moiety in the C (14) side, and also increase our knowledge of structure modification.

2 EXPERIMENTAL

2.1 Reagents and instruments

All reagents were purchased from Aladdin (China) and used without further purification.All compounds were synthesized in our lab and identified by IR, NMR and HRMS.IR spectra were obtained on a Thermo Nicolet NEXUS-670 spectrometer and recorded as KBr thin film and the absorptions were reported in cm^{-1}.HRMS were obtained with a Bruker Daltonics APEX Ⅱ 47e mass spectrometer.^1H NMR spectra were recorded on a Bruker-400MHz spectrometer in appropriate solvents. Chemical shifts (δ) were expressed in parts per million (ppm) relative to the tetramethylsilane. Multiplicities of NMR signals were designated as s (singlet), d (doublet), t (triplet), q (quartet), m (multiplet), br (broad), and so on.^{13}C NMR spectra were recorded on 100MHz spectrometers.The single-crystal structure of the title compound was determined on a Bruker SMART APEX Ⅱ X-diffractometer.All reactions were monitored by TLC on 0.2mm thick silica gel GF254 pre-coated plates.After elution, the plate was visualized under UV illumination at 254nm for UV active materials.Further visualization was achieved by staining with 0.05% KMnO4 aqueous solution. Column chromatography was carried out on silica gel (200~300 mesh).The products were eluted in appropriate solvent mixture under air pressure.Concentration and evaporation of the solvent after reaction or extraction were carried out on a rotary evaporator.

2.2 Synthesis and characterization of the title compound

2.2.1 Synthesis of 14-O- (p-toluene sulfonyloxyacetyl) mutilin 2

5mL of NaOH aqueous solution (2g, 50mmol) was added dropwise to a mixture of pleuromutilin (7.57g, 20mmol) and p-toluenesulfonyl chloride (4.2g, 22mmol) in methyl isobutyl ketone (10mL) and water (5mL).The mixture was vigorously stirred under reflux for 45min at 60℃, and then the reaction mixture was cooled to 10℃ and separated.The organic layer was washed with 5mL water and 5mL saturated sodium carbonate solution.The organic phase was dried overnight with anhydrous sodium sulfate. After filtration, the solvent was concentrated in vacuo to give 10. 56g of yellow oil which was used in the next step without further purification. Yield: 93%. IR (KBr): 3446 (s, OH), 2924 (s, CH_2), 2863 (m, CH_3), 1732 (s, C = O), 1633 (w, C-C), 1456 (m, C-C), 1371 (s, CH_3), 1297 (s, C-O-C), 1233 (m, CH), 1117 [s, C- (C=O) -C], 1035 (s, C-O-C), 832 (s, C-H), 664 (s, =CH_2), 560 (s, = CH_2) cm^{-1}. ^{1}H NMR (400MHz, $CDCl_3$): δ = 7.80 ~ 7.82 (d, J = 4.0Hz, 2H, benzene-H), 7.35 ~ 7.37 (d, J = 4.0Hz, 2H, benzene-H), 6.43 (dd, J = 17.2Hz, 10.8Hz, 1H, CH), 5.75~5.78 (dd, J = 8.8Hz, J = 6.4Hz, 1H, CH), 5.31~5.34 (d, J= 6.4Hz, 1H, CH_2), 5.17 ~ 5.21 (d, J = 8.8Hz, 1H, C = CH_2), 4.48 (s, 2H, OCH_2), 3.34 (d, J= 6.4Hz, 1H, OCH), 2.45 (s, 3H, CH_3), 2.21 ~ 2.29 (m, 3H, CH_2, OH), 2.01 ~ 2.08 (m, 3H, CH_2, CH), 1.63 ~ 1.65 (dd, J = 10Hz, J = 7.2Hz, 2H, CH_2), 1.41 ~ 1.50 (m, 5H, CH_2, CH_3), 1.33 ~ 1.36 (m, 1H, CH), 1.22 ~ 1.26 (s, 5H, CH_2, CH_3), 1.11 ~ 1.15 (m, 1H, CH), 0.87 (d, J = 6.8Hz, 3H, CH_3), 0.63 (d, J = 6.8Hz, 3H, CH_3). ^{13}C NMR (100MHz, CDC_{13}): δ = 216.7 (C=O), 164.8 (C=O), 145.2 (benzene-C), 138.6 (CH=C), 132.5 (benzene-C), 129.9 (benzene-C), 127.9 (benzene-C), 117.2 (C=CH_2), 74.4 (OCH_2), 70.2 (OCH), 64.9 (OCH), 57.9 (CH), 45.3 (C), 44.4 (C), 43.9 (CH_2), 41.7 (C), 36.4 (CH), 35.9 (CH), 34.3 (CH_2), 30.2 (CH_2), 26.7 (CH_2), 26.3 (CH_3), 24.7 (CH_2), 21.6 (CH_3), 16.4 (CH_3), 14.6 (CH_3), 11.4 (CH_3). HRMS (ESI) calcd. $[M+H]^+$ for $C_{29}H_{40}O_7S$: 533.250; found: 533.2507.

2.2.2 Synthesis of the title compound 3

4-Amine-6-hydroxy-2-mercatopyrimidine monohydrate (2.32g, 2.0mmol) was dissolved in 5mL methanol and 30% NaOH aqueous solution (2.2mmol) at room temperature and stirred for 30 minutes. 14-O-(p-toluene sulfonyloxyacetyl) mutilin 2 (1.17g, 2.2mmol) in CH_2Cl_2 (15mL) was added dropwise to the reaction mixture and stirred at 40℃ for 12h. The reaction mixture was concentrated in vacuo and EtOAc was added, followed by washing with brine and water. The organic layer was dried with anhydrous $MgSO_4$, filtered, and concentrated in vacuo to dryness. The crude residue thus obtained was purified by column chromatography using silica gel to give the desired products, white solids (Yield: 89%). IR (KBr): 3368 (NH_2, m), 2930 (CH_2, m), 1729 (C=O, s), 1629 (C=N, s), 1575 (C=N, m), 1542 (C=C, m), 1457 (C-CH_3, m), 1286 (C-O-C, m), 1117 [C- (C=O) -C, m], 1019 (C-O-C, w), 981 (C-H, m), 809 (C-H, m) cm^{-1}. 1H NMR (400MHz, $CDCl_3$): δ = 13.13 (s, OH), 6.48 (dd, J = 17.3, 11.1Hz, 1H, CH), 5.75 (d, J = 8.2Hz, 1H, CH), 5.34 (d, J = 11.0Hz, 1H, CH), 5.21 (s, 1H, CH), 5.16 (s, 1H, CH), 4.83 (s, 2H, NH_2), 3.79 (dd, J = 64.9, 16.5Hz, 2H, CH_2), 3.37 (d, J = 5.7Hz, 1H, CH), 2.30 (dd, J = 13.4, 7.2Hz, 1H, CH), 2.20 (dt, J = 19.6, 9.9Hz, 2H, CH_2), 2.10 (s, 1H, CH), 2.06 ~ 1.97 (m, 1H, CH), 1.76 (d, J = 13.7Hz, 1H, CH), 1.71 ~ 1.58 (m, 2H, CH_2), 1.61 (q, 1H, OH), 1.53 (d, J = 13.3Hz, 1H, CH), 1.47 (s, 1H, CH), 1.43 (s, 3H, CH_3), 1.36 (d, J = 12.8Hz, 1H, CH), 1.28 (d, J = 15.8Hz, 1H, CH), 1.14 (s, 3H, CH_3), 1.09 (s, 1H, CH), 0.87 (d, J = 6.6Hz, 3H, CH_3),

0.73 (d, J = 6.7Hz, 3H, CH_3).13C NMR (101MHz, $CDCl_3$): δ = 217.02 (C=O), 166.94 (C=O), 166.08 (pyrimidine C), 162.74 (pyrimi-dine C), 160.31 (pyrimidine C), 139.25 (CH), 117.15 (CH_2), 84.01 (CH), 74.59 (CH), 70.08 (CH), 58.11 (CH), 45.44 (C), 44.33 (CH_2), 43.94 (C), 41.86 (C), 36.70 (CH), 36.01 (CH), 34.45 (CH_2), 33.24 (CH_2), 30.39 (CH_2), 26.89 (CH_2), 26.38 (CH_3), 24.82 (CH_2), 16.79 (CH_3), 14.90 (CH_3) 11.48 (CH_3).HRMS (ESI) calcd. $[M+H]^+$ for $C_{26}H_{37}N_3O_5S$: 504.2526; found: 504.2520.

2.3 Crystal data and structure determination

The bulk crystals of the title compound suitable for X-ray structure determination were obtained by slow spread of a mixed solvent of methanol, dimethyl sulphoxide and water for about 2 days at room temperature.A colorless single crystal with dimensions of 0.30mm × 0.25mm × 0.20mm was selected and mounted in air onto thin glass fibers.X-ray intensity data were collected at 293 (2) K on a Bruker SMART APEX Ⅱ diffractometer equipped with a mirror-monochromatic MoKa (λ = 0.71073 Å) radiation.A total of 5337 reflections were collected in the range of θ = 2.3°~28.4° (index ranges: $-12\leq h\leq 12$, $-19\leq k\leq 20$, $-16\leq l\leq 20$) by using an ω scan mode with 5337 independent ones (R_{int} = 0 0197), of which 4975 with $I>2\sigma(I)$ were considered as observed and used in the succeeding refinements.The structure was solved by direct methods using SHELXS-97program[6] and refined with SHELXL-97 program[7] by full-matrix least-squares techniques on F^2.The non-hydrogen atoms were refined anisotropically, and hydrogen atoms were determined with theoretical calculations.A full-matrix least-squares refinement gave the final R = 0.0381, wR = 0.1159 $\{1/[\sigma^2(F_O^2)+(0.095P)^2+0.408P]$, where $P=(F_O^2+2F_C^2)/3\}$, $(\Delta/\sigma)_{max}$ = 0.000, S= 1.001, $(\Delta\rho)_{max}$ = 0.592 and $(\Delta\rho)$ min = -0.251 e · $Å^{-3}$.

2.4 Antibacterial activity measurement

The title compound was further tested for its minimum inhibitory concentrations (MICs) by trace broth dilution method (CLSI)[8].The number of bacterium was confirmed using plate count method.Tiamulin fumarate was co-assayed as a reference drug.

3 RESULTS AND DISCUSSION

The synthesis route of the title compound (compound 3) from pleuromutilin by two steps with good yield was depicted in Scheme 1. This compound was isolated by silica gel column chromatography and its structure was elucidated by IR, H RMS, ^{1}H NMR and ^{13}C NMR.The single crystal of 3 was cultured using a mixed solvent of water, methanol and dimethyl sulfoxide at a proper ratio and was determined by X-ray diffraction method to confirm the configuration.The selected bond lengths and bond angles are shown in Table 1 and hydrogen bonding parameters in Table 2.

Table 1 Selected Bond Lengths (Å) and Bond Angles (°)

Bond	Dist	Bond	Dist	Bond	Dist
S(1)-C(22)	1.811(2)	O(2)-C(3)	1.206(4)	N(2)-C(24)	1.417(3)
S(1)-C(23)	1.778(2)	O(3)-C(10)	1.500(3)	N(2)-C(23)	1.365(3)

(continued)

Bond	Dist	Bond	Dist	Bond	Dist
S(2)-O(5)	1.522(2)	O(3)-C(21)	1.324(3)	N(3)-C(23)	1.297(3)
S(2)-C(27)	1.757(4)	O(4)-C(24)	1.254(4)	N(3)-C(26)	1.393(3)
S(2)-C(28)	1.780(4)	O(6)-C(21)	1.203(3)		
O(1)-C(13)	1.443(3)	N(1)-C(26)	1.349(3)		
Angle	**(°)**	**Angle**	**(°)**	**Angle**	**(°)**
C(22)-S(1)-C(23)	99.13(10)	O(3)-C(10)-C(11)	104.41(17)	S(1)-C(23)-N(2)	114.06(16)
C(2)7-S(2)-C(28)	97.79(17)	O(3)-C(10)-C(9)	105.59(16)	S(1)-C(23)-N(3)	121.10(16)
O(5)-S(2)-C(28)	106.95(15)	O(1)-C(13)-C(14)	110.17(18)	O(4)-C(24)-N(2)	117.0(3)
O(5)-S(2)-C(27)	106.24(15)	O(1)-C(13)-C(12)	110.84(19)	N(2)-C(24)-C(25)	115.1(2)
C(10)-O(3)-C(21)	120.99(17)	O(6)-C(21)-C(22)	124.5(2)	O(4)-C(24)-C(25)	127.9(2)
C(23)-N(2)-C(24)	121.1(2)	O(3)-C(21)-O(6)	126.1(2)	N(3)-C(26)-C(25)	122.8(2)
C(23)-N(3)-C(26)	116.26(19)	O(3)-C(21)-C(22)	109.47(18)	N(1)-C(26)-C(25)	122.1(2)
O(2)-C(3)-C(2)	123.2(3)	S(1)-C(22))-C(21)	109.67(15)	N(1)-C(26)-N(3)	115.1(2)
O(2)-C(3)-C(4)	127.6(3)	N(2)-C(23)-N(3)	124.9(2)		

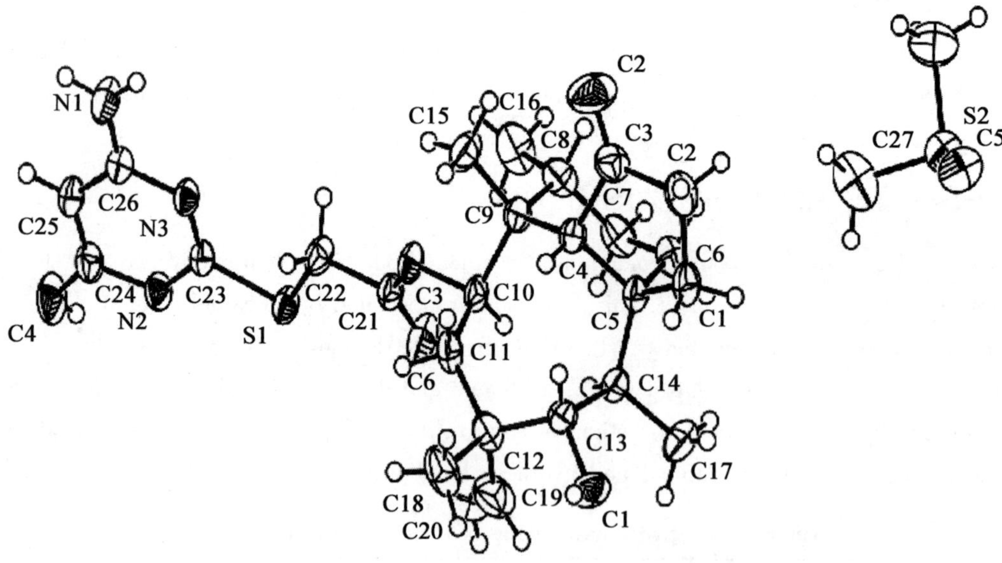

Scheme 1. Synthetic route of the title compound

Table 2 Hydrogen *Bond Lengths* (Å) and *Bond angles* (°)

D-H...A	d(D-H)	d(H···A)	d(D-A)	∠DHA
O(1)-H(1)···O(5)^{i}	0.82	2.04	2.835(3)	163
N(1)-H(1C)...O(4)^{ii}	0.86	2.01	2.837(3)	162
N(1)-H(1D)...O(1)^{iii}	0.86	2.16	2.918(3)	146
C(10)-H(10)...O(6)	0.98	2.45	2.824(3)	102
C(15)-H(15A)···O(2)	0.96	2.26	2.910(4)	124
C(15)-H(15C)...O(3)	0.96	2.25	2.674(4)	106
C(20)-H(20A)···O(6)	0.93	2.41	3.251(4)	151
C(27)-H(27B)...O(6)^{iv}	0.96	2.59	3.317(4)	133
C(27)-H(27C)...O(4)^{v}	0.96	2.57	3.417(5)	148
C(28)-H(28A)...O(4)^{v}	0.96	2.58	3.426(4)	147
C(28)-H(28C)...O(2)^{vi}	0.96	2.46	3.418(5)	176

Symmetry codes: i: −1/2+x, 1/2−y, 1−z; ii: 1/2+x, 5/2−y, 2−z; iii: 1/2+x, 3/2−y, 2−z; iv: 1−x, −1/2+y, 3/2−z; v: 1+x, −1+y, z; vi: 2−x, −1/2+y, 3/2−z

The title compound crystallizes in orthorhombic symmetry, space group $P2_12_12_1$. Its crystal structure (Fig.1) shows that the molecule consists of a 5−6−8 tricyclic carbon skeleton and a 4−amine−6−hydroxy−2−mercatopyrimidine ring, in which all bond lengths fall in normal ranges. The five−membered ring [C (1), C (2), C (3), C (4), C (5)] is a twist envelope−like conformation and the six−membered ring [C (4), C (5), C (6), C (7), C (8), C (9)] takes a chair−like conformation. The eight−membered ring is not planar. C (4) and C (5) were spiro−atom to three rings. The lengths of C (5) −C (4) and C (9) −C (4) are shorter than that of C (4) −C (3), which may result from the effect of carbonyl group on the five−membered ring. The side chain of C (10) exhibits short sharp turns or angles. The O (3) −C (21) −C (22) angel is 109.5 (2) which may be caused by O (3) and O (6). The dimethyl sulfoxide molecule and the title compound have behavior of co−crystallization which is good for the quality of the crystal.

The packing of the title compound is shown in Fig.2, in which eleven kinds of hydrogen bonds are present in the crystal structure. In addition, there are four intramolecular hydrogen bonds in the structure: C (10) −H (10)...O (6), C (15) −H (15A)...O (2), C (15) −H (15C)...O (3) and C (20) −H (20A)...O (6) whose hydrogen bond distances are 2.824 (3), 2.910 (4), 2.674 (4) and 3.251 (4) Å, making the side chains extend to the c axis. The structure is stabilized by O (1) −H (1)...O (5), N (1) −H (1C)...O (4), N (1) −H (1D)...O (1), C (27) −H (27B)...O (6), C (27) −H (27C)...O (4), C (28) −H (28A)...O (4) and C (28) −H (28C) ···O (2) interactions to form a three−dimensional network.

The antibacterial activities of compound 3 and tiamulin fumarate were tested against MRSA, MASE and Staphylococcus aureus by broth microdilution method, with the results listed in Table 3. The antibacterial activity of compound 3 is more active than that of the tiamulin fumarate.

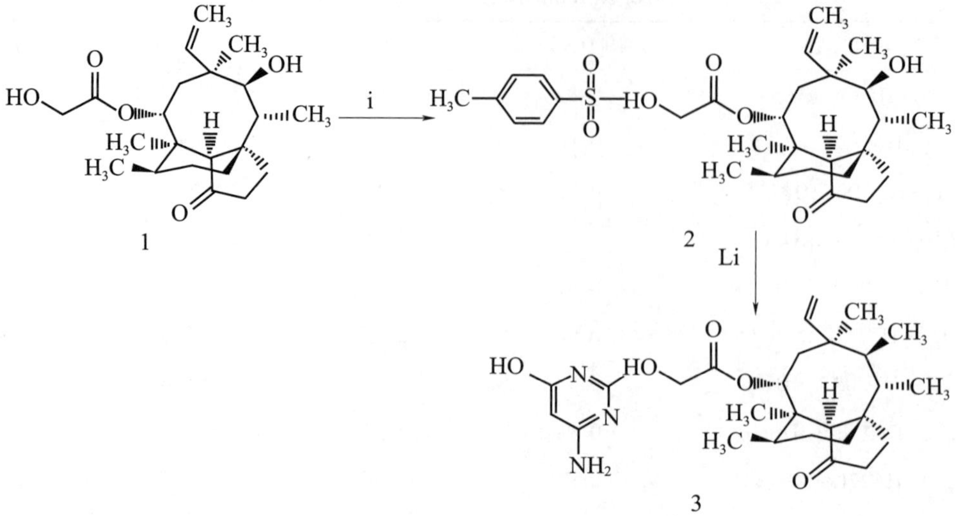

Fig.1 Crystal structure of the title compound

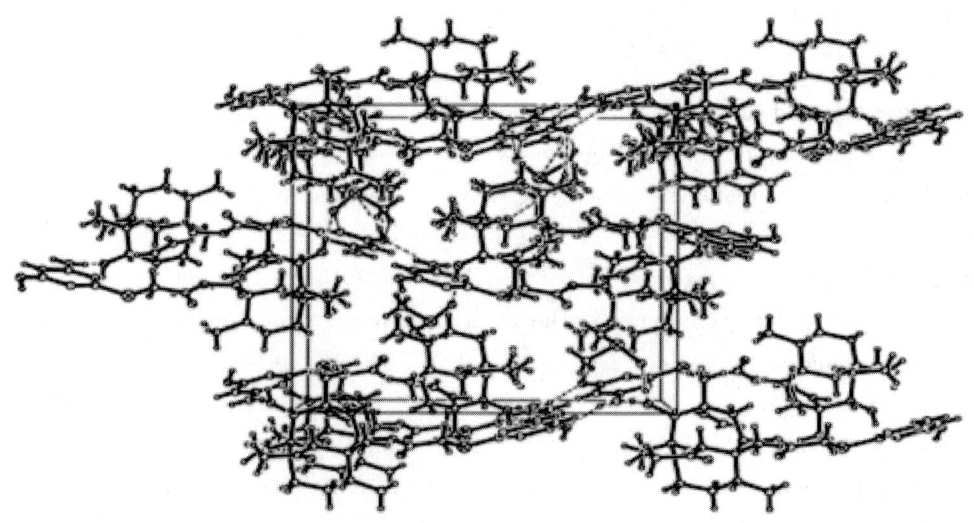

Fig.2 Packing of the title compound

Table 3 Antibacterial Activities of Compound 3 and Tiamulin Fumarate

Compound	MRSA (μg/mL)	MRSE (μg/mL)	Staphylococcus aureus (μg/mL)
3	0. 25	0. 125	0. 0625
Tiamulin fumarate	0. 5	0. 25	0. 125

REFERENCES OMITTED

(发表于《Chinese J. Struct. Chem》，院选 SCI，IF：0. 583)

Efficacy and Safety of Ban Huang Oral Liquid for Treating Bovine Respiratory Diseases

Bing LI, Xu-Zheng ZHOU, Jian-Rong NIU, Xiao-Juan WEI, Jian-Yong LI, Ya-Jun YANG, Xi-Wang LIU, Fu-Sheng CHENG, Ji-Yu ZHANG*

(Key Laboratory of New Animal Drug Project ofGansu Province/Key Laboratory of Veterinary Pharmaceutical Development, Ministry of Agriculture/Lanzhou Institute of Husbandry and Pharmaceutical Sciences of CAAS, Lanzhou, Gansu Province, China)

Abstract: [Background]: Ban Huang oral liquid was developed as a veterinary compound preparation by the Lanzhou Institute of Husbandry and Pharmaceutical Sciences of the Chinese Academy of Agricultural Sciences (CAAS).The purpose of this study was to determine whether the oral liquid preparation of traditional Chinese medicine, Ban Huang, is safe and effective for treating respiratory diseases in cattle.

[Materials and Methods]: Acute oral toxicity experiments were conducted in Wistar rats and Kunming mice via oral administration.The minimum inhibitory concentration of the drug against Mycoplasma bovis in vitro with the double dilution method was 500mg/mL, indicating good sensitivity.The results of laboratory pathogen testing, analysis of clinical symptoms, and analysis of pathological anatomy were combined to diagnose bovine respiratory diseases in 147 Simmental cattle caused by mixed infections of M.bovis, bovine respiratory syncytial virus, bovine parainfluenza virus type 3, and Mannheimia haemolytica.These cattle were randomly divided into three groups: drug treatment group 1 (treated via Tilmicosin injection), drug treatment group 2 (treated with Shuang Huang Lian oral liquid combined with Tilmicosin injection), and drug treatment group 3 (treated with Ban Huang oral liquid combined with Tilmicosin injection).Treatment effects were observed within 7 days.

[Results]: The results showed no toxicity and a maximum tolerated dose greater than 20g/kg BW.For the 87 cattle in drug-treatment group, the cure rate was 90. 80%, whereas the response rate was 94. 25%.The cure rate of drug treatment group was increased by 14. 13% in comparison with that of drug control group 1 and by 7. 47% in comparison with that of drug control group 2 (both $P<0.05$).

[Conclusion]: This study demonstrates that Ban Huang oral liquid is a safe and effective treatment for bovine respiratory diseases, especially for mixed infection caused by M.bovis, bacte-

* Corresponding author, E-mail: infzjy@ sina.com

ria, and viruses.

Key words: Ban Huang oral liquid; Mycoplasma bovis; Minimum inhibitory concentration; safety; Clinical efficacy

1 INTRODUCTION

Bovine respiratory diseases threaten the healthy development of the cattle breeding industry throughout the world. Ranking second only to severe infectious diseases in incidence and mortality, respiratory disease is the most significant and widespread cause of economic loss in the beef cattle industry (Ellis, 2001). The Chinese Ministry of Agriculture has organized relevant experts to conduct in-depth investigations on recent prevalent bovine respiratory diseases in China. By combining laboratory pathogen testing with analysis of clinical symptoms and epidemiological surveys, it was found that the main pathogen inducing respiratory diseases in cattle in China was Mycoplasma bovis, followed by Pasteurella multocida, bovine respiratory syncytial virus, and bovine parainfluenza virus type 3 (BPIV-3), all of which can join in mixed infections. The M.bovis pathogen is a particularly serious concern for cattle breeders because of the relative difficulty in diagnosing respiratory disease caused by the mycoplasma, the lack of an effective antibiotic against it, and its persistence and transmissibility (Caswell, 2007; Fox et al., 2005). The first report on the connection between M. bovis and bovine respiratory diseases was published in 1976 (Nicholas and Ayling, 2003). Although M.bovis can act as a primary pathogen, many cases of disease showed co-infection with other bacteria or viruses (Caswell et al., 2010).

The worldwide spread of M.bovis has recently gained momentum (Cai et al., 2005; van der Burgt et al., 2008; Dyer et al., 2008), causing significant economic losses for the cattle industry. For example, the annual economic loss in the United States due to bovine respiratory diseases reached 500 million US dollars (Wilkinson, 2009; Miles, 2009). In Holland, Laak et al. found that the rate of M.bovis infection was greater than 20% at a cattle-fattening farm, where few cattle were considered completely healthy (Buchvarova and Vesselinova, 1989). In Switzerland, bovine respiratory diseases caused by M.bovis account for more than 50% of all respiratory diseases and reduce the growth rate of latently infected herds by 8% (Poumarat et al., 2001). Nearly all outbreaks of M.bovis infection in China are related to transport, with most cattle showing signs of onset approximately one week after arriving at their destination (Shi et al., 2008). To date, there is no effective vaccine to prevent M.bovis infection (Mulongo et al., 2013). Currently, bovine respiratory disease associated with M. bovis infection is treated primarily with antibiotics. However, antibiotic therapy is rarely effective, the prevalence of antimicrobial resistance is reportedly increasing (Gautier-Bouchardon et al., 2014; Li et al., 2011), and antibiotic residue in cattle poses a serious threat to food safety and public health.

Traditional Chinese medicine (TCM) has played an important role in health protection and disease control for millennia. The therapeutic efficacy of TCM is based on the combined actions of constituents of phytochemical mixtures. TCM is utilized in many countries and districts, and the global market for herbal medicine is growing annually. TCM is widely accepted in Asian countries, and its popularity is also increasing in western countries. For example, the publication "Guidance for Industry: Botanical Drug Products" produced by the US Food and Drug Administration signals basic acceptance of TCM in western countries (Xu, 2008). TCM is useful in the prevention and control of bacterial infections (Wei et al., 2008). Huang et al. (Huang et al., 2005) generated a TCM preparation using the extracts of ten medicinal herbs (including Chrysanthemum indicum, Andrographis paniculata, Radix bupleuri, Houttuynia cordata, and Folium isatidis) and tested its protective effects against avian influenza virus H9N2, Newcastle disease virus, infectious bronchitis virus, and Mycoplasma gallisepticum using challenge experiments, demonstrating that the TCM preparation effectively controlled respiratory disease in poultry caused by these four pathogens. Qingkailing Injection was developed based on the ingredients in the Cow-Bezoar Bolus for Resurrection, the primary TCM preparation used to treat febrile and epidemic diseases. Qingkailing Injection is an outstanding representative of heat-clearing and detoxifying Chinese patent medicines that promotes regaining consciousness and has antibacterial, antiviral, antipyretic, and analgesic actions. Qingkailing Injection contributed greatly to efforts to control SARS, influenza A virus subtype H1N1, human avian influenza, and hand-foot-mouth disease (Guo, 2015; Lu et al., 2006). Zhang et al. (Zhang et al., 2007) controlled nephropathogenic infectious bronchitis in broiler chickens by adding veterinary Hushensuxiaosan Decoction (Shaanxi ShengAo Animal Pharmaceutical Co., Ltd.) and Shenzhongbaidu Decoction (Jiangxi Zhongcheng Medicine Group Co., Ltd.) to a centralized drinking water system. Studies have demonstrated that Chinese medicinal herbs can fully activate the immune system to enhance immunity. Chinese medicinal herbs have the advantages of easy accessibility, low cost, low toxicity, and drug tolerance (Huo and Li, 2002; Shen et al., 2004; Zhang et al., 1998).

The goal of this study was to prepare a compounded Chinese medicinal prescription to treat bovine respiratory diseases, especially pneumonia, caused by mycoplasmic infections. TCM drugs can be used to reduce antibiotic tolerance and residue in cattle. Ban Huang oral liquid was developed as a veterinary compound preparation by the Lanzhou Institute of Husbandry and Pharmaceutical Sciences of Chinese Academy of Agricultural Sciences (CAAS). The main ingredients of Ban Huang oral liquid included Isatidis Radix, Coptidis Rhizoma, Lonicerae Joponicae Flos, Scutellariae Radix, Anemarrhenae Rhizoma, Glycyrrhizae Radix et Rhizoma. Isatidis Radix, Coptidis Rhizoma, Lonicerae Joponicae Flos, and Scutellariae Radix, which have heat-clearing and detoxifying properties, as well as blood-cooling, throat-easing, and antibacterial actions (Li, 2005; Chinese Pharmacopoeia Commission, 2010). Anemarrhenae Rhizoma has heat-clearing, fire-purging, yin-nourishing, dryness-moistening, blood-cooling, throat-easing, and anti-inflammatory actions (Chinese Pharmacopoeia Commission, 2010). Glycyrrhizae Radix et Rhizoma has the effects of invigorating the spleen, replenishing qi, heat-clearing, detoxifying, and expelling phlegm to arrest coughing; therefore, it is used for the coordination of various drugs (Chinese Pharmacopoeia Commission, 2010).

The combined use of the ingredients in Ban Huang oral liquid can achieve the effects of heat-clearing, detoxifying, removing dampness and swelling, arresting bleeding, and easing throat soreness; indeed, these ingredients have been applied to treat many febrile diseases. Coptidis Rhizoma and Scutellariae Radix, the main ingredients in Ban Huang oral liquid intended to treat respiratory diseases, have heat-clearing, detoxifying, and antibacterial actions. Berberine hydrochloride and baicalin were chosen as indices for quality control of Coptidis Rhizoma and Scutellariae Radix, respectively, because the detection methods for these compounds are stable and reliable.

Shuanghuanglian oral liquid (composed of Lonicerae Joponicae Flos, Scutellariae Radix, and Forsythiae Fructus) (Jia, 2013) was compared with Ban Huang oral liquid because Shuanghuanglian oral liquid the former has heat-clearing, detoxifying, and antibacterial actions. It is widely used in the clinical mainly for to treat respiratory infections, including pneumonia., etc. (Jia, 2013; Zhang, 2010). Through the acute toxicity experiments and clinical pharmacodynamic studies presented here, this study demonstrates that Ban Huang oral liquid is a safe and effective treatment for bovine respiratory diseases, especially mixed infection caused by M. bovis, bacteria, and viruses.

2 MATERIALS AND METHODS

2.1 Materials

Isatidis Radix, Coptidis Rhizoma, Lonicerae Joponicae Flos, Scutellariae Radix, Sophorae Tonkinensis Radix et Rhizoma, Arctii Fructus, Platycodonis Radix, Anemarrhenae Rhizoma, and Glycyrrhizae Radix et Rhizoma were purchased from the Lanzhou HuiRenTang Pharmacy and certified by GSP. Ban Huang oral liquid (1g/mL crude drug, batch no. 20090615) was provided by the Lanzhou Institute of Husbandry and Pharmaceutical Sciences, CAAS.

Shuanghuanglian oral liquid (1g/mL crude drug, batch no. 2006 020115030) was provided by Tianjin Zhongsheng TiaoZhan Bioengineering Co., Ltd. Tilmicosin injection (3g/10mL, batch no. 20090717) was provided by Sichuan Zhibang Biological Technology Co., Ltd. Baicalin (batch no. 110715-200815) and berberine hydrochloride (batch No. 110713-200911) were provided by the National Institute for the Control of Pharmaceutical and Biological Products and used as standards. Chromatographically pure acetonitrile and methanol were obtained from Fisher Company (USA). Industrial grade ethanol, 95% ethanol, potassium dihydrogen phosphate, sodium dodecyl sulfate, sodium benzoate, phosphoric acid, agar, thallium acetate, penicillin, sodium hydroxide, and phenolsulfonphthalein were obtained from Sinopharm Group Co. Ltd. PPLO broth base was obtained from BD Co. (USA). Glucose and β-nicotinamide adenine dinucleotide were obtained from Sangon Biotech (Shanghai) Co., Ltd. Fetal bovine serum was obtained from Zhejiang Tianhang Biological Technology Co., Ltd. M. bovis strain gs-1 was isolated, identified, and preserved by the Lanzhou Institute of Husbandry and Pharmaceutical Sciences, CAAS.

2.2 Equipment

A Waters 2695 HPLC System coupled with a UV detector (Waters Co., USA), an AEL-1600 electronic balance (Shimadzu, Japan), and a Millipore water purification system (USA) were used in these experiments.

2.3 Ethics statement

The protocol was approved by the Ethics Committee of Animal Experiments of Lanzhou Institute of Husbandry and Pharmaceutical Sciences of CAAS, which was commissioned by The International Cooperation Committee of Animal Welfare (ICCAW) affiliated with the China Association for the Promotion of International Agricultural Cooperation (CAPIAC). The mouse toxicity study and cattle treatment procedures were subject to the ethical oversight of the Ethics Committee of Animal Experiments of Lanzhou Institute of Husbandry and Pharmaceutical Sciences of CAAS and implemented in accordance with the technical guidelines for clinical trials for veterinary TCM and natural drugs.

2.4 Study protocol

This study used six batches of Ban Huang oral liquid prepared with stable and quality controllable processes. Quality control indices were detected using high performance liquid chromatography (HPLC). Inter-batch differences were tested; relative standard deviation (RSD) was required to be less than 5%. Acute oral toxicity experiments were conducted in rats to evaluate the clinical safety of the drug. The sensitivity of M.bovis to the drug was determined according to the guidelines for therapeutic drug trials in the Collections of Technical Requirements for Veterinary Drug Trials (Ministry of Agriculture, 2001). The method for diagnosing bovine diseases was established along with the criteria for evaluating the treatment effect. The mortality, cure rate, and response rate were calculated after Ban Huang oral liquid was administered. The results were compared with those obtained using Shuang Hang Lian oral liquid.

Intergroup differences were statistically tested, and the efficacy of Ban Huang oral liquid for treating bovine respiratory infectious diseases was evaluated.

2.5 Preparation of Ban Huang oral liquid

The optimal soaking time was determined based on the water absorption of the drug using the concentration of the quality control ingredients as indicators. The water extraction process was optimized using L_9 (3^4) orthogonal array testing. An analysis of variance was performed to determine and verify the process. The concentration and relative density were measured, and the clarity and transport rate of the main quality control ingredients were used as indicators to characterize the alcohol precipitation method.

2.6 Concentration determination for quality control

An HPLC-UV method was used to determine the concentrations ofbaicalin and berberine hydrochloride in Ban Huang oral liquid. The baicalin reference and berberine hydrochloride reference were accurately weighed, after which reference solutions of 5mg/mL and 10mg/mL, respectively, were generated using methanol. Six batches of Ban Huang oral liquid (1mL each) were individually weighed and placed into separate 50mL flasks. The solution was diluted to 50mL with methanol and mixed well. A C18 HPLC column was used with an aqueous solution of methanol and 0.1% phosphoric acid (50 : 50) as the mobile phase. Baicalin was detected at a wavelength of 274nm (1.0mL/min flow rate, 10μL sample size). The number of theoretical plates was calculated using the baicalin peak and was required to be greater than 2,500. Octadecyl silane chemically bonded to silica was used as filler, and an aqueous solution of acetonitrile and potassium dihydrogen phosphate (47 : 53, v/v, 3.2g potassium dihydrogen phosphate and 1.6g sodium dodecyl sulfate in every 1 L of water) was used as the mobile phase. The concentration of berberine hydrochloride was detected at a wavelength of 347nm (10μL sample size). The number of theoretical plates was calculated using the berberine hydrochloride peak and was required to be greater than 5,000.

2.7 Drug safety evaluation in acute oral toxicity experiments in rats and mice

The protocol followed the guidelines recommended in the Guidance of Veterinary Drugs for Technical Study, 2006-2011 (Ministry of Agriculture, 2012). Sixty Kunming mice (30 female and 30 male) weighing 18-22g (license no.scxk [Gansu] 2012-0075) and sixty Wistar rats (30 female and 30 male) weighing 200-250g (license No.scxk [Gansu] 2012-0075) were used in the experiments. All rodents were purchased from the Medical School of Lanzhou University. The rats and mice were randomly divided into six groups (n = 10 per group, 5 males and 5 females) based on body weight. Intragastric irrigation was performed once daily. For control mice, normal saline was gavaged at a dose of 10g/kg BW. Five experimental groups were gavaged at doses of 5, 10, 15, 18, or 20g/kg BW (one dose per group). The mental state, behavior, and food intake of the mice were observed for 14 consecutive days.

2.8 Susceptibility testing

A PPLO broth base was supplemented with fetal bovine serum, 25% yeastleachate, 10% glucose, 10% arginine, 0.4% phenolsulfonphthalein, 1% thallium acetate, and 1% penicillin. The M.bovis strain was aspirated using a micropipettor, inoculated into the PPLO broth, and cultured at 37℃ in a 5% CO_2 incubator for 48-72h. The PPLO broth was added to 12 sterile screw-cap test tubes (1.8mL each) using the double dilution method and cultured at 37℃ in a 5% CO_2 incubator for 48-72h. The color change of the culture media was observed, and the color-changing unit (CCU) was determined as the concentration of the highest dilution that changed the color of the media. For example, if the 5th tube in the dilution series was the highest dilution to show a color change, then the CCU was 1×10^{-5}/mL.

The MIC of Ban Huang oral liquid was determined using themicrodilution method in accordance with the guidelines of the Clinical and Laboratory Standards Institute (CLSI/NCCLS, 2003). The optimal concentration of M. bovis in the drug sensitivity test was 1×10^{-3} to 1×10^{-4} CCU per 0.2mL; a concentration of 1×10^{-4} CCU was used for MIC determination. The MIC was defined as the drug concentration in the tube before the first color change after 48-72h of incubation at 37℃. The experiment was repeated for confirmation when no tubes showed a colour change after 3-6d.

2.9 Study animals

A total of 147 Simmental cattle (aged 6-12 months) affected by respiratory diseases were reared in two-row barns at Fucheng Beef Cattle Farm in Sanhe City (Hebei Province) and Wanhe Beef Cattle Farm in Zhangye City (Gansu Province). All cattle were in the stage of disease onset and showed the following symptoms: increased body temperature (approximately 41℃), continued fever, poor appetite, coarse and disordered hair, emaciation, cough, shortness of breath, diarrhea, and hemafecia. We had the permission of the cattle farm owners to perform the treatment study.

2.10 Clinical diagnosis

Diagnosis was made by combining the results of the clinical observations of symptoms with those of pathological anatomy analysis and laboratory pathogen testing. Laboratory pathogen testing was conducted as follows: a TaqVet Triplex Pasteurella multocida/Mannheimia haemolytica fluorescence qPCR kit and TaqVet Mycoplasma bovis fluorescence qPCR kits were used to examine mucus from the nasal cavities of affected cattle. DNA extraction was performed using a QIAamp DNA mini kit (QIAGEN, 51304). Bovine respiratory syncytial virus and BPIV-3 were detected using the TaqVet Triplex bRSV & PI3 kit with mucus from the trachea and bronchus. A QIAGEN kit (QIAamp Viral RNA Mini Kit: 52904 or 52906) was used for RNA purification and identification. Detections were conducted before and after treatments.

2.11 Efficacy evaluation

Cattle were considered cured when all clinical symptoms of respiratory diseases were absent, appetite and mental state were restored to normal, and laboratory pathogen tests of respiratory secretions were negative. Cattle were considered responsive when clinical symptoms were markedly alleviated, with restoration of appetite and mental state. The treatment was considered ineffective when clinical symptoms of respiratory diseases were not alleviated or were aggravated, appetite declined or was completely lost, mental state was poor, and laboratory pathogen tests of respiratory secretions were positive.

2.12 Treatment

The 147 affected cattle (clinically diagnosed and experimentally confirmed) were randomly divided into three groups. Drug control group 1 was comprised of 30 cattle that were treated with Tilmicosin injection at a dose of 10mg/kg body weight (BW) by subcutaneous injection once daily for 1d. Drug control group 2 was comprised of 30 cattle that were treated with Shuang Hang Lian oral liquid at a dose of 0.4mL/kg BW twice daily for 7d, followed by combined treatment with Shuang Hang Lian oral liquid (0.4mL/kg BW) and Tilmicosin injection (10mg/kg BW) by subcutaneous injection for 1d. The experimental drug treatment group (n = 87 cattle) was treated with Ban Huang oral liquid (0.4mL/kg BW) twice daily for 7d, followed by combined treatment with Ban Huang oral liquid (0.4mL/kg BW) and Tilmicosin injection (10mg/kg BW) by subcutaneous injection once daily for 1d.

3 RESULTS

3.1 Preparation of Ban Huang oral liquid

The optimal preparation process for Ban Huang oral liquid was determined through preliminary trials. Isatidis Radix, Coptidis Rhizoma, Lonicerae Joponicae Flos, Scutellariae Radix, Sophorae Tonkinensis Radix et Rhizoma, Arctii Fructus, Platycodonis Radix, Anemarrhenae Rhizoma, and Glycyrrhizae Radix et Rhizoma were soaked in water for 12h and boiled twice for 1h each time. For the first boiling, 7 times the volume of water was added, and for the second boiling, 5 times the volume of water was added. The resulting liquid was combined and filtered, after which the filtrate was condensed to a relative density of 1.18-1.25 (25℃). Next, ethanol was slowly added until the proportion reached 60%; this solution was mixed well and left to stand for 24h. The supernatant was collected after filtration to recover ethanol until there was no ethanol aroma left in the remaining filtrate. Next, ethanol was added again until the proportion reached 75%, after which the solution was mixed and left to stand for 24h. The supernatant was collected by filtration and ethanol was recovered until no ethanol aroma remained. Sodium benzoate (3g) was added to the resulting solution, which was diluted with water to a volume of 1 000mL, mixed well, sterilized by boiling, and bottled under sterilized conditions.

3.2 Concentration determination

The prescription consisted of the nine herbs listed above. Berberine hydrochloride and baicalin were detected using HPLC to evaluate the preparation process and perform quality control of active ingredients Coptidis Rhizoma and Scutellariae Radix, respectively. The results of these assays demonstrated that the RSD values for baicalin and berberine hydrochloride in all six batches of Ban Huang oral liquid were less than 5%, indicating good reliability and stability of the preparation process (Tables 1, 2 and Fig 1, 2).

Table 1 Baicalin concentration in Ban Huang oral liquid.

Batch no.	Mean content±SD (μg/mL)	RSD (%)
20100122	1032.55±5.37	
20100123	1103.14±2.55	
20100124	983.18±3.18	4.37
20090614	1036.55±6.73	
20090615	1068.97±3.21	
20090616	1100.72±6.42	

Table 2 HPLC chromatograms ofberberine hydrochloride in Ban Huang oral liquid.

Batch no.	Mean content±SD (pg/mL)	RSD (%)
20100122	33.56±1.12	
20100123	30.17±2.03	4.05
20100124	32.67±0.98	
20090614	32.11±1.31	
20090615	31.98±0.83	
20090616	33.80±1.24	

3.3 Safety evaluation

Acute oral toxicity experiments were conducted in rats and mice (n = 10 of each in control and each treatment group). No rodents died following treatment with Ban Huang oral liquid (5, 10, 15, 18, or 20mg/kg BW) for 14 days. During the treatment, rats and mice showed a normal mental state, normal food intake, and normal water intake, without any adverse reactions. The maximum tolerated dose (LD_0) of Ban Huang oral liquid in rodents was greater than 20g/kg BW;

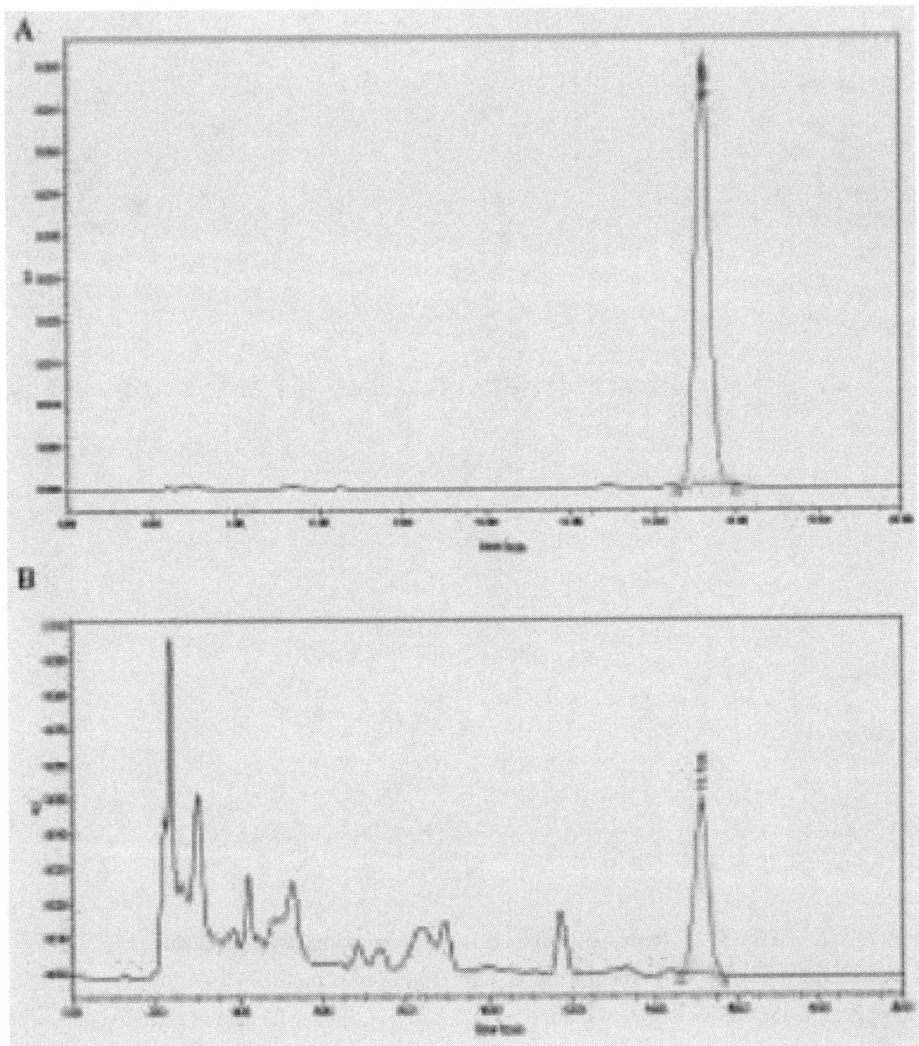

Figure 1 Representative HPLC chromatograms showingbaicalin peaks in Ban Huang oral liquid. (A) is the chromatogram of the reference substance, and (B) is the chromatogram of the sample.

thus, the median lethal dose (LD_{50}) is greater than this value. Therefore, according to the standards for toxicological classification of drugs, Ban Huang oral liquid was considered nontoxic.

No fatalities occurred in any of the groups of rats or mice, and no animal was sacrificed during this study.

3.4 MIC determination

The M.bovis strain was cultured in an incubator for 48h to 72h until the color of the culture medium turned translucent yellow. Proliferating M.bovis cells were subjected to 10-fold serial dilution and cultured at 37℃ until the media did not change color after 48-72h. The results demonstrated that the endpoint dilution was 1×10^{-7}. Thus, the CCU of M.Bovis strain gs-1 was 1×10^{7} CCU/mL. The MIC of Hainosankyuto was greater than 500mg/mL (Fig.3). These results indicate that Ban

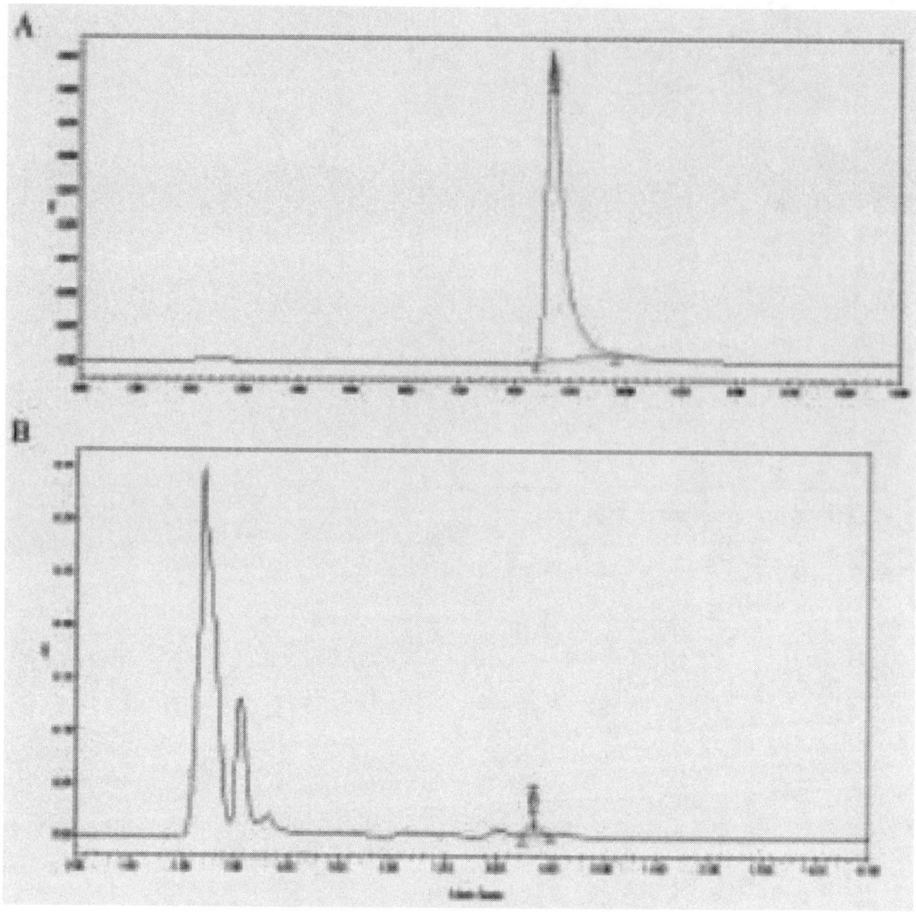

Figure 2 Representative HPLC chromatograms ofberberine hydrochloride in Ban Huang oral liquid.

Note: (A) is the chromatogram of the reference substance, and (B) is the chromatogram of the sample.

Huang oral liquid significantly inhibited visible growth of M.bovis.

3.5 Diagnosis

The cattle showed clinical symptoms consistent with those of bovine respiratory disease. During early onset of respiratory disease, severely affected cattle stopped eating, while their body temperature increased to 41.3±0.2℃. The affected cattle had difficulty breathing, groaned, and presented with abdominal breathing and tenderness upon rib pressing. The cattle refused to crouch and demonstrated pain on coughing, which was aggravated early in the morning and during the night. Clear or purulent nasal discharge was observed with weakened or no vesicular breath sounds during lung auscultation. Death rattles and pleural friction sounds were heard. The stool was loose and contained blood in secondary diarrhea. Lameness and joint abscesses were observed along with secondary arthritis. Few severely affected cattle showed a poor mental state, disordered and lusterless hair, emaciation, or dry cough. In the late stage of infection, the affected cattle were severely anorectic and lay down; they exhibited shortness of breath or weak breathing and eventually died of respiratory failure.

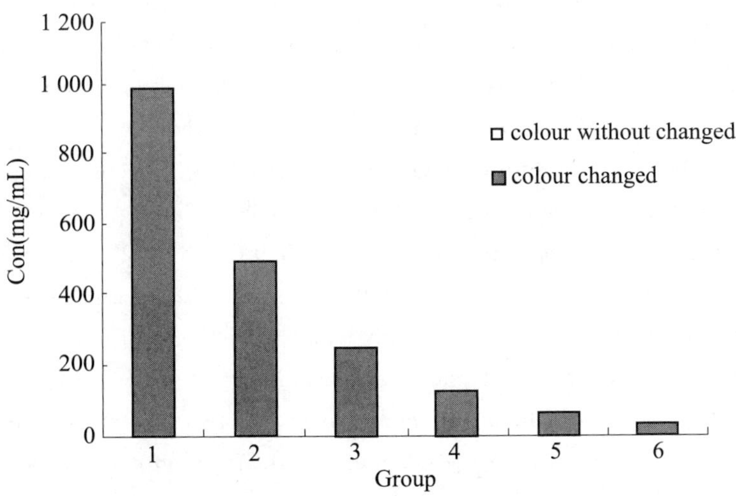

Figure 3 Inhibitory effect of Ban Huang oral liquid on the growth of M.bovis in vitro.

Note: The X-axis values (1-6) represent different concentrations of drug (1 000mg/mL, 500mg/mL, 250mg/mL, 125mg/mL, 62.5mg/mL, 31.25mg/mL); 1 and 2, color changed; 3-6, no color changes; the lowest concentration of drug that inhibited visible growth of *M.bovis* was 500mg/mL.

As shown in Figure 4, the results of laboratory pathogen testing indicated that cattle infection was mainly caused by M.bovis. These results also demonstrated that Ban Huang oral liquid had the strongest inhibitory effect on M.bovis and the weakest inhibitory effect on BPIV-3, among the tested pathogens. After assessing the pathogen testing results with clinical symptoms, 147 cattle were diagnosed as having bovine respiratory diseases caused by M.bovis, bovine respiratory syncytial virus, M.haemolytica, and BPIV-3.

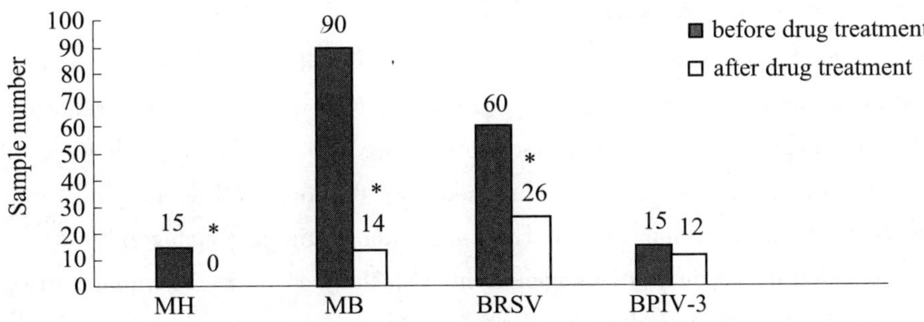

Figure 4 Results of laboratory pathogen testing in 147 cattle.

Note: MH: Mannheimia haemolytica; MB: Mycoplasma bovis; BRSV: bovine respiratory syncytial virus; BPIV-3: Because Pasteurella multocida was not detected before or after drug treatment, results for this organism are not shown. *, significant difference.

3.6 Treatment outcome

After drug treatment for 7 days, clinical symptoms completely disappeared in some cattle, in which the results of laboratory testing for pathogens were negative; moreover, these cattle showed normal body temperature, uniform breathing without any coughing, and no death rattles or diarrhea.

These cattle had a good mental state, glossy hair, and normal appetite; thus, they were considered cured. Some of the cattle tested positive for pathogens, but their symptoms were alleviated; these animals were considered responsive. Other animals showed no alleviation or even aggravation of symptoms; treatment was considered ineffective in these animals.

Variation in the body temperature of drug control group 1, drug control group 2, and the experimental drug treatment group over 7 days of treatment is shown in Fig.5. The average temperature of drug control group 1 declined from 41.1℃ to 38.9℃, a 1.2℃ reduction. The average temperature of drug control group 2 declined from 40.8℃ to 38.8℃, a 2.0℃ reduction. For the experimental drug treatment group, the average body temperature declined from 41.1℃ to 38.6℃, a 2.5℃ reduction. These results show that there was a significant decline in the body temperature of each group within 24h after drug administration. Moreover, there was a significant difference in the decline in the body temperature of the experimental drug treatment group in comparison with that of drug control group 1 and drug control group 2 ($P<0.05$).

After 7 days of treatment, 79 animals in the experimental drug treatment group were cured (90.80% cure rate), whereas 82 were responsive and 4 did not respond to the treatment, which was deemed ineffective. In drug control group 1, 23 animals were cured (76.67% cure rate) and 26 were responsive; the treatment was ineffective in 4 animals. In drug control group 2, 25 animals (83.33% cure rate) and 27 were responsive; the treatment was ineffective in 3 animals. These results are shown in Figure 6. No animal died in the experimental drug group, in which the cure rate was increased by 14.13% compared with that of drug control group 1 and by 7.47% compared with that of drug control group 2 (both $P<0.05$). Thus, Ban Huang oral liquid was effective against bovine respiratory disease, superior to Shuang Hang Lian oral liquid and Tilmicosin alone, and nontoxic.

4 DISCUSSION AND CONCLUSIONS

Berberine hydrochloride and baicalin were chosen as quality control indices. Other components were analyzed using thin layer chromatography (results not shown). The concentrations of berberine hydrochloride and baicalin in the drug were measured using HPLC. In batch production, the concentrations of berberine hydrochloride and baicalin should be the top priorities of quality control testing if the sources of crude drug are not uniform. The application of the chromatographic fingerprint technique to the evaluation of the quality homogeneity and stability of crude medicinal herbs, semi-finished drugs, and finished drug products is currently under study.

M.bovis infection impairs immunity in the respiratory mucosa of cattle. Most bovine respiratory diseases occur in calves during long-distance transportation. Immunity greatly declines after long-distance transportation, when the cattle are likely to show stress responses and be infected by M.bovis and other pathogens. Chinese medicinal herbs destroy the cell membranes and cell walls of pathogens, inhibiting synthesis and expression of proteins. Moreover, TCM herbs may also act directly on DNA structure and interfere with its normal function (Tang et al., 2005), thereby affecting microbial growth and proliferation (Chen et al., 1994), eventually leading to pathogen death (Snyder and Gillies, 2002; Mittra et al., 2000). In this study, the ability of Ban Huang oral liquid to inhibit the growth of M.bovis was confirmed in vitro. However, the effectiveness of the drug in vivo re-

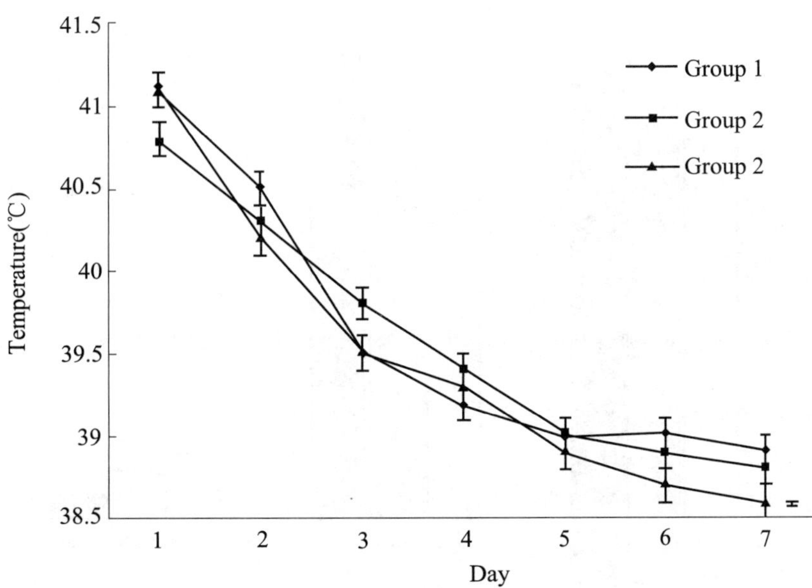

Figure 5 Variation in the body temperature of drug control group 1, drug control group 2, and the experimental drug treatment group over 7 days of treatment.

Note: The average temperature of drug control group 1 declined from 41. 1℃ to 38. 9℃, a 1. 2℃ reduction. The average temperature of drug control group 2 declined from 40. 8℃ to 38. 8℃, a 2. 0℃ reduction. For the experimental drug treatment group, the average body temperature dropped from 41. 1℃ to 38. 6℃, a 2. 5℃ reduction. *, significant difference.

quires further experimental verification.

Simon Frantz, the chief editor of a column on drug discovery in the journal Nature, said that researchers should "forget drugs carefully designed to hit one particular molecule—a better way of treating some complex diseases may be to aim for several targets at once" (Lee et al., 2005). Both humans and animals are complex organisms, which need to be controlled by complex drugs. TCM uses a combination of active components to reach this goal (Kitano, 2007), triggering the immune system and enhancing immunity (Huo and Li, 2002; Shen et al., 2004; Zhang et al., 1998), and thus preventing and treating disease.

Ban Huang oral liquid is easy to prepare, while its quality control methodology is simple. The quality and stability of Ban Huang oral liquid can meet the demands of mass production. Thus, bovine respiratory diseases caused by a mixed infection can be effectively controlled using Ban Huang oral liquid. As a Chinese medicinal product, Ban Huang oral liquid has no toxicity and causes no drug residue or drug resistance problems. Ban Huang oral liquid has great potential for use as a new veterinary drug for respiratory diseases; in addition to therapeutic effects, this preparation has preventive effects against respiratory diseases in livestock and poultry. Because of its good efficacy and safety, Ban Huang oral liquid warrants extensive promotion as a veterinary drug. Moreover, Ban Huang oral liquid provides inspiration for system-oriented drug design and efforts aimed at discovering new drugs to treat and prevent bovine respiratory diseases. Because bovine respiratory diseases have a high rate of incidence and cause significant harm, drug-based prevention is only part of the

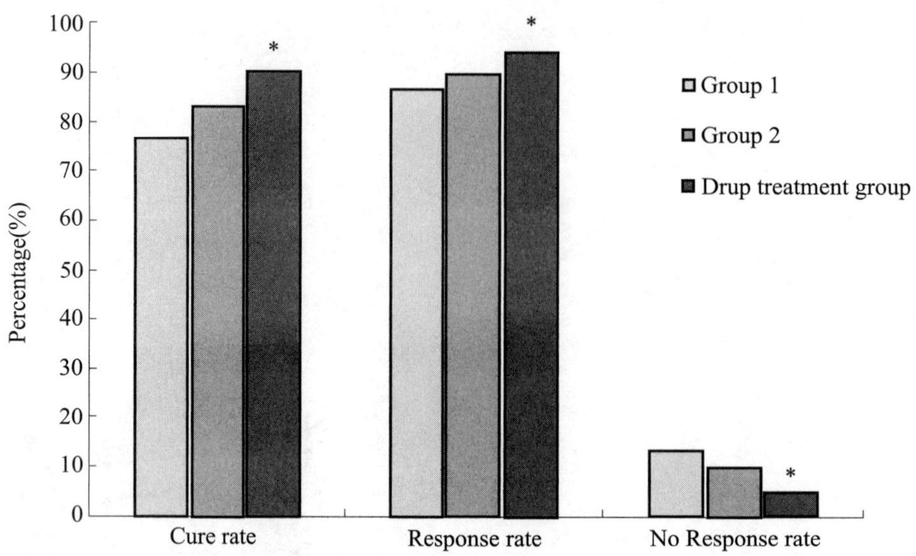

Figure 6 Inhibitory effect of Ban Huang oral liquid against bovine respiratory diseases caused by a mixed infection.

Note: The number of ineffective cases was defined as the number of cattle showing no alleviation of symptoms at the end of treatment. Cure rate (%) = (number cured/number affected) × 100. Response rate (%) = (number cured + number of responsive cattle) /number affected × 100. No response rate (%) = (number of ineffective cattle/number affected) × 100. *, significant difference.

treatment strategy. All necessary control and prevention measures should be adopted to ensure the healthy development of the animal farming industry by optimizing the breeding and rearing environment, as well as transportation, vaccination, safety isolation, and medication methods.

Acknowledgments

This work was supported by the earmarked fund for the China Agriculture Research System (cars-38) and the Special Fund for Agro-scientific Research in the Public Interest (No.201303038-4).

REFERENCES OMITTED

(发表于《Afr J Tradit Complement Altern Med》, 院选 SCI, IF: 0.553)

Characterization of the Complete Mitochondrial Genome of Kunlun Mountain Type Wild Yak (*Bos mutus*)

Xiaoyun WU, Min CHU, Xuezhi DING, Xian GUO, Hongbo WANG, Pengjia BAO, Chunnian LIANG*, Ping YAN*

(Key Laboratory for Yak Breeding Engineering/Gansu Province and Lanzhou Institute of Husbandry and Pharmaceutical Sciences, Chinese Academy of Agricultural Science, Lanzhou 730050, China)

Abstract: The Kunlun Mountain type wild yak (*Bos mutus*) is one of the most charismatic members of the Tibet/Qinghai Plateau, but its population is reduced. In this study, the complete mitochondrial genome of Kunlun Mountain type wild yak has been determined by polymerase chain reaction method for the first time. The circular genome was 16, 325 base pairs (bp), containing 13 protein-coding genes, 2 rRNA genes, 22 tRNA genes and a non-coding control region. The base composition of the genome was A (33.70%), C (25.81%), G (13.22%) and T (27.28%) with an A+T content of 60.98%. A phylogenetic analysis based on complete mitochondrial genome sequences showed that Kunlun Mountain type wild yak (Bos mutus) was closer to domesticated yaks (Bos gruuniens) and bison (Bison bison).

Key words: *Bos mutus*; Mitochondrial genome; Phylogeny

The wild yaks occur on the Tibetan Plateau at elevations of 3 000–5 500m, their range comprising the western edge of Gansu Province, Qinghai Province, the southern rim of the Xinjiang Autonomous Region, and the Tibet Autonomous (Schaller and Liu 1996). In the past 100 years, the population of wild yaks has declined rapidly due to human poaching and habitat destruction (Shi et al., 2016). World population estimates of wild yaks were fewer than 15,000 (Liang et al., 2017). There are two "ecological types" of wild yaks have been described based on body characteristics and geographical location: the smaller, more docile Qilian Mountain type and the massive, aggressive Kunlun Mountain type (Leslie and Schaller 2009). Mitochondrial DNA (mtDNA) was proven one of the most useful molecular tools in population genetics and molecular phylogenetics (Galtier et al., 2009; Wang et al., 2010). The complete mitogenome of Kunlun Mountain type wild yak (GenBank: KY829451) was sequenced, assembled and characterized in this study to provide molecular genetic evidence for the conservation of wild yak.

* Corresponding author, E-mail: chunnian2006@163.com; pingyanlz@163.com

Total genomic DNA was extracted from the individual blood samples using the genomic DNA isolation kit (TIANGENE, Beijing, China) according to the manufacturer's instructions. Six primers were designed to amplify the PCR products for sequencing.The sequencing results were then assembled by DNAStar 5. 01 software (DNASTAR, Inc., Madison, WI).The locations of protein-coding genes were determined by comparing with the corresponding sequences of other yak species.A physical map of the genome was generated by using the web server OGDRAW (http: //ogdraw.mpimp-golm.mpg.de/) (Lohse et al., 2013).

The complete mitochondrial genome of Kunlun Mountain type wild yak was 16, 325 bp, which is consisted of 37 mitochondrial genes, including 13 protein-coding genes, 22 tRNA genes, two rRNA genes (12S and 16S rRNA) and one D-loop region.The base composition of the genome is asymmetric (33. 70% A, 13. 22% G, 25. 81% C, and 27. 28%T) with an overall A+T content

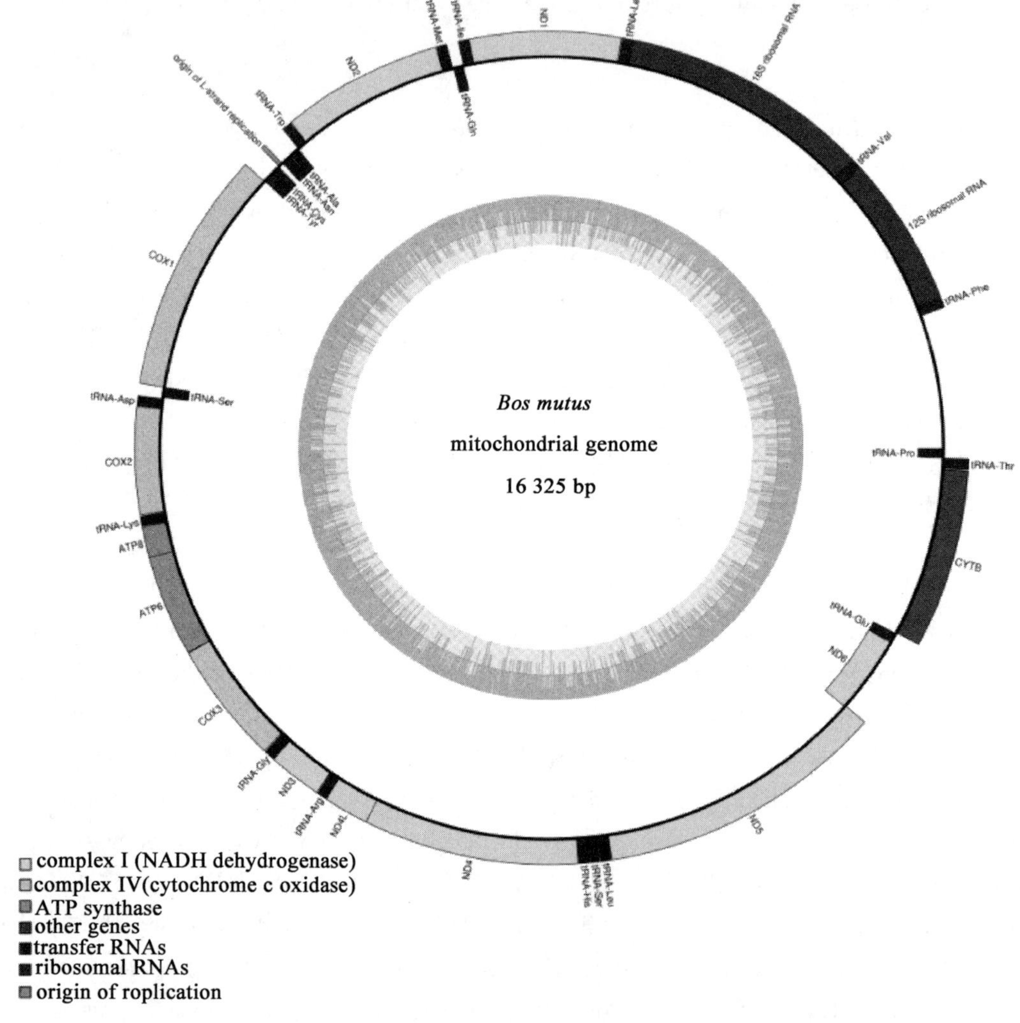

Fig.1 Physical map of the Kunlun Mountain type wild yak mitochondrial genome.

Note: Genes drawn outside of the circle are transcribed in the clockwise direction, whereas those inside are transcribed in the counterclockwise direction.Areas dashed light and darker gray in the inner circle indicates the A+T and G+C contents of the genome, respectively

of 60.97%.

To validate the phylogenetic position of Kunlun Mountain type wild yak, we constructed a neighbor-joining (NJ) tree using 1000 bootstrap replicates using MEGA5 (Tamura et al., 2011) based on 13 complete mitochondrial genomes of Bovidae. The phylogenetic tree showed that the 13 species are clustered into two orders, which is consistent with the previous studies (Ritz et al., 2000; Hassanin et al., 2013). Kunlun Mountain type wild yak was closer to domesticated yaks (Bos gruuniens) and bison (Bison bison) (Figs.1, 2).

In summary, the complete mitochondrial genome and genomic data of Kunlun Mountain type wild yak will provide the genetic basis for protection of germplasm resources of wild yak (Bos mutus).

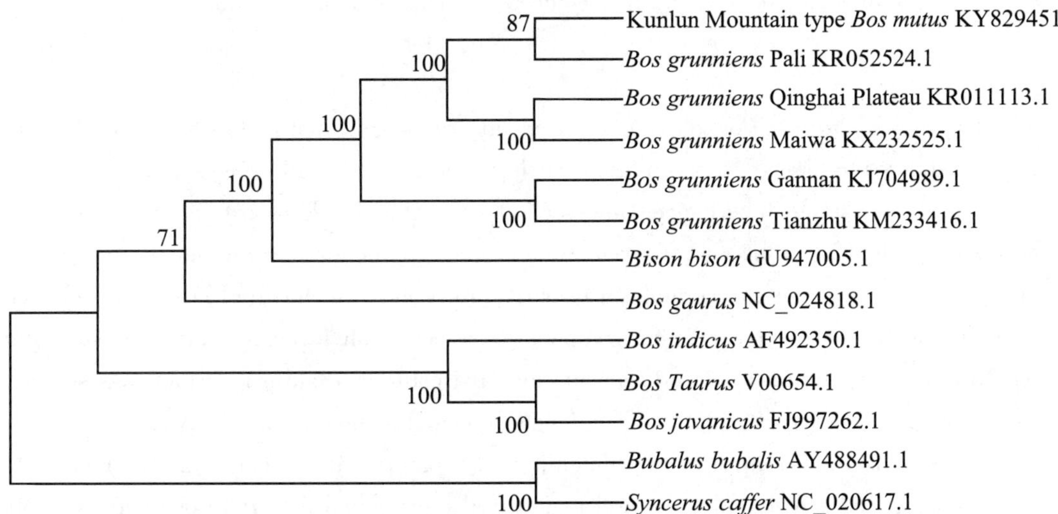

Fig.2 Phylogenetic analysis of selected species from Bovidae based on the neighbor-joining (NJ) analysis of the concatenated coding sequences of 13 mitochondrial genomes. The bootstrap values were based on 1000 replicates, and are shown next to the branches

Acknowledgements This study was supported by grants from the Agricultural Science and Technology Innovation Program (CAAS-ASTIP-2014-LIHPS-01) and National Beef Cattle Industry Technology & System (CARS-38).

REFERENCES OMITTED

(发表于《Conservation Genet Resour》, 院选 SCI, IF: 0.470)

Complete Mitochondrial Genome of Anxi Cattle (*Bos taurus*)

Xian GUO, Xuezhi DING, Xiaoyun WU, Pengjia BAO,
Lin XIONG, Ping YAN, Jie PEI*

(Lanzhou Institute of Husbandry and Pharmaceutical Sciences, Chinese Academy of Agricultural Sciences, Lanzhou 730050, China)

Abstract: Anxi cattle (*Bos taurus*) is an endangered species native to China.It has strong adaptive abilities to harsh desert conditions such as heat, cold, and drought tolerance; disease resistance; and stable inheritance. However, it is on the brink of extinction because of its shrinking population in the past few years due to the introduction and invasion of exotic species and the lack of awareness of germplasm resource protection.Mitochondrial DNA, a well-known genetic marker, has been successfully applied to molecular phylogenetics and population genetics.This study reported a whole mitogenome for Anxi cattle (B.taurus), which was sequenced using Illumina Hiseq 2 500 platform.The length of complete mitogenome of Anxi cattle (B.taurus) breed was 16,341 bp (bp), containing 13 protein-coding genes, 2 rRNA genes, 22 tRNA genes, and a non-coding control region.It also contained the origin of L-strand replication (32 bp long), which was between tRNA-Asn and tRNA-Cys.The base composition of the genome was A (33.4%), C (26.0%), G (13.4%), and T (27.2%) with an A + T content of 60.6%.The molecular phylogenetic analysis revealed four major lineages, and the Anxi cattle breed had a close genetic connection with B.taurus and B.indicus clusters but was still distinctly outside them.This study identified the phylogenetic status of Anxi cattle (B.taurus) breed and offered essential molecular genetic information for the conservation of germplasm resources.

Key words: Anxi cattle (*Bos taurus*); Illumina sequencing; Mitochondrial genome

Anxi cattle (*Bos taurus*), a valuable breeding resource from Anxi county, Gansu province, China, has long been used to transport goods due to its genetic adaptation to the desert environment. It can survive against extremely hot weather, bitter cold, and high winds, indicating that this species is rich with a large collection of excellent genes to adapt to extreme environments.In addition, Anxi cattle is the best original resource to cultivate high-yield breeds and develop animal husbandry

* Corresponding author, E-mail: yanping@caas.cn; peijie@caas.cn

because of the stable pure lines despite difficult survival conditions for exotic species in this area. However, Anxi cattle has been in danger of becoming extinct in recent years as a result of the popularization of agricultural mechanization and transition from transportation to meat production in livestock farming.Also, the lack of awareness of the conservation of Anxi cattle germplasm resources has led to the reduction in its population size.Mitochondrial DNA has been widely applied as a powerful molecular tool in biodiversity conservation and population genetics (Lai et al., 2006; Mannen et al., 2004).In the present study, the complete mitogenome (GenBank: MF281256) of B.taurus was sequenced, assembled, and characterized to provide detailed molecular genetic information for the conservation and research of Anxi cattle.

The total DNA was isolated from the blood of *B.Taurus* collected from Anxi County (Gansu, China) and sheared to construct a genomic library with the insert fragment of approximately 180 bp (bp).Then, the fragments were end-repaired, A-tailed, tagged, and subjected to NGS on the Illumina HiSeq 2 500 Sequencing System (Illumina, CA, USA) according to the standard operating protocol.A total of 13, 145, 432 raw paired reads with an average length of 150 bp were produced.High-quality clean reads were obtained by removing poor-quality sequences and filtering data with CLC Genomics Workbench v10 (CLC Bio, Aarhus, Denmark) and used to assemble the mitochondrial genome using the program MITObim v1. 8 (Hahn et al., 2013), using the sequence of B.taurus (GenBank: AF492351) (Hiendleder et al., 2008) as an initial reference sequence. A total of 3 399 reads were used to assemble with an average coverage of 30. 8×.Then, the consensus sequence produced was proofread.Subsequently, the mitochondrial genome of B.taurus was annotated using GENEIOUS R10 (Biomatters Ltd., Auckland, New Zealand).Finally, the complete mitogenome of Anxi cattle (*B.taurus*) breed was determined, which was 16, 341 bp and consisted of 13 protein-coding genes, 22 tRNAs genes, 2 rRNAs genes, and a non-coding control region (Fig.1).Also, the origin of L-strand replication (32 bp long) was between tRNA-Asn and tRNA-Cys.The content of A, C, T, and G in the mitochondrial genome was 33. 4, 26, 27. 2, and 13. 4%, respectively.An overall A + T content of the whole mitochondrial genome was 60. 6%.

Aphylogenetic tree was constructed by combining ten complete mitogenome sequences previously reported from Bos with a mitogenome of Bubalus bubalis as an outgroup to determine the phylogenetics status of Anxi cattle (*B.taurus*) breed.Based on the maximum likelihood method, a phylogenetic tree was created in Mega 6 (Mega Inc., NJ, USA) (Tamura et al., 2011) with 1 000 supports estimated from 1 000 replicate searches.Both NJ and ML trees (Fig.2) showed an identical topology supported well by bootstrap values and were consistent with a previous study (Hiendleder et al., 2008).Phylogenetic analyses suggested distinct branches.Anxi cattle formed a separated clade (Troy et al., 2001) and was distantly outside *B. taurus* and *B. indicus* groups (Kikkawa et al., 2003).This research produced a compete mitochondrial genome of *B. taurus*, presented significant molecular genetic information for further investigating the population genetic relationships of B.Taurus mitochondrial genome, and provided scientific guidance for protecting and developing Anxi cattle as a resource.

Acknowledgements: The work was supported or partly supported by grants from China Agriculture Research System (CARS-37), Science and Technology Support Projects in Gansu Province

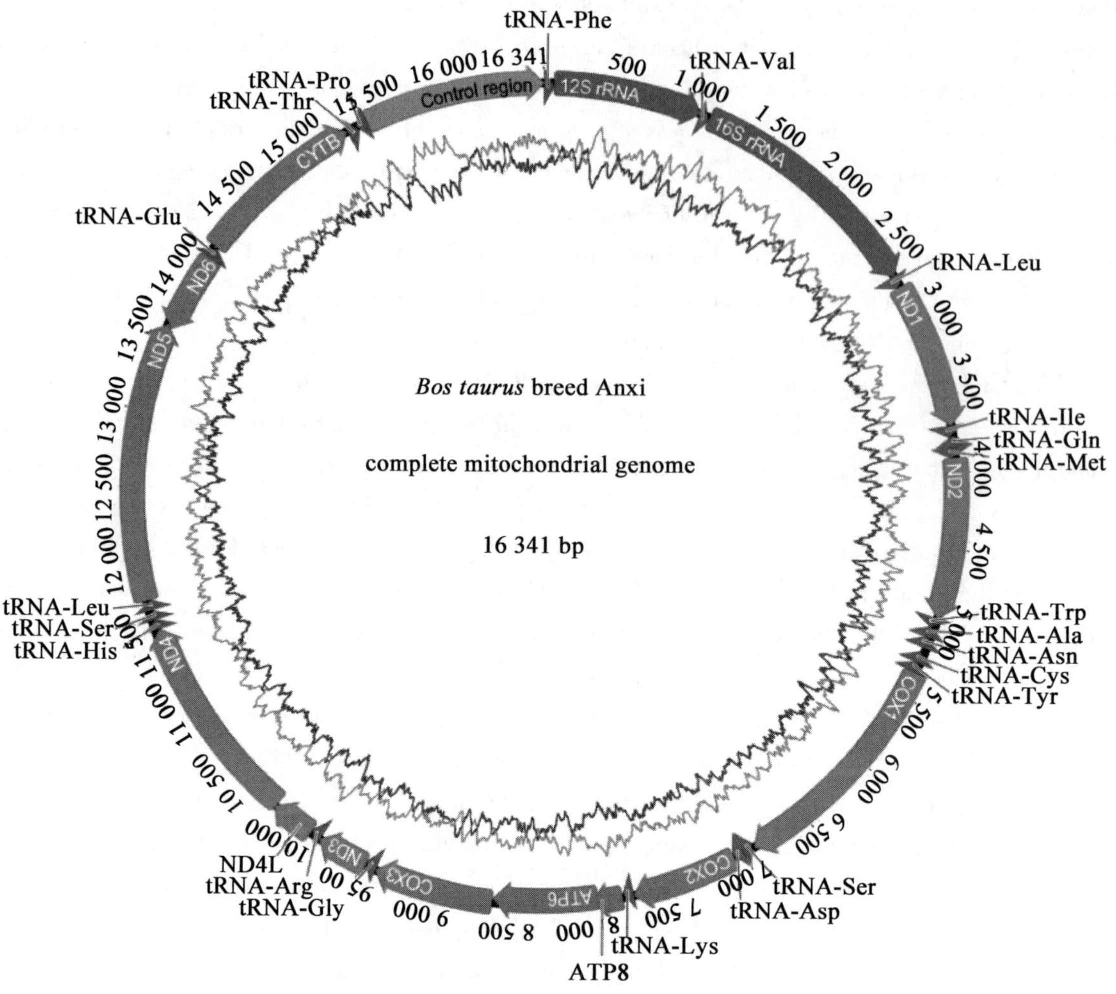

Fig.1 The gene map of the mitochondrial genome for Anxi cattle (*Bos taurus*)

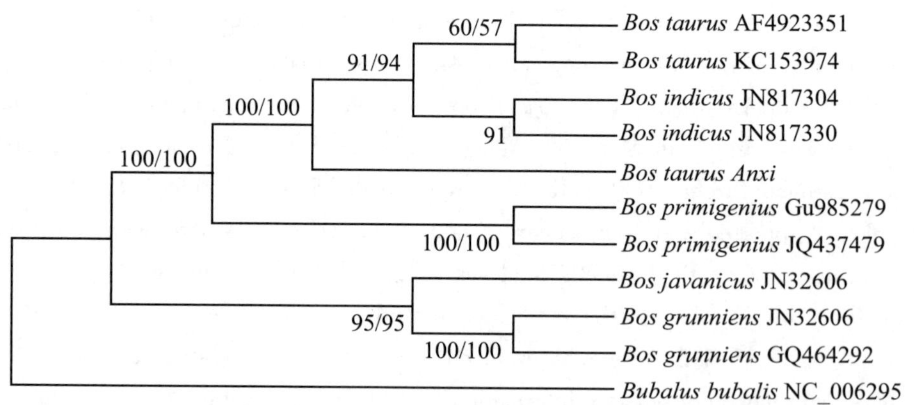

Fig.2 Phylogenetic tree of 5 Bos species based on 13 mitochondrial protein-coding genes. The values at the nodes correspond to bootstrap support in percentages for neighbor-joining method (first) and maximum likelihood method (second)

(1504NKCA052), Innovation Project of Chinese Academy of Agricultural Sciences (CAAS-ASTIP-2014-LIHPS-01), and Central Public-interest Scientific Institution Basal Research Fund (1610322016006).

REFERENCES OMITTED

(发表于《Conservation Genet Resour》，院选 SCI，IF：0.470)

Complete Mitochondrial Genome of Qingyang Gonkey (*Equus asinus*)

Xian GUO[1], Pengjia BAO[1], Jie PEI[1], Xuezhi DING[1], Chunnian LIANG[1], Ping YAN[1], Dengxue LU[2*]

(1. Lanzhou Institute of Husbandry and Pharmaceutical Sciences, Chinese Academy of Agricultural Sciences, Lanzhou 730050, Gansu, China; 2. Institute of Biology, Gansu Academy of Sciences, Lanzhou 730000, China)

Abstract: The present study aimed to report the entire mitochondrial genome of Qingyang donkey (*Equus asinus*). Donkeys play an important role in the daily lives of human beings. The population of Chinese donkey breeds has undergone an enormous decline with the advance of agricultural mechanization, rapid development of transportation and introduction of new different breeds over the past few decades. As a result, a large number of donkey breeds are close to extinction, especially Qingyang donkey. Mitochondrial DNA is used as a molecular tool in research on biodiversity conservation and sustainable utilization of biological resources. The circular genome sequenced using the Illumina Hiseq 2 500 platform was 16,568 base pairs (bp), containing 13 protein-coding genes, 2 rRNA genes, 22 tRNA genes and a non-coding control region. The initiation site of the L-strand replication origin, 32 bp in length, was between tRNA-Asn and tRNA-Cys. The base composition of the genome was A (32.4%), C (28.7%), G (13.2%) and T (25.7%) with an A+T content of 58.1%. Molecular phylogeny demonstrated two separated clades in Equidae and the presence of a distant relationship between the domestic donkey breeds and Asian wild asses, indicating that domestic donkeys might not probably have originated from Asian wild asses. However, African wild asses were grouped as a sister to domestic donkeys. This study presented the complete mitochondrial genome of E. asinus, providing the important genetic background of protection and adequate utilization of domestic donkey breeds.

Key words: Equus asinus; Complete mitogenome; Illumina sequencing

Qingyang donkey (*Equus asinus*) in the family Equidae belongs to a medium-sized domestic donkey breed and is commonly used to transport goods and produce meat. However, unfortunately, its number recently has fallen sharply for the sake of economic benefits and owing to neglect of germ-

* Corresponding author, E-mail: yanping@caas.cn; ludengxue@126.com

plasm conservation.Mitochondrial DNA (mtDNA) has been emerged as a well-accepted molecular tool with extensive applications including origin and evolution, genetic relationships and population genetic structure of livestock (Lippold et al., 2011; Achilli et al., 2012).The complete mitogenome of Qingyang donkey (GenBank: KX683425) was sequenced, assembled and characterized in this study to provide molecular genetic evidence for the conservation of this breed.

The high-quality total genomic DNA was isolated from the blood of Qingyang donkey derived from the Qingyang City of Gansu Province in China (35°10′-37°20′N, 106°45′-108°45′E).The extracted DNA was sheared to produce several fragments of approximately 180 bp, which were end-repaired, A-tailed and tagged. This was followed by the whole genomic sequencing using the Illumina HiSeq 2 500 Sequencing System (Illumina, CA, USA) according to the standard operation protocol.

A total of 16, 285, 492 raw paired reads with an average length of 125 bp were generated. Stringent data quality control was performed to obtain high-quality clean reads by removing poor-quality sequences and filtering using CLC Genomics Workbench v9 (CLC Bio, Aarhus, Denmark).The complete mitochondrial genome of Qingyang donkey was assembled using the program MITObim v1. 8 (Hahn et al., 2013) and then aligned against that of E.ferus przew-alskii (GenBank: JN398402) (Achilli et al., 2012) as an initial reference.A total of 2 335 reads were used to assemble and achieve an average coverage of 17. 4×.Subsequently, the consensus sequence was obtained by proofreading strictly.The final size of the complete mitogenome of the Qingyang donkey was 16,568bp, containing A (32. 4%), C (28. 7%), G (13. 2%) and T (25. 7%) with an A+T content of 58. 1%.This mitochondrial genome was annotated after assembly using Geneious R9 (Biomatters Ltd., Auckland, New Zealand).The present analysis demonstrated that the circular genome comprised 13 protein-coding genes (PCGs), 2 rRNA genes, 22 tRNA genes and a non-coding control region (Fig.1), and the 32-bp long initiation site of L-strand replication origin (OL) was between tRNA-Asn and tRNA-Cys. Simultaneously, the abnormal termination codon and frame shift mutation were not observed, indicating that the reliability of the assembly was verified.

Also, phylogenetic trees were constructed to determine the phylogenetic position of *E. asinus* with the sequences from the Qingyang donkey and other previously published breeds of donkeys (Mendez et al., 2004; Sun et al., 2016).The entire mitogenome of rhinos involved served as an out-group. The concatenated sequences of the 13 PCGs were aligned separately using ClustalW within MEGA 6 (MEGA Inc., NJ, USA) (Tamura et al., 2011; Zhang et al., 2016).The optimal nucleotide substitution model was selected using jModelTest v.0. 1. 1 (Posada 2008) based on Akaike information criterion (AIC) values.A maximum likelihood (ML) analysis was performed using PhyML 3. 0 (Guindon and Gascuel 2003) with the GTR + I + G model.A total of 1 000 bootstrap replicates were run, and bootstrapping frequencies were calculated. Based on the neighbor-joining (NJ) method, the other phylogenetic tree was created in MEGA 6 with 1 000 supports estimated from 1 000 replicate searches.Both ML and NJ trees based on mitochondrial PCG sequences showed identical topology with high boot-strap values, which was consistent with previous reports (Pereira et al., 2004; Chen et al., 2006; Sun et al., 2016) (Fig.2).The phylogenetic tree revealed two distinct clades presented in Equidae. As expected, all the wild asses and domestic donkeys formed a monophyletic group. The Qingyang donkey (E.asinus KX683425) was clustered with the Chinese domestic donkey but distantly related to the Asian wild ass, further supporting the point that Chinese domestic donkeys probably did not originate from Asian wild asses. In contrast, African wild asses and domestic donkeys were grouped as a sister group, indicating that the domestic donkey might have originated from the African wild ass with the long-term domestication (Pereira et al., 2004). This study presented the complete mitogenome of E.asinus and provided the genetic basis for molecular genetic studies on the protection of germplasm resources and breeding variety of Qingyang donkey (*E.asinus*).

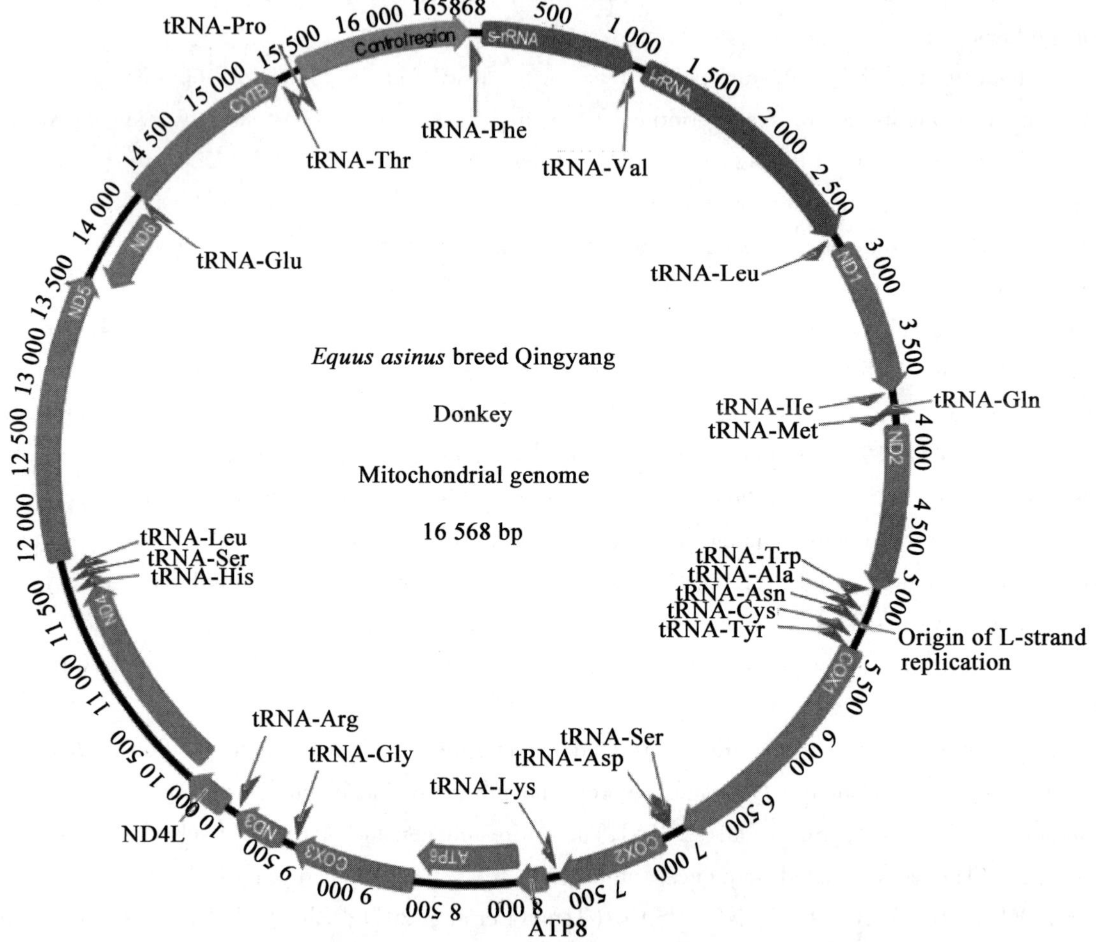

Fig.1 The gene map of the mitochondrial genome for Qingyang donkey (Equus asinus)

Acknowledgements: This study was supported or partly supported by the National Natural Science Foundation of China (31301976), Science and Technology Support Projects in Gansu Province (1504NKCA052) and the Innovation Project of Chinese Academy of Agricultural Sciences (CAAS-ASTIP-2014-LIHPS-01).

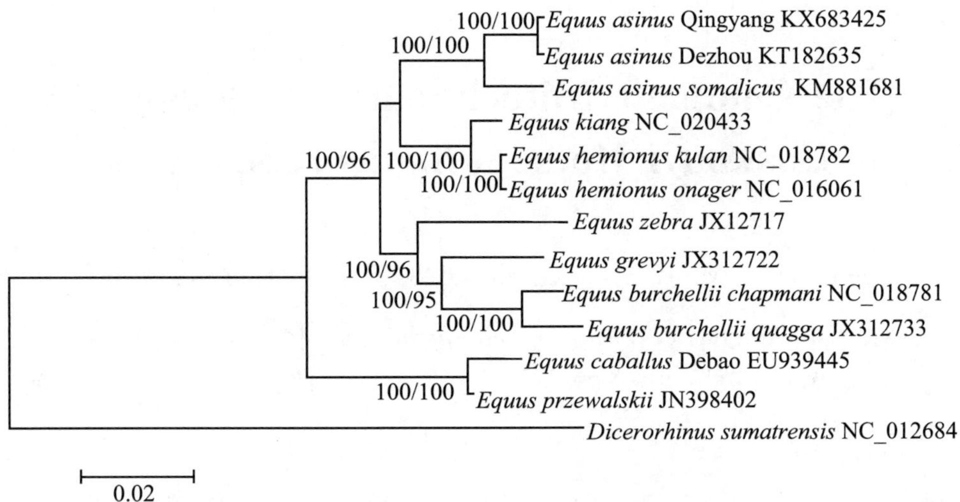

Fig.2 Phylogenetic analysis of selected species from Equidae based on the maximum likelihood (ML) and neighbor-joining (NJ) analysis of the concatenated coding sequences of 13 mitochondrial PCGs. The values at the nodes correspond to ML (right) and NJ (left) boot-strap support in percentages

REFERENCES OMITTED

（发表于《Conservation Genet Resour》，院选 SCI，IF：0.470）

The Complete Mitochondrial Genome of Chakouyi Horse (*Equus caballus*)

Xian GUO, Min CHU, Xuezhi DING, Jie PEI, Ping YAN*

(Lanzhou Institute of Husbandry and Pharmaceutical Sciences, Chinese Academy of Agricultural Sciences, Lanzhou 730050, China)

Abstract: Chakouyi horse (*Equus caballus*) is a native and endangered breed of riding horse in China. This study reports the complete mitochondrial genome of the Chakouyi horse using next-generation sequencing. The full length of the circular genome is 16,656bp. It consists of 37 typical animal mitochondrial genes including 13 protein-coding genes (PCGs), 22 transfer RNA (tRNA) and 2 ribosomal RNA (rRNA) genes and a control region. The orientation and arrangement of these genes is similar to that of the other sequenced horse mitochondrial genomes. The overall nucleotide composition is: 32.2% A, 25.8% T, 28.6% C, and 13.4% G, with a total A + T content of 58.0%. Phylogenetic analysis indicated that the genome of Chakouyi horse clustered within the lineage of most Asian horses and is closely related to Russian riding horse.

Key words: *Equus caballus*; Chakouyi horse; Mitochondrial genome; MITObim

Chakouyi horse (*Equus caballus*) is an old breed of Chinese riding horse characterized by a smooth gait. This breed has a narrow geographical distribution, mainly conined to Tianzhu Tibetan Autonomous County (Gansu province). In recent years, the number of Chakouyi horses decreased (60%) compared with that of the 1980s in Tianzhu. Although the local government protected the breed, the number of purebred Chakouyi horses is less than 3 000 currently (Gao 2011). Although the genetic diversity of Chakouyi horse is based on the mitochondrial control region and Cytb sequences (Yue et al., 2012; Yang et al., 2016), its mitochondrial genome and phylogenetic position are still unknown. Mitochondrial genome has been widely used for phylogeographic studies, population genetics and molecular diagnostics of domestic animals (Lippold et al., 2011). In this work, we sequenced the mitochondrial genome to provide useful molecular markers for conservation biology, population genetics, and evolutionary studies of endangered Chakouyi horse.

Total genomic DNA was extracted from the blood of Chakouyi horse. DNA shotgun library was constructed and sequenced with the Illumina Hiseq 2 500 Sequencing System (Illumina, CA,

* Corresponding author, E-mail: yanping@ caas.cn

USA).A total of 14. 76M of 125 bp raw reads were trimmed and the inal mitochondrial genome was reconstructed using MITObim v1. 8 with default parameters (Hahn et al., 2013).The mitogenome of Equus caballus breed Debao (GenBank: EU939445) was employed as the reference sequence (Jiang et al., 2011).The genome was annotated using Mitos web server (Bernt et al., 2013).The tRNA genes and their secondary structures were also identiied with tRNAscan-SE 1. 21 (Lowe and Eddy 1997).

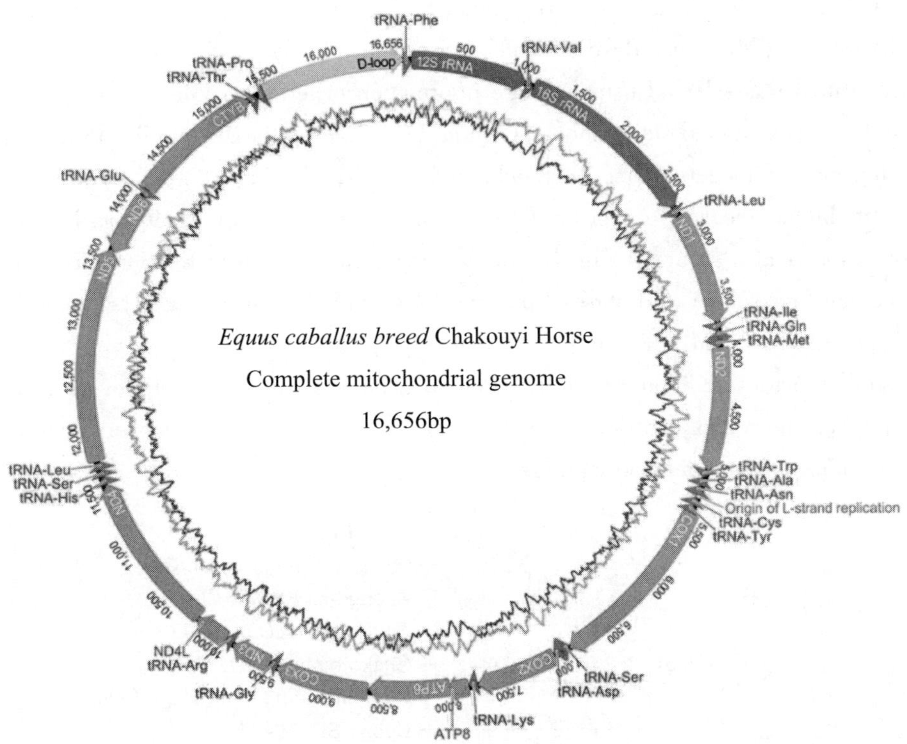

Fig.1 The mitochondrial genome map of Chakouyi horse (*Equus caballus*)

The complete mitochondrial genome of Chakouyi horse (GenBank: KU575247) is 16, 656 bp in size.It contains 37 mitochondrial genes including 13 protein-coding genes (PCGs), 22 transfer RNA (tRNA), 2 ribosomal RNA (rRNA) genes, and a control region (D-Loop) (Fig.1). The gene order was identical to the reported horse mitochondrial genomes.Similar to the other horses, the heavy strand harbors 12 PCGs and 14 tRNA genes, while the light strand carries the ND6 and other eight tRNA genes. The base composition of Chakouyi horse mitochondrial genome is: A (32. 2%), T (25. 8%), C (28. 6%), and G (13. 4%), with a total A + T content of 58. 0%. The average A + T content of PCGs is 57. 5%, ranging from 54. 6% (COIII) to 64. 2% (ATP8). The highest A + T content of the coding region (62. 1%) was found in the tRNA genes. All the PCGs comprise a start codon ATG, except for ND2 and ND3, which begin with ATA codon.In addition, COIII, ND3 and ND4 are terminated with an incomplete stop codon T, while 10 PCGs terminate in a typical stop codon TAA or TAG.All tRNA genes are folded into a common cloverleaf secondary structure except for $tRNA^{Ser(AGY)}$ lacking the dihydro-uridine (DHU) arm.The typical origin of L-strand replication (O_L) is 32 bp between $tRNA^{Ans}$ and $tRNA^{Cys}$, which is folded into a stable stem-loop secondary structure.The 1 192bp D-loop region is located between tRNA^{Pro} and tRNA^{Phe}, which contains several conserved sequences involved in the replication and transcription of mitochondrial genome.

Phylogenetic analysis of selected representative breeds (Xu et al., 2007; Lippold et al., 2011; Der Sarkissian et al., 2015) was conducted using the concatenated coding sequences of 13 mitochondrial PCGs (aligned length 11, 391 bp) with codon partitioning scheme.MrBayes (Ronquist and Huelsenbeck 2003) and RAxML (Stamatakis et al., 2005) were used to reconstruct the maximum likelihood (ML) and Bayesian inference (BI) phylogeny.The best-it model GTR + I + G for each partition was selected using Akaike information criterion in jModelTest (Posada 2008). Both ML and BI trees showed stable topology (Fig.2), which is largely similar to the reported molecular phylogenies of modern horse (Lippold et al., 2011; Ning et al., 2014). Most Chinese breeds exhibit similar lineage, except for Tibetan horse (Deqin), which clustered with Ardennais. Equus przewalskii is also grouped with this branch indicating the close relationship between extant Asian horses and Przewalskii wild horses.Chakouyi horse and Russian riding horse show a sister relationship supported by an ML bootstrap value of 91%, and a Bayesian posterior probability of 0. 99. Our study provides additional data characterizing the structure and evolution of Chakouyi horse mitochondrial genome, and contributes to the molecular genetics and conservation of this rare breed, using mitochondrial genome markers.

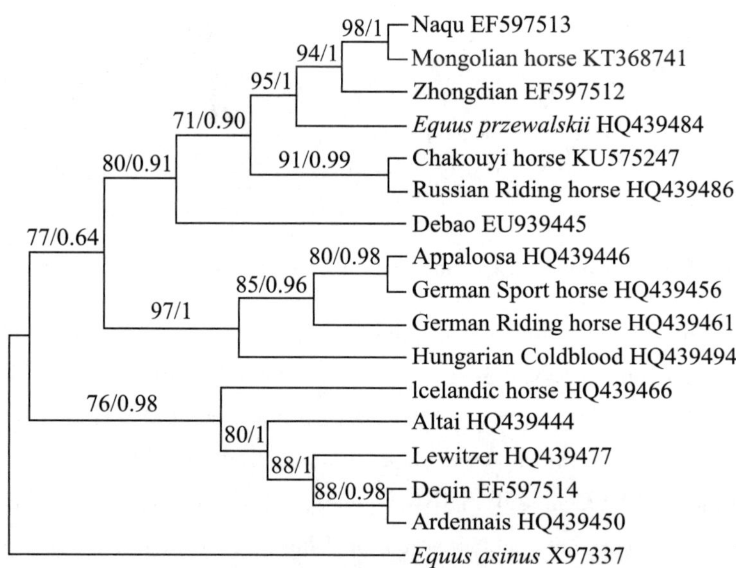

Fig.2 Phylogeny of 15 horses and Equus przewalskii based on the maximum likelihood (ML) and Bayesian inference (BI) analysis of the concatenated coding sequences of 13 mitochondrial PCGs.The nodes correspond to ML bootstrap support values in percentages (left, 1000 resamplings) and Bayesian posterior probabilities (right), respectively

Acknowledgements This study was supported or partly supported by the National Natural Science Foundation of China (31301976), Science and Technology Support Projects in Gansu Province (1504NKCA052) and the Innovation Project of Chinese Academy of Agricultural Sciences

(CAAS-ASTIP-2014-LIHPS-01).

REFERENCES OMITTED

(发表于《Conservation Genet Resour》, 院选 SCI, IF: 0.470)

The Complete Mitochondrial Genome of Zhangmu Cattle (*Bos taurus*)

Xian GUO[1], Pengjia BAO[1], Lin XIONG[1], Yanbin ZHU[2], Wangdui BASANG[2], Xiaoyun WU[1], Xuezhi DING[1], Jie PEI[1*], Ping YAN[1*]

(1. Lanzhou Institute of Husbandry and Pharmaceutical Sciences, Chinese Academy of Agricultural Sciences, Lanzhou 730050, China; 2. Institute of Animal Husbandry and Veterinary Medicine, Tibet Academy of Agriculture and Animal Husbandry Science, Lhasa 850001, China)

Abstract: The Zhangmu cattle (*Bos taurus*) is a rare breed native to the central Himalayas, which has recently suffered drastic reduction in its population. In the present study, the complete mitochondrial genome of Zhangmu cattle was determined.The 16,340 bp circular mitochondrial genome contained a common set of 37 mitochondrial genes, including 13 protein-coding genes, 22 transfer RNA genes, two ribosomal RNA genes, and a control region.The gene order was identical to that of other B.taurus breeds.The overall base composition of this genome was as follows: 33.4% A, 26.0% C, 27.2% T and 13.4% G, with a total A+T content of 60.6%.Both phylogenetic trees based on the maximum likelihood and neighbor joining methods showed that Zhangmu cattle was grouped with other Asian breeds such as Korean cattle.

Key words: Zhangmu cattle; *Bos taurus*; Mitochondrial genome; Himalayas

The Zhangmu cattle (*Bos taurus*) is a rare breed native to the central Himalayas, which has a narrow geographical distribution in Nyalam County.Zhangmu cattle is charactered with the qualities of zebu and yak, which is able to adapt to the plateau and canyon environment. However, until 1998, it was identified as a separate breed.A previous investigation showed that the number of this breed had decreased to 307, including 73 non-breeding individuals.Therefore, Zhangmu cattle was listed as an endangered breed and protected by China's Ministry of Agriculture.There is no molecular information about this cattle breed, which hinders its conservation and reproduction. Mitochondrial genome provides extensive information for species identification, population structure and dynamic research, and has been widely used as a powerful molecular marker for conservation biology and ecology (Boore 1999). This is the first molecular data study on the

* Corresponding author, E-mail: peijie@caas.cn; yanping@caas.cn

Zhangmu cattle, which assessed its phylogenetic position using mitochondrial genome dataset. This mitochondrial genome will provide extensive information for further management and conservation of this vulnerable cattle breed.

The genomic DNA was extracted from the blood of Zhangmu cattle. DNA shotgun library was constructed and sequenced with the Illumina Hiseq 2 500 sequencing system (Illumina, CA, USA). A total of 13.61M of 125 bp raw reads were trimmed using CLC Genomics Workbench (CLC Bio, Aarhus, Denmark). The mitochondrial genome was reconstructed using Geneious (Biomatters Ltd., Auckland, New Zealand) and then annotated with B. taurus (AF492351). Mitos web server (Bernt et al., 2013) and tRNAscan-SE 1.21 (Lowe and Eddy 1997) were used to identified secondary structures of tRNA genes. Furthermore, the phylogenetic relationships of 11 Bos species or breeds were reconstructed based on 13 mitochondrial protein-coding genes (PCGs) with maximum likelihood (ML) and neighbor joining (NJ) methods implemented within MEGA (Kumar et al., 2016). The African buffalo (Syncerus caffer) was selected as the outgroup. The best-fit model (GTR+I+G) was selected using the Akaike information criterion by JmodelTest (Posada 2008).

The complete mitochondrial genome of Zhangmu cattle (GenBank: MF663793) was a double-stranded circular molecule of 16,340bp. It contained a common set of 37 mitochondrial genes including 13 PCGs, 22 tRNA genes, 2 rRNA genes, and a presumed control region (D-loop). The mitochondrial gene arrangement of Zhangmu cattle is identical to that of other cattle breeds. Among the 37 mitochondrial genes, the majority of genes, including 12 PCGs and 14 tRNA genes, were encoded by the heavy strand, while ND6 and eight other tRNA genes were encoded by the light strand (Fig. 1). The base composition of Zhangmu cattle mitochondrial genome showed a slight bias towards A+T, with an overall A+T content of 60.6%. The total length of PCGs was 11, 406 bp, which was comparable to that of other cattle genomes. As expected, ATG was used as the start codon for most PCGs, while ND5, ND3 and ND2 employed ATA as the start codon, respectively. Three typical stop codons were found: TAA, TAG and AGA, as well as the incomplete stop codon T. All 22 tRNAs displayed the stable cloverleaf secondary structure, with the exception of tRNA-Ser^{AGY}, which lost the stable DHU arm. The rrnS and rrnL genes were 956 and 1570 bp in length, respectively. The 912 bp control region of Zhangmu cattle was located between the tRNA-Pro and tRNA-Phe, which harbored conserved blocks involved in replication and transcription of mitochondrial genome.

Fig. 2 shows the identical topology between NJ tree and ML tree, which is similar to previous results (Olivieri et al., 2015). The sister relationship of Bos indicus and B. taurus was confirmed with high support values. The representative of European breeds occupied the basal clade of B. taurus. Our results also strongly supported that Zhangmu cattle was clustered with Korean breeds as sister species. In summary, the mitochondrial genome of Zhangmu cattle identified in this study would facilitate further studies on the genetic structure of Zhangmu cattle population. This could also enrich genetic resources of threatened cattle breeds in China, and provide essential information for investigating their biodiversity and conservation biology.

Acknowledgements The work was supported or partly supported by grants from China

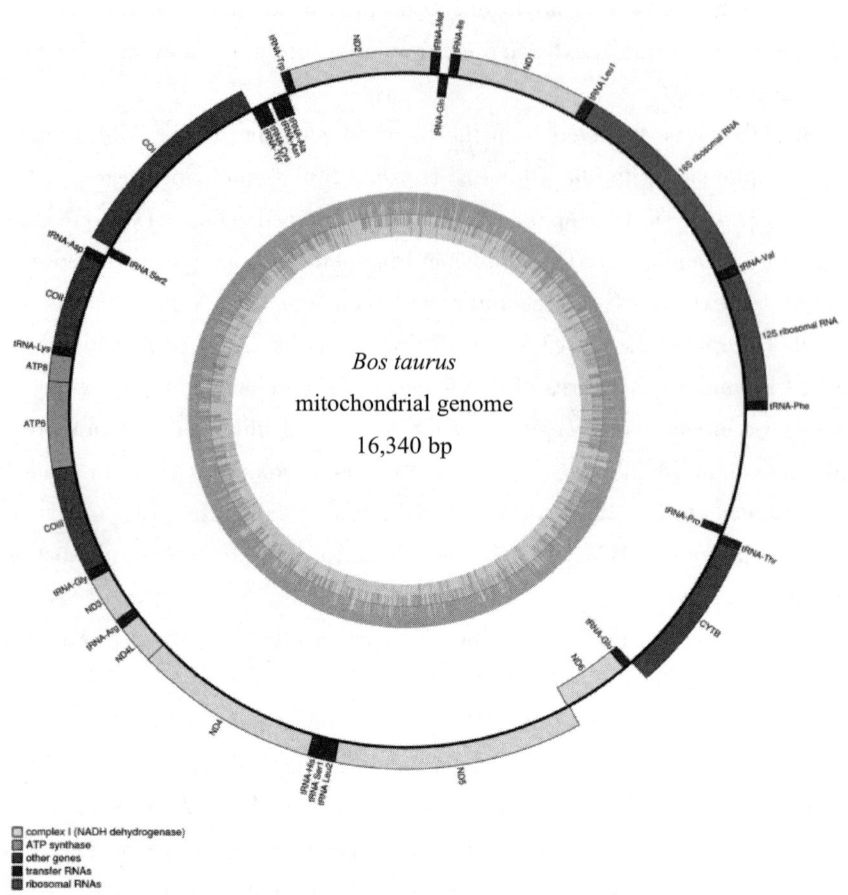

Fig.1 The gene map of the Zhangmu cattle (Bos taurus) mitochondrial genome

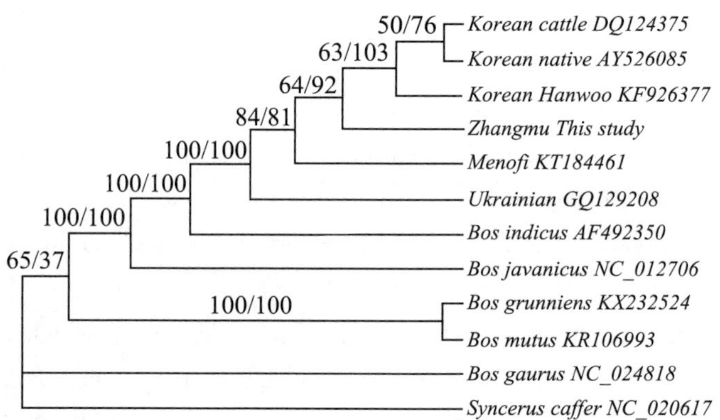

Fig.2 Phylogeny of 11 Bos species and Bos taurus breed based on 13 mitochondrial protein-coding genes. The values at the nodes correspond to neighbor joining (left) and maximum likelihood (right) bootstrap support in percentages, respectively

Agriculture Research System (CARS-37), Science and Technology Support Projects in Gansu Province (1504NKCA052), Innovation Project of Chinese Academy of Agricultural Sciences (CAAS-ASTIP-2014-LIHPS-01), and Central Public-interest Scientific Institution Basal Research Fund (1610322016006).

REFERENCES OMITTED

(发表于《Conservation Genet Resour》，院选 SCI，IF：0.470)

The Complete Mitochondrial Genome of Shigaste Humped Cattle (*Bos taurus*)

Xian GUO[1], Jie PEI[1], Lin XIONG[1], Pengjia BAO[1], Yanbin ZHU[2], Wangdui BASANG[2], Xiaoyun WU[1], Min CHU[1], Ping YAN[1*], Xuezhi DING[1*]

(1. Key Laboratory of Yak Breeding Engineering of Gansu Province, Lanzhou Institute of Husbandry and Pharmaceutical Sciences, Chinese Academy of Agricultural Sciences, Lanzhou 730050, China; 2. Institute of Animal Husbandry and Veterinary Medicine, Tibet Academy of Agriculture and Animal Husbandry Science, Lhasa 850001, China)

Abstract: Shigaste humped cattle (*Bos taurus*), generated through long-term breeding, is a distinctive but threatened species native to China. This species plays an essential role in local stockbreeding and breeding newer species. Its population has been declining rapidly over the past few years owing to poor understanding of the significance of conservation of germplasm resources and the introduction of abundant species. To effectively protect and wisely utilise cattle breed resources, researchers have performed extensive studies on genetic diversity of domestic animals. In this study, the complete mitochondrial genome sequence of Shigaste humped cattle was successfully obtained by Illumina Hiseq Xten platform. The circular genome is 16, 340 bp, which is composed of 13 protein-coding genes, two rRNA genes, 22 tRNA genes and a control region. Base composition of the genome is A (33.4%), C (26.0%), G (13.4%), T (27.2%), with an A + T content of 60.6%. Phylogenetic relationship based on 10 mitochondrial genome sequences indicated that Shigaste humped cattle was clustered with domestic cattle from China, and was closely related to Bos indicus. This study provided detailed molecular genetic information for the conservation and utilization of cattle germplasm resources.

Key words: Shigaste humped cattle (*Bos taurus*); Complete mitochondrial genome; Illumina sequencing

Shigaste humped cattle (*Bos taurus*) is a valuable and endangered domestic breed in high altitude areas of Tibet. They have strong resistance to high heat and cold, extreme drought, roughage and disease as compared to other cattle in Tibet. However, the number of this species has decreased drastically recently due to the introduction of excessive exotic species and low awareness of germplasm resources protection. Mitochondrial DNA (mtDNA) is widely used for genetic research and

* Corresponding author, E-mail: yanping@caas.cn; dingxuezhi@caas.cn

protection of endangered species (Lai et al., 2006; Lei et al., 2006; Olivieri et al., 2015; Sun et al., 2016). In this study, we sequenced, assembled and characterized the complete mitochondrial genome of Bos taurus by next-generation sequencing data generated on Illumina HiSeq Xten sequencing system to provide a scientific basis for the study of cattle breeds.

The total genomic DNA was extracted from muscle tissue of B.taurus, subjected to quality control, and sheared to produce size-selected pieces of approximately 350 bp.These pieces are then end-repaired, A-tailed, tagged and purified to construct a genomic library, followed by whole genomic sequencing on Illumina HiSeq Xten sequencing system (Illumina, CA, USA) according to the standard operation protocol.In total, 13,899,602 raw paired reads with average length of 150 bases was generated. By removing unidentified nucleobases and filtering poor-quality sequences, clean reads were produced and further processed by quality trimming with CLC Genomics Workbench v10 (CLC, Bio, Aarhus, Denmark).Then 5 169 high-quality reads were used to assemble the mitochondrial genome of B.taurus breed Shigatse humped cattle with average coverage of 47. 1X, using the MITObim v1. 9 program (Hahn et al., 2013), with that of B. taurus (Genbank: AF492351) (Hiendleder et al., 2008) as an initial reference sequence.

The complete mitogenome sequence of Shigatse humped cattle breed (GenBank accession No. MF663792) was subsequently annotated with GENEIOUS R10 (Biomatters Ltd., Auckland, New Zealand).It was 16,340bp in length and comprised 13 protein-coding genes (PCGs), two rRNA genes, 22 tRNA genes, a non-coding control region, and initiation site of L-strand replication (OL) with 32 bp in length between tRNA-Asn and tRNA-Cys (Fig.1).The A, C, T and G contents in mitochondrial genome were 33. 4%, 26. 0%, 27. 2% and 13. 4%, respectively.An overall A + T content of whole mitochondrial genome was 60. 6%.Phylogenetic analysis was performed on 10 mitochondrial genome sequences from B.taurus and B.indicus with entire mitogenome of B.grunniens as an outgroup to investigate the status of B.taurus Shigatse humped cattle breed.A phylogenetic tree was constructed in Mega 7 based on the maximum likelihood (ML) method (MEGA Inc., Englewood, NJ) (Kumar et al., 2016) with 1 000 supports estimated from 1 000 replicate searches. Both NJ and ML trees showed identical topology, strongly supported by bootstrap values (Fig.2). The phylogenetic research indicated two major branches, and B. taurus Shi-gatse humped cattle breed was grouped in the cluster with a Chinese cattle breed.In addition, B.taurus Shigatse humped cattle breed had a close genetic relationship with B.indicus, which could be because it is a hybrid breed of the local B.taurus breed from Tibet and the exotic B.indicus breed.This study identified the complete mitochondrial genome and phylogenetic status of B. taurus Shigatse humped cattle breed, and provided essential molecular genetic information for the conservation of germplasm resources of this endangered species.

Acknowledgements: The work was supported or partly supported by grants from China Agriculture Research System (CARS-37), Science and Technology Support Projects in Gansu Province (1504NKCA052), Innovation Project of Chinese Academy of Agricultural Sciences (CAAS-ASTIP-2014-LIHPS-01), and Central Public-interest Scientific Institution Basal Research Fund (1610322016006).

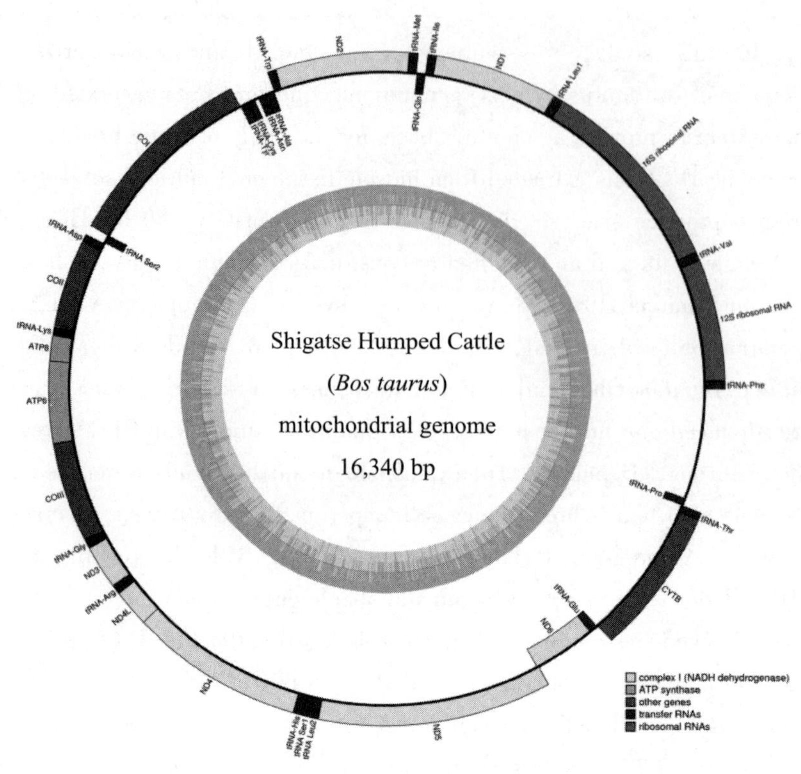

Fig.1 The gene map of Shigaste humped cattle (Bos taurus) mitochondrial genome

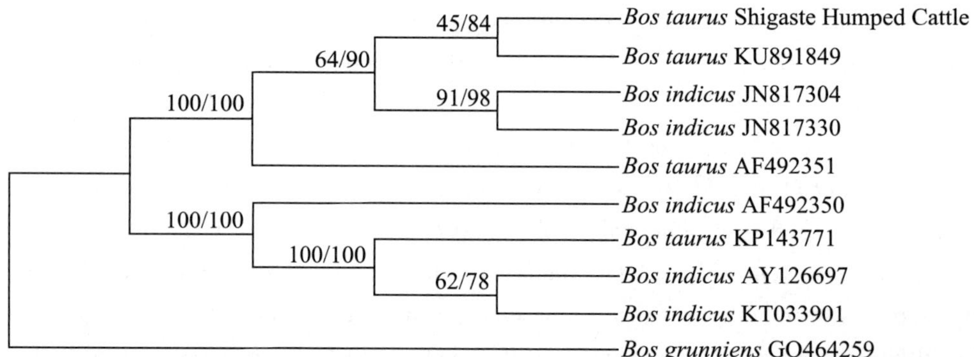

Fig.2 Phylogenetic analyses of 10 Bos species based on 13 mitochondrial protein-coding genes. The values at the nodes correspond to neighbor joining (first) and maximum likelihood (second) bootstrap support in percentages, respectively

REFERENCES OMITTED

（发表于《Conservation Genetics Resources》，院选 SCI，IF：0.470）

The Complete Mitochondrial Genome of Sanhe Horse (*Equus caballus*)

Jie PEI[1], Min CHU[1], Pengjia BAO[1], Zhongcheng SHA[2], Xuezhi DING[1], Ping YAN[1], Xian GUO[1]*

(1. Lanzhou Institute of Husbandry and Pharmaceutical Sciences, Chinese Academy of Agricultural Sciences, Lanzhou 730050, China; 2. Veterinary Station, Jinlai Limited Liability Company of Agriculture and Husbandry, Sanhe Stud-farm, Sanhe Hui Nationality Township, Ergun City 022256, China)

Abstract: The present study reports the whole mitochondrial genome sequence of Sanhe horse (*Equus caballus*).Sanhe horse adapts quickly to severe cold conditions, which might be attributed to the unique genetic characteristics for acclimatizing to the harsh environment.However, the number of Sanhe horses has decreased sharply in recent years.Developing reasonable conservation and utilization measures for Sanhe horse requires a deeper understanding of molecular genetics.Because animal mitochondrial DNA evolves faster than the nuclear genetic markers, it is used as a powerful tool in the study of biodiversity conservation and sustainable utilization of biological resources.In this study, the complete mitochondrial genome of Sanhe horse was determined.The circular mitogenome sequenced by the Illumina HiSeq 2 500 platform retrieved 16,504-bp-long sequences, containing 13 protein-coding genes, 22 tRNA genes, two rRNA genes, and a non-coding control region.The 32-bp initiation site of the L-strand replication origin was located between the tRNA-Asn and tRNA-Cys.The overall base composition of the genome was A (32. 3%), C (28. 5%), G (13. 3%), and T (25. 9%).The phylogenetic tree was constructed using MEGA 7 with the neighbor-joining method, demonstrating that the Sanhe horse was clustered closely with the breeds in northern Europe, although it was relatively independent.

Key words: Sanhe horse; Mitochondrial genome; Illumina sequencing; Phylogenetic tree

Sanhe horse (Equus caballus), is a ride-pull dual-purpose horse breed, native to the Three Rivers Region (Gen River, Deerbuer River, and Hawuer River), Hulunbuir, Inner Mongolia.It is primarily distributed in Ergun City at the west foot of Greater Khingan, which experiences the continental climate of the northern cold temperate zone.Owing to the cold Hulunbuir Plateau climate,

* Corresponding author, E-mail: pingyan63@ 126. com; guoxianlz@ 163. com

and low energy and low carbohydrate diet, Sanhe horse has some unique genetic characteristics that differentiate it from the other horse breeds.

Horses have played a significant role in the daily lives of humans since the early days of human civilization. However, the populations of horse breeds have undergone an enormous decline in the past few years owing to the advances in agricultural mechanization and rapid development of transportation. Therefore, a large number of horse breeds are becoming endangered. The number of Sanhe horses has declined from 17,357 in 1985 to 720 in 2005, and it is predisposed to further decline. Unless protected, Sanhe horse will become extinct in the future. Developing reasonable conservation and utilization measures for Sanhe horse requires a deeper understanding of molecular genetics. In the present study, the complete mitochondrial genome of Sanhe horse was sequenced, assembled, and characterized.

The mitochondrial DNA was extracted using the conventional phenol-chloroform method from the blood of the Sanhe horse fed at the Sanhe Hui Nationality Town-ship (Inner Mongolia, China; 50° 27′N, 120°07′E). DNA shotgun library of the mitogenome was constructed and sequenced with the Illumina Hiseq 2,500 sequencing system (Illumina, CA, USA) (Johns and Paulus-Thomas 1989). A total of 11.79 million raw reads, 150 bp in length, were trimmed using the CLC Genomics Workbench (CLC Bio, Aarhus, Denmark). The mitochondrial genome was reconstructed using MITObim v1.8 (Hahn et al., 2013), followed by annotation with reference to the sequence of Equus caballus breed Russian Riding (HQ439486) using Geneious (Bio-matters Ltd, Auckland, New Zealand). The protein-coding region was determined by comparing the genome sequence with that of the Equus caballus breed Debao (EU939445) as reported previously (Jiang et al., 2011). In addition, the transfer RNA (tRNA) genes were identified using MITOS Web Server (Bernt et al., 2013) and tRNAscan-SE 1.21 (Lowe and Eddy 1997). MEGA 7 was used to construct the phylogenetic tree of 51 equine breeds by the neighbor-joining method (Kumar et al., 2016).

As in other vertebrates, the wholemitogenome of Sanhe horse was 16,504bp in length (GenBank accession No. MF925712), containing 13 protein-coding genes (PCGs), 22 transfer RNA (tRNA) genes, two ribosomal RNA (rRNA) genes, and a control region (Fig.1). The organization and gene order of the mitogenome were similar to those of other Equidae. Among the mitochondrial genes, the majority, including 12 PCGs and 14 tRNAs, were encoded by the light strand, whereas ND6 and the other eight tRNA genes were encoded by the heavy strand (Fig.1). The overall base composition of the H-strand was 32.3% A, 28.5% C, 13.3% G, and 25.9% T, and the A + T content was 58.2%. According to the annotation, ATG was used as the start codon for most PCGs, whereas ND2 and ND3 employed ATA as the start codon. Two standard stop codons (TAA and AGA) and the incomplete stop codon (T) were found. All the 22 tRNAs displayed the stable cloverleaf secondary structure. The 12S rRNA and 16S rRNA genes were 976 and 1 579bp in length, respectively. The 1 040-bp control region of the Sanhe horse mitogenome was located between the tRNA-Pro and tRNA-Phe, which harbored the conserved region involved in the replication and transcription of the mitochondrial genome. According to the annotation, 13 protein-encoding genes were interpreted and no abnormal stop codons or shift in the genomic codes were found, indicating the reliability of the assembly.

Thephylogenetic tree constructed by the neighbor-joining method using the MEGA 7 showed that the Sanhe horse sequence was clustered with most breeds of Equus, which mainly included Asian, European, and American (Fig.2).The phylogenetic tree showed that all the horse breeds have very close relationships; however, the tow breed, Sanhe horse, and Jeju native horse were distinct among these equine breeds. This result indicates that Sanhe horse has unique mitochondrial genetic characteristics that are distinguishable from the other horse breeds. This finding suggests that the mechanism of adaptability to severe cold conditions may be related with the mitochondrial genetic characteristics.

In summary, the mitochondrial genome of Sanhe horse characterized in the present study would facilitate further studies on the genetic constitution of Sanhe horse population, thereby providing an insight into the crucial genetic background for the protection of the breed.It would also enrich essential information for investigating the biodiversity and conservation biology of Sanhe horse.

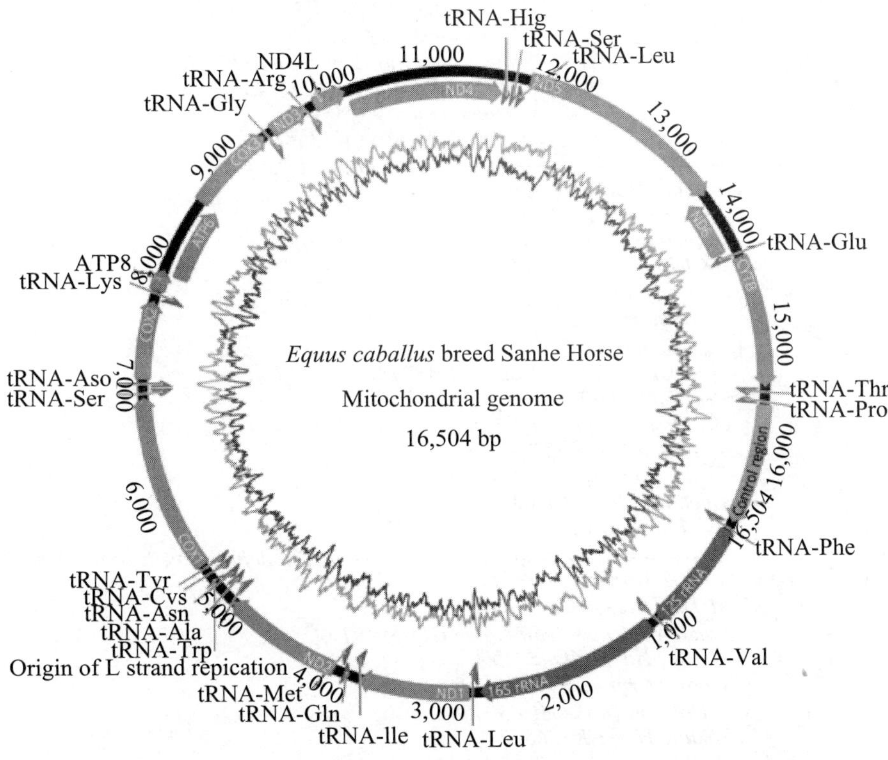

Fig.1 The gene map of Sanhe horse (Equus caballus) mitochondrial genome. (Color figure online)

Acknowledgements: This work was supported by Grants from the National Natural Science Foundation of China (31402034), China Agriculture Research System (CARS-37), and the Innovation Project of Chinese Academy of Agricultural Sciences (CAAS-ASTIP-2014-LIHPS-01).

COMPLIANCE WITH ETHICAL STANDARDS

Conflict of interest The authors declare that they have no conflicts of interest.

Ethical approval The experimental design, and sample collection protocols and procedures were approved by the Ethics Committee of the Research (ECR) of Chinese Academy of Agricultural Sci-

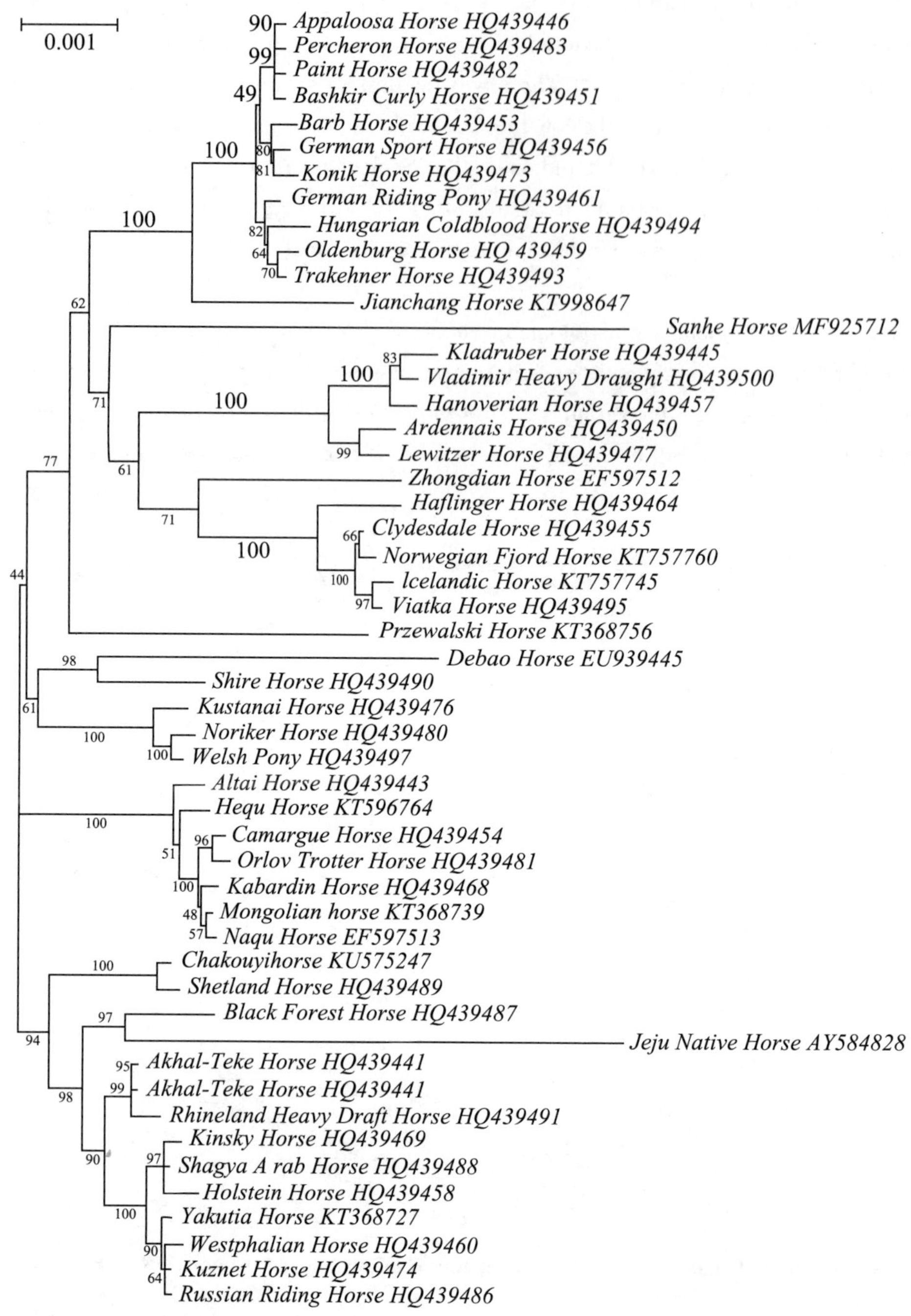

Fig.2 Phylogeny of 51 equine species based on mitochondrial protein-coding regions inferred using the neighbor-joining method in MEGA 7.

Note: The values at the nodes correspond to neighbor-joining bootstrap support in percentages. The tree is drawn to scale, with branch lengths in the same units as those of the evolutionary distances used to infer the phylogenetic tree. The percentage of replicate trees, wherein the associated taxa clustered together in the bootstrap test (1 000 replicates), are shown next to the branches. The Sanhe breed is marked by a filled triangle

ences (CAAS). Furthermore, the research activities were performed in compliance with the local animal welfare laws.

REFERENCES OMITTED

（发表于《Conservation Genetics Resources》，院选 SCI，IF：0.470）

长期饲喂高脂饲料对大鼠血脂、肝及肠道菌群的影响*

马　宁**，刘希望，孔晓军，李世宏，秦　哲，焦增华，杨亚军***，李剑勇***

（中国农业科学院兰州畜牧与兽药研究所/农业部兽用药物创制重点实验室/甘肃省新兽药工程重点实验室，兰州 730950）

摘要：旨在探讨长期饲喂高脂饲料对大鼠血脂、肝及肠道菌群的影响。本研究选择成年健康SD大鼠随机分为两组：空白对照组和高脂饲料组（$n=10$），分别采用维持饲料和高脂饲料饲喂13周，在第8及第13周，采集血清进行血脂分析，并于第13周采集肝组织及盲肠内容物，检测肝病变及肠道菌群变化，并预测与其相关的代谢通路。结果表明，高脂饲料可显著升高TG、TCH及LDL水平（$P<0.01$），并导致严重的肝脂肪变性；相比于第8周结果，延长饲喂时间可显著增高TCH和HDL（$P<0.01$）。高脂饲料可显著降低肠道菌群Shannon及Chao指数（$P<0.001$），降低菌群多样性，并显著改变菌群组成，引起厚壁菌门及放线菌门的增加和拟杆菌门的减少（$P<0.01$）；PICRUSt预测对照组差异代谢通路，主要涉及细菌的生存与繁殖等，而高脂饲料组主要与能量、氨基酸代谢相关，这些通路的改变可能与血脂异常及肠道菌群多样性的降低紧密联系。本研究证实了长期饲喂高脂饲料对大鼠血脂、肝及肠道菌群的危害，为血脂异常成因的阐明提供了依据，也对宠物健康饮食管理有一定的指导作用。

关键词：高脂饲料；肠道菌群；血脂；肝；大鼠

---

* 基金项目：国家自然科学基金（31402254；31572573）；甘肃省青年科技基金计划（1506RJYA148）

** 作者简介：马宁（1990—　），男，河北保定人，博士，主要从事兽医药理与毒理学研究，E-mail：maning9618@163.com

*** 通信作者：杨亚军，博士，副研究员，E-mail：yangyue10224@163.com；李剑勇，博士，研究员，主要从事新兽药研究与开发，E-mail：lijy1971@163.com

Effects of Long-term Feeding of High Fat Diet on Blood Lipids, Liver and GutMicrobiota in Rats

Ning MA, Xi-wang LIU, Xiao-jun KONG, Shi-hong LI, Zhe QIN,
Zeng-hua JIAO, Ya-jun YANG*, Jian-yong LI*

(Key Laboratory of New Animal Drug Project of Gansu Province/Key Laboratory of Veterinary Pharmaceutical Development of Ministry of Agriculture/Lanzhou Institute of Husbandry and Pharmaceutical Sciences of Chinese Academy of Agricultural Sciences, Lanzhou 730050, China)

Abstract: The objective of this study was to explore the effects of long-term feeding with high fat diet (HFD) on blood lipids, liver and gutmicrobiota in rats. Healthy adult SD rats were selected and divided into 2 groups: control group and HFD group ($n = 10$). Rats in different groups were fed with normal diet and HFD for 13 weeks, respectively, and blood samples were collected on 8th and 13th week for blood lipids analysis. Liver tissues and cecal contents were collected to observe the changes of liver and gut microbiota on 13th week. Metabolic pathways related with gut microbiota were also predicted. The results showed that HFD could significantly increase the levels of TG, TCH and LDL ($P<0.01$) and result in serious steatosis in liver. Compared with the results on 8th week, prolonging time of HFD feeding increased the levels of TCH and HDL ($P<0.01$). Shannon and Chao indices were significantly lowered by HFD ($P<0.001$), indicating the reduction of gut microbiota richness. HFD also altered the composition of gut microbiota such as increasing Firmicutes and Actinobacteria and reducing Bacteroidetes ($P<0.01$). Results of PICRUSt showed that the metabolic pathways of control group were mainly associated with bacteria living and reproduction, while the pathways in HFD group were mainly associated with the metabolism of energy and amino acids. These changes of pathways might be related to the dyslip-idemia and the decrease of gut microbiota richness. The study proved the harm of long-term feeding with HFD to blood lipids, liver and gut microbiota, and provided information for dyslipidemia and some guidance for diet management of pets.

Key words: High fat diet; Gutmicrobiota; Blood lipid; Liver; rat

高脂血症（Hyperlipemia）是由先天性的基因缺乏或脂代谢紊乱引起的体内血脂水平过高[1]。伴随着生活水平的提高和生活方式的改变，高脂血症在人群中的发病率逐年升高，并且有年轻化趋势。由于高脂血症与血栓、高血压、动脉粥样硬化等心血管疾病有紧密联系，因此预防高脂血症，降低血脂水平对有效预防心血管疾病的发生，发挥着重要的作用[2]。在伴侣动物中由于缺乏运动，食物结构的改变及食物摄入过多等原因，导致肥胖、血脂异常等的发病率日益增加，已严重威胁宠物健康与动物福利[3-4]。因此，高脂血症引起医学和兽医学科研人员的关注，并成为研究的热点和难点。

高脂血症动物模型，在研究降血脂药物的药效和作用机制方面发挥着重要的作用。目前，啮齿类动物在高脂血症动物模型中使用较为普遍，造模的方法也较多，如：喂饲法、基因敲除法、卵蛋白注射法等[5-6]。相关文献报道，不同配方的高脂饲料在较短时间内（4~8周）即可引起金黄地鼠、大鼠等血脂水平的显著变化[7-9]。然而，长期饲喂高脂饲料对大鼠血脂水平、肝病理组织学变化，以及肠道菌群的影响等的研究尚未见报道。

本研究拟采用血脂水平测定、病理组织学检查和16S rDNA扩增子测序等手段，观察长期（13周）饲喂高脂饲料对大鼠血脂、肝及肠道菌群的影响，并挖掘肠道菌群与血脂异常的潜在联系，以期探究高脂血症的病理机制，为伴侣动物健康饮食管理提供参考。

1 材料与方法

1.1 材料

大鼠维持饲料和高脂饲料均由北京科澳协力饲料有限公司提供。高脂饲料配方：维持料77.8%，蛋黄粉10%，猪油10%，胆固醇2%，胆酸盐0.2%。维持饲料的主要营养成分为12.3%脂类，63.3%碳水化合物和24.4%蛋白质（kcal%）；高脂饲料主要营养成分为41.5%油脂，40.2%碳水化合物和18.3%蛋白质（kcal%）。

高密度脂蛋白（HDL）、低密度脂蛋白（LDL）、总胆固醇（TCH）和甘油三脂（TG）检测试剂盒购自宁波美康生物科技股份有限公司。分析纯伊红、美兰、甲醇等试剂购于国药集团化学试剂有限公司。

1.2 仪器

DHP-9082型电热恒温培养箱（上海朵弗实业有限公司），RM-2235精密转轮半自动螺旋切片机，13395H2X光学显微镜，XP-600E偏振光显微镜（均为德国LEICA公司），XL-640全自动生化分析仪（德国Erba公司），Illumina HiSeq测序平台。

1.3 动物及分组

清洁级健康雄性SD大鼠20只，体重250~280g，由甘肃省中医药大学提供。随机分为对照组（Control）和高脂饲料组（High fat diet，HFD），每组10只，分别连续自由采食维持饲料和高脂饲料13周。

1.4 样品采集与处理

为了探讨血脂水平在试验期间的变化，在给予高脂饲料后第8及第13周末，自大鼠尾尖采血，以检测血脂水平。血液样本采集方法：大鼠腹腔注射戊巴比妥钠（$30mg \cdot kg^{-1}$）进行麻醉，尾部40℃温浴2min后，断尾采集血液1.5mL，$4\,000r \cdot min^{-1}$离心15min制备血清，用于血脂分析。饲喂高脂饲料13周后，采集肝组织及盲肠内容物用于病理组织学和肠道菌群的分析。肝组织切成1cm×1cm×0.6cm大小，10%中性多聚甲醛磷酸盐缓冲液固定，常规石蜡包埋，切片，苏木精-伊红染色，进行病理组织学观察。盲肠内容物分装于离心管中，液氮速冻后，于-80℃保存。

1.5 Illumina HiSeq测序分析

利用Illumina HiSeq测序平台，对细菌的16S rDNA基因V4区进行高通量测序，并对测序结果进行生物信息学分析。

1.5.1 盲肠内容物总 DNA 提取 利用基因组 DNA 提取试剂盒提取大鼠盲肠内容物样品中微生物的总 DNA（PowerFecal™ DNA Isolation kit，USA）。利用 1%琼脂糖凝胶电泳检测 DNA 的纯度和浓度，使用无菌水稀释 DNA 样品至 1 ng · μL^{-1}。所提取的 DNA 于-20℃保存备用。

1.5.2 16SrDNA 的扩增 以稀释后的基因组 DNA 为模板，针对 16S rDNA 基因 V4 区，合成带有 Barcode 的特异性引物 515F（5′-GTTTCGGTGCCAGCMGCCGCGGTAA-3′）和 806R（5′-GATCAGGGACTACHVGGGTWTCTAAT-3′）。PCR 扩增采用 Phusion © High-Fidelity PCR Master Mix with GC Buffer 和高效高保真酶反应体系，以确保扩增效率和准确性。PCR 产物用 2%的琼脂糖凝胶电泳进行检测，目的条带（400~450 bp）用相应的回收试剂盒回收纯化（Qiagen Gel Extraction Kit，Germany）。

1.5.3 文库构建与测序 用建库试剂盒（TruSeq © DNA PCR-Free Sample Preparation Kit）进行文库构建，具体步骤按说明书操作。利用 Qubit@ 2.0 Fluorometer（Thermo Scientific）和 Agilent Bioanalyzer 2100 system 对构建好的文库进行评估。文库合格后采用 Illumina HiSeq2500 PE250 平台进行测序。

1.5.4 生物信息学分析 各样品数据截去 Barcode 和引物序列后使用 FLASH 对数据进行拼接得到原始数据，经 Qiime 软件过滤、UCHIME Algorithm 软件去除嵌合体后，与数据库比对（Gold database），得到有效数据。用 Uparse 软件对有效数据在 97%水平上进行操作分类单元（Operational taxonomic unit，OTU）聚类。采用 Mothur 法与 SILVA 的 SSUrRNA 数据库进行物种注释分析（阈值：0.8~1）。使用 PyNAST 软件与 Green Gene 数据库进行快速多序列比对，得到的所有 OTUs 代表序列的系统发生关系。R 软件绘制稀释曲线以评估测序深度。各样品的数据进行均一化处理后，采用 Qiime 软件进行样品复杂度分析（Alpha diversity）和多样品比较分析（Beta diversity）。运用 PICRUSt（Phylogenetic Investigation of Communities by Reconstruction of Unobserved States），并结合 KEGG 数据库，对肠道菌群的功能进行预测。

1.6 统计分析

血脂检测结果用“Mean±SD”表示，采用 SPSS13.0 对结果进行统计学分析（t 检验），$P<0.05$ 表示差异有统计学意义。

2 结果

2.1 高脂饲料对大鼠血脂水平的影响

大鼠饲喂高脂饲料 8 周后进行血脂水平检测，以评估高脂饲料对血脂的影响（表 1）。相比于对照组，饲喂高脂饲料大鼠的 TG、TCH 和 LDL 水平均显著增高（$P<0.01$），HDL 显著降低（$P<0.01$）。高脂饲料持续饲喂至 13 周，TG，TCH 及 LDL 相对于对照组均显著增高（$P<0.01$），而 HDL 差异不显著。通过比较 8 周与 13 周模型组大鼠的血脂水平，探讨延长高脂饲料饲喂时间对血脂水平的影响。结果表明，延长高脂饲料饲喂时间 5 周，可显著增高 TCH 和 HDL 水平（$P<0.01$）。

2.2 高脂饲料对大鼠肝组织的影响

眼观，对照组肝呈红褐色，肝组织质地柔软（图 1A）；相比于对照组，饲喂高脂饲料 13 周后，大鼠肝肿大，体积明显增大，被膜紧张，肝实质呈土黄色，边缘变钝、质脆，表面光滑且有油腻感（图 1B）。镜检，对照组大鼠肝细胞结构正常，结构清晰（图 1C）；高脂饲料组大鼠肝细胞发生不同程度的变性、坏死，部分细胞核坏死、崩解、淡染；肝细胞发生严重的脂肪变性，形成大小不等的脂滴空泡（图 1D）。

表 1　饲喂高脂饲料 8 周和 13 周后大鼠血脂水平的变化（Mean±SD，$n=10$）

Table 1　Blood lipid changes in rats after feeding with high fat diet for 8 weeks and 13 weeks（Mean±SD，$n=10$）　（$mmol \cdot L^{-1}$）

血脂指标 Blood lipid index	8 周 8 week		13 周 13 week	
	对照组 Control	高脂饲料组 HFD	对照组 Control	高脂饲料组 HFD
甘油三酯 TG	0. 77±0. 42	1. 51±0. 38**	0. 78±0. 20	1. 66±0. 22**
总胆固醇 TCH	1. 04±0. 25	2. 19±0. 31**	1. 10±0. 16	2. 51±0. 14** ##
高密度脂蛋白 HDL	0. 58±0. 06	0. 45±0. 06**	0. 48±0. 04	0. 52±0. 05**
低密度脂蛋白 LDL	0. 13±0. 08	0. 51 ±0. 08**	0. 12±0. 04	0. 54±0. 06**

*.$P<0.05$；**.$P<0.01$：8 周与 13 周高脂饲料组血脂水平相比于对照组；##.$P<0.01$：13 周高脂饲料组血脂水平相比于 8 周高脂饲料组血脂水平

*.$P<0.05$；**.$P<0.01$：significant difference between control and HFD groups at week 8 and week 13，respectively；##.$P<0.01$：significant difference between week 8 and week 13 of the blood lipid results in HFD group

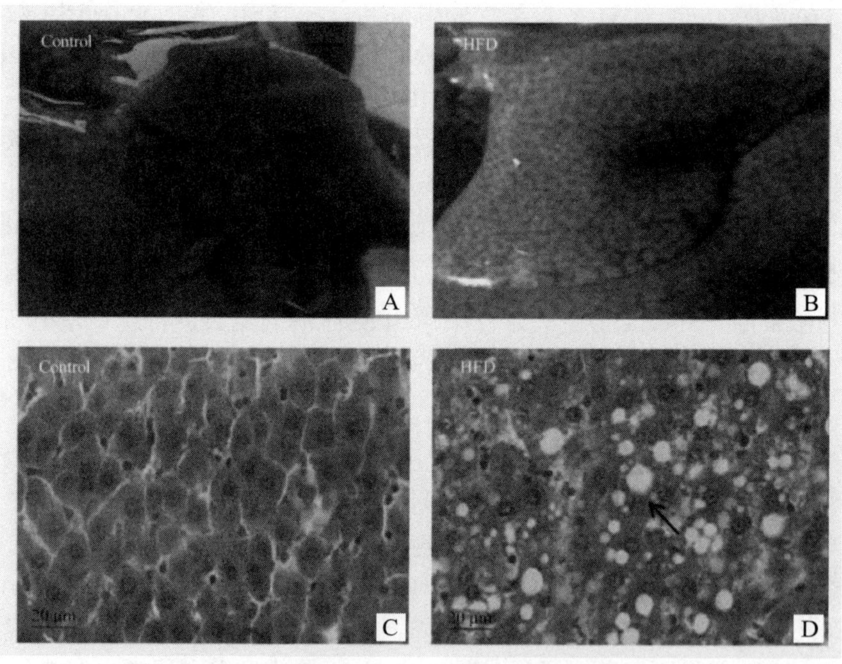

图 1　饲喂高脂饲料 13 周后大鼠肝的病理变化

Fig.1　Pathological changes of liver from the rats fed with high fat diet for 13 weeks

2. 3　高脂饲料对大鼠肠道菌群的影响

2. 3. 1　盲肠微生物测序数据及评价　采用 Illumina HiSeq 测序平台得到原始数据（Raw Data）后，为了使信息分析的结果更加准确、可靠，对原始数据进行拼接、过滤，以得到有效数据（Effective tags）。以 97%的一致性将有效数据聚类成为 OTUs，并进行物种注释用于后续分析（表 2）。稀释曲线趋向平坦说明测序数据量渐进合理（图 2A）。

2.3.2 高脂饲料对肠道菌群组成的影响及聚类分析　连续饲喂高脂饲料 13 周后，相比于正常组，高脂饲料可显著改变盲肠菌群的组成（图 2B）。统计学结果表明，相比于对照组，高脂饲料组的厚壁菌门及放线菌门显著增加（$P<0.01$），拟杆菌门、广古菌门、柔膜菌门和 TM7 门显著减少（$P<0.01$）；同时，高脂饲料也显著降低了其他菌群的丰度（$P<0.05$）；变形菌门、疣微菌门、蓝藻门及螺旋体门在两组之间差异不显著（表 3）。此外，对照组和高脂饲料组之间有 26 个属的细菌，统计学差异显著（$P<0.05$）。相比于对照组，高脂饲料组中 12 个种属的细菌的丰度显著升高（如：Faecalibacterium，Holdemania，Ruminococcus 等），14 个种属的细菌的丰度显著降低（如：Jeotgalicoccus，YRC22，Parabacteroides 等），见表 4。

为了研究样品间的相似性，结合最大相对丰度排名前十的物种，采用 UPGMA（Unweighted Pair-group Method with Arithmetic Mean）的分析方法，对样品进行聚类分析。结果表明，所有的菌群样本被分为两大类，对照组为一类样本，高脂饲料组为一类样本，其聚类分析结果与样本处理的分类结果保持一致（图 2E）。

表 2　肠道菌群测序数据统计

Table 2　Statistical results of sequencing data of gutmicrobiota

样本编号 Sample name	原始标签 Raw tag	有效标签 Effective tag	OTU 数量 OTU number
Con1	68 600	66 438	931
Con2	52 442	50 760	899
Con3	49 712	47 830	887
Con4	64 193	61 921	941
Con5	52 483	50 119	867
Con6	49 345	47 797	928
Con7	60 259	58 415	921
Con8	50 562	48 748	922
Con9	62 811	60 303	940
Con10	57 732	55 493	923
HFD1	61 862	58 603	884
HFD2	61 038	58 039	873
HFD3	63 980	60 758	875
HFD4	54 513	51 622	864
HFD5	63 201	59 838	857
HFD6	60 216	56 798	850
HFD7	59 691	56 665	864
HFD8	61 077	57 644	853

（续表）

样本编号 Sample name	原始标签 Raw tag	有效标签 Effective tag	OTU 数量 OTU number
HFD9	57 636	54 632	892
HFD10	61 834	58 725	877

表 3　对照组与高脂饲料组肠道菌群门水平的统计分析（Mean±SD）

Table 3　Statistical analysis of gutmicrobiota（phylum level）between control and HFD groups（Mean±SD）%

肠道菌群（门水平） Gutmicrobiota（Phylum level）	对照组 Control group	高脂饲料组 HFD group	P 值 P-value
厚壁菌门 Firmicutes	65. 28±10. 11	78. 45±5. 49	0. 002 8
拟杆菌门 Bacteroidetes	22. 59±8. 89	8. 32±3. 55	0. 000 5
变形菌门 Proteobacteria	6. 44 ±2. 17	6. 22±2. 57	0. 833 6
放线菌门 Actinobacteria	0. 72±0. 24	5. 66±3. 18	0. 000 8
疣微菌门 Verrucomicrobia	1. 53±2. 54	0. 02±0. 05	0. 092 3
广古菌门 Euryarchaeota	2. 34±1. 46	0. 31±0. 38	0. 001 6
蓝藻门 Cyanobacteria	0. 22±0. 24	0. 44±0. 70	0. 376 3
螺旋体门 Spirochaetes	0. 03±0. 02	0. 19±0. 34	0. 172 8
柔膜菌门 Tenericutes	0. 45±0. 22	0. 20±0. 06	0. 006 3
TM7 门 TM7	0. 16±0. 13	0. 01±0. 01	0. 004 2
其他 Others	0. 24±0. 09	0. 19±0. 07	0. 012 6

表 4　对照组与高脂饲料组肠道菌群属水平的统计分析（Mean±SD）

Table 4　Statistical analysis of gut microbiota（genus level）between control and HFD groups（Mean±SD）%

肠道菌群（属水平） Gutmicrobiota（Genus level）	对照组 Control group	高脂饲料组 HFD group	P 值 P-value
Bcicteroicles	5. 114±2. 112	1. 289±0. 789	0. 000 2
Dorea	1. 291 ±0. 643	2. 584±0. 784	0. 000 8
Coll inset la	0. 365±0. 150	5. 028±3. 08	0. 001 0
vadinCAW	0. 006±0. 004	0	0. 001 6
Methanobrevibacter	2. 332±1. 462	0. 313±0. 385	0. 001 7
CF231	2. 284±1. 281	0. 620±0. 561	0. 002 6
Ruminococcus	2. 354±1. 178	4. 409±1. 462	0. 002 9
Allobaculuni	2. 882±1. 465	6. 868±3. 111	0. 002 9

（续表）

肠道菌群（属水平） Gutmicrobiota（Genus level）	对照组 Control group	高脂饲料组 HFD group	P 值 P-value
Flexispira	0. 121±0. 065	0. 041±0. 025	0. 003 6
rc4-4	0. 217±0. 089	0. 118±0. 036	0. 006 7
Odoribacter	0. 028±0. 019	0. 007±0. 004	0. 007 5
Blautia	2. 437 ±0. 645	4. 091±1. 504	0. 007 6
Desul fovibrio	0. 736±0. 185	1. 374±0. 605	0. 009 0
Parabacteroides	0. 410±0. 252	0. 145±0. 086	0. 009 1
YRC22	0. 191±0. 101	0. 080±0. 071	0. 011 4
Jeotgalicoccus	0. 306±0. 151	0. 155±0. 066	0. 012 7
Holdemania	0. 056±0. 027	0. 099±0. 043	0. 017 6
Faecalibacterium	0. 084±0. 055	0. 198±0. 127	0. 022 1
Sporosarcina	0. 027±0. 013	0. 015±0. 008	0. 025 9
Coprobacillus	0. 055±0. 037	0. 194±0. 168	0. 028 5
Eubacterium	0. 010±0. 009	0. 002±0. 004	0. 029 5
Coprococcus	1. 121±0. 230	1. 360±0. 229	0. 031 7
Prevotella	0. 193±0. 231	0. 020±0. 011	0. 041 9
Oscillospira	10. 720±1. 605	8. 339±2. 981	0. 043 3
Mogibacteriurn	0. 183±0. 063	0. 257±0. 087	0. 045 2
Anaerofustis	0. 006±0. 004	0. 010±0. 004	0. 046 8

2. 3. 3 高脂饲料对菌群多样性的影响　Chao 和 Shannon 指数用来评估菌群多样性的变化。相比于对照组，高脂饲料组的 Shannon 和 Chao 指数均极显著的降低（图 2C、D），表明高脂饲料可显著降低肠道菌群的多样性。

2. 3. 4 肠道菌群的 PICRUSt 分析　PICRUSt 分析用来挖掘与肠道菌群改变相关的代谢通路（KEGG level 3），来说明菌群和高脂饲料之间的联系。试验结果表明，两组之间共有 35 个代谢通路存在差异（图 3）。高脂饲料组中与能量相关的代谢通路丰度相对较高，其主要包括淀粉蔗糖代谢、果糖-甘露糖代谢、戊糖磷酸盐途径、糖酵解与糖异生途径以及部分氨基酸的代谢等。而对照组中涉及肽酶、细菌运动蛋白、分泌系统核糖体及 DNA 修复等代谢通路的丰度相对较高。

3 讨论

近几年来，诸多原因导致的食物结构不合理，至使伴侣动物长期处于亚健康状态如肥胖、高血脂等，已成为非常突出的问题[10-11]。高脂血症通常表现为 TG、TCH 和 LDL 的升高及 HDL 的降低。TCH 和 TG 是机体重要的组成成分，主要来自肝的合成和食物摄取，且在血液中的浓度恒定，当其浓度过高时则引起相应的疾病。HDL 可将周围组织中的胆固醇运

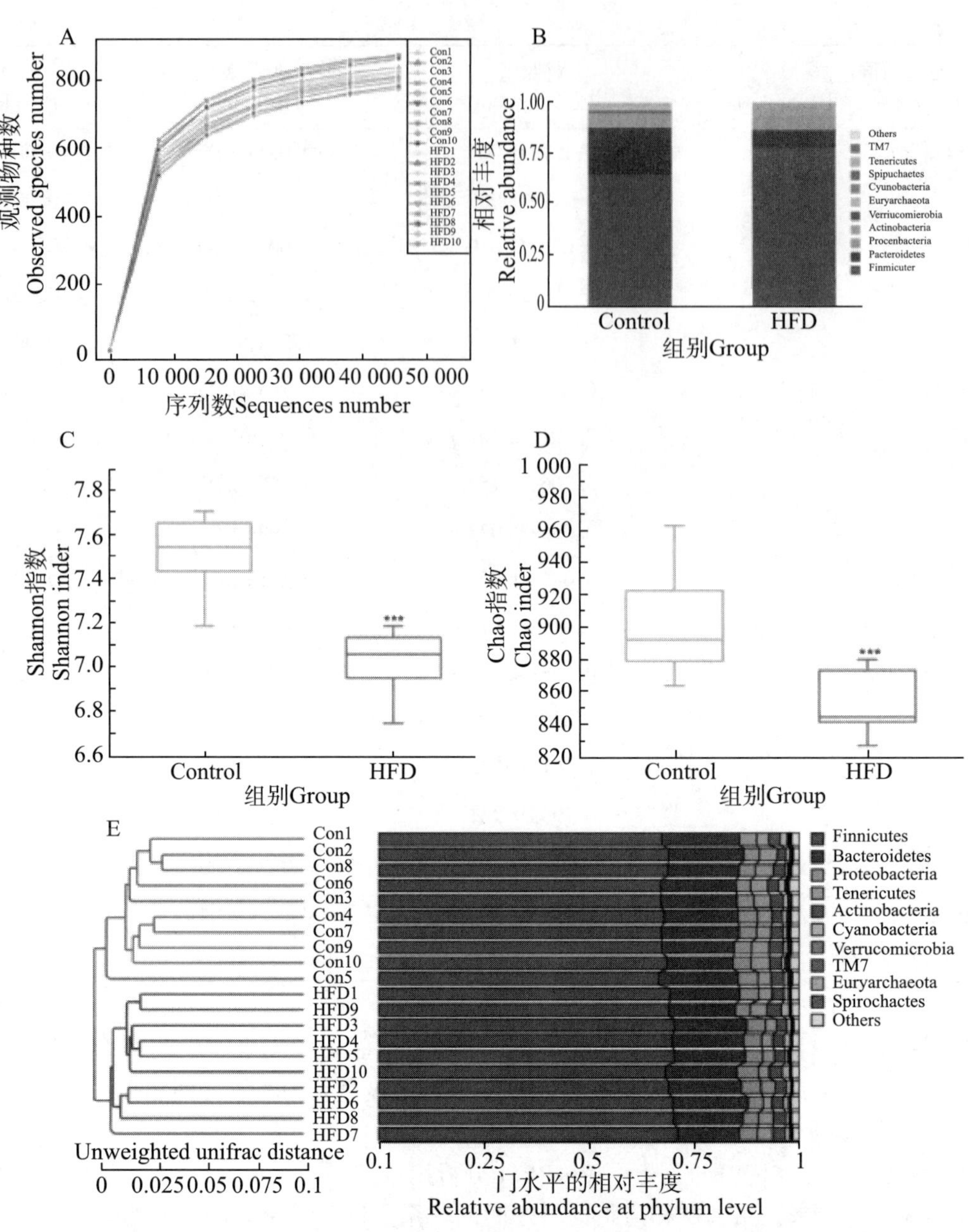

图 2　高脂饲料对大鼠肠道菌群的影响

Fig.2　Effects of HFD on gutmicrobiota in rats

送至肝，将其转化为胆汁酸或者直接通过胆汁从肠道排出[12-13]。本研究中采用高脂饲料诱导大鼠高脂血症的发生，结果表明，饲喂高脂饲料 8 周即可显著升高 TG、TCH 和 LDL，降低 HDL。罗朵生等的研究表明，高脂饮食（30 天）可显著增加大鼠 TG、TCH 和 LDL，并降低 HDL 水平[14]，和本试验结果基本一致。然而，13 周后两组之间 HDL 指标差异不显著，其原因可能是由于高脂饲料饲喂时间及组成成分之间的差异导致（如饱和脂肪酸含量低等）[15]。相比于第 8 周，随着高脂饲料饲喂时间的延长，TCH 及 HDL 水平进一步增加。高脂饲料饲喂时间的延长可能导致肝细胞合成胆固醇增多；伴随着血脂水平的提高，大鼠自身

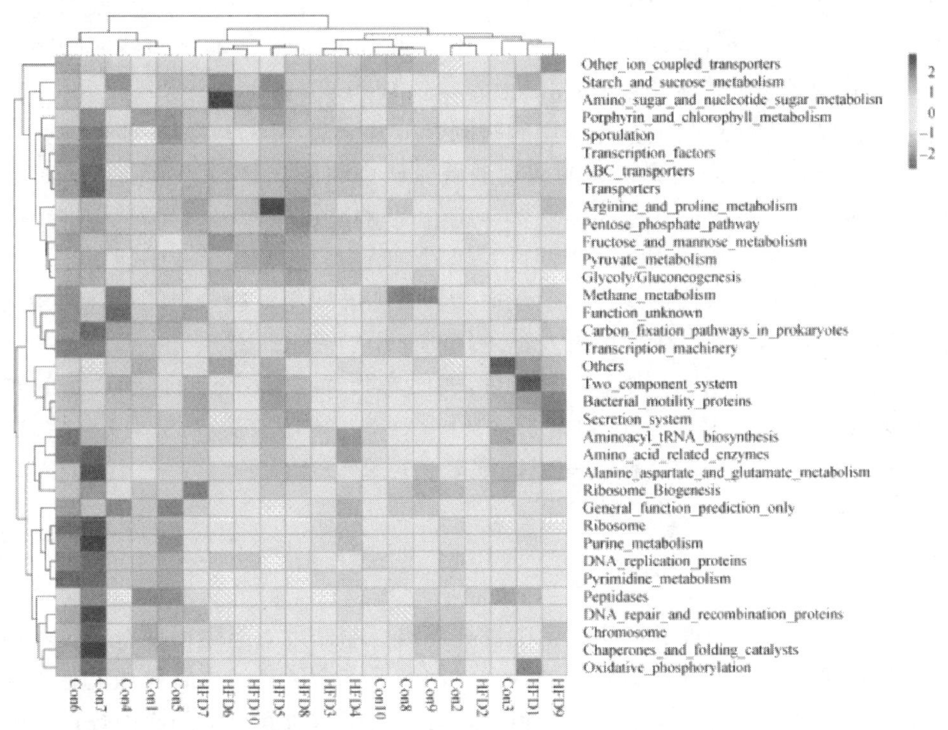

图 3 对照组和高脂饲料组 PICRUSt 分析（KEGG 第三层级代谢通路）

Fig.3 PICRUSt analysis between control and HFD groups (the third level of KEGG pathway)

需要更多的 HDL 来转运增加的胆固醇，从而导致 HDL 水平的升高。因此，HDL 的升高可能是大鼠通过自身调节对高脂饲料的一种适应结果。肝的病理学检查结果与血脂检测结果呈正相关，血脂水平的升高对肝造成了不同程度的损伤，胆固醇合成的增加以及脂类的积累可能是造成肝脂肪变性的主要原因[16-17]。

肠道菌群是一个极其复杂的微生态系统，与宿主的生理、病理、代谢等有着密切的相互作用[18]。近年来相关研究证实，肠道菌群在生殖、营养吸收、肥胖以及免疫等方面发挥着重要作用[19-21]。本研究中，高脂饲料组的 Chao 及 Shannon 指数显著低于对照组，说明肠道菌群的多样性显著降低。黄红丽等通过高脂饮食构建非酒精性脂肪肝动物模型，并分析其肠道菌群的特点[22]，结果证实，高脂饲料可显著降低肠道菌群多样性，与本研究结果一致。E.Le Chatelier 等报道，非肥胖者和肥胖者的肠道微生物的多样性存在显著差别，其结果表明，肠道微生物多样性的降低与肥胖程度、血脂异常及胰岛素抗性呈正相关关系[23]。因此，高脂饲料所导致的菌群多样性的降低对机体构成潜在的威胁。

朱超霞等报道，随着高脂饮食的摄入，大鼠肠道菌群呈现高比例的厚壁菌门和低比例的拟杆菌门[24]。本研究结果也显示，高脂饮食在门水平上对物种组成比例的影响较大，主要表现为厚壁菌门和放线菌门的增加，拟杆菌门的减少。厚壁菌门与拟杆菌门比值的增大，将导致肠道菌群更为有效地从食物中获取能量，增加机体的能量获取，从而促进脂肪、胆固醇等的合成引发血脂升高或脂肪肝病变。因此，高脂饲料可通过影响肠道菌群组成，增加能量摄入而导致血脂异常。D.W.Chen 等报道，高脂饲料可导致大鼠 Dorea 及 Bacteroides 丰度的增加，同时引起 Ruminococcus、Allobaculum 和 Collinsella 丰度的降低[25]。本研究中，Dorea 丰度的变化与其报道一致，而 Bacteroides、Ruminococcus、Allobaculum 及 Collinsella 的丰度变化与报道结果存在差异，这种属水平菌群的差异，可能是由于高脂饲料饲喂时间的不同所导致。不同种属的细菌对宿主发挥着不同的作用，如 Collinsella 能够产生短链脂肪酸，并且有助于维护肠黏膜屏障的完整[26]；Ruminococcus 与能量获取有紧密关系[27]，而高脂饲料对这些菌的影响，以及细菌对宿主发挥作用的机制尚不完全清楚，需要进一步的研究。

本研究通过 PICRUSt 来预测相关的代谢通路，研究高脂饲料对肠道菌群的影响。高脂饲料组中，大多数发生变化的代谢通路主要与能量代谢相关，其结果与肠道菌群组成的改变相符。因此，可以推测，长期饲喂高脂饲料引起菌群结构的改变，导致了能量相关代谢通路的过度富集，进而引发能量代谢异常以至于血脂异常和肝病变。这些与能量代谢相关的通路可作为桥梁来说明高脂饲料、肠道菌群及宿主之间作用的关系。对照组中所涉及的代谢通路主要与细菌的生存与繁殖相关，如 DNA 修复等，而高脂饲料所引起的这些通路的改变可能与肠道菌群多样性的降低相关。虽然 PICRUSt 分析对于肠道菌群功能的解读有很大的帮助，但由于数据库及方法的限制，其存在一定的缺陷，需要结合更多的方法（宏基因组、转录组等）对试验结果加以确认。

4 结论

肠道菌群在高脂饲料引发的血脂异常中发挥着重要的作用，但限于现有的技术条件，高脂饲料对肠道菌群的影响及其血脂异常的具体机制尚不完全清楚。本研究从血脂、肝病理学及肠道菌群的角度说明了长期饲喂高脂饲料的危害，为探讨血脂异常的成因提供了依据，并对伴侣动物饮食管理有一定的指导作用。

参考文献（略）

（发表于《畜牧兽医学报》，院选一级学报）

基于液质平台代谢组学生物样本的采集和制备*

马　宁**，杨亚军，刘希望，李剑勇***

（中国农业科学院兰州畜牧与兽药研究所/农业部兽用药物创制重点实验室/甘肃省新兽药工程重点实验室，兰州　730050）

摘要：代谢组学是利用高通量检测技术对内源性代谢物进行定性和定量分析，从而挖掘代谢物与相应生物学变化的相对关系。目前，代谢组学在药物研发、疾病研究、药物毒理学等领域发挥着越来越大的作用。液相-质谱联用平台，检测的相对分子质量范围大、分辨能力强、扫描速度快，拥有很高的灵敏度和较宽的动态范围，其在代谢组学中的应用得到越来越多研究者的认可。生物样品的采集和制备是代谢组学试验中至关重要的步骤，直接决定着试验结果的可靠性。由于样本自身的复杂性以及对样本采集和制备过程中细节的忽视，往往导致试验结果出现偏差，甚至导致错误的结论。因此充分验证样本的采集和处理方法在试验中尤为重要。本文主要综述了基于质谱平台的代谢组学常用生物样本的采集和制备方法，为代谢组学试验提供相关参考。

关键词：代谢组学；质谱；样本采集；制备

* 基金项目：国家自然科学基金资助项目（31572573，31402254）；甘肃省青年科技基金计划资助项目（1506RJYA148）

** 作者简介：马宁（1990—　），男，博士。

*** 通讯作者，E-mail：lijy1971@163.com

DOI：10.16303/j.cnki.1005-4545.2017.06.38

Biological Sample Collection and Preparation for Metabonomic Study with LC-MS Platform

Ning MA, Ya-jun YANG, Xi-wang LIU, Jian-Yong LI*

(Key Laboratory of New Animal Drug Project of Gansu Province/Key Laboratory of Veterinary Pharmaceutical Development, Lanzhou Institute of Husbandry and Pharmaceutical Science of CAAS, Ministry of Agriculture, Lanzhou 730050, China)

Abstract: Metabonomics could make qualitative and quantitative analysis of endogenous metabolites with the application of high-through test technology, and thus dig out relative relationship between metabolites and the relevant biological change. Currently, metabonomics play an important role in many research fields including drug discovery, disease research, drug toxicology and et al. Liquid chromatography-mass spectrometry (LC-MS) platform has a wide detection range of molecule weight, high resolution ability, fast scan speed, high sensitivity and wide dynamic range. The application of LC-MS platform in metabonomic study has been gradually accepted by many researchers. The collection and preparation of biological sample are crucial steps in metabonomic experiment, which directly determine the reliability of the experimental results. The complexity of sample itself and the ignoring details in sample collection and preparation process frequently cause deviations in experimental results, and even erroneous conclusion. Therefore, it is essential to validate the method of sample collection and preparation. This article mainly reviewed the method of sample collection and preparation for metabonomic experiment with LC-MS platform, which could provide reference for related research.

Key words: Metabonomic; MS; Sample collection; Preparation

在系统生物学中，代谢组学（metabolomics）是继基因组学、转录组学、蛋白组学之后的最终组学分支[1-2]。在体液、细胞、组织或器官中，经转录、翻译、翻译修饰后最终产生的全部小分子物质，最能够体现生物体系对外界干扰所做出的反应[1,3-4]。因此，代谢产物的变化被视为生物体受遗传、疾病、外部环境等干扰后的最终应答。根据研究层次的不同，代谢组学主要分为非靶向代谢组学（untargeted metabolomics）和靶向代谢组学（targeted metabolomics）[5-7]。非靶向代谢组学无偏向性地对所有小分子代谢物同时进行检测分析，因此又称非目标代谢组学或发现代谢组学。靶向代谢组学为目标性代谢组学或定量代谢组学，即通过定量分析对潜在的目标代谢物进行分析，用于验证非靶向代谢组学试验提出的假说。目前，代谢组学作为系统生物学的重要组成和新的技术手段，在营养科学、疾病诊断、微生物代谢、药物毒性评价、药物研发、中药现代化等方面，发挥着不可代替的作用[8-11]。

1　液质平台的优势

测定存在于复杂生物样本中小分子物质需要灵敏的分析检测平台。在代谢组学数据采集中，许多分析技术得到广泛应用，诸如核磁共振（NMR）技术、气相色谱-质谱联用（GC-MS）及液相色谱-质谱联用（LC-MS）等[12-13]。质谱由于其普适性、高灵敏度和特异性，在代谢组学研究领域被广泛地应用[14-15]。样品中中等极性至强极性代谢小分子诸如有机酸、脂肪酸、氨基酸、糖类和次级代谢产物等均可在LC-MS平台中得到有效检测。相比于GC-MS和NMR技术平台而言，LC-MS不需要对样品进行衍生化处理，同时又具有经济实用的特点。通过与不同质量分析器的结合，LC-MS在非靶标代谢组学和靶标代谢组学中均有出色的发挥。液相色谱与飞行时间质谱（LC-TOF-MS）的联用，具有稳定、灵敏的定性分析能力，使之成为非靶标代谢组学中最好的分析技术之一。在靶标代谢组学分析中，液相色谱与串联三重四极杆质谱联用技术（LC-QQQ-MS）可以对不同极性的目标代谢物进行高灵敏度、高选择性、高通量的定量分析。因此，LC-MS平台能够满足代谢组学分析中对于小分子物质的定性与定量要求[16]。此外，随着不同质量分析器技术的完善与发展，高效液相色谱（HPLC）、超高效液相色谱（UPLC）及多维色谱等技术在代谢组学中的不断应用，可显著提高质谱平台的分辨率、灵敏度和通量[17-18]。

2　生物样本的重要性

生物体液是代谢组学的主要研究对象，包括：脑脊液、血液、尿液、唾液、组织和细胞提取液等。作为快速发展的科学研究领域，代谢组学力求分析生物体系中的所有代谢产物，尽可能地保留和反映总的代谢产物信息。由于代谢组学分析对象种类繁多，性质差异大，浓度范围广，为了获得稳定可重复的试验结果，必须对样品的处理方法进行选择，遵循一定的标准化程序[19-21]。因此，样品的采集与制备是代谢组学研究中重要的一步，生物样本的质量品质及前处理方法将直接影响试验结果的重复性与准确性[22]。目前，越来越多的科研工作者关注样本的采集、储存、前处理等环节，然而，在样本处理的标准化问题上仍然存在很多的问题，饱受争议。

3　生物样本的采集与制备

代谢组学研究对实验动物有着严格的要求，确保实验动物在品种、性别、周龄、体质量、饲养环境等方面的相近，以减小个体间的差异[20]。其次，试验过程中的诸多外界因素会对试验结果造成影响，诸如饮食、光照、温度、运输等应激。尿液、血清、血浆、粪便、组织等是代谢组学试验中常用的生物样本。在代谢组学试验中要结合实验动物自身的特征，同时兼顾试验目的、仪器平台、分析方法等的要求，减少生物样本自身差异，提高试验结果的可靠性和重现性。

3.1　尿液

尿液是代谢组学研究中使用最为广泛的生物样本之一，其中含有大量代谢终产物具有丰富的代谢信息，同时具有非破坏性和可重复采集等特点[15]。尿液的主要成分为水，占96%~97%，其他为尿素、尿酸、肌酐、氨、硫酸盐等，其广泛的pH范围、浓度、离子强度等给代谢组学分析带来一定的挑战，需要建立相应的样本制备及技术分析方法[23]。

尿液采集与存储：在动物试验中，代谢笼的使用极大地便利了尿液的收集，然而在样本采集过程中仍存在许多细节需要注意，诸如贸然使用代谢笼对动物的应激；食物残渣、粪便等污染尿液；尿液的蒸发以及尿液收集瓶塑化剂污染等问题。采集尿液要对防腐剂、饮水、食物、采集时间等情况给予充分考虑。叠氮化钠（NaN_3）是一种常用的防腐剂，通过添加到尿液采集瓶中（0.05%~1%），来抑制微生物污染及其造成的产物降解[24]。如 LIU 等[25]在收集的兔子尿液中添加了一定量 NaN_3（0.5%）用于抑制微生物污染，防止尿液中的代谢物发生改变。然而，相关试验证实添加或未添加叠氮化钠（3mmol/L）在代谢轮廓及主成分数据分析上并无明显差异。相关分析认为，不同处理的样本之间可能存在相关差异，但主成分的分析方法可能掩盖相应的试验结果[26]。添加 NaN_3 的尿液样本中钠离子的引入可能会对质谱结果造成一定影响。尿液在低温环境条件下仍存在一定程度的微生物污染及降解，在样本存储时间过长且用于靶标代谢组学分析时，建议添加一定量的 NaN_3。在试验设计中，要根据试验目的具体要求，对动物采取禁食禁水等措施，同时对尿液采集时间给予充分的考虑。禁食禁水等措施可能会导致尿液量、成分以及浓度等的变化；在慢性肾衰竭、高血脂等疾病试验中，要考虑禁水或禁食等措施对尿液采集的影响[27-28]。啮齿动物夜间较为活跃，排尿量大且代谢产物信息丰富，白天采集可能尿量少甚至无尿，不同时间段采集的尿液，其代谢轮廓存在较大差异[29]。GAO 等[30]在研究酵母介导的大鼠发热模型试验中，尿液的采集被设定在不同时间点，同时记录了尿液的采集量、pH 及渗透压，使试验结果更加严谨。采集时间过长（大于 12h）会导致尿液长时间暴露，导致蒸发、污染及产物降解，因此建议将采集管置于冰或干冰中以控制温度[24,29,31]。尿液采集后及时低温高速（4℃）离心除去细胞、微粒等杂质，从样本采集到离心时间间隔一般不超过 2h[32]。离心处理的样品保存于低温冰箱中，短时间可存储于-20℃，长时间需保存在-80℃或液氮中，同时避免样品反复冻融，建议样品保存时间不超过 26 周[33-34]。

尿液样品制备：尿液前处理方法主要分为 2 种。一种是将尿液与水按一定的比例（1∶1~1∶4）进行混合，混合后低温高速离心，取上清液过滤检测；另一种是将尿液与一定比例的有机溶剂（甲醇或乙腈）混合震荡后，低温高速离心以除去蛋白，取上清液过滤检测。人尿液蛋白含量较少，通常无须进行除蛋白处理，啮齿动物存在生理性尿蛋白，建议做除蛋白处理。上述两种方法得到的上清液样品，均要经过滤膜过滤，滤膜孔径一般为 0.22μm。此外，在提取尿液中某些特殊性质的成分时，亲和色谱和固相萃取技术适用于尿液的前期处理[35]。在具体的试验中要结合仪器、流动相、分析柱特征及预试验结果等，对尿液的处理方法进行选择并优化，诸如有机试剂甲醇或乙腈的选择，尿液与水或有机溶剂的比例、离心速度、离心时间等。

3.2 血液样本

血样是临床和病理试验中最常用的样本，通常能够较好的反应机体的整体水平[36-37]。血液的主要成分为血浆、血细胞，同时含有无机盐、氧、激素、酶、抗体及细胞代谢产物等。血浆和血清是代谢组学分析中常用的血液衍生样品，相关研究证实血清中代谢物数量高于血浆并且两者存在一定的差别[38-39]。在采集制备血液样本时，一方面要充分考虑实验动物伦理学，选择合适的麻醉剂、采血时间、采血部位等以减小样本间的个体差异；另一方面要充分考虑试验目的及质谱分析平台的特征，样品的前期制备要具有简便性、重现性等。由于血清和血浆中含有大量的蛋白质，在进行质谱分析前需经过蛋白沉淀处理。血清和血浆样本中，代谢物成分的复杂性，含量差异的显著性，极性跨度大等特点，增加了化合物提取与样品分析的难度，同时对仪器的性能提出更高的要求[40-41]。

血液样本采集与存储：为了提高试验结果的可靠性，在试验操作中应严格控制麻醉剂的使用、采血部位、应激因素、样本放置时间、抗凝剂等因素。基于动物伦理学的考虑，采血过程中要使用麻醉剂以减少动物的痛苦。麻醉剂的使用可以减少动物的不适反应，降低应激对试验结果的影响，非麻醉状态下采血可能会引起代谢轮廓的改变。采血部位是需要考虑的因素之一，试验中要根据不同的实验动物以及疾病相关特征，选择不同的部位采集血样。KIYOKO 等[42]采用主成分分析（PCA）的方法评估了麻醉剂使用及采血部位对代谢轮廓的影响。研究表明，颈静脉与腹主动脉采集的样本存在一定的差异，其说明血液的成分可能与动脉血和静脉血种类及组织的局部代谢有关；通过异氟烷呼吸麻醉同一只大鼠，颈静脉采血，其 PCA 结果能够明显区分，并且内源性代谢物存在差异。因此，在采血过程中应尽量减少应激造成的生理改变，确保麻醉剂的使用剂量、采血位置及麻醉后采血时间的一致性。血样采集到离心处理时间、血清或血浆储存条件及放置时间会影响血液样品的质量。血浆样本采集后一般要尽快与血细胞分离，为避免代谢物的酶解及细胞代谢的影响，样本采集后一般进行冰浴，建议放置时间不宜超多 15min，高速低温离心（2 500×g，4℃），取上清液，超低温（-80℃）保存。血清样本采集后室温放置凝集 30~60min 或 4℃放置凝集 1h，放置过程中，不稳定的血液代谢产物会被降解，随后低温离心分离（2 500×g，4℃），取上清液超低温（-80℃）保存。在试验过程中，为减少样本之间的偏差，制备血清和血浆样品时，要确保样品收集管，处置时间及离心参数保持一致。相关试验证实，样本收集管、处理时间、溶血等因素会导致样本差异的产生，引起代谢轮廓的改变[43-44]。因此，采血过程要执行标准化操作，避免因操作失误引起样本污染、增加样本间差异，甚至溶血。同时根据试验需求，确保采血量相同，一方面在使用抗凝管采集血液时，采血量的不一致将导致抗凝剂浓度发生变化；另一方面，试验中若多次长时间采血，建议采血量不超多大鼠血量的 10%，以免因失血过多导致血液动力学及代谢等发生改变从而影响结果。

抗凝剂的使用是采血过程中不可忽视的重要因素。乙二胺四乙酸（EDTA）、柠檬酸钠和肝素等作为常用的抗凝剂在试验中得到广泛的应用。目前，关于抗凝剂对代谢轮廓的影响，已有广泛的研究。研究结果证实使用不同抗凝剂会导致代谢数据 PCA 分析存在的显著差异，并能够引起特定代谢产物含量的变化[45-47]。综合比较各种抗凝剂的优缺点，相关研究结果建议，在基于质谱平台的代谢组学研究中，肝素钠是首选的抗凝剂。相比于 EDTA 和柠檬酸盐，肝素钠具有较好的重现性，污染较少，基质效应弱等优势[45]。

血液样本制备：相比于尿液，血液样本的前期处理相对复杂，主要体现在两方面：血清和血浆中含有大量的蛋白质成分，在进行分析前需要进行除蛋白处理；样本中成分多，极性跨度大。血液样本前期处理的一般流程为：添加一定比例的有机溶剂—涡旋震荡、静置—低温高速离心—上清液过滤直接上机分析或氮气挥干、复溶、过滤上机分析[16,48]。徐婧等[49]等考察了不同有机溶剂以及固相萃取技术对特定化合物的提取效果，采用 4 种不同的溶剂系统：甲醇、乙腈、甲醇-乙醇（1∶1）、甲醇-乙腈-丙酮（1∶1∶1）和固相萃取方法进行蛋白沉淀，用色谱峰数量、峰强度以及化合物峰面积来考察各溶剂系统的提取效果；结果表明乙腈提取效果最好，固相萃取最差；随后，对乙腈沉淀蛋白法进行优化，确定使用 3 倍体积的冷藏乙腈可达到最佳的处理效果。相关研究表明，低温溶剂（4℃）在代谢物提取和蛋白质沉淀上效果比室温有机溶剂好[50-51]。因此，在基于质谱平台血液样本的制备中，建议使用冰乙腈沉淀蛋白和提取有机物。同时，要充分考虑实验目的与自身试验条件，在预试验的基础上，对有机溶剂比例、涡旋时间、静置时间、离心力、复溶体系等因素进行细致优化。

3.3 粪便样品的采集与制备

粪便作为固体或半固体的生物样本具有采集简便，量大、可重复采集及无侵害性等特点。粪便的主要成分是水，其余大多是蛋白质、维生素、无机物、脂肪、未消化的食物纤维以及从肠道脱落的细胞和细菌。粪便的成分和近期食用的食物有着紧密的联系，同时能够反映宿主和肠道菌群的共同代谢情况。粪便样品中富含丰富的代谢信息，在与消化系统疾病相关的代谢组学研究中有着广泛的应用[52-55]。

实验动物粪便的采集主要有两种采集方法，一种是将动物处死后直肠取样[56]；另一种是使用代谢笼收集排泄的粪便[57]。样本收集后可直接冻存[58]，或者液氮冻干研磨处理后冻存[59]。常用的冻存条件为-80℃，-40℃或-20℃。目前粪便样品的储存方法没有统一的标准，相关试验利用核磁共振技术研究存储条件对粪便代谢轮廓的影响，相关结果并未得出一致的结论[60-61]。在质谱平台代谢组学分析中，考虑到粪便样品自身含有大量微生物菌群，建议样本采集后立即使用液氮淬灭，并冻存于-80℃条件下或液氮中。

粪便的前期处理相对简单，主要包括以下的步骤：称取一定质量的粪便与有机溶剂混合（甲醇）—涡旋震荡、超声提取—高速离心去除杂质沉淀—上清液过滤、上机分析。在基于质谱平台的粪便代谢组学研究中，甲醇是常用的有机体系，将粪便与甲醇按照一定的比例混合[53,62-63]，如粪便：甲醇=1：6 或 3mL/g。黄海军等[64]考察了甲醇/粪便（v/g）的比例、过滤膜孔径的大小对提取粪便的影响，结果确定甲醇/粪便（v/g）的比例为：600μL/0.2g，过滤膜的大小为 0.22μm 时，提取效果较好。建议在振荡或超声提取过程过增加频率或者延长时间，以确保代谢物提取充分。

3.4 组织样品的采集与制备

动物组织样品是一种固体或者半固体样品，能够反映病变组织，器官或者机体的整体代谢状况，是代谢组学生物样品研究中的重要组成部分。动物组织样品在采集过程中一定要快，要在最短的时间内将采集的样品进行淬灭。组织样品离体后，要去除结缔组织，并用生理盐水冲洗干净，去除多余血液。残留的结缔组织和血液会导致后期样品成分的改变，因此在样品采集过程中要引起足够的重视。在离体组织中残留酶的活性或氧化还原反应会导致代谢产物的降解或产生新的代谢产物[65]。因此，组织样本采集后需要立即进行生物反应灭活处理。目前，常用的组织样品快速灭活的方法为-80℃速冻或液氮处理。灭活处理后的样品，分装后液氮或-80℃低温存储[66-67]。

对于组织样品处理，需要增加组织均匀化的步骤，其次涉及萃取溶剂的选择及去除蛋白。组织的均匀化处理能够充分释放代谢物并使之与萃取溶剂充分接触。在代谢组学研究中，常用的组织匀浆方法为机械破碎法主要分为两种，一种是利用研钵或超声进行研磨；另一种则是利用组织匀浆机处理组织样品。张蕾等[68]利用 LC-MS/MS 测定了小鼠脑组织中 7 种神经递质含量，将精密称量的脑组织样品与 4 倍体积的 0.2%甲酸-水进行超声，获得脑匀浆液，然后利用 0.2%的甲酸-乙腈与一定量脑匀浆液混匀放置 30min 提取相关化合物。刘磊等[69]在威尔逊病的研究中，利用组织匀浆仪制备肝匀浆，以水-乙腈（1：1）为溶剂体系，通过离心浓缩和复溶对代谢物进行提取。在使用研砵对组织进行研磨时，通常加入液氮处理组织样品，使研磨更加充分。然而，研砵及组织匀浆机的使用，易引起样品间的交叉污染，应引起足够的重视。在处理不同个体的样本时，应充分清洗研砵、匀浆机与样品接触的部位，避免因残留而导致污染。为了避免样品在均匀化过程中发生代谢物的改变或降解，在相关处理过程中建议低温进行并使用冷萃取溶剂。研究对象的不同对萃取溶剂与方法的选择起着至关重要的作用。在靶向代谢组学中，如果分析对象是特定的某一类或几类代谢物，通常对萃取方法的选择性有较高要求，常用的处理过程诸如固相萃取等。在非靶向代谢组学研究中，关注样品中全部代谢物，对萃取方法的选择性要求较低，尽量能够提取全部的代谢产物。在具体的过程中，结合预试验结果，应充分优化样本与萃取溶液比例、萃取方法、处理时间、温度及离心参数等。

4 展望

伴随着分析化学的进展，近年来基于 LC-MS 的代谢组学技术突飞猛进。代谢组学技术在生物样本采集与制备、数据采集、数据统计分析及生物学解释等方面已逐步成熟。然而，仍存在许多的问题亟待解决，需要进一步的发展与完善，主要包括：由于生物样本自身的复杂性以及个体间的差异，需要稳健的样品处理技术及分析方法确保结果的可靠性和重现性；目前没有分析方法将样本中所有的代谢物进行全面的研究，因此多平台的联合应用将充分发挥互补优势，成为代谢组学的发展趋势；非靶标代谢组学与靶标代谢组学相互结合，在非靶标代谢组学定性筛选的基础上，利用靶标代谢组学手段进行定量分析，从而增强试验结果的说服力；以生物信息学为纽带，对基因组学、转录组学、蛋白质组学及代谢组学进行组学整合分析，可从不同的层次进行生物学解释，从而对试验结果有更加深刻的理解。

参考文献（略）

（发表于《中国兽医学报》，院选一级学报）

抗球虫中药常山口服液的亚急性毒性试验*

王　玲[1**]，郭志廷[1***]，林春全[2]，张晓松[3]，罗小琴[4]，杨　珍[1]，郭爱民[3]

（1. 中国农业科学院兰州畜牧与兽药研究所/农业部兽用药物创制重点实验室/甘肃省新兽药工程重点实验室，兰州 730050；2. 兰州理工大学 生命科学与工程学院，兰州 730050；3. 甘肃农业大学 动物医学院，兰州 730070；4. 兰州市动物卫生监督所畜产品质量安全检测实验室，兰州 730050）

摘要：通过评价常山口服液的安全性，为临床安全用药提供试验依据。在小鼠急性毒性试验基础上，进行常山口服液对大鼠的亚急性毒性试验研究，通过病理组织学观察其毒性损伤的主要靶器官。取 100 只 Wister 大鼠随机分成 4 个给药组和 1 个空白对照组，给药组按照 1.20，0.60，0.30，0.15g/kg 剂量，连续灌胃给药 30d，所有大鼠于给药后分 3 次（10，20，30d）扑杀，观察测定其临床症状、体质量、采食量、脏器系数、血常规、血液生化、病理及组织学等指标。结果显示：常山口服液可显著降低特高、高及中剂量组的平均日采食量和饲料利用率，降低特高剂量组平均日增重（$P<0.01$），但对大鼠体质量的增加及生长发育无抵制作用；随着药物剂量的增加和给药时间的延长，常山口服液会对各剂量组的凝血系统指标造成一定影响，与给药剂量无显著相关性，推测无明显毒性反应；常山口服液对大鼠主要脏器的总体损伤程度为特高剂量组>高剂量组>中剂量组>低剂量组>空白组，高剂量组常山口服液的脏器质量及相对质量与对照组比较差异具有统计学意义（$P<0.01$）；组织病理学检查发现，特高、高剂量组大鼠脏器出现不同程度的肿大、淤血、变性和坏死，中、低剂量组的病变程度在给药 10，20d 时较对照组差异不大，至 30d 时毒性损伤以肝、肾、脾的组织病理变化较为明显。结果表明：常山口服液高剂量、长期给药，其毒性损伤的靶器官主要为肝脏、肾脏和脾脏，呈现“时-毒”和“量-毒”关系。在临床合理剂量及合理用药的条件下，其毒性总体较低，用药安全性较高。

关键词：常山口服液；亚急性毒性；组织学病理学；脏器指数；血液生化指标

---

* 基金项目：中央级公益性科研院所专项资金资助项目（1610322011004）；横向委托资助项目(2015-007)

** 作者简介：王玲（1969—　），女，副研究员，硕士。

*** 通讯作者。

DOI：10.16303/j.cnki.1005-4545.2017.07.22

Subacute Toxicity Test of Anticoccidial Oral Liquid from Dichroa febrifuga in Rats

Ling WANG[1], Zhi-ting GUO[1*], Chun-quan LIN[2], Xiao-song ZHANG[3], Xiao-qin LUO[4], Zhen YANG[1], Ai-min GUO[3]

(1. Gansu Province Key Laboratory of New Animal Drug Project/Ministry of Agriculture Key Laboratory of Veterinary Pharmaceutics Discovery/Lanzhou Institute of Animal Science and Veterinary Pharmaceutics, Chinese Academy of Agricultural Sciences (CAAS), Lanzhou 730050, China; 2. College of Life Science and Engineering, Lanzhou University of Technology, Lanzhou 730050, China; 3. College of Veterinary Medicine, Gansu Agricultural University, Lanzhou 730070, China; 4. Animal Products Quality and Safety Testing Laboratory, Lanzhou Institute of Animal Health Supervision, Lanzhou 730050, China)

Abstract: Thesubacute toxicity test of Dichroa febrifuga oral liquid (DFOL) was established in rat based on acute toxicity test method in mice, and the main target organs of its toxic injury were observed by histopathological examination.100 Wister rats were selected and divided randomly into 4 experimental group (EG) and 1 control group (CG), DFOL was administered in 4 EG by intra-gastric gavage at 4 different doses (1.20, 0.60, 0.30, 0.15g/kg) for 30 days and all rats were killed at 10, 20, 30d.The results showed that daily feed intake and feed utilization of rats in high and medium dose group declined, average daily gain in high dose group declined ($P<0.01$), and had no resist effect to body weight and growth for rats.DFOL had a certain effect on the routine blood index in 4 EG, and no significant relativity and toxicity reaction were found with increasing drug dose and time. Viscera index in high dose group showed very significant difference compared with control group ($P<0.01$).The viscera injury were found in high dose group by histopathological examination, and the liver cell showed severe edema, congestion, kidney showed congestion, interstitial edema, renal tubule epithelial cloudy swelling and degeneration. The injury to liver, kidney and spleen in medium and low dose group were found at 30d with increasing drug dose and time.It is concluded that the main toxic organ of DFOL were liver, kidney and spleen under high dose and long course of medication, and showed time-toxicity and dose-toxicity relationship. The research suggested that the toxicity of DFOL was lower and high security in a conventional dose for clinical anti-coccidiosis application.

Key words: Dichroa febrifuga oral liquid (DFOL); Subacute toxicity test; Histopathological examination; Viscera index; Serum biochemical index

鸡球虫病是由艾美耳属的多种球虫寄生于鸡肠道所引起的一种严重危害养鸡业的原虫病。目前，控制球虫病的主要手段是化学药物防治，因球虫易对化学药物产生耐药性且药物会在肉蛋中残留，使得新药的开发速度已赶不上耐药虫株出现的速度，尤其是多重耐药虫株的出现，加之鸡球虫病的疫苗免疫预防技术因免疫接种后的不良反应以及在我国抗球虫作用的菌药性限制等，短期内还难以推广应用[1-3]。中草药相对于化学合成药具有毒副作用小、药物残留低、不易产生耐药性等优点，具有直接杀死虫体或抑制虫体、虫卵的作用[4-6]。常山（Dichroa febrifuga Lour.）是目前国内常用的抗球虫中药材，具有驱虫杀虫、抗疟解热、止血消炎、促进伤口愈合等功效，其中主要有效成分为常山碱[7-9]。然而常山碱在有良好抗球虫效果的同时，亦存在一定的细胞毒性[10-11]。为了更系统地认知常山的毒性，评价其长期用药的安全性，本试验将通过观察常山口服液连续灌胃对大鼠的长期毒性，对其毒性定性、定位、定量以及与时间、剂量的相关性进行研究，拟明确毒性是其药效的延伸还是中药的自身毒性，旨在进一步正确认识和深入研究常山口服液的抗球虫效果，并制定合理的给药方案，为今后科学指导临床用药提供理论依据。

1 材料与方法

1.1 试验动物

Wistar 大白鼠（SPF 级）100 只，购自兰州大学医学院实验动物中心（GLP 实验室）。试验清洁级颗粒鼠粮（苏饲审〔2008〕01008）购自中国农业科学院兰州兽医研究所。大鼠初始体质量 100.0～130.0g，雌雄各半，雌性大鼠未产无孕。饲养条件：室内温度（22±2)℃，自由采食与饮水，连续观察 1 周，临床健康的大鼠即可用于试验，试验期间按常规饲养。

1.2 试验药物

常山口服液，由中国农业科学院兰州畜牧与兽药研究所农业部兽用药物创制重点实验室提取制备，其中常山碱的含量为 0.15%。制备主要过程为：（1）称取中药常山 1.0kg，粉碎，用 70%乙醇回流提取 2 次（第 1 次加入 7 倍量回流 1.5h，第 2 次加入 6 倍量，回流 1.0h），合并滤液，减压回收乙醇（在 0.09MPa 和 50℃的条件下真空减压浓缩），浓缩液喷雾干燥，即得常山提取物（为褐色粉末状物质）。(2) 试验前，称取适量常山提取物，用纯化水配制成质量浓度为 0.04g/mL 的混悬液，密封保存，备用。

1.3 主要仪器和设备

全自动血细胞分析仪（Coulter-JT，美国 Coulter 公司)；全自动血液生化检测仪（7150，日立公司)。

1.4 试验方法[12-13]

1.4.1 分组与给药

根据“兽用天然药物长期毒性研究技术指导原则”[14]，在进行试验之前对 Wister 大鼠进行 1 周的饲喂观察，每天观察并记录各试验组大鼠的饮食与饮水状况、精神状态、行为活动、皮毛光泽度及其他临床症状的变化，观察期内大鼠自由采食与饮水。挑选临床健康的 100 只 Wister 大鼠用于正式试验。试验前将大鼠随机分为 5 个药物组，每组 20 只，雌雄各半，分别为特高剂量组、高剂量组、中剂量组、低剂量组和空白对照组，各给药物组剂量分别为1.20，0.60，0.30，0.15g/kg，空白对照组不给药，按照每只大鼠的体质量以等质量浓度常山口服液计算给药量。试验期间每天上午同一时间用金属灌胃针经口灌胃给药，连续给药30d。试验期间所有大鼠自由饮水、采食。每周测1次体质量，并随体质量增减调整给药量。记录每周摄食量、饮水量的变化。发现有毒性反应的动物取出单笼饲养，重点观察，对死亡或濒死动物及时尸检，作肉眼大体观察和病理组织学检查。

1.4.2 临床症状、摄食量、增重及饲料消耗量的观察

每天早晚观察并记录大鼠的一般体征（包括行为表现、运动功能、呼吸、口、眼、鼻、耳、粪、尿等）、摄食量、饮水量、精神状态、被毛光泽度、生长发育等临床症状。试验开始及结束前，以每5d为1个周期，对所有大鼠测体质量，计算各药物组各阶段平均日增重，并按照新的体质量重新计算给药量。另外，再对剩余的饲料测质量，并根据投料量计算出每组大鼠每个周期的饲料消耗量，以每个药物组为单位计算平均日采食量、料重比。

1.4.3 血液生理生化指标测定

所有大鼠分3次进行扑杀（给药后10，20，30d）。给药后10，20d，分别从各组随机选取大鼠4只，雌雄各半，测体质量后进行扑杀，并在给药后30d将剩余的60只大鼠全部扑杀。所有大鼠在扑杀之前禁食不禁水12h以上（采血化验前1d晚上开始禁食，次日上午8：30左右采血）。扑杀大鼠经股静脉采血，分别用于血常规检测和血液生化指标检测。其中用于血常规检测的采血管（抗凝管，内含抗凝剂）保证采血量为1~2mL，采完血后立即摇匀以防凝血形成；用于血液生化指标检测的采血管（不加抗凝剂）保证采血量达3mL左右。采好的血样尽快置于冰箱中短时间冷藏保存。血液生化指标检测：包括白细胞数目（WBC）、红细胞数目（RBC）、血红蛋白数目（HGB）、红细胞压积（HCT）、平均血红蛋白浓度（MCHC）和淋巴细胞数目（LY）。血常规检测：包括丙氨酸氨基转移酶（ALT）、天冬氨酸氨基转移酶（AST）、总蛋白（TP）、总胆红素（TBIL）、尿素氮（BUN）和肌酐（Cr）的浓度。

1.4.4 脏器指数

剖检分离大鼠心脏、肝脏、脾脏、肺脏、肾脏主要脏器，测质量，测定脏器指数。并用t检验比较组间差异的显著性，$P<0.05$代表有统计学意义。脏器系数=（脏器质量/体质量）×100%。

1.4.5 剖检和组织病理学观察

剖检观察大鼠心脏、肝脏、肾脏、脾脏及肺脏的大小、形态、色泽、质感，尤其是肝脏、肾脏、肠道、胃的病理变化。同时重点采集肝脏、肾脏、心脏、脾脏、肺脏，置于10mL离心管中，用配制好的10%中性缓冲甲醛溶液固定，常规石蜡包埋切片，HE染色，光镜下观察组织病理学的变化。

1.5 统计学数据处理

采用Microsoft Excel软件将所有的试验数据汇总成表，应用SPSS 22.0软件对各试验原始数据进行方差齐性检验，满足方差齐性要求的数据采用单因素方差分析方法，各给药组分别与对照组进行组间均数的两两比较，以$P<0.05$为差异显著，数据采用$\bar{x}\pm s$的形式表示。

2 结果

2.1 常山口服液对大鼠临床症状、平均摄食量的影响

给药观察期间和停药后，常山口服液特高剂量、高剂量、中剂量、低剂量组大鼠的饮水

量、精神状态、活动状况、被毛光泽度和体温等特征与对照组相比均无明显异常变化，部分高剂量组大鼠出现腹泻及体质量减轻症状。对表1中各剂量组和空白对照组大鼠摄食量进行统计学分析（t检验），结果表明与空白对照组比较，给药后5，10d，特高剂量组、高剂量组大鼠的平均摄食量显著减少（$P<0.05$），中剂量组和低剂量组无显著性影响（$P>0.05$）；而其余时间常山口服液各剂量组大鼠的摄食量均无显著性差异（$P>0.05$）。

表1　常山口服液对大鼠平均摄食量的影响（$\bar{x}\pm s$） g

T给药后/d	特高剂量	高剂量	中剂量	低剂量	对照组
5	533.85±45.89*	622.35±37.26	668.85±14.78	655.20±22.63	566.60±14.97
10	570.70±13.58*	517.35±10.90*	615.40±42.28	646.45±21.99	720.90±30.55
15	654.95±30.80	500.00±41.01	504.05±11.53	559.40±52.04	619.65±10.43
20	580.75±20.04	553.45±78.42	533.75±23.26	555.45±36.27	643.85±12.70
25	389.35±20.15	378.70±71.42	416.30±51.34	408.65±68.52	450.65±11.86
30	416.05±46.88	370.90±50.11	392.95±58.36	424.40±68.17	463.20±65.62

注：* 表示与空白对照组相比 $P<0.05$；** 表示与空白对照组相比 $P<0.01$。下同

2.2 常山口服液对大鼠体质量、日采食量、日增重及饲料消耗率的影响

与空白对照组比较，给药后5d，特高剂量、高剂量、中剂量和低剂量组对大鼠体质量的增长均有明显的抑制作用（$P<0.01$）；给药后20，25，30d，部分给药组大鼠出现腹泻，体质量增长慢于空白对照组，以剂量较高组较为明显。特高剂量组对大鼠体质量增长有明显的抑制作用（$P<0.05$），其余时间无显著影响（$P>0.05$）；高剂量、中剂量和低剂量组对大鼠体质量增长均无显著影响（$P>0.05$）。与空白对照组相比，常山口服液可显著降低特高剂量组、高剂量组和中剂量组的平均日采食量（$P<0.01$），降低特高剂量组平均日增重（$P<0.01$）。料重比结果表明，常山口服液特高剂量组、高剂量组、中剂量组及低剂量组均显著高于空白对照组（$P<0.01$）。常山口服液可降低饲料利用率，这与试验期间观察到的各给药组大鼠腹泻率较高、增重慢等临床症状相一致（表2）。

表2　常山口服液对大鼠体质量的影响（$\bar{x}\pm s$） g/只

体质量/g	特高剂量	高剂量	中剂量	低剂量	对照组
初始	129.17±16.72	129.00±19.94	120.36±21.49	110.10±24.12	100.48±22.03
5d	131.01±14.48**	136.29±18.80**	135.12±17.15**	129.44±17.01**	110.36±20.50
10d	137.95±15.80	139.54±18.11	147.23±20.17	147.45±21.42	142.32±21.13
15d	143.86±15.50	146.53±14.43	154.04±18.66	154.51±19.81	155.23±22.89
20d	155.02±20.27*	163.57±18.71	168.86±21.40	169.67±24.35	176.36±29.04
25d	164.65±19.65*	173.17±16.82	177.27±24.42	177.67±23.33	187.79±37.16
30d	172.38±20.73*	181.75±19.78	185.50±26.08	187.27±27.45	196.83±41.96
日采食量/($g\cdot d^{-1}$)	11.79±2.45**	12.39±1.27**	13.41±1.32	13.62±1.01	14.25±1.77
日增重/($g\cdot d^{-1}$)	1.25±0.29**	1.57±0.31**	1.87±0.10**	2.36±0.17	2.55±0.44
料重比	9.69±1.76**	7.88±0.61**	7.13±2.11**	5.35±0.42**	4.08±0.38

2.3 常山口服液对大鼠血常规的影响

与空白对照组比较，给药 10d 后，常山口服液特高剂量、高剂量、中剂量和低剂量组大鼠血液中 RBC、HGB 和 HCT 的水平显著升高（$P<0.01$ 或 $P<0.05$），WBC、MCHC 和 LY 的水平无显著变化（$P>0.05$）。给药 20d 后，常山口服液特高剂量组大鼠血液中 WBC、HGB 的水平显著降低（$P<0.05$），RBC、HCT、MCHC 和 LY 的水平无显著变化（$P>0.05$）；高剂量组大鼠血液中的上述指标均无显著变化（$P>0.05$）；中剂量组大鼠血液中 RBC 的水平显著升高（$P<0.05$），WBC、HGB、HCT、MCHC 和 LY 的水平无显著变化（$P>0.05$）；低剂量组大鼠血液中 RBC 和 HCT 的水平显著升高（$P<0.05$），WBC、HGB、MCHC 和 LY 的水平无显著变化（$P>0.05$）。给药 30d 后，常山口服液特高剂量组大鼠血液中 HGB 和 HCT 的水平显著降低（$P<0.01$ 或 $P<0.05$），LY 的水平显著升高（$P<0.05$），WBC、RBC 和 MCHC 的水平无显著变化（$P>0.05$）；常山口服液高剂量组大鼠血液中 HGB 的水平显著降低（$P<0.05$），WBC、RBC、HCT、MCHC 和 LY 的水平无显著变化（$P>0.05$）；常山口服液中剂量组和低剂量组大鼠血液中的上述指标均无显著变化（$P>0.05$）（表 3）。

表 3　常山口服液对大鼠血常规的影响（$\bar{x}\pm s$，$n=10$）

t 给药后/d	指标	特高剂量	高剂量	中剂量	低剂量	对照组
10	WBC/($\times10^9\cdot L^{-1}$)	11.68±3.22	13.53±1.44	10.15±0.90	10.75±1.42	11.90±3.73
	RBC/($\times10^{12}\cdot L^{-1}$)	8.19±0.28**	8.29±0.18**	8.28±0.22**	7.95±0.26*	7.41±0.57
	HGB/($g\cdot L^{-1}$)	159.75±5.25*	160.75±3.30*	162.50±2.52**	160.25±6.02*	150.50±8.35
	HCT/%	50.43±2.01	51.85±1.41*	52.28±1.74**	50.98±2.22*	47.70±2.54
	MCHC/($G\cdot L^{-1}$)	317.25±5.12	310.00±4.97	311.00±7.12	314.50±5.92	315.75±6.18
	LY/($\times10^9\cdot L^{-1}$)	8.65±1.77	10.60±1.20	8.25±0.58	8.93±1.16	8.74±1.71
20	WBC/($\times10^9\cdot L^{-1}$)	8.85±1.32*	11.35±0.68	11.15±1.45	10.65±1.38	11.30±2.54
	RBC/($\times10^{12}\cdot L^{-1}$)	7.49±0.47	7.61±0.49	7.89±0.42*	8.02±0.53*	7.08±0.32
	HGB/($g\cdot L^{-1}$)	124.25±27.90*	148.00±7.35	159.00±0.82	160.75±7.27	150.25±5.25
	HCT/%	44.60±3.06	45.63±3.10	48.38±2.33	49.43±3.60*	45.10±1.54
	MCHC/($G\cdot L^{-1}$)	279.75±64.36	326.25±35.37	329.25±18.12	326.00±15.38	332.00±10.65
	LY/($\times10^9\cdot L^{-1}$)	4.93±3.35	8.03±1.16	7.33±1.91	8.18±1.51	7.75±1.08
30	WBC/($\times10^9\cdot L^{-1}$)	11.82±1.56	11.90±1.84	12.25±1.21	17.16±1.36	11.48±1.54
	RBC/($\times10^{12}\cdot L^{-1}$)	7.92±0.51	8.03±0.62	8.00±0.49	8.00±0.33	7.90±0.28
	HGB/($g\cdot L^{-1}$)	148.58±6.97**	155.45±7.72*	157.00±8.39	160.83±4.69	161.50±3.45
	HCT/%	45.18±2.70*	46.37±3.69	46.67±3.03	47.73±2.21	48.12±2.24
	MCHC/($G\cdot L^{-1}$)	329.17±6.63	335.91±11.08	336.64±9.16	337.33±8.94	335.83±13.18
	LY/($\times10^9\cdot L^{-1}$)	9.99±1.43*	9.26±1.27	9.34±1.00	8.73±1.28	8.60±1.03

2.4 常山口服液对大鼠血液生化指标的影响

表4结果显示，给药10d后，与空白对照组比较，常山口服液高剂量、中剂量和低剂量大鼠血液中TBIL的水平显著降低（$P<0.01$ 或 $P<0.05$），ALT、AST、TP、BUN和Cr的水平无显著变化（$P>0.05$）；常山口服液特高剂量组大鼠上述血液生化指标均无显著变化（$P>0.05$）。给药20d后，常山口服液特高、高、中及低剂量组大鼠血液中ALT、AST、TP、TBIL、BUN和Cr的水平均无显著变化（$P>0.05$）。给药30d后，常山口服液特高剂量、高剂量和中剂量组大鼠血液中AST的水平显著降低（$P<0.01$ 或 $P<0.05$），ALT、TP、TBIL、BUN和Cr的水平无显著变化（$P>0.05$）；常山口服液低剂量组大鼠上述血液生化指标均无显著变化（$P>0.05$）。

表4 常山口服液对大鼠血生化的影响（$\bar{x}\pm s$，$n=10$）

t给药后/d	指标	特高剂量	高剂量	中剂量	低剂量	对照组
10	ALT/(U·L^{-1})	69.00±21.83	44.25±9.03	49.50±9.68	46.00±11.17	63.50±13.67
	AST/(U·L^{-1})	312.50±27.61	162.75±10.97	206.00±24.75	189.75±45.40	254.75±86.77
	TP/(g·L^{-1})	68.10±1.08	68.38±2.44	67.80±1.63	64.25±3.70	67.78±3.47
	TBIL/(μmol·L^{-1})	2.05±0.26	1.58±0.17**	1.80±0.22*	1.85±0.17*	2.30±0.36
	BUN/(mmol·L^{-1})	10.72±3.20	10.35±1.33	10.86±0.67	12.60±4.11	11.79±1.70
	Cr/(μmol·L^{-1})	61.55±5.74	49.85±11.05	48.03±6.87	56.75±22.07	48.95±13.43
20	ALT/(U·L^{-1})	47.00±3.16	52.50±5.45	53.50±6.56	52.00±5.72	51.25±3.59
	AST/(U·L^{-1})	197.50±38.68	202.25±25.86	219.25±28.02	201.25±20.34	198.75±15.15
	TP/(g·L^{-1})	69.33±2.60	68.20±3.21	65.73±1.77	67.28±2.24	65.95±1.21
	TBIL/(μmol·L^{-1})	1.30±0.24	1.43±0.39	1.58±0.40	1.20±0.16	1.68±0.54
	BUN/(mmol·L^{-1})	11.56±3.18	10.79±2.01	10.20±1.13	9.93±0.57	9.44±0.41
	Cr/(μmol·L^{-1})	46.68±26.06	44.08±13.01	40.25±8.41	41.90±6.21	35.48±2.52
30	ALT/(U·L^{-1})	53.75±7.16	57.45±5.35	54.91±6.39	54.42±8.80	58.27±5.57
	AST/(U·L^{-1})	199.08±32.82**	194.73±15.92**	218.36±27.71*	238.75±29.36	248.18±27.5
	TP/(g·L^{-1})	68.81±2.56	68.04±2.48	68.47±3.07	67.92±3.13	67.57±1.99
	TBIL/(μmol·L^{-1})	1.32±0.24	1.35±0.32	1.14±0.18	1.33±0.22	1.22±0.23
	BUN/(mmol·L^{-1})	9.36±2.28	9.75±0.95	9.43±0.91	11.44±3.15	10.16±1.27
	Cr/(μmol·L^{-1})	42.03±14.01	33.46±7.88	30.30±4.67	43.83±18.38	36.50±5.98

2.5 常山口服液对大鼠脏器系数的影响

与空白对照组比较，给药10d后，常山口服液特高剂量组大鼠肝的脏器系数有显著性改变（$P<0.01$），心、脾、肺和肾的脏器系数无显著性改变（$P>0.05$）；常山口服液高剂量组大鼠的脾脏器系数有显著性改变（$P<0.05$），心、肝、肺和肾的脏器系数无显著性改变

($P>0.05$)；中剂量和低剂量组大鼠的各脏器系数均无著性改变（$P>0.05$）。给药 20d 后，常山口服液特高剂量、高剂量、中剂量、低剂量组大鼠肝的脏器系数有显著性改变（$P<0.01$ 或 $P<0.05$），心、脾、肺和肾的脏器系数无显著性改变（$P>0.05$）。给药 30d 后，常山口服液特高剂量组大鼠的心、肝、脾、肺和肾的脏器系数均有显著性改变（$P<0.01$ 或 $P<0.05$）；常山口服液高剂量组大鼠的心、肺和肾的脏器系数均有显著性改变（$P<0.01$ 或 $P<0.05$），肝和脾的脏器系数无显著性改变（$P>0.05$）；常山口服液中剂量组大鼠肺的脏器系数均有显著性改变（$P<0.05$），心、肝、脾和肾的脏器系数无显著性改变（$P>0.05$）；常山口服液低剂量组大鼠的各脏器系数均无著性改变（$P>0.05$）（表 5）。

表 5　常山口服液对大鼠脏器系数的影响（$\bar{x}\pm s$，$n=10$）

t 给药后/d	器官	特高剂量	高剂量	中剂量	低剂量	对照组
10	心脏	0.427±0.018	0.466±0.038	0.434±0.014	0.518±0.173	0.432±0.021
	肝脏	3.814±0.395**	4.251±0.203	4.317±0.218	4.571±0.250	4.596±0.395
	脾脏	0.221±0.010	0.246±0.015*	0.217±0.007	0.217±0.040	0.208±0.021
	肺脏	0.781±0.073	0.736±0.257	0.647±0.087	0.532±0.109	0.638±0.064
	肾脏	0.752±0.055	0.792±0.039	0.750±0.032	0.799±0.164	0.757±0.078
20	心脏	0.377±0.023	0.384±0.017	0.414±0.061	0.414±0.052	0.421±0.023
	肝脏	4.112±0.162**	4.209±0.065*	4.309±0.263*	4.289±0.184*	4.753±0.495
	脾脏	0.211±0.012	0.211±0.028	0.221±0.028	0.246±0.061	0.238±0.034
	肺脏	0.609±0.058	0.628±0.074	0.599±0.108	0.583±0.048	0.624±0.048
	肾脏	0.742±0.024	0.740±0.022	0.736±0.065	0.772±0.073	0.784±0.078
30	心脏	0.423±0.056*	0.429±0.063*	0.370±0.043	0.383±0.049	0.373±0.046
	肝脏	3.724±0.288*	3.923±0.149	3.841±0.176	3.942±0.284	3.942±0.238
	脾脏	0.193±0.025*	0.234±0.035	0.216±0.025	0.231±0.041	0.263±0.151
	肺脏	0.736±0.068**	0.682±0.090**	0.659±0.113*	0.614±0.150	0.544±0.133
	肾脏	0.772±0.078**	0.753±0.038*	0.728±0.039	0.704±0.039	0.690±0.093

2.6　剖检及组织病理学观察结果

常山口服液对大鼠各主要脏器的总体损伤程度：特高剂量组>高剂量组>中剂量组>低剂量组>空白组，并呈剂量依赖性。特高剂量组大鼠中有 1 例出现肝脏中度肿胀、边缘变白、脾脏变脆的病理特征，1 例出现肝脏和肾脏均发白的特点；高剂量组大鼠中有极个别出现肝脏轻微肿胀，肝脏和肾脏均轻微发白等病理变化；中剂量组、低剂量组大鼠中有个别出现肝脏肿大、轻微发白，肝脏边缘轻微变圆变钝，肾脏轻微肿大、发白，肺脏上散布出血点等病理特征；胃、十二指肠黏膜完好，未见黏膜上皮细胞坏死、脱落及炎性细胞浸润。

组织学变化发现，随着常山口服液剂量的不断增加及给药时间的延长，低、中、高及特高剂量组的大鼠脏器损伤程度逐渐加重，甚至出现不同程度的肿大、淤血、变性和坏死，其中以特高、高剂量组表现较为明显。肝脏多为颗粒变性，以小叶中央区较明显，亦可见水泡变性肝细胞，光镜下多表现中心性肝坏死，肝小叶内肝板排列欠规则，可见肝小叶中心（即小叶中央静脉周围）的肝细胞肿胀、坏死及崩解，肝窦淤血，小叶周围有少量淋巴细胞、浆细胞等炎性细胞浸润，中央静脉及其周围窦间隙扩张，充满红细胞（图 1A～D）。肾脏病变主要发生在肾小管上皮细胞，表现为细胞肿胀，颗粒变性和透明变性，间质充血、出血，多表现轻度急性增生性肾小球炎（图 2A～D）。脾淋巴细胞出现程度不等坏死，镜检脾髓内充盈大量血液，少量脾实质细胞（淋巴细胞、网状细胞）可见坏死、崩解，红髓中固有的细胞成分大为减少，白髓体积缩小，鞘动脉周围网状细胞坏死及多量浆液和纤维素渗出，血管壁发生坏死，并与周围渗出物连成一片（图 3A～D）。肺脏具间质性肺炎、支气管肺炎的病理组织学变化，镜检可见肺泡间隔、小叶间的间质、支气管周围、血管周围及胸膜下组织发生程度不等的水肿以及淋巴细胞、单核细胞浸润，浸润明显时可见支气管和血管周围有管套形成，肺泡腔中可见少量水肿液，肺泡壁上皮细胞由于肿胀、增生和化生呈立方状，细支气管腔和肺泡腔内可见渗出性的浆液、中性粒细胞和脱落的上皮细胞，炎灶邻近的有些肺泡融合成较大的多囊状空腔，呈代偿性肺气肿（图 4A～D）。心肌轻度颗粒变性，心肌纤维结构基本完整，未见明显肿胀、变粗，横纹、肌原纤维不清楚，原纤维间出现微细的蛋白颗粒，其次间质的渗出变化不显著，可见轻度的毛细血管充血，局灶性集聚炎性细胞（图 5A～D）。镜检中剂量组和低剂量组，各脏器组织学病变与高剂量组比较程度均有所减轻。给药 10d 时，各内脏器官结构基本正常，界限清晰，组织病理学变化较对照组差异不大，但随着给药时间的延长，至 20d 时各内脏器官出现轻度淤血、水肿及程度不等的变性、坏死，以肝、脾、肾的组织病理变化表现较为明显，说明大鼠已经出现了轻微的脏器损伤。给药 30d 后，毒性损伤进一步增强，肝、肾、脾、肺、心除表现轻度淤血变化外，肝细胞出现程度不等的颗粒变性、水泡变性及坏死；肾上皮细胞轻度颗粒变性；脾淋巴细胞则表现为程度不等坏死；肺脏呈间质性肺炎、支气管肺炎变化；心肌纤维多数表现为轻度颗粒变性，与高剂量组组织病变表现相同。

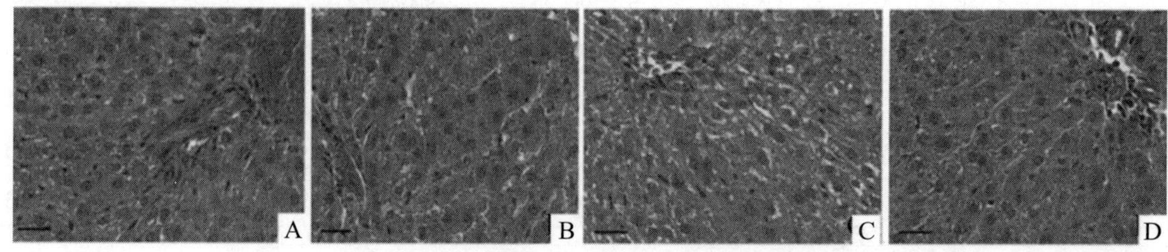

图 1　肝脏病理组织学观察（HE 染色，400×）

A.高剂量组；B.中剂量组；C.低剂量组；D 空白对照组。下同

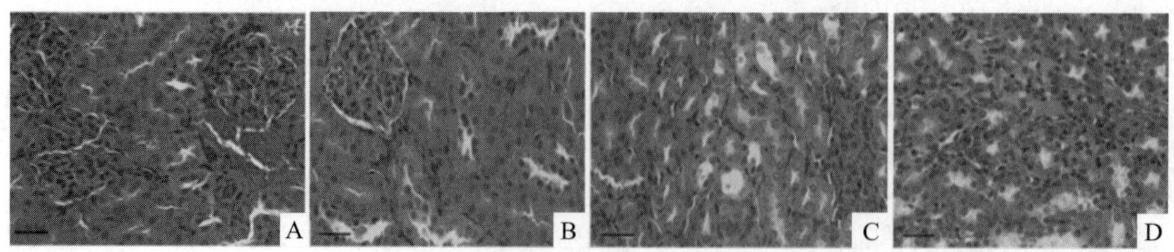

图 2　肾脏病理组织学观察

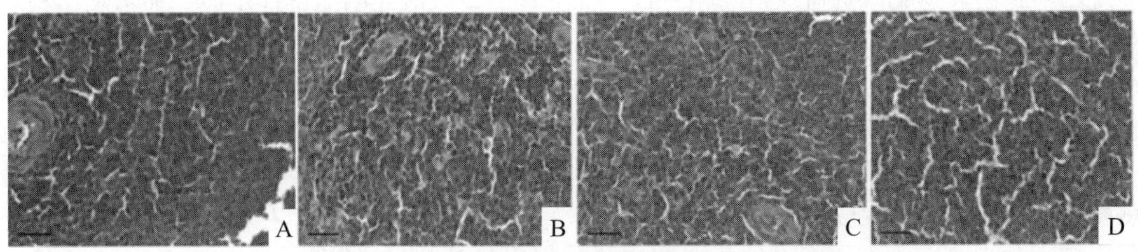

图 3　脾脏病理组织学观察

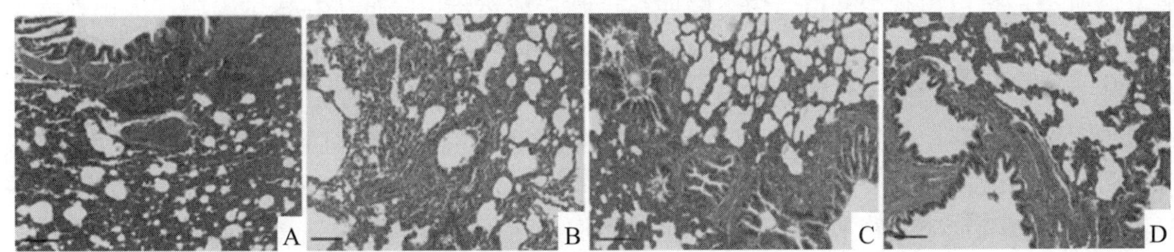

图 4　肺脏病理组织学观察

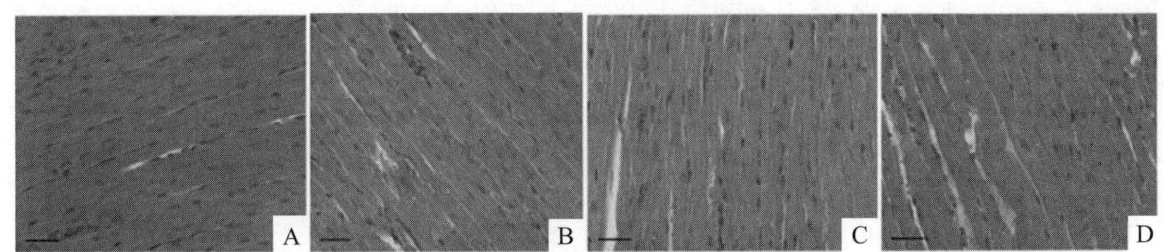

图 5　心脏病理组织学观察

3　讨论

3.1　常山口服液的给药反应及对大鼠行为学、生长的影响

常山口服液各剂量组大鼠的外观体征、行为表现、饮水量、被毛光泽度和体温等特征与对照组相比均无明显的异常变化，各剂量组大鼠对外界声音、触动的刺激均反应正常，眼、口、鼻处无异常分泌物。常山口服液特高、高、中及低剂量组大鼠与空白对照组比较，均出现不同程度的腹泻，其中以特高剂量组最为严重，随着给药剂量的减少，腹泻程度逐渐减轻，这与常山口服液特高剂量组大鼠的体质量增长受到明显的抑制，而高剂量、中剂量和低剂量组均无显著变化的结果基本一致。试验期间，常山口服液对大鼠体质量的增加及生长发育无抵制作用。

3.2　常山口服液对血液学及血液生化指标的影响

药理学研究可通过检测试验动物血液的相关生理、生化指标在受试药物的作用下有无改变，来准确衡量受试药物毒性的大小。在本试验中，与空白对照组比较，随着药物剂量的增加和给药时间的延长，常山口服液会对凝血系统指标造成一定影响，但各剂量组的血液学指标在不同给药时间未见明显规律性，且与给药剂量无显著相关性，这与试验期间各组大鼠未发现因药物引起感染、贫血等情况相一致，推测常山口服液无明显毒性反应。血清酶学检查是用来探查细胞是否受到损害最经济、实用的检测手段，肝损伤后肝细胞内的酶类可释放到血液中，造成血清中酶活性的异常升高，其中 ALT 和 AST 能敏感地反映肝细胞损伤与否及损伤程度。当肝细胞变性、细胞膜的通透性增加时，从细胞中逸出的主要是 ALT，当肝细胞严重病变、坏死时，线粒体内 AST 便释放出来；因此，反映急性肝细胞损伤时以 ALT 为最敏感，而反映其损伤程度时则以 AST 为敏感[15]。总胆红素（TBIL）主要用来诊断是否有肝脏疾病（如黄疸、中毒性肝炎等）或胆道是否发生异常（胆道梗阻）。总蛋白（TP）检查主要是检测肝功能代谢能力的试验，反映肝脏的储备能力，当肝细胞受损，肝功能出现障碍，肝脏合成蛋白质减少，白蛋白下降明显，常造成总蛋白降低。在本试验中，大鼠血液生化指标结果显示，给药 10d 后，常山口服液对高、中、低剂量组大鼠血液生化指标 TBIL 的产生均有明显的抑制作用，特高剂量组无明显的异常变化；给药 20d 后，常山口服液各剂量组对大鼠各项血液生化指标未产生明显影响；给药 30d 后，常山口服液特高剂量、高剂量、中剂量大鼠血液的 ALT、AST、TP 水平与空白对照组比较无显著变化，低剂量组无明显差异。由此可以判定，随着常山口服液剂量的增加及给药时间的延长，高剂量的常山口服液毒性部位对大鼠肝脏未造成严重的器质损伤，肝细胞未发生严重的变性、坏死。但病理组织学观察结果发现，高剂量的常山口服液毒性部位对大鼠肝脏大造成了较严重的损伤，肝细胞具程度不等颗粒变性和水泡变性，肝小叶中心有少量肝细胞坏死，在考虑动物个体差异的前提下，总体来说常山口服液对大鼠肝脏的损伤作用还不够明确，需进一步研究。因此，常山口服液在临床长期用药时，临床用药仍需控制在安全剂量范围内使用。血尿素氮（BUN）、尿肌酐（Cr）通常可作为判断肾功能损伤的指标，当肾脏的排泄功能受到不同程度的抑制，BUN、Cr 会有不同程度的升高。在此试验中，常山毒性部位特高、高、中、低剂量组大鼠的 BUN 和 Cr 值在给药 10~30d 期间，与对照组相比较无显著变化（$P>0.05$），这说明常山毒性部位对肾脏功能无显著影响，与病理组织学观察并未见肾脏损伤结果较一致。

3.3 常山口服液对脏器指数的影响

常山口服液特高剂量组大鼠的肝脏和脾脏的脏器系数较对照组明显减小，心脏、肺脏和肾脏的脏器系数明显增高。常山口服液高剂量组大鼠的心脏、肺脏和肾脏的脏器系数明显增高，中剂量和低剂量组大鼠的各脏器系数基本上无显著变化。试验动物的体质量变化是反映动物机体中毒效应与毒物毒性的最基本指标，而脏器系数则是提供受试化合物作用靶器官的重要指标，脏器质量及相对质量下降意味着该器官萎缩、退行性病变或功能减弱，反之则可能是该器官充血、水肿、增生性肥大或功能增强[16-17]。本试验结果表明，高剂量组常山口服液的脏器质量及相对质量与对照组比较差异具有统计学意义，随给药剂量和时间的增加，常山口服液毒性损伤依次逐渐增强。

3.4 常山口服液对各脏器组织病理学的影响

组织病理学检查可以更加直观地反映出受试药物毒性的大小，试验发现，随着常山口服液剂量的不断增加及用药时间的延长，低、中、高及特高剂量组大鼠各脏器的损伤程度逐渐增强，甚至出现不同程度的肿大、淤血、变性和坏死。中、低剂量组给药 10d 时，各内脏器官的组织病理学变化较对照组差异不大，但随着给药时间的延长，至给药 30d 时毒性损伤逐渐增强，中毒侵犯脏器以肝、肾、脾的的损害较为明显。本试验结果表明：常山口服液长期给药，可对大鼠造成明显的肝、肾、脾毒性，且呈现明显的“时-毒”和“量-毒”关系。

在结果判定当中，肝脏和肾脏比较准确，而肺脏和心脏相对代表性较差，试验中出现个别肺脏散布出血点的症状，可能是由于试验过程中灌药不当造成的损伤所致。试验中高剂量组和中剂量组雌性大鼠各有 1 只死亡，病理学解剖发现死亡的 2 只大鼠的肺脏均出现严重充血的症状，可能是由于灌药不当，使药物误入肺脏而造成的肺脏损伤所致。肝脏是药物在体内最主要代谢场所，绝大多数药物均需经过肝脏的首过效应，在肝脏经过生物转化而被清除。采用小鼠急性毒性试验、大鼠亚急性毒性试验，通过病理组织学观察明确中药常山毒性损伤的靶器官主要为肝脏、肾脏和脾脏，在长期连续、重复多次给药的情况下，大鼠出现了较为严重的肝毒性反应及肾毒性反应。

前期通过对中药常山不同组分进行急性毒性研究发现，常山口服液小鼠经口灌胃的 LD_{50}为 2.0961g/kg，LD_{50}的 95%可信限为 1.7414~2.5429g/kg。给药组中部分小鼠出现中毒反应并死亡，剖检主要脏器未见异常病理变化，表明常山口服液对受试小鼠未造成明显的急性毒性反应，其毒性很低，临床用药安全性较高。结果初步提示：常山口服液在鸡球虫病的防制应用中安全可靠，但并不能说明常山口服液没有毒性。为观察长期用药对大鼠所产生的毒性反应，并为拟订临床抗球虫安全剂量、确定临床给药疗程提供实验依据，本试验通过常山口服液对大鼠的长期毒性的研究，发现常山口服液长期、大剂量给药可造成一定的长期毒性，且随给药剂量增加、给药时间延长，其毒性增强，毒性损伤定位主要在肝、肾。因此，为避免出现上述不良反应，应该建立质量标准，控制毒性成分的含量，加强对此类中药的不良反应监测，建议常山临床用量掌握在 0.15~0.20g/（$kg \cdot d^{-1}$）内为宜，如需加大用量则不宜长期连续服用，用药的剂量过大或不合理用药都可能对动物的健康产生不利的影响。

参考文献（略）

（发表于《中国兽医学报》，院选一级学报）

青海高原牦牛 PRDM1 基因克隆及生物信息学与差异表达分析*

赵生军[1,2]**，张 勇[2]，佘平昌[1]，阎 萍[1]，张利君[3]，郭 宪[1]***

（1. 中国农业科学院 兰州畜牧与兽药研究所，甘肃省牦牛繁育工程重点实验室，兰州 730050；2. 甘肃农业大学 动物医学院，兰州 730070；3. 96351 部队牧场，西宁 810000）

摘要：以青海高原牦牛淋巴结，90 日龄胎儿头部皮肤，牦牛精子以及西门塔尔牛精子为材料，通过 RT-PCR 克隆了包含 PRDM1 基因编码区的 cDNA 序列。将克隆获得的青海高原牦牛 PRDM1 基因的 cDNA 与黄牛相应序列进行比对，对青海高原牦牛 PRDM1 蛋白与其他物种 PRDM1 蛋白进行序列比对和进化树分析，对青海高原牦牛 PRDM1 蛋白的特性和结构进行预测。荧光定量检测青海高原牦牛淋巴结，90 日龄胎儿头部皮肤，牦牛精子以及西门塔尔牛精子中 PRDM1 基因 mRNA 表达量。结果显示，克隆获得青海高原牦牛 PRDM1 基因该部分 cDNA 片段长度为 761bp，其中 PRDM1 基因编码区长 741bp，编码 246 个氨基酸；序列分析显示，克隆获得的牦牛 cDNA 序列与黄牛该序列存在 3 个碱基的变异；青海高原牦牛 PRDM1 蛋白与黄牛、野牦牛、绵羊、马相应序列同源性分别是 99.0%，99.0%，97.0%，92.0%；各物种 PRDM1 蛋白进化树符合物种进化规律。包含卷曲螺旋等典型结构域，人甲基转移酶蛋白结构域 PR 蛋白 1 具有 94% 的相似性，人甲基转移酶蛋白结构域 PR 蛋白 10 具有 36%的相似性，具有 SET 结构域活性。青海高原牦牛 PRDM1 基因在西门塔尔、牦牛精子中表达量极显著高于青海高原牦牛淋巴结（$P<0.05$），90 日龄胎儿头部皮肤中表达量极显著低于青海高原牦牛淋巴结（$P<0.01$）。从青海高原牦牛克隆获得 PRDM1 基因编码区揭示了其分子特征，为青海高原牦牛 PRDM1 蛋白质功能的研究奠定基础。

关键词：牦牛；PRDM1 基因；基因克隆；分子特征

---

* 基金项目：国家自然科学基金资助项目（31301976）；现代农业（肉牛牦牛）产业技术体系建设专项资助项目（CARS-38）；甘肃省科技支撑计划资助项目（1504NKCA052）；中国农业科学院创新工程资助项目（CAAS-ASTIP-2014-LIHPS-01）

** 作者简介：赵生军（1989— ），男，硕士。

*** 通讯作者，E-mail：guoxian@caas.cn

DOI：10.16303/j.cnki.1005-4545.2017.04.30

Gene Cloning and Analysis of Biological Information and Differential Expression of PRDM1 Gene in Qinghai Plateau Yak

Sheng-jun ZHAO[1,2], Yong ZHANG[2], Ping-chang SHE[1], Ping YAN[1], Li-jun ZHANG[3], Xian GUO[1*]

(1. Key Laboratory of Yak Breeding Engineering of Gansu Province, Lanzhou Institute of Husbandry and Pharmaceutical Sciences, Chinese Academy of Agricultural Sciences, Lanzhou 730050, China; 2. College of Veterinary Medicine, Gansu Agricultural University, Lanzhou 730070, China; 3. 96351 Forces Farm, Xining 810000, China)

Abstract: Qinghai plateau yak lymph nodes, 90-day-old fetal head skin, yak sperm, and Simmental sperm were used as experimental materials, cloned cDNA sequence comprising the coding region of the gene PRDM1 by RT-PCR; compared to cattle PRDM1 gene, and forecasted properties and structural proteins of Qinghai plateau yak PRDM1; detected PRDM1 mRNA expression of lymph nodes, 90-day-old fetal head skin, yak sperm, Simmental sperm by real time PCR. The results showed that yak PRDM1 cDNA fragment is 761bp, including coding region 741bp, encoding 246 amino acids. Sequence analysis showed that the cloned cDNA sequence of yak exists variation of three bases compared with cattle, the homology of yak PRDM1 sequence with cattle, yak, sheep, horse were 99.0%, 99.0%, 97.0% and 92.0% respectively, various species PRDM1 protein phylogenetic tree accorded with the evolution law. This fragment contains the typical structure of the coiled-coil domain and so on. methyltransferase protein domain protein 1 PR shared a 94% similarity with human, methyltransferase protein domain proteins PR 10 shared 36% similarity with human, with domain activity. The expression of yak PRDM1 gene in Simmental and yak sperm were significantly higher than that of plateau yak lymph nodes ($P<0.05$), its expression of 90-day-old fetal head skin was significantly lower than in plateau yak lymph nodes ($P<0.01$). Cloned PRDM1 gene coding regions from yak reveal their molecular characteristics, it laid the foundation for the study of yak PRDM1 protein function.

Key words: Yak; PRDM1 gene; Gene clone; Molecular characteristic

青藏高原是世界上最大的连续高海拔生态系统，其面积约有 250 万 km^2，平均海拔在 4 000m 以上。处于高海拔的青藏高原其气候和环境非常严酷，低温、低氧、强紫外线等，但是生活在这一地区的动物对当地的极端环境有着很好的适应性。牦牛一直都是青藏高原牧区的优良家畜，也是牛属动物中能适应高寒高海拔气候的珍稀畜种资源[1]，是青藏高原特有的遗传资源，是唯一能够充分利用高寒高山草原进行动物性生产的优势牛种[2]。哺乳动物种族的延续依赖于其生殖细胞保持了基因的完整性，PRDM1 基因对生殖细胞发育形成具有调控作用，主要是对胚胎发育的影响[3-4]，还表现在对 PGC（原始生殖细胞）的作用[5]。对牦牛 PRDM1 基因的研究还未见报道。本试验对青海高原牦牛 PRDM1 基因部分编码区进行克隆，通过对 PRDM1 基因部分编码区推断获得的 PRDM1 蛋白序列进行蛋白特征分析，为牦牛 PRDM1 蛋白功能的研究奠定基础。

1 材料与方法

1.1 主要试剂与仪器

RNAiso Plus，Prime - Script™ RT Master Mix，Taq PCR Mix（SYBR Green）均购自 TaKaRa 公司；荧光定量试剂盒（FP202-01），胶回收用琼脂糖凝胶回收试剂盒（DP209），pGM-T 连接试剂盒（VT202），TOP10 感受态细胞均购自天根生化科技有限公司。

超微量分光光度计（ND2000 美国），PCR 仪（Bio - RAD，美国），凝胶成像系统（Pland CA91786，美国），Real-time PCR 仪（Bio-Rad 美国）。

1.2 样品采集

在青海省西宁市定点屠宰场，分别选取青海高原牦牛各 5 头，屠宰后 30min 内采集淋巴结，并采集 5 头 90 日龄胎儿头部皮肤组织置于液氮冷冻保存待用。牦牛、西门塔尔牛冷冻精液各 5 支。

1.3 引物设计与合成

参照 GenBank 中已公布的黄牛 PRDM1 基因（登录号：NM_ 001192936. 1）mR-NA 序列，用 Primer Premier 5. 0 软件分别设计 1 对 PRDM1 特异性引物，QRT-PCR 引物及 β-actin 内参引物，引物序列见表 1。

1.4 组织 RNA 提取及反转录

采用 RNAiso Plus 法提取青海高原牦牛淋巴结、90 日龄胎儿头部皮肤、牦牛精子以及西门塔尔牛精子 RNA，用 NanoDrop 仪器检测浓度，并用琼脂糖凝胶电泳检测纯度。以提取的 RNA 为模板，反转录获得相应的 cDNA。反转录体系 10μL：PrimeScrip™ RT Master Mix 2μL，总 RNA 1μL，ddH_2O_7μL。反应程序：37℃反应 15min，85℃灭活 5s，-20℃保存。

表 1　PCR 引物序列

引物	引物序列（5′→3′）	产物长度/bp	退火温度/℃
PRDM1	F：CTGTGAAGGTTCCAGGGTT R：AAGTGAGGGGTGAAATGTTG	759	54.0
PRDM1 （QRT-PCR）	F：CCACTTCATTGATGGCTTTA R：TTGGCAGGGATGGCTTA	144	52.9
β-actin	F：TCCAGCCTTCCTTCCTGGGCAT R：GACAGCACCGTGTTGGCGTAGA	116	56.0

1.5 PCR反应与克隆测序

以cDNA为模板进行PCR反应。PCR反应体系10μL：cDNA模板0.6μL，ddH_2O_4μL，上、下游引物（10μmol/L）各0.2μL，Taq PCR Mix（SYBR Green）5μL。反应程序：94℃预变性5min；95℃变性30s，54℃退火30s，72℃延伸1min，35个循环；72℃延伸5min。PCR产物用1.0%琼脂糖凝胶电泳检测并在紫外灯下切胶回收。PCR产物纯化回收后与pGM-T载体在16℃下连接过夜，然后转化到TOP10感受态细胞中，加入LB（不含抗生素）培养基，150 r/min，37℃振荡培养45min。再涂布在Amp、X-gal和IPTG的LB固体培养基上，37℃培养12~16h。挑取白色阳性菌落，接种于含Amp的LB液体培养基中，过夜振荡培养。筛选阳性转化子进行测序。

1.6 生物学信息分析

使用NCBI ORF Finder对克隆得到的牦牛PRDM1基因序列进行开放阅读框分析，并用BLAST进行序列同源性分析；利用Mega 6.0软件构建PRDM1基因的生物进化树；利用ProtParam分析PRDM1氨基酸序列理化性质；利用ProtScale分析PRDM1氨基酸序列疏水性质；利用Interpro分析PRDM1蛋白结构域；利用Ex-PASy分析PRDM1蛋白二、三级结构。

1.7 QRT-PCR

用反转录所得到的cDNA进行QRT-PCR反应。QRT-PCR反应体系10μL：cDNA模板0.6μL，ddH_2O_4μL，上、下游引物（10μmol/L）各0.2μL，Real Master Mix（SYBR Green）5μL。反应程序：95℃预变性30s；95℃变性5 s，56℃退火30s，40个循环；熔点曲线从65℃升到95℃，0.5℃递增。用内参基因β-actin作为标准对照组，每个试验组设3个重复，用青海高原牦牛淋巴结作为对照品，结果用$2^{-\Delta\Delta Ct}$公式计算，并用SPSS19.0软件进行统计分析。

2 结果

2.1 牦牛PRDM1基因扩增产物

根据普通牛PRDM1基因mRNA设计特异性引物，以牦牛淋巴结等组织cDNA为模板，进行PCR扩增，结果在750~1 000 bp有1条带（图1），经测序后得到预期761 bp长片段。克隆所得青海高原牦牛PRDM1与黄牛PRDM1（登录号：NM_ 001192936.1）mRNA序列进行同源性分析比对，存在3个碱基突变，该cDNA片段位置10T>G，青海高原牦牛在黄牛该cDNA片段位置10与11之间增加1个G，755与756之间增加1个C（图2）。

2.2 核苷酸序列分析结果

使用NCBI ORF Finder对克隆得到的牦牛PRDM1基因序列进行开放阅读框分析，得知本片段含有1个长为741 bp的开放阅读框，编码246个氨基酸，起始密码子为ATG（图3）。用BLAST对不同物种PRDM1基因的核苷酸进行同源性比对，结果发现青海高原牦牛PRDM1基因与黄牛（NM_ 001192936.1）、野牦牛（XM_ 005901607.1）、绵羊（XM_ 012182705.1）、马（XM_ 001501824.3）相应序列同源性分别是99.0%，99.0%，97.0%，92.0%。

2.3 青海高原牦牛PRDM1基因发育树

应用Mega 6.0软件中的NJ法构建生物进化树。构建完成发育树后发现，克隆得到的青

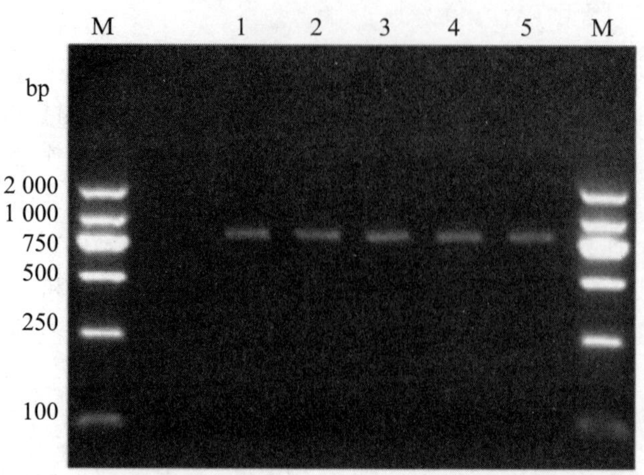

图 1　青海高原牦牛 PRDM1 基因 PCR 扩增产物电泳图 M.DL2000DNA Marker；1~5. PRDM1 基因

```
Bos_mutus_prdm1.seq  CTGTGAAGGGGTCCAGGGTTGGCAGAGGGGACTGAAGGGACCATGAAAAT  50
Bos_taurus_prdm1.seq CTGTGAAGGT  TCCAGGGTTGGCAGAGGGGACTGAAGGGACCATGAAAAT  50
Consensus            CTGTGAAGG   TCCAGGGTTGGCAGAGGGGACTGAAGGGACCATGAAAAT

Bos_mutus_prdm1.seq  GGACATGGAGGACGCGGATATGACTCTGTGGACAGAGGCTGAGTTTGAGG  100
Bos_taurus_prdm1.seq GGACATGGAGGACGCGGATATGACTCTGTGGACAGAGGCTGAGTTTGAGG  100
Consensus            GGACATGGAGGACGCGGATATGACTCTGTGGACAGAGGCTGAGTTTGAGG

Bos_mutus_prdm1.seq  AGAAGTGTACATACATTGTGAATGACCACCCCTGGGATTCTGGTGTGGAA  150
Bos_taurus_prdm1.seq AGAAGTGTACATACATTGTGAATGACCACCCCTGGGATTCTGGTGTGGAA  150
Consensus            AGAAGTGTACATACATTGTGAATGACCACCCCTGGGATTCTGGTGTGGAA

Bos_mutus_prdm1.seq  GGAGGCACTTCGGTTCAGGCGGAGGCTTCCTTACCGAGGAATCTGCTTTT  200
Bos_taurus_prdm1.seq GGAGGCACTTCGGTTCAGGCGGAGGCTTCCTTACCGAGGAATCTGCTTTT  200
Consensus            GGAGGCACTTCGGTTCAGGCGGAGGCTTCCTTACCGAGGAATCTGCTTTT

Bos_mutus_prdm1.seq  CAAATATGCCACAAACAGCAAAGAGATTACTGGAGTGGTGAGTAAAGAAT  250
Bos_taurus_prdm1.seq CAAATATGCCACAAACAGCAAAGAGATTACTGGAGTGGTGAGTAAAGAAT  250
Consensus            CAAATATGCCACAAACAGCAAAGAGATTACTGGAGTGGTGAGTAAAGAAT

Bos_mutus_prdm1.seq  ACATACCAAAGGGAACACGTTTTGGACCTCTAATAGGTGAAATCTACACC  300
Bos_taurus_prdm1.seq ACATACCAAAGGGAACACGTTTTGGACCTCTAATAGGTGAAATCTACACC  300
Consensus            ACATACCAAAGGGAACACGTTTTGGACCTCTAATAGGTGAAATCTACACC

Bos_mutus_prdm1.seq  AGTGATGCAGTTCCCAAGAACGCCAACAGGAAATATTTTTGGCGGATCTA  350
Bos_taurus_prdm1.seq AGTGATGCAGTTCCCAAGAACGCCAACAGGAAATATTTTTGGCGGATCTA  350
Consensus            AGTGATGCAGTTCCCAAGAACGCCAACAGGAAATATTTTTGGCGGATCTA

Bos_mutus_prdm1.seq  TTCCAGAGGGGAGCTTCACCACTTCATTGATGGCTTTAATGAGGAGAAGA  400
Bos_taurus_prdm1.seq TTCCAGAGGGGAGCTTCACCACTTCATTGATGGCTTTAATGAGGAGAAGA  400
Consensus            TTCCAGAGGGGAGCTTCACCACTTCATTGATGGCTTTAATGAGGAGAAGA

Bos_mutus_prdm1.seq  GCAACTGGCTACGCTACGTGAACCCAGCGCACACTGCCCGGGAGCAGAAC  450
Bos_taurus_prdm1.seq GCAACTGGCTACGCTACGTGAACCCAGCGCACACTGCCCGGGAGCAGAAC  450
Consensus            GCAACTGGCTACGCTACGTGAACCCAGCGCACACTGCCCGGGAGCAGAAC

Bos_mutus_prdm1.seq  CTGGCCGCCTGTCAGAACGGCATGAACATCTACTTCTACACCATTAAGCC  500
Bos_taurus_prdm1.seq CTGGCCGCCTGTCAGAACGGCATGAACATCTACTTCTACACCATTAAGCC  500
Consensus            CTGGCCGCCTGTCAGAACGGCATGAACATCTACTTCTACACCATTAAGCC

Bos_mutus_prdm1.seq  CATCCCTGCCAACCAGGAGCTTCTTGTGTGGTACTGCCGGGACTTTGCAG  550
Bos_taurus_prdm1.seq CATCCCTGCCAACCAGGAGCTTCTTGTGTGGTACTGCCGGGACTTTGCAG  550
Consensus            CATCCCTGCCAACCAGGAGCTTCTTGTGTGGTACTGCCGGGACTTTGCAG

Bos_mutus_prdm1.seq  AAAGGCTTCACTACCCTTATTCCGGAGAGCTGACAATGATGAATCTCACA  600
Bos_taurus_prdm1.seq AAAGGCTTCACTACCCTTATTCCGGAGAGCTGACAATGATGAATCTCACA  600
Consensus            AAAGGCTTCACTACCCTTATTCCGGAGAGCTGACAATGATGAATCTCACA

Bos_mutus_prdm1.seq  CAAACCCAGAGCCGTCCAAAGCAGCAGAGCACTGAGAAACATGAACTGTG  650
Bos_taurus_prdm1.seq CAAACCCAGAGCCGTCCAAAGCAGCAGAGCACTGAGAAACATGAACTGTG  650
Consensus            CAAACCCAGAGCCGTCCAAAGCAGCAGAGCACTGAGAAACATGAACTGTG

Bos_mutus_prdm1.seq  CCCAAAGAGTGTCCCGAAGAGAGAGTATAGCGTCAAAGAGATCCTAAAAT  700
Bos_taurus_prdm1.seq CCCAAAGAGTGTCCCGAAGAGAGAGTATAGCGTCAAAGAGATCCTAAAAT  700
Consensus            CCCAAAGAGTGTCCCGAAGAGAGAGTATAGCGTCAAAGAGATCCTAAAAT

Bos_mutus_prdm1.seq  TGGACTCCCACCCTTCCAAAGGGAAGGACTTGTACCGCTCCAACATTTCA  750
Bos_taurus_prdm1.seq TGGACTCCCACCCTTCCAAAGGGAAGGACTTGTACCGCTCCAACATTTCA  750
Consensus            TGGACTCCCACCCTTCCAAAGGGAAGGACTTGTACCGCTCCAACATTTCA

Bos_mutus_prdm1.seq  CCCCCTCACTT                                         761
Bos_taurus_prdm1.seq CCCC TCACTT                                         761
Consensus            CCCC TCACTT
```

图 2　青海高原牦牛 PRDM1 基因部分 CDS 区核苷酸序列同源性分析

```
19  ttggcagaggggactgaagggaccatgaaaatggacatggaggac
    L  A  E  G  T  E  G  T  M  K  M  D  M  E  D
64  gcggatatgactctgtggacagaggctgagtttgaggagaagtgt
    A  D  M  T  L  W  T  E  A  E  F  E  E  K  C
109 acatacattgtgaatgaccacccctgggattctggtgtggaagga
    T  Y  I  V  N  D  H  P  W  D  S  G  V  E  G
154 ggcacttcggttcaggcggaggcttccttaccgaggaatctgctt
    G  T  S  V  Q  A  E  A  S  L  P  R  N  L  L
199 ttcaaatatgccacaaacagcaaagagattactggagtggtgagt
    F  K  Y  A  T  N  S  K  E  I  T  G  V  V  S
244 aaagaatacataccaaagggaacacgttttggacctctaataggt
    K  E  Y  I  P  K  G  T  R  F  G  P  L  I  G
289 gaaatctacaccagtgatgcagttcccaagaacgccaacaggaaa
    E  I  Y  T  S  D  A  V  P  K  N  A  N  R  K
334 tatttttggcggatctattccagaggggagcttcaccacttcatt
    Y  F  W  R  I  Y  S  R  G  E  L  H  H  F  I
379 gatggctttaatgaggagaagagcaactggctacgctacgtgaac
    D  G  F  N  E  E  K  S  N  W  L  R  Y  V  N
424 ccagcgcacactgcccgggagcagaacctggccgcctgtcagaac
    P  A  H  T  A  R  E  Q  N  L  A  A  C  Q  N
469 ggcatgaacatctacttctacaccattaagcccatccctgccaac
    G  M  N  I  Y  F  Y  T  I  K  P  I  P  A  N
514 caggagcttcttgtgtggtactgccgggactttgcagaaaggctt
    Q  E  L  L  V  W  Y  C  R  D  F  A  E  R  L
559 cactacccttattccggagagctgacaatgatgaatctcacacaa
    H  Y  P  Y  S  G  E  L  T  M  M  N  L  T  Q
604 acccagagccgtccaaagcagcagagcactgagaaacatgaactg
    T  Q  S  R  P  K  Q  Q  S  T  E  K  H  E  L
649 tgcccaaagagtgtcccgaagagagagtatagcgtcaaagagatc
    C  P  K  S  V  P  K  R  E  Y  S  V  K  E  I
694 ctaaaattggactcccacccttccaaaggaaggacttgtaccgc
    L  K  L  D  S  H  P  S  K  G  K  D  L  Y  R
739 tccaacatttcacccccctcac 759
    S  N  I  S  P  P  H
```

图 3 PRDM1 基因部分编码区编码的氨基酸

海高原牦牛 PRDM1 片段与黄牛 PRDM1 基因归为一类，藏羚羊、山羊与绵羊归为一类，人与小鼠归为一类（图 4）。

2.4 氨基酸序列

2.4.1 理化性质

利用 ProtParam 软件分析牦牛 PRDM1 基因编码蛋白的基本理化性质，该蛋白相对分子质量为 27 195.5，理论等电点为 11.8。含有 20 种氨基酸，其中含量最高的分别是 Ala（19.9%）和 Thr（14.5%），含量最低的是 Trp（1.0%）（表 2）；含有负电荷的残基为 8 个，正电荷残基 39 个；分子形式 $C_{1197}H_{1940}N_{370}O_{331}S_{12}$，相对分子质量 3 850，其水溶液在 280nm 处的消光系数为 43 345，肽链 N 末端为 C（Cys）。该蛋白不稳定系数为 53.2，属于不稳定蛋白。该蛋白脂溶指数和总平均疏水指数分别为 67.1，-0.507。

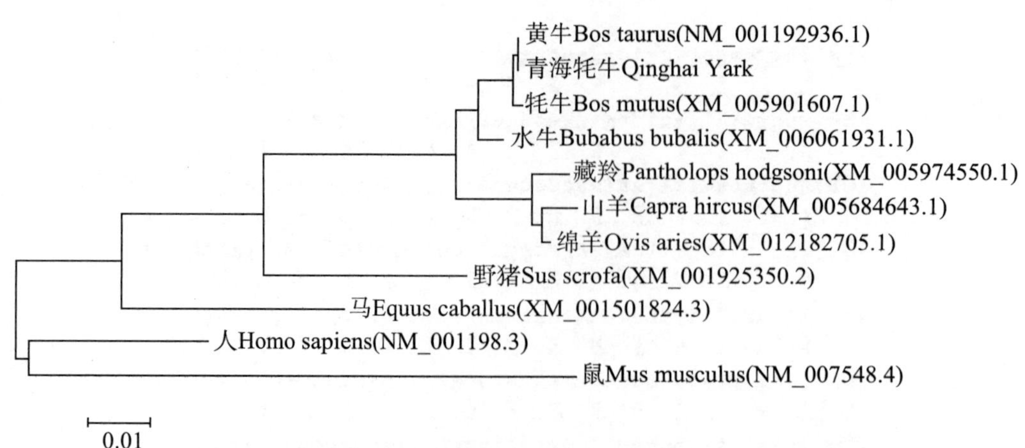

图4 基于PRDM1基因编码区序列构建的发育树

表2 PRDM1基因编码蛋白氨基酸含量

序号	氨 基 酸	含量（%）
1	丙氨酸（Ala）	5.0
2	精氨酸（Arg）	12.0
3	天冬酰胺（Asn）	2.5
4	天冬氨酸（Asp）	0.4
5	半胱氨酸（Cys）	2.5
6	谷氨酰胺（Gln）	4.6
7	谷氨酸（Glu）	2.9
8	甘氨酸（Gly）	7.1
9	组氨酸（His）	1.7
10	异亮氨酸（Ile）	3.3
11	亮氨酸（Leu）	9.5
12	赖氨酸（Lys）	4.1
13	蛋氨酸（Met）	2.5
14	苯丙氨酸（Phe）	3.3
15	脯氨酸（Pro）	8.3
16	丝氨酸（Ser）	8.7
17	苏氨酸（Thr）	13.3
18	色氨酸（Trp）	2.9
19	酪氨酸（Tyr）	1.2
20	缬氨酸（Val）	4.1

2.4.2 疏水性/亲水性预测分析

使用 ProtScale 在线工具对牦牛 PRDM1 基因编码蛋白进行亲水性/疏水性分析。依据氨基酸分值越低亲水性越强或分值越高疏水性越强的规律分析得出，牦牛 PRDM1 基因编码蛋白多肽链大部分区域分值为负值，说明整个多肽链是亲水性（图 5）。

2.4.3 结构域蛋白功能位点预测

利用 Interpro 在线工具预测牦牛 PRDM1 基因编码蛋白结构域显示，该片段属于 PR domain zinc finger protein 家族成员。61～184 位氨基酸具有 SET 蛋白结构域活性。是负调控转录的 RNA 聚合酶Ⅱ发起因子，参与调节细胞发育过程，具有 DNA 结合转录因子活性、蛋白结合活性（图 6）。

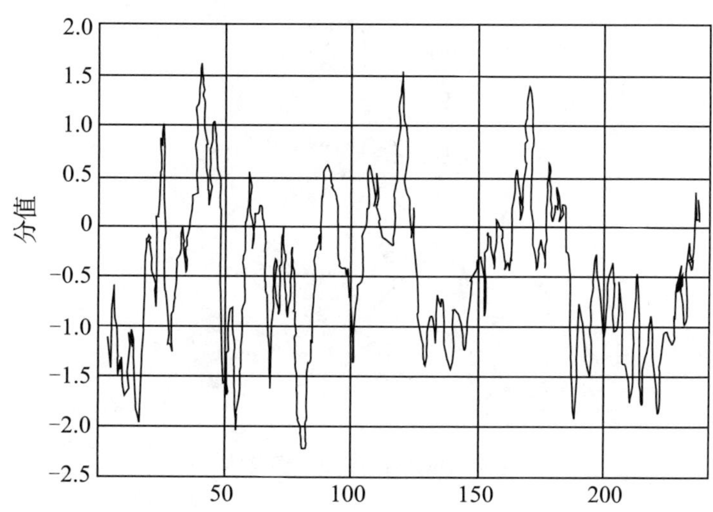

图 5　PRDM1 蛋白亲疏性预测

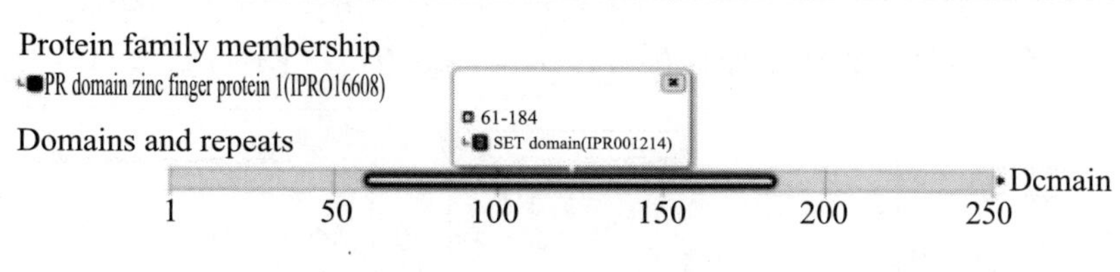

图 6　PRDM1 蛋白结构域

2.4.4 高级结构预测

利用 ExPASy 网站工具预测蛋白二、三级结构，结果见图 7 和图 8。PRDM1 蛋白具有 30.71%的 α 螺旋（Hh），13.28%的延伸连（Ee），10.37%的 β 转角（Tt），45.64%的无规则卷曲（Cc）。与具有组蛋白赖氨酸甲基转移酶活性的 PRDM9 蛋白三级结构具有 37.21%相似性。用 Phyre2 工具预测得出，牦牛 PRDM1 蛋白与人甲基转移酶蛋白结构域 PR 蛋白 1 具有 94%的相似性，100%的确信度，与人甲基转移酶蛋白结构域 PR 蛋白 10 具有 36%的相似性，具有锌指蛋白活性。

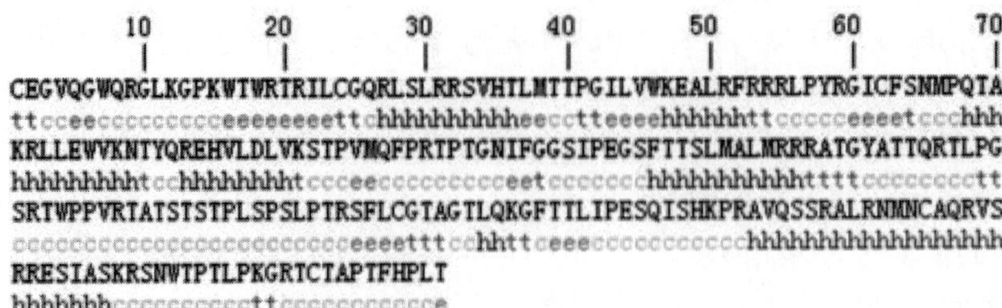

图 7　PRDM1 蛋白二级结构　h.α 螺旋；e.延伸连；t.β 转角；c.无规则卷曲

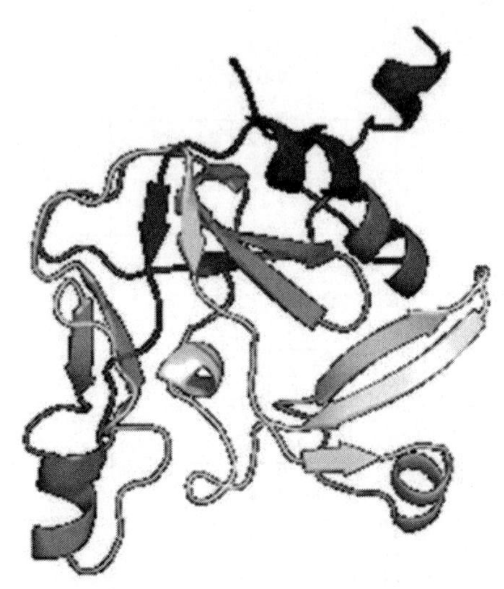

图 8　PRDM1 蛋白三级结构

2.5　PRDM1 基因表达量分析

通过 RT-PCR 对淋巴结、胎儿、西门塔尔精子、牦牛精子中 PRDM1 基因 mRNA 表达水平进行检测，使用 $2^{-\Delta\Delta Ct}$法，计算 PRDM1 基因表达量，结果表明青海高原牦牛 PRDM1 基因在西门塔尔、牦牛精子中表达量极显著高于青海高原牦牛淋巴结（$P<0.05$），90 日龄胎头部皮肤中表达量极显著低于青海高原牦牛淋巴结（$P<0.01$）（图 9）。

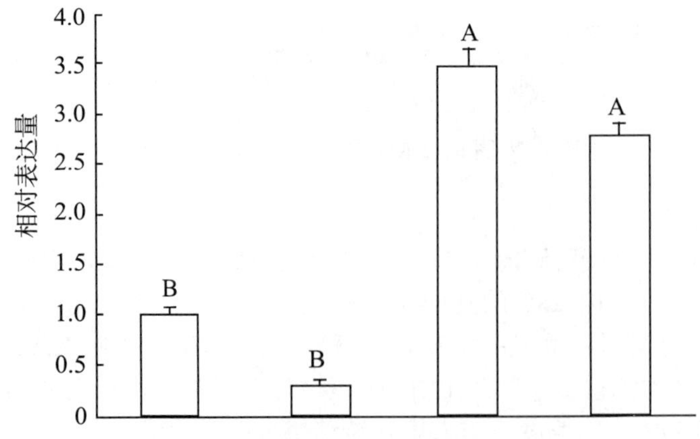

图 9　不同组织中 PRDM1 基因 mRNA 的表达水平

3 讨论

3.1 PRDM1 蛋白的功能分析

PRDM 家族具有一个 PR 结构域和数量不等锌指结构的特点（除了 PRDM11），PRDM 家族蛋白通过甲基转移或募集染色体结构重塑复合物，调控基因表达、修改染色体结构，在细胞特化、疾病调控中起到重要作用。PRDM1 蛋白是一种通过锌指结构绑定特定 DNA 序列的转录抑制因子。作为召集共抑制因子或辅阻遏物的骨架，促进组蛋白修饰。通过与组蛋白 H3k9 乙酰酶抑制剂 HDAC1 和 HDAC2，组蛋白 H3 赖氨酸去甲基化酶 LSD1 相互作用，进而抑制转录[6-7]。

PRDM1 基因 mRNA 有 PRDM1α（5 164 bp）和 PRDM1β（4 675 bp）2 种转录形式，其 N 末端虽然具有 PR 结构域，但并不具有 HMT 活性。人 PRDM1 基因在许多组织中表达，抑制转录。在免疫系统，细胞生长、分化，胚胎发育中有重要作用[6,8-10]。

PRDM1 对生殖细胞发育分化具有一定影响，PRDM1 基因对哺乳动物生殖细胞的调控贯穿于从 PGC 形成初始直至两性生殖细胞的形成早期，具体机理尚不清楚[11-12]。PRDM1 基因表达对小鼠胚胎咽、心脏形成及功能发挥具有影响，对小鼠胎儿出生后的存活也具有重要作用，但具体机理仍不清楚[12-14]。研究表明 PRDM1 基因过表达可以抑制主要组织相容性复合体反式激活因子的转录，诱导小鼠成熟的细胞分化为浆细胞，并不启始浆细胞的分化，而是在后期维持这一进程[15-16]。

3.2 PRDM1 基因（蛋白）序列比对

将克隆获得的牦牛 PRDM1 基因 cDNA 序列与黄牛相应序列进行比对发现，两品种牛该序列存在 3 个碱基的变异。各物种 PRDM1 蛋白质序列比对表明，青海高原牦牛与黄牛的相似度最高为 99.0%。不同物种间 PRDM1 蛋白质存在较高的保守性，说明 PRDM1 蛋白在物种进化过程中起着非常重要的作用。

3.3 PRDM1 蛋白结构分析

通过克隆所得牦牛 PRDM1 基因部分 CDS 区核苷酸序列预测编码蛋白序列，并分析该序列得到基本理化性质，结果表明其编码 246 个氨基酸。肽链 N 末端为半胱氨酸，属于不稳定蛋白，多肽链是亲水性。经过结构域蛋白功能位点预测，该片段属于 PR domain zinc finger protein 家族成员，具有 DNA 结合转录因子活性和蛋白结合活性。对软件预测的牦牛 PRDM1 蛋白三级结构进行比对，克隆的该片段编码蛋白与人 PRDM1 蛋白具有 94% 相似性，与 PRDM9 蛋白具有 37.21% 相似性，与人 PRDM10 蛋白具有 36% 相似性，因此具有 PR 结构域，Zn 指蛋白活性。

3.4 荧光定量分析 PRDM1 表达量

采用 RT-PCR 对不同样品中 PRDM1 基因 mRNA 表达水平进行检测，使用 $2^{-\Delta\Delta Ct}$ 法计算 PRDM1 基因在青海高原牦牛淋巴结、90 日龄胎儿头部皮肤、牦牛精子以及西门塔尔牛精子的表达量。结果发现 PRDM1 基因在青海高原牦牛淋巴结、90 日龄胎儿头部皮肤、牦牛精子以及西门塔尔牛精子有不同程度的表达，在西门塔尔、牦牛精子中表达量极显著高于青海高原牦牛淋巴结（$P<0.05$），90 日龄胎头部皮肤中表达量极显著低于青海高原牦牛淋巴结（$P<0.01$）。有研究表明 PRDM1 对小鼠生殖细胞发育、胚胎发育、淋巴细胞成熟以及胎儿出生

后的存活具有重要作用[13-15]。本试验发现 PRDM1 在青海高原牦牛以上相关组织细胞中都有表达，因此说明 PRDM1 在青海高原牦牛生殖细胞发育、胚胎发育、淋巴细胞成熟以及胎儿出生后的存活等方面发挥作用，但具体机制有待进一步探究。

参考文献（略）

（发表于《中国兽医学报》，院选一级学报）

以锰为代表的过渡金属离子体内吸收及转运机制*

王　慧[1,2]**，张翊华[2]***

（1. 中国农业科学院兰州畜牧与兽药研究所，兰州 730050；
2. 西北农林科技大学 动物医学院，杨凌 712100）

摘要：过渡金属离子；锰内稳态平衡；锰吸收；锰转运；金属离子调控蛋白；分子机制

近年来，金属离子在生命体内的代谢、稳态平衡及其相关疾病是当前国际热点研究领域之一[1]。微量元素是指生物体内含量不足万分之一的元素。在动物体内，微量元素虽然含量微小，但具有极其重要的生理功能，涉及生长发育、新陈代谢、内稳态调节、神经活动、免疫功能、氧化应激、信号转导、基因调控、酶及内分泌等几乎所有的生命活动过程[2]。一方面，金属离子的存在对金属蛋白/金属酶结构的稳定以及功能的发挥是不可缺少的；另一方面，金属离子与某些蛋白的结合又可以调控或影响这些蛋白及其他生物分子的生物学功能[3]。细胞内金属离子的传递、代谢、稳态平衡的失调，是导致多种疾病的关键因素。对细胞内金属离子传递、代谢调控的分子机制研究有助于阐明疾病发生、发展的机制；在此基础上，寻找用于疾病诊断的探针和生物靶标，为疾病的防治和新型药物的设计提供新的思路。

1　过渡金属离子

生物体的生存依赖于大量的化学物质，其中许多不能被合成，必须从外界获得。这些必须的化学物质或营养素所需要的量相对很小，所以被称为微量营养素（micronutrient）。在微量营养素中，过渡金属离子（transition metal ions，tMIs）是非常重要的一类。生物体内最普遍的 tMIs 是 Fe、Zn、Cu 和 Mn[4]。tMIs 通过促进氧化还原反应、化学基团转移反应，通过稳定蛋白结构维持许多蛋白的功能。为了满足对这些 tMIs 的需求，细胞有多种从外环境溶解和摄取 tMIs 的机制。然而，细胞必须同时保护自己免受这些 tMIs 固有的化学特征的危害。如果细胞质基质内 tMIs 浓度不受调控，氧化还原反应会产生有毒自由基。因此，每个生物体通过主动和被动运输系统，从细胞外环境识别、结合 tMIs，转运到必要的细胞区室。细胞质基质内没有游离的 tMIs 存在，表明 tMIs 最终被 tMI 储藏蛋白所储存[5]。

---

* 基金项目：甘肃省青年科技基金资助项目（1606RJYA224）；国家重点研发计划资助项目（2016YFD0501200）；中央级科研院所基本科研业务费资助项目（1610322013003）

** 作者简介：王慧（1985— ），男，博士。

*** 通讯作者。

DOI：10.16303/j.cnki.1005-4545.2017.07.35

Fe、Zn、Mn、Cu 等过渡金属具有多种生物学功能，作为蛋白质结构和催化的辅酶因子对生命至关重要[6]。通过分析蛋白质数据库，发现大约 30%的蛋白质与金属辅酶因子相互作用，凸显了过渡金属对细胞生理学的重要性[7-8]。过渡金属也是脊椎动物维持免疫功能所必须的[9-10]。因此 tMIs 的摄入、有效性、输出等必须在细胞水平上严格调控，这个过程被称为稳态。稳态改变与一些疾病密切相关，如遗传疾病、退化性疾病、癌症、糖尿病等[11-13]。

2 过渡金属离子的运输机制

离子跨膜转运对于生命活动是一个非常重要的过程。金属离子通过体膜进入血液的过程称为吸收。吸收的途径为消化道、呼吸道、皮肤等，其中消化道为离子进入体内的主要途径。因此，金属离子从肠道进入血液需穿过黏膜上皮细胞和毛细血管内皮细胞[14]。在真核细胞中，二价阳离子主要以 2 种状态存在：一种是与蛋白质或其他带阴离子电荷的大分子紧密结合；另一种是以电离状态参与动态化学平衡。游离金属离子是亲水性的，不容易扩散穿过细胞膜的脂质双分子层[15]，需要一种脂溶性化合物或分子载体（图 1）。通道蛋白（channel）被认为是细胞膜中由一类内在蛋白构成的孔道。通道蛋白与所转运物质之间的结合较弱，它能形成亲水的通道，当通道打开时能允许特定大小的溶质通过脂质双分子层。通道蛋白可用化学方式或电学方式激活底物（substrate），控制离子通过细胞膜顺电化学势流动。所有通道蛋白均以自由扩散的方式运输溶质，不消耗能量。离子通道是一类对离子具有选择通透性的跨膜生物大分子，调节机体多种生理活动，如神经肌肉的兴奋、细胞增殖、学习和记忆。离子通道功能紊乱及离子失衡可诱发多种疾病，如癫痫、心律失调和糖尿病等[16]。载体蛋白（carriers）也是一类内在蛋白，由载体转运的物质首先与载体蛋白的活性部位结合，结合后载体蛋白发生构象变化，将被转运物质暴露于膜的另一侧，并释放出去。由载体进行的转运可以是被动的（顺电化学势梯度），也可以是主动的（逆电化学势梯度）。离子泵（pumps）利用 ATP 水解提供能量，驱动底物逆浓度梯度转运；离子泵可以建立膜两侧的浓度梯度，这种浓度梯度为离子通道和其他转运蛋白提供能量以完成营养物质进入细胞、电信号的产生、细胞体积的调节、电解质的分泌等活动[17]。许多金属离子进入细胞的机制是具有相对较低的金属亲和力和特异性[18]。

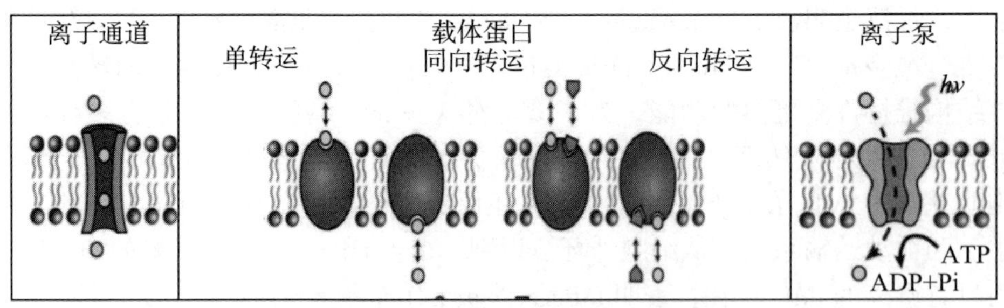

图 1 金属离子穿过生物膜的运输系统模型[17,19]

离子通道为跨膜被动运输，具有高度选择性，取决于离子大小、电荷等因素。载体蛋白根据转运机制可分为 3 种类型：单转运（uniporter）、同向转运（symporter）和反向转运（antiporter）。与转运蛋白一样，离子泵也含有结合位点，可以经过构象改变用于特殊离子或底物的转运，这一过程所需的能量来源于 ATP 转变为 ADP 或吸收光能。

3　过渡金属离子的细胞稳态调节机制

为了满足对不同金属的需要，细胞采取多种机制从细胞外环境中摄取金属离子，并通过一套有效的机制来维持体内的金属离子的质量浓度在生理范围内，同时保护细胞本身免受伤害。由于金属离子可以对细胞产生毒性，不能以自由离子的形式存在，总是被蛋白或生物分子螯合[20]。金属离子体内平衡取决于特定金属蛋白配合物的形成，这些配合物用于影响金属离子的吸收、排泄、细胞内转运及储存（图2）。金属离子往返于金属酶的活性部位或作为生物分子的结构元件从而有利于维持细胞内的金属稳态（图2）。金属转运蛋白将金属离子或小分子金属螯合物以定向的方式转运穿过不透水层；大多数金属转运蛋白为内在膜蛋白，镶嵌在内膜或质膜上（图2）。特定蛋白螯合剂是指金属伴侣蛋白，其将金属离子转运入特定的细胞区室，例如：外周胞质、胞液，易与恰当的蛋白受体结合。这种分子间转运是通过瞬变形成特定的蛋白-蛋白复合物调解协调分子间的金属配位体交换。

所有细胞都拥有一连串的调节蛋白，通过调节金属转运蛋白、细胞内螯合剂及其他解毒酶的编码基因表达来调解 tMIs 的体内平衡[22]。这些蛋白被称为金属传感器蛋白[22-23]或金属调节蛋白[24]。每个金属调节蛋白与金属离子形成特定的配合物，最终抑制或激活 DNA，约束或增强转录活性。细胞用这种方式有效的控制、调节金属离子稳态相关基因的表达。

图2简要介绍了单独的金属离子在不同细胞区室间的转运过程，图中双箭头是为了说明金属离子可以移动和进出靶蛋白响应蛋白质组重构。Zn（Ⅱ）的伴侣蛋白为 YodA/ZinT[25]，Cu（Ⅰ）的伴侣蛋白为 Atx1[26]及 CopZ[27]，Ni（Ⅱ）的伴侣蛋白为脲酶（UreE）[28]或 Ni-Fe 氢化酶（HypA）[28]。自然抗性相关巨噬细胞蛋白（natural resistance associated macrophage proteins，Nramps）是一个高度保守的跨膜蛋白家族，参与大多数有机体的金属离子运输过程。ZIP（ZRT，IRT-like proteins）家族存在于多数真核细胞中，参与 Fe、Zn、Mn 和 Cd 运输，但家族成员的作用底物范围和专一性不同[29-30]。P 型 ATP 酶（P-type ATPases）是与 ATP 结合的蛋白质，其功能在于将无机阳离子转入或排出细胞。ABC 转运蛋白（ATP-binding cassette transporters）是一个具有很强转运功能的膜蛋白家族[31-32]，所有 ABC 转运蛋白都包含4或6个高度疏水跨膜区域，以及外围细胞质的 ATP 结合区域或核苷结合区域[32]。阳离子扩散蛋白（cation diffusion facilitators，CDFs）是存在于所有生命体中的一个重要的蛋白质家族，CDFs 发生突变或是调控变化会改变对细胞功能至关重要的金属离子的浓度，与一些涉及内分泌、神经系统、肝脏和心血管系统的重要疾病有关[33]。

细胞含有外膜（outer membrane，OM）、细胞周质间隙（periplasm）及细胞质膜，金属离子必须通过这些结构才能到达细胞质基质。孔蛋白镶嵌在外膜，允许金属离子通过无选择性的被动扩散穿过外膜（图2）。然而，为了满足细胞对金属离子的需求，细胞质基质必须有效浓缩金属离子[34]，结果外膜的或嵌入原生质膜（plasma membrane，PM）或内膜（inner membrance，IM）的高亲和力的主动转运系统运输及释放金属离子进入细胞质基质。转运系统或者受细胞质膜一侧的 ATP 水解作用的驱动，例如：ABC 转运蛋白、P 型 ATP 酶；或者通过耦合高能量的质子或其他离子穿过双分子层，例如：Nramp 蛋白、CDF 蛋白。胞外脂多糖（LPS）、复合碳水化合物基质及生物膜的存在可能对金属离子的吸收与排出的比率及机制有相当大的影响[35]。ABC 转运蛋白[36-37]及 Nramp 转运蛋白[38-39]调解特定金属离子在细胞质基质中的富集，然而输出这些金属离子主要是依靠 CDFs[40]、P 型 ATP 酶[41-42]及 RND（resistance-nodulation-cell division）转运蛋白[43]。近来又在水稻中发现一种新的 tMIs

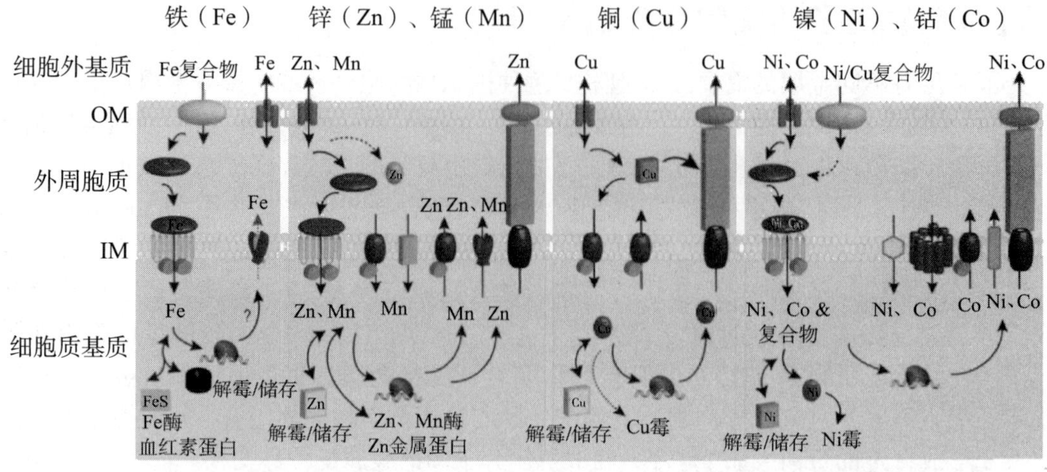

图 2　细胞内 Fe、Zn、Mn、Cu、Ni 和 Co 的金属内稳态模型示意图[21]

1. 外膜受体；2. 孔蛋白；3. ABC 转运体；4. P 型 ATP 酶；5. CDF；6. RND；7. Ni/CoT；8. Nramp；9. CorA；10. RcnA；11. Fe 储存蛋白；12. Fe-S 蛋白；13. DNA 传感器；14. Zn 伴侣；15. 金属硫蛋白；16. 细胞周质 Cu 结合蛋白；17. Cu 伴侣；18. Ni 伴侣；19. Ni-储存

转运蛋白 OsZ-IP6，该蛋白在 Fe^{2+}、Zn^{2+}、Mn^{2+}缺乏时能够被激活，在酸性环境下能增强转运能力；Co^{2+}能竞争性抑制 OsZIP6 对 Fe^{2+}的摄取[44]，但是 OsZIP6 是否在动物细胞中表达目前还不得知。

4　过渡金属 Mn

Mn 被称作“细胞护卫”或“生命体保镖”，在生物体中发挥重要的作用，机体内 Mn 离子的含量必须维持在一个恰当的水平，Mn 缺少或过量都会导致疾病或生物毒性[1]。Mn 有多个化学价态，其中 Mn^{3+}常见于各种酶，Mn^{2+}常见于食物[45]。Mn 的价态能被机体改变，研究表明 Mn^{3+}的细胞毒性大于 Mn^{2+}，原因是具有较高的氧化性[46]。Mn^{3+}增强氧化应激在试验研究中被证实[47]。Mn^{2+}不容易结合巯基（-SH）或胺，其稳定常数不随内源性配体甘氨酸、半胱氨酸、核黄素和鸟苷而改变[45]。在脊椎动物组织内，Mn 的质量浓度范围为 0.3~2.9μg/g，其中代谢活跃的组织浓度较高，例如骨骼、肝脏、胰腺及肾脏[48]。Mn 在细胞内的许多代谢反应过程中具有重要作用，包括脂质、蛋白质、碳水化合物及各种酶。有限的资料显示毒性浓度的 Mn 可能损伤免疫系统[49]。Mn 在酶的激活和抗氧化应激过程中起着重要作用。锰过氧化氢酶（Mn-CAT）和锰超氧化物歧化酶（Mn-SOD）都需要 Mn，所以 Mn 参与氧自由基的解毒和减少氧化应激[50]。Mn 毒性的分子机制目前还不清楚。高浓度的 Mn 可导致神经系统异常，例如行为改变、运动障碍、肌肉痉挛[51]，可能的作用机理为：（1）Mn 优先被脑吸收[50]，增加了氧化磷酸化的几率；（2）Mn 对多巴胺（dopamine，DA）系统的影响[52]；（3）高浓度 Mn 的存在增加了活性氧[53]。

5　Mn 的吸收及转运

动物主要从食物、空气和水中摄入 Mn。Mn 在消化道中吸收缓慢而不完全，吸收率仅

1%~3%，97%以上由粪便排出，因此很容易发生 Mn 的缺乏症[54]。Mn 的吸收部位主要在十二指肠，空肠和回肠吸收很少[54]。Mn^{2+}的吸收分两步进行，首先 Mn^{2+}从肠腔中被摄取，然后经肠黏膜上皮细胞附着缘进入血液，进入血液中的 Mn^{2+}，一部分保持游离状态，一部分与血浆 α2-巨球蛋白或白蛋白相结合进入肝脏，然后经胆汁分泌；一些 Mn^{2+}被血浆铜蓝蛋白（ceruloplasmin，CP）氧化为 Mn^{3+}。Mn^{3+}的血浆载体蛋白为转铁蛋白（transferrin，Tf），Mn^{3+}通过 Tf 受体介导的内吞作用和二价金属离子转运体（divalent metal transporter Ⅰ，DMT1）运输到机体中利用 Mn 的各组织、器官的线粒体和酶的合成部位[55-56]。

Mn 的必需性和毒性之间存在微妙关系，Mn 的稳态对任何生物组织最佳功能的发挥是至关重要的。关于 Mn 转运的确切载体仍然是有争议的。在过去的 20 年里，包括主动运输[57]和易化扩散[58]的各种转运机制已被证明。近年来研究表明，Mn 还可通过高亲和力金属转运蛋白进行转运，如膜 Fe 转运蛋白（FPN1）[59]。FPN1 是哺乳动物肠上皮细胞基底膜上将 Mn 由细胞内转出进入血液循环系统的跨膜转运蛋白。FPN1 主要分布于需要平衡 Fe、Mn 代谢的组织，包括成熟的内质网系统、十二指肠、肝脏、胎盘及中枢神经系统[60-61]。FPN1 为单向转运蛋白[62]。细胞内 Mn 水平提高可上调 FPN1 的表达[63]，进而促进胞内 Mn 进入血液。FPN1 过度表达可减少细胞内 Mn 积累，从而降低可能由 Mn 引起的细胞毒性[64]。人类 FPN1 基因编码 570 个氨基酸，包含 8 个外显子，预测相对分子质量为 62 000[65]。在哺乳动物中 FPN1 是一种高度保守的蛋白，人类、小鼠和大鼠之间有 90%~95%的同源性[65]。已有研究发现，Nramp 家族的 H^+依赖型 DMT1 是位于肠道黏膜细胞顶膜上转运 Mn^{2+}的载体蛋白[39,63]，在动物各组织中广泛表达，尤其在十二指肠和肾脏的表达最高[66-67]。DMT1 基因编码，560~570 个氨基酸，包含 17 个外显子，全长大于 36 kb，主体 DMT1 蛋白相对分子质量约为 65 000[65]。DMT1 在小肠上皮细胞主要定位在肠细胞绒毛膜刷状缘上[66]，从胃部排出的可溶性 Mn^{2+}，通过 DMT1 才能跨越小肠绒毛顶膜进入细胞内部[68-69]。DMT1 突变的大鼠，缺乏吸收和转运 Mn 的能力[70]；向 DMT1 突变的细胞中注入重新构建的 DMT1 基因，细胞对 Mn^{2+}的吸收增加[71]。在十二指肠中，FPN1 主要分布在肠上皮细胞的基底膜，与 DMT1 协同完成 Fe 和 Mn 的跨膜转运[72]。巨噬细胞里的吞噬小体膜上的 Nramp1 也会吸收少量的 Mn，再释放至胞质中[55]。宿主蛋白 Nramp1 在许多细胞中都能表达，例如中性粒细胞和巨噬细胞，且与溶酶体相关膜蛋白 1 有关（lysosomal-associated membrane protein 1，LAMP1）[73]。另外，Nramp1 被研究证实可以将 Fe 和 Mn 转运出溶酶体[73-74]。另外，溶质转运蛋白（solute carrier，SLC）超家族的 ZIP-8[75]、摄取 Fe^{3+}的转铁蛋白受体（transferrin receptor，TfR）[76]、电压门控及钙池操纵钙离子通道[77]、亲离子型谷氨酸受体 Ca^{2+}通道[75]、以 SitAB-CD 和 YfeABCD 为代表的 ABC 转运蛋白被认为是可以转运所有生物体所需的 tMIs 的转运蛋白[24,78-79]。虽然这些转运蛋白对 Mn 的相对转运能力仍未知，但是 Mn 在组织中的富集很可能依赖于这些转运蛋白。

6 Mn 穿过血脑屏障的机理

Mn 可以穿过血脑屏障，而且随着时间的推移，Mn 在脑组织中的相对贮存量大于肝、胰、肾[56]。脑中 Mn 含量过高会形成类似于先天性帕金森症的“多锰病”[80]。对于 Mn 的毒性，DMT1 能促进 Mn 在 DA 含量丰富的组织富集[71]。一些研究表明，Mn 利用胆碱转运体[81]、电压门控及钙池操纵钙离子通道[82]、亲离子型谷氨酸受体 Ca^{2+}通道[75]及一种 Mn 枸橼酸转运蛋白[83]穿过血脑屏障（图 3）。另外，研究表明 Mn 的转运还可通过 Zn 转运蛋白

ZIP8 和 ZIP14[84-85]及 ATP13A2（一种 P 型跨膜的 ATP 酶）[86]。另外高 Mn 还可导致线粒体功能异常[87]，进而影响 DA、谷氨酸（Glu）和 γ-氨基丁酸（GABA）等神经递质的含量[88]。

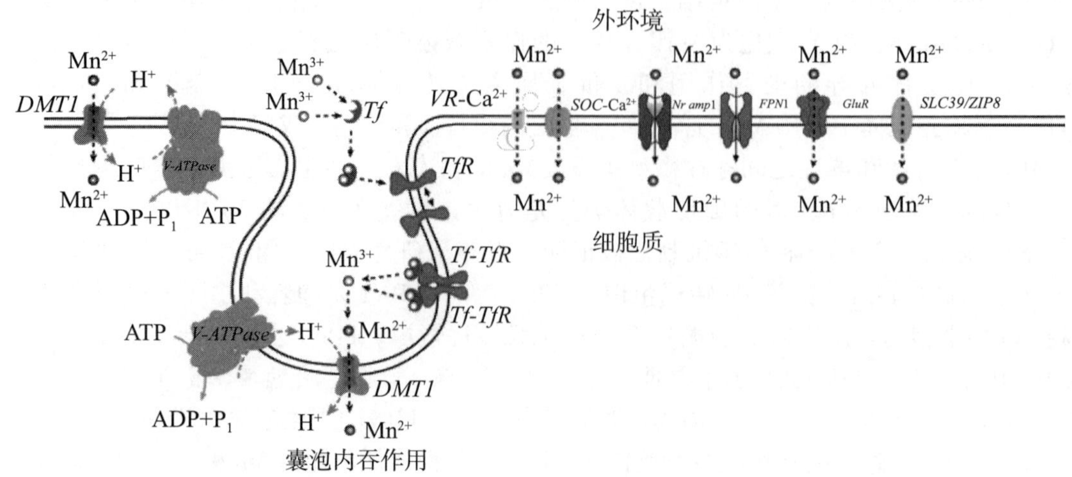

图 3　Mn 穿过血脑屏障分子机制[39,55]　**DMT1. 二价金属离子转运体；V-ATPsae.空泡型 ATP 酶；Tf. 转铁蛋白；TfR.转铁蛋白受体；VR-Ca^{2+}.电子门控钙离子通道；SOC-Ca^{2+}.钙池操纵钙离子通道；Nramp1. 自然抗性相关巨噬细胞蛋白；FPN1. 膜铁转运蛋白；GluR.亲离子型谷氨酸受体；SLC39/ZIP8. 可溶性载体 39 金属转运蛋白；□.酶化的内吞小泡**

DMT1 是一个同向转运体，能被空泡型 ATP 酶（vacuolar-ATPase，V-ATPase）产生的质子激活。从细胞外间隙吸收的质子为运输 Mn^{2+}进入细胞提供了能量。V-ATPase 产生的质子也负责对内吞作用的囊泡进行酸化；酸化后，TfR 系统将 Mn^{3+}转变为 Mn^{2+}，可被 DMT1 转运[65]。电压门控及钙池操纵钙离子通道、亲离子型谷氨酸受体（glutamate ionotropic receptor，GluR）、Nramp1、FPN1、SLC39 家族 ZIP8 蛋白等转运蛋白，被证明在血脑屏障 Mn 离子的吸收过程中扮演重要角色。

7　总结与展望

tMIs 在机体及细胞中的稳态是一个复杂的过程。以 Mn 为重点，介绍了 tMIs 在体内的吸收及转运机制等方面的已有成果，这些数据将为 tMIs 的体内稳态调控机制的深入研究提供重要的线索。人类 30%的基因编码膜蛋白[89]，膜蛋白通过多种方式调节各种离子转运和代谢的信号通路，尽管许多参与过渡金属离子代谢的关键蛋白已被确定，但是转运调节蛋白的细胞及分子转运机调节机制还不是完全清楚。转运调节蛋白的结构及功能还需进一步深入研究。

由于尘肺、肝肾变性、老年痴呆症、帕金森病、各种癌症及其他更复杂的疾病的诊断经常与长期接触金属粉尘相联系，利用基因组、蛋白组、代谢组、脂质组、离子组等技术研究金属离子的抗氧化防御机制、缺乏对大脑细胞的损伤机制、细胞毒性、金属离子配体、生物医学植入材料、癌症等获得更为丰富的生物化学标记物将成为新趋势。在生物体中，蛋白质并不独立行使其功能，而是不同蛋白质相互协调完成一系列生化反应以行使其生物学功能。因此，通路分析有助于更系统、全面地了解细胞的生物学过程、性状或疾病的发生机理、药物作用机制等[90]。随着实验技术的革新，对过渡金属离子的研究会有更多新的切入点，有利于我们更深入认识微量元素的代谢调控网络及其复杂的生物学功能。

参考文献（略）

（发表于《中国兽医学报》，院选一级学报）

我国部分地区牛源金黄色葡萄球菌基因多态性分型研究*

武中庸[1,2**]，倪春霞[1]，赵吴静[1]，徐进强[1,3]，蒲万霞[1***]

（1. 中国农业科学院兰州畜牧与兽药研究所/中国农业科学院新兽药工程重点开放实验室/甘肃省新兽药工程重点实验室，兰州 730050；2. 西北民族大学 生命科学与工程学院，兰州 730030；3. 甘肃农业大学 动物医学学院，兰州 730070）

摘要：为了解我国牛源金黄色葡萄球菌（*S.aureus*）的基因多态性，本研究利用随机引物多态性扩增（RAPD）体系对174株分离自贵州、内蒙古、四川、上海、甘肃等地奶牛乳房炎的*S.aureus*及一株标准菌CVCC2246进行基因分型。结果表明，175株*S.aureus*均得到清晰的RAPD指纹图谱，扩增产物为1~9个片段，产物大小为240~4 500bp。所有菌株共分为8个基因型，其中1型16株；2型和3型各37株；4型15株；5型18株；6型14株；7型13株；8型8株。2型和3型菌株占总菌株42%以上，在贵州、内蒙古、四川、上海、甘肃分布广泛，为流行优势基因型；但5型是内蒙古地区的流行优势基因型。各地区菌株基因型比例有明显差异，可能与奶牛饲养水平和环境差异有关。

关键词：奶牛乳房炎；金黄色葡萄球菌；RAPD；基因型

---

* 基金项目：中央级公益性科研院所基本科研业务费专项资金项目-奶牛养殖环境中耐药数据库及奶牛乳房炎病原菌种库建设（610322016014）；国家重点研发计划“畜禽重要病原耐药性检测与控制技术研究”（2016YFD0501306）。

** 作者简介：武中庸（1990—　），男，河北沧州人，硕士研究生，主要从事兽医微生物方面研究。

*** 通信作者：E-mail：wanxiapu@ yahoo.com.cn

Identification of Genetic Polymorphism of *Staphylococcus aureus* Isolated from Bovines Mastitis in Parts of China

Zhong-yong WU[1,2], Chun-xia NI[1], Wu-jing ZHAO[1], Jin-qiang XU[1,3], Wan-xia PU[1*]

(1. Key Laboratory of New Animal Drug of Gansu Province/Key Laboratory of New Animal Drug/Lanzhou Institute of Animal Sciences and Veterinary Pharmaceutics, Chinese Academy of Agricultural Sciences, Lanzhou 730050, China; 2. College of Life Science and Engineering, Northwest University for Nationalities, Lanzhou 730030, China; 3. College of Veterinary Medicine, Gansu Agricultural University, Lanzhou 730070, China)

Abstract: In order to know the gene polymorphisms of dairy sourced *Staphylococcus aureus*, 174 strains *S. aureus* isolated from dairy cows with mastitis in Guizhou, Inner Mongolia, Sichuan, Shanghai, Gansu and a standard strain were analyzed by random amplified polymorphic DNA (RAPD) in this study.The result showed that the amplified products bands were 1 to 9 fragments and the product sizes was 240 bp to 4 500 bp.Genotyping of these isolates indicated that all isolates were distributed into 8 different genotype. Among 175 isolates, 16 were type 1, 37 were type 2 and type 3 respectively, 15 were type 4, 18 were type 5, 14 were type 6, 13 were type 7 and type 8, 8 were type 8. Both of type 2 and type 3 were the dominant genotypes which were widely distributed in Guizhou, Shanghai, Sichuan, and Gansu provinces.Type 5 was the dominant genotype in Inner Mongolia.The proportion of 8 genotypes in different places were significant diffrence, which could be related to the feeding level and environment difference of dairy cow.

Key words: Bovine mastitis; *Staphylococcus aureus*; RAPD; genotype

金黄色葡萄球菌（*Staphylococcus aureus*）是引起奶牛乳房炎的主要病原菌之一[1]，为影响奶牛养殖业发展的重要因素之一。目前奶牛乳房炎主要依靠抗生素治疗，但研究显示，*S. aureus* 对抗生素呈普遍耐药和多重耐药，耐药菌株数逐年上升，对奶牛养殖业和食品公共安全造成了较强烈的冲击[2]。*S.aureus* 对抗生素的反应主要依赖于菌株自身携带的各种耐药基因[3]，因此对发病地区 *S.aureus* 进行流行菌株的基因分型，从而为分析菌株传播机制，指导抗生素用药，最终采取有效措施降低 *S.aureus* 的感染提供流行病学依据。近年来，各种分型方法已应用到细菌的分型研究中，其中包括随机扩增多态性分析（RAPD）[4]、多位点重复序列分型（MLVA）[5]、高压脉冲场凝胶电泳（PFGE）[6]等技术。RAPD 分型与其他基因分型方法相比，具有分型快速、简便、安全、灵敏、经济，并且不必预先知道待扩增序列、引物通用等特点而被国内外学者广泛应用于微生物的分子分型与鉴定[7]。本研究利用 RAPD-PCR 技术对我国贵州、内蒙古、四川、上海和甘肃地区奶牛乳房炎奶样中 *S.aureus* 分离株进行基因分型，探索不同地区乳房炎 *S.aureus* 分子流行病学特点。

1 材料与方法

1.1 菌株来源

标准菌 CVCC2246 购自中国兽药监察所；174 株 *S.aureus*，分离自临床型奶牛乳房炎奶样，其中四川地区 22 株（CX），内蒙古地区 34 株（NY），甘肃 27 株（HG、QY、KY），贵州地区 37 株（ZY），上海地区 54 株（SH、SX）；表皮葡萄球菌为对照菌株，由本实验室分离鉴定。

1.2 主要试剂

DNA Marker、Premix EX Taq DNA 聚合酶均购自 TaKaRa 公司；细菌染色体 DNA 提取试剂盒（离心柱型）购自天根生化科技（北京）有限公司。

1.3 引物合成

根据文献［8-9］报道进行引物合成，序列 4M：5′-AAGACGCCGT-3′。引物由宝生物工程（大连）有限公司合成。

1.4 RAPD 反应条件

25μL 反应体系为：DNA 模板 2μL，Premix 混合液 12.5μL，去离子水 8.5μL，引物 2μL。PCR 反应条件为：94℃ 5min，37℃ 5min，72℃ 5min，循环 4 次；94℃ 1min，37℃ 1min，72℃ 2min，循环 30 次；72℃延伸 10min。扩增产物经 1.0%琼脂糖凝胶电泳检测。

1.5 数据分析

利用 BioNumerics 软件对上述过程得到的电泳结果进行数据处理，参数如下：最适值、位置公差和阈值分别为 0.5%、1.25%和 80%。图像采用 Dice 相关系数和非加权组算术平均法（Un-weightedpair group method with arithmetic mean，UPG-MA）进行聚类分析。

2 结果与讨论

2.1 RAPD 扩增结果

对 175 株 *S.aureus*（包括标准菌）的 DNA 模板进行 RAPD 扩增，琼脂糖凝胶电泳检测结果显示，所有菌株 DNA 经扩增后均产生了清晰、可分辨的带谱，扩增产物在 1~9 片段之间，片段分子量大小在 240~4 500bp，具有多种带型组成，可以清楚地反映菌株间的基因多态性特征（图 1）。

2.2 RAPD 产物的遗传相关性分析

利用 BioNumerics 分析软件对电泳结果聚类分析结果显示，175 株 *S.aureus* 分为 8 个主要聚类群，各地区基因型分布见表 1。其中 1 型 16 株（9.1%）；2 型和 3 型各 37 株（21.1%）；4 型 15 株（8.5%）；5 型 18 株（10.2%）；6 型 14 株（8%）；7 型 13 株（7.4%）；8 型 8 株（4.6%）。2 型菌株可分为 4 个亚型，其余各型均分为 2 个亚型。2 型和 3 型占总数 42%以上，为主要流行的优势菌群；5 个地区的菌株基因型分布显示，贵州地区菌株主要集中在 2 型，约占总数的 50%，其中 2a 亚型占总数的 33%。其原因可能为奶样采

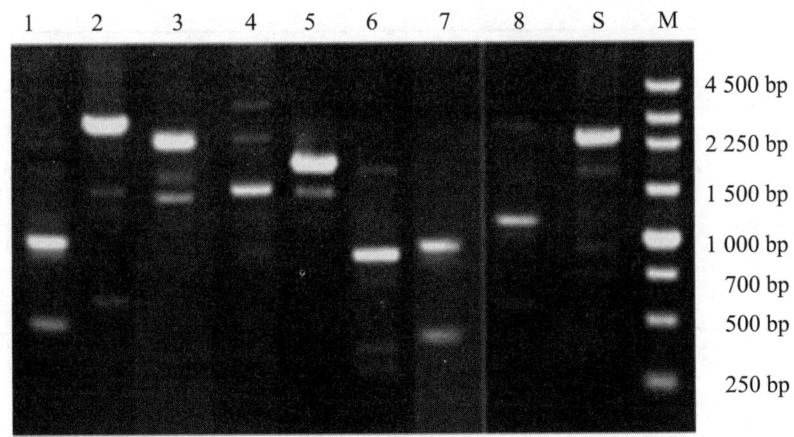

1-8:Genotype 1 to 8，respectively；S：Standard strain；M:DNA Marker

图 1　175 株 S.aureus 基因型的 RAPD 鉴定结果

Fig.1　Identification of the genotypes of 175 S.aureus strains by RAPD

自较偏僻山区，奶牛场规模小、奶牛调动不频繁，病原菌传播范围较小，地区间传播流行的机会较少，因此多数菌株均具有较高的亲缘性。内蒙古地区菌株主要集中在 5 型，约占总数的 33%。内蒙古地区为天然牧场区，奶牛养殖业发达，牛只调动频繁，细菌在牛场、地区间流行传播较频繁，菌株基因易突变，因此内蒙古地区菌株基因型较多且分布较广。四川地区菌株主要集中在 2 型和 3 型，其余各型分布较平均。四川地区菌株分离于大型奶牛场，经常与其他牛场进行牛只调动，因此菌株流行传播较频繁，基因型分布较广。上海地区主要集中在 3 型，占总数的 43%，并且各型均有分布，剩余各型菌株分布数量差距较小。上海地区菌株分离自两个奶牛场，其中一个为独立牧场，与外界接触较少，因此其基因型较集中；另一牧场可能发生了地区间传播。甘肃地区菌株分布较广，且 2d 亚型中仅包括标准菌和分离于甘肃某奶场的 5 株菌，或为同一亲本的克隆株。甘肃地区菌株基因型分布较广，其可能原因为饲养环境及管理水平较差发生了地区间流行。

本实验对我国不同地区的牛源 *S.aureus* 菌株进行了基因分型研究。各奶牛场菌株基因型分布差异较大，表明牛源 *S.aureus* 的感染随着地区、气候及养殖水平的不同而有较大变化。同时大型养殖场因降低饲养成本经常进行不同养殖场间的牛只调动，小型养殖场或散户因扩大规模由异地购牛等均容易引起牛源 *S.aureus* 地区间流行。及时准确地掌握 *S.aureus* 分子流行病学特征，有助于降低其在地区间的传播流行，指导各地区牧场轮换用药，为奶牛乳房炎的防治提供实验依据。

表 1　各地区基因型分布（株）

Table 1　Genotype distribution of *S.aureus* in different regions（strain）

	1 型 Type 1	2 型 Type 2	3 型 Type 3	4 型 Type 4	5 型 Type 5	6 型 Type 6	7 型 Type 7	8 型 Type 8
贵州 GuiZhou	3	18	7	3	0	2	3	0
内蒙古 InnerMongolia	5	5	5	0	11	4	0	1
四川 Sichuan	3	5	2	1	1	2	3	1

（续表）

	1 型 Type 1	2 型 Type 2	3 型 Type 3	4 型 Type 4	5 型 Type 5	6 型 Type 6	7 型 Type 7	8 型 Type 8
上海 Shanghai	4	3	21	5	1	3	6	5
甘肃 Gansu	1	5	2	6	5	3	1	1
标 准 菌 Standard strain（CVCC2246）	0	1	0	0	0	0	0	0
比例% Ratio（%）	9.1	21.1	21.1	8.5	10.2	8	7.4	4.6

参考文献（略）

（发表于《中国预防兽医学报》，院选一级学报）

宫衣净酊中盐酸水苏碱 HPLC-ELSD 测定方法的建立

朱永刚，王　磊，崔东安，王旭荣，张景艳，张　凯，
张　康，王学智，杨志强，李建喜*

（中国农业科学院兰州畜牧与兽药研究所/甘肃省中兽药工程技术研究中心，兰州　730050）

摘要： 为了建立宫衣净酊中盐酸水苏碱的 HPLC 测定方法，采用的色谱条件为色谱柱为赛分 Sepax polar-propylamide（4. 6mm×250mm，5μm），流动相为乙腈-2mL/L 的冰醋酸（65∶35），流速为 0. 5mL/min，进样量 10μL，柱温为 30℃。检测器参数为飘逸管温度为 105℃，雾化室的温度为 60℃，载气体积流量为 1. 0mL/min。结果表明，在此色谱条件下，盐酸水苏碱的线性范围 0. 155～248. 00mg/mL（R2＝0. 999 9，n＝6），平均回收率为 95. 46%（RSD＝1. 08%）。说明本方法简便、准确、重现性好，可以作为宫衣净酊的质量控制方法。

关键词： 高效液相色谱；蒸发光散射检测器；宫衣净酊；盐酸水苏碱；含量测定

（发表于《动物医学进展》，院选中文核心）

某奶牛场乳房炎病原菌的分离鉴定及耐药性分析

王　丹[1]，杨　峰[1]，李新圃[1]，罗金印[1]，刘龙海[1]，
张　哲[1,2]，张亚茹[1,2]，李宏胜[1]*

（1. 中国农业科学院兰州畜牧与兽药研究所/农业部兽用药物创制重点实验室/
甘肃省新兽药工程重点实验室，兰州　730050；2. 甘肃农业大学动物
医学院，兰州　730070）

摘要：为监测甘肃临洮某奶牛场奶牛乳房炎主要致病菌对临床常用治疗药物的耐药情况，指导临床合理用药。从奶牛场采集了45份临床型奶牛乳房炎病样，进行培养分离，并采用16S rDNA对分离纯化的细菌进行鉴定；对主要病原菌，采用牛津杯法测定其对乳房炎常用治疗药物的耐药情况。结果表明，从45份乳房炎奶样中，共分离出以大肠埃希菌和金黄色葡萄球菌为主的7种乳房炎致病菌。耐药性试验表明，金黄色葡萄球菌对无抗头孢、头孢力定、青霉素G钾、强力乳净、乳痈治、乳痈快克、炎乳康、乳肿康和硫酸庆大极其敏感；大肠埃希菌对头孢力定、青霉素G钾、强力乳净、乳痈快克、炎乳康、乳肿康和硫酸庆大极其敏感。金黄色葡萄球菌对强力乳康、无抗乳炎康、乳炎康、十味乳炎清、无抗乳肿消、获特键和强力宫康Ⅱ号耐药作用较强。大肠埃希菌对强力乳康、无抗乳肿消、乳房肿炎立消、获特键和强力宫康Ⅱ号耐药作用较强。研究结果为该奶牛场乳房炎的防治提供了参考。

关键词：奶牛乳房炎；金黄色葡萄球菌；大肠埃希菌；药物敏感性

（发表于《动物医学进展》，院选中文核心）

无乳链球菌荚膜多糖的粗提及多糖含量测定条件优化

张　哲[1,2]，杨　峰[1]，李新圃[1]，罗金印[1]，刘龙海[1]，李宏胜[1*]

（1. 中国农业科学院兰州畜牧与兽药研究所/农业部兽用药物创制重点室/
甘肃省新兽药工程重点实验室，兰州　730050；2. 甘肃农业
大学动物医学院，兰州　730070）

摘要： 为提取无乳链球菌荚膜多糖，并对苯酚-硫酸法测定多糖含量条件进行优化。离心收集发酵上清液，超滤浓缩，经 800mL/L 乙醇沉淀提取无乳链球菌荚膜多糖；采用单因素试验对苯酚硫酸法测定条件进行优化，并对测定方法的可靠性进行验证。结果表明，从 1 L 发酵液中提取到含量为 38. 51%的荚膜多糖 0. 36g；测定的最佳条件为：2mL 样品液中，加入 1mL 50mL/L 的苯酚，5mL 浓硫酸，混匀后 80℃ 作用 20min，在 1. 5h 内测定 485nm 处的光密度。在该条件下，平均加样回收率为 99. 96%，相对标准偏差为 0. 40%。说明从无乳链球菌发酵上清液中成功提取到荚膜多糖，建立了苯酚硫酸法测定荚膜多糖的最优条件。

关键词： 无乳链球菌；荚膜多糖；苯酚-硫酸法

（发表于《动物医学进展》，院选中文核心）

一株驯化啤酒酵母的分离鉴定及扩大培养条件的优化

徐进强[1,2]，梁红雁[1,2]，赵吴静[1]，武中庸[1]，陈　鑫[1,2]，李昱辉[1,2]，蒲万霞[1*]

（1. 中国农业科学院兰州畜牧与兽药研究所/农业部兽用药物创制重点实验室/
甘肃省新兽药工程重点实验室，兰州　730050；2. 甘肃农业大学
动物医学院，兰州　730070）

摘要：为筛选一株驯化啤酒酵母，并对发酵罐扩大培养工艺进行优化，为今后大规模生产提供试验数据。于麦芽汁培养基分离挑选某株具有较好生长力的酵母菌，进行形态学、生化及分子生物学鉴定。应用正交试验设计，从摇瓶温度、pH、取样时间及发酵罐的转速、装液量、接种量等方面进行优化。结果表明，驯化酵母菌株形态学、生化鉴定证明该菌株属于酵母属；发酵罐扩大培养最佳工艺为：转速 200r/min，接种量 50mL/L，装液量 4.5L，此条件下酵母活菌含量为 2.8×10^8CFU/mL。驯化前后表现一致，没有发生变异。优化啤酒酵母菌的发酵罐扩大培养工艺，为下一步规模化生产提供理论参考。

关键词：驯化酵母；生化鉴定；分子鉴定；发酵罐；扩大培养

（发表于《动物医学进展》，院选中文核心）

银翘蓝芩口服液对鸡的安全性试验

许春燕，刘希望，孔晓军，杨亚军，秦　哲，李世宏，李剑勇*

（中国农业科学院兰州畜牧与兽药研究所/农业部兽用药物创制重点实验室/
甘肃省新兽药工程重点实验室，兰州　730050）

摘要：依据兽药研究技术指导原则，开展银翘蓝芩口服液对靶动物鸡的安全性研究，为其临床应用的安全性提供数据资料。按银翘蓝芩口服液临床推荐剂量的1倍、3倍、5倍连续饮水给药7d，给药前后，逐只称重。试验结束后采集血液和组织器官，进行血常规、血液生化指标、脏器指数和病理组织学观察。结果表明，与对照组相比，各给药组鸡只的体重、血常规、血液生化和脏器指数等指标差异均不显著（$P>0.05$），各组织器官未见明显的病理组织学变化。说明银翘蓝芩口服液饮水给药对靶动物鸡是安全的。

关键词：银翘蓝芩口服液；鸡；安全性

（发表于《动物医学进展》，院选中文核心）

酮内酯类抗生素的研究进展

邵莉萍，张继瑜

（中国农业科学院兰州畜牧与兽药研究所/农业部兽药创制重点实验室，兰州　730050）

关键词：大环内酯类抗生素；酮内酯类抗生素；泰利菌素；喹红霉素；solithromycin；进展

Research Progress of Ketolide Antibiotics

Liping SHAO，Jiyu ZHANG

（Key Lahoratory of Veterinary Pharmaceutical Development，Ministry of Agriculture/ Lanzhou Institute of Husbandry and Pharmaceutical Sciences，Chinese Academy of Agricultural Sciences，Lanzhou 730050，China）

Key words：Macrolide antihiotics；Ketolide antibiotics；Telithromycin；Cethromycin；Solithromycin；Progress

（发表于《黑龙江畜牧兽医》，院选中文核心）

不同地区牦牛乳中水解酶活性比较研究

席斌[1,2,3]，高雅琴[1,2,3]，陈　轩[4]，杜天庆[1,2,3]，杨晓玲[1,2,3]

（1. 中国农业科学院兰州畜牧与兽药研究所，兰州　730050；2. 农业部畜产品质量安全风险评估实验室，兰州　730050；3. 农业部动物毛皮及制品质量监督检验测试中心，兰州　730050；4. 兰州大学第一附属医院，兰州　730000）

关键词：地区；胎次；牦牛乳；水解酶；活性

DOI：10. 13881/j.cnki.hljxmsy.2017. 2197

（发表于《黑龙江畜牧兽医》，院选中文核心）

茶树油对5种常见致病菌的体外抑菌作用研究

秦文文，梁剑平，郝宝成，尚若锋，王学红，黄　鑫，衣云鹏，刘　宇

（中国农业科学院兰州畜牧与兽药研究所/农业部兽用药物创制重点实验室/
甘肃省新兽药工程重点实验室，兰州　730050）

关键词：茶树油（TTO）；超临界二氧化碳萃取；牛津杯抑菌圈；最小抑菌浓度（MIC）；悬液定量杀灭

Bacteriostasis of Tea Tree Oil on Five Common Pathogenic Bacteria in Vitro

Wenwen QIN, Jianping LIANG, Baocheng HAO, Ruofeng SHANG, Xuehong WANG, Xin HUANG, Yunpeng YI, Yu LIU

(Key Laboratory of new Animal Drug Project, Gansu province; Key Laboratory of Veternary Pharmaceutics Discovery, Ministry of Agricultural; Lanzhou Institute of Husbandry and Pharmaceutical Sciences of CAAS, Lanzhou 730050, China)

Key words: Tea tree oil (TTO); Supercritical CO_2 extraction; Oxford cup inhibition zone; Minimum inhibitory concentration (MIC); Suspended quantitative kill

（发表于《黑龙江畜牧兽医》，院选中文核心）

畜产品中可能存在的质量安全风险隐患及应对措施

李维红，熊　琳，高雅琴，杨晓玲

（中国农业科学院兰州畜牧与兽药研究所/农业部畜产品质量安全风险评估实验室/甘肃省牦牛繁育工程重点实验室，兰州　730050）

关键词：畜产品；质量安全；兽药残留；禁用物质；风险隐患；措施

Risks of Quality Safety in Animal Products and Countermeasures

Weihong LI，Lin XIONG，Yaqin GAO，Xiaoling YANG

[Lanzhou Institute of Animal Science and Veterinary Pharmaceutics，The Chinese Academy of Agricultural Sciences/Quality Supervising，Inspecting and Testing Center for Animal Fiber，Fur，Leather and Products（Lanzhou），Ministry of Agriculture/Key Laboratory of Yak Breeding Engineering of Gansu Province，Lanzhou 730050，China]

Key words：Livestock products；Quality and safety；Veterinary drug residues；Banned substances；Risk hidden danger；Measures

（发表于《黑龙江畜牧兽医》，院选中文核心）

大鼠口服五氯柳胺的急性毒性研究

张吉丽，张继瑜，朱　阵，程富胜，周绪正，李　冰

（中国农业科学院兰州畜牧与兽药研究所/农业部兽用药物创制重点实验室/
甘肃省新兽药工程重点实验室，兰州　730050）

关键词：五氯柳胺；大鼠；急性毒性；半数致死量（LD_{50}）；95%的可信限；低毒

Study on the Acute Toxicity of Oxyclozanide in Rats by Oral Administration

Jili ZHANG，Jiyu ZHANG，Zhen ZHU，Fusheng CHENG，
Xuzheng ZHOU，Bing LI

（Key Laboratory of New Animal Drug Project of Gansu Province/Key Laboratory of Veterinary Pharmaceutical Development，Ministry of Agriculture/Lanzhou Institute of Husbandry and Pharmaceutical Sciences of CAAS，Lanzhou 730050，China）

Key words：Oxyclozanide；Rat；Acute toxicity；LD_{50}；95% confidence limit；Low toxicity

（发表于《黑龙江畜牧兽医》，院选中文核心）

冬季不同激素处理对高山美利奴羊胚胎移植效果的影响*

冯新宇[1,2]，岳耀敬[1,2]，袁　超[1,2]，王喜军[3]，刘继刚[3]，文亚洲[3]，
郭婷婷[1,2]，罗天照[3]，王天翔[3]，杨博辉[1,2]

（1. 中国农业科学院兰州畜牧与兽药研究所，兰州　730050；2. 中国农业科学院羊育种工程技术研究中心，兰州　730050；3. 甘肃省绵羊繁育技术推广站，甘肃张掖　734031）

关键词：冬季；高山美利奴；卵泡刺激素（FSH）；超数排卵；胚胎移植（ET）

（发表于《黑龙江畜牧兽医》，院选中文核心）

速康解毒口服溶液的稳定性研究

郝宝成[1]，权晓弟[1]，高旭东[1]，黄　鑫[1]，刘建枝[2]，王保海[2]，梁剑平[1]

（1. 中国农业科学院兰州畜牧与兽药研究所/农业部兽用药物创制重点实验室/甘肃省新兽药工程重点实验室，兰州　730050；2. 西藏自治区农牧科学院畜牧兽医研究所，拉萨　850000）

关键词：速康解毒口服溶液；加速试验；性状；稳定性；有效期；贮存

Study on the Stability of Sukang Detoxification Oral Liquid

Baocheng HAO[1]，Xiaodi QUAN[1]，Xudong GAO[1]，Xin HUANG[1]，Jianzhi LIU[2]，Baohai WANG[2]，Jianping LIANG[1]

（1. Key Laboratory of New Animal Drug Project of Gansu Province/Key Laboratory of Veterinary Pharmaceutics Discovery，Ministry of Agriculture/Lanzhou Institute of Husbandry and Pharmaceutical Sciences of CAAS，Lanzhou 730050，China；2. Institute of Animal Husbandry and Veterinary Medicine，Tibet Academy of Agricultural and Animal Husbandry Sciences，Lhasa 850000，China）

Key words：Sukang detoxification oral liquid；Accelerated test；Character；Stability；Validity period；Storage

（发表于《黑龙江畜牧兽医》，院选中文核心）

“慢呼宁”口服液对鸡安全性的试验研究

王贵波[1]，郭妮妮[2]，谢家声[1]，赵小龙[3]，罗超应[1]，
罗永江[1]，郑继方[1]，辛蕊华[1]，李锦宇[1]

（1. 中国农业科学院兰州畜牧与兽药研究所，兰州　730050；2. 湖北省畜禽育种中心，武汉　430070；3. 西安草滩牧业有限公司，华阴　714200）

摘要：为了对“慢呼宁”口服液的安全性进行评价，为临床用药提供安全依据，选取120羽14日龄的三黄肉鸡，适应性饲喂1周后，随机分为治疗剂量的1倍剂量组、3倍剂量组、5倍剂量组和对照组，每组30羽。试验时分别按照每组试验鸡的平均体质量给予不同剂量的药物，连续给药1周。在试验前和试验结束时，各组随机选取10只进行称质量和采血，测定体质量变化、血常规和肝肾功能指标，测定脏器指数，并对鸡的心、肝、脾、肺和肾脏进行病理检查。结果表明，用药后各组鸡的体质量增加量及脏器指数组间差异不显著（P>0.05）。血常规指标则在试验前后组间差异亦不显著（P>0.05）；血清ALT活性、白蛋白、肌酐和尿素含量组间差异不显著（P>0.05）；在用药后，各组鸡的心、肝、脾、肺和肾脏组织变化均无显著差异和异常。结果表明，“慢呼宁”口服液对鸡生理生化指标的影响小，病理剖检未见对心、肝、脾、肺和肾脏产生毒性作用，是一种低毒安全的中兽药制剂。

关键词：“慢呼宁”口服液；中兽药：呼吸道疾病；鸡；安全性评价

（发表于《江苏农业科学》，院选中文核心）

乳铁蛋白抗菌机理研究进展

裴　杰，褚　敏，包鹏甲，阎　萍，郭　宪

（中国农业科学院兰州畜牧与兽药研究所/甘肃省牦牛繁育工程重点实验室，兰州 730050）

摘要：乳铁蛋白（Lactoferrin，LF）是发现于哺乳动物初乳中具有多种生物活性的单体糖蛋白，对幼体初期免疫具有重要作用。前人研究发现 LF 蛋白的酶解后其抗菌活性得到增强，是由于酶解后产生了比其本身抗菌能力更强的肽段。目前关于 LF 蛋白抗菌活性的报道很多，但其抗菌的具体分子机制尚不清楚。以 LF 蛋白与细菌细胞膜的结合能力和 LF 蛋白酶解产物的抗菌活性为切入点，揭示了 LF 蛋白行使抗菌功能的过程，认为 LF 蛋白与细菌表面相结合而发生结构变化，进而暴露出敏感的酶切位点，经酶解而释放抗菌活性肽，这些抗菌肽破坏细菌的细胞膜结构，达到抑菌或杀菌的目的。对 LF 蛋白抗菌机理的阐释将为研制抗菌能力更强的活性蛋白提供理论依据。

关键词：乳铁蛋白；抗菌活性；抗菌肽；膜结合

Research Progress on Antibacterial Mechanism of Lactoferrin

Jie PEI，Min CHU，Peng-jia BAO，Ping YAN，Xian GUO

(Key Laboratory of Yak Breeding Project in Gansu Province，Lanzhou Institute of Husbandry and Pharmaceutical Sciences，Chinese Academy of Agricultural Sciences，Lanzhou 730050)

Abstract：Lactoferrin (LF) is a monomer glycoprotein in mammalian colostrum with multi biological activities，thus it plays a vital role in initial immunologic，reconstitution for young mammals.Previous studies have discovered that enzymatic hydrolysates of lactoferrin have enhanced antihacterial activity compared to lactoferrin because more active antibacterial peptides were produced by the enzymolysis.At present，molecular mechanism of antibacterial function of lactoferrin is still unclear although many antibacterial activities of lactoferrin were reported.In this paper，combination of the bacterial cell membranes and the antibacterial activity of enzymatic hydrolysates were taken as the breakthrough point，and the process of LF acting in antibacterial function was explored. It was considered that the mechanism of the antibacterial

function of lactoferrin was that the structure of lactoferrin changed following binding to bacterial surface, then the sensitive restriction enzyme cutting sites of lactoferrin were exposed, and many antibacterial peptides were released after enzymolysis; subsequently the peptides destroyed the membrane structures of bacterial cells, and thus the purpose of the bacteriostasis or sterilization achieved. In summary, the elucidation of antibacterial mechanism of lactoferrin provides the theoretical basis for developing active proteins with stronger antibacterial ability.
Key words: Lactoferrin; Antibacterial activity; Antimicrobial peptide; Membrane binding

（发表于《生物技术通报》，院选中文核心）

两种毛色牦牛皮肤黑色素组织学分析

高泽成[1,2]，吴晓云[1]，梁春年[1]，李明娜[1]，阎　萍[1*]

（1. 中国农业科学院兰州畜牧与兽药研究所/甘肃省牦牛繁育重点实验室，兰州　730050；
2. 甘肃农业大学动物科学与技术学院，兰州　730050）

摘要：［目的］观察两种不同毛色的牦牛皮肤中黑色素分布。［方法］采用冰冻切片和石蜡切片技术，用甲苯胺蓝和 HE 染色进行组织学分析。［结果］两种切片方法均能展现毛囊结构、皮肤组织分层结构。甲苯胺蓝染色：石蜡切片比冰冻切片着色较浅，易观察到黑色素的分布；冰冻切片通过显微镜观察发现牦牛毛囊由毛鞘、毛乳头和毛球外围的结缔组织形成。HE 染色：冰冻切片着色较深，皮肤结构层次分明；石蜡切片着色均匀，细胞分布明显。［结论］天祝白牦牛表皮的基底层和毛囊周围发现分布着少量的黑色素。大通牦牛的毛囊和表皮中均发现有大量黑色素的存在。大通牦牛毛囊中与毛发相连毛乳头附近的黑色素含量比毛囊周围其他部位含量高，皮肤表皮基底层相对毛囊黑色素含量少。大通牦牛毛囊周围黑色素比天祝白牦牛密集，且更为明显。大通牦牛的毛囊群比天祝牦牛毛囊群密集，毛囊的直径比天祝牦牛的长，毛发较粗。石蜡切片的 HE 染色是比较好的方法。

关键词：牦牛；甲苯胺蓝染色；HE 染色；黑色素

Melanin Histologic Analysis of Yak Skin in Two Different Colors

Ze-cheng GAO[1,2], Xiao-yun WU[1], Chun-nian LIANG[1], Ming-na LI[1], Ping YAN[1*]

(1. Lanzhou Institute of Husbandry and Pharmaceutical Science of CAAS, Lanzhou, Key Laboratory of Yak Breeding Engineering of Gansu Province, Lanzhou 730050, China; 2. Gansu Agricultural University College of Animal Science and Technology, Lanzhou 730050, China)

Abstract:【Objective】To study the melanin distribution in two different coat colors of yak skin.【Methods】The experiment used frozen section and paraffin section technique, Toluidine blue and HE dye were used to histologic analysis.【Results】Two kinds of slice methods could show the hair follicle structure, stratification of the skin tissue structure. Toluidine blue staining: paraffin had shallower color than frozen, and easier for observing the distribution of melanin; frozen section showed that yak hair follicle was made up of wool sheath, hair papilla and the bulb outside the surrounding connective tissue formation through microscope. HE staining: frozen section was deeper colored, and had distinct skin structure; paraffin section colored uniformly, and cell distributed obviously.【Conclusions】A small amount of melanin was found in Tianzhu's skin and hair follicles and a large amount of melanin was found in Datong's skin and hair follicles. The hair dermal papilla near the melanin content in Datong yak hair follicles was higher than other parts around the hair follicles, and skin epidermal basal layer relatively less hair melanin content. Hair follicle melanocytes of Datong yak was denser than Tianzhu yak, and more obvious. Datong yak hair follicle group was denser, the diameter of the hair follicle was longer than Tianzhu and with thicker hair. Paraffin section with HE staining was a better method.

Key words: Yak; Toluidine blue staining; HE staining; Melanin

（发表于《四川农业大学学报》，院选中文核心）

甘肃部分地区致羔羊腹泻大肠埃希菌的分离鉴定及耐药性分析

安　鑫，王胜义，王　慧，崔东安，黄美州，刘永明

（中国农业科学院兰州畜牧与兽药研究所/农业部兽用药物创制重点实验室/甘肃省新兽药工程重点实验室/甘肃省中兽药工程技术研究中心，兰州　730050）

摘要： 为调查甘肃地区大肠埃希菌对新生羔羊腹泻的致病性及其耐药性，从甘肃省8个养殖场采集新生腹泻羔羊粪便92份。采用16S rDNA扩增结合微生物鉴别培养技术对病料中的致病性大肠埃希菌进行鉴定，随机选取分离的6株菌株进行致病性试验，并采用Kirby-Bauer纸片琼脂扩散法对分离得到的菌株进行耐药性分析。结果表明，从92份病料中分离得到21株致病性大肠埃希菌，选取的6株菌株均有一定的致病性，耐药性分析结果显示该地的大肠埃希菌对羊场常用的多种抗生素耐药。表明大肠埃希菌仍是引起甘肃地区新生羔羊腹泻的主要病原菌，而且耐药性严重。

关键词： 新生羔羊；腹泻；大肠埃希菌；致病性；耐药性

（发表于《西北农业学报》，院选中文核心）

主成分分析法与聚类分析法在胎衣不下奶牛血液生化研究中的应用

朱永刚，王　磊，张景艳，王旭荣，崔东安，张　凯，
张　康，王学智，杨志强，李建喜

（中国农业科学院兰州畜牧与兽药研究所/甘肃省中兽药工程技术研究中心，兰州　730050）

摘要：通过分析胎衣不下奶牛血液生化指标的变化特征，为奶牛胎衣不下早期诊断方法的建立提供依据。产犊后立即无菌采集47头奶牛血液，其中25头为胎衣正常排出奶牛，22头为胎衣不下奶牛；血液生化分析仪检测血清生化指标，主成分分析法与聚类分析法分析检测结果。结果表明，16项血清生化指标被聚类成3个主成分，主成分聚类分析的结果与临床采集样品的分类结果一致；与胎衣正常排出奶牛相比，胎衣不下奶牛的血清中乳酸脱氢酶、谷草转氨酶、谷氨酰胺转移酶、肌酸激酶水平显著升高。表明，主成分分析和聚类分析可以有效地区别胎衣不下奶牛血清和胎衣正常排出奶牛血清，因此可以通过此方法对奶牛产后胎衣不下的发生进行提前诊断；还表明奶牛胎衣不下的发生不仅与蛋白质代谢、离子代谢和肌肉代谢紊乱有关，而且肝脏和肾脏的机能损伤也可能导致胎衣不下。

关键词：奶牛胎衣不下；主成分分析；生化指标

（发表于《西北农业学报》，院选中文核心）

阿司匹林丁香酚酯对大鼠的急性毒性研究

赵晓乐，孔晓军，马　宁，杨亚军，刘希望，申栋帅，李剑勇*

（中国农业科学院兰州畜牧与兽药研究所/农业部兽用药物创制重点实验室/
甘肃省新兽药工程重点实验室，兰州　730050）

摘要： 试验旨在评价阿司匹林丁香酚酯（aspirin eugenol ester，AEE）对大鼠的急性毒性，了解AEE毒性特点及可能的毒性靶器官，并全面地分析大鼠的致死原因。试验按照改良寇氏法进行，预试验选取50只大鼠，雌雄各半，分为5组，测定绝对致死量（LD_{100}）和最大非致死量（LD_0）。根据预试验结果，将80只健康大鼠，雌雄各半，随机平均分为7个药物组（剂量分别为2.62、3.25、4.03、5.00、6.20、7.69和9.53g/kg）和1个对照组给予0.5%的羧甲基纤维素钠（CMC-Na）进行正式试验。灌胃给药，观察并记录大鼠给药后的临床症状及体重变化。给药14d后，统计各剂量组大鼠的死亡时间及总死亡数，并计算AEE的半数致死量（LD_{50}）及其95%可信限。同时对死亡大鼠及时进行剖检，将具有病变的组织用10%的甲醛溶液固定，以进行病理学分析。结果显示，AEE对大鼠口服的LD50是5.95g/kg，且LD_{50}的95%可信限为5.30~6.68g/kg。对大鼠的死亡时间进行分析可得出，当口服AEE的剂量大于4.03g/kg时，其死亡时间主要集中在染毒后的48~96h内。大鼠染毒后3d内，体重呈下降趋势，常规饲养第5天开始，大鼠体重可逐渐恢复。病理学检查分析发现，AEE的剂量在5.00g/kg以上时，对大鼠的肝脏、肾脏和胃肠道有明显损伤。根据外源化学物急性毒性分级（WHO）可知，AEE为实际无毒化合物，AEE对大鼠毒性靶器官主要是肝脏、肾脏和胃肠道。

关键词： 阿司匹林丁香酚酯；急性毒性；灌胃；大鼠

Acute Toxicity of Aspirin Eugenol Ester on Rats

Xiao-le ZHAO, Xiao-jun KONG, Ning MA, Ya-jun YANG,
Xi-wang LIU, Dong-shuai SHEN, Jian-yong LI *

(Key Laboratory of New Animal Drug Project of Gansu Province/Key Laboratory of Veterinary Pharmaceutical Development, Ministry of Agriculture/Lanzhou Institute of Husbandy and Pharmaceutical Sciences of CAAS, Lanzhou 730050, China)

Abstract: The acute toxicity study was carried out by intragastric administration in rats to evaluate the toxic characteristics of aspirin eugenol ester (AEE) and target organs and fully analyze the reason of death rats. The experiment was designed in accordance with the method provided by the K? rber arithmetic method. At first, 50 rats (half male and half female) were divided into 5 groups to calculate LD_{100} and LD_0. According to the pre-test results, different doses of AEE (2.62, 3.25, 4.03, 5.00, 6.20, 7.69 and 9.53g/kg) and control group of 0.5% CMC-Na were orally administered to 80 rats, half male and half female, divided into 8 groups. After treatment, clinical signs of rats and the change of body weights were observed and recorded. The microscopic pathology of mild damages on organs was conducted. 14 days after administration, the death time and total deaths of rats were recorded, and the LD_{50} and 95% confidence interual of AEE were calculated according to K? rber. Organs were fixed with 10% formaldehyde solution at the end of the experiment. The LD_{50} of AEE in rats was 5.95g/kg upon oral administration and its 95% confidence interval calculation was 5.30 to 6.68g/kg. The death time of rats were centralized at 48 to 96h after exposing to AEE more than 4.03g/kg by the analysis of the cumulative mortality-time of AEE in each group. Significant decreases in body weights were noted after treatment of AEE for 3 days, after that body weights gradually grew and recovered at the 5th day. Remarkable damages on livers, kidney and gastrointestinal tract were found at AEE dosage of more than 5.00g/kg. From the obtained value of LD_{50}>5.00g/kg, AEE was classified as practically non-toxic compound according to WHO. The target organs of AEE were liver, kidney and gastrointestinal tract.

Key words: Aspirin eugenol ester; Acute toxicity; Intragastric administration; Rats

（发表于《中国畜牧兽医》，院选中文核心）

奶牛乳房炎病原菌诊断技术研究进展

王　丹[1]，杨　峰[1]，李新圃[1]，罗金印[1]，刘龙海[1]，张　哲[1]，
张亚茹[1]，张莉莉[1,2]，李宏胜[1*]

（1. 中国农业科学院兰州畜牧与兽药研究所/农业部兽用药物创制重点实验室/甘肃省新兽药工程重点实验室，兰州　730050；2. 兰州大学公共卫生学院，兰州　730000）

摘要：奶牛乳房炎是影响世界奶牛业发展的最主要疾病之一，不仅给奶牛养殖业带来严重的经济损失，而且还影响牛奶的产量和质量，危及人类健康。引起奶牛乳房炎的病因很复杂，但最主要的病因是病原微生物感染，其中葡萄球菌、链球菌和肠道菌感染是引起奶牛乳房炎的最主要病原菌。为有效治疗奶牛乳房炎，对奶牛乳房炎病原菌的早期快速诊断至关重要。本文对目前用于奶牛乳房炎病原菌诊断的常规微生物生化鉴定法、全自动微生物鉴定系统、刃天青微量板法、免疫诊断法、核酸探针杂交、16S rRNA 基因序列测定、PCR、基因芯片及环介导等温扩增技术等的研究进展进行了综述，并比较分析了这些方法的优缺点，同时对未来奶牛乳房炎病原菌诊断的研究方向和前景进行了展望，以期为进一步开发快速、特异、敏感的奶牛乳房炎病原菌诊断技术提供理论依据。

关键词：奶牛；乳房炎；病原菌；诊断技术

Research Progress on Diagnosis Technology of Dairy Cow Mastitis Pathogens

Dan WANG[1], Feng YANG[1], Xin-pu LI[1], Jin-yin LUO[1], Long-hai LIU[1], Zhe ZHANG[1], Ya-ru ZHANG[1], Li-li ZHANG[1,2], Hong-sheng LI[1*]

(1. Key Laboratory of New Animal Drug Project of Gansu Province, Key Laboratory of Veterinary Pharmaceutical Discovery, Ministry of Agriculture, Lanzhou Institute of Animal Husbandry and Pharmaceutical Sciences, Chinese Academy of Agricultural Sciences, Lanzhou 730050, China; 2. School of Public Health, Lanzhou University, Lanzhou 730000, China)

Abstract: Dairy cow mastitis is one of the most costly diseases affecting the development of the world dairy industry. It causes a serious economic loss to dairy farming, and affects the yield and quality of milk, endangers human health. The cause of dairy cow mastitis is complex, but the main cause is pathogen infection, particularly Staphylococcus, Streptococcus and intestinal bacterial infection. The rapid diagnosis of pathogenic bacteria of mastitis is crucial for the effectively treatment of the dairy cow mastitis. In order to provide theoretical basis for developing rapid, specific and sensitive of dairy cow mastitis pathogens diagnosis, the author summarized the microbiological and biochemical assay method in routine used of dairy cow mastitis pathogen diagnosis, automatic microbial identification system, resazurin trace plate method, immune diagnosis method, the nucleic acid probe hybridization, 16S rRNA gene sequences of sequencing, PCR, gene chip and LAMP etc, and analyzed the advantages and disadvantages of these methods.

Key words: Dairy cow; Mastitis; Pathogenic bacteria; Diagnosis technology

（发表于《中国畜牧兽医》，院选中文核心）

响应面法优化益生菌 FGM 发酵黄芪多糖的提取工艺

边亚彬，张景艳，侯艳华，王旭荣，张　凯，李建喜*

（中国农业科学院兰州畜牧与兽药研究所/甘肃省中兽药工程技术研究中心，兰州 730050）

摘要：为确定益生菌 FGM 发酵黄芪多糖的最佳提取工艺，本试验以发酵黄芪多糖含量为考察指标，采用响应面法对工艺的提取参数进行优化，并建立回归模型。结果显示，提取时间、提取温度和料液比对发酵黄芪多糖的含量影响显著，并且两两因素间存在一定的交互作用；影响发酵黄芪多糖提取含量的工艺因素按主次顺序排列为：料液比>提取温度>提取时间；乳酸菌发酵黄芪液多糖提取最佳工艺为：提取时间 65min、提取温度 80℃、料液比为 1：9，在此条件下，发酵黄芪多糖的含量为 6.72mg/mL。试验证明该工艺可行，可为新兽药开发提供参考。

关键词：非解乳糖链球菌；发酵；黄芪；多糖；响应面法

Optimization of Polysaccharides Extraction from Astragalus Membranaceus Fermented by FGM Strain with Response Surface Method

Ya-bin BIAN, Jing-yan ZHANG, Yan-hua HOU, Xu-rong WANG, Kai ZHANG, Jian-xi LI*

(Engineering & Technology Research Center of Traditional Chinese Veterinary Medicine of Gansu Province, Lanzhou Institute of Husbandry and Pharmaceutical Science of CAAS, Lanzhou 730050, China)

Abstract: Response surface method was used to optimize main process parameters for the polysaccharides extractions from Astragalus membranaceus fermented by probiotic FGM strain. A regression model equation was fitted. The results showed that extraction time, extraction temperature and solid-liquid ratio were significant factors on polysaccharide content fermented from Astragalus membranaceus and there was interactive effects between every two factors. Polysaccharide content was significantly affected by solid-liquid ratio, followed by extraction temperature and extraction time. The optimum conditions were found to be that the extraction time was 65min, extraction temperature was 80℃, solid-liquid ratio was 1 : 9, and the polysaccharide content was 6.72mg/mL at this condition. The results showed that the model was practical that could provide a base for exploitation new medicine.

Key words: *Streptococcus alactolyticus*; Fermentation; *Astragalus membranaceus*; Polysaccharides; Response surface method

（发表于《中国畜牧兽医》，院选中文核心）

伊维菌素微乳中伊维菌素的含量测定方法研究

王嗣涵[1,2]，张继瑜[1]，邢守叶[1,2]，李　冰[1]，周绪正[1*]

（1. 中国农业科学院兰州畜牧与兽药研究所/农业部兽用药物创制重点实验室，兰州　730050；2. 甘肃农业大学动物医学院，兰州　730070）

摘要：为建立伊维菌素微乳中伊维菌素含量的高效液相色谱（HPLC）测定方法，选用 Hypersil ODS2（5μm，4.6mm×250mm）色谱柱，流动相为甲醇：乙腈：水为 35：60：5（V/V/V），检测波长为244nm，柱温为30℃，流速为1mL/min 进行测定。结果显示，伊维菌素在该色谱条件下，系统适应性良好，在80~320μg/mL 浓度范围内线性关系良好，回归方程为：Y=22 700X+2 510，R^2=0.9998，总平均回收率为 101.90%±2.94%，RSD 为 2.88%，对中试生产的 3 批伊维菌素微乳进行含量测定，RSD 为 1.86%。表明该含量测定方法准确可靠，重现性好，可用于伊维菌素微乳中伊维菌素含量的测定，并为该新型制剂的质量标准的制定和质量评价提供依据，也为后期的临床安全应用提供可靠的参考。

关键词：伊维菌素微乳；伊维菌素；高效液相色谱法

Content Determination of Ivermectin in the Ivermectin Microemulsion Injection

Si-han WANG[1,2], Ji-yu ZHANG[1], Shou-ye XING[1,2], Bing LI[1], Xu-zheng ZHOU[1*]

(1. Key Laboratory of New Animal Drug Discovery, Ministry of Agriculture/Lanzhou Institute of Husbandry and Pharmaceutical Sciences, Chinese Academy of Agricultural Sciences, Lanzhou 730050, China; 2. College of Veterinary Medicine, Gausu Agricultural University, Lanzhou 730070, China)

Abstract: In order to establish the determination method of ivermectin (IVM) in the ivermectin microemulsion injection by high performance liquid chromatography (HPLC), Hypersil ODS2 colunm (5μm, 4.6mm×250mm) was used in this study. The mobile phase composed of methanol, acetonitrile and water (35：60：5, V/V/V) at a flow rate of 1mL/

min. The detection wave length was set at 244nm and the column temperature was 30℃. The results showed that the HPLC system suitability of IVM was good. A good linear correlation of IVM was observed within the concentration range 80 to 320μg/mL, and the average recovery rate was 101.90%±2.94% with RSD was 2.88%, the regression equation was Y = 22 700X+2 510 (R^2=0.9998). The RSD of IVM content in ivermectin microemulsion injection was 1.86%. The method was accurate and reliable, reproducible, easy to operate, which could be used in new type of ivermectin microemulsion injection. The method could be used as the basis of quality control and establishing a quality standard, and to provide basis for quality evaluation, also can provide reliable reference for safe veterinary clinical application in the future.

Key words: Ivermectin microemulsion injection; Ivermectin; High performance liquid chromatography

（发表于《中国畜牧兽医》，院选中文核心）

益生菌发酵提高黄芪根、茎、叶活性成分含量的研究

苏贵龙，张景艳，张　凯，王　磊，张　康，王学智，杨志强，李建喜*

（中国农业科学院兰州畜牧与兽药研究所/甘肃省中兽药工程技术研究中心，兰州　730050）

摘要：黄芪药物资源的利用往往仅限于黄芪的根部，而黄芪茎、叶却被大量废弃，造成了药物资源的严重浪费。试验旨在采用益生菌发酵黄芪，研究黄芪根、茎、叶中黄酮类、皂苷类及黄芪多糖等有效成分含量的变化，以期高效利用黄芪。采用从鸡肠道分离保存的一株益生菌（FGM），用于发酵黄芪的根、茎、叶。结果显示，经 FGM 发酵后，黄芪根、一年生茎、两年生茎、一年生叶、两年生叶中粗多糖含量分别提高 177.46%、227.27%、207.11%、170.61%、182.28%；总黄酮含量分别提高 55.67%、33.68%、30.04%、-8.17%、-6.57%；总皂苷含量分别提高 68.50%、55.91%、55.71%、40.93%、46.13%。以上结果表明，利用益生菌发酵可使黄芪各部位中主要活性成分含量提高，这对高效利用黄芪、进一步开发传统中药资源有非常重要的意义。

关键词：黄芪；发酵；黄芪多糖；黄酮；皂苷

Study on Improving Active Ingredients of *Astragalus* Root, Stem and Leaf by Probiotic Fermentation

Gui-long SU, Jing-yan ZHANG, Kai ZHANG, Lei WANG, Kang ZHANG, Xue-zhi WANG, Zhi-qiang YANG, Jian-xi LI*

(Engineering and Technology Research Center of Traditional Chinese Veterinary Medicine of Gansu Province, Lanzhou Institute of Husbandry and Pharmaceutical Sciences of CAAS, Lanzhou 730050, China)

Abstract: The utilization of *Astragalus* resource was often limited to the root, while the stem and leaf had always been discarded, causing serious waste of traditional Chinese medicine resources.This experiment was aimed to study the changes of the content of active ingredients such as *Astragalus* polysaccharides, flavonoids and saponins in *Astragalus* root, stem and leaf by probiotic fermentation.A strain of FGM probiotic isolated from chicken intestines was used in this experiment for the fermentation of *Astragalus* root, stem and leaf.The results showed that, after fermentation, the crude polysaccharide contents of Astragalus root, annual stem, two years stem, annual leaf, two years leaf increased by 177.46%, 227.27%, 207.11%, 170.61% and 182.28%, respectively, the total flavonoids contents increased by 55.67%, 33.68%, 30.04%, −8.17% and−6.57%, respectively, and the total saponins contents increased by 68.50%, 55.91%, 55.71%, 40.93% and 46.13%, respectively.FGM probiotic fermentation made the main component contents of *Astragalus* increased, which would help for the further utilization of different parts of *Astragalus*, and efficient utilization of traditional Chinese medicine resources.

Key words: *Astragalus*; Fermentation; *Astragalus polysaccharide*; Flavonoid; Saponins

（发表于《中国畜牧兽医》，院选中文核心）

中药常山散的体外抑菌作用研究

王　玲[1]，郭志廷[1*]，杨　峰[1]，王文莉[2]，莫亚霞[2]，
郭爱民[2]，罗小琴[3]，魏小娟[1]，吕亚楠[2]

（1. 中国农业科学院兰州畜牧与兽药研究所/农业部兽用药物创制重点实验室/甘肃省新兽药工程重点实验室，兰州　730050；2. 甘肃农业大学动物医学院，兰州　730070；3. 兰州市动物卫生监督所，兰州　730050）

摘要：试验旨在研究中药常山散的体外抑菌活性。分别采用试管二倍稀释法联合琼脂平板稀释法及营养琼脂稀释法对选用的 12 种致病菌进行抑菌试验，测定最低抑菌浓度（MIC），进行抗菌作用量效关系研究，并通过牛津杯法观察药物抑菌效果。试管二倍稀释法联合琼脂平板法结果表明，常山散对链球菌属和芽孢杆菌的抑菌效果较强，MIC 在 15.6~62.5mg/mL 之间；对金黄色葡萄球菌和肠杆菌有较弱的抑菌作用，MIC 在 250~500mg/mL 之间；对真菌的抑菌作用最弱，MIC 值>500mg/mL，其抑菌强度大小依次为：链球菌属、芽孢杆菌、肠杆菌、真菌。营养琼脂稀释法结果表明，常山散对变形杆菌、白色念珠菌及黑曲霉具有一定的抑菌作用，MIC 约为 500mg/mL，而对其他链球菌属、肠杆菌属及芽孢杆菌属细菌的抑菌效果较弱，MIC 值均>500mg/mL。牛津杯法抑菌活性研究结果显示，常山散药液（500mg/mL）对 12 种菌均有一定的抑菌效果，但抑菌效果较弱，绝大多数抑菌环直径≤10mm。牛津杯周围可见明显的药物作用圈，作用圈内细菌数量较其他部位明显减少。综合以上试验结果，中药常山散对常见致病菌均具有一定抑菌效果，但由于药物本身的特性及有效组分含量较低，其抑菌作用效果较弱。

关键词：常山散；最低抑菌浓度；致病菌；抑菌试验

Bacteriostatic Test in Vitro of Traditional Chinese Medicine Radix Dichroa Powder on Common Pathogens

Ling WANG[1], Zhi-ting GUO[1*], Feng YANG[1], Wen-li WANG[2], Ya-xia MO[2], Ai-min GUO[2], Xiao-qin LUO[3], Xiao-juan WEI[1], Ya-nan LV[2]

(1. Key Laboratory of New Animal Drug Project, Gansu Province, Key Laboratory of Veterinary Pharmaceutics Discovery, Ministry of Agriculture, Lanzhou Institute of Animal Science and Veterinary Pharmaceutics, Chinese Academy of Agricultural Sciences, Lanzhou 730050, China; 2. College of Veterinary Medicine, Gansu Agricultural University, Lanzhou 730070, China; 3. Lanzhou Institute of Animal Health Supervision, Lanzhou 730050, China)

Abstract: The purpose of this study was to indicate the antibacterial effect of traditional Chinese medicine Radix dichroa powder (RDP) in vitro.Minimal inhibitory concentration (MIC) of RDP for 12 pathogens were determined by two fold dilution method and agar dilution method, and the dose-effect relationship of antibiotic effect was studied, and the antimicrobial effect was also observed by Oxford-cup method.The results from two fold dilution method indicated that RDP had stronger inhibited effect to Streptococcus and Bacillus subtilis (MIC were from 15. 6mg/mL to 62. 5mg/mL) than that of S.aureus (MIC was about 250mg/mL), Enterobacteria (MIC was about 500mg/mL) and fungi (MIC>500mg/mL); The results of agar dilution method showed that RDP had some antibacterial activities on Proteusbacillus vulgaris, Candida albicans And aspergillus niger, MIC was about 500mg/mL, while it had weak antibacterial effect on other pathogens (MIC > 500mg/mL).The results of Oxford-cup method showed that RDP (500mg/mL) had some degree antimicrobial effect on pathogens, and the bacteria-inhibiting ring diameter of RDP not exceeded 10mm, while drug-action ring was observed obviously around the Oxford-cup, and the number of bacteria was significantly decreased within the ring.In conclusion, RDP had some degree antibacterial effect on common pathogens, but the effects were not strong because of charicteristics of RDP itself and lower effective content.

Key words: Radix dichroa powder (RDP); Minimal inhibitory concentration (MIC); Pathogen; bacteriostatic test

（发表于《中国畜牧兽医》，院选中文核心）

常山散对人工感染的鸡球虫病疗效研究

郭志廷[1]，王　玲[1]，龚振兴[2]，蔡建平[2]，梁剑平[1]

（1. 中国农业科学院兰州畜牧与兽药研究所/农业部兽用药物创制重点实验室/甘肃省新兽药工程重点实验室，兰州　730050）；2. 中国农业科学院兰州兽医研究所/家畜疫病病原生物学国家重点实验室，兰州　730046）

摘要：为了明确常山散治疗鸡球虫病的疗效和最佳给药剂量，通过人工感染鸡柔嫩艾美耳球虫后在饲料中添加一定比例的药物，试验设常山散组（分别为 0.05、0.1、0.3、0.5、0.7、1g/kg 饲料）、妥曲珠利对照组（1.0mL/L 饮水）、感染对照组和健康对照组。结果表明：与感染对照组比较，常山散各剂量组鸡盲肠和十二指肠肿胀明显减轻，血液性内容物明显减少，抗球虫指数分别为 85.5、132.7、141.1、128.8、104.2 和 102.8，均高于感染对照组；常山散按 0.1、0.3 和 0.5g/kg 饲料给药抗球虫指数均高于妥曲珠利对照组（124.8）。结果提示：常山散抗球虫疗效好，且作为中药提取物，绿色环保、低毒、低残留，有望成为新型抗球虫药物。

关键词：常山散；人工感染；鸡球虫病；妥曲珠利

（发表于《中国家禽》，院选中文核心）

奶牛黏合素蛋白的生物信息学分析

李亚娟[1,2]，王　慧[1,3]，张世栋[1]，王东升[1]，何宝祥[2]，
严作廷[1]，杨志强[1]，董书伟[1]，何玉琴[3]

（1. 中国农业科学院兰州畜牧与兽药研究所/农业部兽用药物创制重点实验室/甘肃省新兽药工程重点实验，兰州　730050；2. 广西大学动物科技学院，南宁　530005；3. 甘肃农业大学生命科学院，兰州　730070）

摘要：为进一步了解奶牛CGN蛋白的性质和功能，采用生物信息学方法，利用在线软件和生物学数据库，对其CGN蛋白的氨基酸序列、二级结构、三级结构和生物学功能等进行预测。序列分析可知：CGN蛋白是亲水性不跨膜的蛋白，信号肽存在于20~21位氨基酸之间，二级结构主要以α-螺旋和无规则卷曲为主，并且主要在质膜中发挥生物学作用。生物学功能预测结果表明，CGN蛋白可能在激素、生长因子以及应激反应中发挥重要作用。系统进化分析表明，奶牛CGN与野牛、牦牛、亚洲水牛等物种CGN遗传距离较近，具有高度同源性。本结果为进一步研究CGN蛋白和奶牛的抗病育种提供理论依据。

关键词：黏合素；生物信息学；奶牛

Bioinformatics Analysis of CGN Protein in Dairy Cows

Yajuan LI[1,2], Hui WANG[1,3], Shidong ZHANG[1], Dongsheng WANG[1],
Baoxiang HE[2], Zuoting YAN[1], Zhiqiang YANG[1], Shuwei DONG[1], Yuqin HE[3]

(1. Key Laboratory of Veterinary Pharmaceutical Development, Ministry of Agriculture/Key Laboratory of New Animal Drug Project, Gansu Province/Lanzhou Institute of Husbandry and Pharmaceutical Sciences of CAAS, Lanzhou 730050; 2. College of Animal Science and Technology, Guangxi University, Nanning 530005; 3. College of Life Science of Gansu Agricultural University, Lanzhou 730070)

Abstract: The aim is to explore the property and function of CGN in dairy cows.The bioinformatics was used to predict the amino acid sequence, secondary structure, tertiary structure and biological function of CGN, using online database biological database. Sequence analysis indicated that CGN was hydrophilic untransmembrane protein, and its cleavage position of the signal peptide was located between the 20th and 21st site of amino acid. The secondary structure of CGN was primarily composed of alpha helix and random coil, and the biological effects of CGN were performed in the plasma membrane. Biological function prediction indicated that the function of CGN might be involved in the biological process of hormone, growth factor, stress response and immune response. Phylogenetics analysis showed that the CGN of dairy cows was close to that of Bison, Bos mutus and Bubalus bubalis in the phylogenetic tree, which showed highly homology. This study provides a theoretical basis for further research on CGN and disease-resistant breeding of dairy cows.

Key words: Conglutinin; Bioinformatics; Dairy cows

(发表于《中国农学通报》，院选中文核心)

中兽药临床疗效评价研究及其关键科学问题

崔东安，王胜义，王　磊，王　慧，刘永明*

（中国农业科学院兰州畜牧与兽药研究所/农业部新兽药工程重点实验室/
甘肃省中兽药工程技术研究中心，兰州　730050）

摘要：为探讨中兽药临床疗效评价研究的近期研究任务和创新目标，从加强临床基础研究、遵循国际通则，突出中兽医药特色、多维综合评价、推进中兽医优势病种研究以及科学顶层设计与临床实际执行等方面，综合分析了中兽药临床疗效评价需要关注的科学问题和技术问题。结果显示，中兽医药独特的治疗特点决定了中兽药临床疗效评价需要遵从本身自身特点和发展规律，立足中兽医临床优势，解决好临床实际问题，科学、客观地回答“中兽药的有效性”。

关键词：中兽药；临床疗效评价；优势病种；疗效指标

Key Research Priorities for Clinical Efficacy Evaluation of Veterinary Herbal Medicines

Dong'an CUI, Shengyi WANG, Lei WANG, Hui WANG, Yongming LIU*

(Lanzhou Institute of Husbandry and Pharmaceutical Sciences, Chinese Academy of Agricultural Sciences/Key Laboratory of Veterinary Pharmaceutical Development, Ministry of Agriculture/Engineering/Echnology Research Center of Traditional Chinese Veterinary Medicine of Gansu Province, Lanzhou 730050, China)

Abstract: To investigate the major research objective and priority of the clinical efficacy evaluation of veterinary herbal medicines, a number of science and technology issues are summarized and analyzed to strengthen the clinical basic research, follow general guidelines, highlight the characteristics of herbal veterinary medicines, establish multi-dimensional comprehensive evaluation system, and promote preponderant illness studies, and coordinate top-level design with clinical operation process were analyzed systematically. Due to the unique characteristics of herbal remedies, the clinical efficacy evaluation of veterinary herbal medicines has its own characteristics and developing route. Based on the clinical superiority of traditional Chinese veterinarian medicine, the key research priorities are to improve the capability of resolving clinical problems and respond the effectiveness of veterinary herbal medicines in a scientific objective and reliable way.

Key words: Herbal veterinary medicines; Clinical efficacy evaluation; Preponderant illness; Therapeutic effect index

（发表于《中国农业大学学报》，院选中文核心）

百里香酚对山羊子宫内膜上皮细胞的体外毒活性研究

闫宝琪[1]，张世栋[1*]，董书伟[1]，王东升[1]，那立冬[1,2]，
桑梦琪[1]，杨洪早[1,2]，严作廷[1*]
（1. 中国农业科学院兰州畜牧与兽药研究所/甘肃省新兽药工程重点实验室，
兰州　730050；2. 甘肃农业大学动物医学院，兰州　730070）

摘要：为了探究百里香酚对山羊子宫内膜上皮细胞（goat endometrial epithelial cells，gEECs）的毒性作用，采用不同浓度的百里香酚作用于gEECs后，观测其对细胞生长曲线，细胞形态学变化，以及细胞膜结构完整性的影响。结果表明，百里香酚对gEECs的毒性作用与其浓度呈剂量依赖效应（y=-0.008x+1.2202，R2=0.9767），即随着药物浓度的增加细胞的增殖抑制率增大，药物对细胞的半数增殖抑制率（IC_{50}）为100μg/mL。当50、100μg/mL的百里香酚作用于EECs时，gEECs的细胞核、细胞浆的形态均未产生明显变化，但当150μg/mL的百里香酚作用于gEECs时，细胞膜破裂，细胞死亡；而且药物浓度≥100μg/mL时细胞LDH的释放量显著高于正常细胞组，gEECs细胞膜的完整性受到破坏。百里香酚作用于山羊子宫内膜上皮细胞后呈现一定的毒性作用，且毒性作用和药物的浓度有关。百里香酚作用于gEECs的安全范围在0~100μg/mL，研究结果为百里香酚治疗奶牛等经济动物的炎症性疾病的临床用药浓度提供理论依据。

关键词：百里香酚；子宫内膜上皮细胞；细胞毒性；膜通透性

The Cytotoxic Effect of Thymol on Goat Endometrial Epithelial Cells

Bao-qi YAN[1], Shi-dong ZHANG[1*], Shu-wei DONG[1], Dong-sheng WANG[1], Li-dong NA[1,2], Meng-qi SANG[1], Hong-zao YANG[1,2], Zuo-ting YAN[1*]

(1. Lanzhou Institute of Animal Husbandry and Pharmaceutical Sciences, Chinese Academy of Agricultural Sciences, Lanzhou 730050, China; 2. College of Veterinary Medicine, Gansu Agricultural University, Lanzhou 730070, China)

Abstract: In order to explore the cytotoxicity of thymol on goat endometrial epithelial cells (gEECs), gEECs was treated by different concentrations of thymol. And the cell growth curve, the changes of cell morphology, the integrity of the cell membrane was detected. The results showed that thymol has a role in promoting proliferation of the gEECs. There was a dose-dependent effect between thymol toxicity to gEECs and concentrations ($y=-0.008x+1.2202$, $R2=0.9767$), The inhibitory rate of cell proliferation increased with the increasing of the drug concentration; And 50% inhibitory rate of proliferation (IC_{50}) of this drug on EECs is 100μg/mL. When the 50 and 100g/mL of thymol stimulated gEECs, the nuclear morphology and the cytoplasm of the gEECs cells didn´t change significantly. But when 150g/mL thymol treated gEECs, cell membrane rupture and cell would die; When the concentration was more than 100g/mL, the integrity of cell membrane can be damaged, promoting the release of LDH significantly compared with normal cells group. In summary, there was a certain toxicity of thymol on goat endometrial epithelial cells, and the toxicity related to concentration. The safe concentration of thymol is in the range of 0~100g/mL. Which provided a theoretical basis not only for the clinical treatment on the inflammatory diseases of cows and other economic animal but also for the relevant experiments of thymol at the molecular level.

Key words: Thymol; Endometrial epithelial cells; Cytotoxicity; Membrane permeability

（发表于《中国兽药杂志》，院选中文核心）

不同生长年限紫花苜蓿草地土壤团聚体有机碳分布特征

周　恒[1]，时永杰[1]，胡　宇[1]，陈　璐[2]，路　远[1]，田福平[1*]

（1. 中国农业科学院兰州畜牧与兽药研究所/农业部兰州黄土高原生态环境重点野外科学观测试验站，兰州　730050；2. 甘肃农业大学草业学院，兰州　730070）

摘要：对黄土高原地区不同生长年限紫花苜蓿（*Medicago sativa L.*）草地土壤团聚体0~40cm土层的分布及其有机碳含量特征进行研究。结果表明：土壤总有机碳随着生长年限的增加而呈先增加后减小的趋势，各年限0~10cm土壤有机碳含量均最高，呈表聚现象。不同生长年限紫花苜蓿湿筛处理下土壤团聚体均以<0.106mm粒级团聚体为主，为40%~90%，随土层深度增加而减小；>0.25mm粒级团聚体在4年时含量最高，并随生长年限的增加呈先增加后减小的趋势。不同年限紫花苜蓿土壤团聚体有机碳以<0.106mm粒级团聚体含量最高，随粒级的减小而增加；不同年限紫花苜蓿土壤20~40cm团聚体有机碳含量高于0~20cm土层，并随生长年限增加呈先增加后减小的趋势。

关键词：紫花苜蓿草地；团聚体；有机碳

（发表于《中国土壤与肥料》，院选中文核心）

藏药雪山杜鹃叶挥发油成分的 GC-MS 分析

郭　肖，周绪正，朱　阵，文　豪，张吉丽，张继瑜*

（中国农业科学院兰州畜牧与兽药研究所，兰州　730050）

摘要：[目的]：通过研究雪山杜鹃叶挥发油的化学成分，为雪山杜鹃的药用及开发利用提供依据。[方法]：采用水蒸气蒸馏法提取雪山杜鹃叶挥发油，并运用气相色谱、质谱联用技术（CC-MS）对其挥发油进行了系统的分离和鉴定。[结果]：从雪山杜鹃叶中共鉴定出 71 种挥发性成分，占挥发油总量的 80.51%。雪山杜鹃叶挥发油主要成分为：芳樟醇 10.66%、白菖烯 4.35%、epimanoyl oxide 3.85%、α-松油醇 3.46%、α-杜松醇 3.27%。[结论]：本实验采用 GC-MS 分析方法对雪山杜鹃叶挥发油进行分析，可为合理开发利用雪山杜鹃资源提供科学依据。

关键词：雪山杜鹃；挥发油；GC-MS

（发表于《中药材》，院选中文核心）

Study on Complete Mitochondrial Genome of Oula Sheep (*Ovis aries*)

Xian GUO, Jianbin LIU, Yufeng ZENG, Xuezhi DING, Pengjia BAO, Ping YAN*, Jie PEI*

(Lanzhou Institute of Husbandry and Pharmaceutical Sciences, Chinese Academy of Agricultural Sciences, Lanzhou 730050, China)

Abstract: Mitochondrial genome has been widely used in species identification and gene conservation. In the present study, the complete mitochondrial genome of Oula sheep (*Ovis aries*) was determined using next-generation sequencing. This genome was 16 618 bp (NCBI accession number: KU575248) and contained 13 protein coding genes, 22 transfer RNA genes, two ribosomal RNA genes, and a typical control region. The overall nucleotide composition was 33.7% A, 27.4% T, 25.8% C, and 13.1% G, with a total A+T content of 61.1%. The phylogenetic analysis of selected sheep breeds showed that Oula sheep were clustered within branch A and originated from approximately 6 ka. This mitochondrial genome will provide valuable information for molecular genetic research of Oula sheep.

Key words: Oula sheep; *Ovis aries*; Mitochondrial genom

(发表于《Agricultural Science & Technology》)

Study on the Complete Mitochondrial Genome of Qaidam Cattle (*Bos taurus*)

Xian GUO[1], Lin XIONG[1], Jianbin LIU[1], Shirong QI[2], Zhaxi Zhuoma[3], Gabu Zengcuo[2], Jie PEI[1], Pengjia BAO[1], Ping YAN[1*], Xuezhi DING[1*]

(1. Key Laboratory of Yak Breeding Engineering of Gansu Province, Lanzhou Institute of Husbandry and Pharmaceutical Sciences, Chinese Academy of Agricultural Sciences, Lanzhou 730050, China; 2. Animal Husbandry Station in Zongjia Town of Dulan County, Dulan 816100, China; 3. Haixi Animal Disease Prevention and Control Center of Qinghai Province, Delingha 817000, China)

Abstract: Qaidam cattle (*Bos taurus*) is an important endemic breed in Northwest China, which is mainly distributed in the north west of Qinghai–Tibet Plateau. It has strong adaptive ability to plateau and swamp conditions, such as cold tolerance and insect resistance. In this study, the first complete mitochondrial genome of Qaidam cattle was reported. The circular double-stranded genome is 16,340bp in size, and contains 13 protein-coding genes, 22 transfer RNA genes, two ribosomal RNA genes, and a D-loop region. The overall nucleotide composition is 33.4% A, 27.2% T, 26.0% C and 13.4% G, with a total A+T content of 60.6%. The gene order and composition are similar to those of other B.taurus breeds. Molecular phylogenetic analysis indicated that Qaidam cattle was split as an independent clade and nested within Asian cattle breeds.

Key words: Qaidam cattle; *Bos taurus*; Mitochondrial genome; Qinghai–Tibet Plateau

(发表于《Agricultural Science & Technology》)

Research on Safety of Manhuning Oral Liquid to Chicken

Guibo WANG[1], Nini GUO[2], Jiasheng XIE[1], Xiaolong ZHAO[3], Chaoying LUO[1], Yongjiang LUO[1], Jifang ZHENG[1], Ruihua XIN[1], Jinyu Li[1*]

(1. Lanzhou Institute of Animal Science and Veterinary Medicine, Chinese Academy of Agricultural Sciences, Lanzhou 730050, China; 2. Livestock and Poultry Breeding Center of Hubei Province, Wuhan 430070, China; 3. Xi'an Caotan Animal Husbandry Co., Ltd., Huayin 714200, China)

Abstract: This study aimed to evaluate the safety of Manhuning oral liquid and to provide safety basis for clinical treatment. A total of 120 chickens of 14 days old were selected. After one week of adaptive feeding, they were evenly and randomly divided into four groups: one time the therapeutic dose (T1), three times the therapeutic dose (T2), five times the therapeutic dose (T3) and control (CK). In different groups, the chickens were fed with Manhuning oral liquid according to the set doses for one week. At the beginning and at the end of the experiment, total 10 chickens were selected randomly from each group, respectively. They were weighed, and their blood was sampled for determination of routine blood indexes and examination of liver and kidney function. Organ indexes were determined, and pathological examination was conducted for the heart, liver, spleen, lung and kidney of the chickens. The results showed that the body mass gain and organ indexes differed insignificantly among different groups ($P>0.05$). No significant differences were found in the routine blood indexes, serum ALT activity, albumin content, creatinine content and urea content among different groups before and after the experiment ($P>0.05$). After the administration, the heart, liver, spleen, lung and kidney of the chickens showed no obvious abnormalities. In short, Manhuning oral liquid had little impact on the physiological and biochemical indexes and had no toxic effects on the heart, liver, spleen, lung and kidney of chickens, and it is a low-toxicity safe veterinary drug preparation.

Key words: Manhuning oral liquid; Veterinary medicine; Respiratory disease; Chicken; Safety evaluation

"慢呼宁"口服液对鸡安全性的试验研究

王贵波[1]，郭妮妮[2]，谢家声[1]，赵小龙[3]，罗超应[1]，
罗永江[1]，郑继方[1]，辛蕊华[1]，李锦宇[1*]

（1. 中国农业科学院兰州畜牧与兽药研究所，兰州　730050；2. 湖北省畜禽育种中心，武汉　430070；3. 西安草滩牧业有限公司，陕西华阴　714200）

摘要：为了在对"慢呼宁"口服液的安全性进行评价、为临床用药提供安全依据，选取 120 羽 14 日龄的三黄肉鸡，适应性饲喂 1 周后，随机分为治疗剂量的 1 倍剂量组、3 倍剂量组、5 倍剂量组和对照组，每组 30 羽。试验时分别按照每组试验鸡的平均体质量给予不同剂量的药物，连续给药 1 周。在试验前和试验结束时，各组随机选取 10 只进行称质量和采血，测定体质量变化、血常规和肝肾功能指标，测定脏器指数，并对鸡的心、肝、脾、肺和肾脏进行病理检查。结果表明，用药后各组鸡的体质量增加量及脏器指数组间差异不显著（$P>0.05$）。血常规指标则在试验前后组间差异亦不显著（$P>0.05$）；血清 ALT 活性、白蛋白、肌酐和尿素含量组间差异不显著（$P>0.05$）；在用药后，各组鸡的心、肝、脾、肺和肾脏组织变化均无显著差异和异常。结果表明，"慢呼宁"口服液对鸡生理生化指标的影响小，病理剖检未见对心、肝、脾、肺和肾脏产生毒性作用，是一种低毒安全的中兽药制剂。

关键词："慢呼宁"口服液；中兽药；呼吸道疾病；鸡；安全性评价

（发表于《Agricultural Science & Technology》）

Study on the Complete Mitochondrial Genome of Qaidam Cattle (*Bos taurus*)

Xian GUO[1], Lin XIONG[1], Jianbin LIU[1], Shirong QI[2], Zhaxi Zhuoma[3], Gabu Zengcuo[2], Jie PEI[1], Pengjia BAO[1], Ping YAN[1*], Xuezhi DING[1*]

(1. Key Laboratory of Yak Breeding Engineering of Gansu Province, Lanzhou Institute of Husbandry and Pharmaceutical Sciences, Chinese Academy of Agricultural Sciences, Lanzhou 730050, China; 2. Animal Husbandry Station in Zongjia Town of Dulan County, Dulan 816100, China; 3. Haixi Animal Disease Prevention and Control Center of Qinghai Province, Delingha 817000, China)

Abstract: Qaidam cattle (*Bos taurus*) is an important endemic breed in Northwest China, which is mainly distributed in the north west of Qinghai-Tibet Plateau. It has strong adaptive ability to plateau and swamp conditions, such as cold tolerance and insect resistance. In this study, the first complete mitochondrial genome of Qaidam cattle was reported. The circular double-stranded genome is 16,340 bp in size, and contains 13 protein-coding genes, 22 transfer RNA genes, two ribosomal RNA genes, and a D-loop region. The overall nucleotide composition is 33.4% A, 27.2% T, 26.0% C and 13.4% G, with a total A+T content of 60.6%. The gene order and composition are similar to those of other B.taurus breeds. Molecular phylogenetic analysis indicated that Qaidam cattle was split as an independent clade and nested within Asian cattle breeds.

Key words: Qaidam cattle; *Bos taurus*; Mitochondrial genome; Qinghai-Tibet Plateau

(发表于《Agricultural Science & Technology》)

Test of Feeding Dairy Cows with Chinese Herbal Additive for Improving Production and Quality of Milk

Xiaoping SUN[1*], Jianbin LIU[1], Simin LI[2]

(1. Lanzhou Institute of Husbandry and Pharmaceutical Science of CAAS, Lanzhou 730050, China; 2. Baiyin Livestock Center, Baiyin 730040, China)

Abstract: At Baiyin dairy farm, the Chinese herbal additive was added into feed which was then fed to dairy cows from August to October, 2014, and the changes in milk production and quality were observed.The test showed that the additive added into the feed had obvious milk-increasing effect, the milk production was improved by 12.67%-17.26%, and the milk quality was improved.The additive has the effects of preventing miscarriage, expelling parasite and preventing diseases.The nutritional components in the feed additive were determined, and the results showed that the contents of protein, crude fat, Ca and P in the additive were 12.29%, 2.66%, 1.8% and 0.22%, respectively.

Key words: Chinese herb; Feed additive; Milk production and quality; Test

中草药饲料添加剂饲喂奶牛提高牛奶产量及质量试验

孙晓萍[1*]，刘建斌[1]，李思敏[2]

（1. 中国农业科学院兰州畜牧与兽药研究所，甘肃兰州　730050；
2. 白银畜牧中心，甘肃白银　730040）

摘要：2014 年 8 月至 10 月在白银奶牛场用我们研制加工的中草药饲料添加剂加入奶牛饲料中饲喂奶牛，观察奶牛产奶量及质量的变化。通过一个多月的试验表明：该饲料添加剂饲喂奶牛后有明显的增加产乳量效果，产奶量提高 12.67%~17.26%，同时还改善牛奶的品质。使用该饲料添加剂对奶牛还具有保胎、驱虫、防病等作用，对饲料还可防止其霉变。对中草药饲料添加剂进行营养成分测定，其中蛋白质含量为 12.29%、粗脂肪为 2.66%、钙含量为 1.8%、磷含量为 0.22%。

关键词：中草药；饲料添加剂；牛奶产量及质量；试验

（发表于《Agricultural Science & Technology》）

Whole-Genome Sequence of *Streptococcus parauberis* Strain SP-llh, Isolated from Cows with Mastitis in Western China

Longhai LIU, Feng YANG, Xinpu LI, Jinyin LUO, Zhe ZHANG, Hongsheng LI

(Lanzhou Institute of Husbandry and Pharmaceutical Sciences of Chinese Academy of Agricultural Sciences, Lanzhou, Gansu, 730040 China)

Abstract: *Streptococcus parauberis* strain SP-llh was isolated from cows with mastitis in western China in 2015. The 2522235-bp genome sequence consists of 46 large contigs in 14 scaffolds and contains 2620 predicted protein-coding genes, with a G+C content of 35.3%.

(发表于《Genome Announcements》)

Analysis of Superoxide Dismutase (SOD) Activity of Tianzhu White Yak Milk in Gansu

Bin XI[1,2,3], Ping YAN[1], Wei-hong LI[1,2,3], et al

(1. Lanzhou Institute of Animal Husbandry and Pharmaceutical Science, Chinese Academy of Agricultural Sciences, Lanzhou, Gansu 730050; 2. Animal Product Quality and Safety Risk Assessment Laboratory, Ministry of Agriculture, Lanzhou, Gansu 730050; 3. Lanzhou Quality Supervision, Inspection and Testing Center for Animal Fur and Products, Ministry of Agriculture, Lanzhou, Gansu 730050)

Abstract: [Objective] To analyze the SOD activity of Tianzhu white yak in Gansu. [Method] A total of 9 yak milk samples were collected from 3 yak farms in Tianzhu County, the SOD activity was analyzed. [Result] The SOD activity of Tianzhu white yak milk was about 202U/mL.The SOD activity was different among yak milk samples in Tianzhu County, but the difference was not notable.Birth ank had no significant influence on SOD activity of yak milk. [Conclusion] The study can provide theoretical basis for further development and utilization of white yak milk.

Key words: Tianzhu white yak milk; SOD; Activity; Analysis

甘肃天祝白牦牛乳中超氧化物歧化酶活性分析

席　斌[1,2,3]，阎　萍[1]，李维红[1,2,3]，郭天芬[1,2,3]，熊　琳[1,2,3]

（1. 中国农业科学院兰州畜牧与兽药研究所，兰州　730050；2. 农业部畜产品质量安全风险评估实验室，兰州　730050；3. 农业部动物毛皮及制品质量监督检验测试中心（兰州），兰州　730050）

摘要：[目的] 分析天祝白牦牛乳中超氧化物歧化酶（SOD）活性。[方法] 对采白天祝白牦牛 3 个牧区共 9 头份牦牛乳 SOD 活性进行分析。[结果] 天祝白牦牛乳中 SOD 活性约为 202U/mL。3 个牧区的天祝白牦牛乳中 SOD 活性有所不同，但是差异不显著；SOD 活性随着胎次的变化无显著差异，即胎次对牦牛乳 SOD 活性影响不大。[结论] 研究可为白牦牛乳的进一步开发利用奠定理论基础。

关键词：天祝白牦牛乳；超氧化物歧化酶；活性；分析

（发表于《安徽农业科学》）

甘肃天祝白牦牛与甘南牦牛乳中水解酶活性的比较

席　斌[1,2,3]，高雅琴[1,2,3*]，郭天芬[1,2,3]，李维红[1,2,3]

（1. 中国农业科学院兰州畜牧与兽药研究所，兰州　730050；
2. 农业部畜产品质量安全风险评估实验室，兰州　730050；
3. 农业部动物毛皮及制品质量监督检验测试中心（兰州），兰州　730050）

摘要：［目的］对甘肃天祝白牦牛与甘南牦牛乳中水解酶活性进行比较。［方法］测定并比较甘肃天祝白牦牛和甘南牦牛乳中水解酶（ACP、AKP、AMS）的活性，并比较同一品种不同胎次牦牛乳中水解酶的活性。［结果］天祝白牦牛乳中ACP活性显著高于甘南牦牛（$P<0.05$）；天祝白牦牛乳中AKP活性高于甘南牦牛，但差异不显著（$P>0.05$）；天祝白牦牛乳中AMS活性低于甘南牦牛，但差异不显著（$P>0.05$）。不同胎次牦牛乳中3种水解酶活性均无显著差异（$P>0.05$），说明胎次对牦牛乳中水解酶活性的影响不显著。［结论］该研究结果可为牦牛乳的进一步开发利用提供参考。

关键词：牦牛乳；水解酶；活性；比较

Comparison on the Activity of Hydrolase in the Milk of Tianzhu White Yak and Gannan Yak of Gansu

Bin XI[1,2,3], Ya-qin GAO[1,2,3], Tian-fen GUO[1,2,3], et al

(1. Lanzhou Institute of Animal Husbandry and Pharmaceutical Science, Chinese Academy of Agricultural Sciences, Lanzhou, Gansu 730050;
2. Quality Safety Risk Assessment of Animal Products (Lanzhou), Ministry of Agriculture, Lanzhou, Gansu 730050;
3. Quality Supervising, Inspecting and Testing Center for Animal Fiber, Fur, Leather and Products (Lanzhou), Ministry of Agriculture, Lanzhou, Gansu 730050)

Abstract: [Objective] To compare the activity of hydrolase in the milk of Tianzhu white yak and Gannan yak of Gansu. [Method] The activities of hydrolase (ACP, AKP, AMS) in the milk of Tianzhu white yak and Gannan yak milk of Gansu were determined and compared. And the activities of hydrolase in the milk among different parity of yak were compared. [Result]

ACP activity in the milk of Tianzhu white yak was significantly higher than that of Gannan yak ($P<0.05$). AKP activity in the milk of Tianzhu white yak was higher than that of Gannan yak, but there was no significant difference ($P>0.05$). AMS activity in the milk of Tianzhu white yak was lower than that of Gannan yak, but there was no significant difference ($P>0.05$). The activity of hydrolases in the milk of different parity of yak had no significant difference ($P>0.05$), which inclicated the effects of different parity on the activity of hydrolases in the yak milk were not significant. [Conclusion] The research can provide reference for further development and utilization of yak milk.

Key words: Yak milk; Hydrolase; Activity; Comparison

（发表于《安徽农业科学》）

防治奶牛子宫内膜炎的方法体会

李世宏，刘希望，杨亚军，秦　哲，焦增华，孔晓军，李剑勇

（中国农业科学院兰州畜牧与兽药研究所/农业部兽用药物创制重点实验室/
甘肃省新兽药工程重点实验室，兰州　730050）

摘要：奶牛子宫内膜炎是奶牛三大常见疾病之一，给奶牛养殖业造成巨大的经济损失。文章根据防治奶牛产科疾病的经历与体会，现将防治奶牛子宫内膜炎的方法进行叙述，以供奶牛养殖者进行参考。

关键词：奶牛；子宫内膜炎；防治

（发表于《甘肃畜牧兽医》）

论高山美利奴羊新品种的价值和意义

杨博辉

（中国农业科学院兰州畜牧与兽药研究所，兰州　730050）

摘要：2016 年，高山美利奴羊被农业部列为全国农业主导品种；中国农业科学院、国家绒毛用羊产业技术体系将其作为“十二五”标志性重大成果。高山美利奴羊新品种的培育在推进细毛羊产业转型升级、满足高档精纺羊毛的需求、维护生态平衡等方面均具有巨大作用和重要意义。

关键词：高山美利奴羊；新品种；价值；意义

（发表于《甘肃畜牧兽医》）

基于地表温度-植被指数特征空间的土壤干旱监测

李润林，董鹏程，王　瑜，汪晓斌

（农业部兰州黄土高原生态环境重点野外科学观测实验站/
中国农业科学院兰州畜牧与兽药研究所，兰州　730050）

摘要：以张掖市甘州区绿洲为研究区，采用 5 期遥感影像（2011—2015 年），运用 ENVI 5.2 提取归一化植被指数（NDVI）、改进型土壤调节植被指数（MSAVI）和地表温度（Ts），构建 Ts-NDVI 和 Ts-MSAVI 特征空间，对比分析两种特征空间。结果表明，Ts-MSAVI 特征空间的干边和湿边斜率均小于 0，这与前人的研究干边斜率是负值，湿边斜率是正值的结论有所不同。Ts-NDVI 和 Ts-MSAVI 这两种特征空间具有相同的趋势，其中 2012 年、2013 年、2014 年这 3 年两种特征空间系数产较高，其余 2 年系数 r^2 较低。整体而言，Ts-NDVI 特征空间的干湿边系数相比 Ts-MSAVI 特征空间的干湿边系数要高，稳定性好。从 TVDI 旱情等级分布图上可以得出 2012 年的受旱面积最大，干旱和重旱面积占总面积的 70.39%，2013 年干旱情况最严重，重干旱面积为 1 611.972km^2，重旱面积占到总面积的 43.5%，2014 年干旱程度开始缓解，轻旱、干旱和重旱面积开始降低，湿润和正常面积开始增加，2015 年干旱程度得到全面缓解，湿润和正常面积占到总面积的 21.9%，但是干旱和重旱面积比重依然很大，说明张掖市甘州区绿洲旱情依然很严峻。

关键词：干旱；归一化植被指数（TVDI）；改进型土壤调节植被指数（MSAVI）；地表温度；张掖市甘州区

Soil Drought Monitoring Based on Land Surface Temperature-Vegetation Index Characteristic Space

Run-lin LI[1,2], Peng-cheng DONG[1,2], Yu WANG[1,2], Xiao-bin WANG[1,2]

(1. Lanzhou Scientific Observation and Experiment Field Station of Ministry of Agriculture for Ecological System in Loess Plateau Areas, Lanzhou 730050, China;
2. Lanzhou Institute of Husbandry and Pharmaceutical Sciences, Chinese Academy of Agricultural Sciences, Lanzhou 730050, China)

Abstract: Selecting oasis of Ganzhou district as the study area in Zhangye city, using ENVI 5.2 software to extact normalized difference vegetation (NDVI), modified soil adjusted vegetation index (MSAVI) and temperature of surface (Ts), Ts-NDVI feature space and Ts-MSAVI feature space were built.The two feature spaces were compared and analyzed.The results showed that the slope of dry-edge and wet-edge of Ts-MSAVI feature space was less than 0, which was not consistent with the previous research.The previous research thinked the dry-edge slope was negative and the wet edge slope was positive.The feature space of Ts-NDVI and Ts-MSAVI had the same trend.The r^2 coefficient of two feature spaces was higher in the three years of 2012, 2013 and 2013, and the r^2 coefficient of the other two years was lower.On the whole, the wet-edge coefficient of the Ts-NDVI feature space was higher than that of the Ts-MSAVI feature space, and the stability was good. From the TVDI drought severity map, it could be concluded that the drought area was the largest in 2012, the drought and heavy drought area accounted for 70. 39% of the total area.In 2013, the drought was the most serious, the area of heavy drought was 1 611. 972 km^2, the area of heavy drought occupied 43. 5% of the total area and the degree of drought in 2014 was lightened. And heavy drought area began to decrease, wet and normal area began to increase.In 2015, the degree of drought had been fully relieved, wet and normal area accounted for 21. 9% of the total area. But the proportion of drought and heavy drought area was still great, indicating that the drought of Ganzhou district oasis in Zhangye city was still very serious.

Key words: Drought; Normalized differential vegetation index (TVDI); Modified soil adjusted vegetation index (MSAVI); Temperature of surface; Ganzhou district of Zhangye city

（发表于《湖北农业科学》）

肉品生产中禁用药物西马特罗研究进展

熊　琳*，阎　萍，高雅琴，李维红，杨晓玲

[中国农业科学院兰州畜牧与兽药研究所/农业部畜产品
质量安全风险评估实验室（兰州），兰州　730050]

摘要： 随着肉品中毒事件的不断出现，肉品安全问题越来越受到人们的重视。西马特罗作为一种违禁药物，在畜牧业中的非法使用，会导致严重的后果，严重会导致人员伤亡。因此，加强肉品生产过程中违禁药物西马特罗非法使用的监控，具有十分重要的现实意义。本文对西马特罗的研究进展做了总结综述，主要包括西马特罗作用和危害，以及检测方法研究进展，并提出了防控肉品中西马特罗的措施，以期为畜牧业肉品生产过程中加强西马特罗非法使用的监测提供技术支持，并为国家相关执法部门的监控提供理论依据和指导，保证广大消费者肉品食用安全。

关键词： 西马特罗；肉品；食品安全；检测；危害控制

Review of Forbidden Drug Cimaterol in Meat Production

Lin XIONG[1,2]*, Ping YAN[1,2], Ya-Qin GAO[1,2],
Wei-Hong Li[1,2], Xiao-Ling YANG[1,2]

[1. Lanzhou Institute of Husbandry and Pharmaceutical Sciences,
Chinese Academy of Agricultural Sciences, Lanzhou 730050, China;
2. Laboratory of Quality & Safety Risk Assessment for Livestock Product
(Lanzhou), Ministry of Agriculture, Lanzhou 730050, China]

Abstract: With the growing events of meat poisoning, people pay more and more attention to the meat safety problem.The illegal use of cimaterol as a kind of illicit drugs in animal husbandry can lead to serious consequences and even loss of life, so it is very important to strengthen the monitoring of cimaterol. This paper summarized the research progress of cimaterol, including the effects and hazards of cimaterol and the research progress of its detection

methods, and put forward the measures for prevention and control of cimaterol in meat production. This paper can provide technical support for the monitoring of the illegal use of cimaterol in the process of livestock and theoretical basis and guidance for the monitoring of the relevant national law enforcement departments, and ensure the meat safety for consumers.
Key words: Cimaterol; Meat; Food safety; Detection; Hazard control

（发表于《食品安全质量检测学报》）

天祝白牦牛乳与甘南牦牛乳理化性质比较

席　斌，高雅琴，郭天芬，李维红，熊　琳，杨晓玲

（中国农业科学院兰州畜牧与兽药研究所/农业部畜产品质量安全风险评估实验室/农业部动物毛皮及制品质量监督检验测试中心，兰州　730050）

摘要：对天祝白牦牛（*Bos grunniens* L.cv.*Tianzhu white yak*）与甘南牦牛（*B.grunniens* cv.*Gannan yak*）2个牦牛品种的牛乳理化性质进行了比较分析。结果表明，天祝白牦牛产乳量高于甘南牦牛；甘南牦牛乳中的蛋白质、脂肪、全脂固体含量高于天祝白牦牛乳，天祝白牦牛乳中非乳脂固体、乳糖含量和密度、酸度高于甘南牦牛乳，冰点差别不大；胎次对牦牛乳脂肪、非乳脂固体含量和酸度的影响较大，对蛋白质、全脂固体、乳糖含量和密度、冰点等影响不大。

关键词：牦牛（*Bos grunniens* L.）；牛乳；品种；胎次；理化性质

Comparative Study on Physico-Chenucal Properties of *Bos grunniens* cv.*Tianzhu white yak* Milk and *B.grunniens* cv.*Gannan yak* Milk

Bin XI, Ya-qin GAO, Tian-fen GUO, Wei-hong LI, Lin XIONG, Xiao-ling YANG

(Lanzhou Institute of Animal Husbandry and Pharmaceutical Science, Chinese Academy of Agricultural Sciences/Quality Supervision, Inspection and Testing Center of Agriculture for Animal Fiber, Fur, Leather and Products/Laboratory of the Inistry of Agriculture for Animal Product Safety Risk Assessment, Lanzhou 730050, China)

Abstract: Comparative study on physico-chemical properties of two different yak (*Bos grunniens* L.) breeds of B.grunniens cv.Tianzhu white yak milk and Gannan yak milk was conducted.The results showed that the milk production of *B.grunniens* cv.*Tianzhu white yak* is higher than that of *B.grunniens* cv.*Gannan yak*; *B.grunniens* cv.*Tianzhu white yak* milk was higher in protein, fat, TS, and acidity than *B.grunniens* cv.*Gannan yak milk*, while lower in SNF, lactose content and density; there is no obvious difference on FPD between the two yak milk; parity had great influence on fat, SNF, acidity of yak milk, while had no influence on protein, lactose content, density and FPD.

Key words: Yak (*Bos grunniens* L.); Milk; breeds; Parity; Physico-chemical properties

（发表于《湖北农业科学》）

银翘蓝芩口服液微生物限度检查方法的建立及应用

许春燕，刘希望，杨亚军，孔晓军，秦　哲，李世宏，李剑勇*

（中国农业科学院兰州畜牧与兽药研究所/农业部兽用药物创制重点实验室/
甘肃省新兽药工程重点实验室，甘肃兰州　730050）

摘要：［目的］：建立银翘蓝芩口服液微生物限度的检查方法。［方法］：按照《中国兽药典》2010 年版一部微生物限度检查法进行银翘蓝芩口服液微生物限度检查。结果：银翘蓝芩口服液无明显的抑菌活性，细菌、霉菌及酵母菌计数采用常规法时其回收率在 93.7%~106.8%，大于药典要求的 70%。控制菌检查选用大肠埃希菌，培养基促生长能力、抑制能力、指示能力和控制菌检查方法均符合中国兽药典要求。采用上述建立的方法，3 批银翘蓝芩口服液中细菌、霉菌及酵母菌检查结果均小于 1cfu/mL，大肠埃希菌未检出。［结论］：本研究建立的银翘蓝芩口服液微生物限度检查法简便易行，结果准确，可用于该产品的质量控制。

关键词：银翘蓝芩口服液；方法学验证；微生物限度检查；控制菌

Establishment and Application of Microbial Limit Test Method for Yinqiaolanqin Oral Liquid

Chunyan XU, Xiwang LIU, Yajun YANG, Xiaojun KONG,
Zhe QIN, Shihong LI, Jianyong LI*

(Key Laboratory of New Animal Drug Project of Gansu Province; Key Laboratory of Veterinary Pharmaceutical Development, Ministry of Agriculture; Lanzhou Institute of Husbandry and Pharmaceutical Sciences of CAAS, Lanzhou 730050, China)

Abstract: [Objective]: Establish Yinqiaolanqin oral liquid microbial limit inspection method. [Methods]: According to microbial limit test method in Chinese Veterinary Pharmacopoeia (2010 edition, Version Ⅰ), the microbial limit test of Yinqiaolanqin oral liquid was performed. [Results]: Yinqiaolanqin oral liquid had no significant antibacterial activity, when the count of bacteria, molds and yeasts was investigated by conventional method, the recovery rates were between 93.7% and 106.8%, which were greater than 70%.Escherichia coli was

selected as control bacteria.The growth-promoting ability, inhibition ability, indicating ability of culture medium and the test method of control bacteria were all met the requirements of Chinese Veterinary Medicine. Using the method mentioned above, the results of bacteria, molds and yeasts in three batches were less than 1 cfu/mL, at the same time Escherichia coli could not be detected. [Conclusion]: The microbial limit test method of Yinqiaolanqin oral liquid was simple and accuracy, which was available for quality control of Yinqiaolanqin oral liquid.
Key words: Yinqiaolanqin oral liquid; Verification test; Microbial limit test; Control bacteria

（发表于《亚太传统医药》）

PCR 诊断技术用于奶牛乳房炎病原菌快速检测的研究进展

王　丹[1]，杨　峰[1]，李新圃[1]，罗金印[1]，张　哲[1]，刘龙海[1]，
张亚茹[1]，张莉莉[2]，李宏胜[1]

（1. 中国农业科学院兰州畜牧与兽药研究所/农业部兽用药物创制重点实验室/
甘肃省新兽药工程重点实验室，兰州　730050；
2. 兰州大学公共卫生学院，兰州　730000）

摘要：奶牛乳房炎是一种严重影响奶牛业发展的常见多发病，不仅给乳品业造成巨大的经济损失，而且还影响乳品质量，危及人类健康。目前，对于奶牛乳房炎的治疗大部分奶牛场基本都是在不明确病原菌的情况下滥用抗生素，造成病原菌耐药性增强，治疗效果下降。为了有效治疗奶牛乳房炎，对奶牛乳房炎病原菌的早期快速诊断至关重要。由于传统的乳房炎病原菌的检测方法存在操作烦琐、耗时长、灵敏度低及无法检测抗生素乳样中病原菌等缺陷，因此，临床上急需一种快速、灵敏、简便的检测方法。作者对目前用于奶牛乳房炎病原菌诊断的普通 PCR 法、多重 PCR 法、实时荧光定量 PCR 法、巢氏 PCR 诊断法以及 PCR 检测试剂盒等进行了综述，并分析了这些方法的优缺点，同时对未来奶牛乳房炎病原菌诊断的研究方向和前景进行了展望，以期为进一步开发快速、特异、敏感的奶牛乳房炎病原菌诊断技术提供理论参考。

关键词：PCR；奶牛；乳房炎；病原菌

Research Progress in PCR Diagnosis Technology of Bovine Mastitis Pathogens

Dan WANG，Feng YANG，Hongsheng LI，et al

(Lanzhou Institute of Animal Husbandry and Pharmaceutical Sciences of Chinese Academy of Agricultural Sciences；Key Laboratory of Veterinary Pharmaceutical Discovery，Ministry of Agriculture；Key Laboratory of New Animal Drug Project of Gansu Province，Lanzhou 730050，China)

Abstract：Dairy cow mastitis is a common disease affecting the development of dairy industry. It causes serious economic loss to dairy farming，and affects the yield and quality of milk，endangers human health. At present, the antibiotics were abused in the treatments of dairy cow mastitis with unclearing pathogens caused, resulting in increasement of pathogen resistance and decreasement of the treatment effects. There are a few shortcomings in traditional detection method for mastitis pathogen dignosis, such as complicated operation, time-consuming, low sensitivity and no effect on pathogens in milk samples of antibiotics, so that a rapid, sensitive and simple detection method is urgently needed. The rapid diagnosis of mastitis are crucial for the effectively treatment of the dairy cow mastitis. In order to provide theoretical basis for developing rapid, specific and sensitive diagnosis of dairy cow mastitis pathogens, the author summarized the traditional PCR, multiplex PCR, real time PCR, nested PCR and detection kit etc, and analyzed the advantages and disadvantages of these methods.

Key words: PCR; Cow; Mastitis; Pathogenic bacteria

（发表于《中国草食动物科学》）

大通牦牛体尺与体重性状的多元线性回归与通径分析

裴　杰[1,2]，褚　敏[1,2]，包鹏甲[1,2]，扎西卓玛[3]，骆正杰[4]，武甫德[4]，梁春年[1,2]，丁学智[1,2]，王宏博[1,2]，吴晓云[1,2]，阎　萍[1,2]，郭　宪[1,2]

（1. 中国农业科学院兰州畜牧与兽药研究，兰州　730050；
2. 甘肃省牦牛繁育工程重点实验室，兰州　730050；
3. 青海省海西州动物疫病预防控制中心，德令哈　817099；
4. 青海省大通种牛场，西宁　810100）

摘要：试验旨在通过大通牦牛体尺性状对其体重做较为准确的估计，剖分牦牛各体尺性状对体重的影响。随机选取了 6～12 月龄的 88 头大通牦牛为研究对象，其中公牛 48 头，母牛 40 头。测量各牦牛的体长、体高、胸围 3 个体尺指标，并称量其体重；采用逐步线性回归的方法建立大通牦牛体重与体尺的多元线性回归方程；利用通径分析方法计算各体尺性状对体重的直接作用和间接作用。结果表明：各性状之间的表型相关均达到了极显著水平（$P<0.01$）；大通牦牛公牛和母牛的多元线性回归方程分别为 $Y=-194.708+0.766X_1+0.782X_2+1.229X_3$ 和 $y=-118.056+0.910X_1+1.106X_3$，其中 y 为体重（kg），$X_1$ 为体高（cm），X_2 为体长（cm），X_3 为胸围（cm），体重估计值与实际观察值差异不显著；胸围对体重的直接作用大于通过其他性状影响体重的间接作用，且胸围的间接作用在体高和体长对体重影响中做了主要贡献。通过研究结果可知，研究所得多元线性方程可以应用于大通牦牛良种选育实践，大通牦牛体重的主要影响因素来自其胸围指标。

关键词：大通牦牛；体尺；体重；多元线性回归；通径分析

Multiple Linear Regression and Path Analysis between Body Measurement and Weight on Datong Yak

Jie PEI[1,2], Ping YAN[1,2], Xian GUO[1,2], et al

(1. Lanzhou Institute of Husbandry and Pharmaceutical Sciences, Chinese Academy of Agricultural Sciences, Lanzhou 730050, China;
2. Key Laboratory of Yak Breeding Project in Gansu Province, Lanzhou 730050, China)

Abstract: The aim of this study was to estimate weight of Datong yak through its body measurements, and subdivided effects of body measurement on the weight. In the experiment, 88 Datong yaks at the age of 6 to 12 months, 48 male yaks and 40 female yaks, were choosed randomly as objects of the study. Standing height, body length, chest circumference and weight of the yaks were measured. Multiple linear regression equations of body measurements and weight was established by the means of stepwise multiple linear regression method. The effects of body measurements on weight were split into direct effects and indirect effects. The results indicated that correlation coefficients between phenotypic traits reached significant difference level ($P<0.01$). Multiple linear regression equations of male and female Datong yak were $Y=-194.708+0.766X_1+0.782X_2+1.229X_3$ and $Y=-118.056+0.910X_1+1.106X_3$, and " Y"," X_1"," X_2" and " X_3" were weght, standing height, body length and chest circumference respectively. There was no significant difference between the estimated values and the observed values of weight. The direct effect of chest circumference on the weight was greater than the indirect effects of other body measurement on the weight through chest circumference, and the indirect effects of chest circumference contributed mainly to effects of standing height and body length on weight, suggesting that. The results suggested that the multiple linear regression equations obtained could be used for selective breeding practice, and the major influence factors of Datong yak weight was from chest circumference.

Key words: Datong yak; Body measurement; Weight; multiple linear regression; Path analysis

（发表于《中国草食动物科学》）

甘肃牦牛乳的研究进展

席　斌[1,2,3]，高雅琴[1,2,3]，郭天芬[1,2,3]，李维红[1,2,3]，熊　琳[1,2,3]

（1. 中国农业科学院兰州畜牧与兽药研究所，兰州　730050；
2. 农业部畜产品质量安全风险评估实验室，兰州　730050；
3. 农业部动物毛皮及制品质量监督检验测试中心，兰州　730050）

摘要：牦牛乳是一种天然的浓缩乳，其乳中的脂肪、蛋白质、乳糖、干物质等含量均高于其他牛乳，是加工奶油系列制品的最优原料乳之一。文章对甘肃天祝白牦牛和甘南牦牛乳的物理特性和化学特性的研究现状进行了阐述，并对其进行了展望，以期为甘肃牦牛乳及牦牛乳业的进一步开发利用提供理论参考。

关键词：甘肃；牦牛乳

Research Progress of Yak Milk in Gansu

Bin XI[1,2,3], Yaqin GAO[1,2,3], Tianfen GUO[1,2,3], et al

(1. Lanzhou Institute of Animal Husbandry and Pharmaceutical Science, Chinese Academy of Agricultural Sciences, Lanzhou 730050, China;
2. Lanzhou Laboratory of the Ministry of Agriculture for Animal Product Safety Risk Assessment, Lanzhou 730050, China;
3. Lanzhou Quality Supervision, Inspection and Testing Center of the Ministry of Agriculture for Animal Fiber, Fur, Leather and Products, Lanzhou 730050, China)

Abstract: Yak milk is a kind of natural concentrated milk, and it is one of the most optimal raw milk for processing cream series producucts because its fat, protein, lactose, dry matter, etc are higher than other cattle.This article had discussed the two varieties yak milk´s physical properties, chemical properties, the situation of research and development of yak milk products, and the outlook for the development of yak milk in Gansu, which would provide a theoretical basis for further development and utilization of yak milk in Gansu.

Key words: Gansu; Yak milk

（发表于《中国草食动物科学》）

甘肃永登灌区四翅滨藜根际土壤盐分的变化

张怀山[1]，代立兰[2]，杨世柱[1]，王　平[2]，王国宇[2]

（1. 中国农业科学院兰州畜牧与兽药研究所，农业部兰州黄土高原生态环境重点野外科学观测试验站，兰州　730050；
2. 甘肃省兰州市农业科技研究推广中心，兰州　730000）

摘要：为研究四翅滨藜的耐盐性和对盐渍化土壤的改良作用，选择甘肃永登灌溉农业区次生盐渍化土壤 4 个试验点，测定不同种植年限四翅滨藜根际土壤的盐离子浓度、pH 值和电导率（EC 值）变化。结果表明：1 年期四翅滨藜根际土壤盐离子浓度略有下降，2 年期和 3 年期四翅滨藜根际土壤盐离子浓度明显下降；根际土壤 pH 值逐年下降；四翅滨藜叶片电导率逐年增加；0~80cm 土层电导率逐年下降。说明四翅滨藜对于根际土壤的盐离子具有吸收分化作用，可以明显改善土壤的生态环境（pH 值、EC 值），是一种绿色安全的盐渍化土壤生物改良植物。

关键词：四翅滨藜；灌溉农业区；盐渍化土壤；盐分变化

Changes of Rhizosphere Soil Salinity Around *Atriplex canescens* in Yongdeng Irrigation Area of Gansu Province

Huaishan ZHANG[1]，Lilan DAI[2]，Shizhu YANG[1]，et al

（1. Lanzhou Institute of Husbandry and Pharmaceutical Sciences of CAAS，The Lanzhou Scientific Observation and Experiment Field Station of Ministry of Agriculture for Ecological System in Loess Plateau Areas，Lanzhou 730050，China；
2. Lanzhou Agriculture Science Research Center，Lanzhou 730000，China）

Abstract：In order to study the salt tolerance and salinity soil improvement of *Atriplex canescens*，four experimental sites with secondary salinized soil in Yongdeng irrigation agricultural area of Gansu province were selected.The thizosphere soil salt concentration，pH and EC of *Aiplex canescens* in different planting years were determined.The results showed that the rhizosphere soil salt concentration decreased slightly after having been planted for one year，and de-

creased significantly after having been planted for two and three years, the rhizosphere soil pH and EC decreased and leaf EC of Aiplex canescens increased with planting years. It showed that Aiplex canescens could absorb and differentiate salinity in rhizosphere soil and improve soil ecological environment (pH, EC). Aiplex canescens is green and safe to improve salinized soil.
Key words: *Atriplex canescens*; Irrigated agricultural area; Soil salinization; Salinity change

（发表于《中国草食动物科学》）

黄土高原半干旱荒漠地区盐碱地优良豆科牧草适应性评价

路　远，田福平，胡　宇，张　茜，崔光欣

（中国农业科学院兰州畜牧与兽药研究所，兰州　730050）

摘要：筛选适宜黄土高原半干旱荒漠地区盐碱地生长的优良豆科牧草草种，为盐碱化草地植被恢复与重建提供适宜的豆科牧草资源，对该区草地生态功能恢复和畜牧业可持续发展具有重要的现实意义。文章在典型盐碱地上对引进和采集的13个耐寒性较强的豆科牧草品种的种子发芽特性和生长特性等进行了研究。结果表明：中兰2号紫花苜蓿、毛苕子和沙打旺3个耐盐牧草品种，对黄土高原半干旱荒漠地区盐碱地气候和土壤环境适应性较好，其中中兰2号表现最佳。

关键词：黄土高原；半干旱荒漠；盐碱地；适应性评价

Adaptability Evaluation of Fine Legume Forage in Saline and Alkaline Land in Semi-arid Desert Area on Loess Plateau

Yuan LU，Fuping TIAN，Yu HU，et al

（Lanzhou Institute of Husbandry and Pharmaceutics Sciences of CAAS，Lanzhou 730050，China）

Abstract：Screening fine legume forage adapting to saline and alkaline land in semi－arid desert area on Loess Plateau is good for salinized grassland restoration and reconstruction and also is important for grassland ecological function regain and animal husbandry development.13 legume varieties living in typical saline and alkaline land with strong cold tolerance were chosen and their seed germination and growth characteristics were studied.Results showed that three of the legumes，Medicago sativa L.cv.Zhonglan No.2，Vicia villosa Roth.and A stragalus albus Pall.，were salt tolerant varieties and could adapt to the saline and alkaline soil environment in semi-arid desert area on Loess Plateau，and Medicago sativa L.cv.Zhonglan No.2 was the best.

Key words：Loess plateau；Semi－arid desert area；Saline and alkaline land；Adaptability evaluation

（发表于《中国草食动物科学》）

牦牛剪毛量及部分绒毛特性研究

包鹏甲，梁春年，吴晓云，王宏博，郭　宪，
丁学智，褚　敏，裴　杰，阎　萍

（中国农业科学院兰州畜牧与兽药研究所，甘肃省牦牛繁育
工程重点实验室，兰州　730050）

摘要：试验测定了成年及2岁大通牦牛和无角牦牛的剪毛量及部分绒毛特性。结果表明：大通牦牛和无角牦牛成年公牛分别可剪毛1.82kg±0.13kg和2.10kg±0.43kg，均极显著高于本品种母牦牛（$P<0.01$）；成年无角牦牛公牛体侧毛长极显著长于其母牦牛及大通牦牛（$P<0.01$），但大通牦牛成年公牛裙毛长极显著长于其母牛及无角牦牛（$P<0.01$），背部毛长在两类群中均无显著性差异（$P>0.05$）；成年大通牦牛母牛的绒毛细度显著低于大通牦牛公牛及无角牦牛（$P<0.05$），粗毛及裙毛细度在两类群公母牛之间均无显著性差异（$P>0.05$）。两类群2岁牦牛公母牛剪毛量、体侧毛长及裙毛长均无显著性差异（$P>0.05$），粗毛、绒、裙毛的细度在群体及性别间也均无显著性差异（$P>0.05$）。绒毛伸长率及断裂强力在不同牦牛类群、年龄及性别之间均无显著性差异（$P>0.05$）。

关键词：牦牛；剪毛量；断裂强力；伸长率

（发表于《中国草食动物科学》）

奶牛腐蹄病综合防治措施

李世宏，杨亚军，孔晓军，刘希望，秦　哲，焦增华，李剑勇

（中国农业科学院兰州畜牧与兽药研究所/农业部兽用药物创制重点实验室/
甘肃省新兽药工程重点实验室，兰州　730050）

摘要：奶牛腐蹄病是奶牛肢蹄病中较严重的蹄病之一，严重危害奶牛业的健康发展，给奶牛养殖业带来很大的经济损失。文章结合作者自身对奶牛腐蹄病的防治体会，简述了奶牛腐蹄病的发病原因、临床症状、治疗和预防方法，为奶牛养殖管理者和从业者提供奶牛腐蹄病的防治措施，保障奶牛健康，促进奶牛业持续健康发展。

关键词：奶牛；腐蹄病；防治

（发表于《中国草食动物科学》）

奶牛乳房炎大肠杆菌的青霉素耐药特征研究

张莉莉[1,2]，杨　峰[1]，李宏胜[1]，王益民[2]，刘龙海[1]，
张　哲[1]，王　丹[1]，张亚茹[1]，李新圃[1]，罗金印[1]

（1. 中国农业科学院兰州畜牧与兽药研究所/农业部兽用药物创制重点实验室/
甘肃省新兽药工程重点实验室，兰州　730050；
2. 兰州大学公共卫生学院，兰州　730000）

摘要：为了研究奶牛乳房炎大肠杆菌的青霉素耐药特征，采用药敏纸片法检测了123株大肠杆菌对青霉素的耐药性，并利用β-内酰胺酶试剂盒测定耐青霉素株β-内酰胺酶活性，同时采用刚果红法和改良结晶紫半定量方法检测生物被膜的形成能力。结果表明：123株受试大肠杆菌中，对青霉素耐药的菌株有122株，耐药率99.19%。122株耐青霉素大肠杆菌中，生物被膜检测为阳性的菌株有92株（75.41%），β-内酰胺酶检测为阳性的有72株（59.02%）；β-内酰胺酶阳性而生物被膜阴性的菌株有15株（12.30%），β-内酰胺酶和生物被膜同为阳性的有57株（46.72%）；β-内酰胺酶为阴性但生物被膜为阳性的菌株有35株（28.69%）。92株生物被膜阳性菌株中，29株（31.52%）具有强成膜能力，61株（66.30%）具有中等成膜能力，2株（2.17%）成膜能力较弱。说明奶牛乳房炎大肠杆菌对青霉素的耐药率较高；大肠杆菌青霉素耐药性受β-内酰胺酶和生物被膜多种因子的共同调控。

关键词：奶牛乳房炎；大肠杆菌；青霉素；耐药性；β-内酰胺酶；生物被膜

Penicillin-resistant Characterization of *Escherichia coli* Isolated from Bovine Mastitis

Lili ZHANG[1,2], Feng YANG[1], Hongsheng LI[1], et al

(1. Lanzhou Institute of Husbandry and Pharmaceutical Science of CAAS/Key Laboratory of Veterinary Pharmaceutics Discovery, Ministry of Agriculture/Key Laboratory of New Animal Drug Project of Gansu Province, Lanzhou 730050, China;
2. School of Public Health, Lanzhou University, Lanzhou 730000, China)

Abstract: The aim of this study was to investigate the penicillin resistant characterization of

Escherichia coli isolated from bovine mastitis.The penicillin resistance was determined by disk diffusion method, β–lactamase test kit was used to detect the β–lactamase activity, biofilm formation was tested with Congo red Agar and sami–quantitative adherence assay method.The results showed that, in all tested E. coli, 122 isolates were resistant to penicillin, the resistance rate was 99. 19%.Among the 122 penicillin–resistant E.coli strains, 72 (59. 02%) isolates were identified as β–lactamase producers, and 92 (75. 41%) isolates produced biofilm.In addition, 15 (12. 30%) biofilm–negative strain was identified as β–lactamase producer.Meanwhile, 57 (46. 72%) isolates were positive for both of β–lactamase and biofilm, 35 (28. 69%) β–lactmase–negative strains were identified as biofilm producers.The results of sami–quantitative adherence assay method showed that, among the 92 isolates biofilm producer, 29 (31. 52%) were detected as strong biofilm producer, 61 (66. 30%) were detected asmoderate biofilm producer, and 2 (2. 17%) were detected as weak biofilm producer. The study indicated that *E.coli* exhibited high resistance rates to penicillin, and implied that E. coli resistant to penicillin was attributed to multiple mechanisms.

Key words: Bovine mastitis; *Escherichia coli*; Penicillin; resistance, β–lactamase; Biofilm

（发表于《中国草食动物科学》）

耐盐苜蓿新品种在永登盐渍土区的品比试验

张怀山[1]，代立兰[2]，王春梅[1]，杨世柱[1]，王　平[2]，王国宇[2]

（1. 中国农业科学院兰州畜牧与兽药研究所/农业部兰州黄土高原生态环境重点野外科学观测试验站，兰州　730050；
2. 甘肃省兰州市农业科技研究推广中心，兰州 730000）

摘要：试验以陇中苜蓿为对照，对 3 个耐盐苜蓿新品系的物候期、株高、茎叶比和产草量等指标进行观测比较，并对其营养成分进行检测分析。结果表明：3 号和 1 号品系生长较快，比对照陇中苜蓿早 5~11d，在无霜期较短的甘肃中部干旱、半干旱地区种植具有明显生长优势；1 号品系在现蕾期、开花期和结荚期株高均高于其他苜蓿材料，与当地陇中苜蓿对比具有显著生长优势；1 号、2 号和 3 号品系平均比当地陇中苜蓿增产 37.6%、29.5%和 34.5%，粗蛋白含量分别为 18.35%、17.62%和 18.09%。综合评价结果，1 号苜蓿新品系的品质最好，综合性状突出，产量潜力大，再生性强，适宜在甘肃盐渍土区大面积推广种植。

关键词：耐盐碱；紫花苜蓿；新品种；永登

The Variety Comparison Test to New Salt-tolerant Alfalfa Varieties in Saline Soil of Gansu Yongdeng Region

Huaishan ZHANG[1]，Lilan DAI[2]，Chunmei WANG[1]，et al

(1. Lanzhou Institute of Animal Sciences and Veterinary Pharmaceutics，Chinese Academy of Agriculture Science/The Lanzhou Scientific Observation and Experiment Field Station of Ministry of Agriculture for Ecological System in Loess Plateau Areas，Lanzhou 730050，China；
2. Lanzhou Agriculture Science Research Center，Lanzhou 730000，China)

Abstract：In this experiment，with Alfalfa of Longzhong as contrast，the indexes of the phenological period，plant height，stem-leaf ratio and yield of grass etc.of three salt-tolerant varieties were observed and compared，and the testing analysis of its nutrients.Result showed that the 1 and 3 strains grow faster，earlier than Alfalfa of LongZhong in the controlled 5-11 days, the cultivation in the short frost-free period of arid and semi-arid area of central Gansu had the obvious growth advantage; the plant height of the 1 strain were higher than other alfalfa in budding stage, flowering and podding stage, compared with the local Alfalfa of Longzhong, it had significant growth advantage; the production of 1, 2 and 3 strains than local alfalfa was an average 37.6%, 29.5% and 34.5%, crude protein content was 18.35%, 17.62% and 18.09% respectively.Comprehensive evaluation results: the 1 strain had the best quality, and the comprehensive character was outstanding, a big potential, strong regeneration, so it was favorable in planting on saline soil in Gansu as large area.

Key words: Salt tolerant; Alfalfa; New varieties; Yongdeng

（发表于《中国草食动物科学》）

无角牦牛产肉性能研究

梁春年[1,3]，吴晓云[1,3]，王宏博[1,3]，张国模[2]，
拉　环[2]，包鹏甲[1,3]，冯宇诚[2]，王　伟[2]，阎　萍[1,3]

（1. 中国农业科学院兰州畜牧与兽药研究所，兰州　730050；
2. 青海省大通牛场，大通　810102；
3. 甘肃省牦牛繁育重点实验室，兰州　730050）

摘要：选择6月龄和4周岁的无角牦牛进行屠宰性状测定，并以同年龄的有角牦牛作为对照，比较分析了无角牦牛和当地有角牦牛的产肉性能，以期为今后无角牦牛新品种的培育提供科学的基础数据。结果表明：6月龄无角公牦牛和母牦牛的宰前活重、胴体重和肉骨比均显著高于同龄的有角牦牛（$P<0.05$），4周岁的无角牦牛的宰前活重和胴体重也均显著高于同龄有角牦牛（$P<0.05$）。以上结果说明无角牦牛具有较高的产肉性能。

关键词：无角牦牛；产肉性能；胴体重

（发表于《中国草食动物科学》）

西藏亚东牦牛线粒体 DNA 分析及系统进化研究

郭　宪[1]，包鹏甲[1]，胡显忠[2]，丁学智[1]，吴晓云[1]，阎　萍[1]，裴　杰[1]

（1. 中国农业科学院兰州畜牧与兽药研究所，兰州　730050；
2. 甘肃省科学院，兰州　730000）

摘要：应用基因测序技术对西藏亚东牦牛线粒体基因组进行测序，并对测序数据质控与修剪及基因组组装、精细注释、功能注释与分析。结果表明：亚东牦牛线粒体基因组大小约为 15 389 bp，包含 13 个 PCGs、21 个 tRNAs、2 个 rRNAs 及 1 个 D-loop，其 A+T 碱基含量为 60. 86%，G+C 碱基含量为 39. 14%；编码基因（PCGs、tRNAs & rRNAs）片段总长度为 14 495 bp，占基因组总长度的 94. 19%；同时，分析了亚东牦牛与其他 11 个物种间的进化关系，发现亚东牦牛与甘南牦牛归为一类，亲缘关系最近。西藏亚东牦牛线粒体 DNA 分析为研究牦牛群体遗传、起源与分子进化提供了技术参考。

关键词：牦牛；线粒体 DNA；系统进化

Analysis on Phyletic Evolution and Mitochondrial DNA of Yadong Yak (*Bos grunniens*) in Tibet

Xian GUO, Pengjia BAO, Xianzhong HU, et al

(Lanzhou Institute of Husbandry and Pharmaceutical Sciences Chinese Academy of Agricultural Sciences, Lanzhou 730050, China)

Abstract: The complete mitochondrial genome sequence of Yadong yak (Bos grunniens) in Tibet has been sequenced by gene sequencing technique, the sequencing data were analyzed by quality control and pruning genome assembly, fine annotation, functional annotation and analysis.The results showed that the complete mitochondrial genome of Yadong yak in Tibet was 15 389 bp, which contained 13 protein-coding genes, 21 tRNA genes, 2 rRNA genes and a non-coding control region (D-loop region).The overall base pair composition were 60. 86% A+T and 39. 14% G+C, the total length of coding genes (PCGs, tRNAs & rRNA) were 14 495 bp, accounting for 94. 19% of the total length of genome.At the same time, them tDNA of Yadong yak were analyzed with other evolutionary relationships between 11 species, the result showed that Yadong yak and Gannan yak were classified as a class. The mitogenome sequence of Yadong yak would contribute better population genetics protection and evolution understanding of Bos grunniens.

Key words: Yak; mitochondrial DNA; Phyletic evolution

（发表于《中国草食动物科学》）

复方中药偶蹄康对奶牛口蹄疫疫苗抗体水平、血清细胞因子及生化指标的影响

刘　艳[1]，彭文静[1]，辛蕊华[1]，郑继方[1]，梁　歌[2]，罗永江[1]*

（1. 中国农业科学院兰州畜牧与兽药研究所，兰州　730050；
2. 四川省畜牧科学研究院，成都　610966）

摘要： 为研究复方中药偶蹄康对奶牛口蹄疫O型、A型、亚洲Ⅰ型3个血清型疫苗抗体水平、10种血清细胞因子含量及生化指标的影响，本试验将健康成年奶牛分为药物组和空白对照组，每组100头，连续拌料给药7d后注射口蹄疫三联苗，注射疫苗后0、14、28、56d分别采血，检测血清疫苗效价、血清细胞因子含量及生化指标。结果显示，注射疫苗后0、28和56d药物组与空白对照组抗体效价无显著差异（$P>0.05$）；注射疫苗后14d药物对口蹄疫3种血清型抗体效价的提升作用均极显著（$P<0.01$）。药物组血清总胆红素（T-Bil）降低，0、28d差异显著（$P<0.05$），56d差异极显著（$P<0.01$）；28、56d药物组与空白对照组葡萄糖含量差异显著（$P<0.05$）；0d血磷含量差异极显著（$P<0.01$），但未超出正常生理值范围，且后续检测无显著差异（$P>0.05$）。注射疫苗后14d药物组血清CD4含量显著高于空白对照组（$P<0.05$）；注射疫苗后28d药物组血清CD4、IL-2、IL-6、IL-10、IL-12、IL-18、IFN-γ含量极显著高于空白对照组（$P<0.01$）；注射疫苗后56d药物组血清IL-6、IL-12含量极显著高于空白对照组（$P<0.01$），CD4、IL-2、IL-4、IL-18含量显著高于空白对照组（$P<0.05$）。综上可知，偶蹄康对奶牛口蹄疫O型、A型、亚洲Ⅰ型3个血清型疫苗效价有显著提升，对奶牛肝脏和肾脏功能无明显的损伤，且有一定的利胆作用，对奶牛细胞免疫和体液免疫机能有明显的提升作用。

关键词： 偶蹄康；奶牛；口蹄疫；抗体效价；细胞因子；血清生化指标

Effect of Herbal Compound Outikang on Antibody Titer of Foot and Mouth Disease Vaccine, Cytokines and Biochemical Indices in Cows

Yan LIU[1], Wen-jing PENG[1], Rui-hua XIN[1], Ji-fang ZHENG[1], Ge LIANG[2], Yong-jiang LUO[1*]

(1. Lanzhou Institute of Husbandry and Pharmaceutical Sciences of CAAS, Lanzhou 730050, China; 2. Sichuan Animal Science Academy, Chengdu 610066, China)

Abstract: The experiment was aimed to study the effects of herbal compound Outikang on antibody titer of foot and mouth disease vaccine type O, type A and type Asia Ⅰ, serum levels of ten cytokines and biochemical indices. 200 healthy adult cows were averagely divided into drug group and blank control group. Vaccinating the cows with FMD triplex vaccine when Outikang was given to the cows of drug group for 7d running and then got the blood of all the cows on 0, 14, 28 and 56d, respectively. Then blood biochemical examination, vaccine effectiveness and serum levels of cytokines were performed. The results showed that there were no significant differences in antibody titer of FMDV between blank control group and drug group on 0, 28 and 56d after vaccination ($P>0.05$), while on 14d the effect of Outikang on antibody titer of FMDV reached extremely significant level ($P<0.01$). Compared with blank control group, the contents of T-Bil in drug group was significantly decreased on 0 and 28d ($P<0.05$), while it was extremely significantly decreased on 56d ($P<0.01$). Compared with blank control group, the contents of Glu had significant difference on 28 and 56d ($P<0.05$). The contents of P had extremely significant difference on 0d ($P<0.01$), which was within the normal range. On 14d after vaccination, the content of CD4 significantly increased in the drug group comparing to those in blank control group ($P<0.05$). On 28d after vaccination, the content of CD4, IL-2, IL-6, IL-10, IL-12, IL-18 and IFN-γ extremely significantly increased in the drug group comparing to those in blank control group ($P<0.01$). On 56d after vaccination, the contents of IL-6 and IL-12 extremely significantly increased in the drug group comparing to those in blank control group ($P<0.01$), while the content of CD4, IL-2, IL-4 and IL-18 significantly increased ($P<0.05$). Therefore, Outikang could significantly rise the antibody titer of foot and mouth disease vaccine type O, type A and type Asia Ⅰ. While it had no damage to the liver and kidney functions and had cholagogue effect, and it could enhance cellular immunity and humoral immunity function ob-

viously.
Key words: Outikang; cows; Foot and mouth disease; Antibody titer; Cytokines; serum biochemical indices

(发表于《中国畜牧兽医》)

中药常山口服液微生物限度检查方法的验证

王　玲，郭志廷*，杨　峰，李宏胜，魏小娟，周绪正

（中国农业科学院兰州畜牧与兽药研究所/农业部兽用药物创制重点实验室，兰州　730050）

摘要：本研究旨在验证抗球虫中药常山口服液的微生物限度检查法，为其质量控制、研发新兽药奠定基础。试验通过直接接种法考察常山口服液的抑菌活性，并采用验证过的试验条件和方法（常规法）进行细菌计数、回收率测定及控制菌的检查，建立常山口服液微生物限度检查方法，并进行验证试验。结果证实常山口服液无抑菌作用或抑菌作用极其微弱，可采用常规法建立检查方法。常山口服液对大肠埃希菌、金黄色葡萄球菌、枯草芽孢杆菌、白色念珠菌及黑曲霉的加菌回收率均大于70%，且供试品和所用稀释剂对此结果无干扰。按常规方法对大肠埃希菌、乙型副伤寒沙门氏菌、金黄色葡萄球菌、铜绿假单胞菌进行控制菌验证检查发现，试验组、阳性对照组均检出相应的试验菌，阴性对照组和供试液组均未检出相应的试验菌。综上所述，常规法可用于常山口服液细菌、霉菌和酵母菌的计数检验，控制菌检查及微生物限度检查，试验过程符合微生物限度检查法规定，且方法简便，可操作性强，结果准确可靠。

关键词：常山口服液；微生物限度；方法验证；常规法

Validation of Microbial Limit Test Method for Traditional Chinese Medicine Dichroa febrifuga Oral Liquid（DFOL）

Ling WANG，Zhi-ting GUO*，Feng YANG，Hong-sheng LI，Xiao-juan WEI，Xu-zheng ZHOU

（Key Laboratory of Veterinary Pharmaceutics Discovery，Ministry of Agriculture，Lanzhou Institute of Animal Science and Veterinary Pharmaceutics，Chinese Academy of Agricultural Sciences（CAAS），Lanzhou 730050，China）

Abstract：The purpose of this study was to validate the microbial limit test method for anticoccidial drug Dichroa febrifuga oral liquid（DFOL），and provide data base for its quality control and researching the new veterinary drug.The direct inoculation method was used to validate the antibacterial activity for DFOL, and the bacterial counting, determination of recovery rate and inspection of control bacteria were measured using a validated test condition and method (conventional method).The results showed that DFOL had no or faint bacteriostasis verified by validated test method, the determination and inspection of bacteria could be determined by conventional method.The normal plate counting were used to detect the amount of Escherichia coli, Staphylococcus aureus, Bacillus subtilis, Candida albicans and Aspergillus niger in DFOL in which the recovery rate were all more than 70%, and the results were not disturbed by sample and diluents.The control bacteria including Escherichia coli, Salmonella paratyphi B, Staphylococcus aureus, and Pseudomonas aeruginosa were tested according to conventional method, whose growth were detected in the test group and positive control group, while no growth were detected in the negative control group and sample group.In conclusion, the microbial limit test method for DFOL had been validated in this study, and the conventional method could be applied to DFOL for bacteria, moulds and yeast counting as well as the inspection for control bacteria.The process of test was in accordance with the regulations of microbial limits test method, the method was convenient and simple, and the results was reliable.

Key words: Dichroa febrifuga oral liquid (DFOL); Microbial limit; method validation; Conventional method

(发表于《中国畜牧兽医》)

胎衣不下奶牛血清和胎盘中激素含量变化的研究

朱永刚，王　磊，王旭荣，张景艳，崔东安，张　凯，
张　康，王学智，李建喜，杨志强

（中国农业科学院兰州畜牧与兽药研究所/甘肃省中兽药
工程技术研究中心，兰州　730050）

摘要：本试验在产犊后立即无菌采集32头奶牛血液，其中17头为胎衣正常排出奶牛，15头为胎衣不下奶牛。胎衣正常排出奶牛在胎衣排出后立即采集胎盘，胎衣不下奶牛在确诊后再次静脉采血并采集其胎盘；用酶联免疫吸附测定法对胎衣不下奶牛和胎衣正常排出奶牛血清和胎盘中的激素含量进行检测，并对检测结果进行分析。结果表明，胎衣不下奶牛血清和胎盘中孕酮、纤维蛋白溶酶原和前列腺素（$PCJF_{2\alpha}$）的含量与胎衣正常排出组相比差异显著降低（$P<0.05$），而雌二醇含量则显著升高（$P<0.05$）。本试验通过对胎衣不下和胎衣正常排出奶牛血清和胎盘中激素的含量进行检测并比较，结果发现奶牛胎衣不下的发生与血清和胎盘组织中的孕酮、雌二醇、纤维蛋白溶酶原和前列腺素（$PCJF_{2\alpha}$）的含量变化有密切相关的联系，表明奶牛胎衣不下的发生可能与机体内激素分泌紊乱具有很大的关联。

关键词：胎衣不下；孕酮；雌二醇；前列腺素（$PGF_{2\alpha}$）；纤维蛋白溶酶原
DOI：10.19305/j.cnki.11-3009/s.2017.04.012

（发表于《中国奶牛》）

不同品种、胎次对牦牛乳中超氧化物歧化酶活性的影响

席　斌[1,2]，高雅琴[1,2]，李维红[1,2]，郭天芬[1,2]，杜天庆[1,2]

（1. 中国农业科学院兰州畜牧与兽药研究所，兰州　730050；
2. 农业部畜产品质量安全风险评估实验室，兰州　730050）

摘要：对采自青海天峻、甘肃甘南各30份牦牛乳样以及天祝抓喜秀龙乡红疙瘩村、岱乾村和碳山岭镇四台沟村3个天祝白牦牛主产区的90份乳样，进行超氧化物歧化酶活力检测分析，比较青海高山牦牛与甘肃甘南高原牦牛、甘肃天祝白牦牛乳中SOD活性，为牦牛乳中活性物质的开发利用提供理论基础。结果表明：3个品种牦牛乳的SOD活性存在一定差异，但是差异性不显著（$P>0.05$）；天祝白牦牛3个牧区的牦牛乳中SOD活性差异不显著（$P>0.05$）；不同胎次条件下牦牛乳中SOD活性差异不显著（$P>0.05$）。即品种、胎次对牦牛乳中SOD活性影响均不显著。

关键词：品种；胎次；牦牛乳；超氧化物歧化酶；活性

Impact Study on SOD Activity of Yak Milk under the Condition of Different Variety and Parity

Bin XI[1,2], Yaqin GAO[1,2], Weihong LI[1,2],
Tianfen GUO[1,2], Tianqing DU[1,2]

(1. Lanzhou Institute of Animal Husbandry and Pharmaceutical Science, Chinese Academy of Agricultural Sciences, Lanzhou 730050, China; 2. Quality Safety Risk Assessment of Animal Products (Lanzhou), Ministry of Agriculture, Lanzhou 730050, China)

Abstract: Each 30 yak milk samples were collected from Gannan County and Tianjun County, and 90 yak milk samples were collected from three different yak farms in Tianzhu County, in order to determine and compare the SOD activity to provide a theoretical basis for further development and utilization of enzyme activity of yak milk.The result shows that no significant differences among the SOD activity in three yak milks from Tianzhu County, Gannan County and

Tianjun County, while there was no outstanding difference among the SOD activity in three yak milks from Tianzhu County also.In all, the variety has no obvious effect on the SOD activity in yak milk, so did parity.
Key words: Variety; Parity; Yak milk; SOD; Activity

（发表于《中国乳品工业》）

奶牛三大常见疾病的防治体会

李世宏，孔晓军，刘希望，杨亚军，秦　哲，焦增华，李剑勇

（中国农业科学院兰州畜牧与兽药研究所/农业部兽用药物创制重点实验室/
甘肃省新兽药工程重点实验室）

摘要：奶牛临床型乳房炎、子宫内膜炎和腐蹄病是影响奶牛养殖业的三大常见疾病，给奶牛养殖业造成了巨大的经济损失。结合笔者在实际生产中防治奶牛疾病的经验与体会，现将防治奶牛临床型乳房炎、子宫内膜炎和腐蹄病的方法进行阐述，以供奶牛养殖者进行参考借鉴。

关键词：奶牛；临床型乳房炎；子宫内膜炎；腐蹄病；防治

（发表于《中国乳业》）

医学模式转变之困惑及其复杂性探讨

罗超应[1]，罗磐真[2]，李锦宇[1]，王贵波[1]，谢家声[1]

摘要：针对现代生物医学所面临的日益严峻的慢性疾病防治挑战，以及“生物-心理-社会医学模式”等新医学模式所面临的“多因素非线性分析”的困惑，在对其进行复杂性科学理论分析与讨论的基础上，论证了以复杂性科学为指导，重视整体相互联系与作用，以“状态分析与处理”为认识方法的“复杂整体-状态医学模式”，要比简单地增加与并列不同因素的新医学模式更加科学与可行。它必将促进生物医学的进一步发展与完善。

关键词：慢性疾病；医学模式；复杂性科学；状态分析与处理

The Discussion of Puzzlement and Complexity of Medical Model Transformation

Chaoying LUO, Panzhen LUO, Jinyu LI, et al.

(Engineering and Technology Research Center of Traditional Chinese Veterinary Medicine of Gansu Province, Lanzhou Institute of Husbandry & Pharmaceutics Science, CAAS, Lanzhou, 730050, China)

Abstract: In view of more severe challenges of the modern biomedical at prevention and cure of chronic diseases, and the puzzlement of the nonlinear analysis for multiple factors of the bio-psycho-social model etc., it was discussed from complexity science that the complexity-whole-condition model, in which it was directed by complexity science, attached importance to the relations and interactions of the whole, and regarded the " analysis and treatment at the conditions" as its cognitive method, is more scientific and feasible than the bio-psycho-social model etc., in which new factors were increased and listed simply.It will certainly promote bio-medical development and improvement.

Key words: Chronic diseases; Medical model; Complexity; Analysis and treatment at the conditions

（发表于《中国社会医学杂志》）

阿司匹林丁香酚酯的一代繁殖毒性研究

赵晓乐，孔晓军，杨亚军，刘希望，李世宏，秦 哲，焦增华，李剑勇*

（中国农业科学院兰州畜牧与兽药研究所/农业部兽用药物创制重点实验室/
甘肃省新兽药工程重点实验室，兰州 730050）

摘要：为了初步评价阿司匹林丁香酚酯（AEE）对亲代大鼠的生殖功能和子代早期发育的影响，将360只Wistar大鼠随机分成6组，分别为受试药物AEE高、中、低剂量组剂量分别为498.0、124.5和31.1mg·kg^{-1}·d^{-1}），阿司匹林组（剂量为68.4mg·kg^{-1}·d^{-1}），丁香酚组（剂量为246.0mg·kg^{-1}·d^{-1}），空白对照组（给予常规维持饲料）。各实验组均按设定剂量连续给药10周后，测定亲代大鼠主要的脏器系数、繁殖指数和生产指标以及子代（F_1）仔鼠的出生指标和生长指标。结果显示，与空白对照组相比，高剂量的AEE影响胎鼠的生长（$P<0.05$），并且高剂量AEE组大鼠的妊娠率、窝产仔数、出生活仔率和出生存活率都偏低，其中雄性仔鼠比例偏高；阿司匹林组妊娠大鼠和哺乳期胎鼠生长缓慢（$P<0.05$）；高剂量组AEE亲代大鼠的睾丸和附睾系数偏高（$P<0.05$），阿司匹林和丁香酚组子代大鼠的附睾系数偏高（$P<0.05$）。结果表明，在本试验条件下，AEE对大鼠繁殖毒性的最大无损害作用剂量为31.1mg·kg^{-1}·d^{-1}，对大鼠的最小有损害作用剂量为124.5mg·kg^{-1}·d^{-1}。

关键词：阿司匹林丁香酚酯；生殖毒性；发育毒性；繁殖指标

Reproductive Toxicity in Parental Generation of Aspirin Eugenol Ester on Rats

Xiao-le ZHAO, Xiao-jun KONG, Ya-jun YANG, Xi-wang LIU,
Shi-hong LI, Zhe QIN, Zeng-hua JIAO, Jian-yong LI

(Key Lab of New Animal Drug Project, Gansu Province; Key Lab of Veterinary Pharmaceutical Development, Ministry of Agriculture; Lanzhou Institute of Husbandry and Pharmaceutical Sciences of CAAS, Lanzhou 730050, China)

Abstract: The reproductive and developmental toxicity of aspirin eugenol ester (AEE) in rats were investigated. Parental generation were exposed to AEE at doses of 498. 0、124. 5 and 31. 1mg · kg^{-1} · d^{-1} and aspirin at dose of 68. 4mg · kg^{-1} · d^{-1} and eugenol at dose of 246. 0mg · kg^{-1} · d^{-1} for 10 weeks before breeding.The weights of main organs and breeding indexes were observed in P and first filial generation (F_1).Results showed that the body weights of P and F_1 were decreased ($P<0.05$), and pregnancy rate, litter size, live birth index and viability index on day 4 in high AEE dose group were also lower than those of control group.In addition, high male proportion in sex ratio of F_1 was very abnormal.Pregnant rats and weaning index in aspirin group grew slowly ($P < 0.05$). Compared to control group, testis and epididymis indexes of P rats in high dose AEE group were high ($P<0.05$), and epididymis indexes of F_1 rats in aspirin and eugenol groups were also high ($p<0.05$).Under the experimental condition, AEE had reproductive toxicity in rats.No observed adverse effect level was 31. 1mg · kg^{-1} · d^{-1}.Low observed adverse effect leve was 124. 5mg · kg^{-1} · d^{-1}.

Key words: Aspirin eugenol ester; Reproductive toxicity; Developmental toxicity; Reproductive indexes

（发表于《中国兽药杂志》）

藿芪灌注液中淫羊藿苷和黄芪甲苷稳定性试验研究

杨洪早[1,2]，苗小楼[1]，张世栋[1]，董书伟[1]，闫宝琪[1]，桑梦琪[1]，
那立冬[1,2]，王东升[1*]，严作廷[1*]

（1. 中国农业科学院兰州畜牧与兽药研究所/农业部兽用药物创制重点实验室，兰州 730050；2. 甘肃农业大学动物医学院，兰州 730070）

摘要：为了研究藿芪灌注液的稳定性，采用 HPLC 法测定藿芪灌注液中淫羊藿苷与黄芪甲苷含量，使用 WondaSil C18 柱（4.6mm×250mm，5μm），流动相：乙腈-水 28：72，柱温 40℃，检测波长 270nm，进样量 10μL，流速 1.0mL/min 测定淫羊藿苷的含量；使用 ZORBAX SB-C18 柱（4.6mm×150mm，5μm），漂移管温度为 90℃，载气流速为 2.5L/min，流动相：甲醇-水 65：35，柱温：40℃，流速：1mL/min，用蒸发光散射检测器检测测定黄芪甲苷的含量。藿芪灌注液于高温 60℃放置 10d，高湿度 90%±5%放置 10d，经强光照射 4 500LX±5 00LX 放置 10d，加速试验在温度 40℃±2℃、相对湿度 75%±5%的条件下放置 6 个月，长期试验在温度 25℃±2℃、相对湿度 60%±10%的条件下放置 12 个月，考察藿芪灌注液在不同条件存放后的性状、鉴别、pH、淫羊藿苷含量、黄芪甲苷含量以及无菌检查等项目的变化。结果表明，藿芪灌注液性质稳定，在不同条件存放后所有指标均未有明显变化，符合质量标准要求。

关键词：藿芪灌注液；稳定性试验；高效液相色谱法；淫羊藿苷；黄芪甲苷

Study on Stability of Icariin and Astragaloside in Huoqi Perfusion

Hong-zao YANG[1,2], Xiao-lou MIAO[1], Shi-dong ZHANG[1], Shu-wei DONG[1], Bao-qi YAN[1], Meng-qi SANG[1], Li-dong NA[1,2], Dong-sheng WANG[1*], Zuo-ting YAN[1*]

(1. Lanzhou Institute of Animal Husbandry and Pharmaceutical Sciences, Chinese Academy of Agricultural Sciences, Key Laboratory of Veterinary Pharmaceutical Discovery, Ministry of Agriculture, Lanzhou 730050, China; 2. College of Veterinary Medicine, Gansu Agricultural University, Lanzhou 730070, China)

Abstract: To investigate the stability of Huoqi perfusion, the contents of icariin and astragaloside were determined by HPLC.The separation of the icariin was achieved by a WondaSil-C18 (4. 6mm×250mm, 5μm) and mobile phase composed of acetonitrile-water (28 : 72), the detection wavelength was 270nm, the column temperature was 40℃, the injection volume of 10μL, the flow rate was 1. 0mL/min; The separation of the astragaloside was achieved by a ZORBAX SB - C18 (4. 6mm × 150mm, 5μm), The drift tube temperature is 90℃, The carrier gas flow rate is 2. 5 L/min and mobile phase composed of methanol-water (65 : 35), the flow rate was 1. 0mL/min; the injection volume of 10μL; the column temperature was 40℃, with evaporative light scattering detector. Huoqi Perfusion was treated respectively with high temperature (60℃) for 10d, high wet (RH95% ± 5%) for 10 days, high light (4500LX±500LX) for 10 days, accelerating testing (40℃ ± 2℃, RH75% ± 5%) for six months, and a long-term testing (25℃ ±2℃, RH60%±10%) for twelve months, and then the changes of characters, identification, pH, icariin, astragaloside and sterility test under different storage conditions were observed. All of the tested indexes of Huoqi Perfusion were consistent with the quality standards.The tested of Huoqi perfusion is stable and reliable.

Key words: Huoqi perfusion; Stability test; HPLC; Icariin; Astragaloside

（发表于《中国兽药杂志》）

常山口服液制备工艺研究

郭志廷[1]，罗晓琴[2]，王　玲[1]，梁剑平[1]

（1. 中国农业科学院兰州畜牧与兽药研究所/农业部兽用药物创制重点实验室甘肃省新兽药工程重点实验室，兰州　730050；2. 兰州市动物卫生监督所畜产品质量安全检测实验室，兰州　730050）

摘要：为满足禽病临床防治的需要，在常山散剂的基础上研制常山口服液。从助溶剂、防腐剂、pH 值以及口服液澄清度和常山碱含量等方面综合评价，确定常山口服液的最佳生产工艺条件。结果表明，常山口服液的最佳牛产工艺条件为：1.5%吐温－80、0.5%苯甲酸钠和 pH 值 5~6；在此条件下，口服液澄清度较好、无沉淀和分层、常山碱含量较高。结果提示，以上口服液的制备方法具有工艺简单、成本较低、有效成分含量高等优点，为今后常山口服液的制备和工业化生产奠定了基础。

关键词：常山；口服液；制备；常山碱

（发表于《中兽医医药杂志》）

鸡骨常山和普通常山中常山乙素含量比较

郭志廷[1]，王　玲[1]，潘志忠[2]，梁剑平[1]，罗永江[1]

（1. 中国农业科学院兰州畜牧与兽药研究所/农业部兽用药物创制重点实验室/
甘肃省新兽药工程重点实验室，兰州　730050；
2. 松原职业技术学院，松原　138000）

摘要：目的：比较鸡骨常山和普通常山中常山乙素的含量。方法：应用高效液相色谱法测定常山乙素含量。结果：鸡骨常山和普通常山中常山乙素的平均含量分别为 0.36% 和 0.05%，普通常山仅为鸡骨常山的 1/7。结论：本实验为常山药材的 HPLC 检测和常山碱的工业化生产提供了技术支持。

关键词：HPLC 法；鸡骨常山；普通常山；常山乙素

（发表于《中兽医医药杂志》）

蒲黄药理作用研究进展

焦增华，杨亚军，刘希望，李世宏，李剑勇

（中国农业科学院兰州畜牧与兽药研究所，农业部兽用药物创制重点实验室，
甘肃省中兽药工程技术研究中心，甘肃省新兽药工程重点实验室，甘肃兰州　730050）

摘要：蒲黄为香蒲科植物水浊香蒲、东方香蒲或同属植物的干燥花粉，具有镇痛、抗凝促凝（与浓度有关）、促进血液循环、降低血脂、防止动脉硬化、保护高脂血症所致的血管内皮损伤、兴奋子宫收缩、增强免疫力等作用，还有促进肠蠕动、抗炎、抗低压低氧、抗微生物等药理作用。近年来，随着对蒲黄的深入研究，其药理活性越来越受到重视，就蒲黄化学成分和药理活性进行综述，为蒲黄的进一步开发应用提供参考。

关键词：蒲黄；化学成分；药理活性；研究进展

（发表于《中兽医医药杂志》）

西藏砂生槐主要成分苦参碱类药理学研究进展

牛建荣，张继瑜，周绪正，李　冰，魏小娟，程富胜

（中国农业科学院兰州畜牧与兽药研究所/农业部兽用药物创制重点实验室/
甘肃省新兽药工程重点实验室，甘肃兰州　730050）

摘要：西藏砂生槐是西藏特有植物，具有极强的抗旱、耐瘠薄、抗风沙等生态适应性和很好的防风固沙、保持水土功能。现代药理学研究证明，西藏砂生槐具有很高药用价值，其有效成分主要为苦参碱类生物碱，具有抗炎、抗病毒、抗肿瘤、抑制中枢神经系统和强心、抗心律失常等作用，同时又是植物性杀虫剂的重要原料。西藏砂生槐在西藏等地分布广泛，药源丰富，对其药理学的深入研究和进一步的开发应用十分重要。笔者等就西藏砂生槐近年来的药理学研究进展做一综述，以期对西藏砂生槐的开发利用有所裨益。

关键词：西藏砂生槐；药理学；防治；开发应用

（发表于《中兽医医药杂志》）

银翘蓝芩口服液中黄芩苷 HPLC 含量测定方法的耐用性研究

杨亚军，刘希望，孔晓军，许春燕，秦　哲，焦增华，李世宏，李剑勇

（中国农业科学院兰州畜牧与兽药研究所/农业部兽用药物创制重点实验室/甘肃省新兽药工程重点实验室，甘肃兰州　730050）

摘要：[目的]：考察银翘蓝芩口服液中黄芩苷 HPLC 含量测定方法的耐用性。[方法]：采用单因素法，考察色谱柱、柱温、流速、流动相比例和 pH 值等对分离度、理论板数的影响。[结果]：常用的不同品牌及规格的 C_{18} 色谱柱以及柱温、流速和 pH 值发生微小变动时，均能实现样品中黄芩苷的有效分离，符合系统适用性的要求；但当流动相中甲醇的比例提高至 50% 时，不能实现样品中黄芩苷与其他成分的有效分离，在检测过程中应予以注意。[结论]：该方法稳定性好，可用于银翘蓝芩口服液中黄芩苷的含量测定。

关键词：银翘蓝芩口服液；黄芩苷；HPLC；含量测定；方法耐用性

（发表于《中兽医医药杂志》）

作者简介

杨志强（1957—），中共党员，学士，二级研究员，博士生导师。甘肃省优秀专家，甘肃省第一层次领军人才，中国农业科学院跨世纪学科带头人，甘肃省“555”创新人才，《中兽医医药杂志》主编。兼任中国毒理学兽医毒理学分会会长，中国畜牧兽医学会常务理事，中国兽医协会常务理事，中国畜牧兽医学会动物药品学分会副会长，中国畜牧兽医学会毒物学分会副会长，中国畜牧兽医学会中兽医学分会副会长，中国畜牧兽医学会西北地区中兽医学会理事长，农业部兽药评审委员会委员。长期从事中兽医药学、兽医药理毒理、动物营养代谢与中毒病等研究工作，是该领域内的知名专家，先后主持和参加国家、省、部级科研课题33项，其中主持20项，获奖9项，自主和参与研发新产品8个，获授权专利3项。先后培养硕士研究生20名，培养博士研究生10名。在国内和国际学术刊物上共发表学术论文100余篇，其中主笔发表论文80篇。主编和参与编写《微量元素与动物疾病》等学术专著13部。

刘永明（1957—），中共党员，大学文化程度，三级研究员，硕士研究生导师。先后担任中国农业科学院中兽医研究所党委办公室副主任、主任，中国农业科学院兰州畜牧与兽药研究所人事处处长、副所长、党委副书记和纪委书记等职务，2001年7月至今任中国农业科学院兰州畜牧与兽药研究所党委书记、副所长、工会主席，兼任《中兽医医药杂志》和《中国草食动物科学》杂志编委会主任、中国农业科学院思想政治工作研究会理事、中国兽医协会会员和兰州市科学技术奖励委员会委员等职务。主要从事动物营养与代谢病研究工作。先后主持国家科技支撑计划、公益性行业专项、科技成果转化基金项目、948项目以及省级科研课题或子专题12项，主持基本建设项目4项，获授权专利8项，取得新兽药证书1个、添加剂预混料生产文号5个；主编（主审）、副主编著作6部，参与编写著作6部。

张继瑜（1967—），中共党员，博士研究生，三级研究员，博（硕）士生导师，中国农业科学院三级岗位杰出人才，中国农业科学院兽用药物研究创新团队首席专家，国家现代农业产业技术体系岗位科学家。兼任中国兽医协会中兽医分会副会长，中国畜牧兽医学会兽医药理毒理学分会副秘书长，农业部兽药评审委员会委员，农业部兽用药物创制重点实验室常务副主任，甘肃省新兽药工程重点实验室常务副主任，中国农业科学院学术委员会委员，黑龙江八一农垦大学和甘肃农业大学兼职博导。主要从事兽用药物及相关基础研究工作，重点方向包括兽用化学药物的研制、药物作用机理与新药设计、细菌耐药性研究。带领的研究团队在动物寄生虫病、动物呼吸道综合症防治药物研究上取得了显著进展。在肠杆菌耐药机理、血液原虫药物作用靶标筛选的研究处于领先地位。先后主持完成国家、省部重点科研项目20多项，获得科技奖励6项，研制成功4个兽药新产品，其中国家一类新药一个，以第一完成人申报10项国家发明专利，取得专利授权5项。培养研究生21名，发表论文170余篇，主编出版《动物专用新化学药物》和《畜牧业科研优先序》等著作2部。

阎萍（1963—）中共党员，博士，三级研究员，博士生导师。2012年享受国务院特殊津贴，是中国农业科学院三级岗位杰出人才，甘肃省优秀专家，甘肃省“555”创新人才，甘肃省领军人才。曾任畜牧研究室副主任、主任等职务，2013年3月任研究所副所长职务。兼任国家畜禽资源管理委员会牛马驼品种审定委员会委员，中国畜牧兽医学会牛业分会副理事长，全国牦牛育种协作组常务副理事长兼秘书长，中国畜牧兽医学会动物繁殖学分会常务理事和养牛学分会常务理事等。阎萍研究员主要从事动物遗传育种与繁殖研究，特别是在牦牛领域的研究成绩卓越，先后主持和参加完成了科技部支撑计划、科技部基础性研究项目、科技部“863”计划、“948”计划、农业部行业科技项目、国家肉牛产业技术体系岗位专家、人事部回国留学基金项目、科技部成果转化项目、甘肃省科技重大专项计划、甘肃省农业生物技术项目等20余项课题。现为国家肉牛牦牛产业技术体系牦牛选育岗位专家，甘肃省牦牛繁育工程重点实验室主任。作为高级访问学者多次到国外科研机构进行学术交流。培育国家牦牛新品种1个，填补了世界上牦牛没有培育品种的空白。获国家科技进步奖1项，省部级科技进步奖5项及其他科技奖励3项。培养研究生15名，发表论文180余篇，出版《反刍动物营养与饲料利用》《现代动物繁殖技术》《牦牛养殖实用技术问答》《Recend Advances in Yak Reproduction》《中国畜禽遗传资源志—牛志》等著作。

董书伟（1980—），男，汉族，在读博士，助理研究员，毕业于西北农林科技大学。主要从事奶牛营养代谢病与中毒病的蛋白质组学研究。先后主持参加国家科技支撑计划、中央级公益性科研院所基本科研业务费专项资金项目、中国科学院西部之光项目、国家自然科学青年基金，发表学术论文20余篇，申请专利18项。

郭宪（1978—），中国农业科学院硕士生导师，博士，副研究员。中国畜牧兽医学会养牛学分会理事，中国畜牧业协会牛业分会理事，全国牦牛育种协作组理事。主要从事牛羊繁育研究工作。先后主持国家自然科学基金、国家支撑计划子课题、中央级公益性科研院所基本科研业务费专项资金项目等5项。科研成果获奖3项，参与制定农业行业标准3项，授权专利3项。主编著作2部，副主编著作5部，参编著作2部。主笔发表论文30余篇，其中SCI收录5篇。

郭志廷（1979—），男，内蒙古人，助理研究员，执业兽医师，九三学社兰州市青年委员会委员，中国畜牧兽医学会中兽医分会理事。2007年毕业于吉林大学，获中兽医硕士学位。近年主要从事中药抗球虫、免疫学和药理学研究。先后主持或参加国家、省部级科研项目5项，包括中央级公益性科研院所专项基金。作为参加人获得兰州市科技进步一等奖1项，兰州市技术发明一等奖1项，完成甘肃省科技成果鉴定4项，授权国家发明专利5项（1项为第一完成人），参编国家级著作2部。在国内核心期刊上发表学术论文80余篇（第一作者30篇）。

郝宝成（1983—），甘肃古浪人，硕士研究生，助理研究员，研究方向为新型天然兽用药物研究与创制。先后主持和参与了中央级公益性科研院所基本科研业务费专项资金、国家支撑计划、863项目子课题等项目6项，以第一作者发表论文23篇（其中SCI论文2篇，一级学报2篇），参与编写著作3部《中兽药学》《兽医中药学及试验技术》《天然药用植物有效成分提取分离与纯化技术》，以第一发明人获得国家发明专利3项，荣获2012年度兰州市九三学社参政议政先进工作者。

李宏胜（1964—），九三学社社员，博士，研究员，硕士生导师，甘肃省“555”创新人才。中国畜牧兽医学会家畜内科学分会常务理事。多年来主要从事兽医微生物及免疫学工作，尤其在奶牛乳房炎免疫及预防方面有比较深入的研究。先后主持和参加完成了国家自然基金、国家科技支撑计划、国际合作、甘肃省、兰州市及企业横向合作等20多个项目。先后获得农业部科技进步三等奖1项；甘肃省科技进步二等奖2项、三等奖2项；中国农业科学院技术成果二等奖3项；兰州市科技进步一等奖2项、二等奖2项。获得发明专利4项，实用新型专利14项，培养硕士研究生5名，在国内外核心期刊上发表论文160余篇，其中主笔论文60余篇。

李剑勇（1971—），研究员，博士学位，硕士和博士研究生导师，国家百千万人才工程国家级人选，国家有突出贡献中青年专家。现任中国农业科学院科技创新工程兽用化学药物创新团队首席专家，农业部兽用药物创制重点实验室副主任，甘肃省新兽药工程重点实验室副主任，甘肃省新兽药工程研究中心副主任，农业部兽药评审专家，甘肃省化学会色谱专业委员会副主任委员，中国畜牧兽医学会兽医药理毒理学分会理事，国家自然基金项目同行评议专家，《黑龙江畜牧兽医杂志》常务编委，《PLOS ONE》《Medicinal Chemistry Research》等SCI杂志审稿专家。多年来一直从事兽用药物创制及与之相关的基础和应用基础研究工作。曾先后完成药物研究项目40多项，主持16项。获省部级以上奖励10项，2011年度获第十二届中国青年科技奖；2011年度获第八届甘肃青年科技奖；获2009年度兰州市职工技术创新带头人称号。获国家一类新兽药证书，均为第2完成人。申请国家发明专利22项，获授权9项。发表科技论文200余篇，其中SCI收录22篇，第一作者和通讯作者15篇。出版著作4部，培养研究生15名。

王学智（1969—），研究员、博士，硕士研究生导师，科技管理处处长。主要从事科技管理工作和兽医临床科研工作。主要在中兽药、动物营养代谢病等兽医临床学方面开展基础应用研究，先后主持“国家公益性行业专项中兽药生产关键技术研究与应用课题防治螨病和痢疾藏中兽药制剂制备”“科技基础性工作专项—传统中兽医药资源抢救和整理”和“甘肃省科技重大专项——防治奶牛繁殖病中药研究与应用”、甘肃省科技支撑计划项目等科研课题10项。参与完成“948项目”“国家自然基金项目”“国家科技支撑计划”等各类科研项目16项。先后获得省部院级科技成果奖励7项：“奶牛乳房炎联合诊断和防控新技术研究及示范”获甘肃省农牧渔业丰收一等奖，“重金属镉/铅与喹乙醇抗原合成、单克隆抗体制备及ELLSA检测技术研究”和“新型中兽药饲料添加剂“参芪散”的研制与应用”获中国农业科学院科学技术成果二等奖，“归蒲方中草药饲料添加剂的研制与应用”“禽用复合营养素的研制及其应用”和“非解乳糖链球菌发酵黄芪转化多糖的研究与应用”获甘肃省科技进步三等奖。在国家级学术刊物上先后发表学术论文15篇，其中SCI收录7篇，参编著作17部，主编5部，获得专利22个。

李维红（1978—），博士，副研究员。主要从事畜产品质量评价技术体系、畜产品检测新方法及其产品开发利用研究等。先后主持和参加了中央级公益性科研院所基本科研业务费专项资金项目、甘肃省自然基金项目等。主笔发表论文20余篇。主编《动物纤维超微结构图谱》，副主编《绒山羊》，参加编写《动物纤维组织学彩色谱》和《甘肃高山细毛羊的育成和发展》著作2部。获得专利4项，作为参加人获甘肃省科技进步一等奖一项、三等奖一项。

李新圃（1962—），博士，副研究员。主要从事兽医药理学研究工作。已主持完成省、部、市级科研项目7项。参加完成国际合作、省、部、市级科研项目三十余。参加项目“奶牛重大疾病防控新技术的研究与应用”获2010年甘肃省科技进步二等奖；“奶牛乳房炎主要病原菌免疫生物学特性的研究”在2008年获兰州市科技进步一等奖、中国农科院科学技术成果二等奖和甘肃科技进步三等奖；“绿色高效饲料添加剂多糖和寡糖的应用研究”获2005年兰州市科技进步一等奖；“奶牛乳房炎综合防治配套技术的研究及应用”获2004年甘肃省科技进步二等奖。已发表研究论文40余篇，其中5篇被SCI收录。获实用新型专利授权3个。

梁春年（1973—），硕士生导师，博士，副研究员，副主任，兼任全国牦牛育种协作组常务理事，副秘书长，中国畜牧兽医学会牛学会理事，中国畜牧业协会牛业分会理事，中国畜牧业协会养羊学分会理事等职。主要从事动物遗传育种与繁殖方面的研究工作。现主持国家科技支撑计划子课题“甘肃甘南草原牧区牦牛选育改良及健康养殖集成与示范”和国家星火计划项目子课题“牦牛高效育肥技术集成示范”等课题4项；参加完成国家及省部级科研项目30余项，获得省部级科技奖励7项。参与制定农业行业标准5项。参加国内各类学术会议30余次，国际学术会议4次。主编著作2本，副主编著作5本，参编著作6本，发表论文90余篇，其中SCI文章5篇。

孙晓萍（1962—），硕士生导师，学士，副研究员。主要从事绵羊遗传育种工作，先后主持的项目有：甘肃省自然科学基金项目——绵羊毛生长机理研究；甘肃省农委推广项目——绵羊双高素推广应用研究；甘肃省支撑计划项目——奶牛产奶量的季节性变化规律研究；甘肃省星火项目——肉羊高效繁殖技术研究；甘肃省支撑计划项目——肉用绵羊高效饲养技术研究。先后发表学术论文40余篇，主编参编著作11部，实用新型专利3个。获甘肃省畜牧厅科技进步二等奖1项，甘肃省科技进步二等奖1项，中国农业科学院科技进步一等奖1项，中华农业科技二等奖1项。

王东升（1979—），农学硕士，助理研究员。主要从事奶牛繁殖疾病的研究。主持“奶牛子宫内膜中天然抗菌肽的分离、鉴定及其生物学活性研究”和“狗经穴靶标通道及其生物效应的研究”2个课题，参加“十二五”国家科技支撑计划项目“奶牛健康养殖重要疾病防控关键技术研究”和“十一五”国家科技支撑计划项目“奶牛主要繁殖障碍疾病防治药物研制”等10多个项目。参与申请并获得发明专利4项，实用性新专利5项，取得三类新兽药证书1个，参加的成果获甘肃省科技进步二等奖2项和兰州市科技进步二等奖2项。参编著作《兽医中药配伍技巧》《兽医中药学》《奶牛围产期饲养与管理》和《奶牛常见病综合防治技术》等5部，主笔发表论文20余篇。

王宏博（1977—），博士，助理研究员，博士。主要从事动物营养与饲料科学。先后主持甘肃省科技厅项目2项，中央级公益性科研院所科研业务费专项资金项目3项，先后参加农业部公益性行业（农业）专项3项，“948”项目1项，获得国家发明专利2项，实用新型专利1项，参与完成国家发明专利和实用新型专利总计10余项。参与制定国家和农业部标准10余项。主编图书1部，副主编图书1部，参编图书3部。发表学术论文30余篇。

王磊（1985—），硕士，研究实习员。主要从事中兽医药理学、奶牛疾病防治和药物残留研究。先后参加各类研究课题8项，主持课题1项。2013年参加的“非解乳糖链球菌发酵黄芪转化多糖的研究与应用”获甘肃省科技进步三等奖、2014年参加的“重金属镉/铅与喹乙醇抗原合成、单克隆抗体制备及ELISA检测技术研究”获中国农业科学院科学技术成果二等奖；主笔发表科技论文4篇、其中SCI收录1篇，获得授权专利2项。

王晓力（1965—），副研究员。主要从事牧草资源开发利用方面的研究工作。近年来参加省部级课题多项，主持农业行业科研专项子课题2项，获全国农牧渔业丰收二等奖1项、内蒙古自治区农牧业丰收一等奖1项、2011年贵州省科学技术成果转化二等奖1项。发表论文30余篇，出版著作4部，参编4部，授权专利5项。

王旭荣（1980—），博士，助理研究员。主要从事分子病毒学和分子细菌学方面的研究。主持完成“犬瘟热病毒”的基本科研业务费项目1项；现主持“奶牛乳房炎病原菌“方面的农业行业标准项目1项和甘肃省农业生物技术项目1项；参与国家奶牛产业技术体系项目、948项目、国家科技支撑项目、国家自然基金、科技基础性工作专项等10余项。参与的研究项目在2011—2014年期间获得奖项3个，其中甘肃省科技进步三等奖1项，甘肃省农牧渔丰收奖一等奖1项、中国农业科学院科学技术成果二等奖1项。发表科技论文40余篇，主笔15篇；参编著作1部；实用新型授权20余项，第一发明人授权5项；申请发明专利（第一发明人）3项。

王学红（1975—），硕士，高级兽医师。主要从事天然药物提取研究。先后主持及参加国家级省部级药物研究项目20余项，包括“十二五”农村领域国家科技计划项目、国家支撑计划项目、甘肃省科技支撑计划项目及中央级公益性科研院所基本科研业务费专项资金项目等。获省部级科技进步奖4项，院厅级奖2项，国家发明专利10余个。在国内外学术刊物上发表论文20余篇，出版著作4部。

吴晓云（1986—），博士，研究实习员，主要从事牦牛低氧适应机制和牦牛肉品质形成遗传机制的研究。目前以第一作者发表论文11篇，其中7篇被SCI收录。

严作廷（1962—），硕导，博士，研究员。现为第五届农业部兽药评审专家、中国畜牧兽医学会家畜内科学分会常务理事、中国畜牧兽医学会中兽医学分会理事。主要从事工作奶牛疾病和中兽医学研究（先后主持国家自然科学基金、国家科技支撑计划等项目10多项，现主持“十二五”国家科技支撑计划课题“奶牛健康养殖重要疾病防控关键技术研究”课题）。主编《家畜脉诊》《奶牛围产期饲养与管理》等著作3部，参编《中兽医学》《奶牛高效养殖技术及疾病防治》和《犬猫病诊疗技术及典型医案》等著作6部。发表论文70余篇，获省部级科技进步奖6项，院厅级奖7项。获得国家新兽药证书2个，国家发明专利10个，实用新型专利5个。

张世栋（1983—），硕士，助理研究员。主要从事奶牛疾病与中兽医药研究工作。近年来，主持中央级基本科研业务费及其增量项目共3项，国家自然基金1项；参与国家“十一五”“十二五”课题3项，甘肃省、兰州市科技项目多项。参与获得兰州市科技进步二等奖1项。发表论文10篇，获得专利6项，参与申请专利10多项。

李建喜（1971—），研究员，博士学位，硕士研究生导师。现任中国农业科学院兰州畜牧与兽药研究所中兽医（兽医）研究室主任，中国农业科学院科技创新工程中兽医与临床创新团队首席专家，甘肃省中兽药工程技术研究中心副主任，农业部新兽药中药组评审专家，国家自然基金项目同行评议专家，国家现代农业（奶牛）产业技术体系后备人选等。从事兽医病理学、动物营养代谢病与中毒病、兽医药理与毒理、奶牛疾病防治、中兽医药现代化等研究工作。完成国家和省部级科研项目40余项，其中主持24项，包括国家自然基金面上项目、国家科技支撑计划课题、国家公益性行业（农业）科研专项课题、国家科技基础性工作专项课题、农业部948计划项目、甘肃省农业科技重大专项、甘肃省生物技术专项、中国西班牙科技合作项目、中泰科技合作项目等。先后获科技奖励6项，其中获省部级奖励2项，院厅级奖励4项。研发新产品6个，获授权发明专利9项，以第一导师培养硕士研究生6名，共同培养硕士研究生17名，参与培养博士研究生8名。在国内和国际学术刊物上共发表学术论文99篇，其中以第一作者和通讯作者发表论文54篇，SCI收录6篇，编写著作7部。

梁剑平（1962—），博士，三级研究员，博士生导师。2005年获“农业部有突出贡献的中青年专家”称号。2005年分别获中国科协“西部开发突出贡献奖”、中央统战部“为全面建设小康社会做出贡献的先进个人”、甘肃省“陇上骄子”、九三学社甘肃省委、“十佳青年”称号。现任兰州畜牧与兽药研究所兽药研究室副主任、农业部兽用药物创制重点实验室副主任、中国农业科学院新兽药工程重点开放实验室副主任。是中国农业科学院二级岗位杰出人才，甘肃“555”创新人才。兼任中国毒理学会兽医毒理学分会及中国兽医药理学分会理事，农业部新兽药评审委员会委员，农业部兽药残留委员会委员，中国兽药典委员会委员，中国农业科学院学术委员会委员，中国农科院研究生院教学委员会委员、政协兰州市委常委，九三学社兰州市七里河区主任委员，九三学社兰州市副主委。梁剑平研究员主要从事兽药化学合成和中草药的提取及药理研究，先后主持和参加国家和省部级重大科研项目20余项，获奖8项，其中获国家科技进步二、三等奖各1项，省（部）级二等奖2项、三等奖2项，发明专利10项。

蒲万霞（1964—）博士后，四级研究员，硕士生导师，甘肃省微生物学会理事，政协兰州市七里河区第六届委员。从事兽医微生物与微生物制药研究，重点方向为兽用微生态制剂的研制及细菌耐药性研究，在金黄色葡萄球菌耐药性研究方面取得一定进展。先后主持农业部公益性行业科研专项“新型动物专用药物的研制与应用”子专题，国家科技支撑计划“奶牛主要疾病综合防控技术研究及开发”子专题，中央级公益性科研院所基本科研业务费专项资金项目、甘肃省科技成果重点推广计划项目、甘肃省农业科技创新项目、兰州市院地校企合作项目、兰州市科技局项目、农业部畜禽病毒学开放实验室基金项目等各级各类项目15项。获得各级政府奖7项。取得国家发明专利2项。主编《食品安全与质量控制技术》等著作5部，副主编著作1部，发表论文70多篇，培养硕士生11名。

岳耀敬（1980—），在读博士。主要从事绵羊繁殖、羊毛和高原适应性等重要经济、抗逆性状的分子调控机制研究。国家绒毛用羊产业技术体系分子育种岗位团队成员，兼任中国畜牧兽医学会养羊学分会副秘书长、世界美利奴育种者联盟成员，曾先后到法国、澳大利亚、新西兰等国学习考察细毛羊育种工作。现主持国家自然青年基金、甘肃省青年基金等项目3项；申报发明专利5项，授权发明专利2项、实用新型2项；参编著作5部，发表学术论文38篇，其中SCI收录5篇。

尚小飞（1986—），硕士，助理研究员。主要从事藏兽医药的现代化研究。目前主持中央级公益性院所基本科研业务费项目一项，参与多项国家及省部级课题。在*Journal of Ethnopharmacology*，*Veterinary Parasitology*等SCI杂志发表文章8篇，参与编写著作1部。

杨红善（1981—），硕士，助理研究员。主要从事牧草种质资源搜集与新品种选育研究工作，主持在研项目3项，其中甘肃省青年基金项目1项、甘肃省农业生物技术研究与应用开发项目1项、中央级公益性科研院所基本科研业务费专项资金项目1项，参加国家支撑计划子课题等各类项目共计3项。工作期间以第一完成人或参加人审定登记甘肃省牧草新品种4个。获中国农业科学院科技进步二等奖1项（第三完成人）。参加编写《高速公路绿化》著作1本（副主编）。在各类期刊发表论文20篇，其中主笔13篇。

刘宇（1981—），硕士，助理研究员。2007年7月来中国农业科学院兰州畜牧与兽药研究所工作至今，主要从事有机及天然药物化学研究。先后主持及参与国家级省部级药物研究项目10余项，包括“十二五”农村领域国家科技计划项目、国家支撑计划项目、甘肃省科技支撑计划项目、甘肃省自然科学基金项目及中央级公益性科研院所基本科研业务费专项资金项目等。获省部级科技进步奖1项，院厅级奖2项，国家发明专利4个。在国内外学术刊物上发表论文20余篇，出版著作2部。

程富胜（1971—）博士，副研究员，硕士生导师。主要从事天然药物活性成分免疫药理学研究及其制剂开发研制。先后主持和参加国家、省、部级科研课题20项。参加完成的国家支撑项目“中草药饲料添加剂‘敌球灵’的研制”获农业部科技进步二等奖；国家自然基金项目“免疫增强剂-8301多糖的研究与应用”成果获中国农业科学院科技进步二等奖；“蕨麻多糖免疫调节作用及其机理研究与临床应用”成果已分别获中国农科院、兰州市科技进步二等奖。主持完成的“富含活性态微量元素酵母制剂的研究”获兰州市科技进步二等奖。在国内和国际学术刊物上共发表学术论文50余篇，其中主笔发表论文30余篇。

李冰（1981—），女，硕士研究生、助理研究员，研究方向为兽药新制剂研制与安全评价。主要从事新兽药制剂质量标准制定、体内药代动力学研究与残留研究。中国畜牧兽医学会兽医药理毒理学分会理事。主持在研项目2项，其中公益性行业专项子课题1项、中央级公益性科研院所基本科研业务费专项资金项目1项；参加国家支撑计划项目、国家高技术研究发展计划（863计划）、甘肃省科技重大专项、国家自然基金项目、国家基础专项等各类项目十余项。以主要完成人申报国家新兽药1个。前三以前作者发表论文30余篇，主笔核心期刊3篇，主笔

SCI收录2篇，参编著作2部。获2011年兰州市科学技术进步奖一等奖，第6完成人；2012年中国农业科学院科技进步二等奖，第3完成人；2013年甘肃省科学技术发明奖一等奖，第3完成人；2013年兰州市科学技术发明奖一等奖，第3完成人。

杨博辉（1964—）博士，四级研究员，博士生导师。现为"国家绒毛用羊产业技术体系"分子育种岗位科学家，中国农科院兰州畜牧与兽药研究所"绵羊新品种（系）培育"科技创新团队首席专家。兼任中国畜牧兽医学会养羊学分会副理事长兼秘书长，中国农科院学位委员会委员，中国博士后基金评审专家，中国畜牧业协会羊业分会特聘专家，中国农业科学院三级岗位杰出人才。主要从事绵羊新品种（系）培育、分子育种及产业化研发。先后主持完成国家863计划、国家支撑计划、国家基础专项及国家（部）级标准等项目20余项。制定国颁标准6项，部颁标准5项。获国家发明专利2项，实用新型专利8项。已培育出"大通牦牛"新品种、西北肉羊新品群，基本培育出高山美利奴羊新品种。发表论文120篇，其中7篇SCI，获国际论文一等奖1篇。主编《甘肃高山细毛羊的育成和发展》与《中国野生偶奇蹄目动物遗传资源》著作2部，副主编著作4部，参编著作4部。培养博、硕士研究生21名，其中国际留学生1名。

王慧（1985—），硕士，研究实习员。主要从事动物营养代谢病研究。现主持"中央级科研院所基本科研业务费专项资金项目（NO. 1610322013003）。2013年获农业部全国农牧渔业丰收二等奖（第7完成人）；兰州市科学技术进步二等奖（第9完成人）。目前第一作者发表论文16篇，其中SCI收录8篇。申请专利2项。

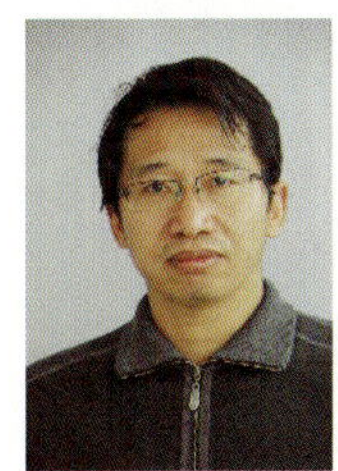

尚若锋（1974—），博士，副研究员。主要从事兽用药物研发工作。主持或参与国家支撑计划、"863"计划以及其他国家级和省部级的科研项目20余项。获得省部级科研奖励4项，国家发明专利12项，以第一作者或通讯作者发表文章30余篇，其中SCI文章12篇。

刘建斌（1977—），甘肃农业大学动物生产系统与工程专业博士，副研究员，主要从事绵羊现代遗传学理论与育种，动物种质资源优良基因发掘和重要经济性状分子遗传机理研究。先后主持省部级科研项目5项。获甘肃省科技进步二等奖1项、中国农业科学院科技进步二等奖1项、中华神农科技三等奖1项。主编出版《现代肉羊生产实用技术》和《绒山羊》著作2部，副主编出版《甘肃高山细毛羊的育成和发展》《中国野生偶奇蹄目动物遗传资源》《羊繁殖与双羔免疫技术》《适度规模肉羊场高效生产技术》和《优质羊毛生产技术》著作5部，参编出版著作5部；主笔发表专业论文30余篇，其中SCI论文7篇；申报国家发明专利6项，授权发明专利3项，合作授权国家实用新型专利30项。

辛蕊华（1981—），硕士，助理研究员。主要从事中兽医药物学工作，先后主持和参与过中央级公益性科研院所基本科研业务费专项资金项目、公益性行业（农业）科研专项、国家科技支撑项目等；发表论文12篇，参与著作两部。参加的“富含活性态微量元素酵母制剂的研究”科研成果获得兰州市科学技术二等奖。

杨峰（1985—），助理研究员，长期从事奶牛乳房炎的诊断、预防和治疗工作。主持2项中央级公益性科研院所基本科研业务费专项资金和1项甘肃省科技计划项目，先后参与国家自然科学基金、国家支撑计划、省部级等项目10多项。以第一完成人获得授权实用新型专利8项。

罗超应（1960—），学士，研究员，中西兽医药学结合研究（主持科技部科研院所开发研究专项“新型中兽药射干地龙颗粒的研究与开发”、科技部基础工作专项“华东区传统中兽医学资源抢救与整理”与国家“十一五”科技支撑子项目“中兽药中试及其生产工艺研究”等）。主编、副主编与参编出版《牛病中西医结合治疗》等著作16部，共计679余万字；主笔发表“Variability of the Dosage，Effects and Toxicity of Fu Zi （Aconite）From a Complexity Science Perspective”“以复杂科学理念指导中西医药学结合”“奶牛乳房炎的复杂性及其对传统科学观念的挑战”等学术论文近100篇，其中英文期刊文章5篇；专利12项，成果奖励5项。

冯瑞林（1959—），本科毕业，助理研究员，研究方向为羊繁殖育种。主要从事羊双羔免疫技术研究工作，参与甘肃省重大专项“甘肃超细毛羊新品种培育示范与推广”、中国农业科学院“羊绿色增产增效技术集成与示范”等十余项。工作期间参与申报细毛羊新品种1个。获国家科技进步三等奖1项（第九完成人），甘肃省科技进步一等奖1项（第九完成人）。主编出版著作《羊繁殖与双羔免疫技术》1部，参加编写《甘肃高山细毛羊的育成与发展》（副主编）、《绒山羊》（副主编）等著作5部。在各类期刊发表论文50余篇，其中以第一作者发表论文12篇。

郑继方（1958—），研究员，硕士生导师。现任甘肃省中兽药工程技术研究中心主任，中国农业科学院兰州畜牧与兽药研究所中兽医药创新团队首席科学家，中国农业科学院兰州畜牧与兽药研究所学术委员会委员，《中兽医医药杂志》编委，亚洲传统兽医学会常务理事，中国畜牧兽医学会中兽医分会理事，中国生理学会甘肃分会理事，西北地区中兽医学术研究会常务理事，中国畜牧兽医学会高级会员，农业部项目评审专家，农业部新兽药评审委员会委员，科技部国际合作计划评价专家，西南大学客座教授。从事中兽医药学的研究工作。先后主持国家自然科学基金、国家科技攻关、国家支撑计划、省部级等项目20多项。获省部级科技进步奖3项，院厅级奖4项，国家级新兽药证书两个，国家发明专利十余项。主编出版了《兽医中药学》《中兽医诊疗手册》《兽医药物临床配伍与禁忌》《常用兽药临床新用》《中草药饲料添加剂配制与应用》《甲型H1N1流感防控100问》和《兽医中药临床配伍技巧》等专著10部，先后在国内外各学术刊物发表论文80余篇。

刘希望（1986—），硕士，助理研究员。2007年毕业西北农林科技大学环境科学专业，2010年获西北农林科技大学化学生物学专业硕士学位，同年参加工作至今。曾主持农业部兽用药物创制重点实验室开放基金项目1项，现主持中央公益性科研院所基本科研业务费1项。以第一作者发表SCI论文4篇，主要从事新兽药研发，药物合成方面的研究工作。文章1篇。

熊琳（1984—），硕士，助理研究。主要从事农产品质量安全的研究。发表相关科研论文4篇，其中SCI收录1篇，授权专利8项，参与制定国家标准1项。

魏小娟（1976年），女，硕士研究生、助理研究员，研究方向为分子药理学。曾先后主持国家自然科学基金1项、甘肃省青年基金项目1项、中央级公益性科研院所基本科研业务费专项资金项目1项；先后参加国家自然科学基金项目，863项目、国家支撑计划项目、公益性行业专项、甘肃省重大专项等项目10余项。工作期间获甘肃省科技进步一等奖1项，中国农业科学院科技进步二等奖2项，兰州市科技进步二等奖2项。参加编写著作5部。在各类期刊发表论文30篇，其中SCI 5篇。

崔东安（1981—），博士，主要从事奶牛胎衣不下方证代谢组学研究（奶牛胎衣不下血瘀证的代谢组学研究No. 1610322015006），发表文章4篇，其中3篇SCI文章。

路远（1980—），硕士，副研究员。主要从事牧草新品种选育及植物组培研究。自毕业以来，主持院所长基金项目“美国杂交早熟禾引进驯化及种子繁育技术研究”、和“黄花矶松驯化栽培及园林绿化开发应用研究”，曾参与并完成了973合作子课题项目“气候变化对西北春小麦单季玉米区粮食生产资源要素的影响机理研究”，全球环境基金（GEF）项目“野生牧草种质资源应用研究”“放牧利用与草原退化关系研究”和国家科技基础条件平台工作项目子课题“牧草种质资源的实物共享及标志性数据采集”“牧草种质资源的标准化整理和整合”、国家科技支撑计划项目子课题“西北优势和特色牧草生产加工关键技术研究与示范”等项目20余项。发表论文20余篇，主编著作1部，或甘肃省科技进步二等奖1项，选育牧草新品种1个，获专利5项。

高雅琴（1964—），研究员，硕士研究生导师，主要从事畜产品加工及动物毛皮质量评价研究工作。现主持农业部畜产品质量安全风险评估项目中牛羊肉质量安全风险评估子项目。曾主持的国家公益类科研项目“动物纤维毛皮产品质量评价技术的研究”，2009年获甘肃省科技进步三等奖；主持国家标准制定项目“GB/T 25885—2010羊毛纤维直径试验方法——激光扫描仪法”“GB/T 25880—2010毛皮掉毛测试方法——掉毛测试仪法”“GB/T 26616—2011裘皮獭兔皮”均已颁布并实施；主持的农业行业标准制定项目“NY 1164—2006裘皮蓝狐皮”“NY 2222—2012动物纤维直径及成分分析方法——显微图像分析仪法”已颁布实施，“动物毛皮各类鉴别方法——显微镜法”上报农业部审定。参与多项国家标准和农业行业标准“GB/T 25243—2010甘肃高山细毛羊”“河西绒山羊”和“大通牦牛”“天祝白牦牛”“NY 1173—2006动物毛皮检验技术规范”等。主编的《动物毛纤维组织学彩色图谱》《动物毛皮质量鉴定技术》；参与出版著作5部。发表科技论文150余篇，其中主笔发表40余篇。